D0516487

INTRODUCTION TO CIRCUIT ANALYSIS and DESIGN

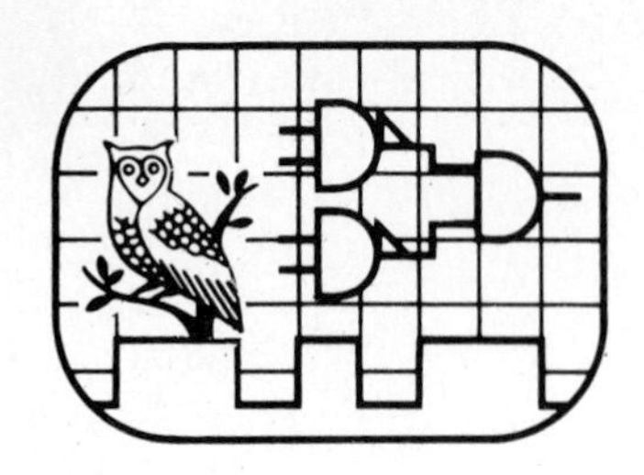

HRW
Series in
Computer Engineering

Michael R. Lightner, Series Editor Computer Engineering

P.E.Allen and D.R.Holberg CMOS ANALOG CIRCUIT DESIGN
M.D.Ciletti INTRODUCTION TO CIRCUIT ANALYSIS AND DESIGN
L.A.Leventhal MICROCOMPUTER EXPERIMENTATION WITH THE SDK-86
L.A.Leventhal MICROCOMPUTER EXPERIMENTATION WITH THE IBM PC
W.A.Wolovich ROBOTICS: BASIC ANALYSIS AND DESIGN

Michael D. Ciletti
University of Colorado at Colorado Springs

INTRODUCTION TO CIRCUIT ANALYSIS and DESIGN

Holt, Rinehart and Winston, Inc.
New York Chicago San Francisco Philadelphia
Montreal Toronto London Sydney Tokyo

THIS BOOK IS DEDICATED TO JERILYNN.
SHINING BRIGHTLY, LOVING DEEPLY.

Publisher Ted Buchholz
Acquisitions Editor Deborah Moore
Senior Project Manager Chuck Wahrhaftig
Production Manager Paul Nardi
Design Supervisor Bob Kopelman
Text Design Photo Plus Art/Celine Brandes
Text Art Scientific Illustrators
Cover Photo Tom Adams

Library of Congress Cataloging-in-Publication Data
Ciletti, Michael D.
Introduction to circuit analysis and design.

ISBN 0-03-070563-0

Includes index.
1. Electric circuit analysis. 2. Electronic
circuit design. I. Title.
TK454.C6 1987 621.381'042 87-12108

PRINTED IN THE UNITED STATES OF AMERICA
Published simultaneously in Canada

8 9 016 9 8 7 6 5 4 3 2 1

Holt, Rinehart and Winston, Inc.
The Dryden Press
Saunders College Publishing

PREFACE

The study of circuits is the foundation on which most other courses in the electrical engineering curriculum rest. For this reason a student's first courses in circuits must be informative, technically rigorous, stimulating, and challenging.

This book was written to help meet these objectives, and to help students develop a physical insight into the behavior of circuits while giving them an early introduction to circuit design. The amount of material and the technical detail in a first course on circuits can sometimes lead students to settle for learning to manipulate the mathematical equations that describe a circuit, without appreciating its physical operation. This is especially true of circuits that contain inductors, capacitors, and op amps, where the task of solving a differential equation rears its head, (usually before the student has had a formal course in the subject). Even if a student has some initial background in differential equations, it will not likely include their use in circuits. In either case, studying the physical behavior of the circuit can be overshadowed by learning the mathematical manipulations required to solve the circuit's differential equation. This text hopes to challenge students to probe more deeply into the physical operation of circuits, so the selection, sequence, and treatment of topics is intended to combine physical insight with a solid foundation in circuit theory.

Each chapter has an introduction that presents the chapter objectives and discusses their relationship to material in previous chapters. This helps students understand the rationale behind the effort asked of them. Skill Exercises are placed at strategic locations within each chapter to let students evaluate their comprehension of the subject matter before proceeding to new material. Several worked examples demonstrate the application of the material to problem solving and develop additional physical insight. Abundant figures and graphs show circuit responses for a variety of input signals and parameter values. Finally, numerous problems are given at the ends of the chapters; some require only direct application of material in the text, others require deeper reflection.

The presentation of the basic properties of circuits relies on nodal analysis more than mesh or loop analysis because nodal equations, rather than mesh or loop equations, are used almost exclusively in practice to describe circuits. Matrix methods are used to formulate and solve the nodal and mesh models of circuits for several reasons: engineering students have commonly been introduced to matrix algebra in high school or in their first year of university-level mathematics; circuits are compactly summarized by matrix models; and fundamental circuit properties (such as superposition and Thevenin or Norton equivalent circuits) evolve directly from a circuit's matrix model. Operational amplifiers are introduced early and used throughout the text, rather than in only one or a few chapters.

Electrical engineering students are often introduced to differential equations at least a semester before their mathematics curriculum covers the material. The usual approach is to show the student how to guess solutions and then solve boundary-value problems. In contrast, this text emphasizes the importance of *constructing* rather than *guessing* the solution to the differential-equation model of a circuit for a variety of input signals, circuit parameters, and initial conditions. The treatment of differential equations is self-contained, and does not presume that a student has taken a course in the subject. This approach de-emphasizes ad-hoc solutions that rely on memorization and tricks, and reduces the confusion that such an approach can create. The text further distinguishes between the mathematical problem—solving a differential equation for its natural- and particular-solution components—and the circuit problem—finding the zero-input, zero-state, and initial-state responses for a variety of sources and initial-stored-energy conditions.

Exponential signals are pivotal in the text because they play a pivotal role in circuits, and because most signals of interest at this level can be represented by real or complex exponentials. When coupled with the concept of the transfer function, exponentials afford an opportunity to present a unified treatment of the response of first-, second- and higher-order linear circuits to steps, sinusoids, pulses, periodic signals (Fourier series), and more general non-periodic signals (Fourier and Laplace transforms). Thus, we introduce the transfer function of a circuit immediately and use it to construct the circuit's response. We strive to have students see the relationship between time-domain and frequency-domain models of a circuit, and to appreciate the progression that unfolds as they learn how to construct the circuit response to a real exponential, a step, a complex exponential, a sinusoid, a periodic signal, a Fourier-transformable signal, and ultimately, a Laplace-transformable signal.

Laplace transforms are covered last because students can develop a deeper appreciation for their significance when they see them placed in an evolutionary perspective with exponential response, superposition, Fourier series and Fourier transforms. Thus, there is no need to mention the so-called "Laplace operator" approach or to ask students to accept the Laplace transform on faith. This chapter is a foundation for a course in linear systems that follows the circuits sequence at many universities.

Chapter 17 specifically addresses the need to provide more balance between analysis and design in the introductory circuits sequence. The dilemma is that the technology for designing modern digital and analog integrated circuits requires additional background in solid-state devices and is far too dynamic to be covered at this level. The compromise taken here is to introduce students to the design of analog filters without addressing fabrication and manufacturing issues. The text develops design equations for analog filters and uses them to determine filter-component values that assure filter performance. This chapter can be abbreviated, sampled, or supplemented to suit the objectives of the instructor. For example, Bode analysis is presented here because of its close relationship to the filter-design problem, but this topic can be covered sooner.

The book contains more material than can be covered in a typical two-semester or three-quarter course sequence for electrical engineering majors. The intent is to provide enough flexibility for instructors to deviate in a variety of ways from a fairly standard core of material, and to encourage students to use additional material for outside reading and reference. For example, the chapter on signals can be covered as a unit or in piecemeal fashion (as signals are used in the overall development). The time-domain response of first-order circuits to sinusoids (Chapter 9) is presented before second-order circuits, but it can also be placed immediately after Chapter 12, so that it leads into Chapter 13, which deals with sinusoidal steady-state analysis. Instructors wishing to spend more time on design issues can skim or skip some of the material on power and three-phase circuits. Conversely, the chapter on frequency response and filter design admits to selected or truncated coverage of the basic filter types. Two-port networks may be skipped entirely, or assigned for outside reading. And Laplace transforms can be covered before Fourier series and Fourier transforms, but with some additional effort to make the transition.

The symbol * is used in the Table of Contents and at section headers in the text to suggest material that the instructor can minimize, skip, or assign for outside reading in a two-semester course. These topics might also be covered in a recitation section.

The material in Chapters 1–10 can serve as the basis for the first semester in a sequence while Chapters 11–18 can serve fo the second semester, or Chapter 10 can be placed in the latter group to accomodate the pace of the first semester. The topic coverage sequence (TCS) can be modified to create a desired emphasis. Some possibilities are given below (* indicates partial coverage).

TCS #1 (Design emphasis) Chapters 1–9*, 10–13*, 14–17
TCS #2 (Design emphasis), Chapters 1–5, 7–8, 10–13*, 14–17
TCS #3 (Traditional) Chapters 1–5, 7, 8, 10–16
TCS #4 (Traditional) Chapters 1–5, 7, 8, 11, 12, 10, 13–16

Circuit Master™ Software

Circuit design requires examining a circuit's behavior and performance over a range of conditions. But lack of time and the volume of computation required to support that effort can be a barrier to working examples or assigning design problems in a circuits course. Even attempts to use computer programs can be unsatisfactory, because they typically require the user to extract useful information from volumes of data. To address this need, the text has an optional software package, called **Circuit Master™**, that students can purchase and use to create and solve a variety of circuit problems over a wide range of circuit parameters, input signals, and initial conditions.

Circuit Master is a courseware aid to help students master the fundamentals of circuit analysis and introduce them to computer-aided design. It runs on an IBM personal computer. A menu-driven turnkey program, it does not require that either student or instructor actually write programs. The text does not require the software, but does suggest ways that students can go beyond the classroom and use a personal computer to gain additional background in circuits. Circuit Master lets students select a circuit from a library, select its component values, and select a source signal. The software has features that automatically generate displays of the circuit's pole-zero patterns, source and response waveforms, transient and steady-state responses, source and response spectra, Bode plots, and phasor diagrams. It lets students study the effects of changing parameter values and initial conditions or input signals. One feature also supports design of a variety of filters.

The software and the text anticipate changes in pedagogy that will place increased emphasis on presenting graphical images. The software is not intended to duplicate the software tools available on modern CAD workstations commonly used to design integrated circuits, but it does acquaint the student with the underlying philosophy of designing with a computer—to simulate and investigate many possibilities and to reduce the chances for error—and it is consistent with the movement toward interactive graphical tools. Such tools encourage the student to think creatively, without being overburdened by having to carry out a multitude of calculations. Their evolution is exciting and imminent.

Many friends have been instrumental and inspirational in the process creating this text. Foremost, my wife and best friend, Jerilynn, encouraged me to write, and held fast for the duration. Our children Monica, Lucy, Rebecca, Christine, and Michael patiently endured my absorption, and, with Jerilynn, pitched in at key points to cut and paste, proofread, and manage the workload. Typists, especially Lynn Scott, helped conquer a monumental morass of manuscript. Students helped by demanding clarity and insight. Many reviewers shaped and ultimately improved the text with constructive evaluations, among them, Bennet Basore of Oklahoma State University, Ross K. Johnson of Michigan Technical University, Bernhard Schmidt of the University of Dayton, Alan Marshak of Louisiana State University, Ron Holzman of the University of Pittsburgh, Susan Reidel of Marquette University, Robert H. Miller of Virginia Polytechnic Institute, John Fleming of Texas A & M, Darrel Vines of Texas Tech University, Robert E. Yantorno of Temple University, Louis W. Eggers of California State University, K.S.P. Kumar of the University of Minnesota, Dennis Herr of Ohio Northern University, Benjamin Nichols of Cornell University, Donald Farris of General Dynamics, Jack Kurzweil of San Jose State University, Mike Lightner of the University of Colorado, and Mac Van Valkenburg of the University of Illinois. Deborah Moore, my editor from the initial proposal to publication, always seemed to know when to challenge, urge, or cajole me to push farther than I thought reasonable. Chuck Wahrhaftig, senior project manager, then took my manuscript and turned it into the book you now hold. I thank you all.

Michael D. Ciletti

CONTENTS

INTRODUCTION TO CIRCUIT ANALYSIS and DESIGN

ELEMENTARY CIRCUIT RELATIONSHIPS

INTRODUCTION

Humanity's fascination with electricity has progressed from fearful observation of the effects of lightning to a deeper understanding of the physical properties of charged matter. While electricity remains a mysterious and potentially destructive force, our ability to harness its effects has enabled us to propel images across a color video screen, provide vital signals to monitor and control high-performance automobile engines, and pulse the human heart with a pacemaker.

We have progressed to the submicroscopic dimensions of integrated circuits, or "electronic engines"—irresistibly exciting, inescapably complex, cheaper, more powerful, and ever smaller. In the world of microcircuits, trial-and-error circuit design has gone by the wayside, doomed by its inherent cost and by an ever-shortening product market window. Long before they are fabricated, circuits must be built and tested on paper. Production costs, especially time, dictate that an integrated circuit must work on its first trial. So circuit designers must have a thorough understanding of how a circuit will behave *before* it is built. Now, our confidence that a circuit will work is based on our confidence that our model of its components and their interconnected behavior is accurate. We need, therefore, to begin with a clear understanding of how circuits behave.

1.1 CIRCUIT MODELS AND ASSUMPTIONS

Our aim in this text is to provide you with the technical tools needed to understand circuits. We intend to combine a quantitative with a qualitative appreciation for circuits, so we will be working in a mathematical context. This approach should be familiar, because you should already be accustomed to the use of point-mass mathematical models in physics (such as Newton's Law which describes the relationship between physical force, mass, and acceleration). Like these familiar models from physics, circuit models are incomplete descriptions of circuits, but they still provide useful results.

Our work will only consider "lumped" circuits—those whose physical geometry is assumed not to affect their electrical behavior. This assumption is generally valid when the physical dimensions of the circuit are much smaller than the electrical wavelength of the signals propagating through the circuit. Nonetheless, many important circuits do not satisfy this assumption, so later courses will treat circuit models that account for the spatial distribution and propagation of the electromagnetic signals within a circuit. Furthermore, the models in this text treat circuits as though they have no physical dimensions, and as though electrical signals propagate instantly across them. In reality, this is not the case. The propagation delays for signals within the integrated circuits of a computer, for example, determine its performance limits at high clock frequencies and ultimately limit our ability to manufacture faster computers. In this text we will only consider "time invariant" circuits—those whose characteristics are fixed for all time of interest.

Electrical engineering is a quantitative discipline that models and measures physical quantities and determines relationships between them. The physical quantities used in our work will be expressed in the International System (SI) of Units, which has standard measures for length (meter), mass (kilogram), time (second), current (Ampere), and temperature (degree Kelvin). The six fundamental physical quantities, along with their symbols and their SI units, are shown in Table 1.1. Since physical quantities may range in value over several orders of magnitude it is convenient to use the standard prefixes shown in Table 1.2 to reduce the amount of notation needed to describe relatively large or small numbers.

TABLE 1.1

QUANTITY	UNIT	SYMBOL
Length	Meter	m
Mass	Kilogram	kg
Time	Second	s
Current	Ampere	A
Temperature	Degree Kelvin	K
Luminous Intensity	Candela	cd

TABLE 1.2

PREFIX	POWER OF 10	SYMBOL
atto	10^{-18}	a
femto	10^{-15}	f
pico	10^{-12}	p
nano	10^{-9}	n
micro	10^{-6}	μ
milli	10^{-3}	m
centi	10^{-2}	c
deci	10^{-1}	d
deka	10^{1}	da
hecto	10^{2}	h
kilo	10^{3}	k
mega	10^{6}	M
giga	10^{9}	G
tera	10^{12}	T

Our study of circuits will proceed logically: first, examine models for the electrical characteristics of *individual* circuit components; next, examine physical laws that describe the electrical behavior of *interconnected* components; and last, examine how a circuit of interconnected components behaves when connected to electrical energy sources. This sequence is easy to state, but difficult to carry out. You will be repeatedly challenged to look beyond the limited mathematical models of circuits and acquire a physical appreciation or "feel" for their behavior. A feel for circuits does not come easily for everyone, so don't falter at your initial discouragement. Continue to ask yourself why a circuit behaves the way it does, and look for logical cause-effect reasons to support your insights.

1.2 VOLTAGE AND CURRENT

Any interconnection of electrical devices such as batteries, resistors, and current sources, can be called a **circuit.** Here, we will be concerned primarily with two measurements: the voltage at the nodes of a circuit, and the current in the circuit branches. **Voltage** is a measure of the work required to move charge between two points in space, and **current** is a measure of the amount of charge that is moving in a unit of time.

The boxes in Fig. 1.1 represent elements of a circuit; the lines connecting the boxes represent allowable paths for current. The heavy dots, called the **nodes** of the circuit, symbolize the physical interconnection of the devices represented by the boxes. They are also the points at which we will measure voltages.

The voltage at the nodes of a circuit will be denoted by the labels attached to the nodes. This voltage is the *difference* between the electrical potential at the node and the electrical potential at the reference node, which is denoted by the ground symbol.

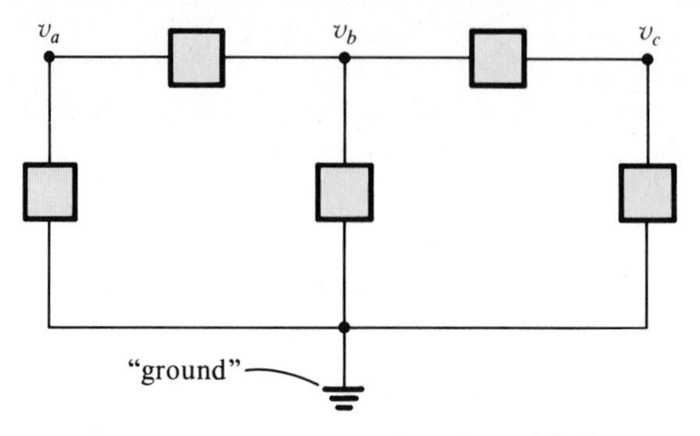

Figure 1.1 Circuit with node voltage labels.

The **voltage drop** between a pair of nodes (a, b) is calculated as

$$v_{ab} = v_a - v_b$$

where v_a and v_b are the electrical potentials of nodes a and b. This measurement convention assigns a positive polarity to the voltage between node a and nobe b when node a is at a higher electrical potential than node b; in this case $v_{ab} > 0$. Consequently, the **voltage rise** between nodes a and b is

$$v_{ba} = v_b - v_a$$

and therefore $v_{ba} = -v_{ab}$, with a positive voltage rise corresponding to a negative voltage drop.

Skill Exercise 1.1

If the nodes in Figure 1.1 have $v_a = 20$ V, $v_b = 35$ V and $v_c = -45$ V, determine the values of v_{ab}, v_{ac}, v_{ba}, and v_{ca}. What voltage drop would be measured from ground to node b? What voltage rise would be measured from node a to node c?

Answer: $v_{ab} = -15$ V $\quad v_{ac} = 65$ V $\quad v_{ba} = 15$ V
$v_{ca} = -65$ V $\quad -v_b = -35$ V $\quad -v_{ac} = -65$ V

Note that a **negative rise corresponds to a positive drop, and a negative drop in potential corresponds to positive rise.**

Charge moves in the presence of an electric field. In conductors the valence electrons are loosely bound to a particular atom, so they can be considered to be moving randomly in the atomic lattice. If an electric field is created by an external source the free electrons in a conductor will move in a direction opposite to the polarity of the field. By convention, however, we will define current to be the motion of an equivalent positive charge in the same direction as the field. If a conductor has a cross sectional area S, the average current can be measured by counting the charge Δq moving through the cross section in a measured time interval Δt. The average current I is

$$I = \frac{\Delta q}{\Delta t}$$

On an instantaneous basis, the current is obtained by shrinking the measurement interval:

$$i(t) = \lim_{\Delta t \to 0} \frac{\Delta q}{\Delta t} = \frac{dq}{dt} \text{ Amps.}$$

Currents in the paths of a circuit will be denoted by a label and by arrows beside the paths. Labels distinguish individual currents and may be signed to indicate a negative quantity (a current in the opposite direction). The direction of the arrow indicates the positive algebraic reference direction for the labeled current. Figure 1.2 illustrates this notation for two equivalent sets of current labels. If $i_1 = 5$ A, then $-i_1 = -5$ A, and if $i_2 = -10$ A, then $-i_2 = 10$ A. A positive value of current corresponds to the flow of positive charge in the direction of the arrow for that current.

This text treats only circuits that can be characterized by the voltages measured between nodes and the currents in the conducting paths between nodes. The assumption that any conductor (line) attached to a node is at the same electrical potential as the node justifies using dots to represent nodes and lines to represent current paths. The voltage measured anywhere on a line is the same as the voltage measured at the node to which it is attached.

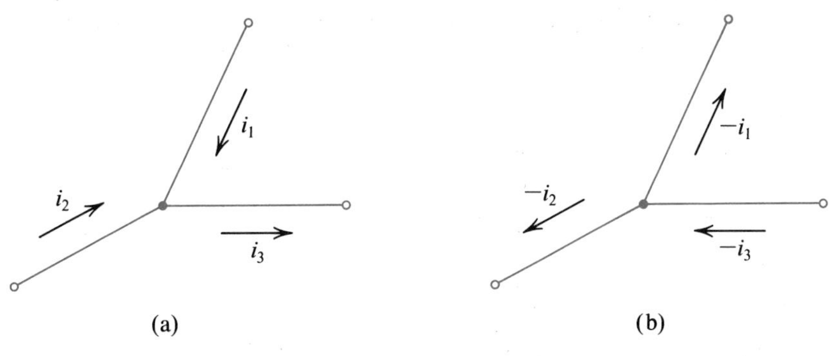

Figure 1.2 Equivalent current polarities and labels.

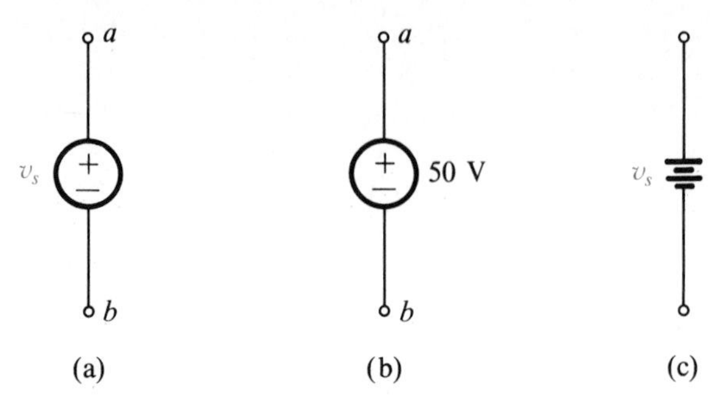

Figure 1.3 Circuit symbols for ideal independent voltage sources.

1.3 IDEAL VOLTAGE AND CURRENT SOURCES

Ideal independent voltage sources establish a voltage between a pair of nodes in a circuit. We use the circuit symbol shown in Figure 1.3a for such sources, with the source label distinguishing individual sources, and the ± symbol specifying the polarity of the source. Sometimes the source label also indicates the value of the voltage assigned to the source, as shown in Fig. 1.3b where $v_s = 50$ V. When a voltage source is time-varying its label will indicate the time dependency. A fixed source is sometimes represented by the symbol for the battery shown in Fig. 1.3c. By convention, the wider bar denotes the battery's positive terminal. The algebraic value of the assigned source voltage corresponds to the measurement polarity indicated by the source symbol. For example, in Fig. 1.4 $v_{ab} = -10$ V and $v_{cd} = 5$ V. The voltage established by an independent voltage source is **independent** of the current in the source path.

A source's ***v-i* characteristic** is a graph showing the voltage established by the source for each value of the current in the source path. In general, the voltage and current are not independent of each other, but for ideal voltage sources the *v-i* characteristic has the shape shown in Fig. 1.5 where the source has the same voltage regardless of the current.

The current established by an ideal independent current source is independent of the voltage across the terminals of the source. This prop-

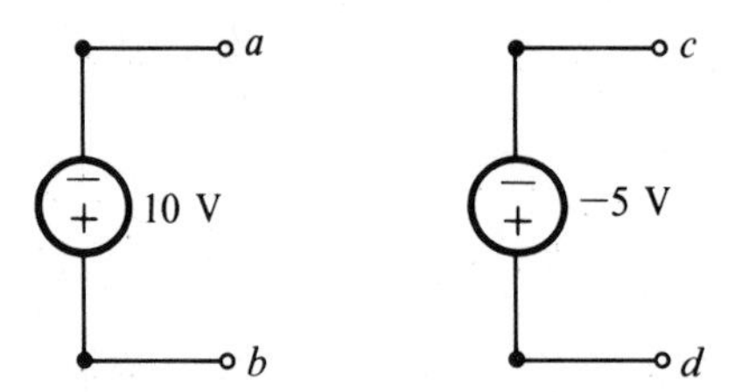

Figure 1.4 Voltage sources establish a voltage between the nodes of a circuit. Polarity symbols and signed voltage labels determine the voltage of a source.

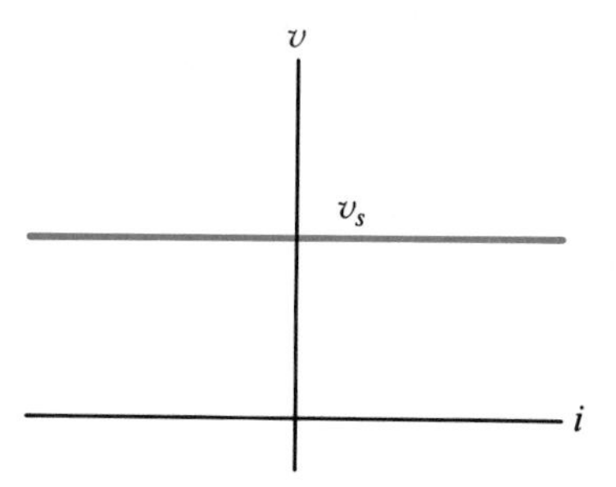

Figure 1.5 The v-i characteristic for an ideal voltage source.

erty is represented by the i-v characteristic shown in Fig. 1.6, with the circuit symbol for an ideal independent current source. The arrow indicates the positive direction of the assigned source current.

1.4 RESISTORS AND OHM'S LAW

The electrical properties of circuit components are determined by the materials and processes used to fabricate them, so the kinds of devices that can be built have been dramatically influenced by the major technological advances of the last decade. Circuits now commonly contain several hundred thousand devices, and even denser circuits are expected to become available over the next decade. This text does not cover the physical fabrication of circuit components, but the basic concepts developed here provide a foundation for understanding the operation of these more advanced circuits.

The simplest circuit components can be represented as two-terminal devices, characterized by the relationship between the voltage across their terminals and the current through the device. One such device is a resistor, denoted by R, whose circuit symbol is shown in Fig. 1.7(a). We will use a measurement convention in which the voltage *across* a resistor and

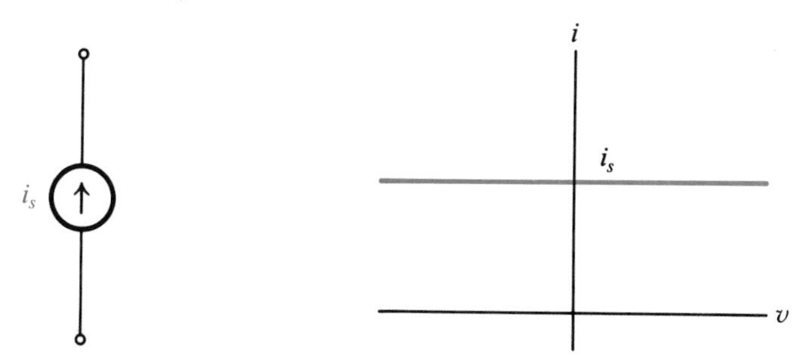

Figure 1.6 The (a) circuit symbol and (b) i-v characteristic for an ideal current source.

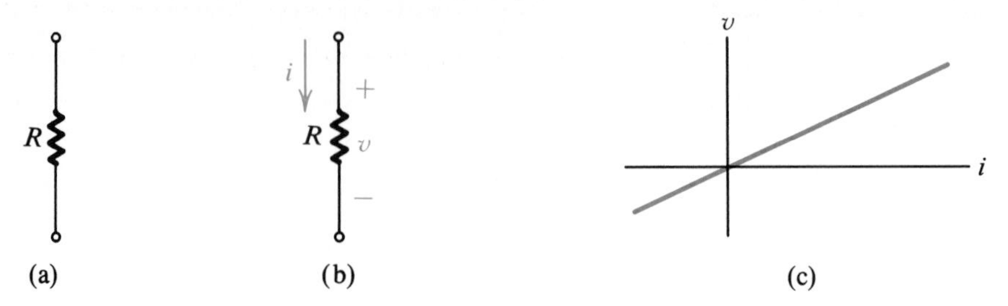

Figure 1.7 The (a) circuit symbol, (b) conventional measurement polarities for v, and (c) v-i characteristic for a resistor.

the current *into* the resistor are measured with the direction for positive current entering the positive terminal for the measurement of voltage, as shown in Fig. 1.7(b). Linear resistors exhibit the linear relationship shown in Fig. 1.7(c). This important property is described by **Ohm's Law:**

$$v(t) = i(t)R$$

or $R = v(t)/i(t)$. The **resistance** of a resistor is the ratio of the resistor voltage and the resistor current.

The unit of measure for resistance is Ohms, or volts/amps, and is symbolized by Ω, upper case Greek omega. We will only consider circuits in which R, the resistance, is a constant, is independent of voltage and current, and does not vary with time.

It is convenient to think of circuits in terms of their input/output properties. That is, we apply a signal to a circuit and we cause an output signal. If we apply a voltage to the terminals of a resistor we cause a current to flow in an amount specified by Ohm's Law, and vice-versa. Thus, Ohm's Law is a model for the resistor's input/output behavior.

1.5 CONDUCTANCE

Ohm's Law also describes the ratio of current to voltage: $i(t)/v(t) = G$, where

$$G = \frac{1}{R}$$

with G having units of siemans (mhos) and the symbol Ω. A high resistance value implies a low conductance value, and vice versa. Given the same applied voltage, a resistor with a high conductance value will have a higher current than a resistor with a lower conductance.

Skill Exercise 1.2

Find i_1 and i_2 in Fig. SE 1.2.

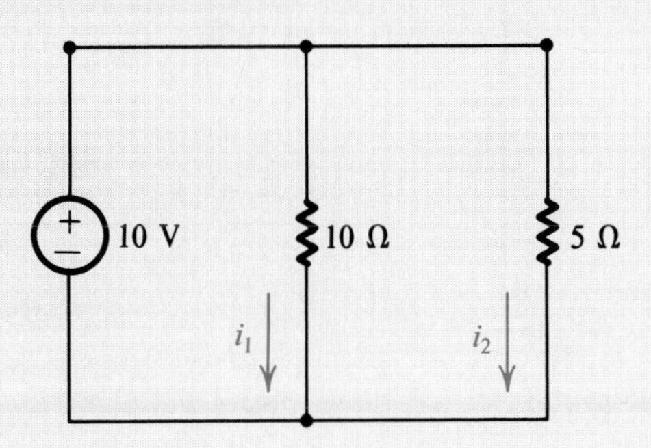

Figure SE1.2

Answer: $i_1 = 1$ A, $i_2 = 2$ A.

Example 1.1

Find v_1, v_2, v_3 and v_4 in Fig. 1.8 if the currents in the circuit are as shown.

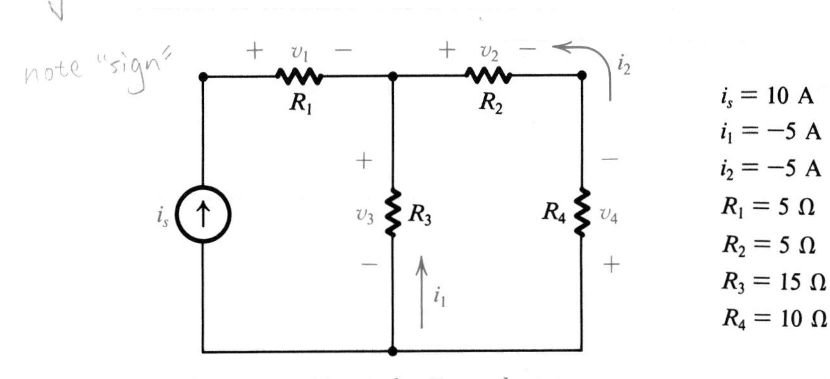

Figure 1.8 Circuit for Example 1.1.

Solution: From Ohm's Law:

$$v_1 = i_s R_1 = 50 \text{ V} \qquad v_3 = -i_1 R_3 = 75 \text{ V}$$

$$v_2 = -i_2 R_2 = 25 \text{ V} \qquad v_4 = i_2 R_4 = -50 \text{ V}$$

Example 1.2

Find R_1, R_2, R_3 and R_4 in Fig. 1.9.

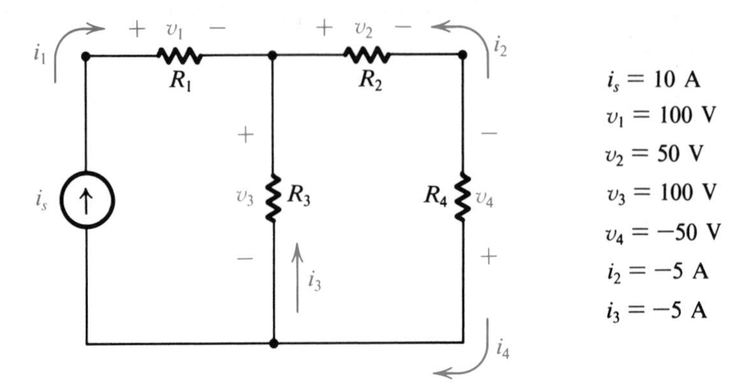

Figure 1.9 Circuit for Example 1.2.

Solution: Noting that $i_1 = i_s = 10$ A and $i_4 = -i_2$ we get

$$R_1 = \frac{v_1}{i_1} = \frac{100}{10} = 10\ \Omega \qquad R_3 = \frac{v_3}{-i_3} = \frac{100}{-(-5)} = 20\ \Omega$$

$$R_2 = \frac{v_2}{-i_2} = \frac{50}{-(-5)} = 10\ \Omega \qquad R_4 = \frac{v_4}{-i_4} = \frac{-50}{-5} = 10\ \Omega$$

Two nodes of a circuit are said to be connected by a **short circuit** if the voltage between them is zero for any current in their connecting path. Since wires that connect components together are assumed to have negligible resistance we model them as short circuits. A short circuit can be modeled as a wire or as a resistor having $R = 0$. When nodes are "shorted" together they have the same voltage.

A branch between two nodes of a circuit is said to be an **open circuit** if the current in the branch is zero for any voltage between the nodes. An open circuit can be modeled by a broken wire or by a resistor having $R = \infty$. For example, the spark plug in an automobile engine behaves like an open circuit for low voltages. However, once the voltage is high enough to ionize the surrounding air the spark plug behaves like a low resistance path for current.

In using ideal source models we will avoid situations that would connect a voltage source across a path of zero resistance (short circuit) since this would contradict the *v-i* characteristic of the source. Even though laboratory instruments are usually protected by fuses, it is generally recommended that you do not short-circuit a physical source because it could lead to very high currents in the shorting path, with subsequent vaporization of the conductor. Such practices could be a hazard to your health (and your wallet). Likewise, we will not attempt to connect an open circuit to an ideal current source. The two mathematical models are contradictory and an attempt to use them would lead to a meaningless description of the actual physical result.

1.6 KIRCHHOFF'S CURRENT LAW (KCL)

Kirchhoff's laws describe relationships between the voltages and currents in a circuit that contains several resistors and sources. These laws are especially important because they enable us to analyze systematically, and thereby understand, the behavior of large circuits.

Kirchhoff's current law expresses the physical fact that charge is always conserved in a circuit. It states that charge cannot accumulate at a node, and therefore the branch currents entering and leaving a node behave like traffic at an ideal intersection—all the cars move through, none of them stall, stack up, or otherwise accumulate. Unless there is a leak, a connection of water pipes behaves the same way. More formally:

KCL-1

> **At any time, the sum of the currents entering any node equals the sum of the currents leaving it.**

A current is said to be entering a node if its reference direction arrow is pointed toward the node, and is said to be leaving a node if the opposite is true. In Fig. 1.10 i_1 and i_3 are *entering* node N, and i_2 and i_4 are *leaving* it. Applying KCL-1 gives

$$i_1 + i_3 = i_2 + i_4$$

We don't know the value of these currents, but we do know that they must satisfy this constraint.

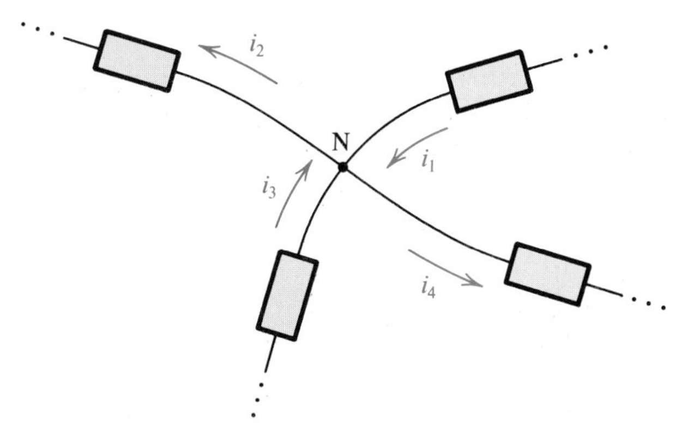

Figure 1.10 Currents entering and leaving a circuit node.

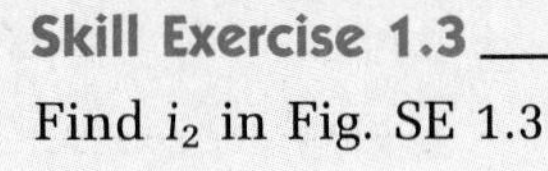

Skill Exercise 1.3

Find i_2 in Fig. SE 1.3

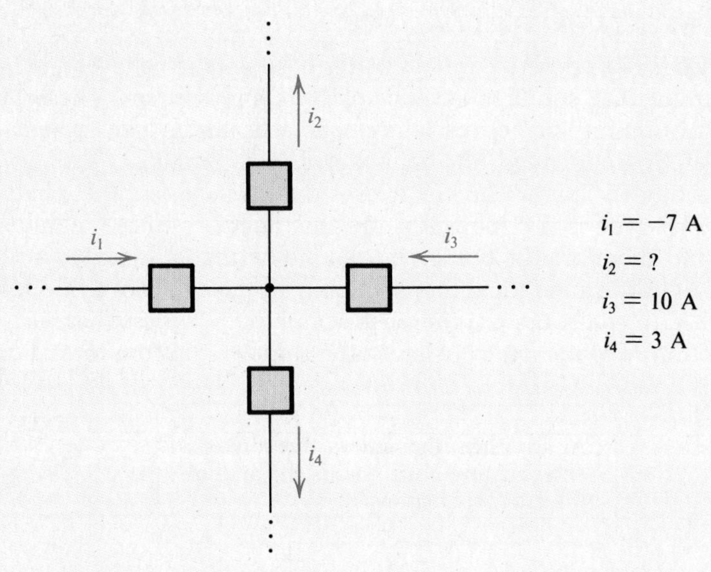

Figure SE1.3

Answer: $i_2 = 0$ A.

Writing KCL in terms of the current labels helps avoid confusion about current polarity. If the current labels are signed, we apply KCL the same way. We use numeric values *after* writing KCL.

Example 1.3

Find i_5 in Fig. 1.11.

Solution: Equating the current entering the node to the current leaving the node gives

$$i_2 + i_4 = i_1 - i_3 - i_5$$

Solving for i_5 with the known values of $i_1, \ldots, i_4$ gives: $i_5 = 1$ A.

When branch currents are known we can use KCL with Ohm's Law to find voltages across the resistors in a circuit.

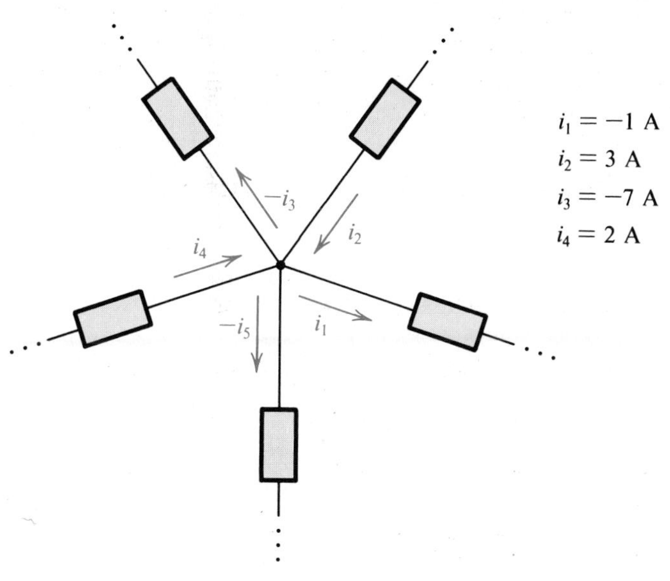

Figure 1.11 Circuit for Example 1.3.

Example 1.4

Find v_4 in Fig. 1.12

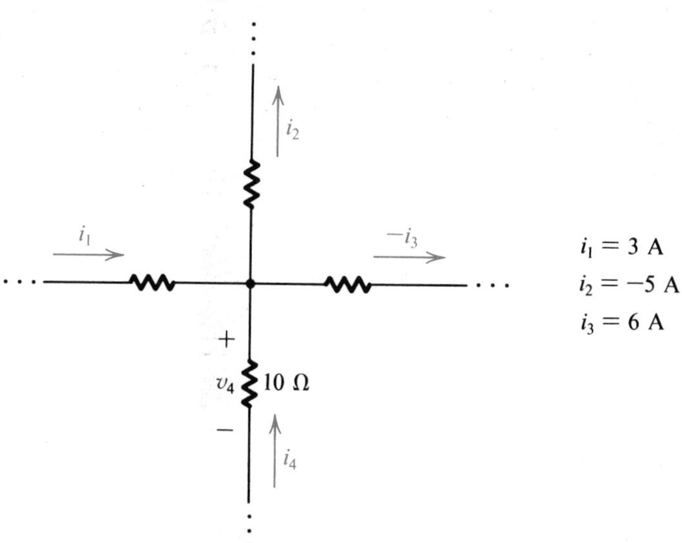

Figure 1.12 Circuit for Example 1.4.

Solution: First apply KCL to find i_4, then use Ohm's Law to find v_4

$$i_1 + i_4 = i_2 - i_3$$

and so $i_4 = -14$ A. Then

$$v_4 = (-i_4)10 = (-(-14))10 = 140 \text{ V}.$$

Sometimes it is convenient to use other forms of KCL.

KCL-2

> **"At every instant of time, the algebraic sum of the current *entering* any node equals zero."**

KCL-3

> **"At every instant of time, the algebraic sum of the current *leaving* any node equals zero."**

In applying these forms of KCL we must be careful to sign each current so that it conforms to an entering current (KCL-2) or a leaving current (KCL-3). Initially, you may want to redraw a circuit diagram to orient all currents as entering currents or as leaving currents, and relabel the currents where needed to reflect the change of orientation.

Example 1.5

Write KCL-2 for the circuit in Fig. 1.13a

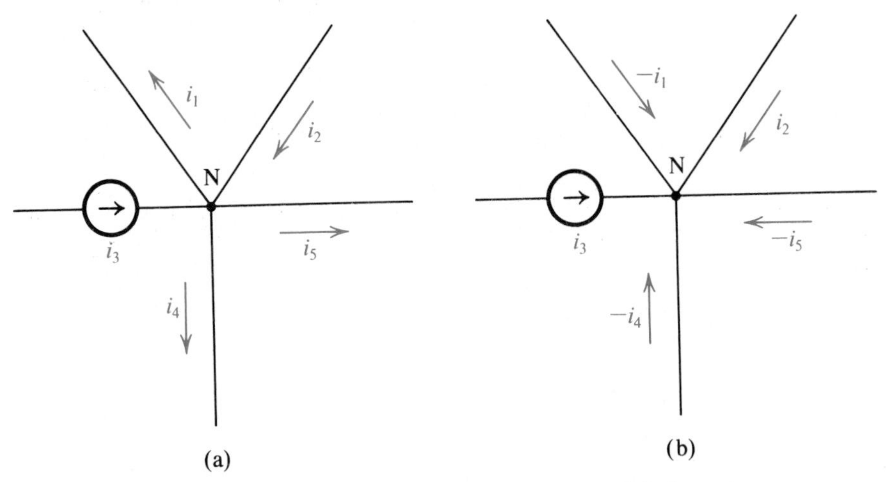

Figure 1.13 Circuit (a) before and (b) after the current labels and arrows are changed to be "entering" the node.

Solution: First, we create Fig. 1.13b by (a) re-orienting currents to enter N, and (b) re-signing the labels affected by (a). Then we apply KCL

$$-i_1 + i_2 + i_3 - i_4 - i_5 = 0$$

Note that the solution includes changing the orientation *and* the sign of i_1, i_4, and i_5. KCL can also be written directly, without the intermediate step of making a new diagram. Just be careful not to drop a minus sign in the wrong place! KCL-2 and KCL-3 are equivalent to KCL-1 (Explore this.)

Example 1.6

Find v, i_1, i_2 and i_3 in Fig. 1.14.

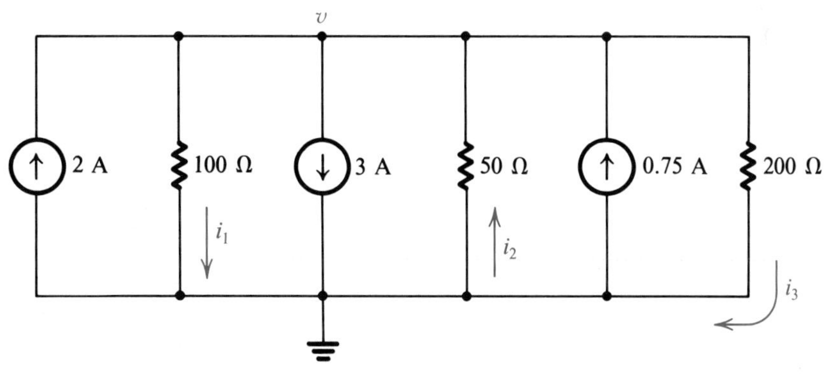

Figure 1.14 Circuit for Example 1.6.

Solution: The node voltage measured to ground is v. The parallel components all have the same voltage across their terminals. The currents at the node must satisfy KCL, so

$$2 + i_2 + \frac{3}{4} = i_1 + 3 + i_3$$

Expressing i_1, i_2 and i_3 in terms of Ohm's law gives:

$$i_1 = \frac{v}{100},\ i_2 = \frac{-v}{50},\ i_3 = \frac{v}{200}$$

so KCL becomes

$$2 + \frac{(-v)}{50} + \frac{3}{4} = \frac{v}{100} + 3 + \frac{v}{200}$$

and we solve for $v = -50/7$ V. The current in each branch is found by using Ohm's Law again

$$i_1 = \frac{v}{100} = \frac{-50}{7(100)} = \frac{-1}{14}\text{ A}$$

$$i_2 = \frac{-v}{50} = \frac{50}{7(50)} = \frac{1}{7}\text{ A}$$

$$i_3 = \frac{v}{200} = \frac{-50}{7(200)} = \frac{-1}{28}\text{ A}$$

As an exercise, confirm that the calculated currents are consistent with KCL. If this check fails we know the calculations are incorrect (the check is made by equating currents entering the nodes to currents leaving the nodes). Verifying that i_1, i_2 and i_3 satisfy KCL is not a guarantee that all of the calculations are correct, since two or more errors could cancel. But it increases our confidence. For certainty, we must verify that the calculated currents satisfy Ohm's Law and have $v = -50/7$.

Kirchhoff's current law also applies to any closed region or surface. If we partition a circuit into parts N_1 and N_2, a region enclosing N_1 can be viewed as a **generalized node,** or a "super node," for KCL. The subcircuits N_1 and N_2 in Fig. 1.15 have connecting current paths. We draw a boundary to enclose N_1 and apply KCL to the current paths that cross the boundary. Whatever the values of i_1, i_2, i_3 and i_g we know they must satisfy $i_1 - i_2 - i_3 + i_g = 0$.

Sometimes a partitioned circuit is connected by a single wire, like the hypothetical circuits shown in Fig. 1.16. Regardless of the contents of N_1

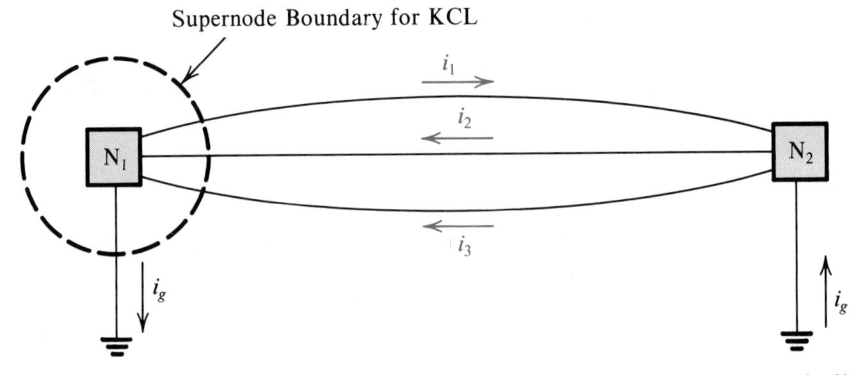

Figure 1.15 A super node surface creates a boundary for applying Kirchoff's current law.

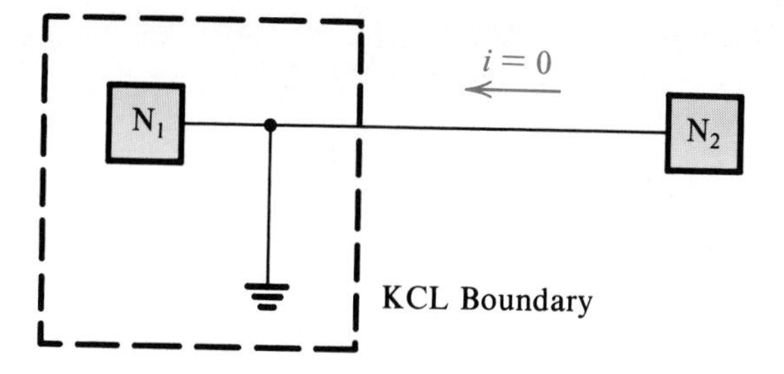

Figure 1.16 The current in the wire connecting the subcircuits must be zero.

and N_2, KCL assures us that the current in the connecting path is zero, even though the connecting wire causes a node in each sub-circuit to have the same potential. We'll use this result when we study the behavior of circuit models for transistor amplifiers.

1.7 CURRENT DIVISION AND PARALLEL RESISTANCE

Two-terminal devices that are connected between the same pair of nodes are said to be **in parallel.** If components are in parallel they must have the same voltage across their terminals, but they need not have the same current. Two circuits are said to be **equivalent** if they have the same v-i characteristics at their terminals. If two or more resistors are connected in parallel they may be replaced by a single resistor whose circuit behavior is equivalent to that of the original resistors. In Fig. 1.17, we consider the case of two resistors in parallel, R_1 and R_2, and our objective is to find the value of R such that the measurement of v at the terminals is the same for a given current source i_s.

Using KCL and Ohm's Law for the original circuit we have

$$i_s = i_1 + i_2 = v\left[\frac{1}{R_1} + \frac{1}{R_2}\right]$$

and

$$\frac{v}{i_s} = \frac{R_1 R_2}{R_1 + R_2}$$

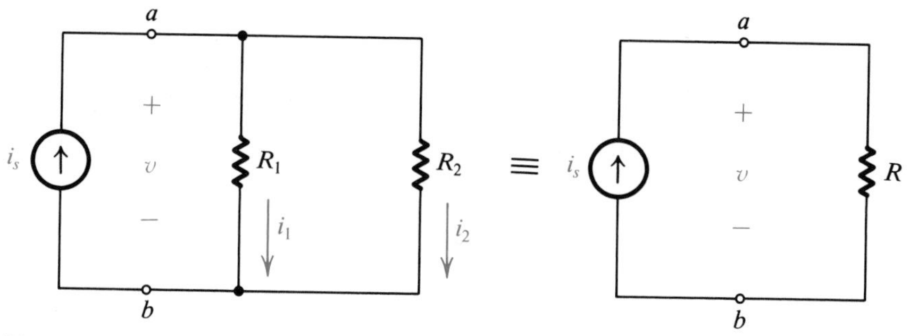

Figure 1.17 A parallel connection of two resistors and its single-resistor equivalent circuit.

The equivalent circuit must have $v/i = R$, and we conclude that

$$\boxed{R = \frac{R_1 R_2}{R_1 + R_2}}$$

Alternately:

$$R = \frac{1}{\dfrac{1}{R_1} + \dfrac{1}{R_2}} = \frac{1}{G_1 + G_2}$$

Note that $R < R_1$ and $R < R_2$. Adding parallel conducting paths reduces the resistance between a pair of nodes. Placing two identical resistors in parallel creates an equivalent resistor of half their size. Other parallel combinations can be used to obtain resistors whose values differ from standard production values.

Referring again to Fig. 1.17, we now want to develop a simple algorithm for determining how i_s divides to form i_1 and i_2. We would expect that i_2 will be larger than i_1 if R_1 is larger than R_2. The result is

$$i_1 = \frac{v}{R_1} = \frac{1}{R_1} \cdot \frac{R_1 R_2}{R_1 + R_2} i_s = \frac{R_2}{R_1 + R_2} i_s$$

Likewise

$$i_2 = \frac{R_1}{R_1 + R_2} i_s$$

In summary

$$\boxed{\begin{aligned} i_1 &= \frac{R_2}{R_1 + R_2} i_s \\ i_2 &= \frac{R_1}{R_1 + R_2} i_s \end{aligned}} \quad \text{and} \quad \boxed{\begin{aligned} \frac{i_1}{i_s} &= \frac{R_2}{R_1 + R_2} \\ \frac{i_2}{i_s} &= \frac{R_1}{R_1 + R_2} \end{aligned}}$$

The circuit in Fig. 1.17 is called a **current divider.** The division of current occurs as the ratio of the resistor in the other path to the sum of the two resistors. In terms of conductance

$$\frac{i_1}{i_s} = \frac{G_1}{G_1 + G_2}, \qquad \frac{i_2}{i_s} = \frac{G_2}{G_1 + G_2}$$

Here we see that current divides as the ratio of path conductance to the sum of the path conductances. Thus, a high conductance path will have a proportionately larger share of i_s.

Skill Exercise 1.4

Find i_1 and i_2 in Fig. SE 1.4

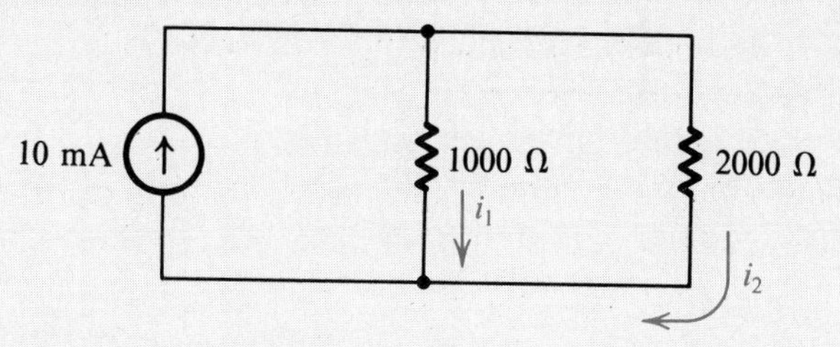

Figure SE1.4

Answer: $i_1 = 6.67$ mA, $i_2 = 3.33$ mA. Note: $i_1 + i_2 = 10$ mA.

The algorithm for combining parallel resistors can be extended to the case shown in Fig. 1.18, where *n* resistors are connected in parallel. We could pair-wise combine these parallel resistors to find their equivalent, R, but it is easier to use KCL and conductance to get

$$i_s = i_1 + i_2 + \cdots + i_n$$
$$i_s = G_1 v + G_2 v + \cdots + G_n v$$
$$i_s = (G_1 + G_2 + \cdots + G_n)v$$
$$\frac{i_s}{v} = G_1 + G_2 + \cdots + G_n$$

and

$$G = \frac{1}{R} = G_1 + \cdots + G_n$$

so

$$R = \frac{1}{G_1 + G_2 + \cdots + G_n}$$

$$R = \frac{1}{\frac{1}{R_1} + \frac{1}{R_2} + \cdots + \frac{1}{R_n}}$$

Note that $G_j \leq G$, $j = 1 \cdots n$. The current division algorithm follows directly:

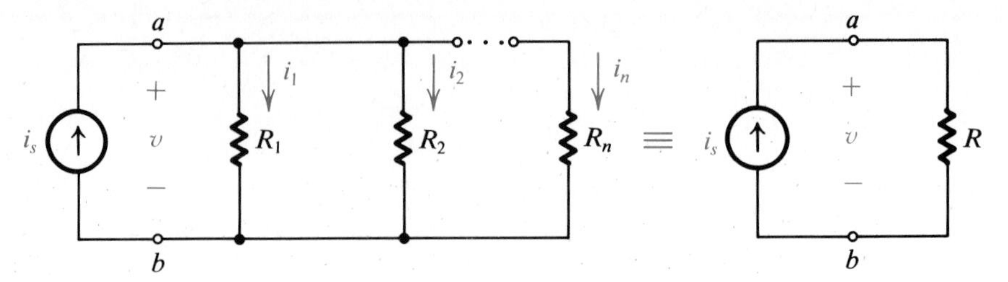

Figure 1.18 A parallel connection of n resistors, and its single-resistor equivalent circuit.

$$i_k = G_k v = G_k \frac{i_s}{G}$$

and

$$\frac{i_k}{i_s} = \frac{G_k}{G}$$

The current share in R_k is the ratio of G_k to G.

Example 1.7

Find the indicated currents and voltages in Fig. 1.19.

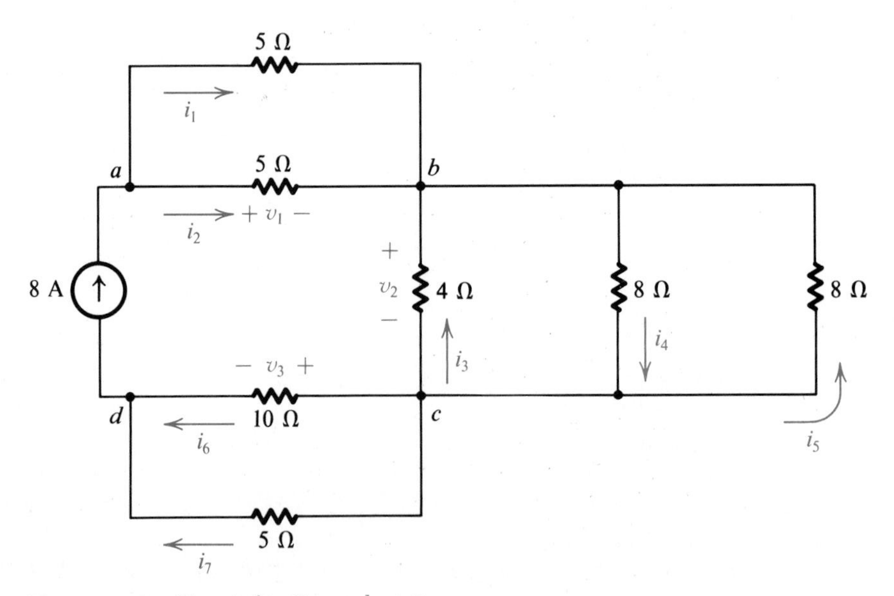

Figure 1.19 Circuit for Example 1.7.

Solution: The source current is known, so we can divide it at each node to find the path currents. At node a

$$i_1 = \frac{5(8)}{5+5} = 4\text{ A}, \qquad i_2 = \frac{5(8)}{5+5} = 4\text{ A}$$

The current entering node b is $i_1 + i_2 = 8$ A. We divide this current according to the ratio of conductances.

$$-i_3 = \frac{\frac{1}{4}(8)}{\frac{1}{4}+\frac{1}{8}+\frac{1}{8}} = 4\text{ A}$$

$$i_4 = \frac{\frac{1}{8}(8)}{\frac{1}{4}+\frac{1}{8}+\frac{1}{8}} = 2\text{ A}$$

$$i_5 = -i_4 = -2\text{ A}$$

Applying KCL enables us to calculate the current to be divided at node c

$$-i_3 + i_4 - i_5 = 4 + 2 + 2 = 8\text{ A}$$

$$i_6 = \frac{5(8)}{10+5} = 2.67\text{ A}$$

$$i_7 = \frac{10(8)}{10+5} = 5.33\text{ A}$$

The voltages across the elements are calculated using Ohm's Law

$$v_1 = 5i_2 = 20\text{ V}$$

$$v_2 = 8i_4 = 16\text{ V}$$

$$v_3 = 10i_6 = 26.7\text{ V}$$

1.8 KIRCHHOFF'S VOLTAGE LAW (KVL)

Kirchhoff's voltage law (KVL) is the second (and last) important physical law that we'll need for our study of circuits. Before stating KVL, let's consider a simple physical analogy. Imagine yourself a mountaineer descending after a climb to the summit and stopping to rest at prearranged locations (nodes). If you kept a log of the altitude during your descent you would find that the sum of the altitude drops from the summit to the first resting place, from there to the next resting place, and so on, would add up to the altitude climbed from base camp to the summit. Or, if you began a hike at point A and then wandered around the mountain and eventually returned to point A, you would be at the same altitude (potential) as you were when you began your journey. Each point on the mountain has a unique altitude relative to a conventional reference—sea level.

In the context of circuits, let's imagine ourselves walking from node to node with voltmeter in hand, taking the voltage measurement between each node and ground. Furthermore, let's agree to travel a closed path, one that begins and ends at the same node. As we travel between a given pair of nodes we calculate the voltage drop across the branch traveled, and subtract the second node voltage from the first node voltage. In Fig. 1.20 we measure a voltage drop if the + terminal of the device is entered, and the − terminal is exited.

Referring to Fig. 1.20(a) we see that in going from a to b we have a drop of $+v_1$. In going from b to a we have a drop of $-v_1$. We have a voltage rise if the − terminal is entered and the + terminal is exited. In going from b to a in Fig. 1.20(b) we have a rise of v_1 volts, and in going from a to b we have a rise of $-v_1$ volts. Algebraically, a negative drop is a rise, and vice versa.

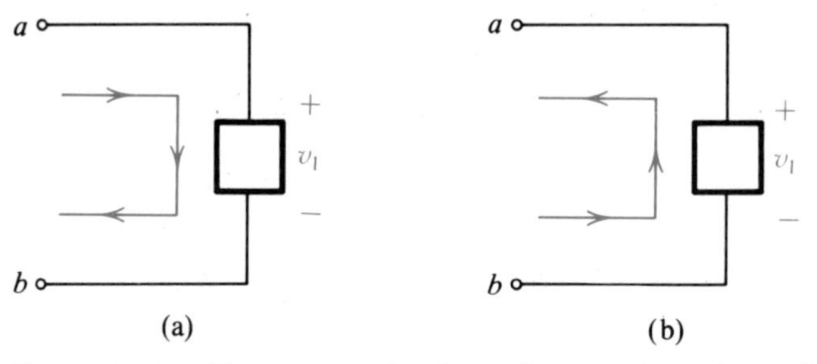

Figure 1.20a The measured voltage drop v_1 along the path from node a to node b.
Figure 1.20b The measured voltage drop -v_1 along the path from node b to node a.

Kirchhoff's voltage laws (KVL) express the physical fact that the node voltages in a circuit are uniquely defined. As with KCL, there are alternate forms of the law:

KVL-1

> **"At every instant of time the sum of the voltage drops equals the sum of the voltage rises around any closed path in a circuit."**

KVL-2

> **"At every instant of time the algebraic sum of the *voltage drops* around any closed path in a circuit equals zero."**

KVL-3

> **"At every instant of time the algebraic sum of the *voltage rises* around any closed path in a circuit equals zero."**

Example 1.8

Using KVL-1, with $v_{s1} = 36$ V and $v_{s2} = 18$ V, find i in Fig. 1.21(a)

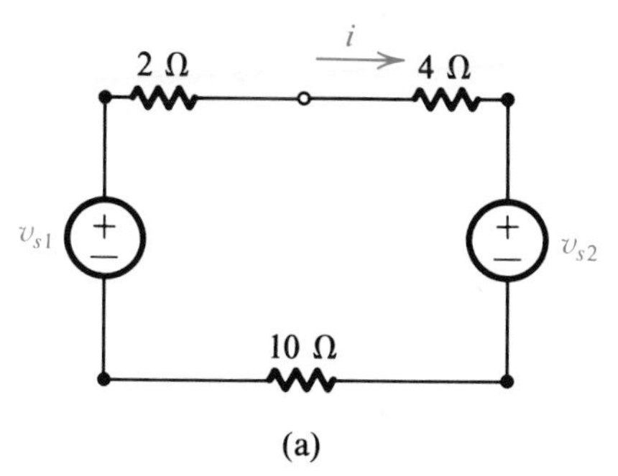

Figure 1.21(a) Circuit for Example 1.8.

Solution: We begin by assigning voltage labels and polarities to the branches of the circuit, as shown in Fig. 1.21(b)

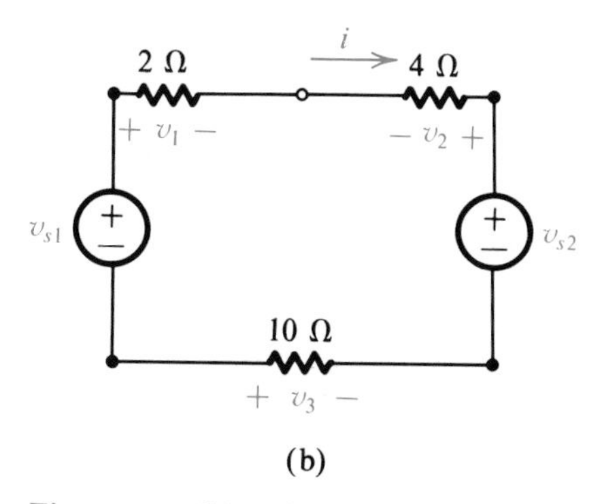

Figure 1.21(b) The circuit of Fig. 1.21(a) after branch voltage labels are defined.

(other labels and polarities are possible). Equating the sum of the drops to the sum of the rises produces

$$v_1 + v_{s2} = v_2 + v_3 + v_{s1}$$

Next, use Ohm's Law to express the resistor voltages in terms of the current:

$$v_1 = 2i$$
$$v_2 = 4(-i) = -4i$$
$$v_3 = 10(-i) = -10i$$

For the given source voltages we have from KVL, $2i + 18 = -4i - 10i + 36$ or $i = 9/8$ A. Now that i is known we can find the branch voltages

$$v_1 = 2\left(\frac{9}{8}\right) = \frac{9}{4}\text{ V}$$

$$v_2 = -4\left(\frac{9}{8}\right) = \frac{-9}{2}\text{ V}$$

$$v_3 = -10\left(\frac{9}{8}\right) = \frac{-45}{4}\text{ V}$$

As an exercise verify that the calculated voltages are consistent with KVL:

1.9 VOLTAGE DIVISION AND SERIES RESISTANCE

Two-terminal devices connected together with only one common node are said to be **in series.** Components connected in series must have the same current, but not necessarily the same voltage.

An algorithm for replacing two resistors in series by a single equivalent resistor can be obtained from Fig. 1.22. For the two circuits to be

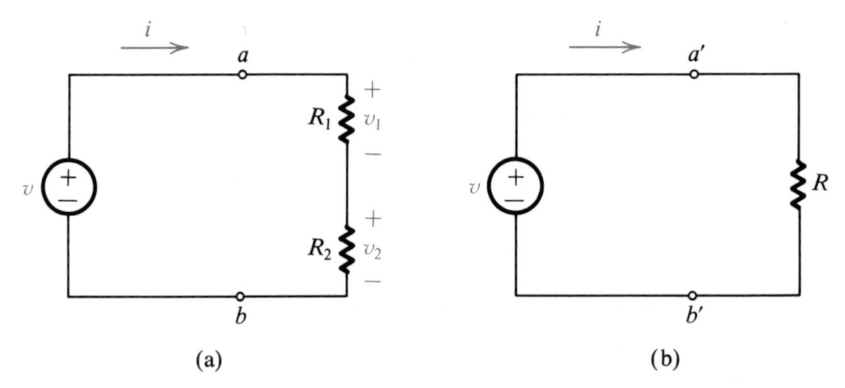

Figure 1.22 A series connection of two resistors, and its single-resistor equivalent circuit.

electrically equivalent they must have the same v-i relationship at their terminals. Thus, an observer could not distinguish measurements made at ab from those made at $a'b'$. Applying a voltage v to either terminal pair will cause the same current i, and vice versa. Here we show only a voltage source, but we could have chosen to measure the v caused by a current source. In either case the circuits will look the same to your instruments.

Using KVL and Ohm's Law we have

$$v = v_1 + v_2 = i(R_1 + R_2)$$

so

$$\frac{v}{i} = R_1 + R_2$$

and

$$\boxed{R = R_1 + R_2}$$

Note that two 10 Ω resistors in series act like a 20 Ω resistor, but two 10 Ω resistors in parallel act like a 5 Ω resistor. In terms of conductance:

$$G = 1/R = 1/(R_1 + R_2)$$

Multiplying numerator and denominator by $1/(R_1R_2)$ gives

$$G = \frac{1/R_1R_2}{R_1/(R_1R_2) + R_2/(R_1R_2)} = \frac{G_1G_2}{G_2 + G_1}$$

and

$$\boxed{G = \frac{G_1G_2}{G_1 + G_2}}$$

The rule for calculating the equivalent conductance for two resistors in series has the same form as that for combining two resistors in parallel. In the case of two parallel resistors we saw that their combined resistance was less than either of their individual resistances. For two (non-zero) resistors in series we have $R = R_1 + R_2$ and $R \geq R_1$, $R \geq R_2$.

Skill Exercise 1.5

Find the equivalent resistance seen by the source in Fig. SE1.5 and find i, v_1 and v_2.

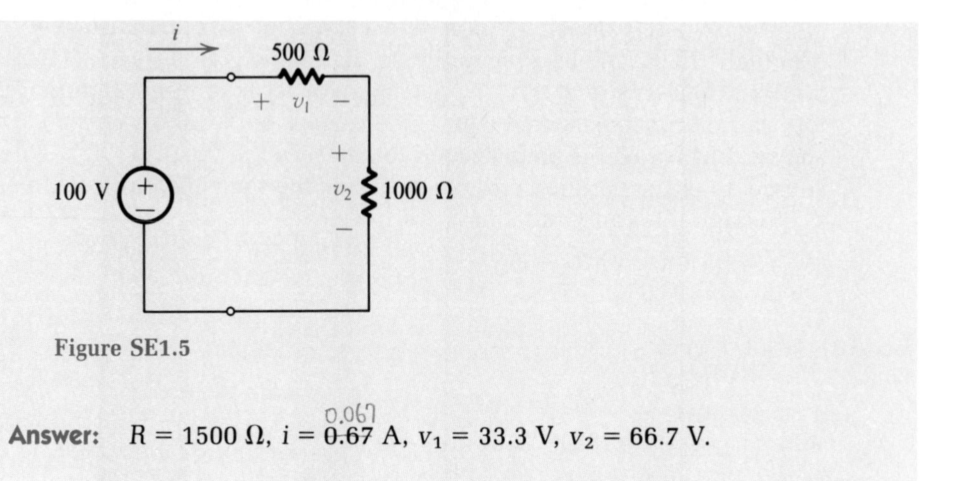

Figure SE1.5

Answer: $R = 1500\ \Omega$, $i = 0.67$ A, $v_1 = 33.3$ V, $v_2 = 66.7$ V.

Next, we derive a **voltage division algorithm** for series resistors. In Fig. 1.23 $v_1 = iR_1$ and $v_2 = iR_2$ so

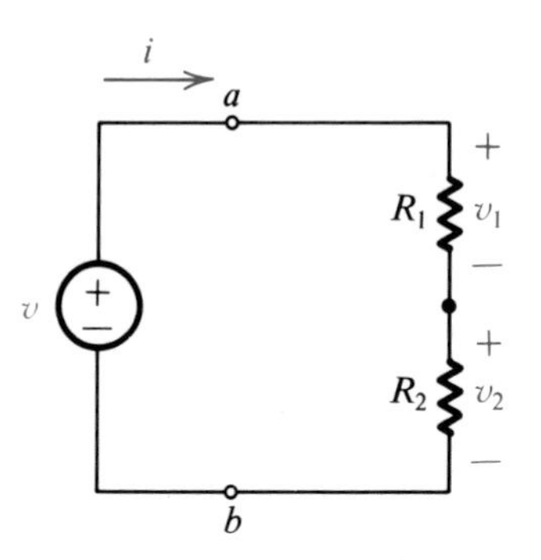

Figure 1.23 Voltage division across resistors.

$$v_1 = \frac{R_1}{R_1 + R_2} v \qquad v_2 = \frac{R_2}{R_1 + R_2} v$$

and

$$\frac{v_1}{v} = \frac{R_1}{R_1 + R_2} \qquad \frac{v_2}{v} = \frac{R_2}{R_1 + R_2}$$

The *division of voltage* is determined by the ratio of the resistor to the sum of the two resistors. The larger resistor will have the greater portion of v. This algorithm provides an alternate approach to SE1.5.

Example 1.9

Use the voltage division algorithm to find v_1, v_2 and i in Fig. SE1.5

Solution: First we find v_1 and v_2:

$$v_1 = \frac{500}{500 + 1000}(100) = 33.33 \text{ V}.$$

$$v_2 = \frac{1000}{500 + 1000}(100) = 66.67 \text{ V}.$$

As a check, note that $v_1 + v_2 = 100.00$ V. Now, either v_1 or v_2 can be used to calculate i from Ohm's Law. The voltage division algorithm produces the same answers as direct application of KVL, but it simplifies the procedure and appeals to our intuition. You should become familiar with both approaches.

Voltage division also applies to n resistors connected in series, as shown in Fig. 1.24. Applying KVL and Ohm's Law, we get

$$v = v_1 + v_2 + \cdots + v_n$$
$$v = iR_1 + iR_2 + \cdots + iR_n$$
$$v = i(R_1 + R_2 + \cdots + R_n)$$

and

$$R = R_1 + R_2 + \cdots + R_n$$

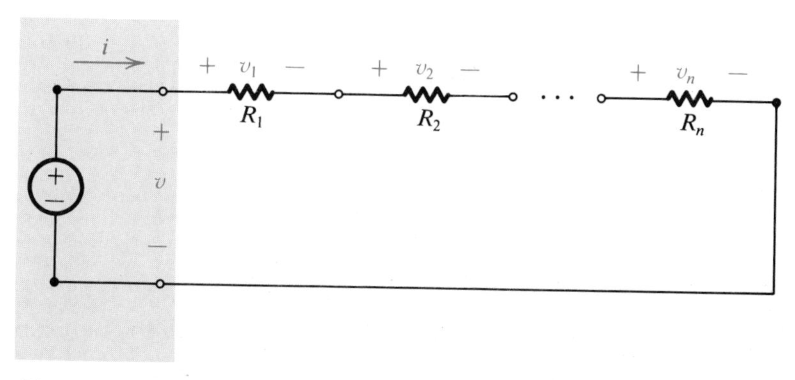

Figure 1.24 Series connection of n resistors.

Note too that $R > R_j$ for each index j. The equivalent series resistance exceeds each of the resistances. The voltage division rule extends to this case too. Consider the voltage across R_k:

$$v_k = i_k R_k$$

$$v_k = \frac{R_k v}{R_1 + R_2 + \cdots + R_n}$$

and

$$\boxed{\frac{v_k}{v} = \frac{R_k}{R_1 + R_2 + \cdots + R_n}}$$

1.10 SERIES/PARALLEL CIRCUITS

Resistive circuits that combine the topological features of a series circuit with those of a parallel circuit can be reduced to an equivalent resistor. The circuit shown in Fig. 1.25 contains sub-circuits that are either parallel or series. The 4 Ω and 2 Ω resistors form a parallel sub-circuit whose equivalent resistance is 1.33 Ω. This equivalent resistor and the 5 Ω resistor are in series, and together they are equivalent to a 6.33 Ω resistor. So, the circuit has a parallel connection of 8 Ω, 3 Ω and 6.33 Ω resistors (equivalent to a single 1.62 Ω resistor) at cd, and the whole circuit reduces to a single 7.62 Ω resistor.

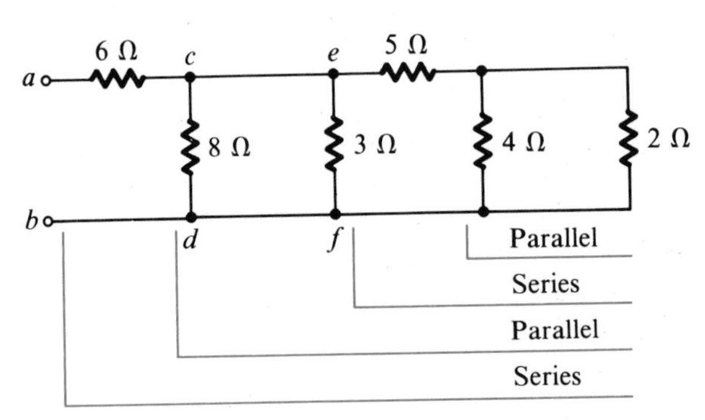

Figure 1.25 Reduction of a series/parallel circuit.

Now suppose we want to find the currents shown in Fig. 1.26 when a 10 V source is applied. One approach is to begin with the reduced circuit and work backwards through it while applying the current division algorithm: $i_1 = 10/7.62 = 1.31$ A. This current enters node c and divides

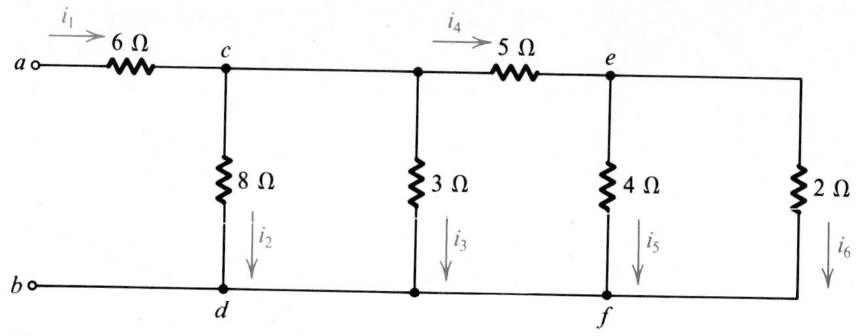

Figure 1.26

among the three parallel paths. The current division rule (with conductances) gives

$$i_2 = \frac{\frac{1}{8}}{\frac{1}{8} + \frac{1}{3} + \frac{1}{6.33}} i_1 = .27 \text{ A}$$

$$i_3 = \frac{\frac{1}{3}}{\frac{1}{8} + \frac{1}{3} + \frac{1}{6.33}} i_1 = .70 \text{ A}$$

$$i_4 = \frac{\frac{1}{6.33}}{\frac{1}{8} + \frac{1}{4} + \frac{1}{6.33}} i_1 = .34 \text{ A}$$

Then i_5 and i_6 can be obtained by dividing i_4:

$$i_5 = \frac{\frac{1}{4}}{\frac{1}{4} + \frac{1}{2}} i_4 = .11 \text{ A}, \qquad i_6 = \frac{\frac{1}{2}}{\frac{1}{4} + \frac{1}{2}} i_4 = .23 \text{ A}$$

1.11 SOURCE COMBINATIONS

Voltage sources connected in series can be added to form a single equivalent source. This result is somewhat intuitive, but follows from KVL. In Fig. 1.27 v_{s1} and v_{s2} are equivalent to a single source whose value is $v_{s1} + v_{s2}$. Series voltage sources are combined according to their polarity. Thus, v_{s3} and v_{s4} are equivalent to a source of value $v_{s3} - v_{s4}$.

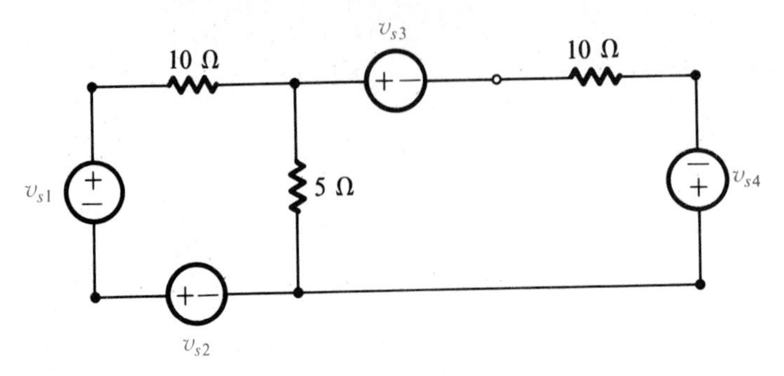

Figure 1.27

When combining voltage sources we must observe the rule that different ideal voltage sources may never be placed in a parallel combination because to do so will contradict KVL. Thus, in Fig. 1.28 KVL will be satisfied only if $v_{s1} = v_{s2}$. An attempt to violate KVL by simultaneously forcing a node to two different potentials usually results in a hazardous display of sparks and molten copper.

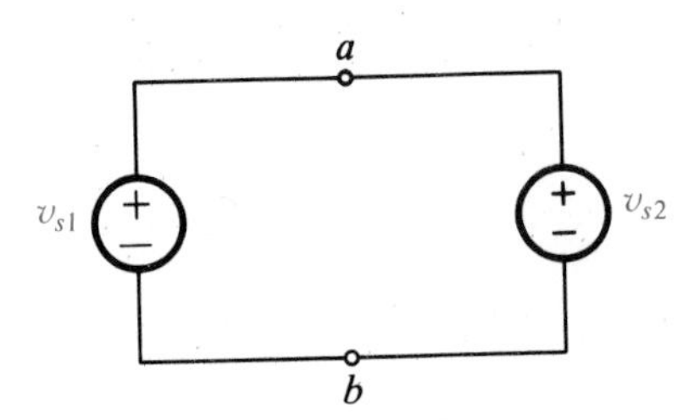

Figure 1.28 A hazardous connection of ideal voltage sources.

Earlier we mentioned that limitations would be imposed on the use of ideal voltage source models. For physical sources, their voltage depends on the current they supply. Such sources can be modeled by the series connection of an ideal voltage source and resistor shown in Figure 1.29.

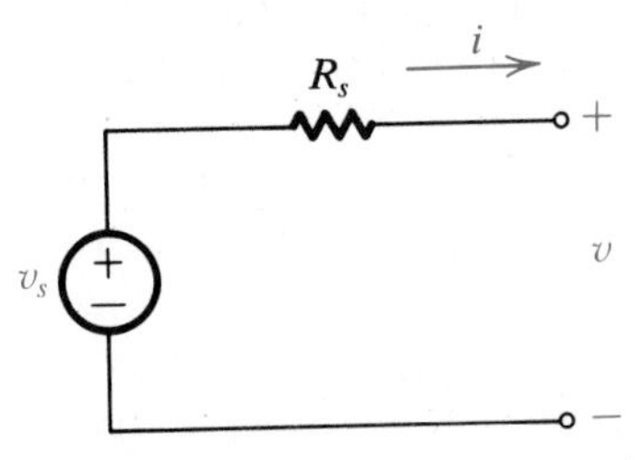

Figure 1.29 A model for a realistic voltage source.

To find a circuit's terminal v-i characteristic, connect a hypothetical circuit to the source and determine the relationship between v and i at the connecting nodes. Applying KCL to the circuit gives

$$v = v_s - iR_s$$

Graphing this characteristic in Fig. 1.30 reveals a fundamental difference between an ideal voltage source, whose terminal voltage would be v_s regardless of i, and the physical source, whose terminal voltage would decrease linearly with current. Ideally, the internal resistance of the source should be as small as possible to decrease the internal voltage drop of the source, and thereby reduce the magnitude of the slope of the v-i characteristic.

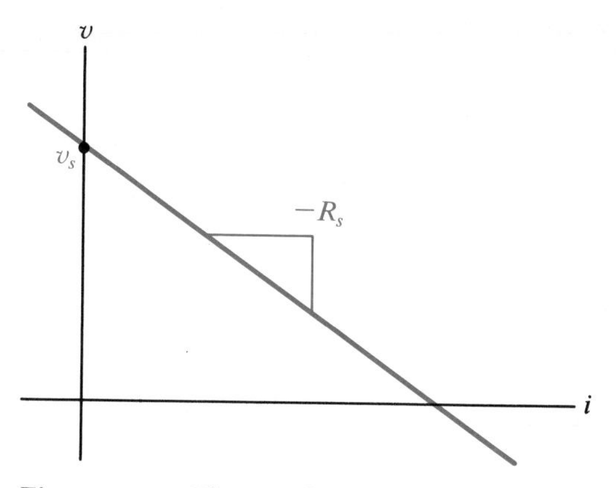

Figure 1.30 The v-i characteristic of a physical voltage source.

The effect of the internal source resistance is to reduce the voltage seen by the attached circuit from the "no-load" value of v_s (when $i = 0$) to a lower voltage, assuming that i has the polarity shown. Thus, a defective automobile battery will have a measured terminal voltage that is close to 12 V under no load, but cannot maintain that voltage under the current load condition imposed by a starter motor. Also, as a battery ages its internal resistance increases until the battery is unable to deliver current to an external circuit. The electronic power supplies that are commonly available as laboratory instruments will operate as ideal voltage sources over a specified range of current loads.

Realistic sources may be combined in series by adding their resistors and by combining their ideal sources,

$$R_s = R_1 + R_2 + \cdots + R_n$$
$$v_s = v_{s1} + v_{s2} + \cdots v_{sn}$$

Example 1.10

Find the equivalent circuit for the sources in Fig. 1.31

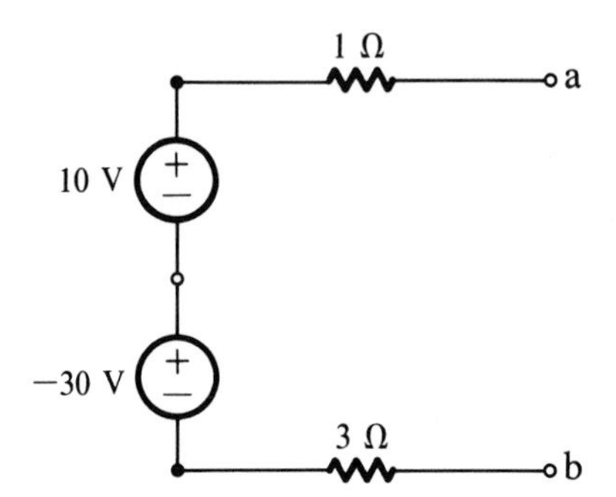

Figure 1.31 Circuit for Example 1.10.

Solution: The series resistors are equivalent to a 4 Ω resistor, and the series sources are equivalent to a 20 V source whose polarity has $v_{ab} = -20$ V.

The source combination rule for current sources is the dual of the rule for voltage sources, i.e., we combine parallel current sources, and we prohibit series configurations of current sources.

The equivalent circuit for a physical current source shown in Fig. 1.32 consists of an ideal current source connected in parallel with a resistor. This corresponds to an internal path for current, and consequently reduces the current available to an attached load. Applying KCL to the circuit configuration gives $i = i_s - v/R_s$ as the i-v model for the circuit. The graph of the physical current source i-v characteristic shown in Fig. 1.33 has a slope that is negative, with a magnitude equal to the internal leakage

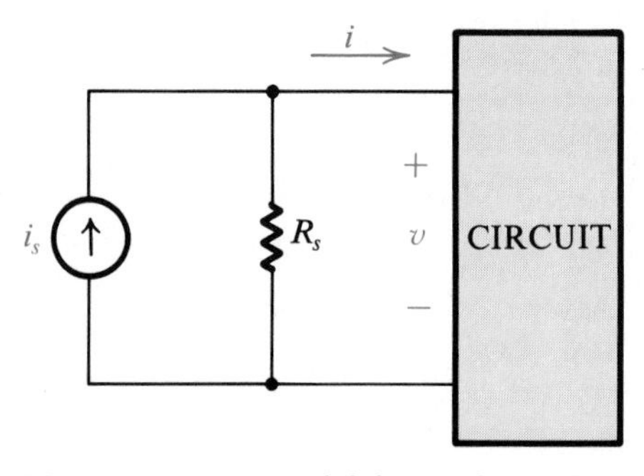

Figure 1.32 A model for a physical current source.

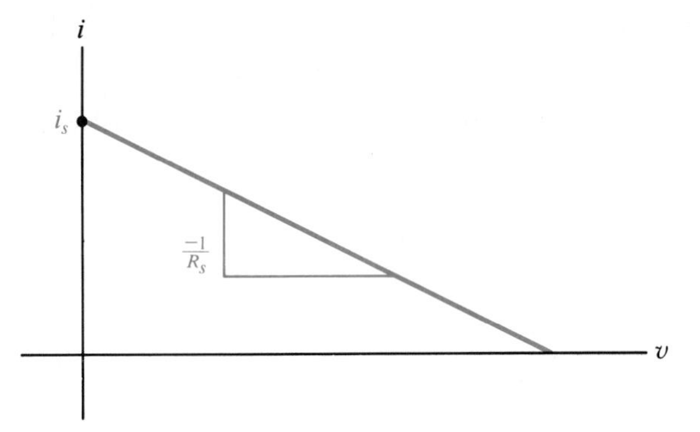

Figure 1.33 The v-i characteristic for a physical current source.

conductance of the source. The current available to the load decreases from the "no-load" value of i_s as the load voltage increases from zero because proportionately more of the fixed current from the ideal source circulates internally through R_s. Finally, n realistic sources in parallel combine according to

$$i_s = i_{s1} + i_{s2} + \cdots + i_{sn}$$

$$R_s = \frac{1}{\frac{1}{R_1} + \frac{1}{R_2} + \cdots + \frac{1}{R_n}}$$

Example 1.11

Find the equivalent current source to replace the sources shown in Fig. 1.34.

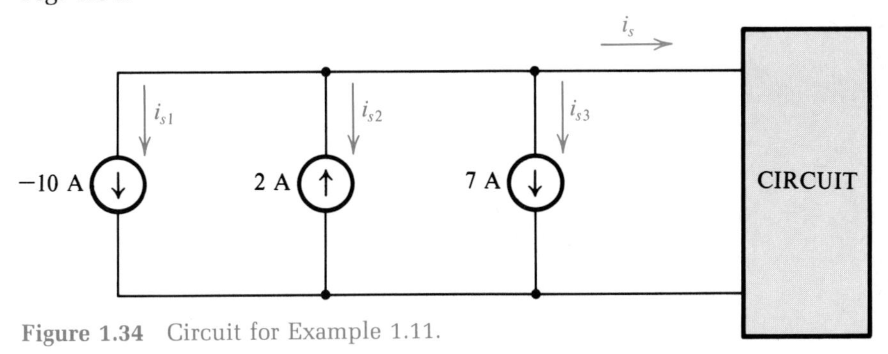

Figure 1.34 Circuit for Example 1.11.

Solution: Using the source combination rule gives

$$i_s = -i_{s1} - i_{s2} - i_{s3} = -(-10) + 2 - (7) = 5 \text{ A}.$$

1.12 CONTROLLED SOURCES

Models for the electrical behavior of some circuit elements require the introduction of another kind of voltage and current source—one that can be controlled by a voltage or current in the circuit. Figure 1.35 shows controlled-current sources and controlled-voltage sources. These establish a voltage between a pair of nodes or a current in a branch, but the value of the established voltage (or current) is controlled either by a current in some branch of the circuit, or by a voltage between two circuit nodes. Controlled sources are also called dependent sources because the value of the source depends on another voltage or current. We will assume that the control mechanism is linear, i.e. the value of the controlled variable is a linear function of the controlling variable.

When a circuit contains a controlled source it is usually difficult to use series/parallel combinations of resistors to solve for the currents and voltages. For example, the circuit in Fig. 1.36 contains a current-controlled current source and has a series/parallel structure, but we cannot combine the resistors in the parallel path since the parallel combination is only valid if the voltage across each resistor is the same. The controlled source forces the current in the path of the 3 Ω resistor to be twice

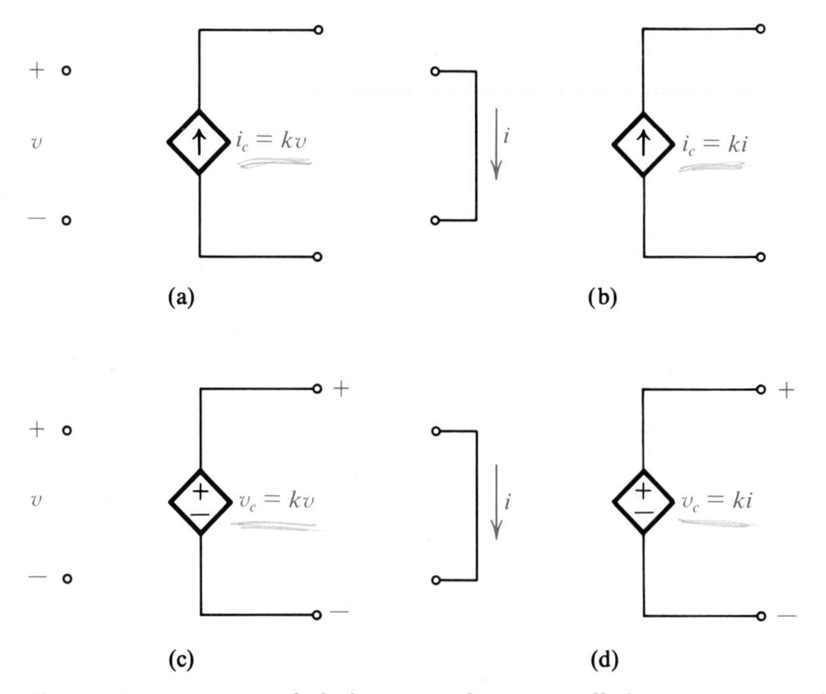

Figure 1.35 Circuit symbols for (a) a voltage-controlled current source, (b) a current-controlled current source, (c) a voltage-controlled voltage source, and (d) a current-controlled voltage source.

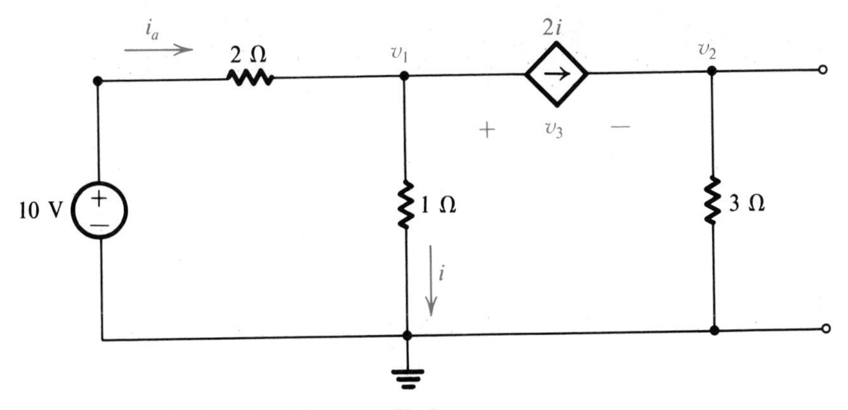

Figure 1.36 A circuit with controlled sources.

the current in the 1 Ω resistor, making it impossible for them to have the same voltage. Now note that the use of KCL at the v_1 node gives

$$i_a = i + 2i = 3i$$

so KVL for the path containing the source and the 1 Ω resistor can be used to solve for i

$$10 = 3i(2) + i(1) = 7i$$
$$i = 10/7 \text{ A}$$

Having found i, we can find v_1, v_2 and v_3 as

$$v_1 = i(1) = \frac{10}{7} \text{ V}$$

$$v_2 = 2i(3) = 6i = \frac{60}{7} \text{ V}$$

$$v_3 = v_1 - v_2 = \frac{10}{7} - \frac{60}{7} = -\frac{50}{7} \text{ V}$$

Several examples of circuits with controlled sources will be discussed in the next chapter.

1.13 ENERGY AND POWER

Because it is costly, the energy required to operate a circuit is an important design factor. The rate at which energy is dissipated in a circuit is defined to be the **power dissipated.** Power is especially important because the energy dissipated is in the form of heat. If the rate at which heat is dissipated is sufficiently high, the temperature of the physical components may rise enough either to destroy them or to alter their operating characteristics. If necessary, measures must be taken to remove the heat, incurring an additional cost.

The power dissipated in the two-terminal device shown in Fig. 1.37 can be calculated by determining the rate at which energy is expended as charge is moved through the electric field applied across it. In terms of the applied voltage, the energy/charge relationship is $v = dw/dq$, and the current through the device determines the rate at which charge moves through the electric field ($i = dq/dt$). Taking the product of v and i gives

$$v(t)i(t) = \frac{dw}{dq}\frac{dq}{dt} = \frac{dw}{dt}$$

Therefore, the power dissipated is

$$p(t) = v(t)i(t)$$

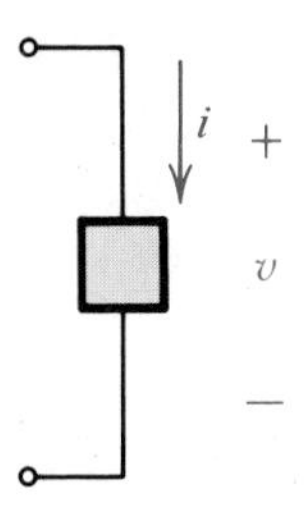

Figure 1.37 Measurement polarity for calculating the power dissipation in a two-terminal circuit element.

For the above polarities, if $p(t) > 0$, energy is being absorbed by the device. On the other hand, if $p(t) < 0$, energy is being supplied by the device. Physically, energy is dissipated when the direction of positive current at the device terminals has the same orientation as a voltage drop across the device, i.e. positive current enters the + terminal.

As a special case of this result, the power dissipated in a resistor can be obtained by using Ohm's law to express the voltage:

$$p(t) = i(t)Ri(t)$$

or

$$\boxed{p(t) = i^2(t)R}$$

Alternately,

$$\boxed{p(t) = \frac{v^2(t)}{R}}$$

Power is expressed in units of Joule/sec, or watts, (W).

SUMMARY

This chapter presented fundamental concepts of current, voltage, and power for electrical circuits, and developed Kirchhoff's voltage and current laws for describing circuit behavior. Voltage and current division algorithms were presented for series and parallel resistive circuits. Independent and dependent sources and source combinations were described, and Ohm's law was developed.

Problems - Chapter 1

1.1 Obtain an equivalent resistor.

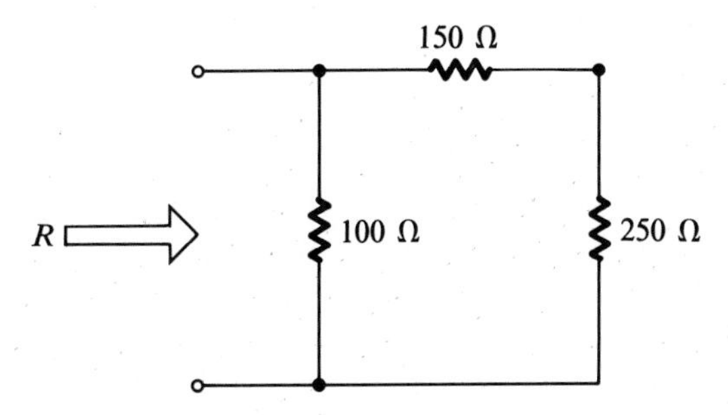

Figure P1.1

1.2 Find i_1 and i_2.

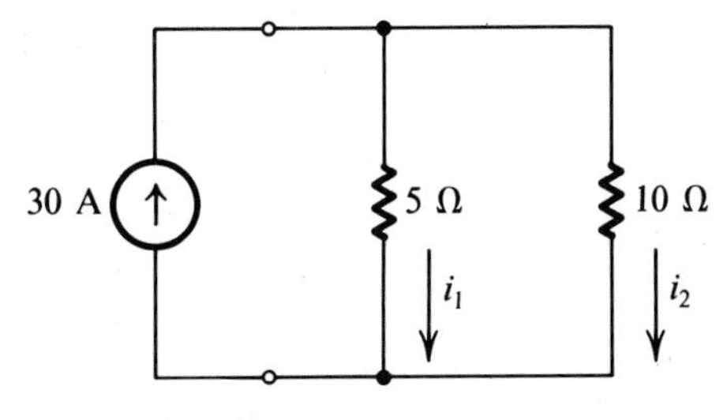

Figure P1.2

1.3 Find v_1 and v_2.

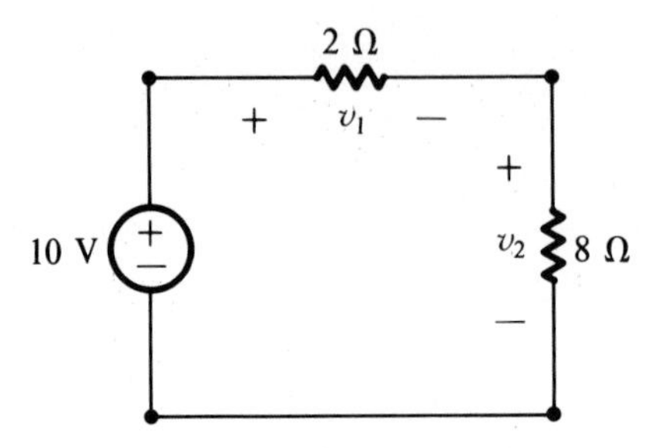

Figure P1.3

1.4 Find v_{12}, v_{23}, and v_{13} if $v_1 = 10$ V, $v_2 = 2$ V and $v_3 = -5$ V.

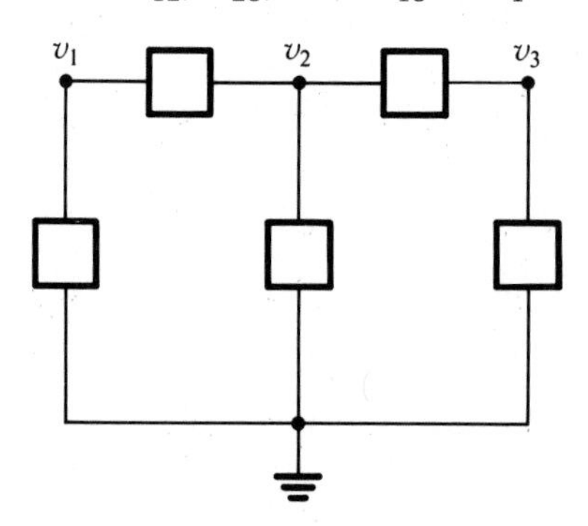

Figure P1.4

1.5 Find i_2 if $i_1 = 3$ A, $i_3 = 7$ A, $i_4 = -5$ A, $i_5 = 1$ A, $i_6 = 4$ A, $i_7 =$ 1 A, and $i_8 = -2$ A.

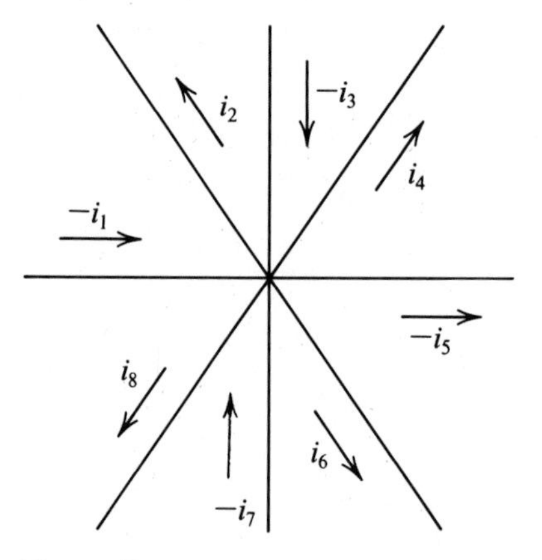

Figure P1.5

1.6 Find i_1 and i_2.

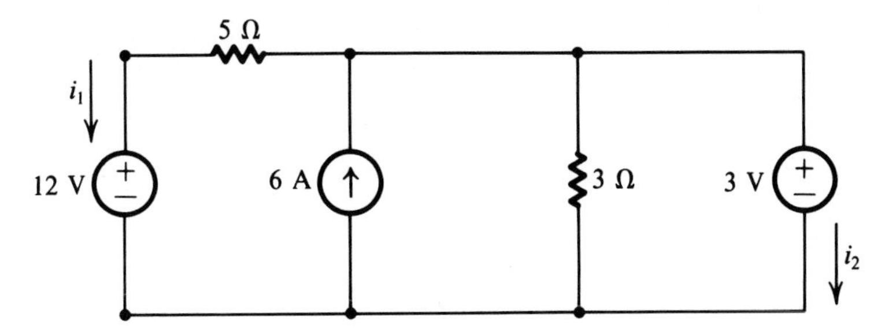

Figure P1.6

1.7 Find v_1, v_2, v_3, v_a, v_b, and v_c if $v_4 = 10$ V, $i_2 = 6$ A, and $i_3 = 5$ A.

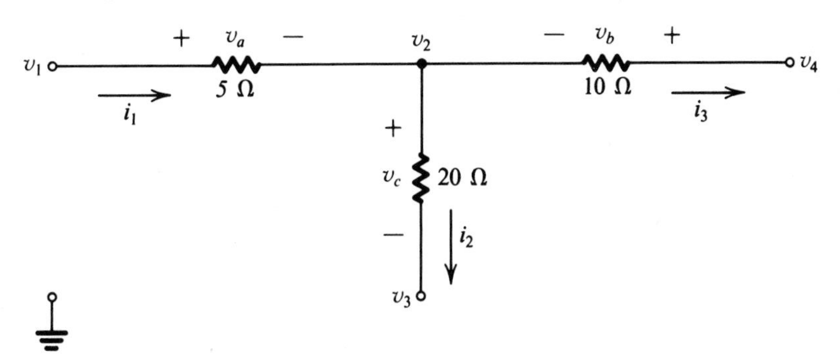

Figure P1.7

1.8 Write KVL for the paths shown in Fig. P1.8.

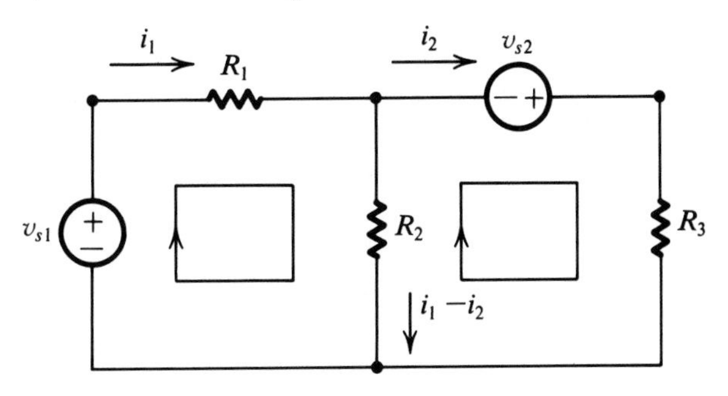

Figure P1.8

1.9 Use KVL and Ohm's law to find i, and v_{ab}.

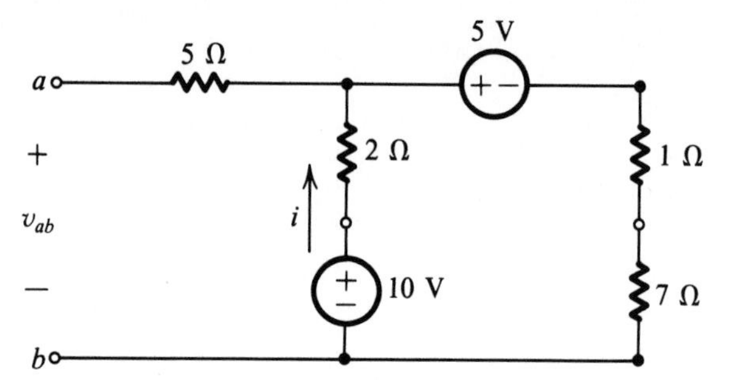

Figure P1.9

1.10 Calculate i.

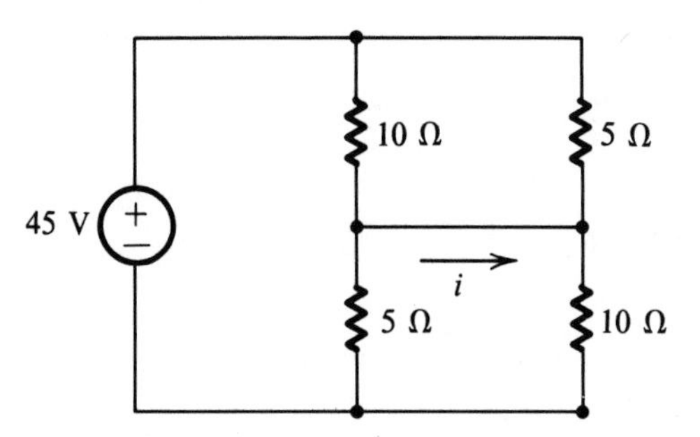

Figure P1.10

1.11 Find i and the power supplied by each source.

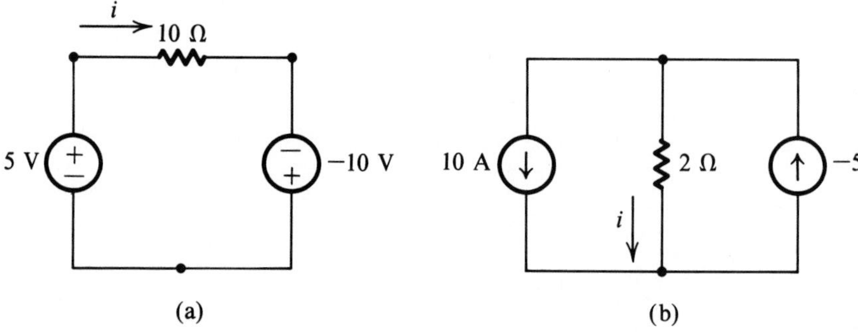

Figure P1.11

1.12 Calculate the power dissipated in R_1, and calculate the power supplied by v_{s1} and v_{s2} when $v_{s1} = 10$ V, $v_{s2} = 3$ V, $i_1 = -5$ A, $i_2 = -4$ A, $i_3 = 5$ A, $v_1 = -10$ V.

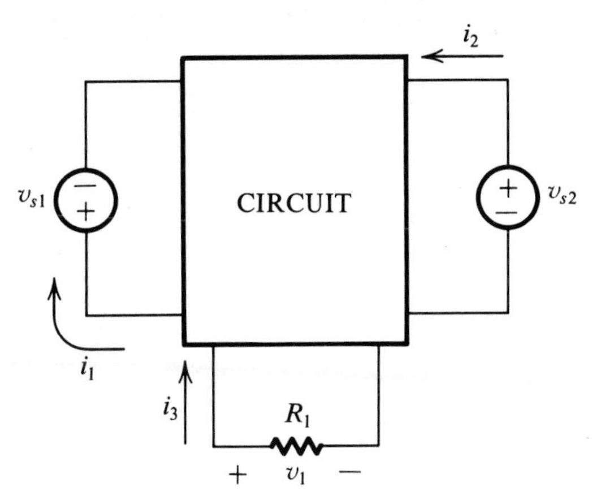

Figure P1.12

1.13 Find i_1, i_2, the power supplied by each source, and the power dissipated in the 5 Ω resistor.

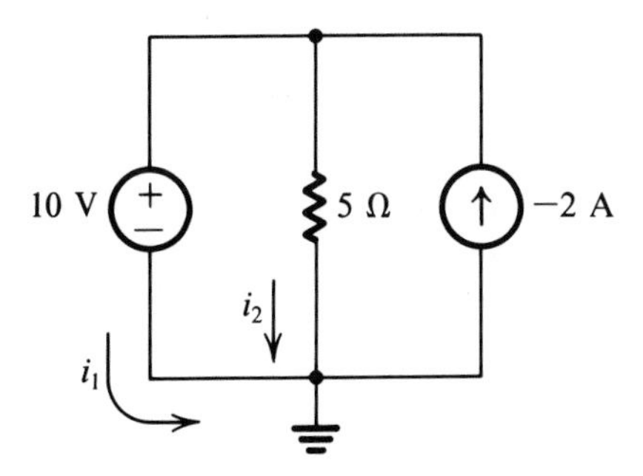

Figure P1.13

1.14 Find i_1, i_2, v_1 and the power supplied by each source.

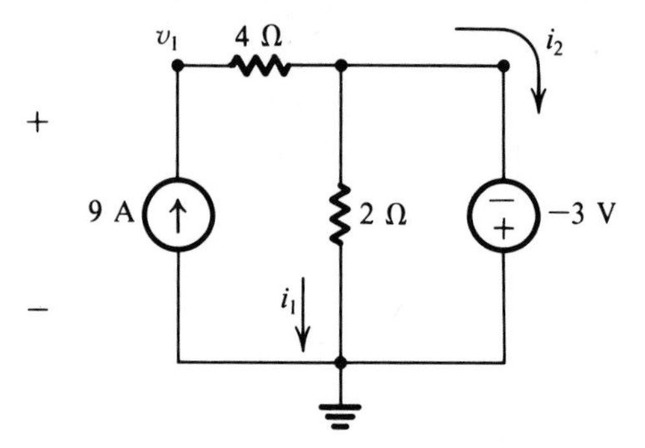

Figure P1.14

1.15 Find i_1, i_2, i_3, v_1 and v_2.

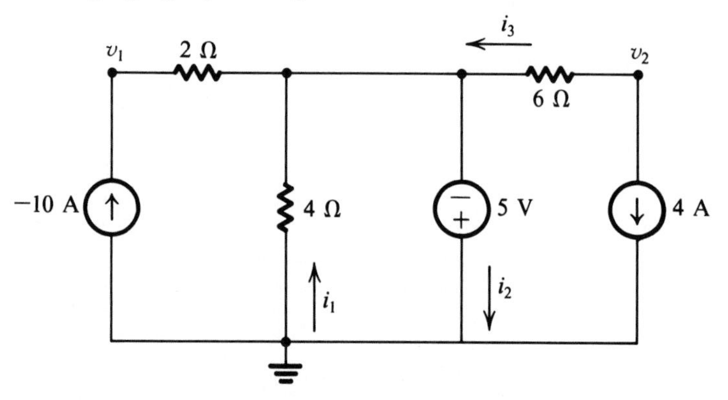

Figure P1.15

1.16 Find $i_1, \cdots, i_5$, v_1 and v_2.

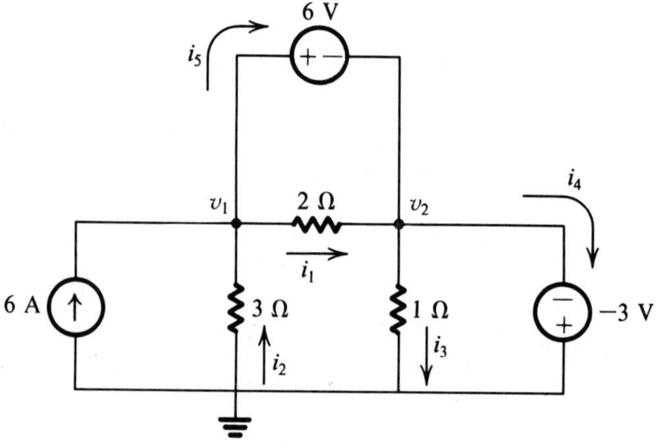

Figure P1.16

1.17 Find i.

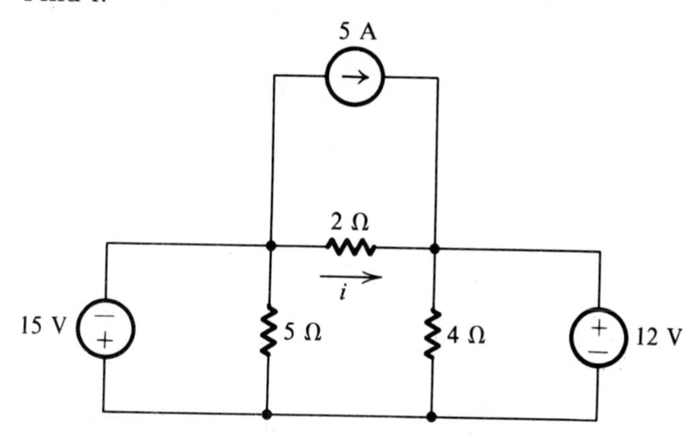

Figure P1.17

1.18 Find $i_1, \cdots, i_6$.

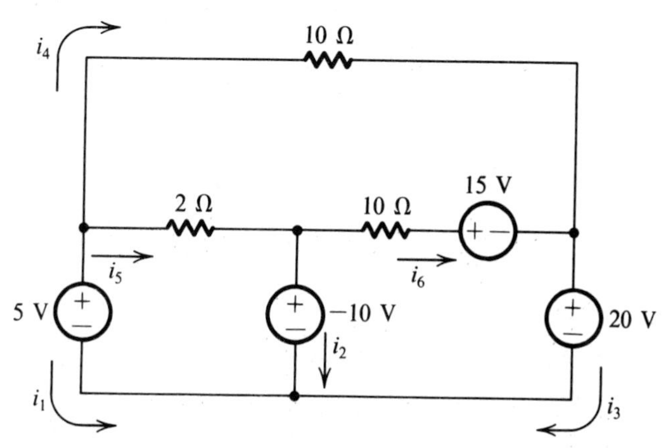

Figure P1.18

1.19 For the circuit in Fig. P1.19:

a. Find expressions for v and i.

b. Sketch a graph of the power dissipated in R_L as a function of R_L, for $0 \leq R_L < \infty$.

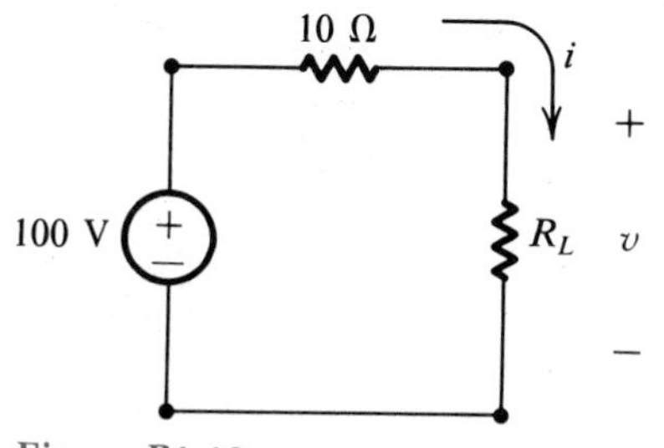

Figure P1.19

1.20 Find R.

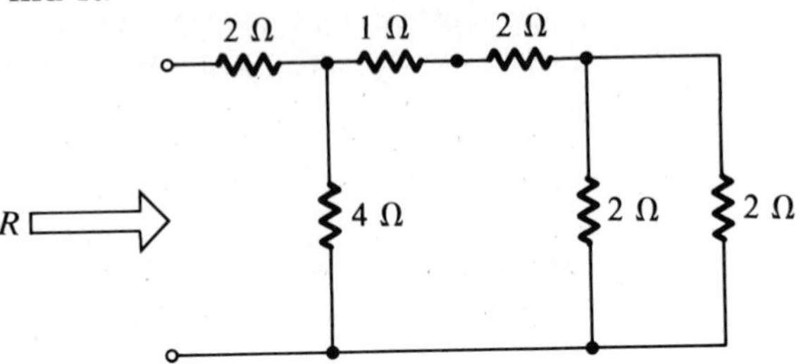

Figure P1.20

1.21 Find $i_1, \cdots, i_5$, v_1 and v_2.

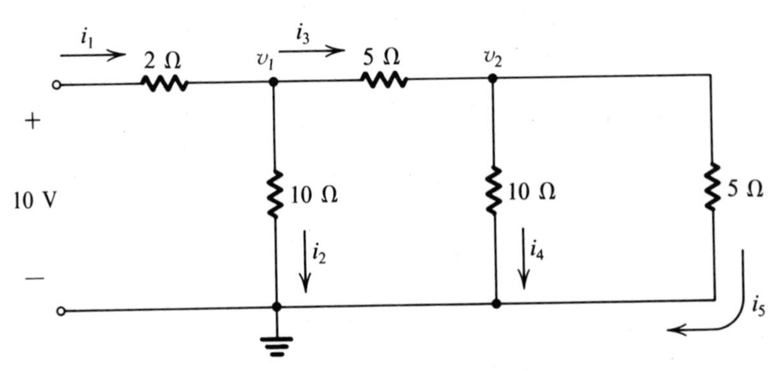

Figure P1.21

1.22 If $v = 100$ V, find i, i_1, i_2, i_3 and v_1.

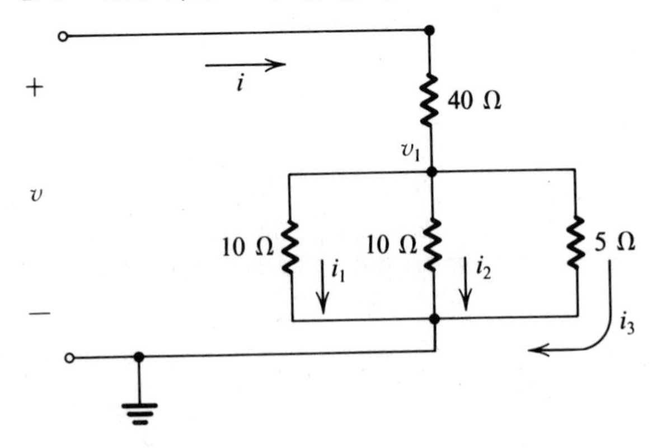

Figure P1.22

1.23 For the circuit in Figure P1.23:

a. Find v, i.

b. Find the power dissipated in the 10 Ω resistor.

c. Find the power supplied by each source.

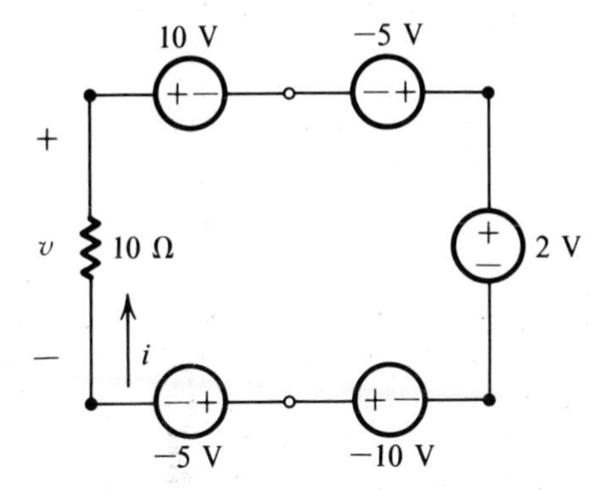

Figure P1.23

1.24 Find v.

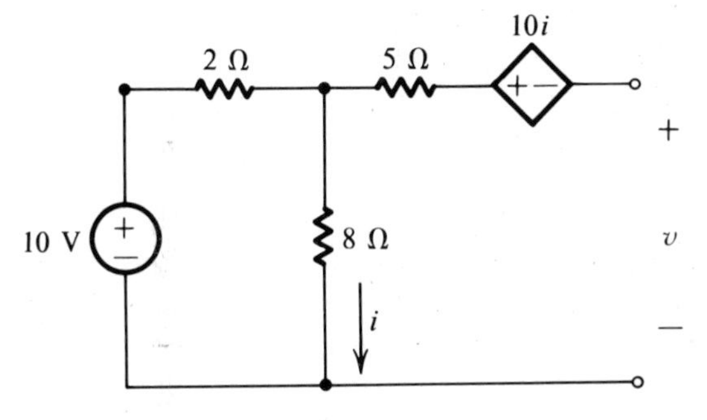

Figure P1.24

1.25 Find v and the power dissipated by the controlled source.

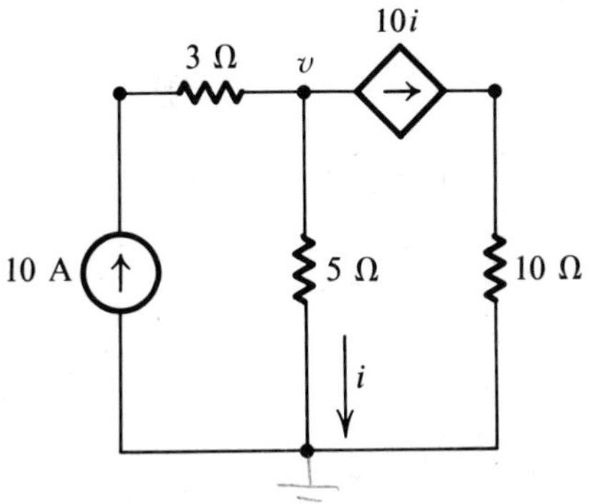

Figure P1.25

1.26 Find i_1 and i_2.

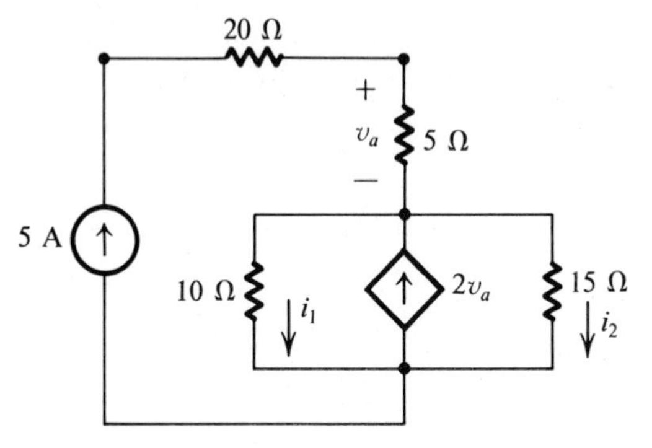

Figure P1.26

1.27 Find i, v_1 and v_2.

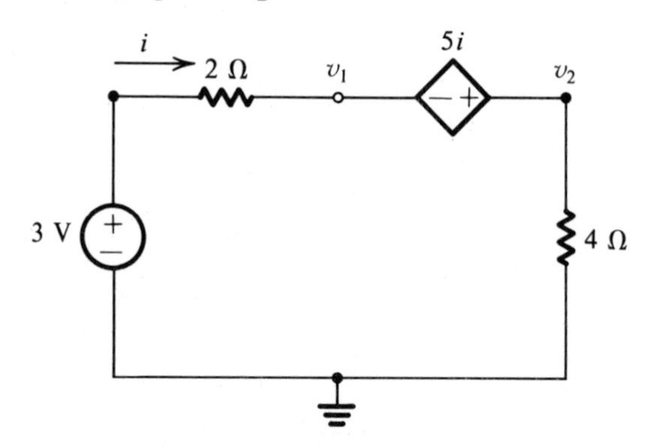

Figure P1.27

1.28 Find i, the current delivered to the load resistor.

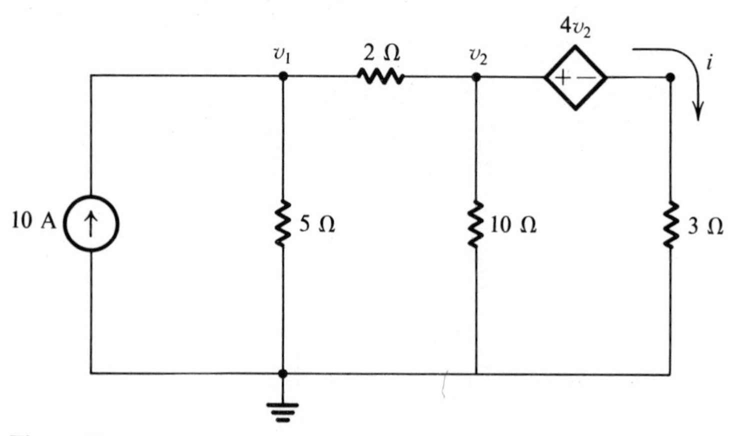

Figure P1.28

1.29 Find i_1, and i_2. in terms of k

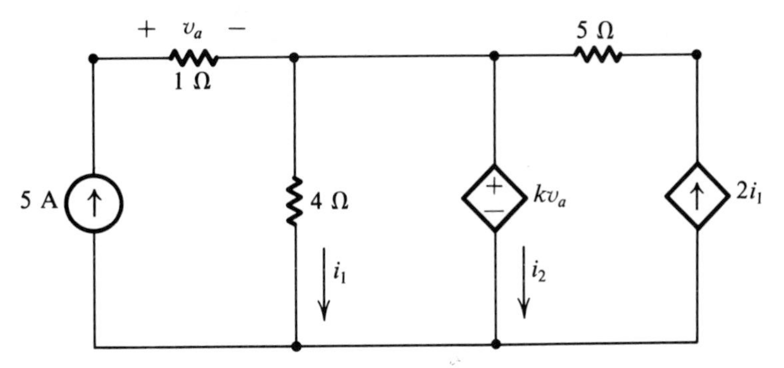

Figure P1.29

1.30 A simplified model of a field-effect transistor amplifier is shown in the shaded part of Fig. P1.30. Find the input/output voltage gain, v_o/v_{in}.

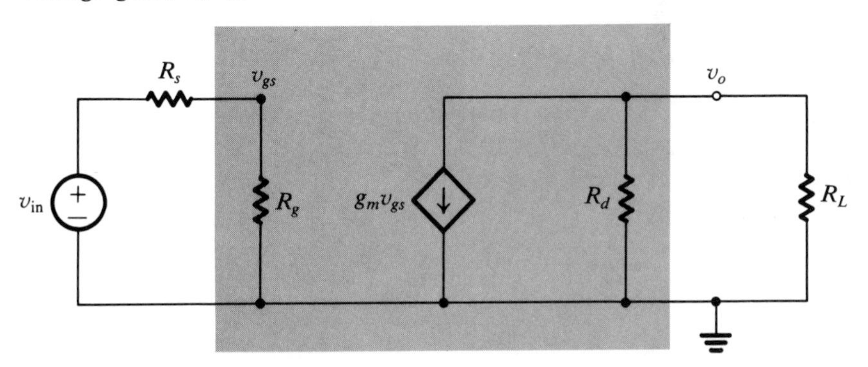

Figure P1.30

1.31 Find i_1 and i_2.

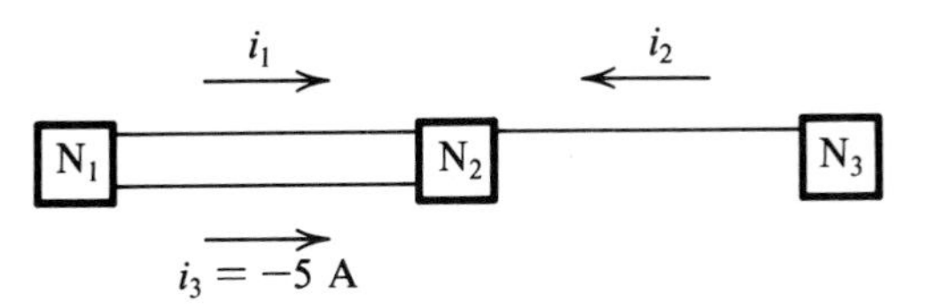

Figure P1.31

1.32 Find i_a and i_b.

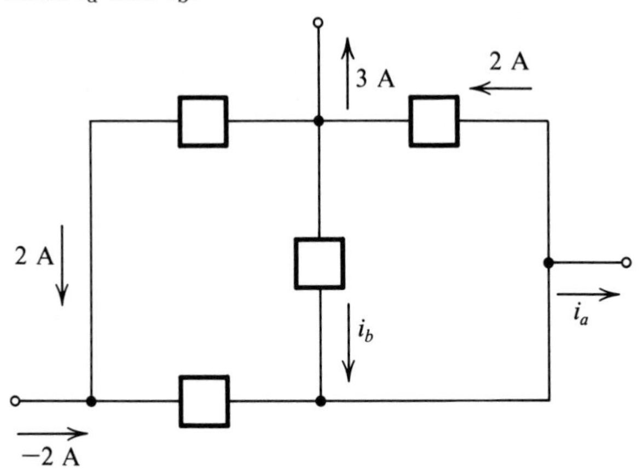

Figure P1.32

1.33 Find v_1 and v_2.

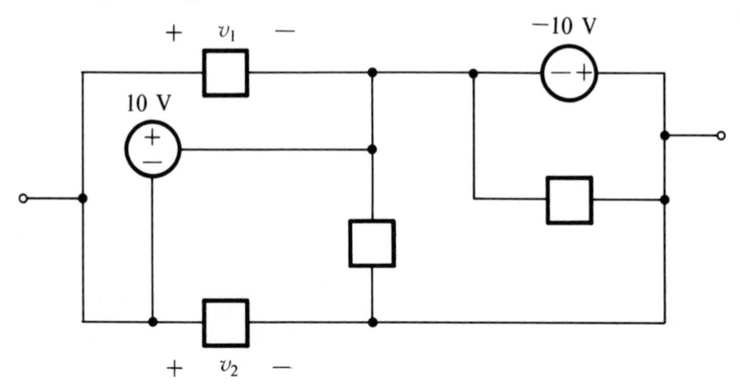

Figure P1.33

1.34 Calculate the power delivered by each source.

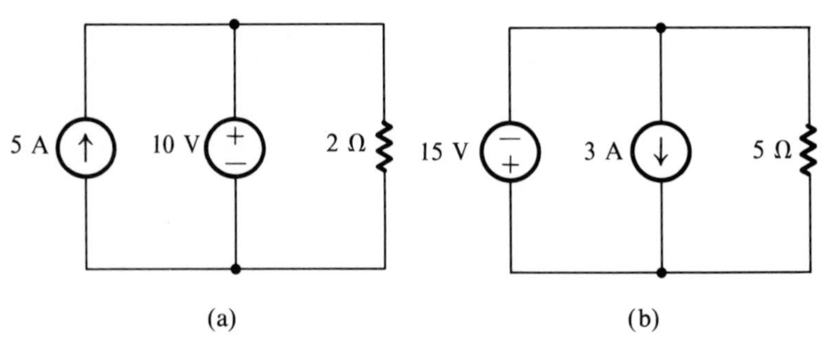

Figure P1.34

CIRCUIT ANALYSIS METHODS

INTRODUCTION

Many modern electrical circuits are very large from the standpoint of the number of components they contain, but very small from the standpoint of their physical size. Integrated circuits containing over two hundred thousand components on a single chip already exist, and even larger circuits will be designed in the near future. The complexity and size of these circuits require a systematic, computer-based method for calculating the voltages, currents, and power dissipations throughout a circuit. Calculated values of current and voltage are used to provide assurance that a circuit will operate correctly from an electrical standpoint, and power calculations are used to determine that a circuit will operate without being destroyed by its own heat dissipation. This is especially important in the design of high-density integrated circuits.

Nodal analysis is a systematic method for analyzing the behavior of large (multi-node) circuits. It uses KCL to determine efficiently and correctly the value of all of the voltages and currents in a circuit, and consequently the power dissipations, in a direct way. It is the cornerstone of modern software used for circuit design.

2.1 NODAL ANALYSIS (NA) AND KCL

Many circuits are complex enough to require more than series/parallel reduction techniques for their analysis. These circuits may be complex because they contain many nodes, and therefore many elements, or because they have a topology that is not series/parallel. Let's examine Fig. 2.1 to see how we can apply Kirchhoff's current law at a circuit's nodes to obtain a set of equations whose solution gives the value of each node voltage with respect to the ground node.

First choose a reference node (denoted by the ground symbol), then assign labels to the other nodes. The label v_j represents the voltage drop from node j to ground. Subscripts will distinguish the nodes from each other, and we may take the liberty of refering to the jth node as "node v_j." The circuit has two nodes and two node voltages: v_1 and v_2. It is easier to apply KCL if we also label the currents in all branches not containing current sources. (Once we have some practice with nodal analysis we'll skip this step.)

Now we apply KCL at the v_1 and v_2 nodes by equating the current *leaving* a node via resistive/conductive paths to the current *entering* the node from current sources. At the nodes of the circuit:

$$\Sigma\ R\text{-path currents leaving} = \Sigma\ \text{source currents entering}$$

and so

$$\text{Node 1:} \quad i_1 - i_2 = -i_{s1} + i_{s2} \tag{2.1}$$

$$\text{Node 2:} \quad i_2 + i_3 = i_{s1} + i_{s3} \tag{2.2}$$

$$\text{Ground:} \quad -i_1 - i_3 = -i_{s2} - i_{s3}$$

Observe that the equation we wrote for the ground node is (and always will be) an extra equation. In fact, the KCL equation at the ground node can be obtained by taking the negative of the sum of the equations for nodes 1 and 2. Thus, we really only need to use (2.1) and (2.2) to describe the circuit. A circuit having a total of N nodes will be completely described by the KCL equations written for at most N − 1 of its nodes.

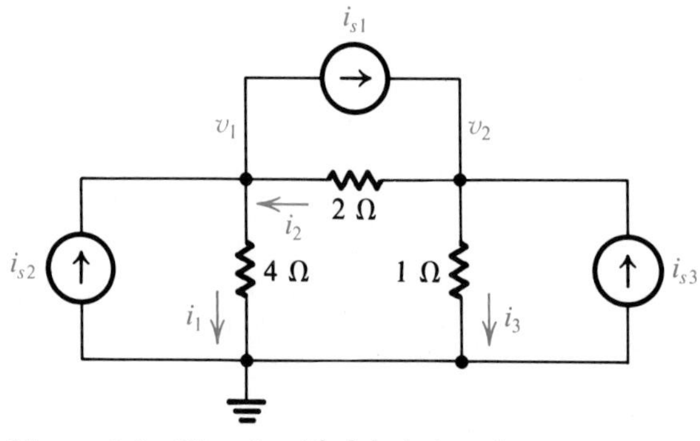

Figure 2.1 Circuit with labeled nodes.

Once KCL has been written for the nodes of a circuit we rewrite the KCL equations by expressing the resistive branch currents in terms of the node voltages. The key is to use Ohm's law

$$i_1 = \frac{v_1}{4} \tag{2.3}$$

$$i_2 = \frac{v_2 - v_1}{2} \tag{2.4}$$

$$i_3 = \frac{v_2}{1} \tag{2.5}$$

The current in any branch between a node and ground can be expressed simply in terms of the respective node voltage, but i_2, which is in the branch between nodes 1 and 2, must be written as the voltage *drop* across the connecting resistor divided by the resistor value. The voltage drop consistent with the direction of i_2 is $v_2 - v_1$. Conversely, the voltage drop consistent with $-i_2$ is $v_1 - v_2$. Now use these expressions for i_1, i_2 and i_3 in the first two KCL equations to get

$$\frac{v_1}{4} - \frac{v_2 - v_1}{2} = -i_{s1} + i_{s2}$$

and

$$\frac{v_2 - v_1}{2} + v_2 = i_{s1} + i_{s3}$$

Regrouping, but not combining coefficients gives

$$v_1\left(\frac{1}{4} + \frac{1}{2}\right) - \frac{1}{2}v_2 = -i_{s1} + i_{s2}$$

$$-\frac{1}{2}v_1 + v_2\left(\frac{1}{2} + 1\right) = i_{s1} + i_{s3}$$

and so

$$\frac{3}{4}v_1 - \frac{1}{2}v_2 = -i_{s1} + i_{s2} \tag{2.6}$$

$$-\frac{1}{2}v_1 + \frac{3}{2}v_2 = i_{s1} + i_{s3} \tag{2.7}$$

Equations (2.6) and (2.7) are called the **nodal equations** of the circuit. They are an independent set of linear, algebraic equations that determine values of v_1 and v_2 that are consistent with given values for the current sources, and they model the behavior of the circuit. If we can solve these equations for v_1 and v_2, we can determine the branch currents from (2.3)–(2.5), and we can get the power dissipation for each resistor as $p = vi$. The power supplied or dissipated by each source can be obtained in the same way. All this information is available from the solution of the nodal equations, so let's examine some methods for solving a pair of linear algebraic equations.

One approach is to solve the nodal equations by using one of the equations to eliminate one of the variables from the set by expressing it in terms of the other, and then substituting the expression back into the unused equation in the set. For example, we can use (2.6) to get

$$v_1 = \frac{2}{3}v_2 - \frac{4}{3}i_{s1} + \frac{4}{3}i_{s2} \tag{2.8}$$

This equation expresses v_1 in terms of v_2. We can substitute it into (2.7) to obtain

$$-\frac{1}{2}\left[\frac{2}{3}v_2 - \frac{4}{3}i_{s1} + \frac{4}{3}i_{s2}\right] + \frac{3}{2}v_2 = i_{s1} + i_{s3}$$

Rearranging gives

$$v_2 = \frac{2}{7}i_{s1} + \frac{4}{7}i_{s2} + \frac{6}{7}i_{s3} \tag{2.9}$$

Using (2.9) in (2.8) give the solution for v_1

$$v_1 = -\frac{8}{7}i_{s1} + \frac{12}{7}i_{s2} + \frac{4}{7}i_{s3} \tag{2.10}$$

Once the values of the sources are specified we use (2.9) and (2.10) to obtain the values of the node voltages. For example, if $i_{s1} = 10$ A, $i_{s2} = -5$ A and $i_{s3} = 3$ A, we get $v_1 = -18.29$ V and $v_2 = 2.57$ V.

A second method for solving nodal equations is the method of **determinants,** although determinants also become cumbersome when the dimension of the system equations exceeds three. Both methods require several steps of hand computation and are prone to human error. So, as an alternative, we will introduce and use the **matrix** method for representing and solving the KCL model. This method produces equivalent results, and has attractive features—it gives a compact representation of the circuit structure, and is especially amenable to numerical solution with a computer or hand-held calculator. Now is a good time to review the material on matrices in Appendix I.

The matrix equation model for Fig. 2.1 is an array of the coefficients that appear in the nodal equations given by (2.6) and (2.7). Take care to arrange the equations in consistent sequential order (v_1, v_2, i_{s1}, i_{s2}, i_{s3} in this case). We place the coefficients in a rectangular array, or matrix, having as many columns as there are node voltage variables, and as many rows as we have nodal equations. The node voltages are arranged in sequential order in the column vector $\mathbf{v}$; likewise the current sources are arranged in the vector $\mathbf{i}_s$ as shown in (2.11). (Notice that the nodal equations can be reconstructed easily by multiplying the elements of the column vectors onto the elements in the rows of the coefficient matrices.)

$$\begin{bmatrix} \left(\frac{1}{4}+\frac{1}{2}\right) & -\frac{1}{2} \\ -\frac{1}{2} & \left(\frac{1}{2}+1\right) \end{bmatrix} \begin{bmatrix} v_1 \\ v_2 \end{bmatrix} = \begin{bmatrix} -1 & 1 & 0 \\ 1 & 0 & 1 \end{bmatrix} \begin{bmatrix} i_{s1} \\ i_{s2} \\ i_{s3} \end{bmatrix} \tag{2.11}$$

This matrix equation has the form of

$$\boxed{\mathbf{Gv} = \mathbf{Bi}_s} \tag{2.12}$$

where **G** is a **conductance matrix**

$$\mathbf{G} = \begin{bmatrix} \frac{3}{4} & \frac{-1}{2} \\ -\frac{1}{2} & \frac{3}{2} \end{bmatrix}$$

and **B** is a **source-coupling matrix**

$$\mathbf{B} = \begin{bmatrix} -1 & 1 & 0 \\ 1 & 0 & 1 \end{bmatrix}$$

Notice that the main diagonal elements of **G** are the sums of the conductances attached to each node, that the off-diagonal term consists of the negative of the conductance connecting the nodes, and that a zero entry in **G** corresponds to the absence of a connecting resistor between nodes i and j. These observations let us write (2.12) by inspection in cases where a circuit consists of resistors and independent current sources. Be careful, though, for **G** must be modified when the circuit has controlled current sources and/or voltage sources.

We should also note that in the case we are considering the **G** matrix is (and always will be) symmetric, i.e. $\mathbf{G} = \mathbf{G}^T$. This provides a quick visual check for mistakes in forming **G**. The source-*coupling* matrix will only contain a 1, a -1, or a 0, depending on whether the current source connects with orientation towards the node, connects with orientation away from the node, or does not connect. The matrix **B** can be written by inspection for this class of circuits.

The association of the minus sign with the off-diagonal entries is **G** is a simple consequence of our choice to write KCL in terms of R-path currents *leaving* the nodes. Thus, the current leaving node i through a resistor to node j has $i = (v_i - v_j)/R_{ij}$, where R_{ij} is the connecting resistor. The $-1/R_{ij}$ entry in **G** accounts for $-v_j/R_{ij}$ in the set of algebraic equations represented by (2.12). Likewise, the main diagonal entry in **G** accounts for all of the resistive paths from the node.

In a more general case of this class of circuits, **G** will have dimension $(n \times n)$, and **B** will have dimension $(n \times m)$, where n is the number of independent nodes, and m is the number of independent current sources. Regardless of the complexity, this class of circuits is still modeled by (2.12), which expresses a generalized form of Ohm's Law.

2.1.1 Nodal Equation Solution

The nodal model for a resistive circuit with independent current sources can be solved by using the matrix inverse of **G**, $\mathbf{G}^{-1}$ where $\mathbf{G}^{-1}\mathbf{G} = \mathbf{I}$. Multiplying (2.12) from the left by $\mathbf{G}^{-1}$ on both sides, gives

$$\boxed{\mathbf{v} = \mathbf{G}^{-1}\mathbf{B}\mathbf{i}_s} \qquad (2.13)$$

For this example $\mathbf{G}^{-1}$ can be computed by hand.

$$\mathbf{G}^{-1} = \begin{bmatrix} \frac{3}{2} & \frac{1}{2} \\ \frac{1}{2} & \frac{3}{4} \end{bmatrix} \times \frac{8}{7} = \begin{bmatrix} \frac{12}{7} & \frac{4}{7} \\ \frac{4}{7} & \frac{6}{7} \end{bmatrix}$$

The inverse of **G** was calculated by first interchanging the elements on its main diagonal and changing the sign of both the elements off the diagonal, then dividing each element by the determinant of **G**. *If the determinant of* **G** *is zero, the equations are not independent, and the node voltages cannot be uniquely specified.* (As an exercise, verify that $\mathbf{G}^{-1}\mathbf{G} = \mathbf{I}$.)

Using (2.13) gives

$$\mathbf{v} = \begin{bmatrix} \frac{12}{7} & \frac{4}{7} \\ \frac{4}{7} & \frac{6}{7} \end{bmatrix} \begin{bmatrix} -1 & 1 & 0 \\ 1 & 0 & 1 \end{bmatrix} \begin{bmatrix} i_{s1} \\ i_{s2} \\ i_{s3} \end{bmatrix} = \begin{bmatrix} -\frac{8}{7} & \frac{12}{7} & \frac{4}{7} \\ \frac{2}{7} & \frac{4}{7} & \frac{6}{7} \end{bmatrix} \begin{bmatrix} i_{s1} \\ i_{s2} \\ i_{s3} \end{bmatrix}$$

Hitherto, we have deliberately worked with generic current sources, and it's a good idea to use them even if specific values are given, if only to preclude a premature evaluation of $\mathbf{B}\mathbf{i}_s$.

Example 2.1

Find v_1 and v_2 when $i_{s1} = 21$ A, $i_{s2} = -7$ A and $i_{s3} = 7$ A in Fig. 2.1

Solution:

$$\begin{bmatrix} v_1 \\ v_2 \end{bmatrix} = \begin{bmatrix} -\frac{8}{7} & \frac{12}{7} & \frac{4}{7} \\ \frac{2}{7} & \frac{4}{7} & \frac{6}{7} \end{bmatrix} \begin{bmatrix} 21 \\ -7 \\ 7 \end{bmatrix} = \begin{bmatrix} -32 \\ 8 \end{bmatrix} \text{ volts}$$

As an exercise, find the branch currents and verify that they are consistent with KCL at *each* node.

The nodal model and its solution are summarized as

$$\mathbf{Gv} = \mathbf{Bi}_s$$
$$\mathbf{v} = \mathbf{G}^{-1}\mathbf{Bi}_s$$

Expressing the model solution in matrix form rather than using Cramer's rule or other algebraic methods displays the circuit structure, and has the benefit that $\mathbf{G}^{-1}\mathbf{B}$ need only be computed once for a given circuit topology. This is easily done with either a calculator or a personal computer. Once $\mathbf{G}^{-1}\mathbf{B}$ is computed, a solution can be generated for any choice of the current sources. In contrast, algebraic methods require several steps of recomputation each time the source vector is varied. The effort is considerable, and the chances for error are greater. We urge you to adopt the matrix method of attacking the problem.

Skill Exercise 2.1

Obtain the nodal model in matrix form for the circuit in Fig. SE 2.1, and solve it for the node voltages where $i_{s1} = 2$ A and $i_{s2} = 1$ A.

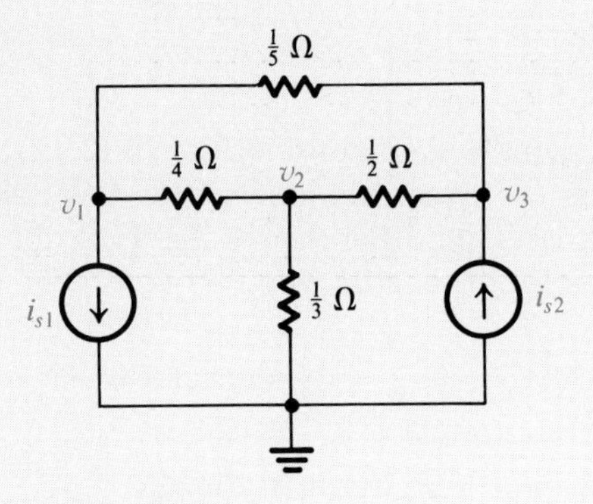

Figure SE 2.1

Answer:

$$\begin{bmatrix} 9 & -4 & -5 \\ -4 & 9 & -2 \\ -5 & -2 & 7 \end{bmatrix} \mathbf{v} = \begin{bmatrix} -1 & 0 \\ 0 & 0 \\ 0 & 1 \end{bmatrix} \mathbf{i}_S, \text{ and } \mathbf{v} = \begin{bmatrix} -0.58 \\ -0.33 \\ -0.35 \end{bmatrix}$$

2.1.2 NA–Controlled Current Sources

Nodal analysis also treats circuits consisting of resistors, independent current sources, and controlled current sources. The circuit shown in Fig. 2.2 contains a current-controlled current source. For this class of circuits we write the left-hand side (LHS) of the matrix model exactly as we did before, but the right hand side (RHS) is modified to include current entering the nodes from independent sources, plus current entering the nodes from dependent sources. If all sources are labeled the matrix model becomes

$$\tilde{\mathbf{G}}\mathbf{v} = \mathbf{B}\mathbf{i}_S + \mathbf{B}_C\mathbf{i}_C$$

and

$$\tilde{\mathbf{G}}\mathbf{v} - \mathbf{B}_C\mathbf{i}_C = \mathbf{B}\mathbf{i}_S$$

where $\mathbf{i}_C$ is, in general, a vector of the controlled sources, and $\mathbf{B}_C$ is the coupling matrix for the controlled sources.

Each controlled source will be specified by its dependency on current or voltage. If the source is controlled by current we can use Ohm's Law to write the current in terms of two node voltages, and thus can re-specify the control to depend on node voltages. If the controlled source is already specified in terms of voltage nothing more needs to be done. In both cases the control mechanism for each controlled source can be described by a dependency on at most two node voltages. The vector of controlled sources depends on the vector of node voltages, and is written explicitly as

$$\tilde{\mathbf{G}}\mathbf{v} - \mathbf{B}_C\mathbf{i}_C = \tilde{\mathbf{G}}\mathbf{v} - \mathbf{B}_C\mathbf{i}_C(\mathbf{v}) = \mathbf{G}\mathbf{v}$$

The model becomes $\mathbf{G}\mathbf{v} = \mathbf{B}\mathbf{i}_S$, which was found earlier in (2.12).

This class of problems *has the same model form* as the case with resistors and independent sources. One difference is that $\mathbf{G}$ is not a symmetric matrix, but the model solution is still given by (2.13).

Example 2.2

Use nodal analysis to find $\mathbf{v}$ for the circuit in Fig. 2.2.

Solution: The RHS of the KCL equations must include the dependent current source. It is advisable first to de-

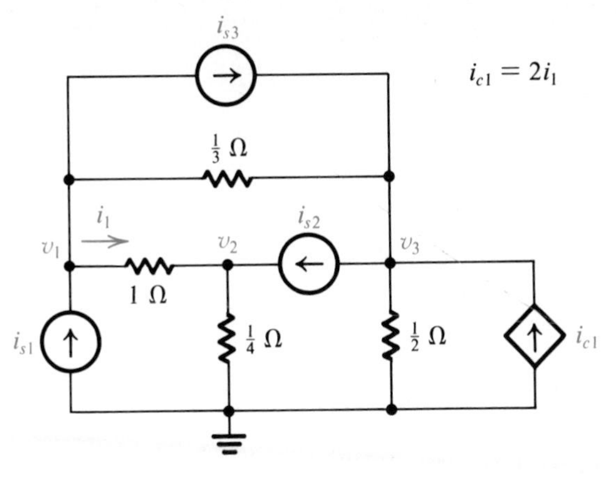

Figure 2.2 Circuit with a controlled current source.

scribe the source in terms of its generic current source label, and then to use its explicit definition

$$\begin{bmatrix} 4 & -1 & -3 \\ -1 & 5 & 0 \\ -3 & 0 & 5 \end{bmatrix} \mathbf{v} = \begin{bmatrix} 1 & 0 & -1 \\ 0 & 1 & 0 \\ 0 & -1 & 1 \end{bmatrix} \begin{bmatrix} i_{s1} \\ i_{s2} \\ i_{s3} \end{bmatrix} + \begin{bmatrix} 0 \\ 0 \\ 1 \end{bmatrix} [i_{c1}]$$

where $i_{c1} = 2i_1 = 2(v_1 - v_2)$. Thus, Ohm's Law allows us to express i_{c1} in terms of the node voltages. Forming $\tilde{\mathbf{G}}\mathbf{v} - \mathbf{B}_c\mathbf{i}_c$ we get

$$\begin{bmatrix} 4 & -1 & -3 \\ -1 & 5 & 0 \\ -3 & 0 & 5 \end{bmatrix} \mathbf{v} - \begin{bmatrix} 0 \\ 0 \\ 2v_1 - 2v_2 \end{bmatrix} = \begin{bmatrix} 1 & 0 & -1 \\ 0 & 1 & 0 \\ 0 & -1 & 1 \end{bmatrix} \begin{bmatrix} i_{s1} \\ i_{s2} \\ i_{s3} \end{bmatrix}$$

which reduces to

$$\begin{bmatrix} 4 & -1 & -3 \\ -1 & 5 & 0 \\ -5 & 2 & 5 \end{bmatrix} \mathbf{v} = \begin{bmatrix} 1 & 0 & -1 \\ 0 & 1 & 0 \\ 0 & -1 & 1 \end{bmatrix} \begin{bmatrix} i_{s1} \\ i_{s2} \\ i_{s3} \end{bmatrix}$$

We now have the nodal model expressed in the form of (2.12).

Notice that $\mathbf{G}$ is formed from the original $\tilde{\mathbf{G}}$ matrix by modifying the latter to include the effects of the controlled source. Applying (2.13), we get

$$\mathbf{G}^{-1} = \frac{1}{26} \begin{bmatrix} 25 & -1 & 15 \\ 5 & 5 & 3 \\ 23 & -3 & 19 \end{bmatrix}$$

and with $i_{s1} = 1$ A, $i_{s2} = 2$ A and $i_{s3} = 4$ A we have

$$\mathbf{v} = \frac{1}{26}\begin{bmatrix} 25 & -1 & 15 \\ 5 & 5 & 3 \\ 23 & -3 & 19 \end{bmatrix}\begin{bmatrix} 1 & 0 & -1 \\ 0 & 1 & 0 \\ 0 & -1 & 1 \end{bmatrix}\begin{bmatrix} i_{s1} \\ i_{s2} \\ i_{s3} \end{bmatrix}$$

Some errors

$$= \frac{1}{26}\begin{bmatrix} -25 & -16 & -10 \\ -5 & 2 & -2 \\ -23 & -22 & -4 \end{bmatrix}\begin{bmatrix} i_{s1} \\ i_{s2} \\ i_{s3} \end{bmatrix} = \begin{bmatrix} -97 \\ -9 \\ -83 \end{bmatrix} \text{A}$$

Skill Exercise 2.2

Solve for the node voltages in Fig. SE 2.2, and calculate the power dissipated in each resistor. Compare the power dissipation in the resistors with the power supplied by the sources. $i_{s1} = 2A$, $i_{s2} = 4A$

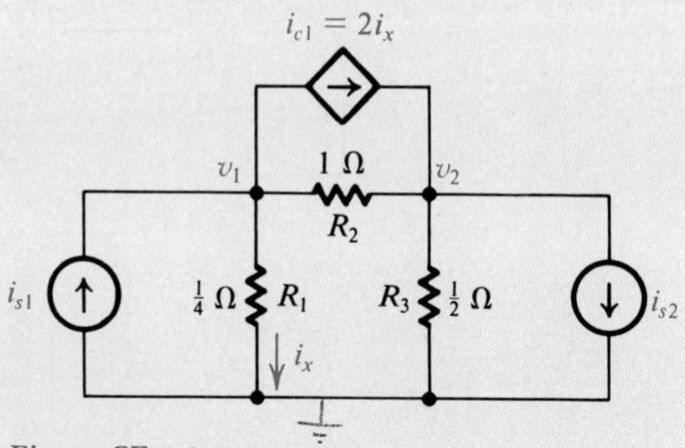

Figure SE 2.2

Answer: $v_1 = 0.067$ V, $v_2 = -1.133$ V

$p_{R1} =$	0.018 W	$p_{s1} =$	0.132 W
$p_{R2} =$	1.439 W	$p_{s2} =$	4.532 W
$p_{R3} =$	2.567 W	$p_{c1} =$	0.640 W
Total:	4.024 W		4.024 W

Skill Exercise 2.3

Find the power supplied by the current source and the power dissipated in the 1000-ohm resistor in the circuit in Fig. SE 2.3.

Answer: $p_R = 100$ kW, $p_{s1} = -2.5$ kW

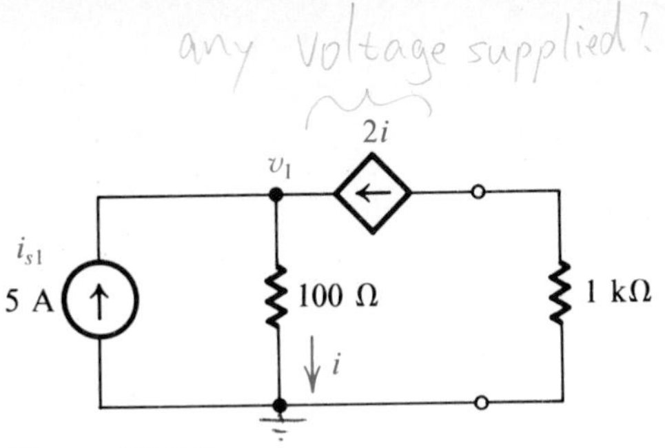

Figure SE 2.3

2.1.3 NA–Voltage Sources

When a circuit contains one or more voltage sources KCL must be modified at the nodes of those sources. A voltage source will be connected either with one of its terminals at ground, as in Fig. 2.3 or with both of its terminals at nodes that are not at ground, as in Fig. 2.4.

Case I

If the voltage source has a node attached to ground, we replace KCL at the other (non-ground) node of the source by a *constraint equation*.

Example 2.3

Find the nodal analysis model for the circuit in Fig. 2.3.

Solution: At node 1 the physical constraint imposed by the voltage source gives

CONSTRAINT EQ. $v_1 = v_{s1}$

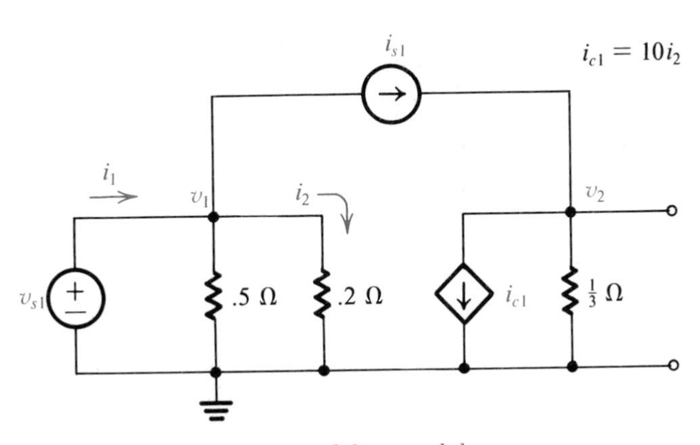

Figure 2.3 Transistor amplifier model.

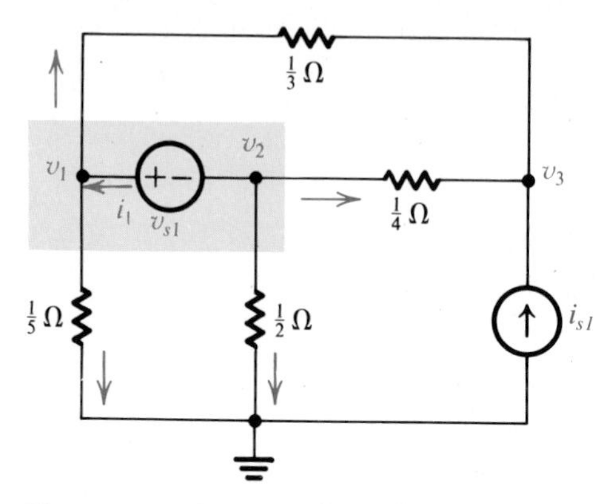

Figure 2.4 Super node surface.

There is no need to write KCL at this node. In fact, if we attempt to write it we face the dilemma of not knowing how to express i_1 in terms of the node voltages. (Try it.) The second node in Fig. 2.3 can be treated in the conventional way:

KCL EQ. $\quad 3v_2 = i_{s1} - i_{c1}$

These two equations can be expressed in matrix form as

$$\begin{bmatrix} 1 & 0 \\ 0 & 3 \end{bmatrix} \begin{bmatrix} v_1 \\ v_2 \end{bmatrix} = \begin{bmatrix} 1 & 0 \\ 0 & 1 \end{bmatrix} \begin{bmatrix} v_{s1} \\ i_{s1} \end{bmatrix} + \begin{bmatrix} 0 \\ -1 \end{bmatrix} [i_{c1}]$$

Using the expression for i_{c1} and rearranging the matrix equation leads to the nodal model:

$$\begin{bmatrix} 1 & 0 \\ 50 & 3 \end{bmatrix} \begin{bmatrix} v_1 \\ v_2 \end{bmatrix} = \begin{bmatrix} 1 & 0 \\ 0 & 1 \end{bmatrix} \begin{bmatrix} v_{s1} \\ i_{s1} \end{bmatrix}$$

The nodal analysis model for a circuit containing a voltage source has the same form as (2.12), but the source vector contains a voltage source as well as a current source. Since circuits may contain independent current and voltage sources, we write matrix model equations using a **generalized source vector $\mathbf{u}_s$** which includes current source components and voltage source components. For these circuits the nodal model is

$$\boxed{\mathbf{G}\mathbf{v} = \mathbf{B}\mathbf{u}_s} \qquad (2.14)$$

with solution

$$\boxed{\mathbf{v} = \mathbf{G}^{-1}\mathbf{B}\mathbf{u}_s} \qquad (2.15)$$

Skill Exercise 2.4

Solve for the nodal voltages in Fig. SE 2.4.

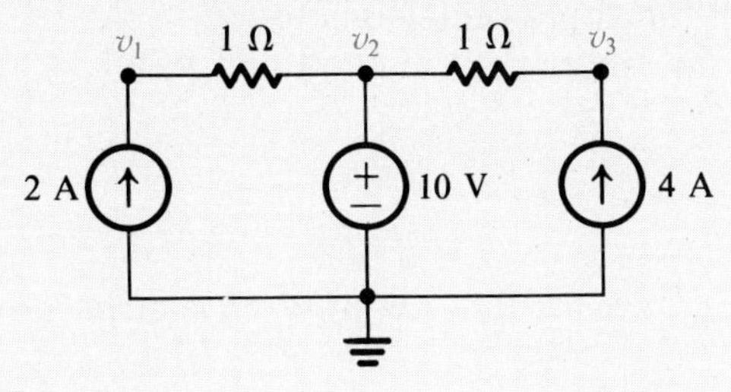

Figure SE 2.4

Answer: $v_1 = 12$ V, $v_2 = 10$ V, $v_3 = 14$ V

Case II

If neither node of a voltage source is at ground, as in Fig. 2.4, an equation can still be written for the constraint imposed by the source:

CONSTRAINT EQ. $v_1 - v_2 = v_{s1}$

However, we face the dilemma of not being able to write KCL at either of the source nodes in terms of the node voltages. KCL at node 1 in Figure 2.4 becomes

$$5v_1 + 3(v_1 - v_3) = i_1$$

At node 2 we would get

$$2v_2 + 4(v_2 - v_3) = -i_1$$

We don't have a way to specify the branch current i_1 in terms of the node voltages. Notice, however, that if we add these two equations we get

$$5v_1 + 3(v_1 - v_3) + 2v_2 + 4(v_2 - v_3) = 0$$

which reduces to a super node equation.

SUPER NODE EQ. $8v_1 + 6v_2 - 7v_3 = 0$

This is exactly the equation we would get if we applied KCL to the surface enclosing the source nodes in Fig. 2.4 (dashed lines). This surface is called a **generalized node,** or a **super node.** We use it to exploit a more general form of KCL which states that **the sum of the current leaving a closed surface by resistive paths must equal the current entering the surface from current sources.** (In this example, there are no current sources with paths entering the super node.) A voltage source can be treated by first drawing a super node to *enclose* both nodes of the source, and then writing KCL for the super node surface. In this example, we would add the currents leaving the surface on the four resistive paths shown with arrows in Figure 2.4 to get the super node equation.

At node 3 we apply the "short cut" method and write KCL in the conventional way:

KCL EQ. $-3v_1 - 4v_2 + 7v_3 = i_{s1}$

This completes the equation set for this example. In matrix format we have

$$\begin{bmatrix} 1 & -1 & 0 \\ 8 & 6 & -7 \\ -3 & -4 & 7 \end{bmatrix} \begin{bmatrix} v_1 \\ v_2 \\ v_3 \end{bmatrix} = \begin{bmatrix} 1 & 0 \\ 0 & 0 \\ 0 & 1 \end{bmatrix} \begin{bmatrix} v_{s1} \\ i_{s1} \end{bmatrix}$$

The model has the form of (2.14) and has the generic solution given by (2.15).

In sum, when a voltage source is connected between two nodes not at ground, the two KCL equations are replaced by a **constraint equation** and a **super node equation.** This preserves the number of equations used to describe the circuit.

Skill Exercise 2.5

Write the constraint equation and the super node equation for the circuit shown in Fig. SE 2.5.

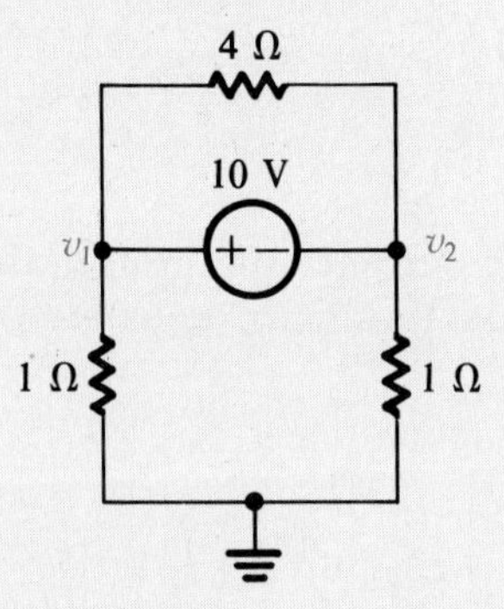

Figure SE 2.5

Answer: CE: $v_1 - v_2 = 10$
SN: $v_1 + v_2 = 0$

Example 2.4

Find the nodal analysis model for the circuit in Figure 2.5

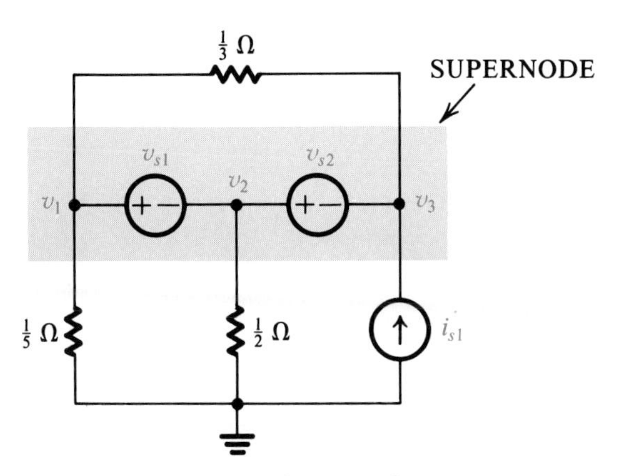

Figure 2.5 Super node example.

Solution: The voltage sources provide two constraint equations:

$$v_1 - v_2 = v_{s1}$$
$$v_3 - v_2 = -v_{s2}$$

The super node must be drawn to enclose both of the voltage sources, since if we draw it to enclose v_{s1} we still would have to know the current in the branch of v_{s2}. In general, when voltage sources are connected to a common node we must form a super node that encloses all of them.

Writing KCL for the super node gives

$$5v_1 + 3(v_1 - v_3) + 2v_2 + 3(v_3 - v_1) = i_{s1}$$

Cancelling the currents in the $\frac{1}{3}\,\Omega$ path gives

$$5v_1 + 2v_2 = i_{s1}$$

These equations can be arranged in matrix format as

$$\begin{bmatrix} 1 & -1 & 0 \\ 0 & -1 & 1 \\ 5 & 2 & 0 \end{bmatrix} \begin{bmatrix} v_1 \\ v_2 \\ v_3 \end{bmatrix} = \begin{bmatrix} 1 & 0 & 0 \\ 0 & -1 & 0 \\ 0 & 0 & 1 \end{bmatrix} \begin{bmatrix} v_{s1} \\ v_{s2} \\ i_{s1} \end{bmatrix}$$

Skill Exercise 2.6

Write the constraint equations and the super node equation for the circuit shown in Fig. SE 2.6.

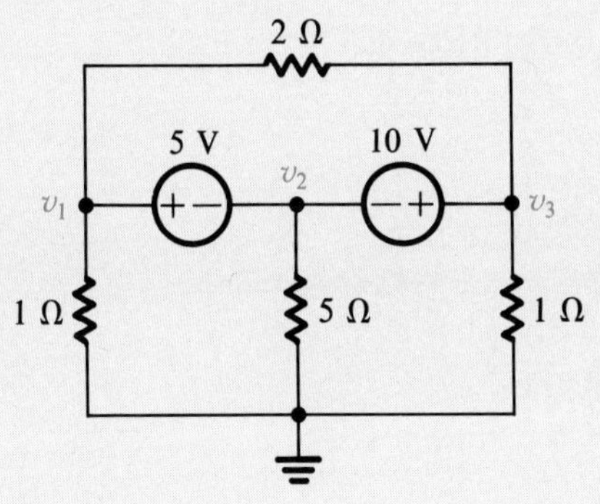

Figure SE 2.6

Answer:

$$\text{CE:} \quad v_1 - v_2 = 5$$
$$\text{CE:} \quad v_2 - v_3 = -10$$
$$\text{SN:} \quad v_1 + \frac{v_2}{5} + v_3 = 0$$

Example 2.5

Find the matrix model for the circuit of Fig. 2.6.

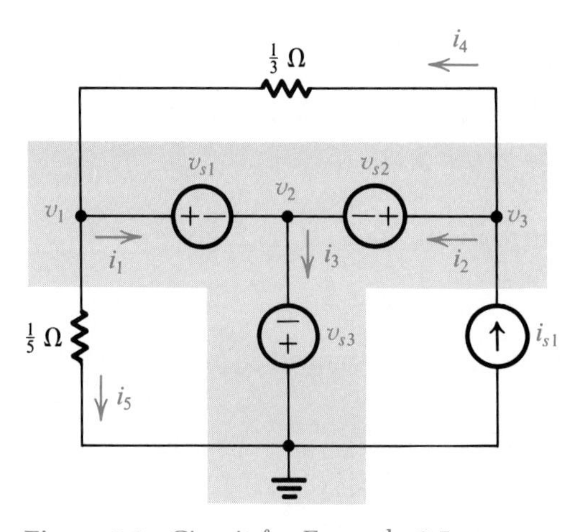

Figure 2.6 Circuit for Example 2.5.

Solution: The circuit has three constraint equations:

$$v_1 = v_{s1} - v_{s3}$$
$$v_2 = -v_{s3}$$
$$v_3 = v_{s2} - v_{s3}$$

In matrix format:

$$\begin{bmatrix} 1 & 0 & 0 \\ 0 & 1 & 0 \\ 0 & 0 & 1 \end{bmatrix} \begin{bmatrix} v_1 \\ v_2 \\ v_3 \end{bmatrix} = \begin{bmatrix} 1 & 0 & -1 \\ 0 & 0 & -1 \\ 0 & 1 & -1 \end{bmatrix} \begin{bmatrix} v_{s1} \\ v_{s2} \\ v_{s3} \end{bmatrix}$$

The constraint equations completely characterize the circuit.

Voltage sources simplify nodal analysis because they either determine the voltage at a node, or the voltage between a pair of nodes. This reduces the number of equations that have to be solved for the node voltages.

Example 2.6

Find the node voltages for the circuit shown in Fig. 2.7.

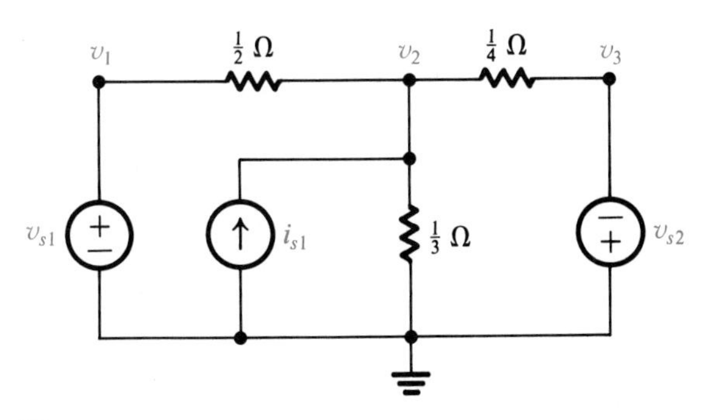

Figure 2.7

Solution: The nodal model for the circuit has two constraint equations and a KCL equation

CE1: $v_1 = v_{s1}$
KCL: $2(v_2 - v_1) + 4(v_2 - v_3) + 3v_2 = i_{s1}$
CE2: $v_3 = -v_{s2}$

v_1 and v_3 are known, so instead of solving the matrix equation directly, we substitute the two constraint equation values for v_1 and v_3 into the KCL equation and then solve for $v_2 = \frac{1}{9}[2v_{s1} - 4v_{s2} + i_{s1}]$.

The preceding problems showed how to modify the nodal analysis method to treat circuits with voltage sources. Computer programs for nodal analysis do not use super node equations or constraint equations. Instead, they obtain a sufficient number of equations by using a more generalized model of a circuit element, one that can include, simultaneously and systematically, the effect of current and voltage sources and resistors. We will not address those methods here.

2.2 MESH ANALYSIS (MA) AND KVL

Mesh analysis characterizes a circuit by invoking Kirchhoff's voltage law for suitably chosen closed paths in the circuit. This leads to a set of equations which can be solved for the **mesh currents** of the circuit.

2.2.1 Mesh Equations

Planar circuits have a schematic that can be drawn in a plane without having crossing paths. The circuit in Fig. 2.8a is planar, while that in Fig. 2.8b is not. The circuit in Fig. 2.8c appears to be non-planar, but actually has the planar form shown in Fig. 2.8d.

A closed path formed by the connected branches of a planar circuit is a **mesh** if it does not enclose another distinct closed path. There are many

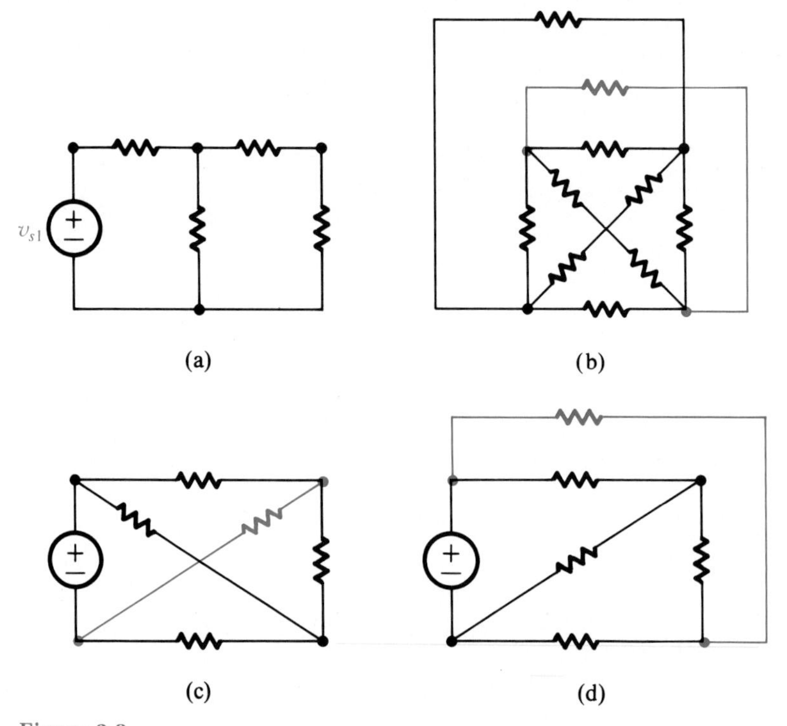

Figure 2.8

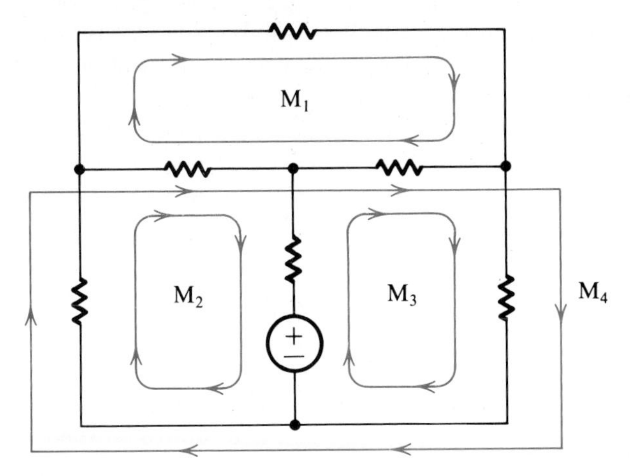

Figure 2.9 Circuit meshes.

different closed paths in a planar circuit, but not all of them are meshes. For example, M1, M2, and M3 are meshes for the planar circuit shown in Fig. 2.9, but M4 is not a mesh because it contains M2 and M3. A branch of the circuit that is in two meshes is said to be a coupling branch.

In mesh analysis (**MA**) we associate currents with the meshes of the circuit, as depicted in Fig. 2.10. For convenience, mesh currents are chosen to flow clockwise. The important analytical results that we will use (but not prove) are that the equations describing KVL around the meshes are an independent set, and that the branch currents are specified by the mesh currents. There are many ways to choose a set of currents corresponding to closed paths of the circuit. We will restrict our choice to the closed paths corresponding to the meshes. Once we have found the mesh currents we can then find the branch currents, branch voltages, node voltages, and power.

The circuit shown in Fig. 2.10 will serve to introduce us to mesh analysis. Its two clockwise mesh currents are shown, along with the

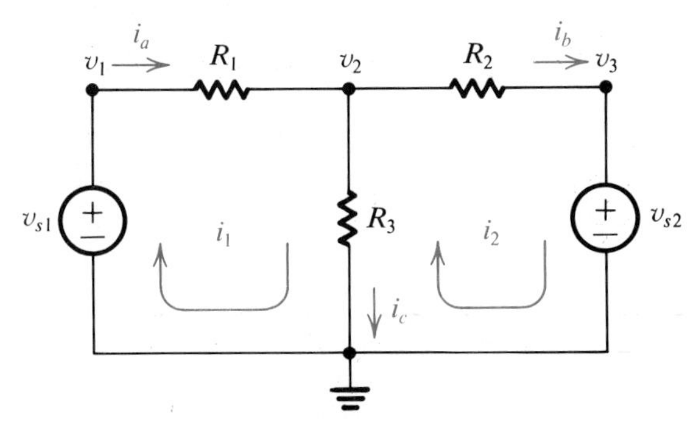

Figure 2.10 Mesh analysis example.

branch currents. We can gain an intuitive feel for mesh current by noting that $i_a = i_1$ and $i_b = i_2$ but $i_c = i_1 - i_2$. Thus, the circulating mesh currents subtract to form i_c, a natural consequence of KCL at Node 2.

The first step in mesh analysis is to label the mesh currents, as we have done. The next step is to write KVL around each mesh by equating the voltage drop across the resistors to the voltage rise of the sources in the clockwise mesh path. We have

$$\text{M1:} \quad i_1R_1 + (i_1 - i_2)R_3 = v_{s1}$$
$$\text{M2:} \quad (i_2 - i_1)R_3 + i_2R_2 = -v_{s2}$$

Note that $i_1 - i_2$ is the current through R_3 in the clockwise direction of M1, and $(i_2 - i_1)$ is the current through R_3 in the clockwise direction of M2. **Failure to observe current polarity is a common mistake.** A way to avoid making this mistake is to note that the current in a mesh-coupling resistor such as R_3 will consist of the difference of two mesh currents, with the positive current being the one corresponding to the mesh equation we are writing, and the other being the negative of the current in the coupled mesh. The mesh equations can be rewritten in matrix format as

$$\begin{bmatrix} (R_1 + R_3) & -R_3 \\ -R_3 & (R_2 + R_3) \end{bmatrix} \begin{bmatrix} i_1 \\ i_2 \end{bmatrix} = \begin{bmatrix} 1 & 0 \\ 0 & -1 \end{bmatrix} \begin{bmatrix} v_{s1} \\ v_{s2} \end{bmatrix}$$

or

$$\boxed{\mathbf{Ri} = \mathbf{Bv}_s} \qquad (2.16)$$

The matrix model has the solution given by

$$\boxed{\mathbf{i} = \mathbf{R}^{-1}\mathbf{Bv}_s} \qquad (2.17)$$

Notice that the elements on the main diagonal of matrix **R** are formed by taking the sum of the resistance values in the mesh, while the off-diagonal terms are formed from the negative of the coupling resistance between meshes. Thus, R_{ij} is the coupling resistor between mesh i and mesh j. If there is no coupling, the entry is zero. These observations allow us to form **R** by inspection when doing mesh analysis for this family of circuits, and to bypass the step of writing the detailed KVL equations. Likewise, the source coupling matrix **B** is found by inspection by noting whether a voltage source is in the mesh, and if so, whether it is a rise or a drop in the clockwise direction for the mesh. A rise has a coupling coefficient of +1, a drop has a coupling coefficient of −1.

Example 2.7

Find the **R** and **B** matrices for mesh analysis of the circuit of Fig. 2.11.

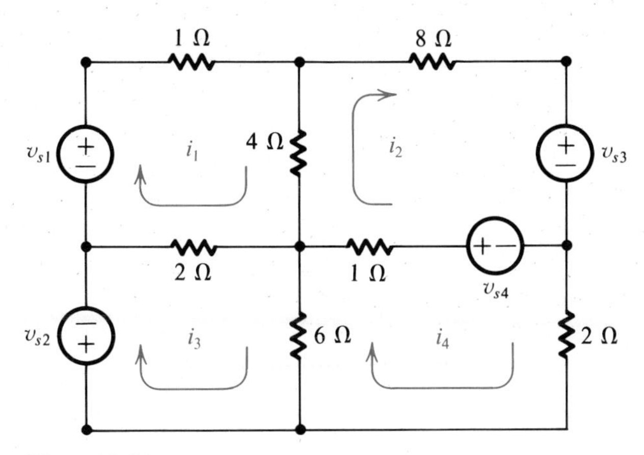

Figure 2.11

Solution: The circuit is planar with 4 meshes, and contains only independent voltage sources and resistors. We form **R** and **B** by inspection:

$$\begin{bmatrix} 7 & -4 & -2 & 0 \\ -4 & 13 & 0 & -1 \\ -2 & 0 & 8 & -6 \\ 0 & -1 & -6 & 9 \end{bmatrix} \mathbf{i} = \begin{bmatrix} 1 & 0 & 0 & 0 \\ 0 & 0 & -1 & 1 \\ 0 & -1 & 0 & 0 \\ 0 & 0 & 0 & -1 \end{bmatrix} \mathbf{v}_S$$

2.2.2 MA–Controlled Voltage Sources

If a circuit contains linear controlled voltage sources the mesh analysis model must be modified to include the effects of these additional sources. For the circuit of Fig. 2.12 we write mesh equations as before, but we form a separate voltage source vector containing the independent sources, and another for the controlled sources

$$\mathbf{v}_S = \begin{bmatrix} v_{S1} \\ v_{S2} \end{bmatrix} \qquad \mathbf{v}_C = \begin{bmatrix} v_{C1} \\ v_{C2} \end{bmatrix}$$

For this circuit the mesh equation is

$$\begin{bmatrix} 3 & -2 \\ -2 & 5 \end{bmatrix} \begin{bmatrix} i_1 \\ i_2 \end{bmatrix} = \begin{bmatrix} 1 & 0 \\ 0 & -1 \end{bmatrix} \begin{bmatrix} v_{S1} \\ v_{S2} \end{bmatrix} + \begin{bmatrix} 1 & 0 \\ -1 & -1 \end{bmatrix} \begin{bmatrix} v_{C1} \\ v_{C2} \end{bmatrix} \tag{2.18}$$

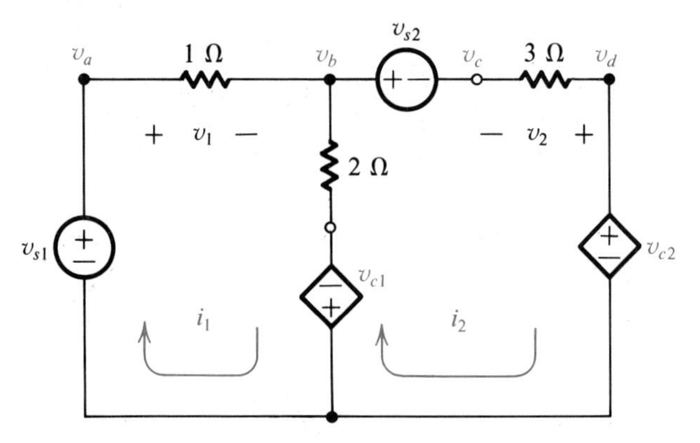

Figure 2.12 Mesh analysis with controlled voltage sources.

In general, a circuit with controlled voltage sources will have a mesh equation model of the form

$$\tilde{\mathbf{R}}\mathbf{i} = \mathbf{B}\mathbf{v}_s + \mathbf{B}_c\mathbf{v}_c$$

This is just (2.16) with an additional term to account for the controlled sources. The vector of controlled sources ultimately depends on the mesh current, so we must be careful to group all of the dependent variables on the LHS as

$$\tilde{\mathbf{R}}\mathbf{i} - \mathbf{B}_c\mathbf{v}_c = \mathbf{B}\mathbf{v}_s$$

Since $\mathbf{v}_c$ can be expressed as a function of $\mathbf{i}$, e.g. $\mathbf{v}_c = \mathbf{f}(\mathbf{i})$, and $\mathbf{f}(\cdot)$ is linear the equation can be rearranged in the form of (2.16). So the methodology is fundamentally the same as for the case without the controlled sources.

2.2.3 MA–Current Sources

Current sources simplify mesh analysis by reducing the number of meshes to which we apply KVL. The current source will constrain either the value of one mesh current or the difference between two mesh currents, depending on whether it appears in a branch that couples two meshes or in a branch that does not. The branch containing the current source in Fig. 2.13 is part of M3, but it is not part of M1 or M2. To model this circuit we expand the usual source vector to include i_{s1}, and use $\mathbf{u}_s$ to represent a generalized source vector containing independent current and voltage sources. The current source in M3 introduces a constraint equation:

$$\text{CE:} \quad i_3 = -i_{s1}$$

so there is no need to write KVL for M3. In fact, we can't—because the voltage across i_{s1} cannot be determined yet. (We could introduce v as another variable, but that would have the effect of increasing the number of unknowns.) The equations for M1 and M2 are modified to include the

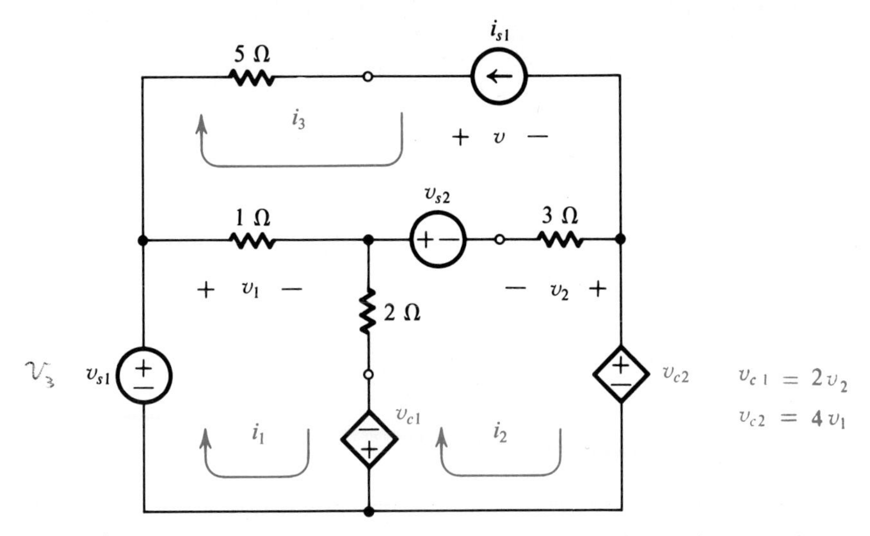

Figure 2.13

voltage drops caused by i_3 in the coupling resistors. The circuit model then yields two mesh equations and one constraint equation:

$$\begin{array}{l}\text{M1:}\\ \text{M2:}\\ \text{CE:}\end{array}\begin{bmatrix}3 & -2 & -1\\ -2 & 5 & -3\\ 0 & 0 & 1\end{bmatrix}\mathbf{i} = \begin{bmatrix}1 & 0 & 0\\ 0 & -1 & 0\\ 0 & 0 & -1\end{bmatrix}\begin{bmatrix}v_{s1}\\ v_{s2}\\ i_{s1}\end{bmatrix} + \begin{bmatrix}1 & 0\\ -1 & -1\\ 0 & 0\end{bmatrix}\begin{bmatrix}v_{c1}\\ v_{c2}\end{bmatrix}$$

From the specification of the controlled sources we obtain:

$$v_{c1} = 2v_2 = 2[3(i_3 - i_2)] = -6i_2 + 6i_3$$
$$v_{c2} = 4v_1 = 4[1(i_1 - i_3)] = 4i_1 - 4i_3$$

which we use in the matrix model to get:

$$\begin{bmatrix}3 & 4 & -7\\ 2 & -1 & -1\\ 0 & 0 & 1\end{bmatrix}\mathbf{i} = \begin{bmatrix}1 & 0 & 0\\ 0 & -1 & 0\\ 0 & 0 & -1\end{bmatrix}\begin{bmatrix}v_{s1}\\ v_{s2}\\ i_{s1}\end{bmatrix}$$

This model can be solved for **i** in the usual way. (Note that the model could be further simplified by eliminating v_3)

Skill Exercise 2.7

Find the mesh current in Fig. SE 2.7.

Answer: $i_1 = 2$ A, $i_2 = -4$ A, $i_3 = -0.857$ A

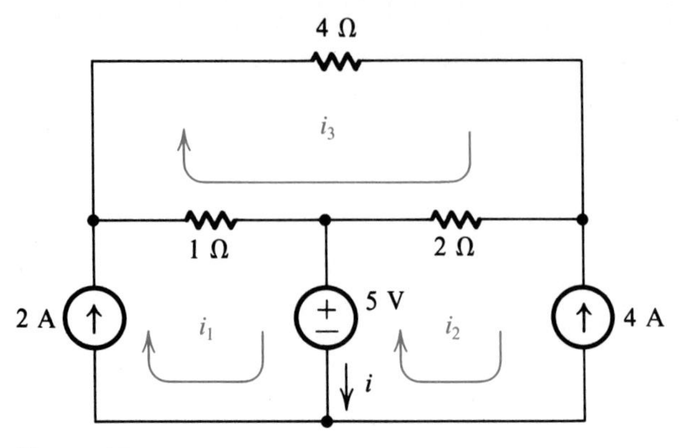

Figure SE 2.7

The KVL mesh equations for current sources that are in noncoupling branches will be replaced by constraint equations, even when the sources are controlled.

Skill Exercise 2.8

Write the mesh equations for the circuit of Fig. SE 2.8 with $i_{c1} = 2i_3$ and $i_{c2} = 4i$.

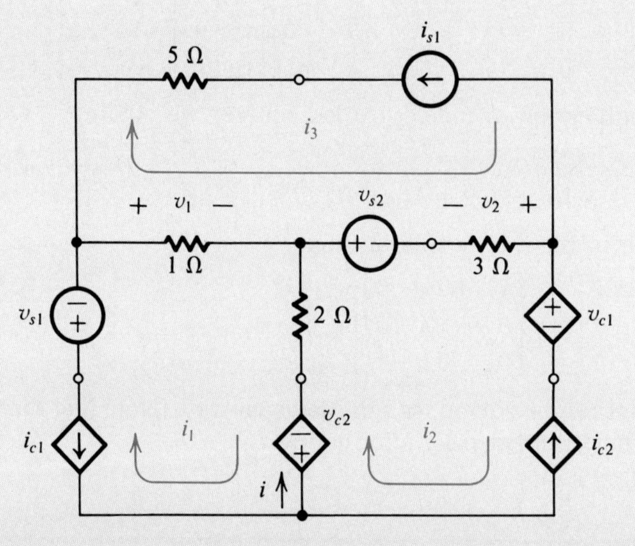

Figure SE 2.8

Answer: M1/CE: $i_1 = -i_{c1} = -2i_3$

M2/CE: $i_2 = -i_{c2} = -4i = -4(i_2 - i_1)$

M3/CE: $i_3 = -i_{s1}$

When a current source (independent or controlled) is located in a branch that couples two meshes (can it ever couple more?) we face the dilemma of not knowing how to represent the voltage across the circuit source in terms of the mesh currents.

Example 2.13

Because the current source in Fig. 2.14(a) is between two meshes, we have to find another way to write two equations for our model. The first equation is just the constraint equation imposed on the mesh currents by the current source:

$$-i_1 + i_2 = i_{s1}$$

The second equation is found by writing KVL on a super mesh that we create by removing the branch containing the current source and identifying the enlarged mesh formed by combining the two meshes that were coupled by the current source. In Fig. 2.14(a) we remove the source to form the super mesh shown in Fig. 2.14(b). The super mesh equation is just KVL around the super mesh. It's easy—just remember to use the *actual* mesh currents in the resistors on this path as they are defined in Fig. 2.14(a).

Beginning at "a" and going clockwise we get

$$2(i_1 - i_3) + 2(i_2 - i_3) + 1(i_2) + 1(i_1) = -v_{s2} + v_{s1}$$

Rearranging

$$3i_1 + 3i_2 - 4i_3 = v_{s1} - v_{s2}$$

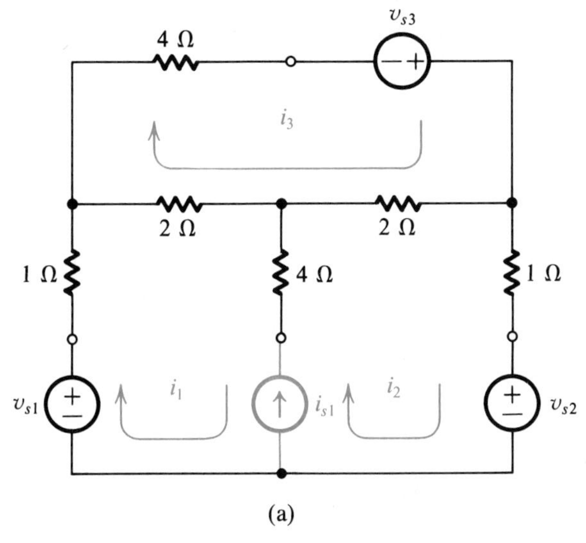

Figure 2.14(a)

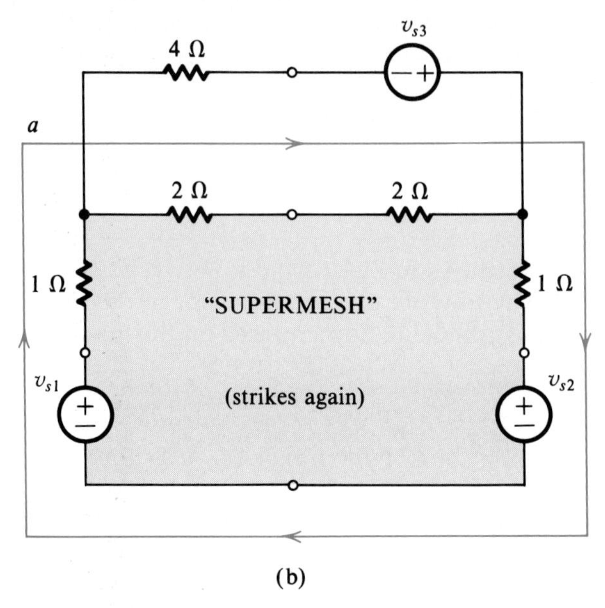

Figure 2.14(b)

The third equation is from M3

$$-2i_1 - 2i_2 + 8i_3 = v_{s3}$$

In matrix format

$$\begin{matrix} \text{M1,2/CE:} \\ \text{SM1,2:} \\ \text{M3:} \end{matrix} \quad \begin{bmatrix} -1 & 1 & 0 \\ 3 & 3 & -4 \\ -2 & -2 & 8 \end{bmatrix} \begin{bmatrix} i_1 \\ i_2 \\ i_3 \end{bmatrix} = \begin{bmatrix} 0 & 0 & 0 & 1 \\ 1 & -1 & 0 & 0 \\ 0 & 0 & 1 & 0 \end{bmatrix} \begin{bmatrix} v_{s1} \\ v_{s2} \\ v_{s3} \\ i_{s1} \end{bmatrix}$$

The solution of this model for given sources provides the mesh current vector i.

SUMMARY

This chapter showed how to use Kirchhoff's current law (KCL) to develop models for the behavior of the node voltages in a circuit driven by independent sources. Matrix methods were used to represent and solve the circuit's mathematical model for the values of the $N - 1$ node voltages measured relative to a chosen ground node. These voltages uniquely determine all of the branch voltages in the circuit. Kirchhoff's voltage law (KVL) was used to develop models for the behavior of the mesh or loop currents of a circuit. Finally, we found that the mesh currents of a planar circuit uniquely determine all of the branch currents in that circuit.

Problems - Chapter 2

2.1 For the circuit shown in Fig. P2.1:

a. Write KCL at node 1; repeat for node 2.

b. Write the nodal model in matrix form.

c. Find the node voltage vector.

d. Find v_1, v_2, i_1, i_2 and i_3 when $i_{s1} = 2$ A and $i_{s2} = -3$ A.

e. Calculate the power supplied by the current sources.

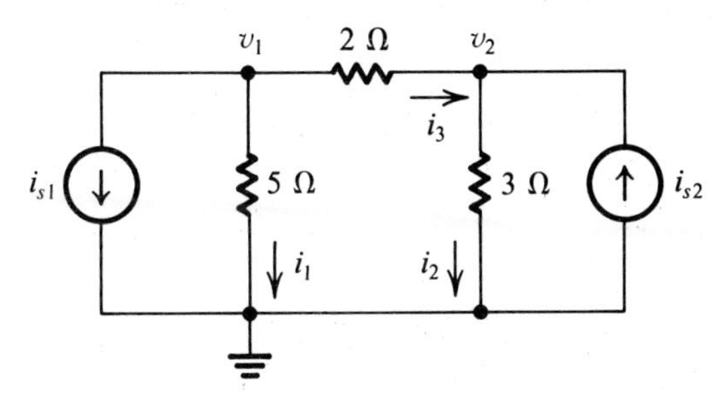

Figure P2.1

2.2 Use the short-cut method to write the nodal model for the circuit in Fig. P2.1.

2.3 Use nodal analysis to find i when $i_{s1} = 3$ A and $i_{s2} = 8$ A.

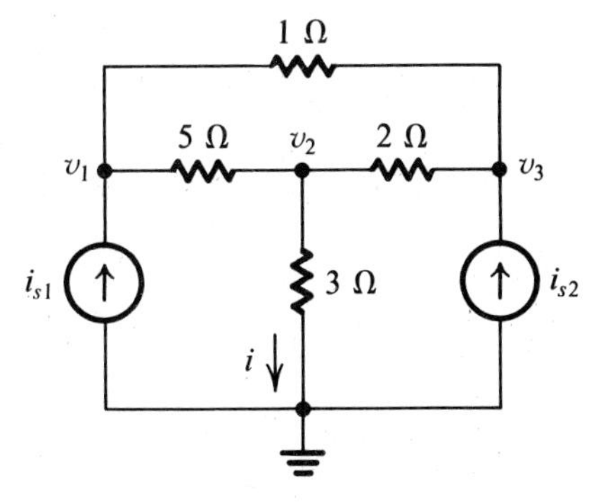

Figure P2.3

2.4 Find v_1, v_2, and v_3.

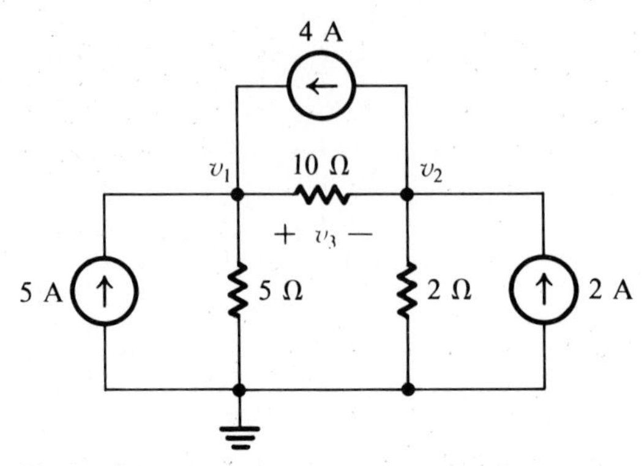

Figure P2.4

2.5 For the circuit shown in Fig. P2.5:

a. Write the constraint equation.

b. Write the nodal model in matrix form.

c. Find the node voltage vector when $v_s = 100$ V and $i_s = 0$ A.

d. Repeat c with $v_s = 0$ V and $i_s = 5$ A.

e. Repeat c with $v_s = 100$ V and $i_s = 5$ A.

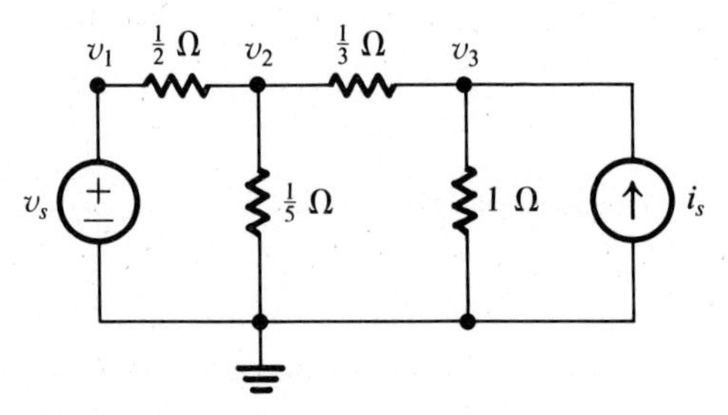

Figure P2.5

2.6 Develop the nodal model in matrix form for the circuit in Fig. P2.6.

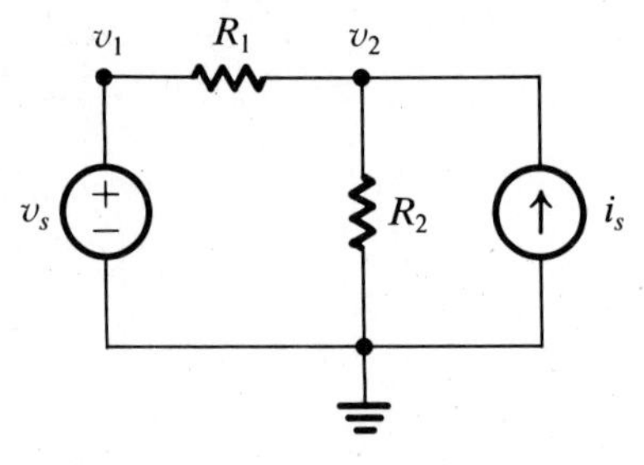

Figure P2.6

2.7 Find v when $v_s = -5$ V and $i_s = 4$ A.

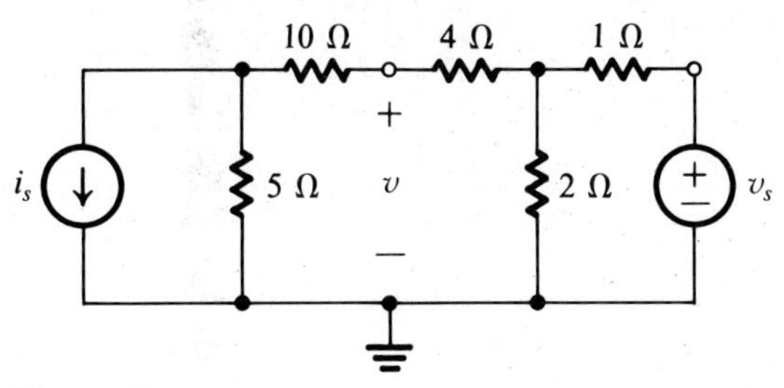

Figure P2.7

2.8 Find v_o when $v_s = 10$ V and $i_s = 5$ A.

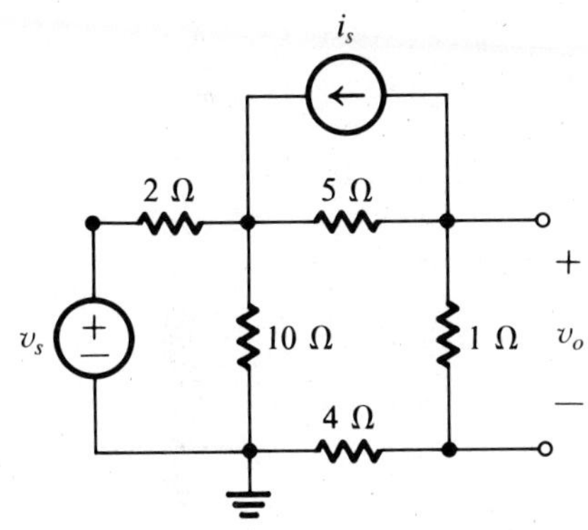

Figure P2.8

2.9 If $v_{s1} = -5$ V and $i_{s1} = 4$ A, find v_1, v_2, and v_3.

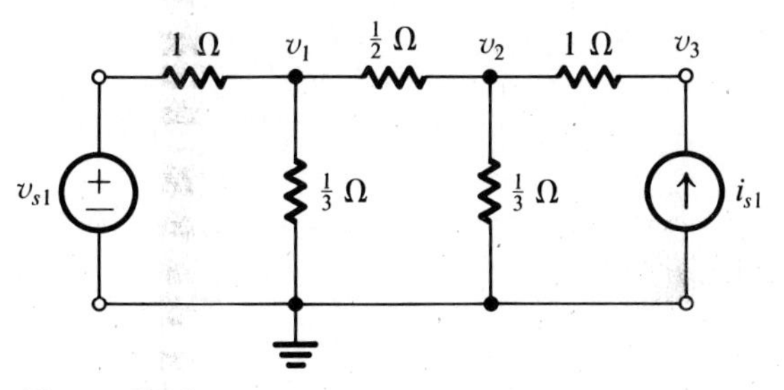

Figure P2.9

2.10 For the circuit in Fig. P2.10:

a. Find v_o when $v_{s1} = 10$ V.

b. Find v_o when $v_{s1} = 20$ V.

c. Find the power delivered by the source in both cases.

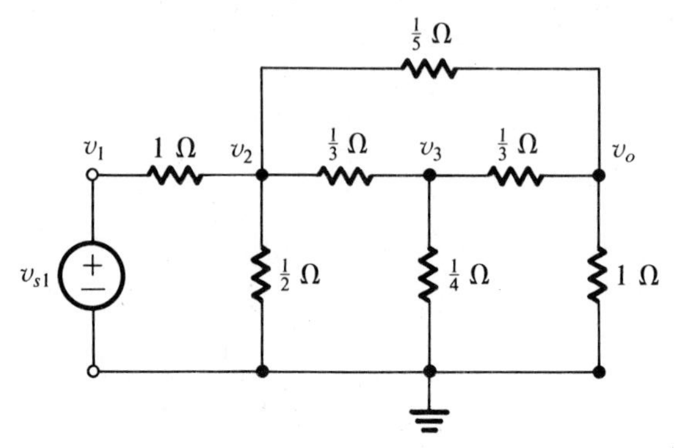

Figure P2.10

2.11 Find the node voltage vector when

a. $v_{s1} = 5$ V, $v_{s2} = 10$ V, and $i_{s1} = 2$ A.

b. $v_{s1} = 5$ V, $v_{s2} = 0$ V, and $i_{s1} = 0$ V.

c. $v_{s1} = 0$ V, $v_{s2} = 10$ V, and $i_{s1} = 0$ V.

d. $v_{s1} = 0$ V, $v_{s2} = 0$ V, and $i_{s1} = 2$ A.

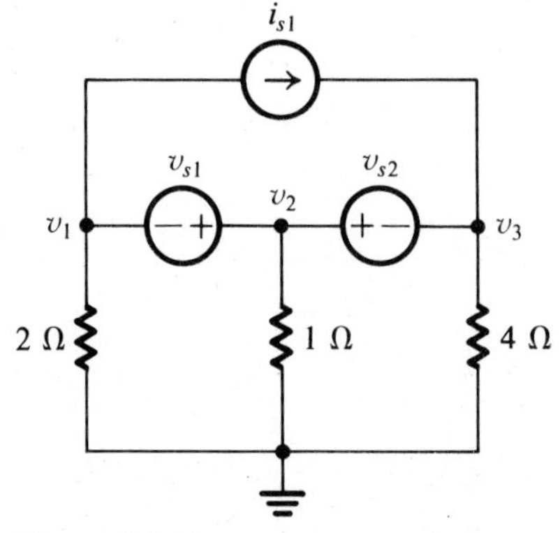

Figure P2.11

2.12 For the circuit in Fig. P2.12, find the current i and the power supplied by the voltage source if $v_s = 50$ V, $i_s = 10$ A, and $i_{c1} = 3i_a$

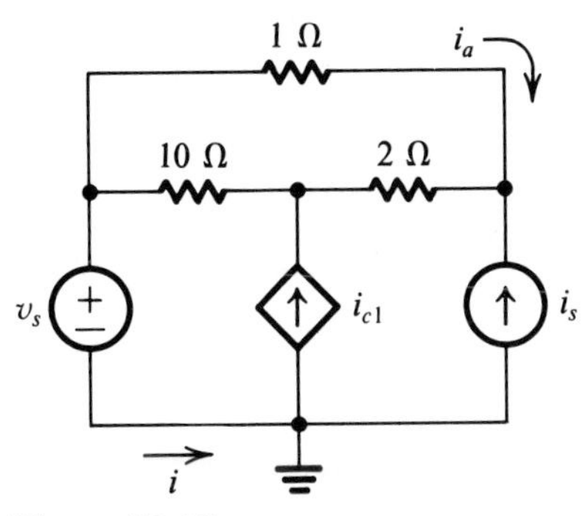

Figure P2.12

2.13 The circuit in the shaded portion of Fig. P2.13 is a small-signal model of a bipolar transistor connected in a common-emitter configuration. It can be used to describe the transistor in its linear region of operation, such as when it is used as a signal amplifier. In this example, the input node and the output node are connected by a feedback resistor, and the output signal v_o depends on the input signal v_{in} in a manner that is a function of the circuit and transistor parameters.

a. Find an expression for v_o

b. Find an expression for the input/output voltage gain, v_o/v_{in}.

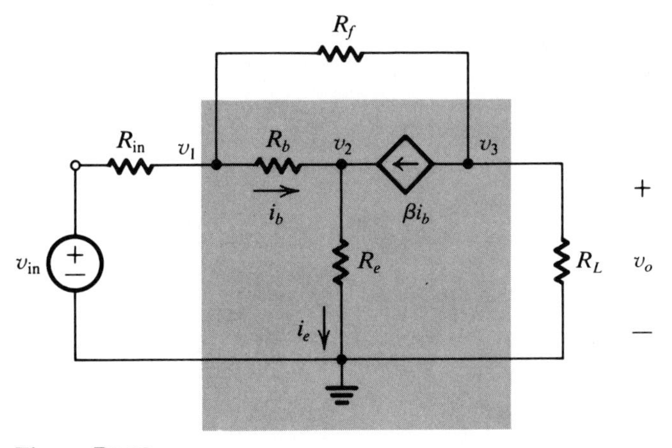

Figure P2.13

2.14 Find the power supplied by the controlled current source.

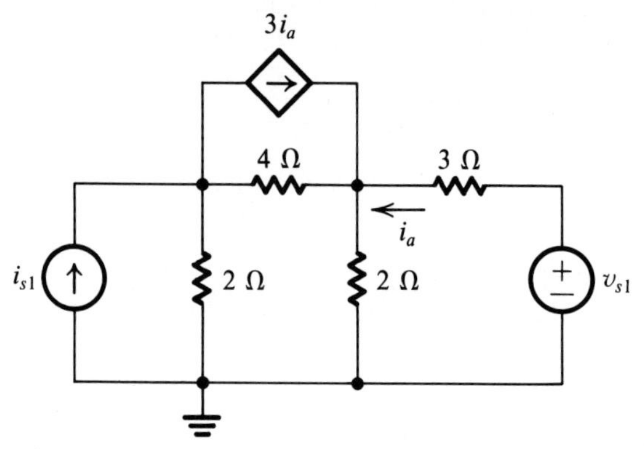

Figure P2.14

2.15 Find v_o when $i_s = 20$ A and $v_s = 15$ V.

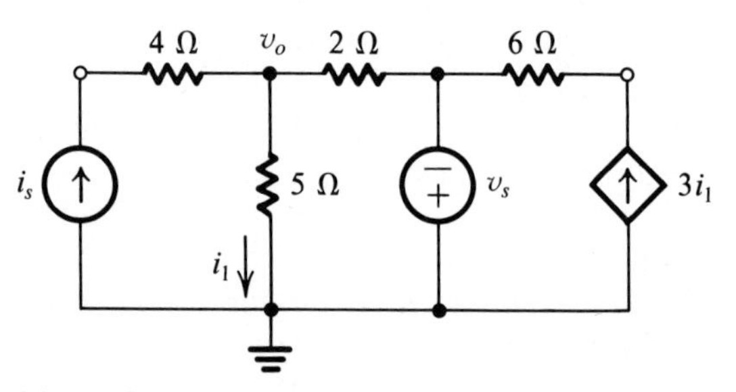

Figure P2.15

2.16 Find i_a when $v_s = 50$ V.

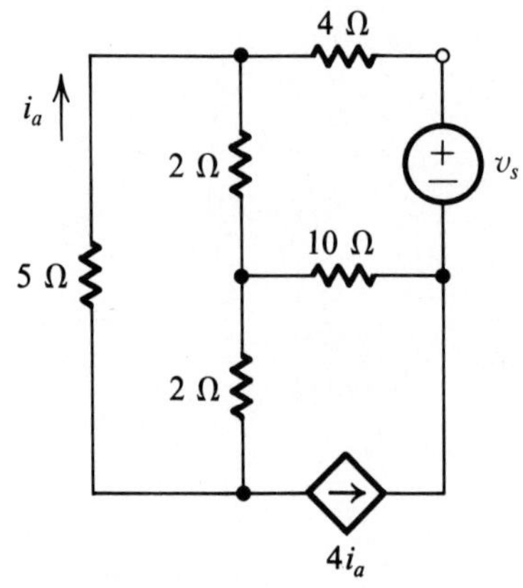

Figure P2.16

2.17 Find the power delivered by the controlled sources when $v_{s1} = 10$ V and $i_{s1} = 4$ A.

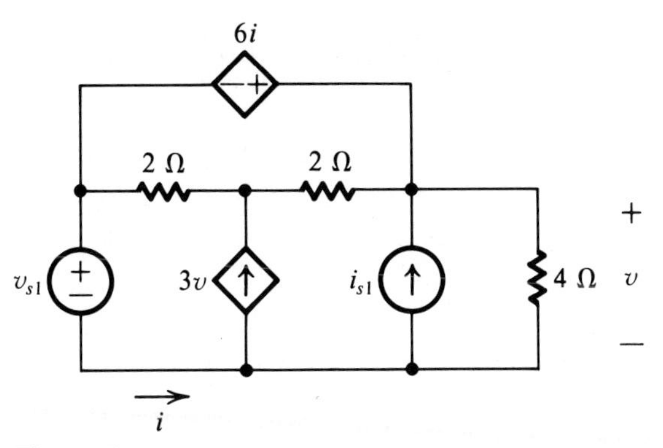

Figure P2.17

2.18 Find v_o/v_s, the gain of the transistor amplifier in Fig. P2.18.

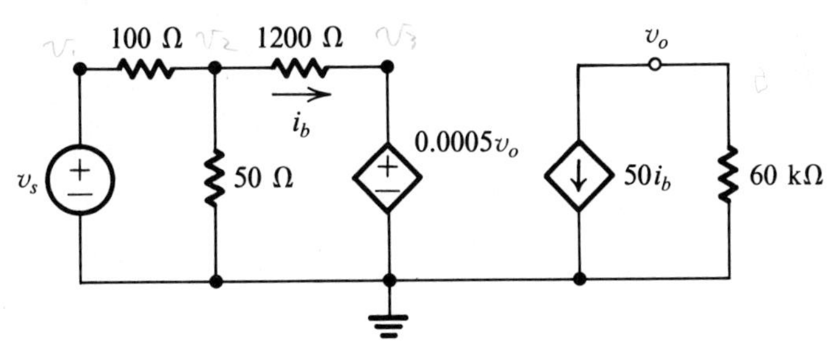

Figure P2.18

2.19 Find the power dissipated by the 100 Ω resistor.

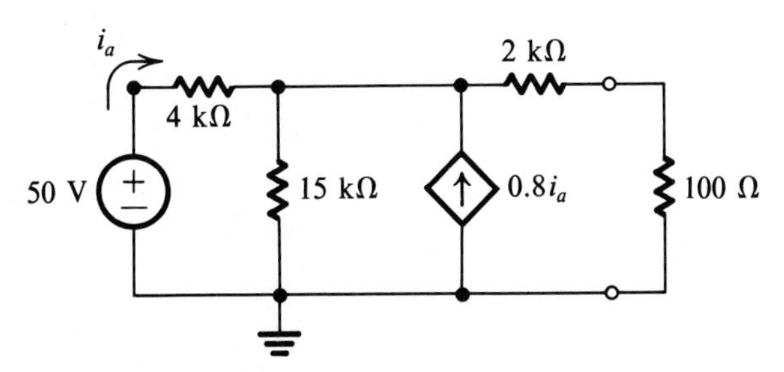

Figure P2.19

2.20 Find i_a when $i_{s1} = 3$ A and $i_{s2} = 5$ A.

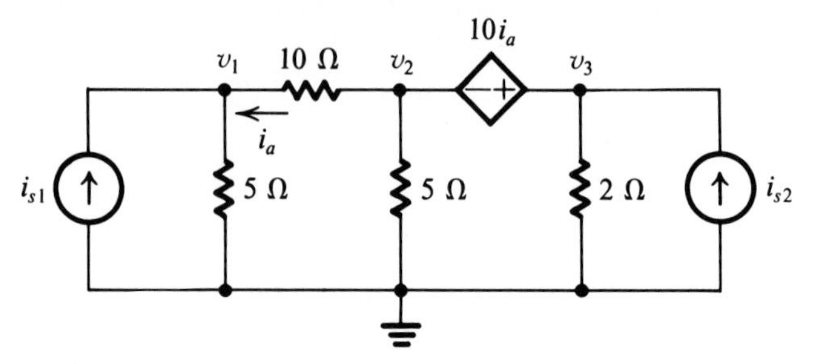

Figure P2.20

2.21 Find the input/output gain, v_o/v_s.

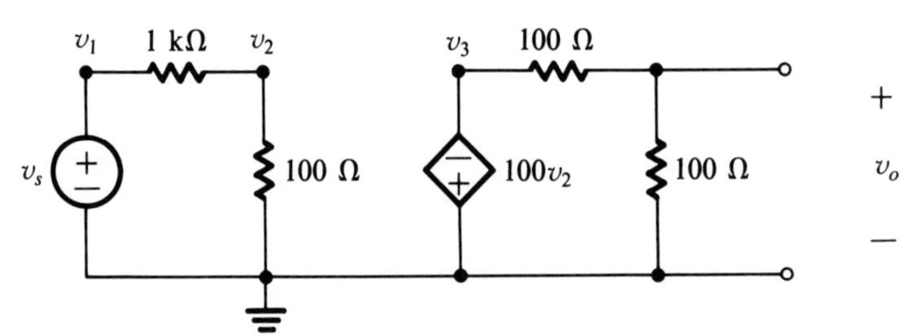

Figure P2.21

2.22 For the circuit in Fig. P2.22,

a. Write the super node equation

b. Find i.

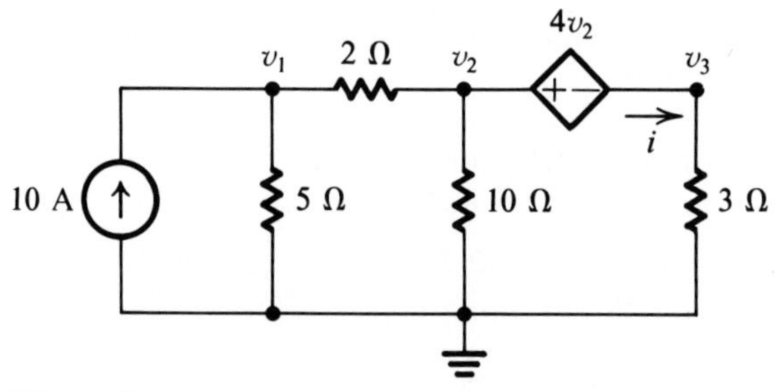

Figure P2.22

2.23 For the circuit in Fig. P2.23:

a. Write the super node equation.

b. Develop the matrix nodal model.

c. Find the node voltages when $i_{s1} = 5$ A and $i_{s2} = 8$ A.

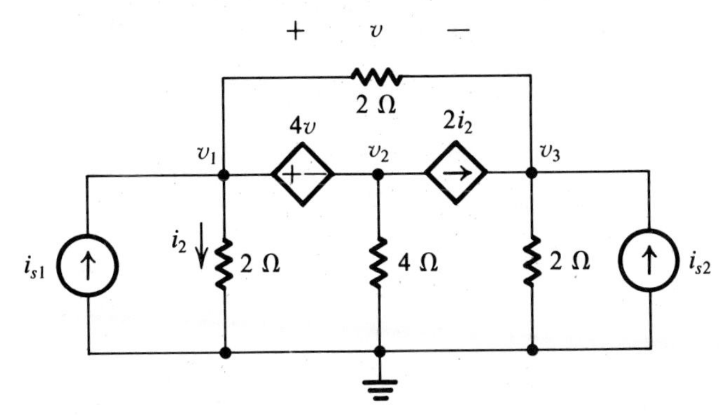

Figure P2.23

2.24 Find v_a when $v_s = 20$ V.

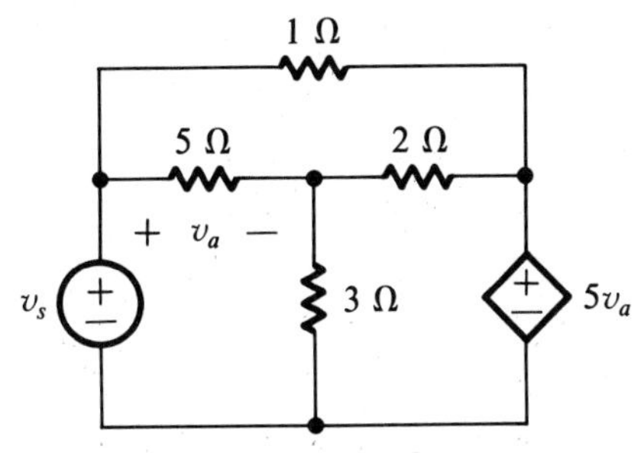

Figure P2.24

2.25 For the circuit in Fig. 2.25, write the constraint equations in terms of the node voltages.

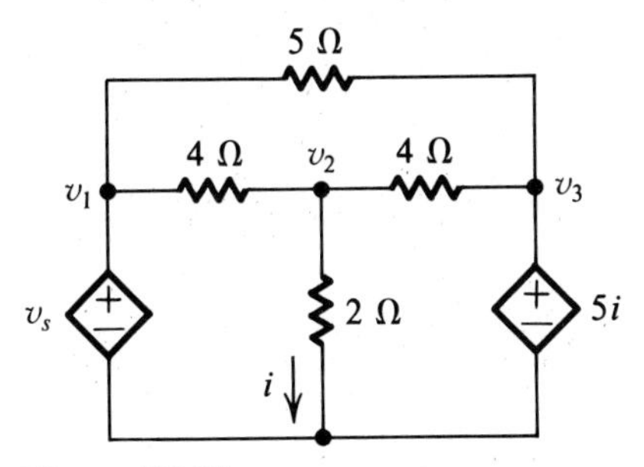

Figure P2.25

2.26 Find v_1, v_2, and v_3.

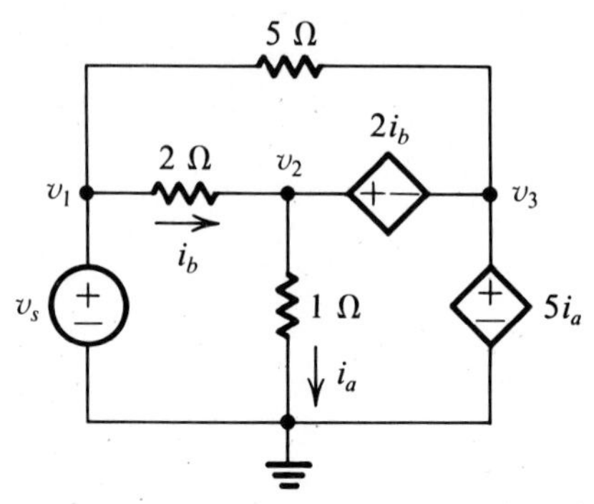

Figure P2.26

2.27 Find the power dissipated in the 2 Ω resistor when $v_s = 10$ V, $v_c = 3i_a$ and $i_s = 5$ A.

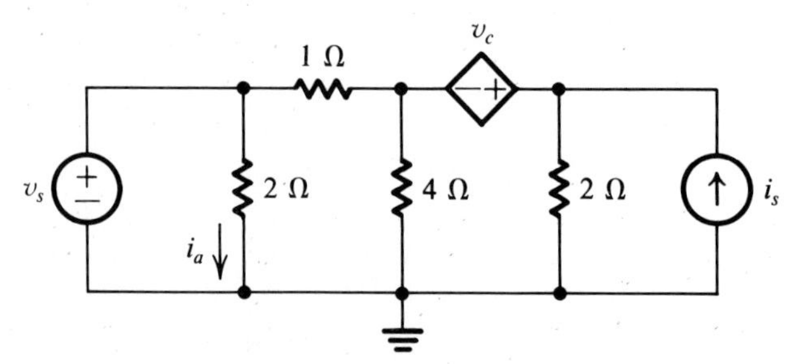

Figure P2.27

2.28 If $v_{s1} = 5$ V and $v_{s2} = 2$ V, use mesh analysis to

a. Find i_1, i_2, and i_3.

b. Find the branch voltages, v_a, v_b and v_c.

c. Find the node voltages, v_1 and v_2.

d. Find the power supplied by each source.

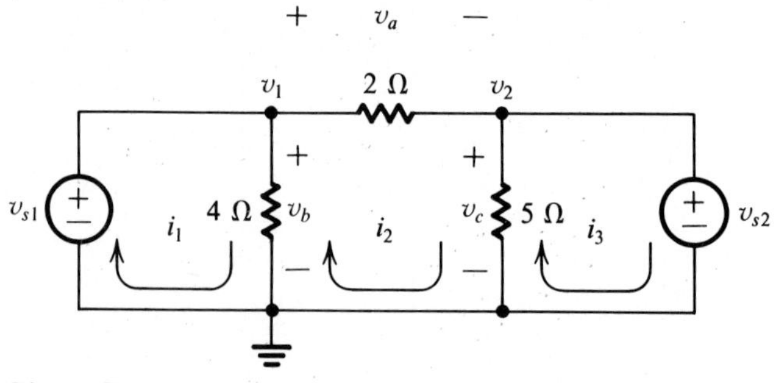

Figure P2.28

2.29 Using mesh analysis with $v_{s1} = 24$ V and $v_{s2} = 15$ V,

a. Find the branch currents, i_a, i_b, and i_c.

b. Find v.

c. Find the power supplied by each source.

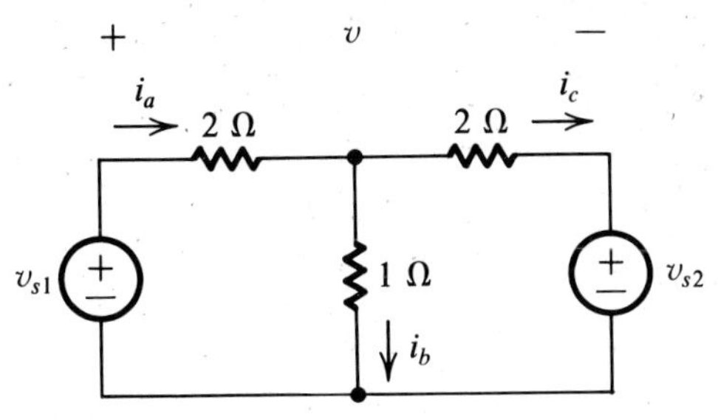

Figure P2.29

2.30 Draw the super mesh for the circuit in Fig. P2.30.

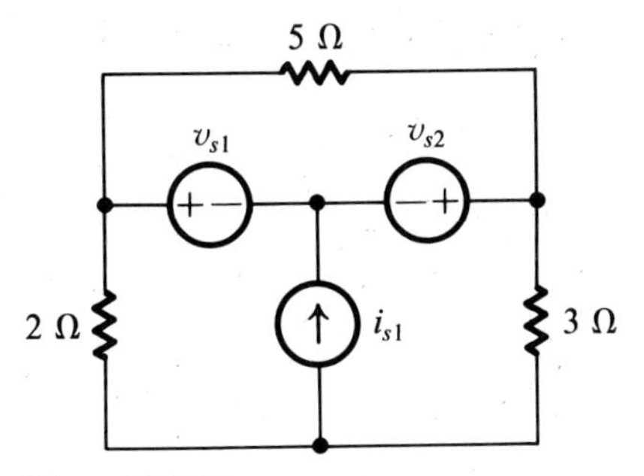

Figure P2.30

2.31 Solve for the mesh currents with $v_{s1} = 20$ V and $v_{s2} = -5$ V.

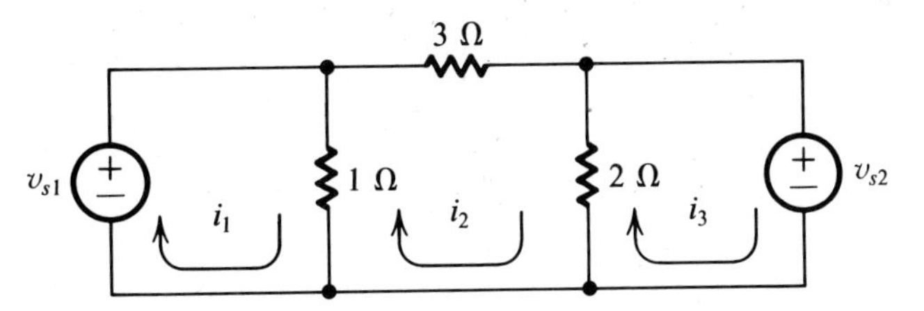

Figure P2.31

2.32 Develop the mesh model for the circuit in Fig. P2.32.

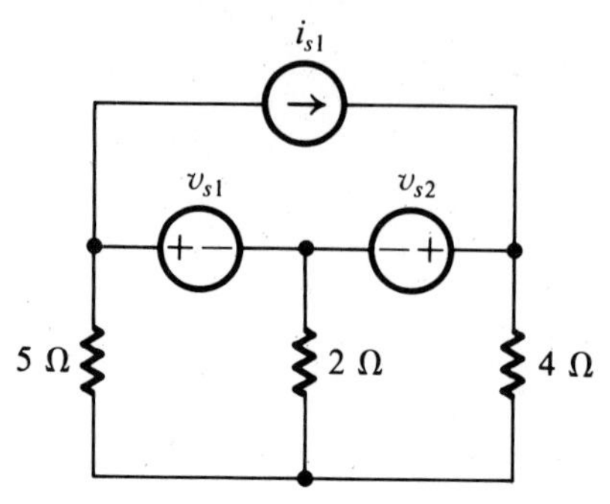

Figure P2.32

2.33 Draw the supermesh and write KVL for the super mesh in Fig. P2.33.

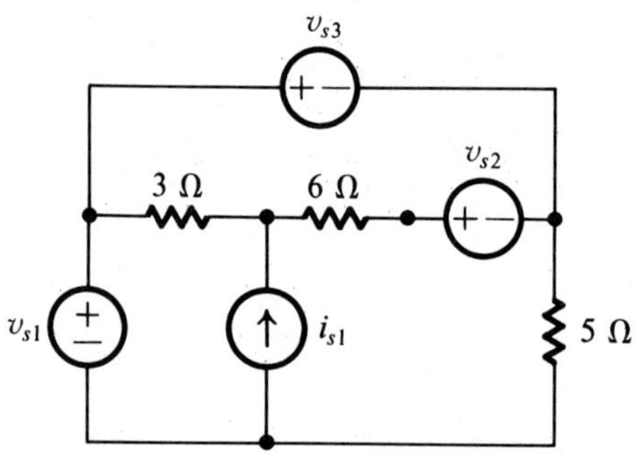

Figure P2.33

2.34 If $i_c = Kv_a$,

a. Draw the super mesh.
b. Write KVL for the super mesh.
c. Develop the mesh model of the circuit.
d. Find the voltage gain, v_o/v_s when $K = 2$.

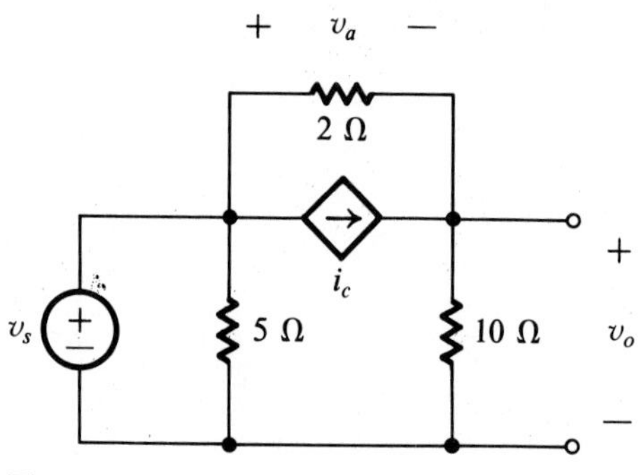

Figure P2.34

2.35 For the circuit in Fig. P2.35, find i_a when $v_{s1} = 10$ V, $v_{s2} = 5$ V and $i_c = 3i_a$.

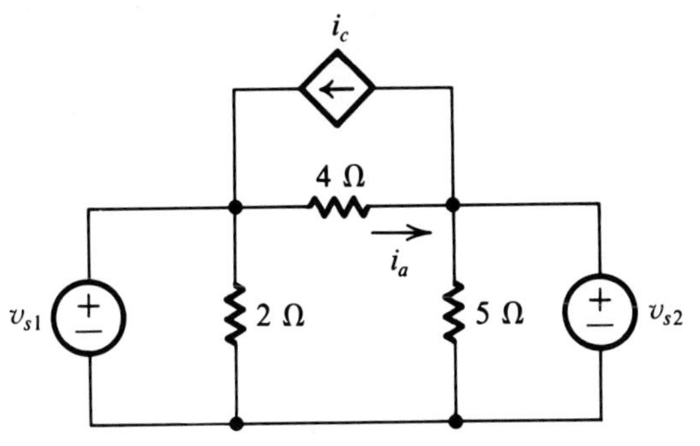

Figure P2.35

2.36 Using mesh analysis, find v_o in Fig. P2.36 when $v_{s1} = 12$ V and $v_{s2} = 18$ V.

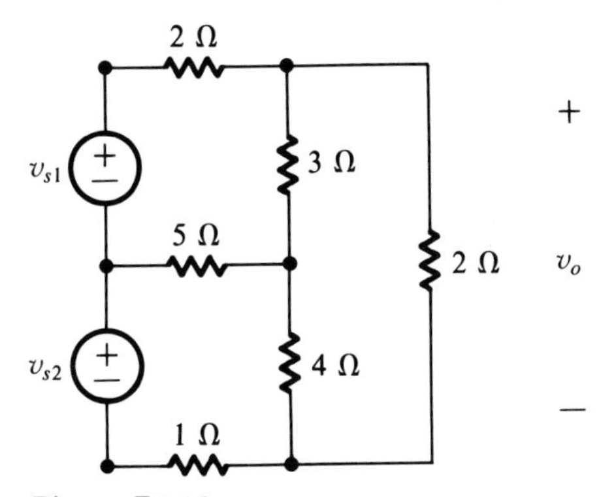

Figure P2.36

2.37 Find the power dissipated in each resistor in Fig. P2.37 when $v_s = 10$ V.

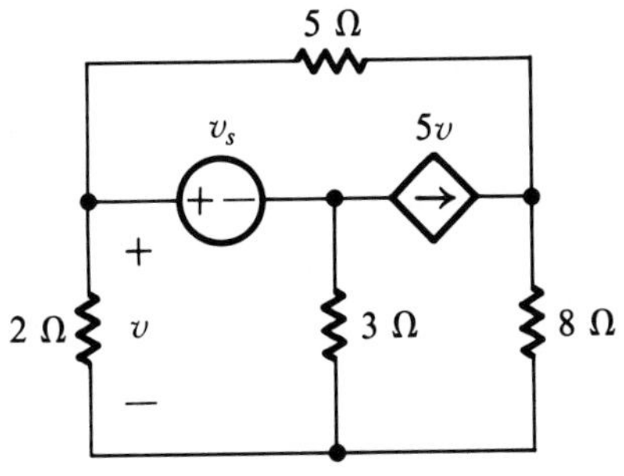

Figure P2.37

2.38 Using mesh analysis, find the node voltages when $v_s = 100$ V.

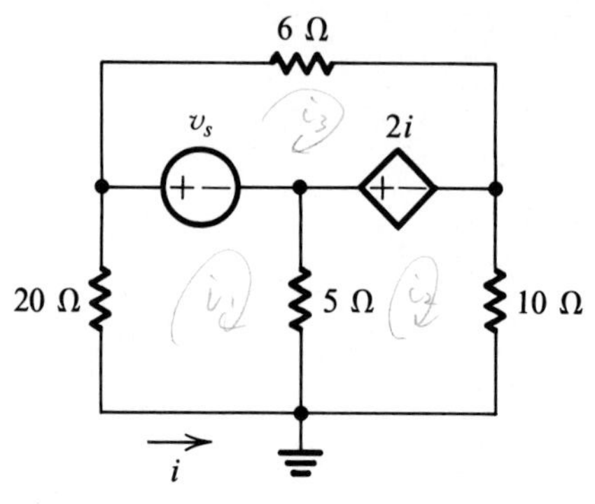

Figure P2.38

2.39 Find v_1, v_2, v_3 and v in the bridge circuit of Fig. P2.39.

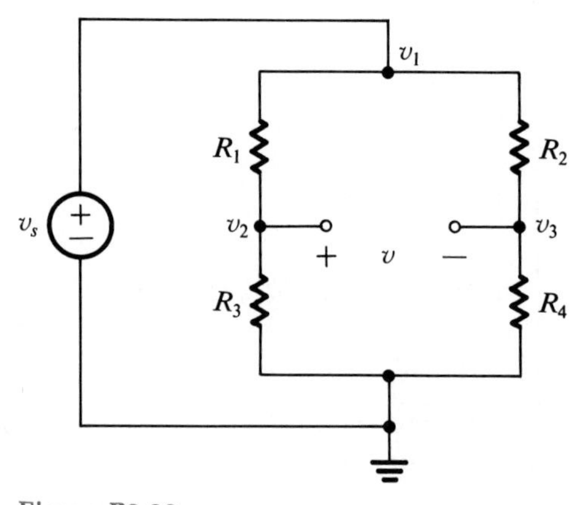

Figure P2.39

2.40 Find an expression for v_o in terms of the source and resistors in Fig. P2.40.

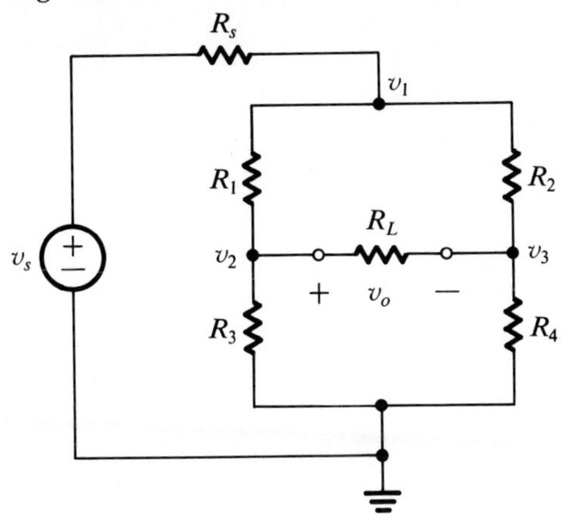

Figure P2.40

USEFUL CIRCUIT PROPERTIES

INTRODUCTION

Chapter 2 presented the fundamental physical laws for electrical circuits and used those laws to develop nodal and mesh analysis models describing the circuits' behavior. This chapter will show that the models have a linear structure—which leads to several important properties that can simplify circuit analysis. The first property, superposition, is one we are already familiar with. From the experience of hearing two radio stations at once, we have an intuitive understanding that the response of a circuit to many voltage and current sources acting simultaneously can be related to its response to each source acting separately. This property has a far-reaching effect, providing as it does a basic framework for understanding how circuits respond to very complex signal shapes, such as a radar echo or a musical ensemble. Another important property that follows from linearity is that a complex circuit can be replaced by a very simple equivalent circuit.

3.1 LINEARITY AND SUPERPOSITION

The nodal model for a circuit provides a set of algebraic equations that summarize the physically constrained behavior of the circuit. The model expresses a relationship between the node voltage vector and the independent current and voltage sources that cause the node voltages and branch currents. In general, the node voltage vector is a function of the source vector, which we express as

$$\mathbf{v} = \mathbf{f}(\mathbf{u}_s)$$

We say that the function $\mathbf{f}(\cdot)$ is **linear,** if

i. $\mathbf{f}(\mathbf{u}_{s1} + \mathbf{u}_{s2}) = \mathbf{f}(\mathbf{u}_{s1}) + \mathbf{f}(\mathbf{u}_{s2})$

and

ii. $\mathbf{f}(k\mathbf{u}_s) = k\mathbf{f}(\mathbf{u}_s)$

for any scalar k and any source vectors $\mathbf{u}_{s1}$ and $\mathbf{u}_{s2}$. The first property is called **additivity,** and the second is called **scaling.** The property of additivity says that the node voltage vector of a circuit with two or more sources is the sum of the node voltage vectors caused by each source separately. The property of scaling implies that if the source *vector* is scaled by a factor of k the node voltage *vector* will also be scaled by the same factor, i.e. each node voltage is scaled by k. If the model of a circuit is linear, the circuit is said to obey the **principle of superposition.**

The nodal equation model of a circuit is linear. Each equation expresses a linear relationship between the node voltages and the independent sources, even when the circuit includes controlled sources—provided that the control mechanism is linear, e.g. we do not allow a controlled current source to be of the form $i = kv^2$. The linearity of the model is easily demonstrated by noting that

$$\mathbf{v} = \mathbf{G}^{-1}\mathbf{B}(\mathbf{u}_{s1} + \mathbf{u}_{s2}) = \mathbf{G}^{-1}\mathbf{B}\mathbf{u}_{s1} + \mathbf{G}^{-1}\mathbf{B}\mathbf{u}_{s2}$$

for any source vectors $\mathbf{u}_{s1}$ and $\mathbf{u}_{s2}$ and that

$$\mathbf{v} = \mathbf{G}\mathbf{B}\quad(k\mathbf{u}_s) = k\mathbf{G}^{-1}\mathbf{B}\mathbf{u}_s$$

for any scalar k. The component of $\mathbf{v}$ due to $\mathbf{u}_{s1}$ is $\mathbf{G}^{-1}\mathbf{B}\mathbf{u}_{s1}$ and the component due to $\mathbf{u}_{s2}$ is $\mathbf{G}^{-1}\mathbf{B}\mathbf{u}_{s2}$. Consequently, **the node voltages due to two or more independent sources acting simultaneously is the sum of the node voltages when the sources act separately**. Circuits constructed from resistors, independent sources, and controlled sources obey the principle of superposition.

The mesh current model of a circuit is linear too, since

$$\mathbf{i} = \mathbf{R}^{-1}\mathbf{B}(\mathbf{u}_{s1} + \mathbf{u}_{s2}) = \mathbf{R}^{-1}\mathbf{B}\mathbf{u}_{s1} + \mathbf{R}^{-1}\mathbf{B}\mathbf{u}_{s2}$$

for any source vectors $\mathbf{u}_{s1}$ and $\mathbf{u}_{s2}$ and

$$\mathbf{i} = \mathbf{R}^{-1}\mathbf{B}(k\mathbf{u}_s) = k\mathbf{R}^{-1}\mathbf{B}\mathbf{u}_s$$

for any scalar k.

Example 3.1

The circuit in Fig. 3.1 is described by

$$\begin{bmatrix} 1 & 0 & 0 \\ -1 & 7 & -4 \\ 0 & -4 & 4 \end{bmatrix} \begin{bmatrix} v_1 \\ v_2 \\ v_3 \end{bmatrix} = \begin{bmatrix} 1 & 0 \\ 0 & 0 \\ 0 & 1 \end{bmatrix} \begin{bmatrix} v_{s1} \\ \\ i_{s1} \end{bmatrix}$$

and so

$$\mathbf{v} = \frac{1}{12} \begin{bmatrix} 12 & 0 & 0 \\ 4 & 4 & 4 \\ 4 & 4 & 7 \end{bmatrix} \begin{bmatrix} 1 & 0 \\ 0 & 0 \\ 0 & 1 \end{bmatrix} \begin{bmatrix} v_{s1} \\ \\ i_{s1} \end{bmatrix} = \frac{1}{12} \begin{bmatrix} 12 & 0 \\ 4 & 4 \\ 4 & 7 \end{bmatrix} \begin{bmatrix} v_{s1} \\ \\ i_{s1} \end{bmatrix}$$

If $v_{s1} = 3$ and $i_{s1} = 2$ we have

$$\mathbf{u}_s = \begin{bmatrix} 3 \\ \\ 2 \end{bmatrix} \qquad \mathbf{v} = \begin{bmatrix} 3 \\ \frac{5}{3} \\ \frac{13}{6} \end{bmatrix}$$

Now consider the two nodal vectors that result when v_{s1} and i_{s1} act separately. Let $\mathbf{u}_1$ symbolize the source *vector* when $v_{s1} = 3$ and $i_{s1} = 0$. If $\mathbf{u}_1$ is applied to the circuit, the node voltage vector is obtained by solving the model with $\mathbf{u}_1$ to get

$$\mathbf{u}_1 = \begin{bmatrix} 3 \\ \\ 0 \end{bmatrix} \qquad \mathbf{v} = \begin{bmatrix} 3 \\ 1 \\ 1 \end{bmatrix}$$

Likewise, let $\mathbf{u}_2$ denote the source vector when $v_{s1} = 0$ and $i_{s1} = 2$. If $\mathbf{u}_2$ is applied, solving the nodal model gives

$$\mathbf{u}_2 = \begin{bmatrix} 0 \\ \\ 2 \end{bmatrix} \qquad \mathbf{v} = \begin{bmatrix} 0 \\ \frac{2}{3} \\ \frac{7}{6} \end{bmatrix}$$

Note that $\mathbf{u}_s = \mathbf{u}_1 + \mathbf{u}_2$—the node voltage vector that results from $\mathbf{u}_s$ is the sum of the node voltage vectors that result from $\mathbf{u}_1$ and $\mathbf{u}_2$.

Superposition of the effects of the separate sources can also be seen in Fig. 3.2a, where the current source is turned off, and Fig. 3.2b, where the voltage source is turned off. With $i_{s1} = 0$ in Fig. 3.2a, $v_3 = v_2$ because no current flows in the 1/4 Ω resistor. Then, by voltage division, $v_2 = 1$, and

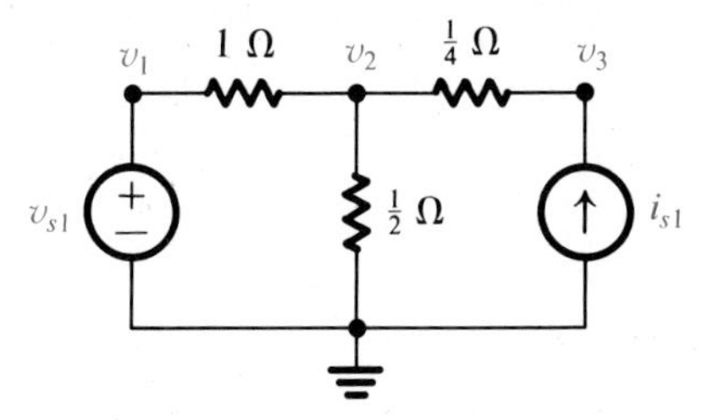

Figure 3.1 The separate voltages caused by v_{s1} and i_{s1} are super imposed.

the node voltage vector is $\mathbf{v} = [3 \quad 1 \quad 1]^T$, where $[\ ]^T$ denotes the transpose of the enclosed row vector. This agrees with the result obtained from solving the nodal model. Similarly, setting $v_{s1} = 0$ in Fig. 3.2b forces the condition that $v_1 = 0$. Combining the parallel resistors leads to $v_2 = 2/3$ and $v_3 = 7/6$, so $\mathbf{v} = [0 \quad 2/3 \quad 7/6]^T$. Again, note that the node voltage vectors due to each source acting separately *add* to give the node voltage vector that results when both sources act simultaneously.

The previous example showed that the effect of the voltage source can be obtained by inspecting the circuit when the current source is set to 0 (open circuit), and the effect of the current source can be obtained by inspecting the circuit when the voltage source is set to 0 (short circuit). The example also demonstrates the important fact that **a current source acts like an open circuit to the other independent sources,** i.e. no other source can cause a current in the branch containing an independent current source. For example, the branch currents in Fig. 3.3a when $i_{s1} = 0$ are $i_a = i_b = 2$, and $i_c = 0$, so i_{s1} behaves like an open-circuit. Likewise, the branch currents in Fig. 3.3b when $v_{s1} = 0$ are $i_a = -2/3$, $i_b = 4/3$, and $i_c = -i_{s1} = -2$. In this case v_{s1} has no effect on the distribution of i_{s1} among the branches, and it behaves like a short circuit *for the current caused by* i_{s1}. **In general, an independent voltage source behaves like a short circuit**

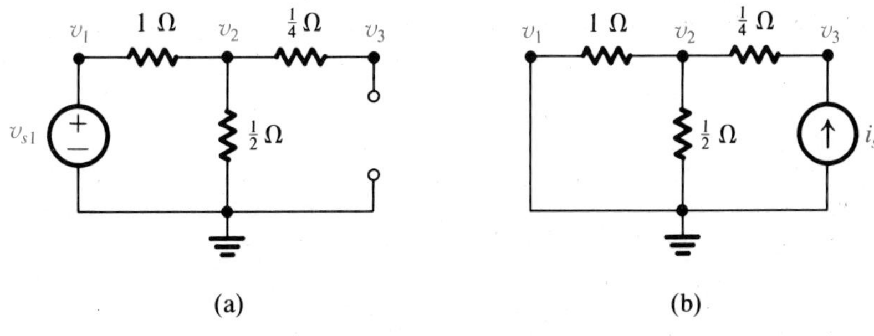

Figure 3.2(a) Circuit of Fig. 3.1 with current source turned off.
Figure 3.2(b) Circuit of Fig. 3.1 with voltage source turned off.

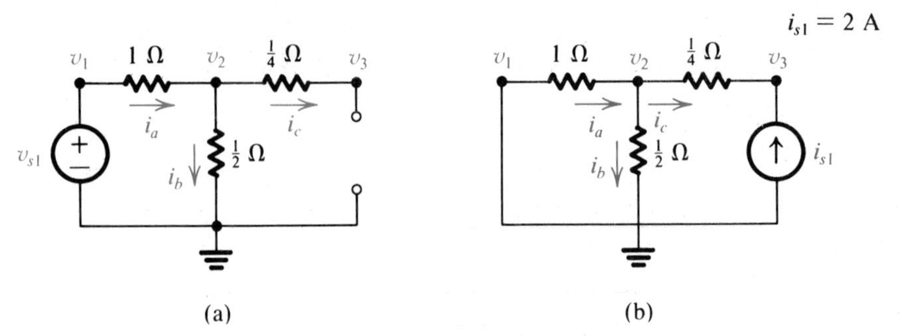

Figure 3.3(a) Branch currents when the current source is turned off in Fig. 3.1. **Figure 3.3(b)** Branch currents when the voltage source is turned off in Fig. 3.1.

to the other independent sources. No other independent source can cause a voltage at the node of an independent voltage source (forbidding a pathological connection of ideal sources).

Skill Exercise 3.1

Use superposition to find the current i_a in Fig. SE3.1.

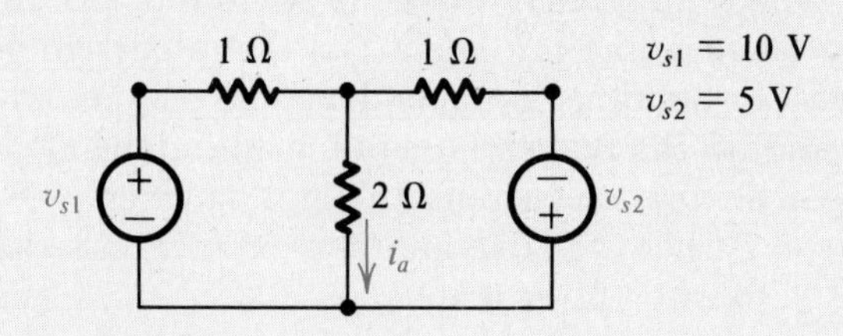

Figure SE3.1

Answer: $i_a = 1$ A.

Skill Exercise 3.2

Find the voltage at node 2 in Fig. SE3.2 when only i_{s2} is applied to the circuit and $i_{s2} = -4$.

Answer: $v_2 = -2$ V.

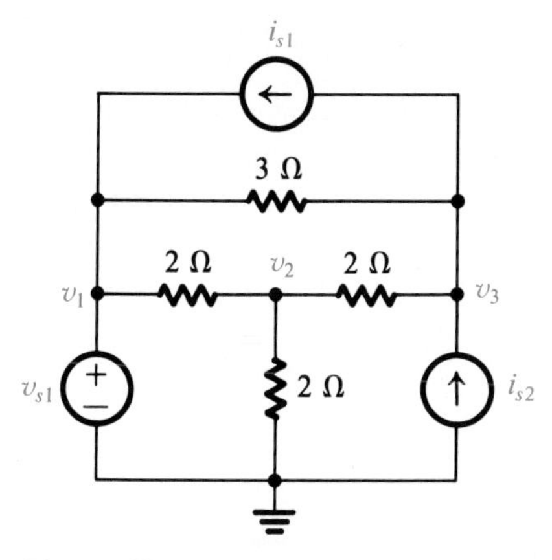

Figure SE3.2

Many circuits are more easily analyzed with superposition because their complexity is reduced when each source acts separately. There is a tradeoff though, because superposition requires solution of as many separate problems as there are independent sources. If a circuit has m independent voltage sources and q independent current sources, superposition of the sources produces:

$$\begin{aligned}\mathbf{v} &= \mathbf{G}^{-1}\mathbf{B}[\mathbf{u}_1 + \mathbf{u}_2 + \cdots + \mathbf{u}_n] \\ &= \mathbf{G}^{-1}\mathbf{B}\mathbf{u}_1 + \mathbf{G}^{-1}\mathbf{B}\mathbf{u}_2 + \cdots + \mathbf{G}^{-1}\mathbf{B}\mathbf{u}_n\end{aligned}$$

where $n = m + q$ and $\mathbf{u}_j$, for $j = 1, \cdots, n$, is a column vector containing zeros everywhere but in the jth row, where it has an entry corresponding to the value of the jth independent source. Writing the nodal vector in this form displays the fact that **each node voltage has a contribution from each of the independent sources.** The contribution is the voltage that would be caused by that source acting separately.

Skill Exercise 3.3

Find the voltage at node 2 due to each source in Fig. SE3.2 when $v_{s1} = 1.5$, $i_{s1} = 2$, and $i_{s2} = 1$. Also find the voltage at node 2 when all three sources are operative.

Answer: $v_{s1} = 1.5$ causes $v_2 = 7/8$, $i_{s1} = 2$ causes $v_2 = -1$, and $i_{s2} = 1$ causes $v_2 = 1/2$. The total voltage at node 2 is $v_2 = 3/8$.

A word of caution. The node voltage vector will be scaled by the same factor as the sources if and only if the entire source vector is scaled. Scaling only one of the sources will scale only that source's contribution to the node voltage vector.

Superposition is a direct consequence of the matrix model of a circuit. In contrast, the superposition property is not readily apparent when Cramer's rule or other algebraic methods are used to solve the nodal equations. Therefore, we believe it is worthwhile to master the matrix format for nodal analysis and mesh analysis.

Superposition applies to circuits containing controlled sources. The matrix **G** includes the effects of controlled sources, but the node voltage vector **v** is solely a function of the independent source vector $\mathbf{u}_s$. It is a common misapplication of superposition to scale the controlled sources or to remove them from the circuit as though they were independent sources. Since their output values depend on the voltages and currents within the circuit, we cannot control them directly; attempts to do so will lead to erroneous results.

Example 3.2

Using superposition, find **v** for the circuit in Fig. 3.4, with $v_{s1} = 2$, $i_{s1} = 6$ and $v_{d1} = 4v_2$.

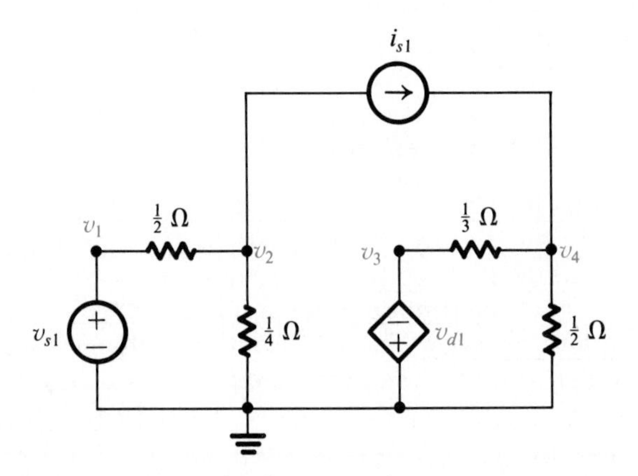

Figure 3.4 Circuit for Example 3.3.

Solution: With v_{s1} acting separately, the path for i_{s1} is open. Thus, $v_1 = v_{s1} = 2$, and voltage division leads to $v_2 = 1/3\ v_{s1} = 2/3$. Consequently, $v_3 = -v_{d1} = -8/3$. Voltage division gives $v_4 = -8/5$. When only i_{s1} is operative (the path containing v_{s1} is shorted to ground), placing the 1/2 Ω and 1/4 Ω resistors in parallel at node 2 and leads to $v_2 = -1/6\ i_{s1} = -1$. Then $v_3 = 4$. At node 4 we write KCL (why not use voltage division?):

$$3(v_4 - v_3) + 2v_4 = -3v_3 + 5v_4 = i_{s1}$$

and with $v_3 = 4$ we get $v_4 = 18/5$. Finally, the node voltages due to both sources are the sums of the voltages when each acts separately:

$$v_1 = 2 + 0 = 2 \qquad v_3 = -8/3 + 4 = 4/3$$
$$v_2 = 2/3 - 1 = -1/3 \qquad v_4 = -8/5 + 18/5 = 2$$

Thoughtless application of current division at node 4 would give the incorrect currents and voltages shown in Fig. 3.5. Ignoring the voltage source attached to node 3 leads to currents that satisfy KCL at node 4, and voltages that do *not* satisfy KVL around the mesh formed by node 3, node 4 and ground. (Verify this statement).

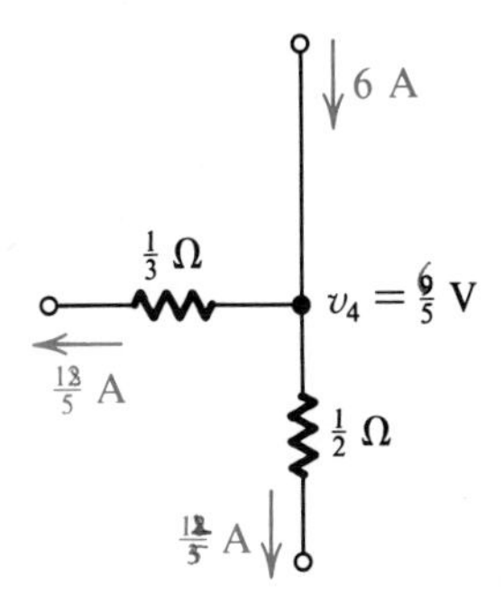

Figure 3.5 Current division? Beware!

3.2 THEVENIN EQUIVALENT CIRCUIT

In many practical problems a circuit provides power to a resistive load, as shown in Fig. 3.6. The current in the load resistor depends on the voltage across the pair of nodes connected to it. Since, by superposition, the node voltages depend on all of the internal independent sources of the circuit, we could determine the current in the load resistor by adding its components from each of the internal sources. As an alternate approach we could formulate and solve a nodal analysis problem for the loaded circuit to obtain the voltages at the load nodes, and then use Ohm's law to find

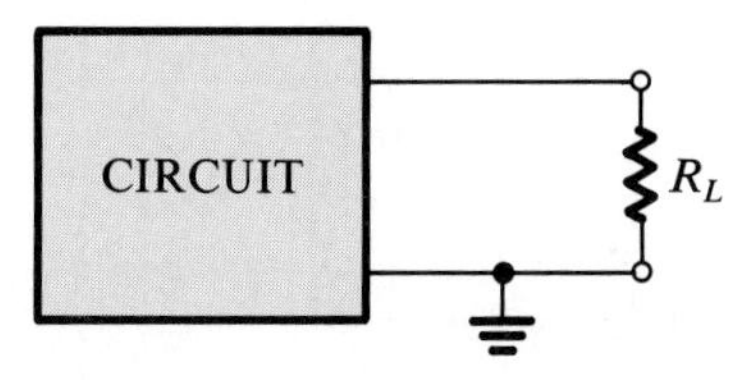

Figure 3.6 Circuit with load resistor.

the current. This task could be tedious and complicated if the circuit is very complex, with several internal nodes, sources, and resistors. For example, the signal voltage at the terminals of a speaker depends on all the internal circuitry of the amplifier and the preceding stages of the stereo system. Certainly, we would not want to have to analyze the complete stereo system as a prelude to finding the speaker response. We also know from experience that several different stereo systems are "equivalent," in the sense that they create the same terminal conditions for the speaker input.

Two circuits are equivalent if their electrical behavior at a pair of nodes is equivalent, i.e. they deliver the same voltage and current to *any* given load. Thevenin's theorem states that for a specified pair of nodes the Thevenin equivalent of a given circuit consists of a voltage source connected in series with a resistor, as shown in Fig. 3.7. The two circuits are **electrically equivalent** if they cannot be distinguished on the basis of current and voltage measurements at the terminal pair (a, b). Suppose we measure the open-circuit voltage at the terminals of the circuits. For equivalence, we require $v_{oc} = v_{ocT}$. But the open-circuit voltage of the equivalent circuit is given by $v_{ocT} = v_{TH}$, so **the Thevenin equivalent voltage source equals the open-circuit voltage of the circuit:**

$$\boxed{v_{TH} = v_{oc}} \tag{3.1}$$

The Thevenin equivalent voltage of a circuit is the open-circuit voltage due to its internal independent sources.

Next, suppose we create "short" circuits by connecting the load terminals of both circuits as in Fig. 3.8. If the circuits are equivalent $i_{sc} = i_{scT}$. But $i_{scT} = v_{TH}/R_{eq}$, so the Thevenin equivalent resistance is

$$\boxed{R_{eq} = v_{oc}/i_{sc}} \tag{3.2}$$

This indicates how to compute R_{eq} from terminal pair measurements of the circuit. (Note that the polarity of i_{sc} is chosen to be consistent with v_{oc} and Ohm's law for R_{eq}.)

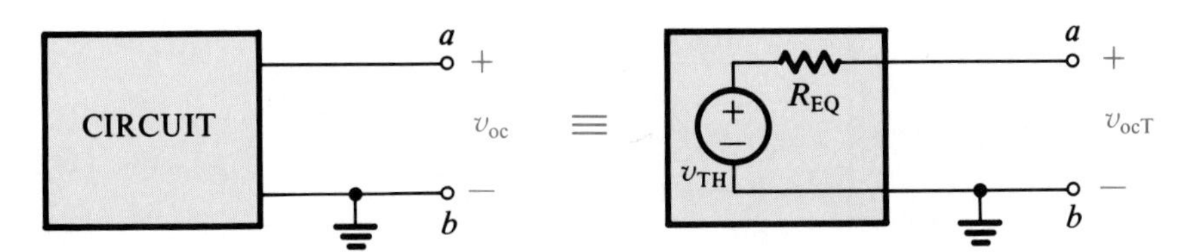

Figure 3.7 Open-circuit voltage equivalence.

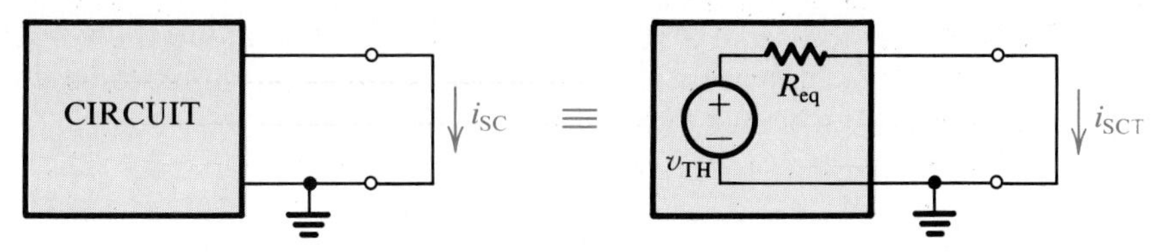

Figure 3.8 Short-circuit current equivalence.

It is important to realize that we have not shown that the two circuits in Fig. 3.7 are equivalent for all values of the load, only that (3.1) and (3.2) must be satisfied if they are. These constraints assure that the circuits are equivalent when $R_L = \infty$ and $R_L = 0$. Next, we connect a current source to (a, b) with the polarity shown in Fig. 3.9. For equivalence, the measured voltage at (a, b) must be the same for any value of i_{in}, so

$$v_{in} = i_{in}R_{eq} + v_{oc} \tag{3.3}$$

This is a **terminal model** of the circuit. Since this must be true for any value of i_{in}, it must hold when $i_{in} = 0$, so we conclude that

$$v_{oc} = v_{in}\Big|_{i_{in} = 0} \tag{3.4}$$

which corresponds to the open-circuit voltage, and

$$R_{eq} = \frac{v_{in}}{i_{in}}\Big|_{v_{oc} = 0} \tag{3.5}$$

which is the input resistance when the internal sources are set to zero. In general, $v_{oc} = 0$ when all of the *internal* independent sources are turned off (i.e. set to 0). So we write

$$R_{eq} = \frac{v_{in}}{i_{in}}\Big|_{\text{Int. Indep. Sources Off}} \tag{3.6}$$

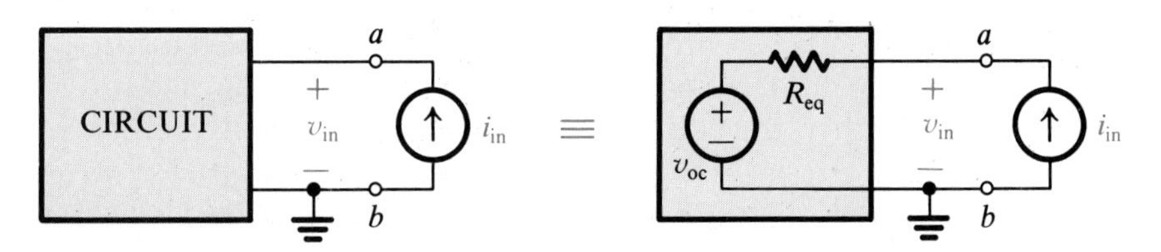

Figure 3.9 Circuit with external current source.

If the circuits of Fig. 3.9 are to be electrically equivalent, the values of the equivalent resistor and the equivalent voltage source must be chosen to satisfy (3.1), (3.2) and (3.6).

Example 3.3

Find the Thevenin equivalent circuit at (a, b) in Fig. 3.10(a).

Solution: We'll find the equivalent circuit using three methods.

Method #1

Set the independent sources to 0 and calculate R_{eq} at (a, b). With the sources turned off the circuit reduces to that shown in Fig. 3.10(b), and the resistors combine to give $R_{eq} = 2/3\ \Omega$. Next, obtain v_{OC} by superposition. With $v_{s1} = 0$ in Fig. 3.10(a) the current source produces $v_3 = 0$ because the 2 Ω and 1 Ω resistors are shorted together. On the other hand, with only v_{s1} turned on, the open-circuit voltage is $v_3 = 1/3\ v_{s1}$. With both sources turned on $v_{OC} = 1/3\ v_{s1}$. (The current source has no effect on v_{OC}.)

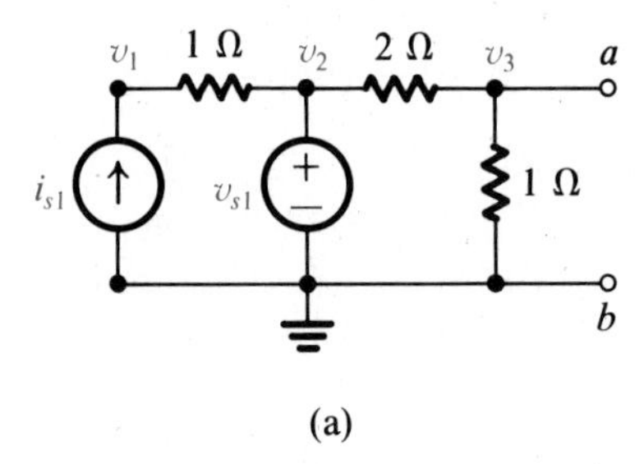

Figure 3.10(a) Circuit for Example 3.3

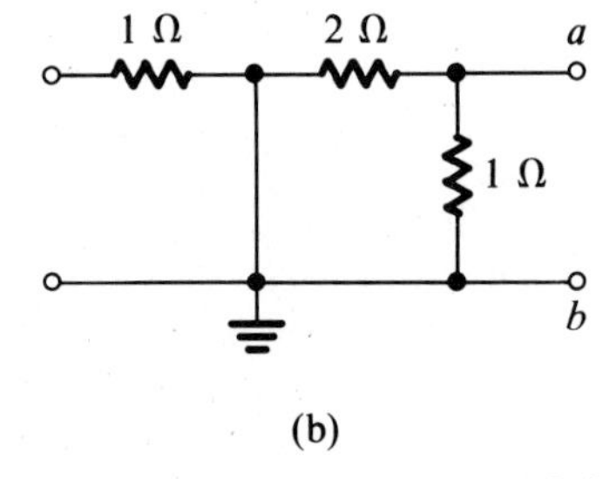

Figure 3.10(b) Circuit with both sources off.

Method #2

The second method finds i_{SC} for the circuit in Fig. 3.11. Since the terminal pair (a, b) is shorted together, the current i_1 must be 0. Consequently, all of i_{SC} passes through the 2 Ω resistor, and $i_{SC} = 1/2\ v_{s1}$. With $v_{OC} = 1/3\ v_{s1}$ already known, we use (3.2) to get $R_{eq} = v_{OC}/i_{SC} = 2/3\ \Omega$.

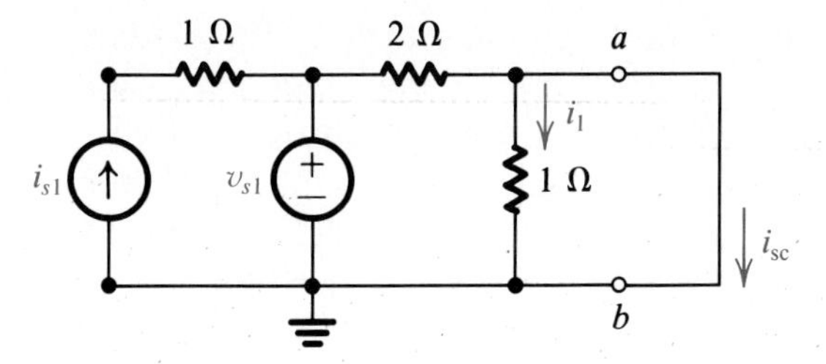

Figure 3.11 The circuit of Fig. 3.10 with the output terminals shorted.

Method #3

With this method a current source i_{in} is attached to the circuit, as shown in Fig. 3.12, and the nodal model is developed:

$$\begin{bmatrix} 1 & -1 & 0 \\ 0 & 1 & 0 \\ 0 & -\frac{1}{2} & \frac{3}{2} \end{bmatrix} \mathbf{v} = \begin{bmatrix} 0 & 1 & 0 \\ 0 & 0 & 1 \\ 1 & 0 & 0 \end{bmatrix} \begin{bmatrix} i_{in} \\ i_{s1} \\ v_{s1} \end{bmatrix}$$

Solving for **v** gives

$$\mathbf{v} = \begin{bmatrix} 0 & 1 & 1 \\ 0 & 0 & 1 \\ \frac{2}{3} & 0 & \frac{1}{3} \end{bmatrix} \begin{bmatrix} i_{in} \\ i_{s1} \\ v_{s1} \end{bmatrix}$$

Selecting the equation corresponding to v_3, with $v_{in} = v_3$, leads to $v_{in} = 2/3\ i_{in} + 1/3\ v_{s1}$. Comparing this expression to the equivalent circuit model in (3.3), we conclude that for equivalence we must have $R_{eq} = 2/3\ \Omega$ and $v_{oc} = 1/3\ v_{s1}$.

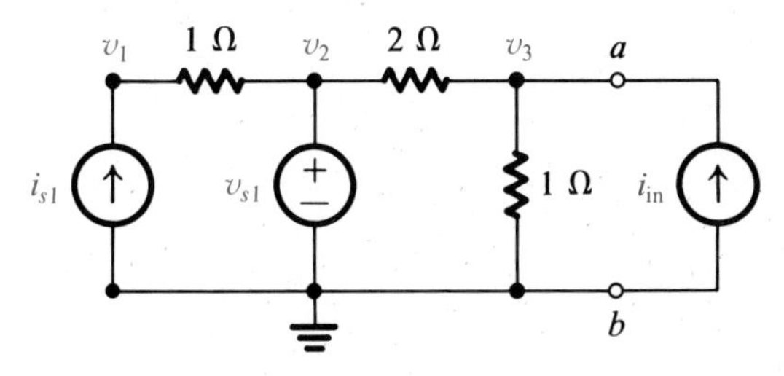

Figure 3.12 The circuit of Fig. 3.10 with an external current source added.

Note that **the three methods lead to the same equivalent circuit. The first method can only be used with circuits that do not contain controlled sources.** When the circuit contains controlled sources we cannot find R_{eq} by series/parallel reduction, so either (3.1) or (3.3) with (3.6) must be used. It's a good idea to be familiar with each approach.

Example 3.4

Find the Thevenin equivalent circuit at (a, b) for the circuit in Fig. 3.13.

Solution:

Method #1 (Find v_{oc} and R_{eq})

To find v_{oc} note that from KCL and Ohm's law we have

$$v_{oc} = v_3 = (i + 4i)\left(\frac{1}{4}\right) = \frac{5}{4}i$$

and from KVL we get

$$v_{s1} = \frac{1}{2}i + (i + 4i)\left(\frac{1}{2} + \frac{1}{4}\right) = \frac{17}{4}i$$

so that $i = 4/17 v_{s1}$ and $v_{oc} = 5/4(4/17 v_{s1}) = 5/17\ v_{s1}$. Note that R_{eq} can't be found by inspection. (Why?)

Method #2 (Find i_{sc} and R_{eq})

To find i_{sc} note that if the shorting path is connected to (a, b) in Fig. 3.13 the current in the 1/4 Ω resistor must be zero (Why?). So $i_{sc} = 5i$, and from KVL we conclude that

$$v_{s1} = \frac{1}{2}i + \frac{1}{2}(5i) = 3i$$

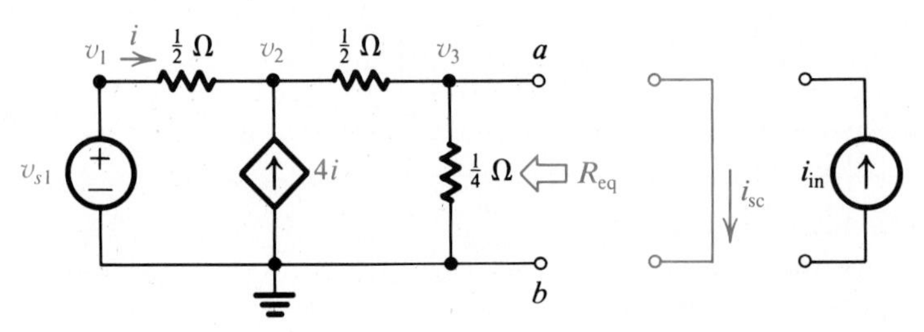

Figure 3.13 Circuit for Example 3.4.

and $i = 1/3\, v_{s1}$. Therefore

$$i_{sc} = \frac{5}{3} v_{s1}$$

and with v_{oc} from the first method we get

$$R_{eq} = v_{oc}/i_{sc} = \frac{3}{17}\Omega.$$

Method #3 (Terminal model)

If the current source is attached to (a, b) in Fig. 3.13 the nodal model of the circuit is

$$\begin{bmatrix} 1 & 0 & 0 \\ -2 & 4 & -2 \\ 0 & -2 & 6 \end{bmatrix} \begin{bmatrix} v_1 \\ v_2 \\ v_3 \end{bmatrix} = \begin{bmatrix} 1 & 0 \\ 0 & 0 \\ 0 & 1 \end{bmatrix} \begin{bmatrix} v_{s1} \\ \\ i_{in} \end{bmatrix} + \begin{bmatrix} 0 \\ 1 \\ 0 \end{bmatrix} [i_{d1}]$$

With $i_{d1} = 4i = 4(2(v_1 - v_2))$ we have

$$\mathbf{v} = \frac{1}{68} \begin{bmatrix} 68 & 0 \\ 60 & 2 \\ 20 & 12 \end{bmatrix} \begin{bmatrix} v_{s1} \\ \\ i_{in} \end{bmatrix}$$

Removing the equation for v_3 gives the terminal model at (a, b):

$$v_3 = v_{in} = \frac{20}{68} v_{s1} + \frac{12}{68} i_{in} = v_{oc} + R_{eq} i_{in}$$

and we conclude that $R_{eq} = 3/17\ \Omega$ and $v_{oc} = 5/17\, v_{s1}$.

We should become familiar with each method for finding a Thevenin equivalent circuit. In general, it is usually easier to find R_{eq} and v_{oc} by inspection, unless the circuit contains controlled circuits or is large. Also, care must be taken in applying the nodal method when the Thevenin nodal pair does not include the ground node.

Skill Exercise 3.4

Find the Thevenin equivalent at (a, b) for the circuit shown in Fig. SE3.4.

Answer: $v_{TH} = 0$ V, $R_{eq} = 1\ \Omega$

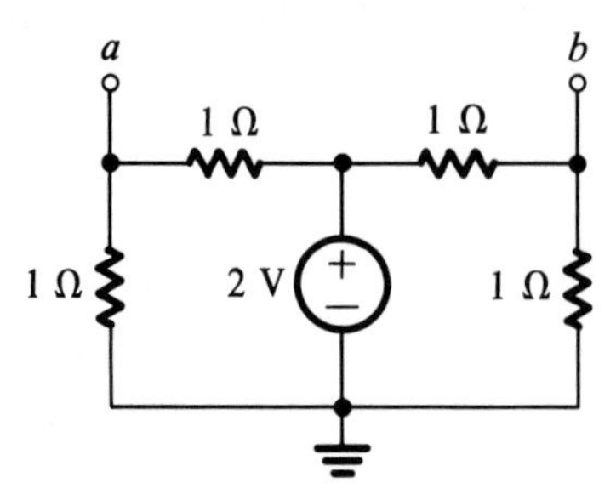

Figure SE3.4

3.3 THEVENIN EQUIVALENT CIRCUITS—SUPERPOSITION*

This *optional* section will show that the Thevenin equivalent circuit is a direct consequence of superposition, and will establish a relationship between a circuit's nodal analysis model and its Thevenin equivalent circuit.

The circuit in Fig. 3.14 is assumed to contain resistors, independent sources and controlled sources. We have access to a pair of terminals (a, b) which are also connected to internal nodes. Without a loss of generality we assume that node b is ground. If it is not, we are free to choose it as the reference node and to relabel the remaining internal nodes. We arbitrarily label node a as the first node so that $v_1 = v_{in}$, and we connect a current source to (a, b) as shown. The NA model can be written as

$$\mathbf{Gv} = \mathbf{Bu}_s + \mathbf{B}_{in}[i_{in}]$$

where $\mathbf{B}$ is the source coupling matrix of the original circuit, $\mathbf{u}_s$ is the vector of the circuit's internal independent sources, and $\mathbf{B}_{in}$ is a column vector having a 1 in its first row and zeroes in all of the remaining rows. Solving for $\mathbf{v}$ gives

$$\mathbf{v} = \mathbf{G}^{-1}\mathbf{Bu}_s + \mathbf{G}^{-1}\mathbf{B}_{in}[i_{in}]$$

The first row of this relationship is just the equation

$$v_1 = [\mathbf{G}^{-1}\mathbf{Bu}_s]_1 + [\mathbf{G}^{-1}\mathbf{B}_{in}]_1[i_{in}]$$

or, equivalently

$$v_{in} = v_{oc} + R_{eq}i_{in}$$

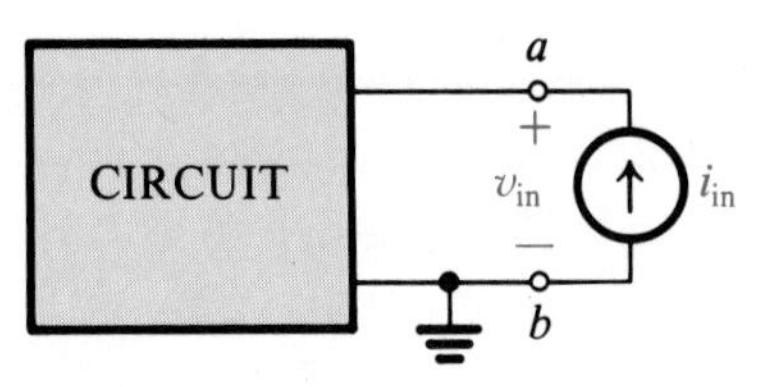

Figure 3.14 Circuit with external current source added.

where

$$v_{oc} = [\mathbf{G}^{-1}\mathbf{B}\mathbf{u}_s]_1$$

and

$$R_{eq} = [\mathbf{G}^{-1}\mathbf{B}_{in}]_1$$

and $[\ ]_1$ denotes the first row of the enclosed column vector. The expression for v_{oc} contains the effect of all of the internal independent voltage and current sources. If the external applied source i_{in} is set to zero, v_{in} is the open-circuit voltage at (a, b). On the other hand, if the internal source vector $\mathbf{u}_s$ is set to zero the expression gives $R_{eq} = v_{in}/i_{in}$. Thus, **the circuit's Thevenin equivalent can be extracted from its nodal analysis model** when the circuit is driven by i_{in}. Note that this can be accomplished without having to re-solve the nodal model, assuming $\mathbf{G}^{-1}$ has already been calculated, with the exception of a possible re-labeling of the nodes to correspond to the choices made for the pair (a, b).

Example 3.5

Find the input resistance and the forward voltage gain v_3/v_{in} for the loaded small-signal hybrid-parameter model of a bipolar-transistor amplifier shown in Fig. 3.15.

Solution: Applying nodal analysis gives

$$v_{in} = \left[h_i - \frac{h_f h_r R_L}{1 + h_o R_L}\right] i_{in}$$

and

$$v_3 = \frac{-h_f R_L i_{in}}{1 + h_o R_L}$$

The input resistance is

$$R_{eq} = \frac{v_{in}}{i_{in}} = h_i - \frac{h_f h_r R_L}{1 + h_o R_L}$$

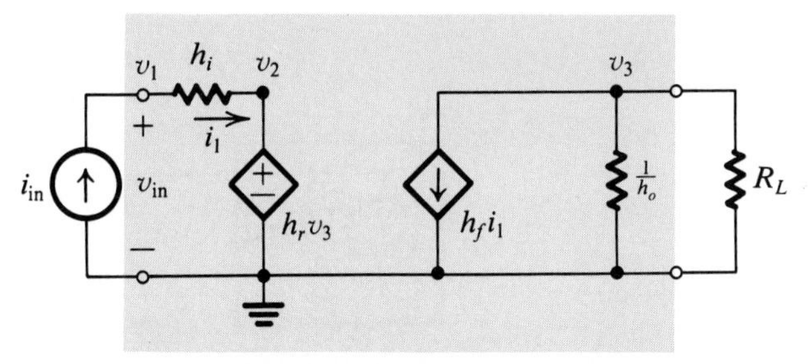

Figure 3.15 Hybrid parameter model of a bipolar-transistor amplifier.

and the forward voltage gain is

$$\frac{v_3}{v_{in}} = \frac{-h_f}{h_i/R_L + h_i h_o - h_f h_r}$$

Typical values for the transistor parameters are $h_i = 1\ \text{k}\Omega$, $h_f = 50$, $h_r = 2.4 \times 10^{-4}$ and $h_o = 24\ \mu\text{mhos}$. With a load resistor $R_L = 20\ \text{k}\Omega$ we have $R_{eq} = 837.84\ \Omega$ and $v_3/v_{in} = -568.18$. The output signal v_3 is an inverted, amplified copy of the input voltage, with over two orders of magnitude amplification.

3.4 NORTON EQUIVALENT CIRCUITS

A Norton equivalent circuit replaces a given circuit by a two-terminal model consisting of a parallel configuration of a current source and a resistor, as shown in Fig. 3.16. The current source i_N is referred to as the Norton equivalent current source for the original circuit, and R_N is the "Norton resistor."

A simple way to find the values of i_N and R_N is to choose them so that the circuit's behavior is consistent with the Thevenin equivalent circuit, as shown in Fig. 3.17. If a short circuit is connected across (a, b) the circuits must have

$$i_N = i_{sc}$$

and

$$i_N = v_{oc}/R_{eq}$$

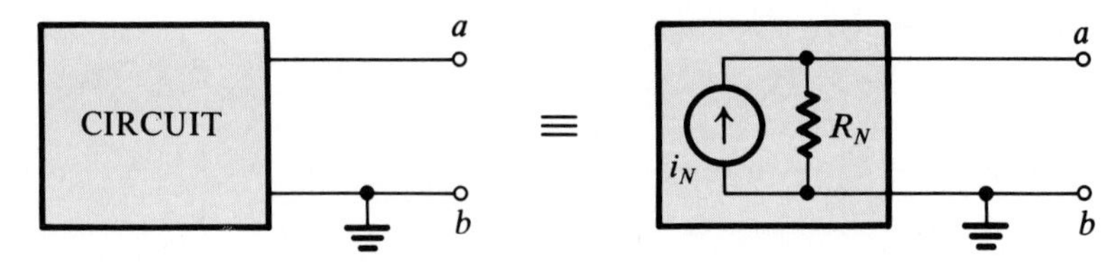

Figure 3.16 Norton equivalent circuit model.

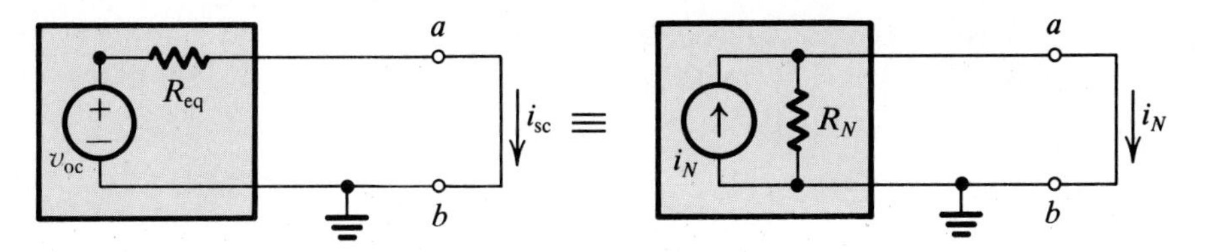

Figure 3.17 Thevenin and Norton equivalent circuits.

The Norton equivalent current source has a value equal to the short-circuit current of the original circuit. If the circuits are equivalent their open-circuit voltages must also be equal, so

$$v_{oc} = i_N R_N$$

and

$$R_N = v_{oc}/i_{sc}$$

Consequently, **the Norton equivalent resistance is the same as the Thevenin equivalent resistance of the circuit.**

Skill Exercise 3.5

Find the Norton equivalent circuit in Fig. SE3.5.

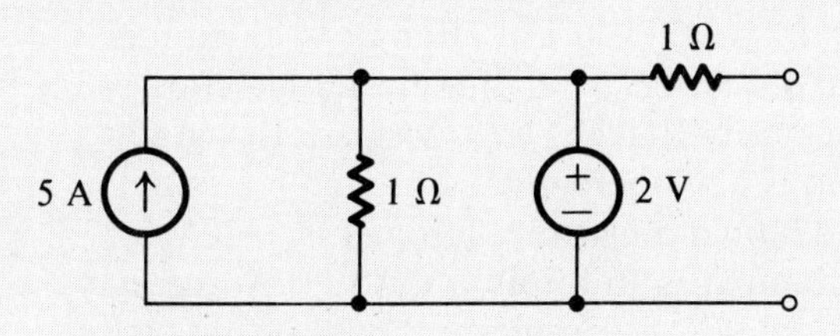

Figure SE3.5

Answer: $i_N = 2$ A, $R_{eq} = 1\ \Omega$.

An alternate approach to finding a Norton equivalent circuit is to drive the circuit by a voltage source, as shown in Fig. 3.18. In this configuration the terminal model of the circuit model is

$$i_{in} = v_{in}/R_{eq} - i_{sc}$$

Then the values of i_{sc} and R_{eq} can be found from measurements at (a, b) according to

$$R_{eq} = v_{in}/i_{in} \Big|_{i_{sc} = 0}$$

which corresponds to the input impedance when the internal sources are set to 0, and

$$i_{sc} = -i_{in} \Big|_{v_{in} = 0}$$

which gives the short-circuit current at (a, b).

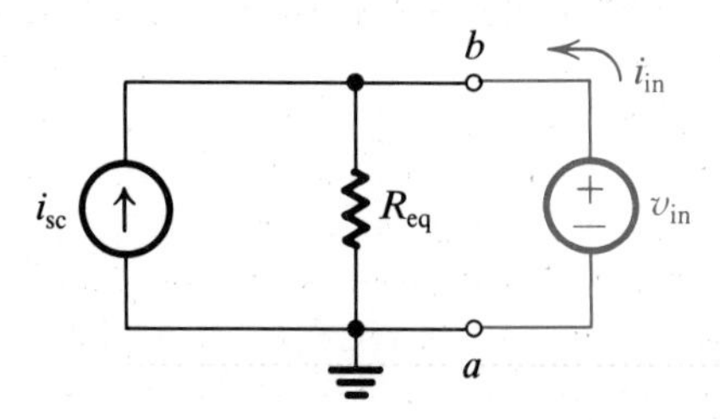

Figure 3.18 Circuit with external voltage source attached.

Example 3.6

Find the Norton equivalent circuit at (a, b) in Fig. 3.19.

Solution: Since the circuit has no internal independent sources, $v_{oc} = 0$ and $i_{sc} = 0$. An attempt to calculate $R_{eq} = v_{oc}/i_{sc} = 0/0$ leads to an indeterminate form. This does not imply that $R_{eq} = 0$. Instead, the indeterminate form is a mathematical clue that R_{eq} is not defined by the ratio of these measurements. One way out of this dilemma is to apply a current source at (a, b) and then obtain the terminal model of the circuit. Nodal analysis gives

$$\begin{bmatrix} \frac{1}{8} & -\frac{1}{2} \\ -\frac{1}{2} & \frac{3}{4} \end{bmatrix} \begin{bmatrix} v_1 \\ v_{in} \end{bmatrix} = \begin{bmatrix} 0 \\ 1 \end{bmatrix} [i_{in}]$$

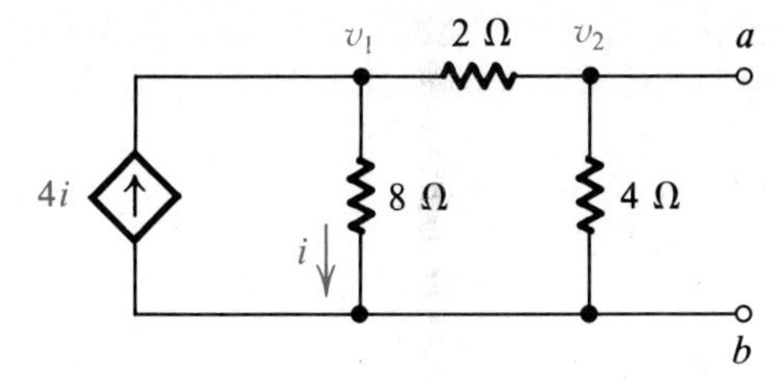

Figure 3.19 Circuit for Example 3.6

and so

$$\begin{bmatrix} v_1 \\ v_{in} \end{bmatrix} = - \begin{bmatrix} \frac{3}{4} & \frac{1}{2} \\ \frac{1}{2} & \frac{1}{8} \end{bmatrix} \begin{bmatrix} 0 \\ 1 \end{bmatrix} [i_{in}]$$

The terminal relationship is $v_{in} = -i_{in}$ and we conclude that $R_{eq} = -4\ \Omega$. When connected to an external circuit at (a, b) the original circuit behaves like a -4-Ω resistor. It will still have the linear v-i relationship characteristic of resistors, but with a negative slope. The circuit will supply power rather than absorb power. Such models are useful in describing the behavior of transistors.

3.5 NORTON EQUIVALENT CIRCUITS—SUPERPOSITION*

The Norton equivalent circuit will now be developed from the mesh model of a circuit. Suppose that a circuit is connected to a voltage source as in Fig. 3.20, and, without loss of generality, suppose that the first labeled mesh of the circuit contains the attached source. Then superposi-

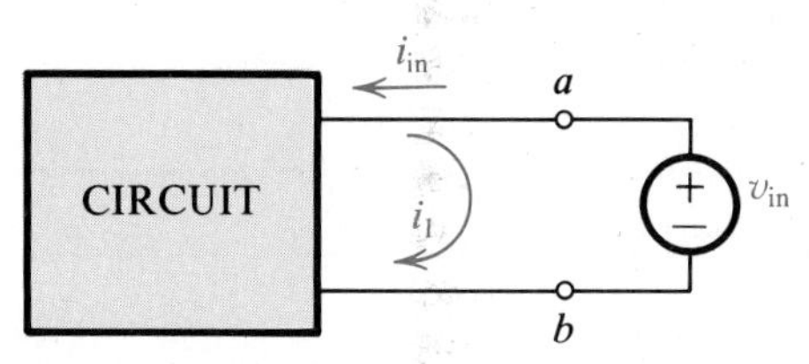

Figure 3.20 Mesh current definition for the Norton equivalent circuit.

tion allows us to write the mesh current vector in terms of the internal independent sources and the attached voltage source as

$$\mathbf{i} = \mathbf{R}^{-1}\mathbf{B}\mathbf{u}_s + \mathbf{R}^{-1}\mathbf{B}_{\text{in}}[-v_{\text{in}}]$$

where $\mathbf{B}_{\text{in}}$ is a *column* vector having a 1 in the first *row* and 0 elsewhere. (Note that $-v_{\text{in}}$ is the voltage *rise* in the mesh for i). Taking the first row of this matrix relationship gives the equation:

$$-i_{\text{in}} = [\mathbf{R}^{-1}\mathbf{B}\mathbf{u}_s]_1 - [\mathbf{R}^{-1}\mathbf{B}_{\text{in}}[v_{\text{in}}]]_1$$

where $[\ ]_1$ denotes the first row of the enclosed vector, and $-i_{\text{in}}$ corresponds to the conventional clockwise direction of mesh currents. Then, by inspection, we write

$$i_{\text{in}} = -[\mathbf{R}^{-1}\mathbf{B}\mathbf{u}_s]_1 + [\mathbf{R}^{-1}\mathbf{B}_{\text{in}}[v_{\text{in}}]]_1$$

and the Norton terminal model of the circuit is

$$i_{\text{in}} = -i_{\text{sc}} + v_{\text{in}}/R_{\text{eq}}$$

where the short-circuit current is

$$i_{\text{sc}} = -i_{\text{in}} \Big|_{v_{\text{in}} = 0}$$

and the input resistance is

$$R_{\text{eq}} = \frac{v_{\text{in}}}{i_{\text{in}}} \Big|_{i_{\text{sc}} = 0}$$

To measure R_{eq} the circuit's internal independent sources must be set to 0 to force i_{sc} to 0.

3.6 SOURCE TRANSFORMATIONS

Thevenin and Norton equivalent circuits simplify circuit analysis by reducing a complex circuit to an equivalent one having a single source and a single resistor. A Thevenin equivalent circuit can be transformed to its Norton equivalent circuit, and vice versa, as shown in Fig. 3.21. Note, however, that if $R = 0$ the voltage source cannot be replaced by a current source because such a source would have to deliver infinite current. Likewise, an ideal current source with $R = \infty$ cannot be replaced by a finite voltage source. Source transformations are useful because they sometimes eliminate the need to develop a nodal or mesh model.

Example 3.7.

Find i in Fig. 3.22(a).

Solution: Instead of using nodal or mesh analysis, we transform the sources as shown in Fig. 3.22(b). Then the

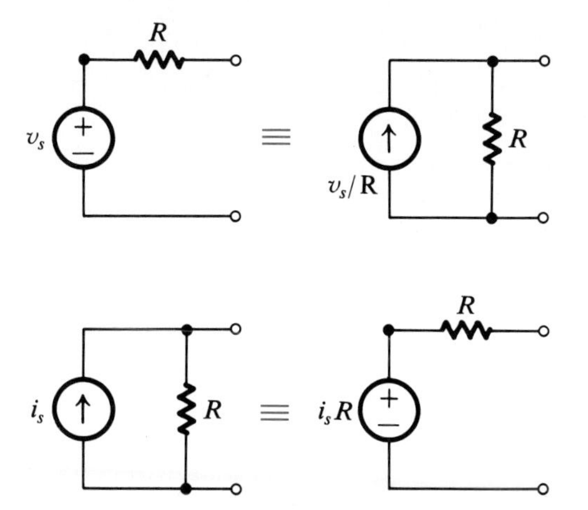

Figure 3.21 Equivalent voltage and current source models.

parallel resistors are combined and the current-source model is transformed back into the voltage-source model in Fig. 3.22(c). The current can be found by inspection: $-i = 1/8[3 + 20(i + 3)]$, which can be solved to give $i = -2.25$ A. Note that the transformation of the controlled source was made without eliminating the controlling variable v.

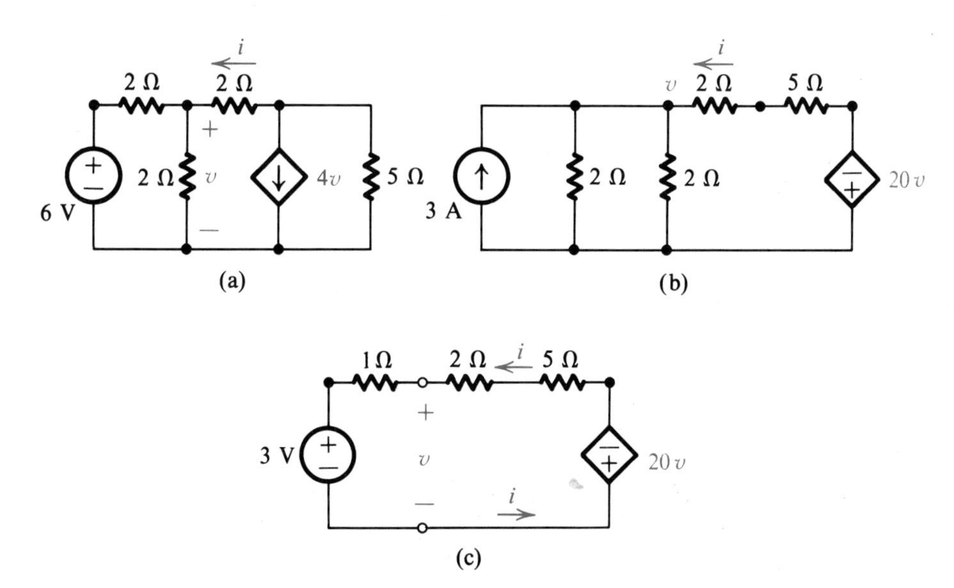

Figure 3.22 Source transformation with a controlled source.

3.7 MAXIMUM POWER TRANSFER

Thevenin and Norton equivalent circuits are useful when a load resistor R_L is to be selected so that the power P_L delivered to R_L by a given circuit is maximized. For example, R_L might be the resistance of a speaker that is to be driven by a given stereo amplifier. The driving circuit can be represented by its Thevenin equivalent circuit, as shown in Fig. 3.23, where a variable resistor is shown, with $0 \leq R_L \leq \infty$. The extreme allowed values of R_L correspond to a short circuit or an open circuit. In both cases, the power delivered to the load is 0. In the first case, $P_L = i_L{}^2R_L = 0$ because $R_L = 0$. In the second case, $P_L = 0$ because $i_L = 0$. (As an exercise, sketch a graph of P_L vs. R_L.) We assume that the driving circuit corresponds to a physical source, i.e. $R_s \neq 0$. To find the value of R_L for which P_L is a maximum, describe P_L in terms of the voltage and current:

$$P_L = i_L v_L = \frac{v_s}{R_s + R_L} \frac{R_L}{R_s + R_L} v_s = \frac{R_L}{(R_s + R_L)^2} v_s{}^2$$

For a fixed v_s and R_s the power delivered to R_L is a function of R_L. Taking the derivative dP_L/dR_L and setting it to 0 indicates that the maximum value of P_L occurs for

$$\boxed{R_L = R_s}$$

The power delivered to the load is maximized when the load resistor equals the source resistor. We say that R_L is **matched** to R_s when $R_L = R_s$. Under this condition, the maximum power P_{max} is obtained with $i_L = v_s/(2R_L)$, and $P_{max} = i_L{}^2R_L$, or

$$\boxed{P_{max} = \frac{v_s{}^2}{4R_L}}$$

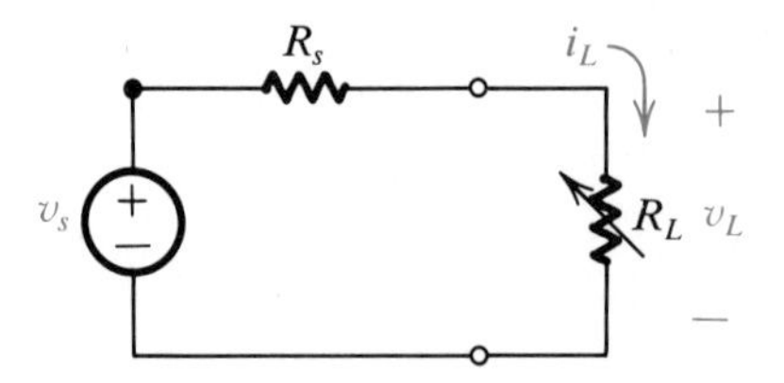

Figure 3.23 Source with a variable-load resistor.

Example 3.8

The circuit in Fig. 3.24 is an h-parameter model of a transistor amplifier connected to (a) a load resistor, (b) an open circuit, and (c) a short circuit. Find R_L to maximize the power transfer to the load.

Solution: First, we remove R_L and find the equivalent circuit for the amplifier in Fig. 3.24(b). By inspection,

$$v_{oc} = v_4 = \frac{-h_f}{h_o} i$$

and

$$i = (v_{s1} - h_r v_4)/h_i$$

Under open-circuit conditions with $v_3 = v_4$, the current becomes

$$i = (v_{s1} - h_r v_3)/h_i$$

The controlled current source is then used to give

$$i = \frac{v_{s1}}{h_i} - \left(\frac{h_r}{h_i}\right)\left(\frac{-h_f}{h_o}\right) i$$

which rearranges to

$$i = \frac{h_o v_{s1}}{h_i\, h_o - h_r h_f}$$

Therefore

$$v_{oc} = \frac{-h_f[h_o v_{s1}]}{h_o[h_i h_o - h_r h_f]} = \frac{-h_f v_{s1}}{h_i h_o - h_r h_f}$$

Next, we find i_{sc} in Fig. 3.24(c), where v_4 is constrained by $v_4 = v_{sc} = 0$:

$$i_{sc} = \frac{-h_f i}{h_o} h_o = -h_f i$$

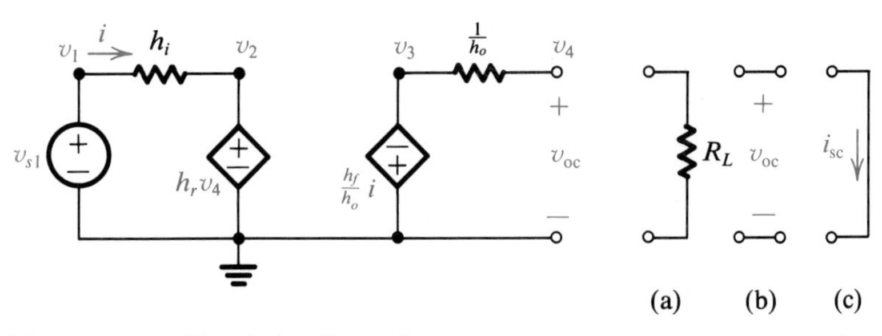

Figure 3.24 Circuit for Example 3.8

where

$$i = \frac{v_{s1} - h_r v_4}{h_i} = \frac{v_{s1}}{h_i}$$

Therefore

$$i_{sc} = \frac{-h_f}{h_i} v_{s1}$$

The Thevenin resistance is

$$R_{eq} = \frac{v_{oc}}{i_{sc}} = \frac{h_i}{h_i h_o - h_r h_f}$$

Using the parameters that were given for Example 3.5, we get $R_{eq} = 83.33$ kΩ, and $v_{oc} = -4166.67\ v_{s1}$. For maximum power transfer choose $R_L = 83.33$ kΩ. The maximum power delivered to the load is $P_L = 52.1\ \mu$W when $v_{s1} = 1$ mV.

SUMMARY

Circuits obey the principle of superposition, which states that the response of a circuit to several independent sources is the sum of the circuit's responses to each individual source when the others are turned off. A circuit can be replaced at a pair of terminals by a Thevenin equivalent circuit consisting of a voltage source connected in series with a resistance. The Thevenin equivalent voltage is due to the circuit's independent internal sources and equals the open-circuit voltage at the terminals. At the same pair of terminals a circuit has a Norton equivalent circuit consisting of a current source connected in parallel with a resistor. The value of this current source is equal to the short-circuit current at the terminals of the circuit, and is due to the internal independent sources. The resistance in the equivalent circuits can be obtained by computing v_{in}/i_{in} at the terminals after all of the internal independent sources are turned off, or, in most cases, by calculating v_{oc}/i_{sc}. The Thevenin model of a circuit can also be obtained from its nodal model, as the Norton model can be obtained from its mesh model. Physical voltage sources can be transformed into physical current sources, and vice versa. Finally, the power transferred from a source to a load is maximized when the load resistance equals the internal resistance of the source.

Problems - Chapter 3

3.1 Using superposition, find i when $v_{s1} = 10$ V and $v_{s2} = 20$ V.

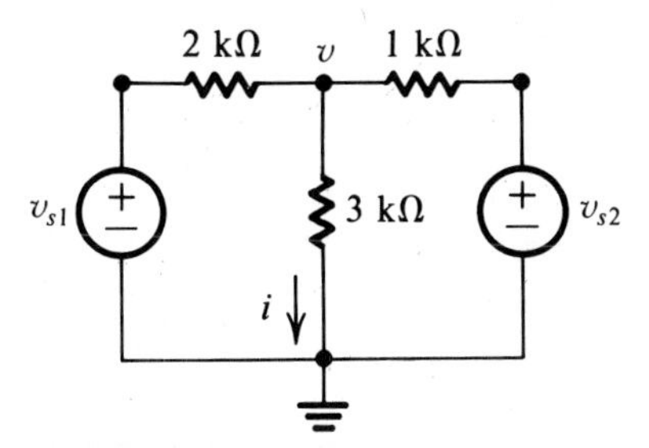

Figure P3.1

3.2 For the circuit of Fig. P3.1:

a. Solve for v when $v_{s1} = 20$ V and $v_{s2} = 0$ V.
b. Solve for v when $v_{s1} = 0$ V and $v_{s2} = 45$ V.
c. Solve for v when $v_{s1} = 20$ V and $v_{s2} = 45$ V.
d. Compare your answers from parts a, b and c.

3.3 For the circuit in Fig. P.3.3:

a. Find the component of i_1, i_2 and i_3 due to i_{s1} alone.
b. Find the component of i_1, i_2 and i_3 due to i_{s2} alone.
c. Find the power dissipated in the 10 Ω resistor when i_{s1} is operative and i_{s2} is turned off.
d. Repeat c when i_{s2} alone is operative.
e. Repeat c when both sources are operative.
f. Discuss your answers for a–e.

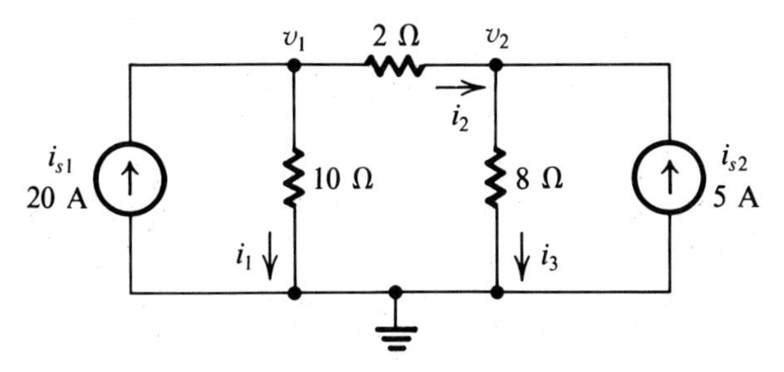

Figure P3.3

3.4 Use superposition to find i.

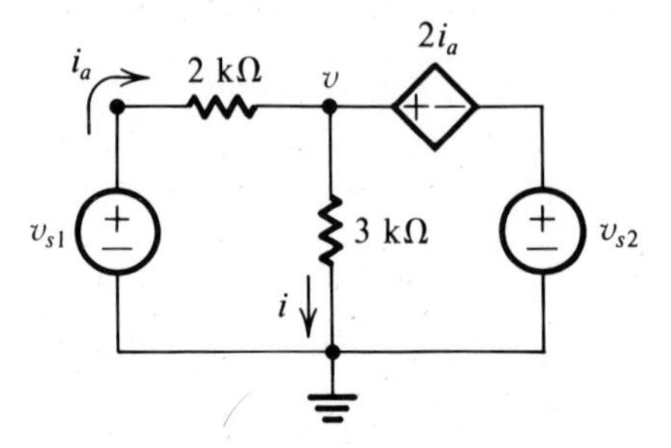

Figure P3.4

3.5 Use superposition to find v_{ab}.

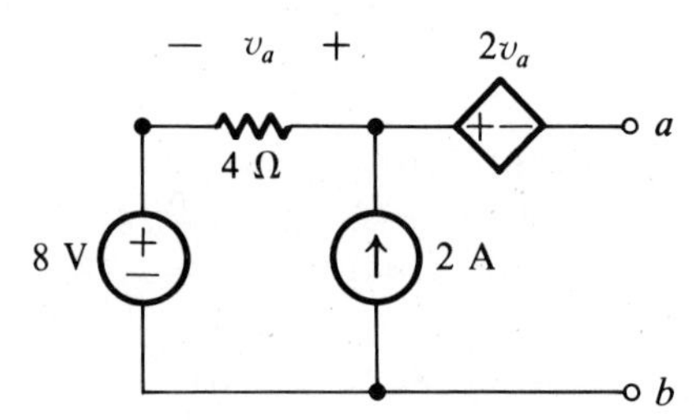

Figure P3.5

3.6 For the circuit in Fig. P3.6,

a. Develop the matrix nodal model.

b. Using the solution to the nodal model, find the component of i_o due to v_s.

c. Repeat b to find the component of i_o due to i_s.

d. By inspection, find i_o when v_s alone is operative.

e. By inspection, find i_o when i_s alone is operative.

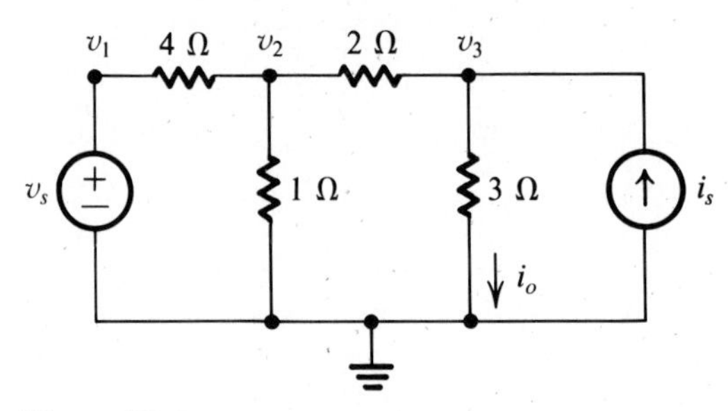

Figure P3.6

3.7 For the circuit in Fig. P3.7,

a. Find the component of **v** due to each source when $i_{s1} =$ 10 A and $i_{s2} = 25$ A.

b. Find **v** when $i_{s1} = 20$ A and $i_{s2} = 75$ A.

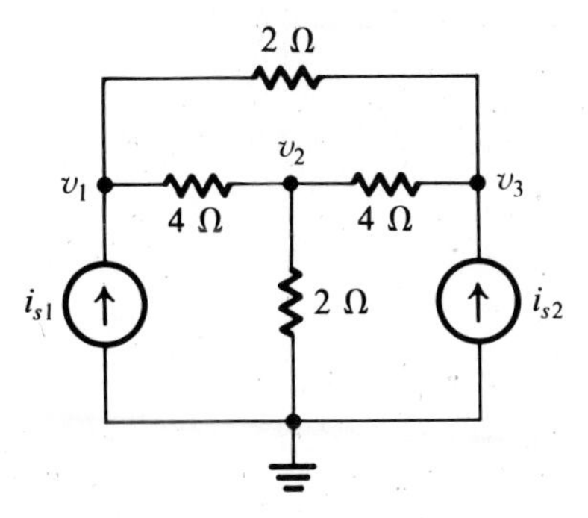

Figure P3.7

3.8 By inspection, construct $\mathbf{G}^{-1}\mathbf{B}$ for the nodal model of the circuit in Fig. P3.8.

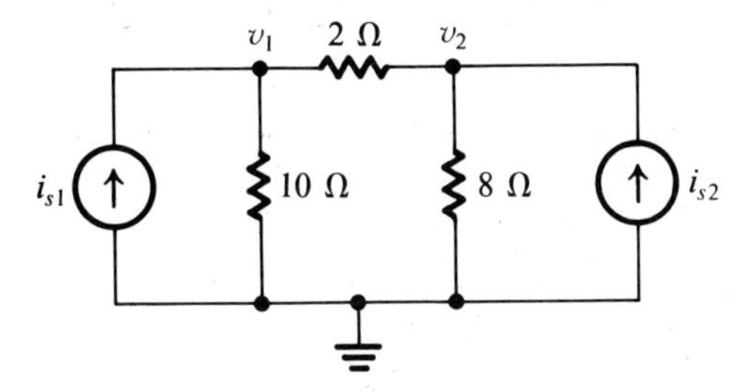

Figure P3.8

3.9 Find v_o by inspection, with $v_s = 20$ V, and $i_s = 5$ A.

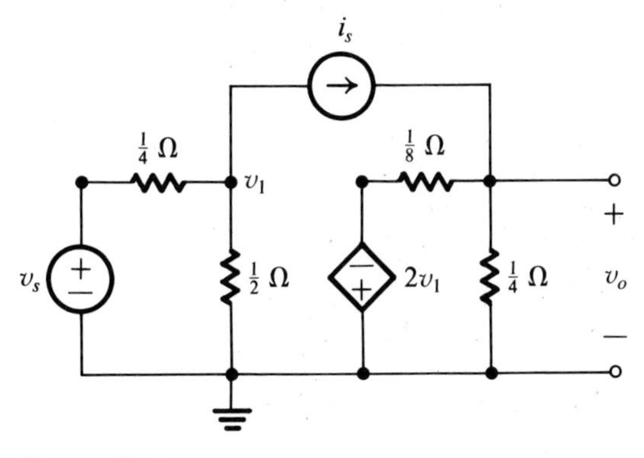

Figure P3.9

3.10 For the circuit in Fig. P3.10, with $v_{s1} = 10$ V and $v_{s2} = 4$ V,
a. Find the component of i due to v_{s1}.
b. Find the component of i due to v_{s2}.

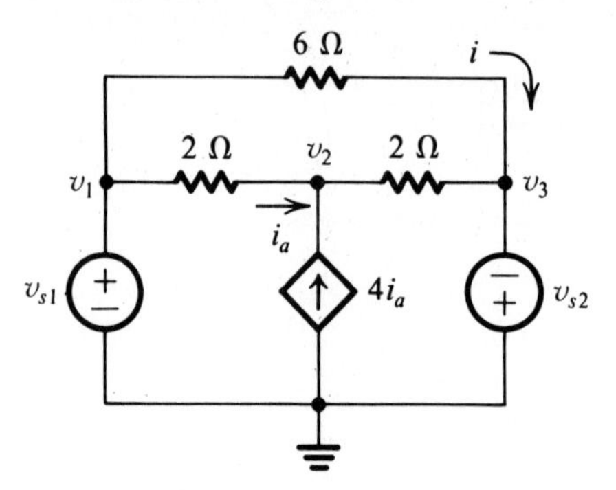

Figure P3.10

3.11 Find the components of **v** due to v_s and i_s, with $v_s = -2$ V and $i_s = 10$A.

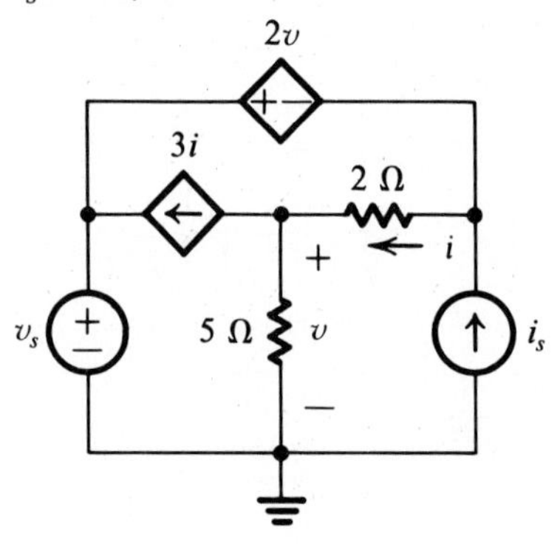

Figure P3.11

3.12 Find the Thevenin equivalent circuit at (a, b).

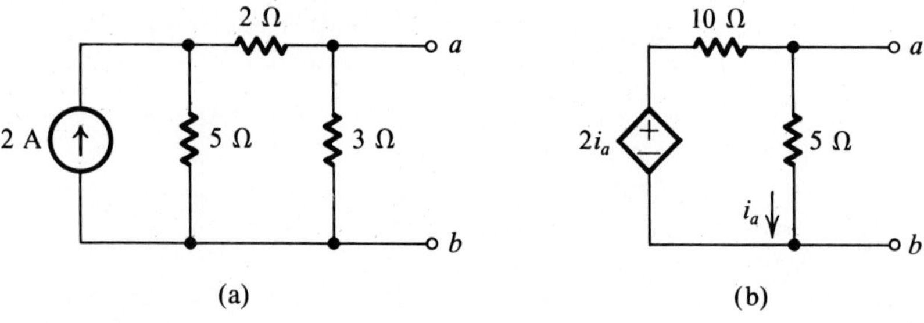

Figure P3.12(a) (b)

3.13 Find the Thevenin resistance seen by R_L.

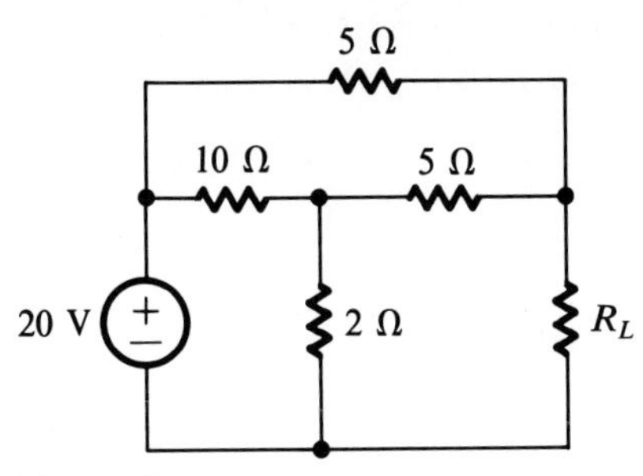

Figure P3.13

3.14 The circuit in Fig. P3.14 has the following nodal model:

$$\mathbf{v} = \begin{bmatrix} 6 & 1 & 2 & 8 \\ 3 & 2 & 4 & -2 \\ 0 & 1 & -3 & 6 \\ 12 & 7 & 1 & 4 \end{bmatrix} \begin{bmatrix} v_{s1} \\ v_{s2} \\ v_{s3} \\ i_{s1} \end{bmatrix}$$

If v_{s1}, v_{s2} and v_{s3} are internal sources and i_{s1} is connected as shown, find v_{oc}, R_{TH} and i_{sc} for the Thevenin and Norton equivalent circuits at (a, b), with $v_{s1} = 2$, $v_{s2} = -1$ and $v_{s3} = 5$.

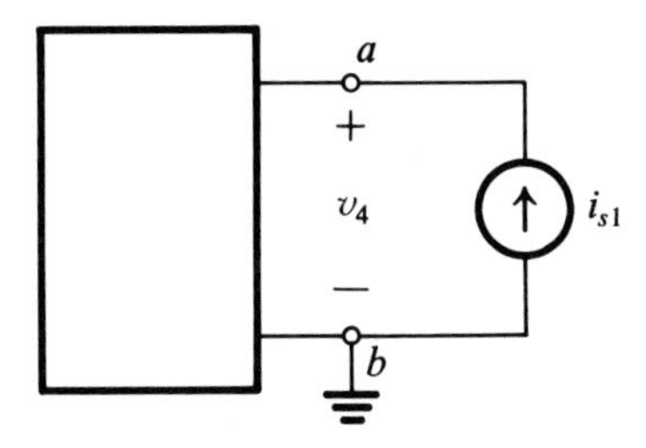

Figure P3.14

3.15 For the bridge circuit shown in Fig. P3.15, find the Thevenin equivalent circuit at (a, b).

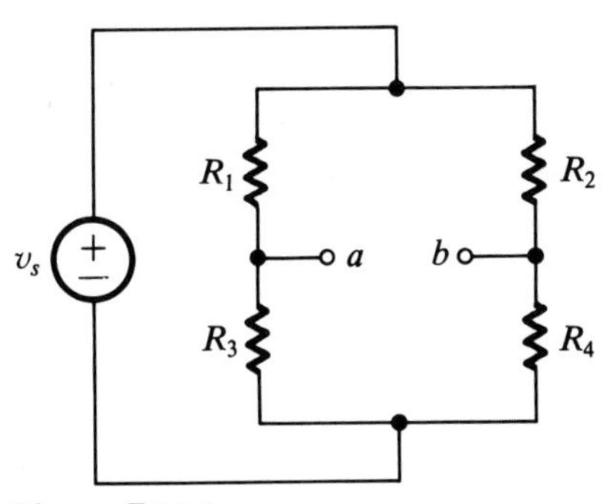

Figure P3.15

3.16 Find the Thevenin equivalent circuit seen by the 1 Ω resistor.

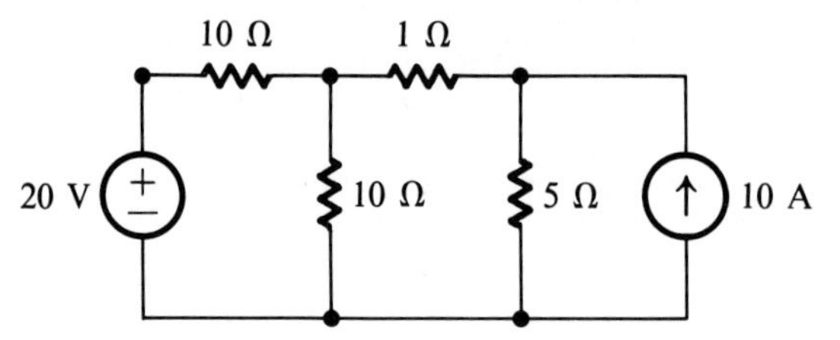

Figure P3.16

3.17 Find the Thevenin equivalent circuits for (a) two circuits connected in series, and (b) two circuits connected in parallel.

3.18 For the circuit in Fig. P3.18:

a. Find the open-circuit voltage.

b. Find the short-circuit current.

c. Use the Thevenin model of the circuit to obtain the equation describing the $v - i$ characteristic of the circuit.

d. Draw the $v - i$ characteristic.

e. For a given value of i_s, find the value of v_s that causes $v = 0$.

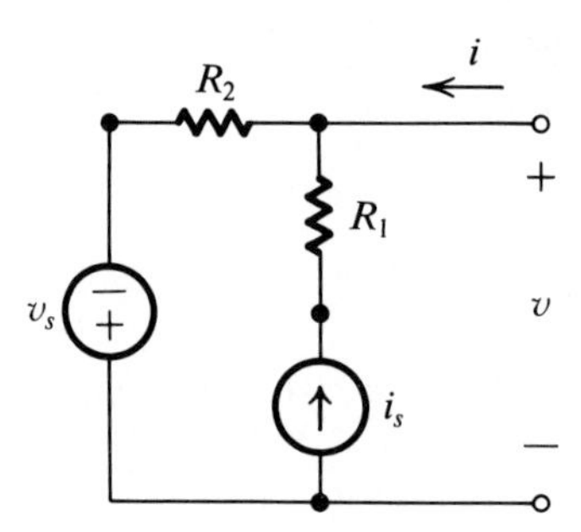

Figure P3.18

3.19 If $v_o = 100$ V when $R_L = 10\ \Omega$ in the circuit shown in Fig. P3.19, and if $v_o = 60$ V when $R_L = 40\ \Omega$, find v_s and R_s.

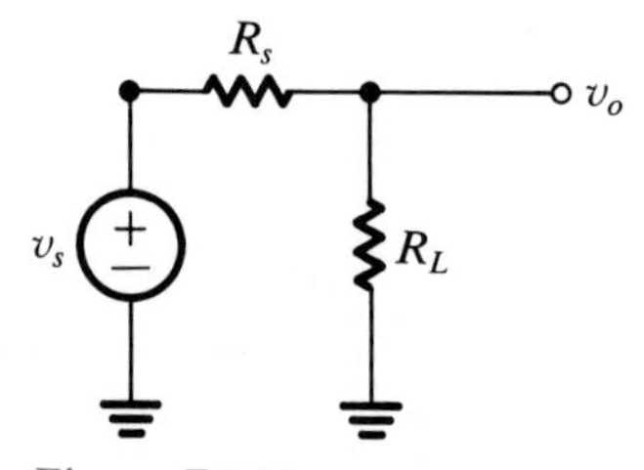

Figure P3.19

3.20 Find v_o for the potentiometer circuit in Fig. P3.20, with $0 \leq k \leq 1$, and find the Thevenin equivalent circuit that would be seen by a circuit attached to v_o.

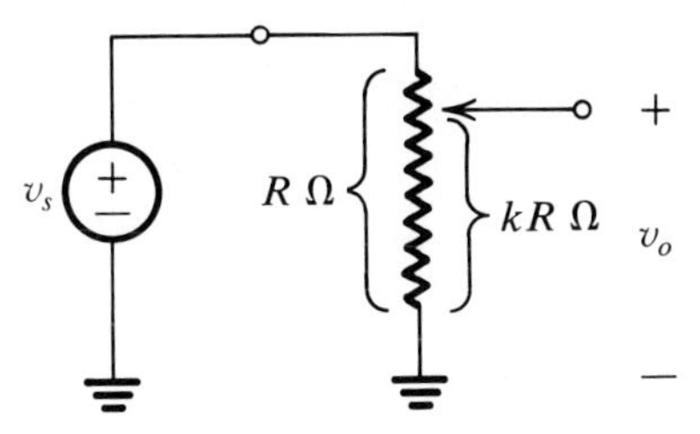

Figure P3.20

3.21 Draw the v − i curve for the circuit shown in Fig. P3.21.

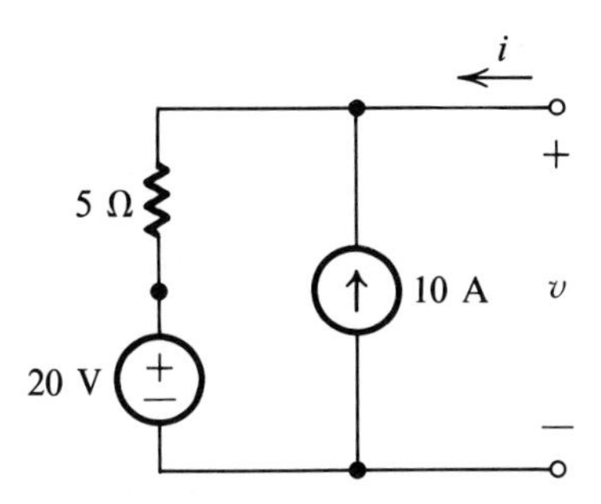

Figure P3.21

3.22 Find the Norton equivalent circuit for the circuit shown in Fig. P3.22.

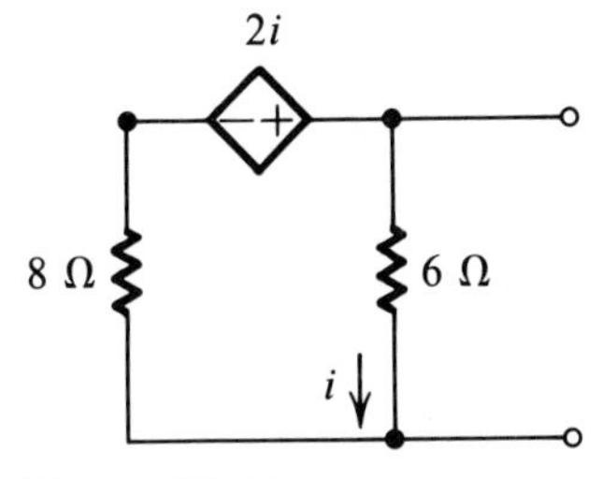

Figure P3.22

3.23 Find the Norton equivalent circuits for (a) two circuits connected in series, and (b) two circuits connected in parallel.

3.24 Find the equivalent resistance seen at (a, b) in Fig. P3.24 by attaching a source at (a, b) and computing v_{in}/i_{in}.

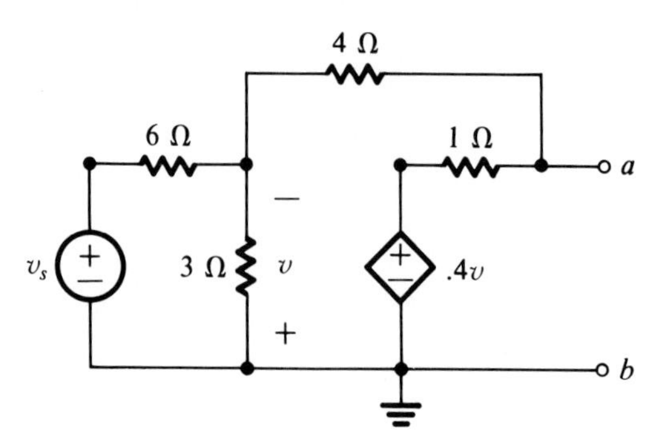

Figure P3.24

3.25 For the circuit of Fig. P3.24, find v_{oc} and i_{sc}.

3.26 Find v by the following methods:

a. Use nodal analysis.
b. Find the Thevenin equivalent seen by the 40 Ω resistor.
c. Use superposition.
d. Use mesh analysis.

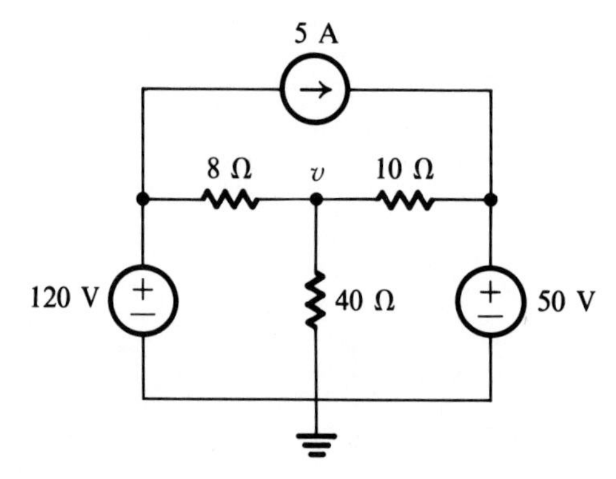

Figure P3.26

3.27 Find the Thevenin equivalent circuit seen by the 1 Ω resistor, and then calculate the power dissipated by the resistor.

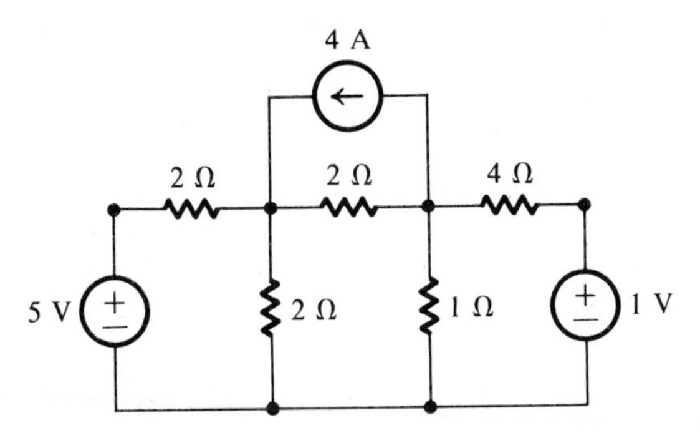

Figure P3.27

3.28 Find the power delivered by the 20 V source.

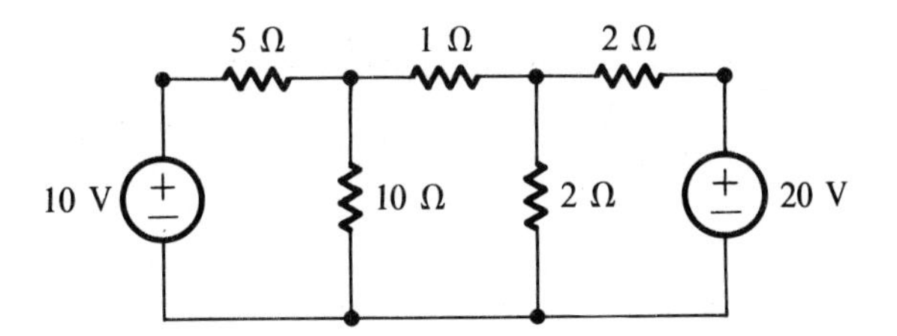

Figure P3.28

3.29 Find i.

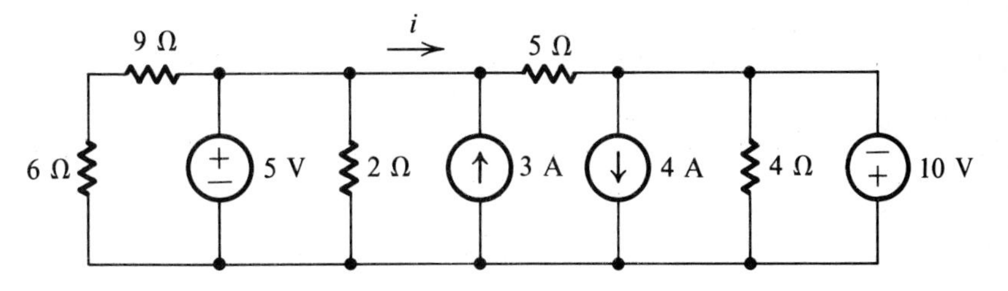

Figure P3.29

3.30 Find i_1, i_2, i_3, i_4, v_1 and v_2 by the following methods:
a. Superposition.
b. Nodal analysis.
c. Source transformation.

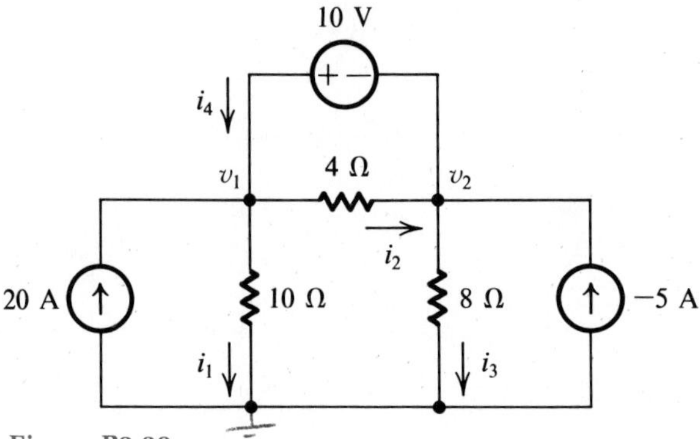

Figure P3.30

3.31 What value of R_L will absorb maximum power from v_s in Fig. P3.31?

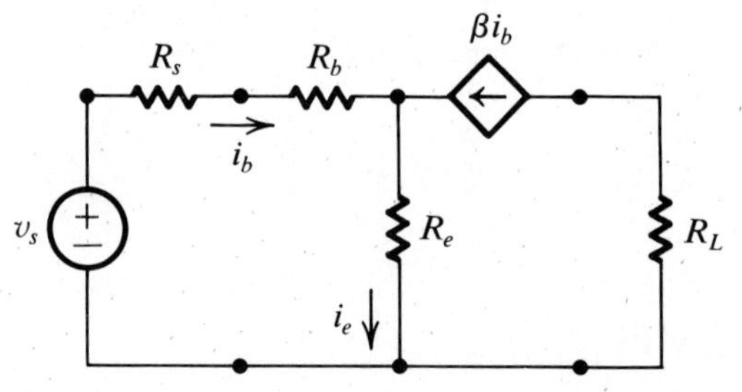

Figure P3.31

3.32 Solve for v_2 by the following methods:
a. Nodal analysis.
b. Mesh analysis.
c. Superposition.
d. Find the Thevenin equivalent circuit seen by each source.

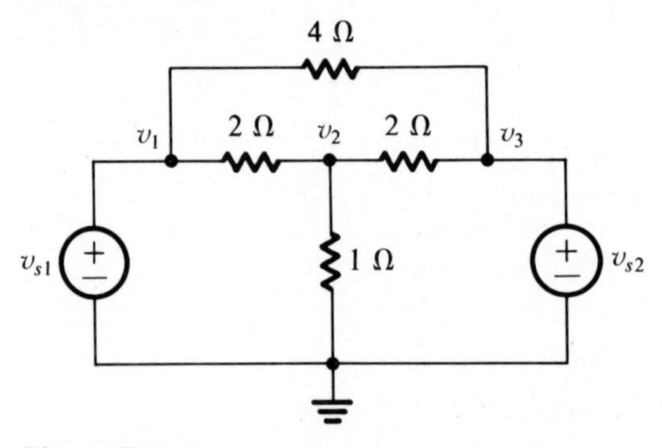

Figure P3.32

3.33 For the circuit in Fig. P3.33:

a. Find the input resistance v_1/i_1.

b. Find the voltage gain v_3/v_1.

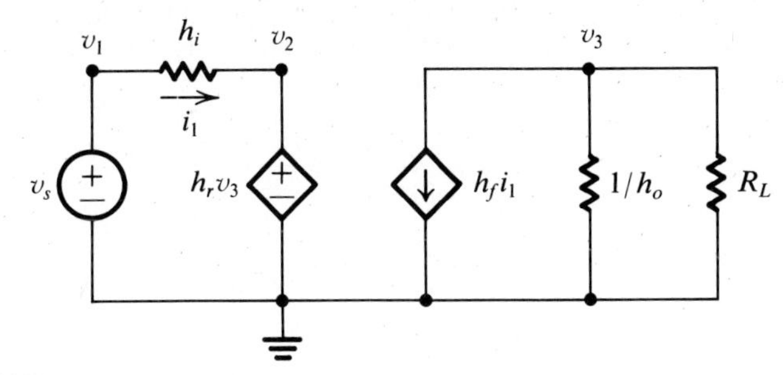

Figure P3.33

3.34 Find the Thevenin resistance seen by v_S.

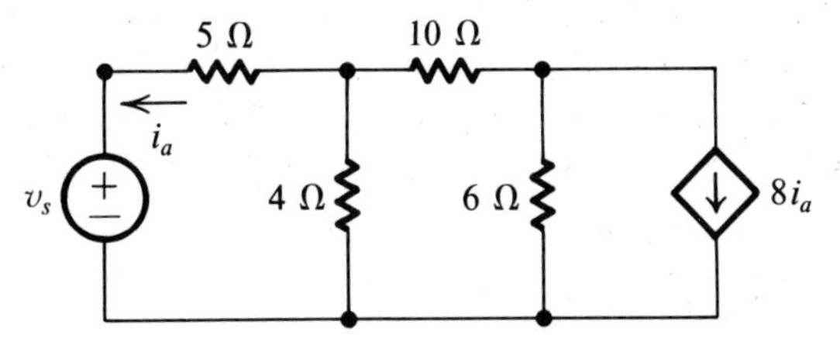

Figure P3.34

3.35 Find the Norton equivalent of the circuit in Fig. P3.35.

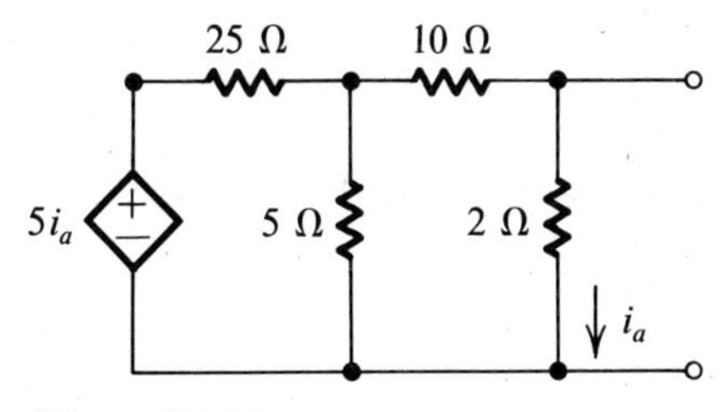

Figure P3.35

3.36 Find the power delivered to the 1 Ω resistor.

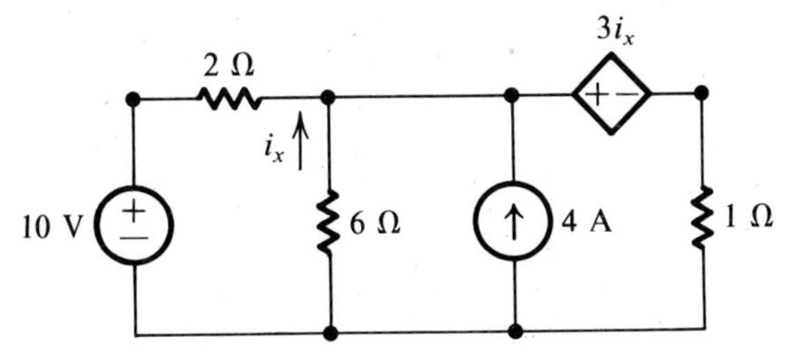

Figure P3.36

3.37 Using nodal analysis, find the Thevenin equivalent circuit seen by the 5 A source.

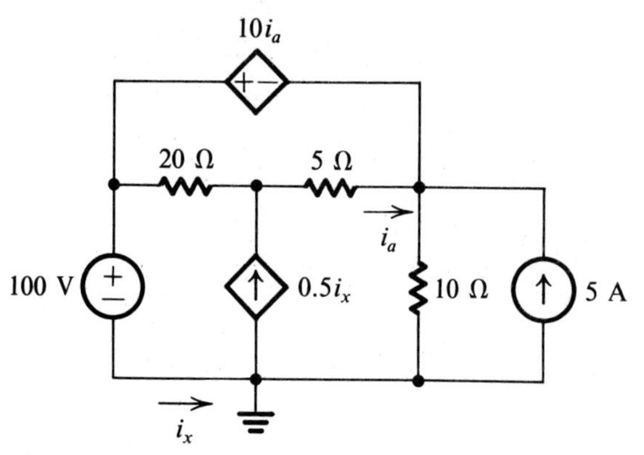

Figure P3.37

3.38 Find the output resistance of the circuit in Fig. P3.38.

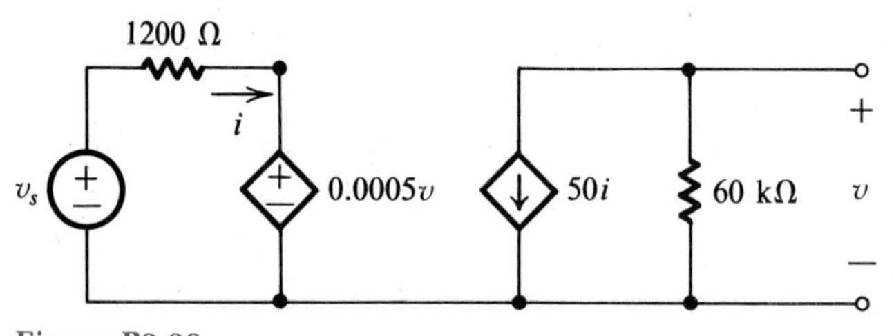

Figure P3.38

OPERATIONAL AMPLIFIERS

INTRODUCTION

Operational amplifiers play an important role in circuits used for data acquisition, timers, and other applications where signals must be amplified to useful levels. Although the internal implementation of operational amplifiers (op amps) can be complex, their actual behavior is simple enough to be modeled and understood with the tools that have been presented in the first three chapters. Here we will develop the input/output model for several op amp circuits and examine their Thevenin equivalents.

4.1 FINITE-GAIN FEEDBACK AMPLIFIERS

The basic building block of the operational amplifier is the differential voltage amplifier shown in Fig. 4.1. This circuit amplifies the difference

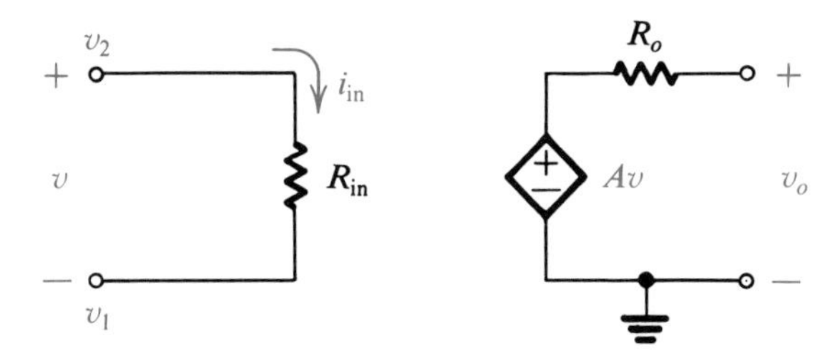

Figure 4.1 A voltage amplifier circuit.

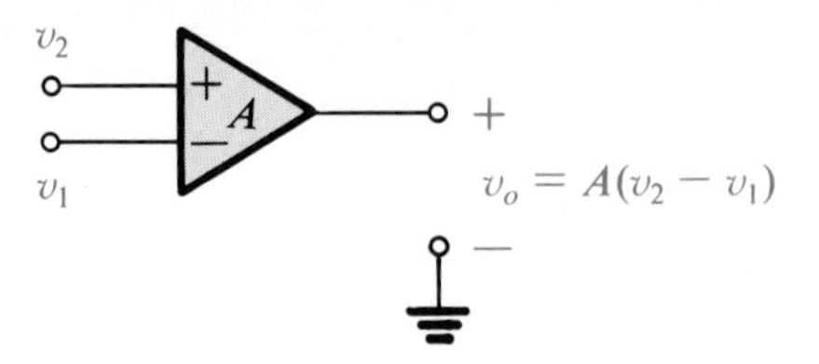

Figure 4.2 Circuit symbol for a voltage amplifier.

of the voltage applied to its two input terminals, with the open-circuit output voltage

$$v_o = Av = A(v_2 - v_1)$$

and $A > 0$. The terminal marked with "+" is called the noninverting input, and the terminal marked with "−" is called the inverting input because if $v_1 = 0$, $v_o = Av_2$, and if $v_2 = 0$, $v_o = -Av_1$.

The circuit in Fig. 4.1 is called an **amplifier** because its output voltage is a scaled copy of the differential input voltage, and because the voltages have $|v_o| > |v|$ if $A > 1$. The gain G of the circuit is the ratio of the output voltage to the input voltage:

$$G = \frac{v_o}{v_{in}}$$

with $v_{in} = v$. Since there is no conductive path between the output node and the input nodes, A is called the **open-loop gain** of the circuit. The circuit symbol for a voltage amplifier having an open-loop gain of A is shown in Fig. 4.2, and its internal circuitry is displayed in Fig. 4.3.

A: open-loop gain

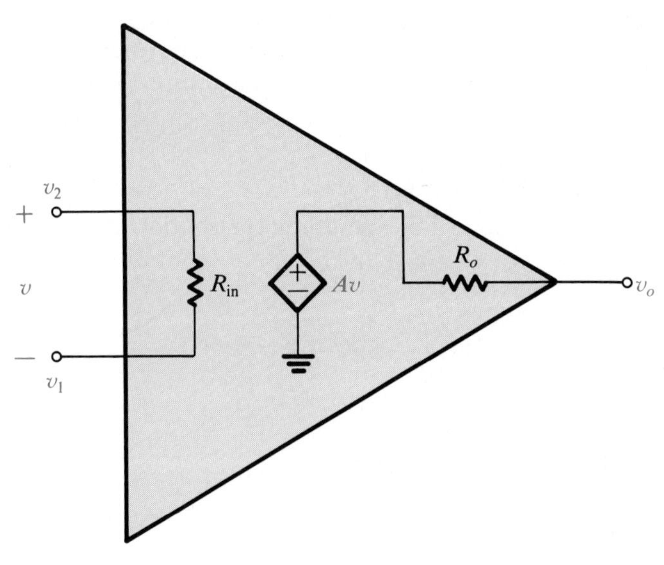

Figure 4.3 Internal circuitry of the voltage amplifier model.

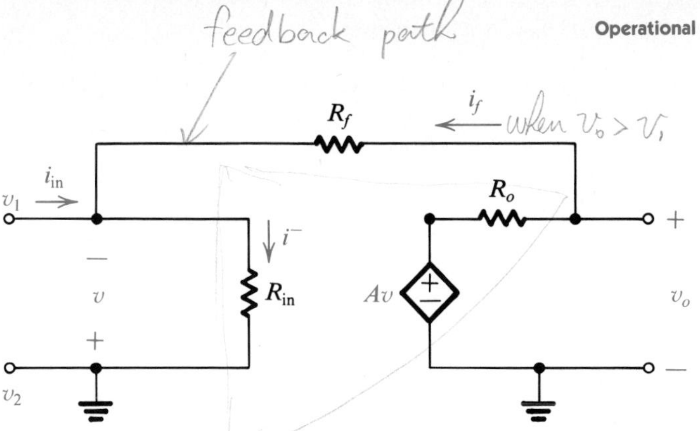

Figure 4.4 Voltage amplifier with a feedback resistor.

Next, we connect the noninverting input node of the amplifier to ground and connect a resistor R_f between the output node and the inverting input node, as shown in Fig. 4.4. The conductive path formed by R_f is called a **feedback path** because, in this configuration, the current in the feedback path i_f will be nonzero whenever the output voltage and the input voltage are unequal. This current is fed back, or oriented, towards the input portion of the circuit from its output, whenever $v_o > v_1$.

Placing a feedback resistor around the voltage amplifier leads to the basic form of a feedback amplifier shown in Fig. 4.5, where the circuit drawing is simplified by not showing the internal components of the amplifier. This amplifier is referred to as a feedback amplifier, or an amplifier with negative feedback, since the feedback resistor connects the output node to the inverting input terminal of the open-loop amplifier.

Feedback amplifiers have many important applications in which signals are modified to convey or represent information in an electrical circuit. In a typical application, a signal voltage from a circuit is the input to a feedback amplifier whose output is connected to a load resistor. Since such a circuit can be represented by its Thevenin equivalent, we need only consider the configuration shown in Fig. 4.6, where v_s and R_s are the equivalent of the circuit that is attached to the amplifier. Our objective is

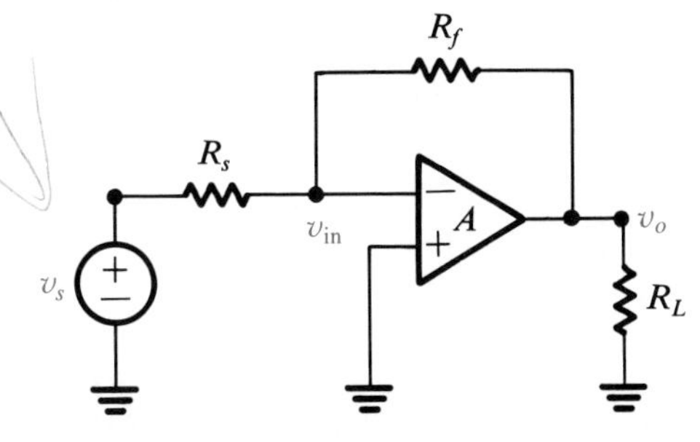

Figure 4.5 A feedback amplifier circuit.

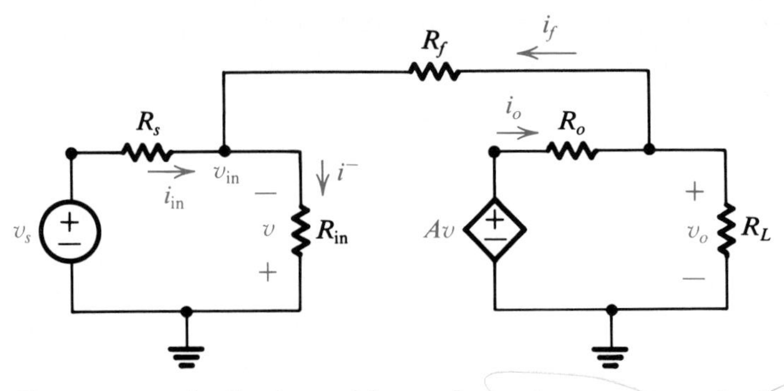

Figure 4.6 A feedback amplifier with a voltage source and a load resistor.

to develop a model for the input/output relationship between v_s and v_o in the circuit with feedback. Such a model will enable us to understand how the output signal at R_L depends on the input signal and the circuit components.

The nodal model for the loaded amplifier circuit can be formed for the voltage v_{in} at the amplifier input node and the voltage v_o at the amplifier output node:

$$\begin{bmatrix} \frac{1}{R_s} + \frac{1}{R_{in}} + \frac{1}{R_f} & -\frac{1}{R_f} \\ -\frac{1}{R_f} + \frac{A}{R_o} & \frac{1}{R_o} + \frac{1}{R_f} + \frac{1}{R_L} \end{bmatrix} \begin{bmatrix} v_{in} \\ v_o \end{bmatrix} = \begin{bmatrix} \frac{1}{R_s} \\ 0 \end{bmatrix} [v_s]$$

with solution

$$\begin{bmatrix} v_{in} \\ v_o \end{bmatrix} = \frac{1}{\Delta} \begin{bmatrix} \frac{1}{R_o} + \frac{1}{R_f} + \frac{1}{R_L} & \frac{1}{R_f} \\ \frac{1}{R_f} - \frac{A}{R_o} & \frac{1}{R_s} + \frac{1}{R_{in}} + \frac{1}{R_f} \end{bmatrix} \begin{bmatrix} \frac{1}{R_s} \\ 0 \end{bmatrix} [v_s] \tag{4.1}$$

where

$$\Delta = \left[\frac{1}{R_s} + \frac{1}{R_{in}} + \frac{1}{R_f}\right]\left[\frac{1}{R_o} + \frac{1}{R_f} + \frac{1}{R_L}\right] + \frac{1}{R_f}\left[-\frac{1}{R_f} + \frac{A}{R_o}\right]$$

Taking the row corresponding to v_o gives

$$v_o = \frac{\frac{1}{R_s}\left[\frac{1}{R_f} - \frac{A}{R_o}\right] v_s}{\left[\frac{1}{R_s} + \frac{1}{R_{in}} + \frac{1}{R_f}\right]\left[\frac{1}{R_o} + \frac{1}{R_f} + \frac{1}{R_L}\right] + \frac{1}{R_f}\left[-\frac{1}{R_f} + \frac{A}{R_o}\right]} \tag{4.2}$$

Multiplying the numerator and denominator of v_o in (4.2) by the factor R_oR_f/A and solving for v_o/v_s gives the gain as

$$G = \frac{v_o}{v_s} = \frac{-\dfrac{R_f}{R_s}\left[1 - \dfrac{R_o}{AR_f}\right]}{\dfrac{1}{A}\left[\dfrac{1}{R_s} + \dfrac{1}{R_{in}} + \dfrac{1}{R_f}\right]\left[R_f + R_o + \dfrac{R_oR_f}{R_L}\right] + 1 - \dfrac{R_o}{AR_f}} \tag{4.3}$$

This gain is called the **closed-loop gain** of the circuit because the feedback resistor closes the loop between the output and input nodes of the circuit. The model for the amplifier circuit depends on each of the resistors and the amplifier gain A.

Having derived the gain model of the circuit, let's explore it to see if it fits our intuition about the behavior of the circuit for various choices of the component values. First, note that removing the feedback resistor forces i_f to 0. Letting $R_f \to \infty$ in (4.2) and rearranging coefficients gives

$$\frac{v_o}{v_s} = \frac{-AR_{in}R_L}{(R_{in} + R_s)(R_o + R_L)} \tag{4.4}$$

The gain expression in (4.4) is just the product of the voltage divider factors that determine v from v_s, and v_o from Av. If $R_{in} = 0$ the gain is 0 because the controlling voltage is 0. Thus, the loaded amplifier model is consistent with the expected behavior of the circuit for these choices of resistors.

The Thevenin equivalent circuits at the input and output terminals of the feedback amplifier provide additional insight about the amplifier's operation. Since the Thevenin resistance seen by v_s in Fig. 4.6 is the sum of R_s and the Thevenin resistance that would be seen by an ideal source attached at node v_{in}, we remove v_s and R_s, attach an ideal current source

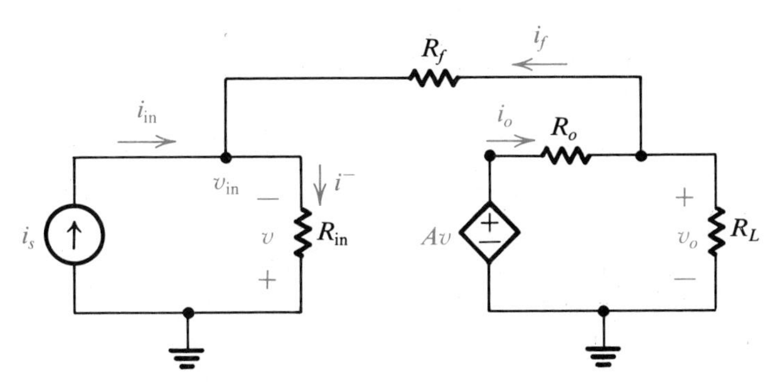

Figure 4.7 Circuit for determining the amplifier input resistance.

to the inverting input node, and develop the nodal model for the circuit shown in Fig. 4.7. The nodal model has

$$\begin{bmatrix} \dfrac{1}{R_{in}} + \dfrac{1}{R_f} & -\dfrac{1}{R_f} \\ -\dfrac{1}{R_f} + \dfrac{A}{R_o} & \dfrac{1}{R_o} + \dfrac{1}{R_f} + \dfrac{1}{R_L} \end{bmatrix} \begin{bmatrix} v_{in} \\ v_o \end{bmatrix} = \begin{bmatrix} 1 \\ 0 \end{bmatrix} [i_{in}]$$

with v_{in} and v_o given by

$$\begin{bmatrix} v_{in} \\ v_o \end{bmatrix} = \frac{1}{\Delta} \begin{bmatrix} \dfrac{1}{R_o} + \dfrac{1}{R_f} + \dfrac{1}{R_L} & \dfrac{1}{R_f} \\ \dfrac{1}{R_f} - \dfrac{A}{R_o} & \dfrac{1}{R_{in}} + \dfrac{1}{R_f} \end{bmatrix} \begin{bmatrix} 1 \\ 0 \end{bmatrix} [i_{in}] \tag{4.5}$$

and

$$\Delta = \left[\frac{1}{R_{in}} + \frac{1}{R_f}\right]\left[\frac{1}{R_o} + \frac{1}{R_f} + \frac{1}{R_L}\right] + \frac{1}{R_f}\left[\frac{A}{R_o} - \frac{1}{R_f}\right]$$

Solving (4.5) for v_{in} gives the Thevenin equivalent resistance $R_{TH/in} = v_{in}/i_{in}$ at the *input* terminals of the feedback amplifier:

$$R_{TH/in} = \frac{\dfrac{1}{R_o} + \dfrac{1}{R_f} + \dfrac{1}{R_L}}{\left[\dfrac{1}{R_{in}} + \dfrac{1}{R_f}\right]\left[\dfrac{1}{R_o} + \dfrac{1}{R_f} + \dfrac{1}{R_L}\right] + \dfrac{1}{R_f}\left[\dfrac{A}{R_o} - \dfrac{1}{R_f}\right]}$$

This expression is easy to evaluate, but it does not give us much insight into the behavior of the circuit. So we first factor $1/R_f$ from the numerator and denominator and then cancel it to get

$$R_{TH/in} = \frac{1 + R_f\left[\dfrac{1}{R_o} + \dfrac{1}{R_L}\right]}{\left[\dfrac{1}{R_{in}} + \dfrac{1}{R_f}\right]\left\{1 + R_f\left[\dfrac{1}{R_o} \quad \dfrac{1}{R_L}\right]\right\} + \dfrac{A}{R_o} - \dfrac{1}{R_f}}$$

Noting that the parallel equivalent of R_o and R_L is

$$R_{OL} = \frac{1}{\dfrac{1}{R_o} + \dfrac{1}{R_L}} \tag{4.6}$$

we multiply the numerator and denominator by R_{OL} to create a simpler expression for the Thevenin resistance:

$$R_{TH/in} = \frac{R_{OL} + R_f}{\left[\dfrac{1}{R_{in}} + \dfrac{1}{R_f}\right]\left[R_{OL} + R_f\right] + R_{OL}\left[\dfrac{A}{R_o} - \dfrac{1}{R_f}\right]} \tag{4.7}$$

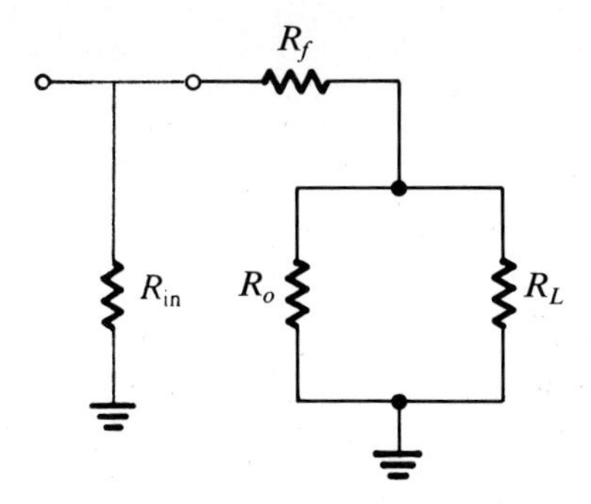

Figure 4.8 The Thevenin input resistance when the amplifier gain has $A = 0$.

Expanding the denominator and cancelling terms reduces (4.7) to

$$R_{TH/in} = \frac{R_f + R_{OL}}{1 + \dfrac{1}{R_{in}}[R_f + R_{OL}] + \dfrac{AR_{OL}}{R_o}} \tag{4.8}$$

Lastly, we multiply the numerator and denominator of (4.8) by the input resistor R_{in} and rewrite $R_{TH/in}$ as:

$$R_{TH/in} = \frac{R_{in}[R_f + R_{OL}]}{R_{in} + [R_f + R_{OL}] + AR_{in}R_{OL}/R_o} \tag{4.9}$$

Inspection of equation (4.9) reveals that the algorithm for calculating the Thevenin input resistance has a form corresponding to the resistance of R_{in} in parallel with the series connection of R_f and R_{OL}, but with the denominator modified by an additional term that depends on the resistors and the open-loop gain. When A is 0 the expression gives the equivalent resistance of the series/parallel circuit shown in Fig. 4.8. For $A > 0$ the value is reduced, and as $A \to \infty$ the value approaches 0.

The Thevenin resistance seen by the voltage source v_s in Fig. 4.6 is the sum of the Thevenin resistance at the amplifier input terminals, as given by (4.9), and the source resistance R_s:

$$R_{TH/s} = R_s + R_{TH/in}$$

Next, we determine the Thevenin equivalent circuit that is seen by the *load* resistor of the feedback amplifier. With the input source voltage at zero, $R_{TH/out} = v_L/i_L$ can be gotten from the matrix model of the circuit in Fig. 4.9:

$$\begin{bmatrix} \dfrac{1}{R_s} + \dfrac{1}{R_{in}} + \dfrac{1}{R_f} & -\dfrac{1}{R_f} \\ -\dfrac{1}{R_f} + \dfrac{A}{R_o} & \dfrac{1}{R_f} + \dfrac{1}{R_o} \end{bmatrix} \begin{bmatrix} v_{in} \\ v_L \end{bmatrix} = \begin{bmatrix} 0 \\ 1 \end{bmatrix} [i_L]$$

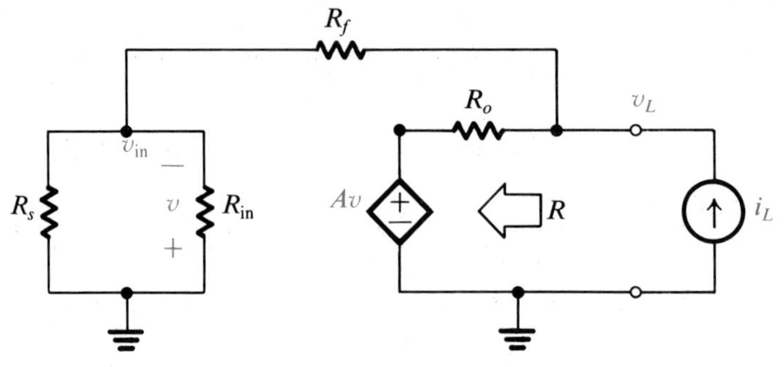

Figure 4.9 Circuit for determining the amplifier output resistance.

and

$$v_L = \frac{\left[\frac{1}{R_s} + \frac{1}{R_{in}} + \frac{1}{R_f}\right] i_L}{\left[\frac{1}{R_s} + \frac{1}{R_{in}} + \frac{1}{R_f}\right]\left[\frac{1}{R_f} + \frac{1}{R_o}\right] + \frac{1}{R_f}\left[-\frac{1}{R_f} + \frac{A}{R_o}\right]}$$

so

$$R_{TH/out} = \frac{\frac{1}{R_s} + \frac{1}{R_{in}} + \frac{1}{R_f}}{\left[\frac{1}{R_s} + \frac{1}{R_{in}} + \frac{1}{R_f}\right]\left[\frac{1}{R_f} + \frac{1}{R_o}\right] + \frac{1}{R_f}\left[-\frac{1}{R_f} + \frac{A}{R_o}\right]} \tag{4.10}$$

This expression can be used to calculate the *output* resistance of the amplifier for finite open-loop gain, and given values of the resistors.

Example 4.1

Remove R_L and find the open-circuit voltage of the feedback amplifier shown in Fig. 4.5.

Solution: The open-circuit voltage can be obtained by solving (4.1) for v_o and then letting $R_L \to \infty$, so

$$v_o = \frac{\frac{1}{R_s}\left[\frac{1}{R_f} - \frac{A}{R_o}\right] v_s}{\left[\frac{1}{R_s} + \frac{1}{R_{in}} + \frac{1}{R_f}\right]\left[\frac{1}{R_o} + \frac{1}{R_f} + \frac{1}{R_L}\right] + \frac{1}{R_f}\left[-\frac{1}{R_f} + \frac{A}{R_o}\right]}$$

Letting $R_L \to \infty$ gives

$$v_{oc} = \frac{\frac{1}{R_s}\left[\frac{1}{R_f} - \frac{A}{R_o}\right] v_s}{\left[\frac{1}{R_s} + \frac{1}{R_{in}} + \frac{1}{R_o}\right]\left[\frac{1}{R_o} + \frac{1}{R_f}\right] + \frac{1}{R_f}\left[-\frac{1}{R_f} + \frac{A}{R_o}\right]} \tag{4.11}$$

The open circuit voltage can be calculated from (4.11) when no load is connected to the op amp, or when a load is connected in a manner that forces i_L to 0.

4.2 IDEAL OPERATIONAL AMPLIFIERS

The circuit model for an ideal operational amplifier can be determined from the properties of the finite-gain amplifier when the open-loop gain A is made arbitrarily large. Using the expression that was developed for the gain of the feedback amplifier, we let $A \to \infty$ in (4.3) to get

$$G = \frac{v_o}{v_s} = -\frac{R_f}{R_s} \tag{4.12}$$

This gain is called the **ideal closed-loop gain** of the circuit because the gain of the open-loop amplifier is taken to be arbitrarily large. Since G is negative the output voltage waveform will be an inverted copy of the input waveform when the feedback and source resistors are equal. Otherwise, the gain is established by the *ratio* of the feedback and source resistors, rather than by their absolute values. This makes it possible to fabricate op amps as integrated circuits with accurately determined gains that are either less than, equal to, or greater than unity. Furthermore, G does not depend on R_L, which enables the gain of the amplifier to be fixed independently of its load—thereby making possible the design of "nearly ideal" voltage sources.

The closed-loop op amp circuit has two important, interesting, and intriguing properties. The first property to note is that **the differential voltage at the input nodes of the amplifier becomes vanishingly small as the open-loop gain is made arbitrarily large.** Some texts refer to this condition as a "virtual short," or a "virtual ground," as illustrated in Fig. 4.10. The op amp's virtual ground property can be verified by inspecting v_{in} in (4.1) and observing that the differential input voltage v becomes 0 as $A \to \infty$ because $v = -v_{in}$ and $v_{in} \to 0$ as $A \to \infty$:

$$v_{in} = \frac{\left[\frac{1}{R_o} + \frac{1}{R_f} + \frac{1}{R_L}\right]\frac{1}{R_s}v_s}{\left[\frac{1}{R_s} + \frac{1}{R_{in}} + \frac{1}{R_f}\right]\left[\frac{1}{R_o} + \frac{1}{R_f} + \frac{1}{R_L}\right] + \frac{1}{R_f}\left[-\frac{1}{R_f} + \frac{A}{R_o}\right]} \to 0$$

Consequently, a second property is that the inverting and noninverting input branch currents i^- and i^+ approach 0, or **the circuit behaves as though the op amp input resistor R_{in} is infinite.** Thus, even though the open-loop gain is made very large, the closed-loop gain remains finite

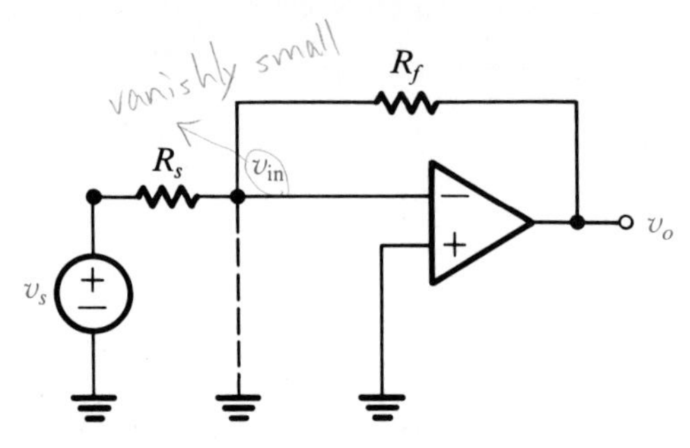

Figure 4.10 The op amp virtual ground.

because the input current is sufficiently small, and the input node is virtually at ground. In reality, the node is not actually shorted to ground, but is at the same potential. *Beware!* There is no short-circuit path for current between the input nodes of the op amp.

Next, we note that when $A \to \infty$ in (4.9) and (4.10) the feedback amplifier circuit's Thevenin-equivalent-input resistance and output resistance become zero. Thus, the input to the ideal feedback amplifier looks like a short circuit to an external source, which accords with the fact that the differential input voltage goes to zero as the open-loop gain is made arbitrarily large. In effect, the current supplied to the ideal feedback amplifier will be limited only by the resistance of the source itself. Similarly, the output resistance is zero and the circuit looks like an ideal voltage source to an external load. In practice, the load current cannot be arbitrarily large and will be limited by the current-carrying capacity of the internal op amp components.

Skill Exercise 4.1

Find the input/output gain of the amplifier circuit shown in Fig. SE4.1 for $R_L = 1000\ \Omega$ with (a) $A = 3000$ and (b) $A = 30000$, and again for $R_L = \infty\ \Omega$ with (c) $A = 3000$, and (d) $A = 30000$.

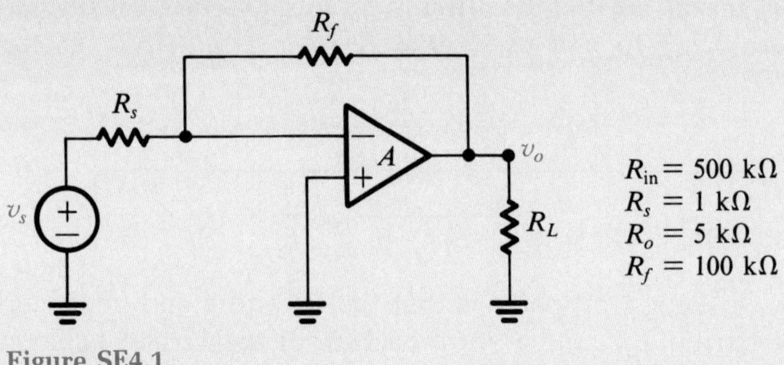

Figure SE4.1

Answer:

R_L \ A	3000	30000
1000	−80.38	−97.62
∞	−95.64	−99.58

Notice that the circuit with $R_L = 1000\ \Omega$ in the previous example has a greater drop in gain as the open-loop gain is decreased, suggesting a need for careful design to examine the behavior of the circuit over the range of load conditions that might be possible. For example, how large does A have to be made to assure that the closed-loop gain is nearly ideal? If we can choose, what values of resistors are most desirable? These and other questions will be explored in the problems at the end of the chapter.

The amplifier with $A = \infty$ is called an ideal operational amplifier because an infinite gain cannot be realized. However, physically realizable gains are large enough to warrant use of an ideal op amp model without a compromise of accuracy. The circuit symbol of an ideal op amp is shown in Figure 4.11. If we restrict its use to circuits in which the op amp is connected with a negative feedback path its properties are as summarized in Table 4.1.

TABLE 4.1 IDEAL OP AMP PROPERTIES

Differential Input Voltage	$v_2 - v_1 = 0$
Input Current	$i^- = i^+ = 0$
Input Resistance	$R_{in} = \infty\ \Omega$
Output Resistance	$R_o = 0\ \Omega$

Ideally, the voltage across the op amp input terminals is zero, and the current into either input terminal is zero.

Operational amplifiers have been used as components in circuits since the 1940s, when they were used in the first analog computers and feedback-control systems. The performance of early vacuum-tube op amps depended strongly on temperature, and in some cases computers

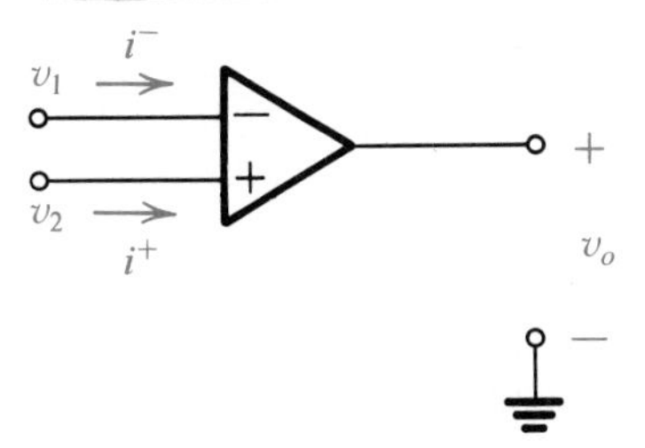

Figure 4.11 Ideal operational amplifier.

using them took several hours to reach a stable operating condition. Today, operational amplifiers are available as inexpensive, packaged integrated circuits. Table 4.2 describes the LM741 op amp, introduced by National Semiconductor in 1972, which remains very popular.

Although "off the shelf" op amps contain several transistors, we can ignore the internal complexity of the device and exploit the simplicity of its model.

TABLE 4.2 LM741 OP AMP PARAMETERS

Input Resistance	R_{in}	2M Ω
Open-loop Gain	A	50,000
Output Resistance	R_o	100 Ω

Example 4.2

Find an expression for the gain of the ideal op amp circuit in Fig. 4.12

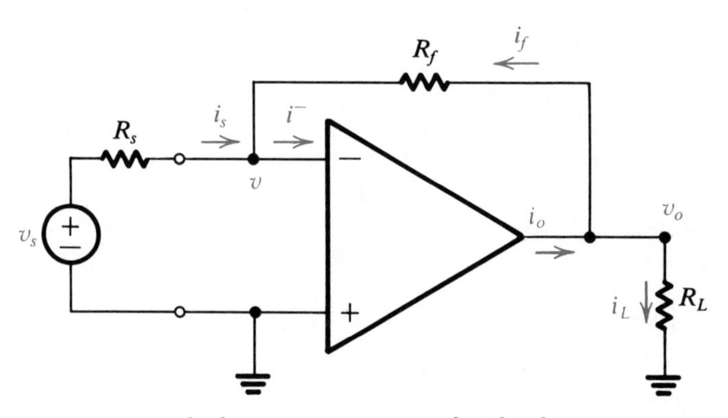

Figure 4.12 Ideal op amp circuit with a load resistor.

Solution: At the op amp output node $v_o = i_f R_f + v$, but for an ideal op amp $v = 0$, so $v_o = i_f R_f$ and

$$i_f = v_o/R_f \tag{4.13}$$

Similarly, the input circuit behavior is described by

$$i_s = v_s/R_s \tag{4.14}$$

Applying KCL at the input node gives $i_s + i_f = i^-$, but $i^- = 0$ for an ideal op amp, so

$$i_f = -i_s \tag{4.15}$$

Using (4.15) and (4.14) in (4.13) gives the output voltage

$$v_o = i_f R_f = -i_s R_f = (-v_s/R_s)R_f$$

and the closed-loop gain

$$G = \frac{v_o}{v_s} = -\frac{R_f}{R_s}$$

Using the ideal op amp model to obtain the closed-loop gain leads to the same expression for the gain that was found in (4.12), but the analysis was greatly simplified. In fact, (4.12) can be obtained directly from KCL at the inverting input node by using $v = 0$ and $i^+ = 0$:

$$\frac{v_s}{R_s} + \frac{v_o}{R_f} = 0$$

Example 4.3

If $R_s = 100\ \Omega$ and $R_L = 1000\ \Omega$ in Fig. 4.12, find v_o, i_s, i_f, i_L, i_o and the power dissipated in the load when the feedback resistor is $R_f = 200\ \Omega$ and the source voltage is 5 V.

Solution: Using (4.14) to determine the source current gives $i_s = v_s/R_s = 5/100 = 50$ mA and from (4.15) the feedback current is $i_f = -i_s = -50$ mA. The closed-loop gain of the op amp is $G = -R_F/R_s = -200/100 = -2$, so the output voltage becomes $v_o = -2(5) = -10$ V. The load current i_L is determined by Ohm's law: $i_L = v_o/R_L = -10/1000 = -10$ mA. The op amp output current must satisfy KCL at the output node for the values of i_f and i_L found above: $i_o = i_f + i_L = -50 - 10 = -60$ mA. The power dissipated in the load resistor is $P_L = (-.01)^2(1000) = 100$ mW.

Note that we did not write mesh equations to describe the op amp circuit. In general, we cannot use mesh analysis to solve an op amp circuit because we do not have a way to determine the voltage across the op amp itself. Instead, we use **nodal analysis.**

Example 4.4

Using the parameters of the LM741 in Table 4.1, find the gain of the amplifier in Fig. 4.13 and compare it to the ideal op amp gain found in

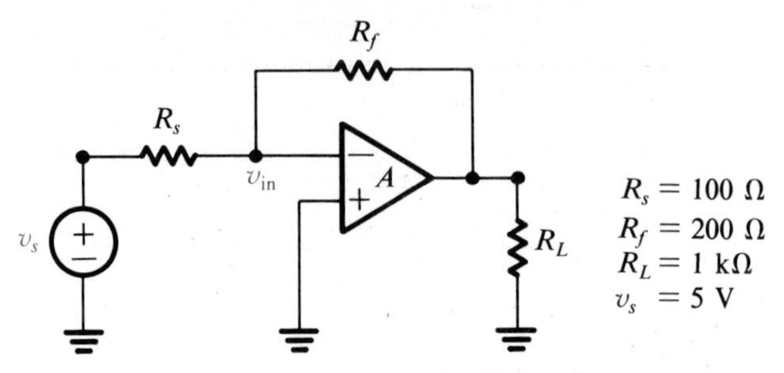

Figure 4.13 Amplifier circuit for Example 4.4.

Example 4.3. Also calculate the voltage v_{in} at the inverting node of the amplifier.

Solution: The gain of the amplifier is given by (4.7)

$$\frac{v_o}{v_s} = -1.999808$$

and the voltage at the inverting input node can be determined with the use of (4.5)

$$v^- = v_{in} = 0.320 \text{ mV}$$

The current in the inverting input terminal of the amplifier is

$$i^- = (0.00032)/(2 \times 10^6) = 1.6 \times 10^{-10} \text{ A}$$

Comparing the values of the voltages and currents with those for an ideal op amp model shows that the ideal op amp model provides a very close approximation to the finite-gain amplifier model, with significantly less computation.

4.3 A CLOSER LOOK AT THE OP AMP

The ideal op amp behaves in a seemingly mysterious manner. First, it is essential to note that **the ideal op amp model does not satisfy Kirchhoff's current law!** The op amp from Example 4.4 is shown in Fig. 4.14 with the values of i_s and i_L calculated. A total of 60 mA *enters* a KCL surface drawn around the op amp, not the 0 mA that is predicted by KCL! The explanation for this anomaly is that the *ideal* op amp model does not show the internal path to ground in the *actual* feedback amplifier circuit. The internal path exists in all physical op amps, and it provides a return path for the current in v_s and R_L. The internal ground path carries 60 mA leaving the KCL surface, so that KCL is ultimately satisfied for the physical de-

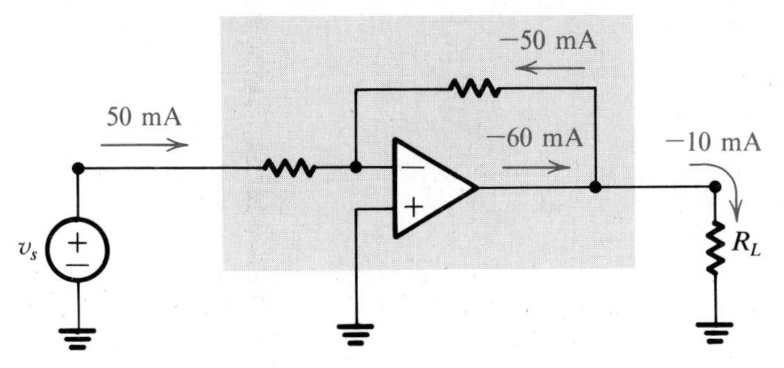

Figure 4.14 Branch currents for an op amp.

vice, even though the ideal op amp model fails to satisfy it. The rule of thumb to remember is that it is inappropriate to apply KCL to the output node of the op amp, unless the currents in all but the op amp return branch are known. Likewise, it is impossible to apply KVL to a path that includes the output branch of the op amp because there is no way to specify the voltage drop across the op amp itself.

4.4 VOLTAGE FOLLOWER

The feedback amplifier developed in Section 4.2 has negative gain. In some applications it is desirable to use the op amp circuit of Fig. 4.15 to realize a positive gain. Since the op amp differential input voltage $v_1 - v_2$ must be zero, the voltage across R_1 is the same as the source voltage, $v = v_s$ and the current through R_1 is $i_1 = v_s/R_1$. Next, observe that $i_2 = i_1$ because the current into the noninverting input must be zero. Therefore, the output voltage is

$$v_o = i_1(R_1 + R_2)$$

Substituting the expression for i_1 into this equation gives

$$v_o = v_s[1 + R_2/R_1]$$

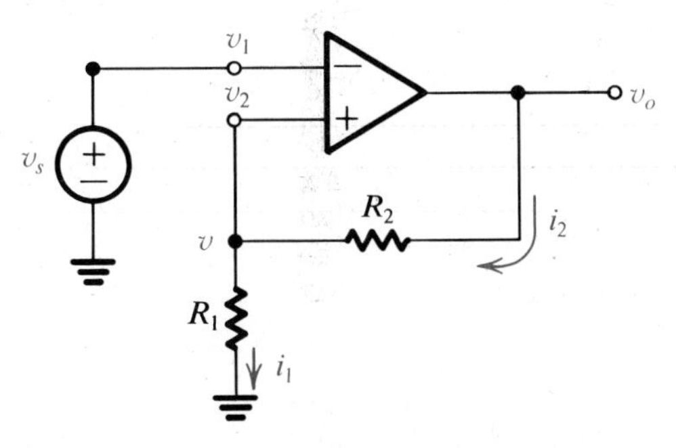

Figure 4.15 A "voltage follower" circuit.

which can be rearranged to display the input/output gain:

$$G = v_o/v_s = 1 + R_2/R_1$$

The gain of the circuit is always positive and greater than one, i.e. the amplitude of the output signal will always exceed the amplitude of the input signal. This circuit is sometimes referred to as a "noninverting amplifier" because the output signal always has the same polarity as the input signal.

4.5 BUFFER CIRCUIT

The op amp circuit and the short circuit in Fig. 4.16 both have the property that $v_o = v_s$, and for both circuits $i_L = v_o/R_L = v_s/R_L$. This might lead you to conclude that the op amp realization of a short circuit is a needless waste of hardware. However, the circuits differ in a major way, because the op amp must have $i_s = 0$, while for the short circuit $i_s = v_s/R_L$. In effect, the op amp acts as a buffer between the source and the load, because it establishes a load voltage without causing a source current. This property is extremely useful in practical circuits having a source resistor, like those shown in Fig. 4.17. In this case, the op amp circuit has $v_o = v_s$ but the short circuit (direct connection) has $v_o = [R_L/(R_s + R_L)]v_s$. In the direct connection, the source voltage is split between the internal resistance of the source and the external load resistance. Consequently, the power that can be delivered to the load is less than it would be for the op amp circuit.

Figure 4.16 A load driven (a) by an *ideal* source with an op amp, and (b) by a directly connected source.

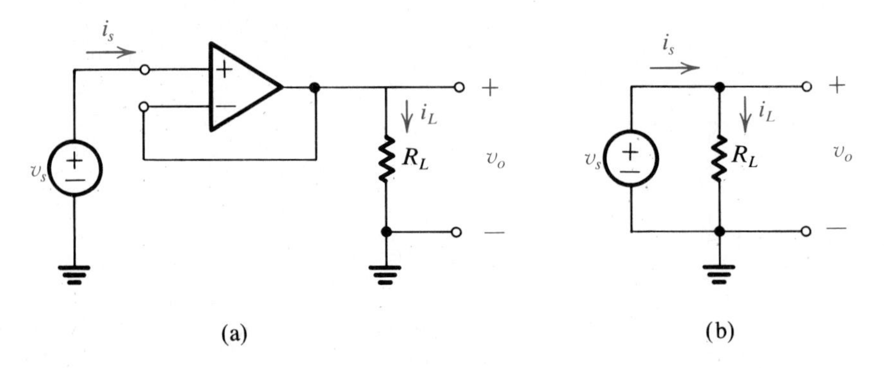

Skill Exercise 4.2

What is the power delivered to R_L in the op amp circuits for Fig. 4.17?

Answer: a. $P_L = (v_s)^2/R_L$ b. $P_L = [v_s/(R_s + R_L)]^2 R_L$.

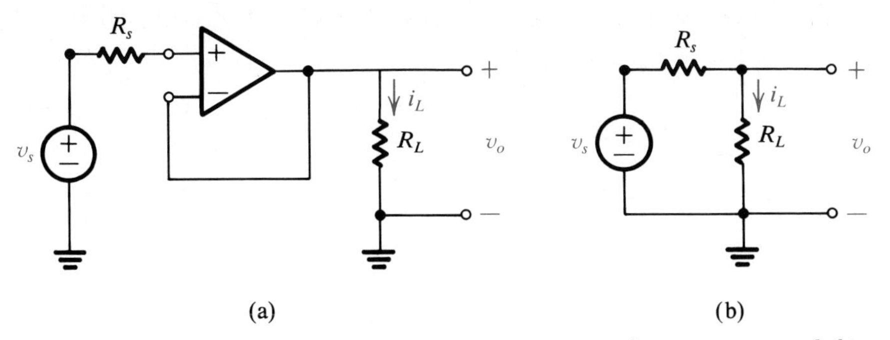

Figure 4.17 A load driven (a) by a *physical* source with an op amp, and (b) by a directly connected source.

Buffer circuits are used when several circuits are to be driven from the same source, for example, when many logic gates fan out from a single gate, as shown in Fig. 4.18. The fanout of a gate is a measure of how many gates can be driven by its output while maintaining satisfactory signal levels. If each of the parallel gates presents the same effective load resistance R_L to the driving gate, their parallel combination has $R_{eq} = R_L/N$, with the result that the power delivered to the aggregate load is much less than the power that would have been delivered to a single load, and the voltage applied to the parallel combination of loads is also reduced. Since logic circuits are typically implemented as threshold devices whose voltage levels represent either 0 or 1, degrading signal levels can change a 1 to a 0 (or vice versa), thereby degrading the circuit's behavior.

Loading the output of the direct-connection circuit with additional resistors has the effect of reducing the equivalent resistance seen by the source (driving gate). Unfortunately, this causes most of the power supplied by the source to be dissipated in its internal resistor. The load in Fig. 4.17(b) has a power dissipation given by

$$P_L = [v_s/(R_s + R_L)]^2 R_L$$

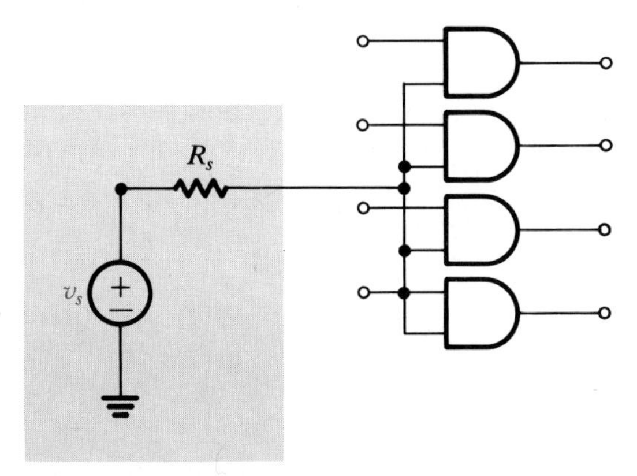

Figure 4.18 A voltage source loaded by several logic gates.

On the other hand, each of the N parallel resistors attached to a physical source dissipates power according to

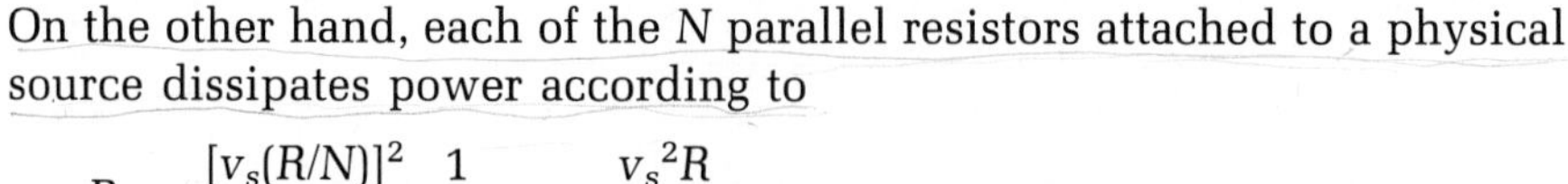

$$P = \frac{[v_s(R/N)]^2}{[R_s + R/N]^2}\frac{1}{R} = \frac{v_s^2 R}{[NR_s + R]^2}$$

The power delivered to one of N parallel resistors decreases as the square of N. On the other hand, if the loading gates are driven by a buffer, the driving gate would only see an effective resistance of an open circuit and the op amp buffer would maintain a constant source voltage, its loading being limited only by the current-carrying capacity of its internal components.

4.6 PRACTICAL CONSIDERATIONS

The model for the ideal operational amplifier is a simplified description of actual op amps. Physical op amps have additional input terminals to supply bias-voltage levels for the internal transistors, with ±15 V being typical bias voltages. Sometimes these additional terminals are shown on the op amp circuit symbol (see Fig. 4.19). Recently, op amps have become available that require only a single bias voltage, e.g. + 1 V, thereby simplifying their power-supply requirements. When specified supply voltages are needed, the circuit symbol shown in Fig. 4.19 can be used.

The magnitude of the output voltage of a physical op amp cannot exceed the magnitude of the supply voltage—a limitation imposed by the current saturation of the transistors used to implement the high-gain voltage amplifier. Thus, a more realistic I/O model of the high-gain amplifier is given by

$$v_o = \begin{cases} +V_{cc} & \text{if } (v_2 - v_1) > +V_{cc}/A \\ A(v_2 - v_1) & \text{if } -V_{cc}/A < v_2 - v_1 < +V_{cc}/A \\ -V_{cc} & \text{if } (v_2 - v_1) < -V_{cc}/A \end{cases}$$

This results in the input/output characteristic for the inverting amplifier shown in Fig. 4.20, with the output voltage saturating at values determined by the supply voltages.

In addition to having a saturation characteristic, a real op amp may have a nonlinear input/output characteristic, rather than a linear one. This can cause nonlinear distortion of the input signal, so that the output sig-

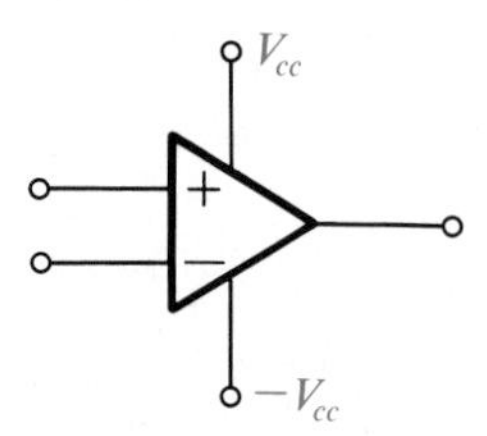

Figure 4.19 Op amp symbol with supply voltages shown.

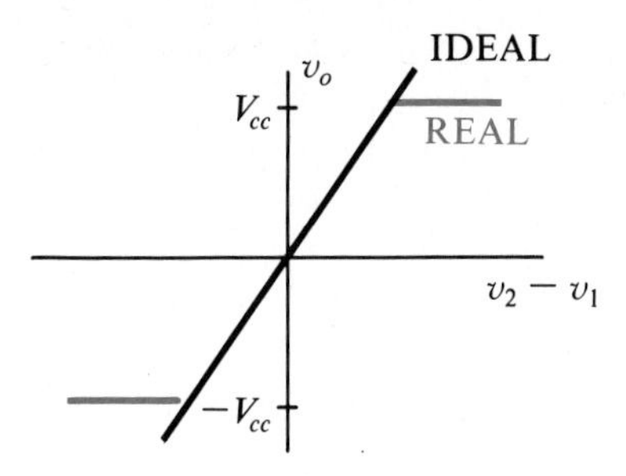

Figure 4.20 Input/output characteristic for the inverting amplifier with saturation.

nal may not be a uniformly scaled copy of the input signal—the scaling will depend on the signal level. A second characteristic of real op amps is that their amplification of a sinusoidal signal is dependent on the frequency of the signal. Typically, high-frequency signals are amplified less than low-frequency signals. This can cause the output signal to be a severely distorted copy of the input signal. Lastly, the manufacturing processes used to fabricate op amps cannot be controlled accurately enough to guarantee that both channels of the op amp are identical. Thus, a signal may be amplified differently if it is applied to the inverting rather than the noninverting input terminal. This causes the op amp to have an output voltage even though the differential input voltage is zero. This output signal is called a **bias.** Packaged op amps contain a bias-offset terminal to permit compensation for the effect of this asymmetry.

SUMMARY

This chapter developed the model for a voltage amplifier and then showed how increasing the gain of the amplifier leads to a circuit having the properties of an ideal operational amplifier. An ideal operational amplifier can be used to simplify circuit analysis containing high-gain op amps. Several examples of op amp circuits were discussed, to illustrate their broad utility in circuits.

Problems - Chapter 4

4.1 Find the input/output gain v_o/v_s for the cascaded op amp circuit shown in Fig. P4.1.

4.2 Find the gain of the voltage follower in Fig. 4.15 when $R_1 = R$ and $R_2 = 2R$.

4.3 For the feedback amplifier shown in Fig. 4.6, find v_{in}, i_{in}, i^-, i_f, i_o, v_o, and the input/output gain, with $R_{in} = 1.5\ \text{M}\Omega$, $A = 20{,}000$, $R_o = 200\ \Omega$, $R_s = 100\ \Omega$, $v_s = 10$ V, and $R_f = 100\ \Omega$.

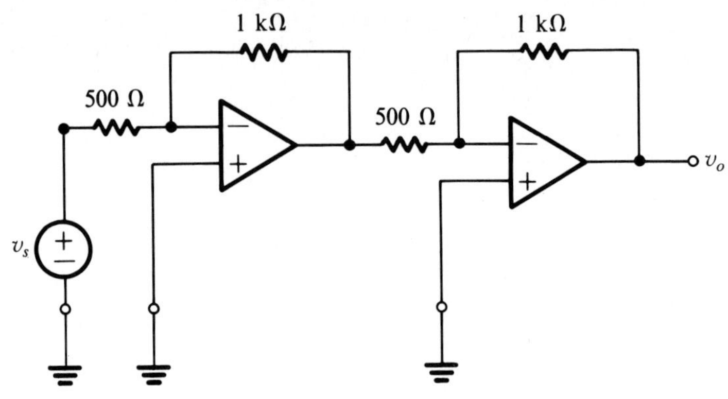

Figure P4.1

4.4 Find the power dissipated in R_L in Fig. 4.13 when $R_s = 1000\ \Omega$, $R_f = 2000\ \Omega$, $R_L = 1000\ \Omega$, and $v_s = 2$ V.

4.5 Analog adders are used in analog computers and other electronic circuits that implement the mathematical operations of addition, subtraction, and scaling. Derive an expression for v_o in Fig. P4.5, and determine the conditions under which $-v_o = v_{s1} + v_{s2}$.

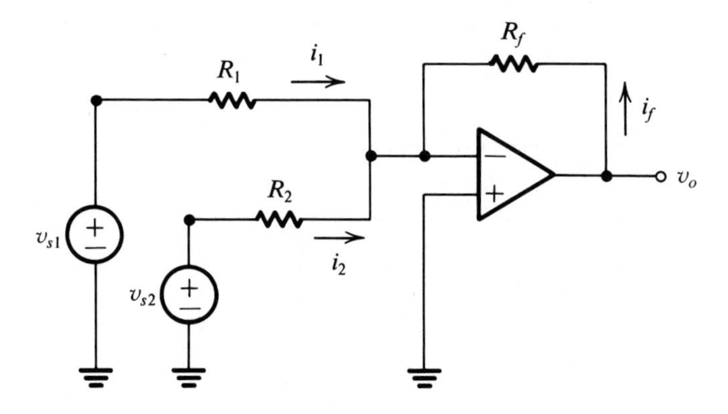

Figure P4.5

4.6 If a buffer is connected to 10 parallel resistors, each having a resistance of 10 kΩ, calculate the total power absorbed by the resistors when the circuit is driven by a source having $v_s = 5$ V and $R_s = 2$ kΩ. Repeat the calculations with $R_s = 0$.

4.7 Find v_o, i_f, i_1, and i_2 in Fig. P4.5, using $R_f = 10{,}000\ \Omega$, $R_1 = 5{,}000\ \Omega$, $R_2 = 10{,}000\ \Omega$, $v_{s1} = 1$ V and $v_{s2} = 2$ V.

4.8 ~~Find the input/output relation of the circuit~~ in Figure P4.8. What kind of functional behavior does it exhibit?

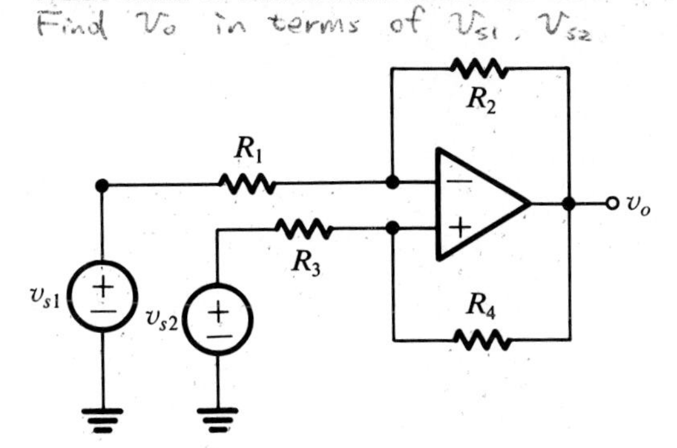

Figure P4.8

4.9 Find an expression for v_o in the differential amplifier circuit in Fig. P4.9.

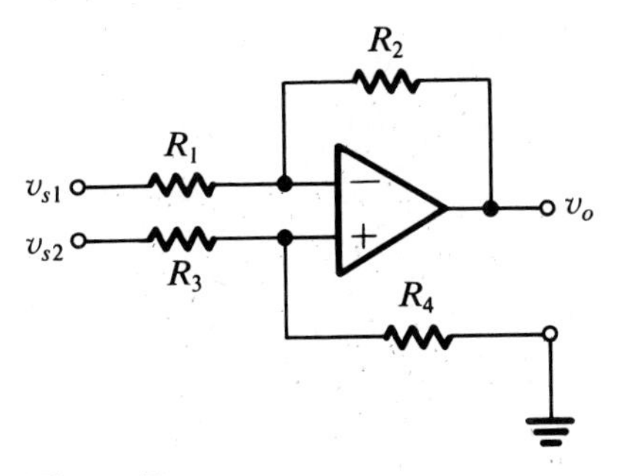

Figure P4.9

4.10 Find v_o and determine the Thevenin equivalent circuit seen by the current source in Fig. P4.10.

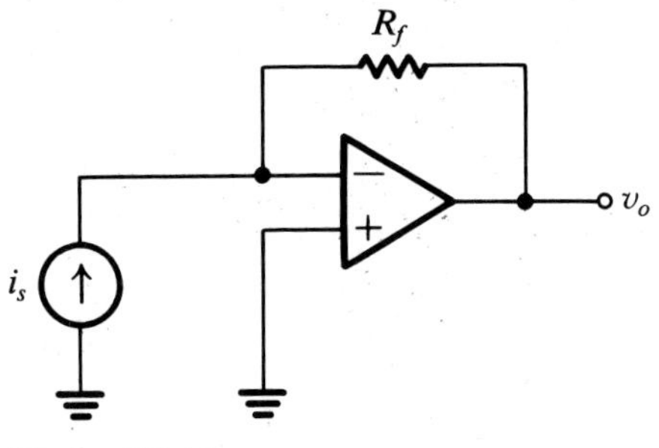

Figure P4.10

4.11 Explain why the resistor R_s can be removed from the circuit in Fig. P4.11 without affecting the input/output relationship of the circuit.

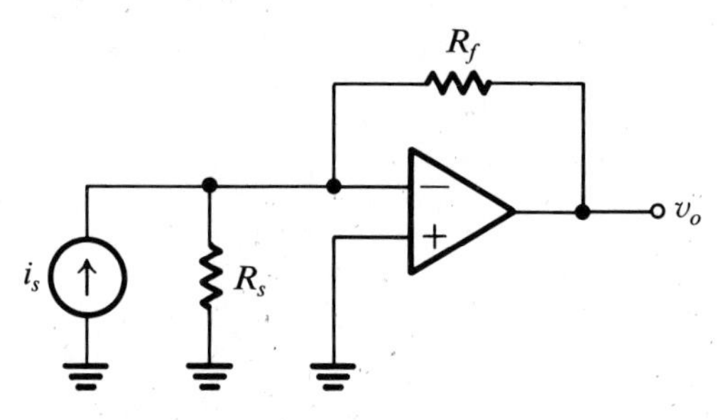

Figure P4.11

4.12 Find the input/output gain of the op amp circuit in Fig. P4.12.

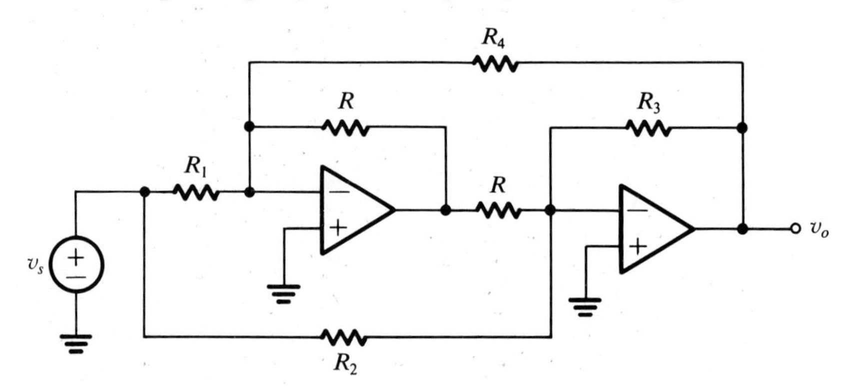

Figure P4.12

4.13 Find the Thevenin input and output resistances of the circuit in Fig. 4.15.

4.14 Find the resistances "seen" by v_s and by R_L in Fig. 4.13.

4.15 Find the input resistance "seen" by i_s, and the output resistance "seen" by R_L in Fig. P4.15.

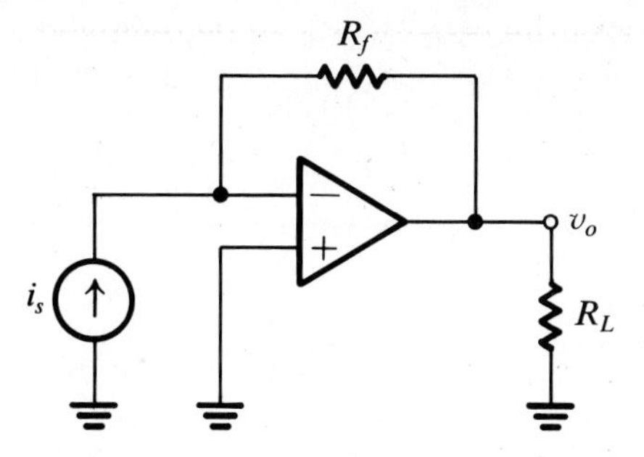

Figure P4.15

4.16 Find the Thevenin equivalent circuit at (a, b) in Fig. P4.16.

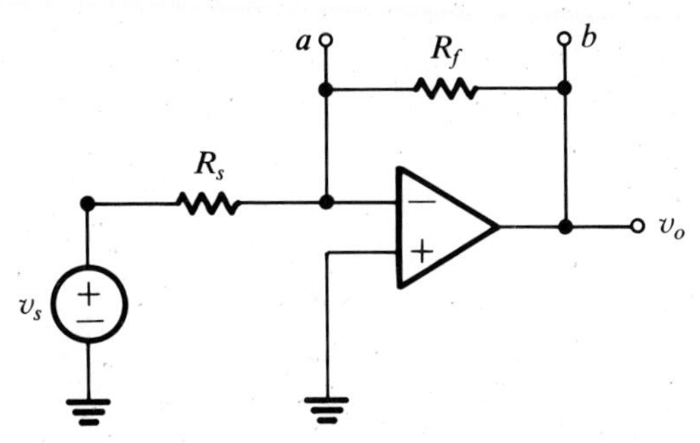

Figure P4.16

4.17 Find the input resistance "seen" by v_s in Fig. P4.17.

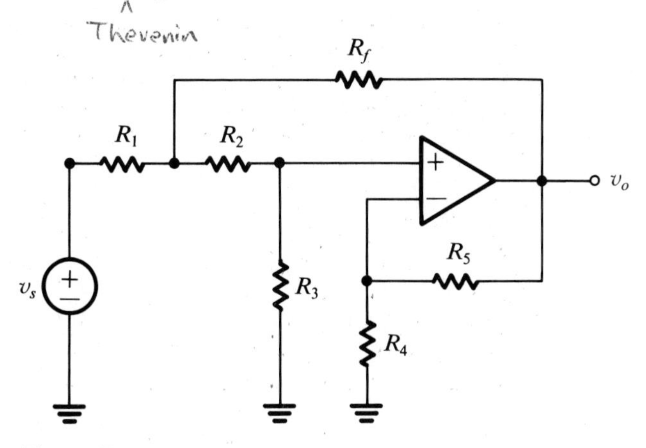

Figure P4.17

4.18 Find v_o in Fig. P4.18.

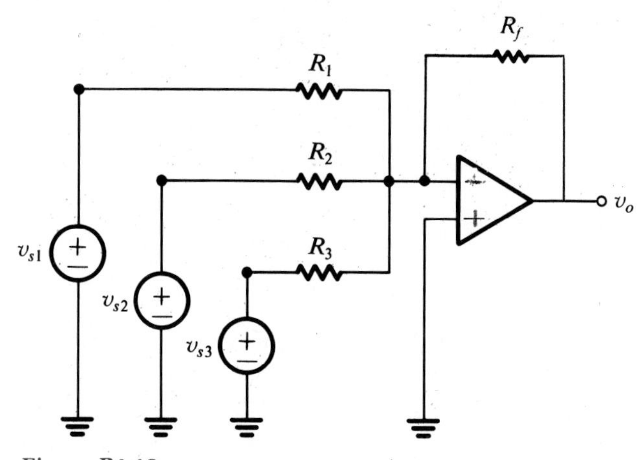

Figure P4.18

4.19 Find the input/output gain of the op amp circuit in Fig. P4.19.

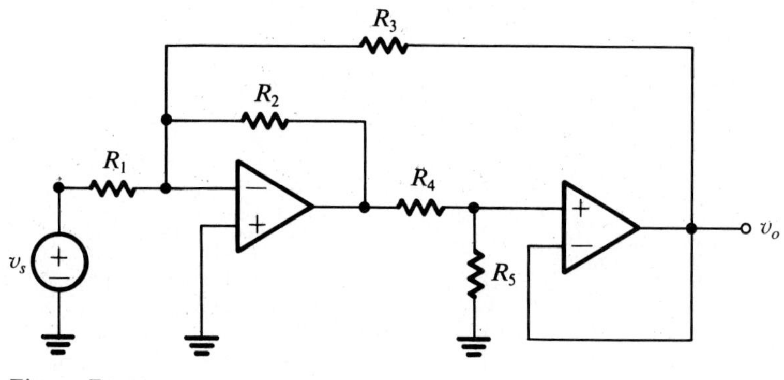

Figure P4.19

4.20 Find the power supplied by the voltage source in Fig. P4.20.

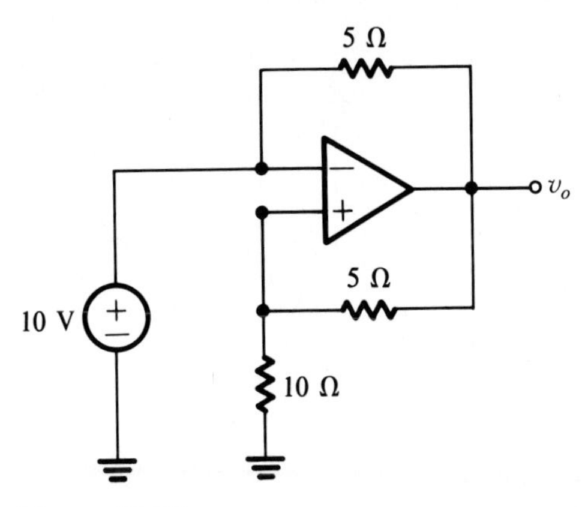

Figure P4.20

4.21 Find the gain of the circuits and find the power supplied by v_s in Figs. P4.21, a, b, and compare.

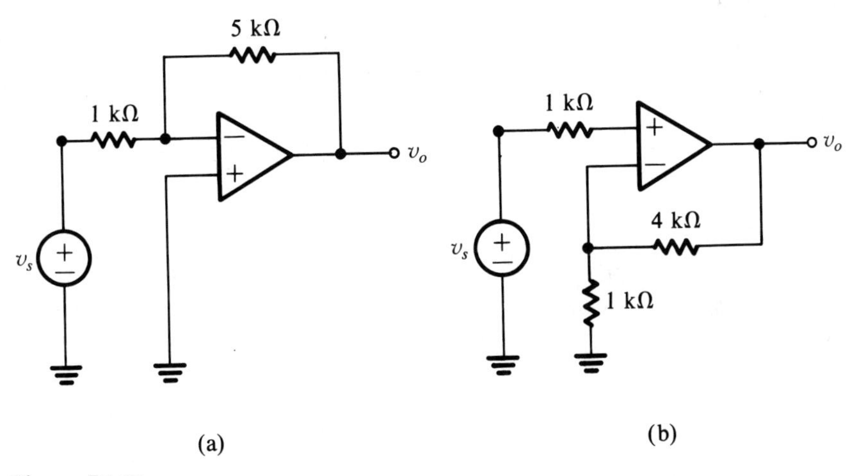

Figure P4.21

4.22 Find the input/output gain of the circuit in Fig. P4.22.

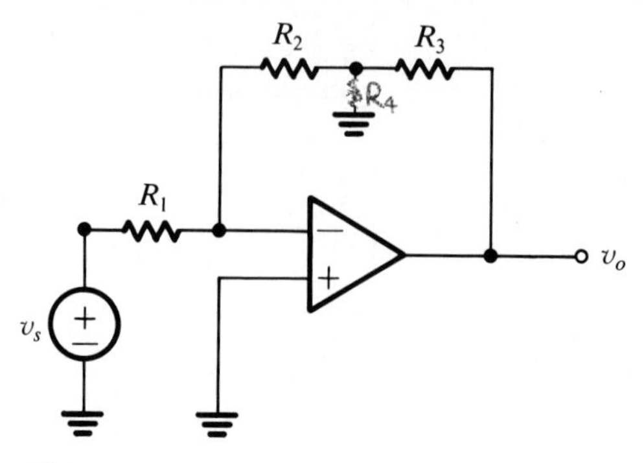

Figure P4.22

4.23 Verify that both of the circuits in Fig. P4.23 are op amp adders, and discuss the relative advantages and disadvantages of each.

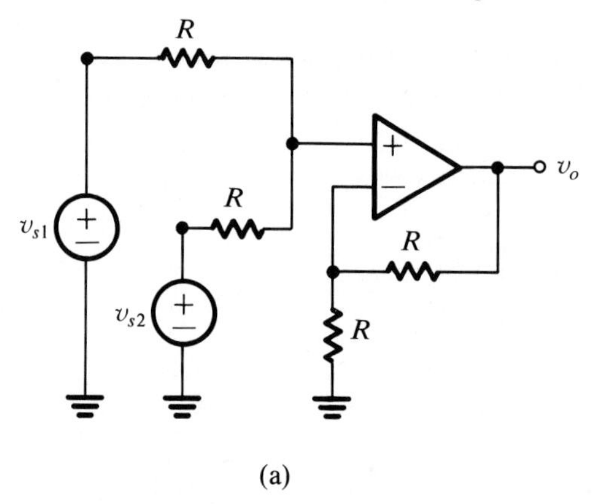

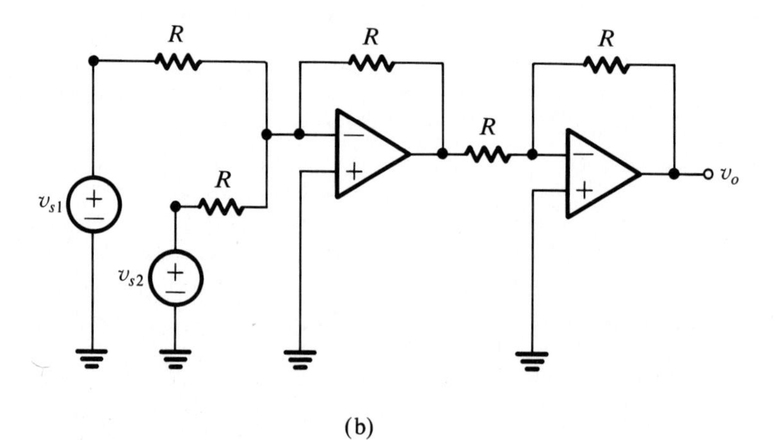

Figure P4.23

ENERGY STORAGE ELEMENTS

INTRODUCTION

The previous chapters have shown how to apply Kirchhoff's voltage and current laws to analyze circuits containing resistors, controlled sources and operational amplifiers. This family of circuits has the property that the output voltage or current depends instantaneously on the input signal. For example, in the feedback operational amplifier the output voltage is given by $v_o(t) = -(R_2/R_1)v_{in}(t)$. The value of the input signal at each time t uniquely determines the value of the output voltage. Circuits and systems having this property are said to be static or memoryless, in the sense that the output depends only on the present input, and not on the past history of the input.

Two elements, capacitors and inductors, exhibit a response behavior in which the output depends on the past waveform of the input, not just on its present value. Circuits containing capacitors and/or inductors are called dynamic circuits, and they are said to have memory. Understanding the dynamic response of individual capacitors and inductors is the first step towards developing an intuitive feel for how more complex circuits behave.

5.1 ENERGY STORAGE ELEMENTS

Capacitors store energy in their electric field, and inductors store energy in their magnetic field. Unlike resistors, these devices do not consume or dissipate energy.

5.1.1 The Capacitor

A capacitor is formed by placing two conducting plates in proximity to each other but separated by an insulating material or dielectric. If charges of equal magnitude but opposite polarity are somehow placed on the conductors, a voltage will be induced between them. The amount of voltage depends on the quantity of charge on the plates, and the model of an ideal capacitor is

$$Cv = q$$

where C is the capacitance of the plate/dielectric configuration. For a given C the voltage across the plates of the capacitor increases linearly with charge. Also, a given charge will create a greater voltage on the smaller of two capacitors. Conversely, it takes more charge to induce a given voltage level on a big capacitor compared to a smaller one.

We will only consider circuits in which C is a constant having units of farads (F). For the parallel plate capacitor shown in Fig. 5.1 the constant C that relates v to q is given by $C = \epsilon(A/d)$ where ϵ is the dielectric constant, or permitivity of the insulating material.

Capacitors store charge just as dams store water. A given amount of water will cause a higher water level in a smaller dam. In our model, charge stored on the plates of a capacitor will remain there indefinitely. In reality, a small leakage current exists between the plates of the capacitor, and the voltage will decay to zero, just as the water level in a dam drops because of evaporation. In Chapter 8 we will see how to model the leakage current for a capacitor.

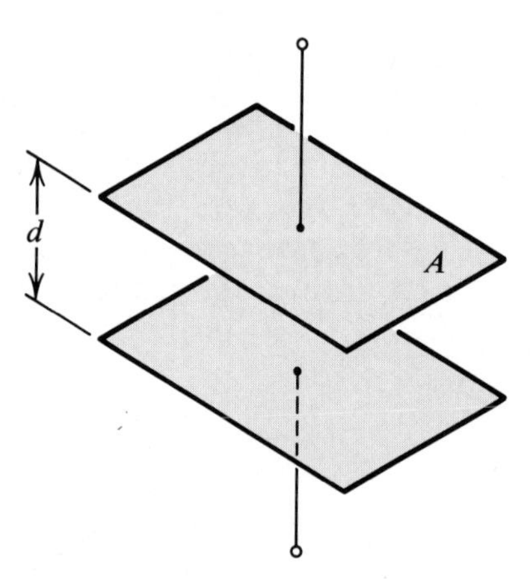

Figure 5.1 Geometry of a parallel plate capacitor.

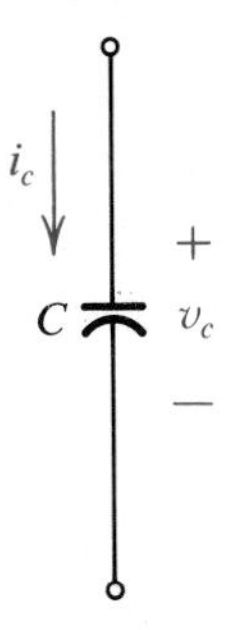

Figure 5.2 The circuit symbol for a capacitor, and its current and voltage polarities.

The circuit symbol and polarity notation for a capacitor are shown in Figure 5.2. The voltage v_C is measured across the terminals of the capacitor, and the current i_C is oriented towards the + reference terminal for voltage. If an external circuit provides a current i_C to a capacitor, the resulting accumulated charge is

$$q(t) = \int_{-\infty}^{t} i_c(\alpha)d\alpha$$

since $i_c(t) = dq/dt$. If the charge on a capacitor is known at time t_0, the charge for $t \geq t_0$ is

$$q(t) = q(t_0) + \int_{t_0}^{t} i_c(\alpha)d\alpha$$

The resulting voltage is

$$v_c(t) = v_c(t_0) + \frac{1}{C}\int_{t_0}^{t} i_c(\alpha)d\alpha$$

This is a **charge storage model** for the behavior of a capacitor; it allows for the possibility that a voltage was present when i_c began, and we summarize the history of the charging current by writing

$$v_c(t_0) = v_c(-\infty) + \frac{1}{C}\int_{-\infty}^{t_0} i_c(\alpha)d\alpha$$

Notice how the voltage for $t \geq t_0$ depends on the initial voltage $v_c(t_0)$ and the charging current waveform from t_0 to time t. The incremental change in voltage from t_0 to t caused by the capacitor current is given by

$$\Delta v_c = v_c(t) - v_c(t_0) = \frac{1}{C}\int_{t_0}^{t} i_c(\alpha)d\alpha$$

It is important to have a careful interpretation of the capacitor model. The presence of *charge causes* the observed *voltage* v_c.

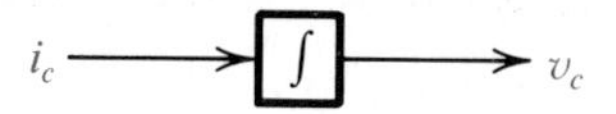

Figure 5.3 Symbolic representation of the current/voltage model for a capacitor.

Voltage and charge have a static relationship. If there is charge on the plates of a capacitor there will be a voltage across its terminals, and conversely, charge will always be present if there is a voltage. In circuits we cannot control capacitor charge directly; instead, we control the *accumulated* charge by controlling the capacitor current. Consequently, if we want to establish a voltage on a capacitor, we must first provide a current. **A change in capacitor voltage is always preceded by capacitor current** because a change in voltage requires a change in the capacitor charge level. This input/output model is represented by the integration symbol shown in Fig. 5.3.

A circuit is said to be in the **DC condition** (direct current) when all the currents and voltages are constant. If a circuit is in DC the capacitors must be acting like open circuits, i.e. have zero current—otherwise the capacitor voltage would be changing.*

Example 5.1

Find the DC voltage v_c in Fig. 5.4.

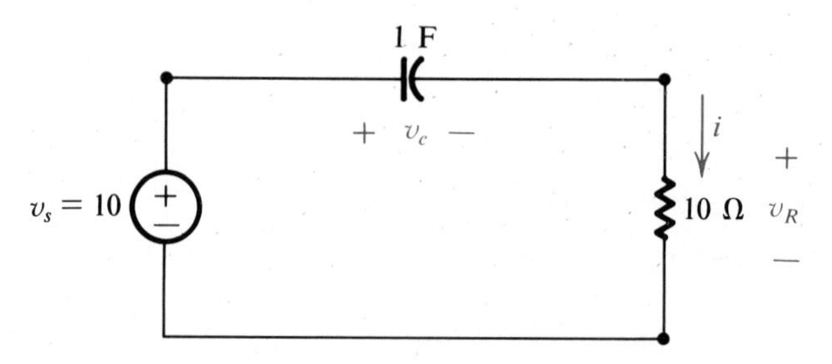

Figure 5.4 Circuit for Example 5.1.

Solution: From KVL we get $v_C = v_S - v_R$ but if the circuit is in the DC condition, the capacitor voltage is a constant and $i = 0$, so $v_C = 10$ V.

*Recall that a branch of a circuit is said to be open if the current in the branch is zero for any applied voltage.

Skill Exercise 5.1

Find the dc voltage across the capacitor in Fig. SE 5.1.

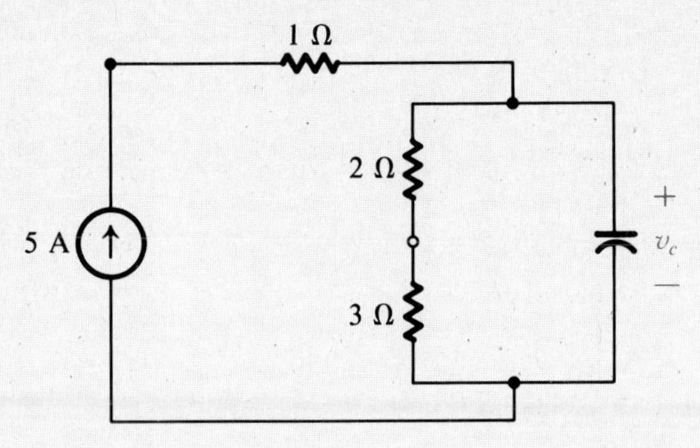

Figure SE5.1

Answer: $v_C = 25$ V.

Example 5.2

A 2 F capacitor is charged by $i_s(t)$ given in Fig. 5.5(a). Find $v_c(t)$.

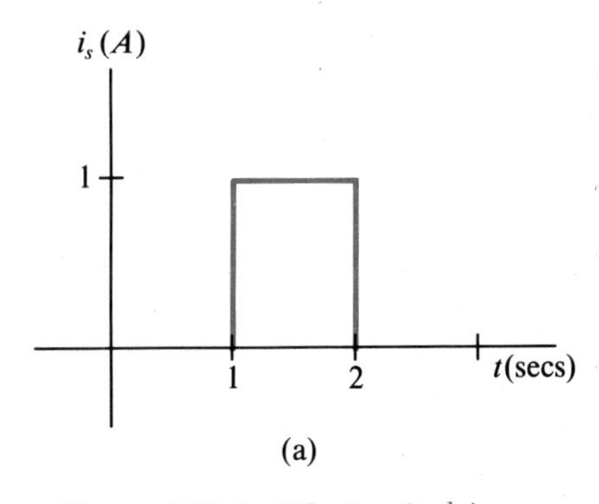

Figure 5.5(a) RC circuit driven by a delayed current pulse.

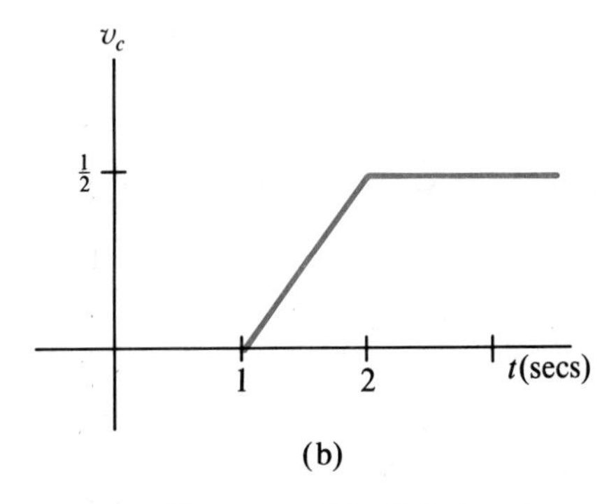

Figure 5.5(b) RC circuit response to a delayed current pulse.

Solution: The capacitor current is described by

$$i_c(t) = \begin{cases} 0 & -\infty < t < 1 \\ 1 & 1 \leq t \leq 2 \\ 0 & t > 2 \end{cases}$$

Since i_c is not described by a single analytic expression, equation must be applied in a piecewise fashion:
For $-\infty < t < 1$

$$v_C(t) = 0 + \frac{1}{2}\int_{-\infty}^{t} 0 d\alpha = 0$$

For $1 \le t \le 2$

$$v_C(t) = v_C(0) + \frac{1}{2}\int_{1}^{t} 1 d\alpha = 0 + \frac{1}{2}\alpha \Big|_1^t = \frac{1}{2}(t-1)$$

For $t \ge 2$

$$v_C(t) = v_C(2) + \frac{1}{2}\int_{2}^{t} 0 d\alpha = v_C(2) = \frac{1}{2}.$$

The graph of v_c is shown in Fig. 5.5(b). Notice how the voltage increases linearly, while the current is constant for $1 \le t \le 2$, and how the voltage remains constant after the current becomes zero.

Example 5.3

Repeat Example 5.2 with i_c as given in Fig. 5.6(a).

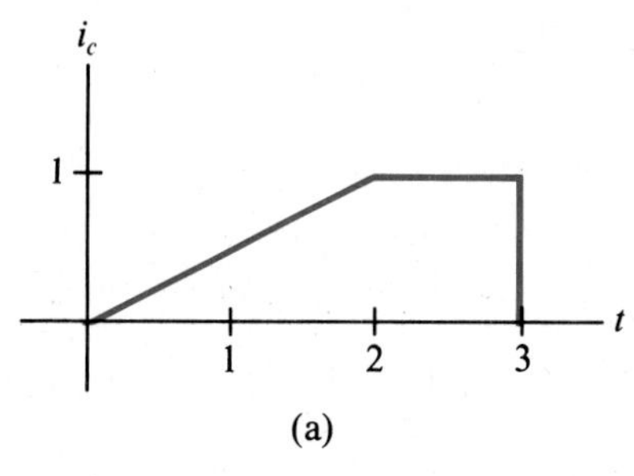

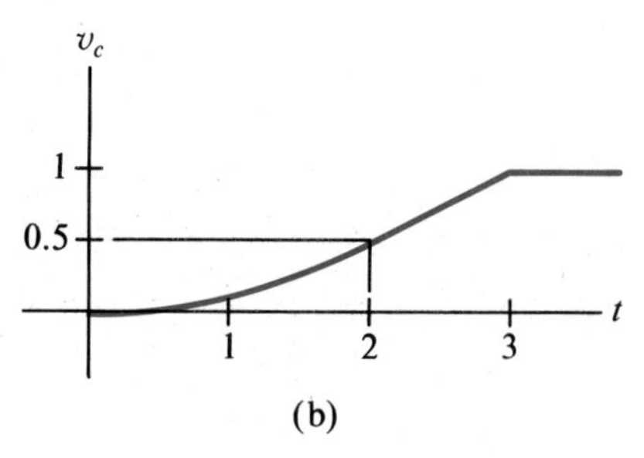

Figure 5.6 Waveforms for capacitor (a) current and (b) voltage.

Solution: The capacitor current is described by

$$i_C(t) = \begin{cases} 0 & -\infty < t \le 0 \\ \frac{1}{2}t & 0 \le t \le 2 \\ 1 & 2 \le t \le 3 \\ 0 & t > 3 \end{cases}$$

Since the capacitor is uncharged at $t = 0$, its voltage for $0 \le t \le 2$ is obtained by integrating $i_C(t)$:

$$v_C(t) = v_C(0) + \frac{1}{2}\int_{0}^{t} \frac{1}{2}\alpha d\alpha = 0 + \frac{1}{4}\frac{\alpha^2}{2}\Big|_0^t = \frac{1}{8}t^2 \text{ V}$$

Likewise, for $2 \leq t \leq 3$ we use the value of $v_C(2)$ and again integrate $i_C(t)$:

$$v_C(t) = v_C(2) + \frac{1}{2}\int_2^t 1\,d\alpha = \frac{1}{2} + \frac{\alpha}{2}\bigg|_2^t = \frac{1}{2} + \frac{1}{2}(t-2)\text{ V}$$

After $t = 3$ the current is zero and the capacitor remains at the voltage $v_C(3) = 1$ V. The graph of $v_C(t)$ is shown in Fig. 5.6(b).

Skill Exercise 5.2

A 2F capacitor is charged by $i_s(t)$ in Fig. SE 5.2(a). Draw $v_C(t)$.

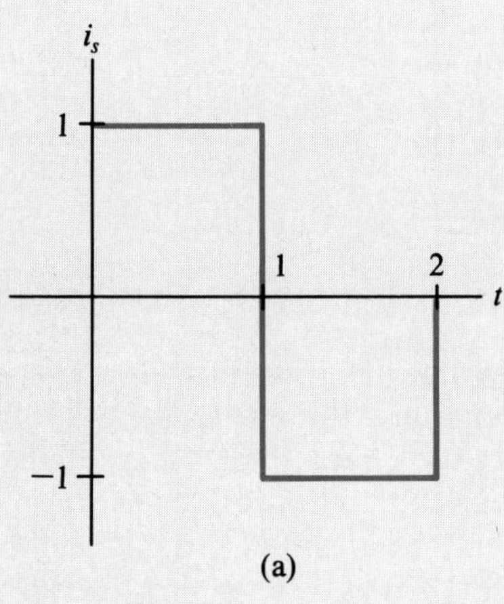

v_c
.5
1
2
t
(b)

Figure SE5.2

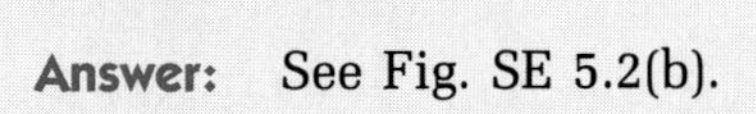

Answer: See Fig. SE 5.2(b).

Example 5.4

Find an expression for $v_o(t)$ in the circuit of Fig. 5.7.

Solution: At the inverting input node $i_R(t) = i_c(t)$. From Ohm's law

$$i_R(t) = \frac{1}{R}v_{in}(t) = i_c(t)$$

so

$$v_c(t) = v_c(t_0) + \frac{1}{RC}\int_{t_0}^t v_{in}(\alpha)d\alpha$$

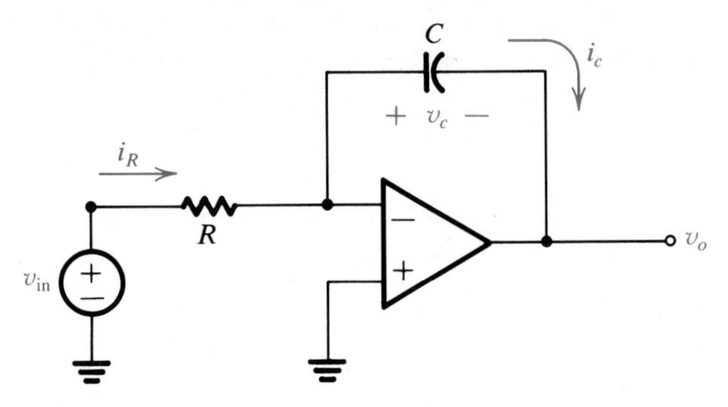

Figure 5.7 Op amp circuit with a feedback capacitor.

Since the voltage drop across the input to an ideal op amp is zero

$$v_o(t) = -v_c(t)$$

and

$$v_o(t) = v_o(t_0) - \frac{1}{RC}\int_{t_0}^{t} v_{in}(\alpha)d\alpha$$

The circuit in Fig. 5.7 behaves like an integrator. If $v_o(t_0) = 0$, the output voltage is a scaled, inverted copy of the integral of the input voltage. That is, $v_o(t)$ depends only on the charge that accumulates on the capacitor up to time t. This is determined by the area under the curve for $v_{in}(t)$ up to time t because the capacitor current is proportional to $v_{in}(t)$. The area can be obtained by plotting $v_{in}(\alpha)$ vs α to display the input waveform and then calculating the area under the curve for a given value of t as shown in Fig. 5.8.

The circuit operates as an integrator because the input source establishes the input current through R. This same current charges the capaci-

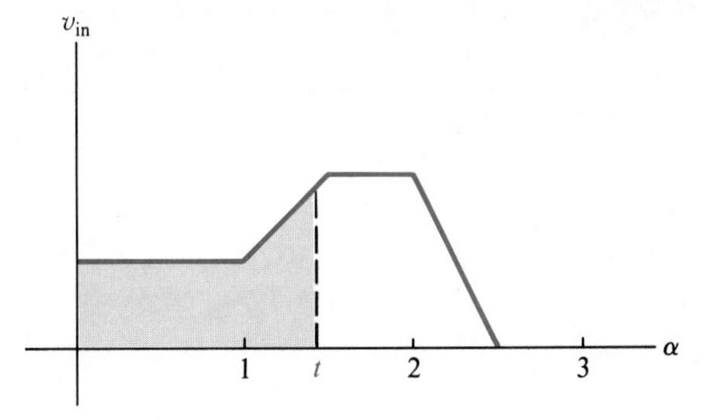

Figure 5.8 The shaded region is the area under the curve of v_{in} from $\alpha = 0$ to $\alpha = t$.

tor, with the capacitor voltage being proportional to the accumulated charge. If C is large, a relatively large amount of charge is required to raise the output voltage to a specified level. (What accounts for the minus sign?)

Skill Exercise 5.3

If an integrator with $R = 1\ \mathrm{M\Omega}$ and $C = 5\ \mu\mathrm{F}$ is driven by the signal in Fig. SE 5.3(a), find $v_o(t)$ (assuming $v_c(0) = 0$).

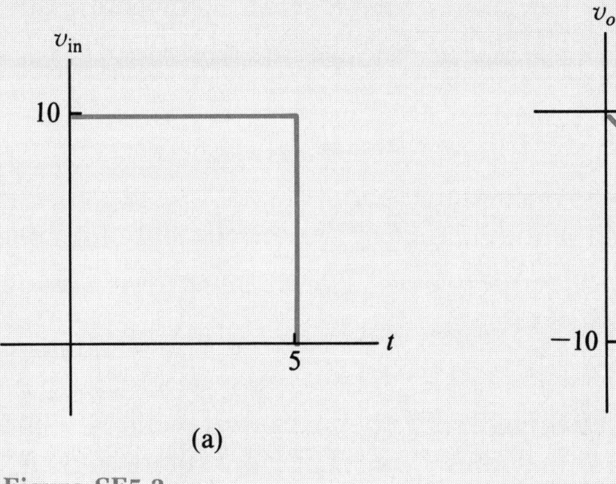

Figure SE5.3

Answer: See Fig. SE 5.3(b)

A differentiator circuit can be built by interchanging the resistor and capacitor in Fig. 5.7 to get the circuit shown in Fig. 5.9. Since $v_c(t) = v_{in}(t)$ and $i_c(t) = C(dv_{in}/dt)$ we have $v_o(t) = -i_R(t)R = -RC(dv_{in}/dt)$. Differentiator circuits are susceptible to spurious noise signals and consequently

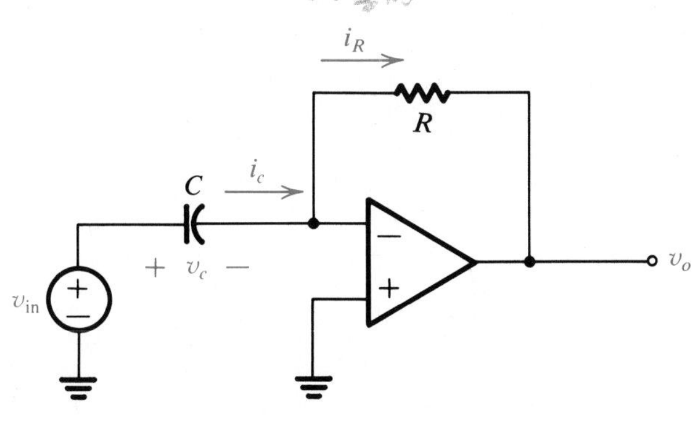

Figure 5.9 A differentiator circuit.

have very limited utility in practical circuits, where noise cannot be eliminated. Noise signals typically have waveforms that vary rapidly, i.e. their derivatives have very large values, which would cause signals to drive op amps out of their linear range of operation.

5.1.2 The Inductor

An inductor is formed by coiling a wire, which greatly increases the magnetic field around the wire when it carries current, as shown in Fig. 5.10 (along with the inductor circuit symbol and reference polarities).

Recall from your background in physics that the magnetic flux linkage λ is related to the voltage across the coiled wire by

$$\lambda(t) = \lambda(-\infty) + \int_{-\infty}^{t} v_L(\alpha)\, d\alpha$$

and the flux linkage is instantaneously related to the coil current by $\lambda(t) = Li(t)$, where L is the inductance of the coil measured in henries (H). It follows that

$$i(t) = i(-\infty) + \frac{1}{L}\int_{-\infty}^{t} v_L(\alpha)\, d\alpha$$

Inductor current may be thought of as being caused by the applied voltage via the accumulation of magnetic flux. That is, inductor voltage controls inductor current by governing the accumulation of magnetic flux. We can also write for $t \geq t_0$.

$$i(t) = i(t_0) + \frac{1}{L}\int_{t_0}^{t} v_L(\alpha)\, d\alpha$$

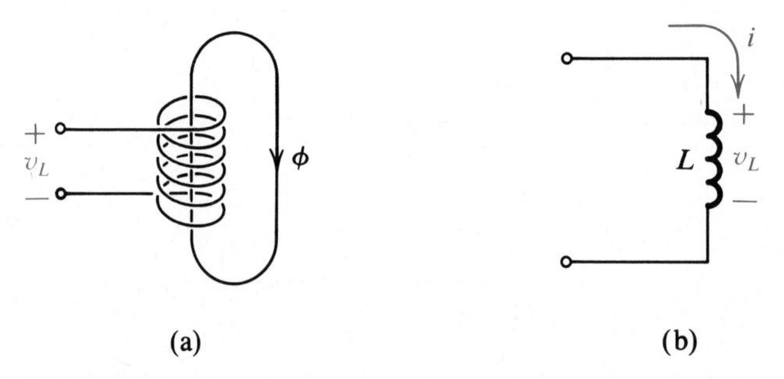

Figure 5.10 Diagram of (a) coiled-wire inductor with (b) its circuit symbol and its current and voltage polarities.

where

$$i(t_0) = i(-\infty) + \frac{1}{L}\int_{-\infty}^{t_0} v_L(\alpha)\,d\alpha$$

This expression for $i(t)$ is a **flux storage model** for the inductor, since $i(t) = \lambda(t)/L$.

Inductor current and the magnetic flux linkages have a static relationship. Inductor voltage causes flux accumulation just as capacitor current causes charge accumulation. However, magnetic flux cannot be established instantly, just as charge cannot accumulate instantly on a capacitor. **A change in inductor current is always preceded by inductor voltage** because a change in current requires a change in the magnetic flux level. Since a voltage across the terminals of an inductor causes flux linkages to accumulate, the current in an inductor will change unless the applied voltage is held to zero. Thus, DC conditions can exist for an inductor if and only if its voltage is zero.

Example 5.5

Find i, v_c and i_L under DC conditions in Fig. 5.11.

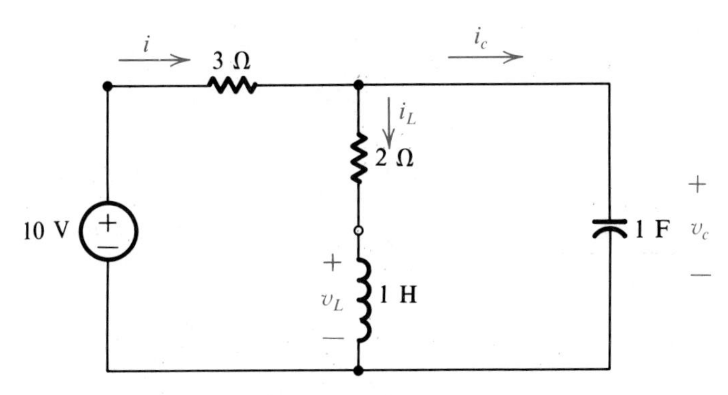

Figure 5.11 Circuit for Example 5.8.

Solution: Under DC conditions we have $i_c = 0$ and $v_L = 0$, so by voltage division $v_c = 4$ V, and by Ohm's law $i = i_L = 2$ A.

Skill Exercise 5.4

Find v_c and i_L under DC conditions in Fig. SE 5.4.

Answer: $i_L = 5$ A, $v_c = 75$ V

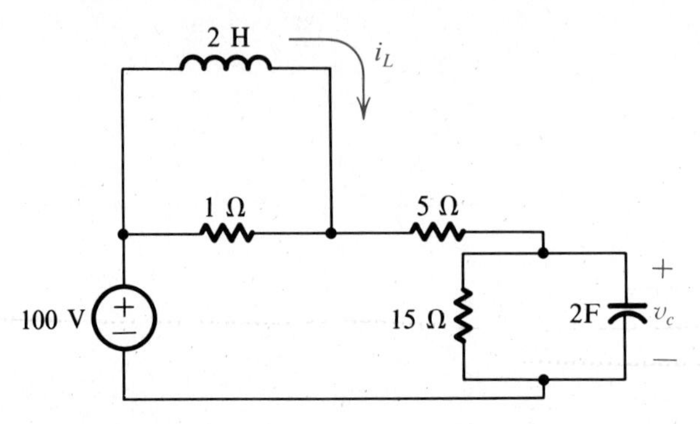

Figure SE5.4

5.1.3 Derivative Models

The derivative models for capacitors and inductors are defined by differentiating the integral models to get

$$i_c(t) = C\frac{dv_c}{dt}$$

$$v_L(t) = L\frac{di_L}{dt}$$

These relationships are appealing because they are easy to memorize. However, we must be careful to interpret them correctly. The capacitor model lets us infer the value of the capacitor current that is causing an observed change in capacitor voltage. Current causes a change in capacitor voltage, not vice versa. Likewise, the inductor model allows us to infer

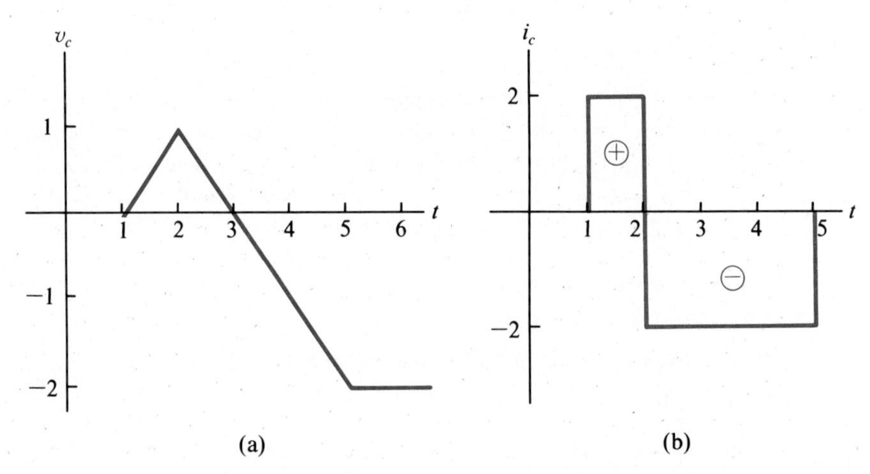

Figure 5.12 Capacitor (a) voltage and (b) current for Example 5.6.

the value of the voltage that is causing a change in inductor current. Understanding the causality of these relationships will make it easier to develop a physical "feel" for circuits because we can use KVL and KCL to determine i_c and v_L. In contrast, we find it difficult to look at a circuit and "see" dv_c/dt or di_L/dt. The derivative model for a capacitor can be misinterpreted to mean that a changing capacitor voltage causes capacitor current. Rather it specifies the amount of current that must be in the C-path to cause a change in v_C. Even when a voltage source is connected directly across a capacitor, the capacitor voltage is caused by the current that causes charge to accumulate on its plates. In this extreme case the current is exactly the amount required to cause the capacitor voltage to equal the applied source voltage.

Example 5.6

The voltage across a 2 F capacitor is observed to have a waveform as shown in Fig. 5.12(a). Find the capacitor current for $t \geq 0$.

Solution: Since the slope of the curve for the capacitor voltage determines the current according to $i_c(t) = C dv_c(t)/dt$, and from Fig. 5.12.

$$\frac{dv_C}{dt} = \begin{cases} 0 & 0 < t < 1 \\ 1 & 1 < t < 2 \\ -1 & 2 < t < 5 \\ 0 & 5 < t \end{cases} \quad \text{and} \quad i_c(t) = \begin{cases} 0 & 0 < t < 1 \\ 2\text{A} & 1 < t < 2 \\ -2\text{A} & 2 < t < 5 \\ 0 & 5 < t \end{cases}$$

The curve for i_c in Fig. 5.12(b) has been labeled to illustrate its polarity, and this can be seen to correspond to the slope of the curve for v_c in Fig. 5.12(a).

5.2 ENERGY STORAGE

We saw in Chapter 1 that the instantaneous power dissipated by a resistor is given by $p_R(t) = i_R^2(t)R$. Resistors always dissipate power in the form of heat. To compare the behavior of capacitors and inductors with that of resistors, we next consider the instantaneous power supplied to a capacitor by whatever source is providing its charging current

$$p(t) = v(t)i(t) = Cv\frac{dv}{dt}$$

The incremental change in energy stored in the electric field of the capacitor is

$$\Delta W(t) = \int_{t_0}^{t} p(\alpha)d\alpha = C\int_{t_0}^{t} v\frac{dv}{d\alpha}d\alpha = \frac{1}{2}Cv^2(t) - \frac{1}{2}Cv^2(t_0)$$

Likewise, for an inductor

$$p(t) = i(t)v(t) = Li(t)\frac{di}{dt}$$

and the incremental energy stored in the magnetic field of the inductor is

$$\Delta W(t) = L\int_{t_0}^{t} i(\alpha)\frac{di}{d\alpha}d\alpha$$

$$= \frac{1}{2}Li^2(t) - \frac{1}{2}Li^2(t_0)$$

If $v_c(t_0) = 0$ and $i_L(t_0) = 0$ the elements are said to be initially de-energized. Then the energy stored at any time t is given by

$$W_c(t) = \frac{1}{2}Cv_c{}^2(t) \geq 0$$

$$W_L(t) = \frac{1}{2}Li_L{}^2(t) \geq 0$$

The energy stored in a capacitor or an inductor is a non-negative quantity. For a capacitor, W_c is just the energy absorbed from the charging current.

For a resistor $p(t)$ and $W(t)$ are non-negative, so the device always absorbs energy, which it dissipates as heat. This heat can never be recovered. On the other hand, the energy absorbed by a capacitor is always non-negative, but $p(t) = Cvdv/dt$ can be positive, negative or zero. If p is negative, the polarity of the capacitor current is directed away from the positive side of C, and the capacitor is actually delivering power back to the external circuit. This reduces W by reducing v. A similar line of reasoning applies to inductor energy.

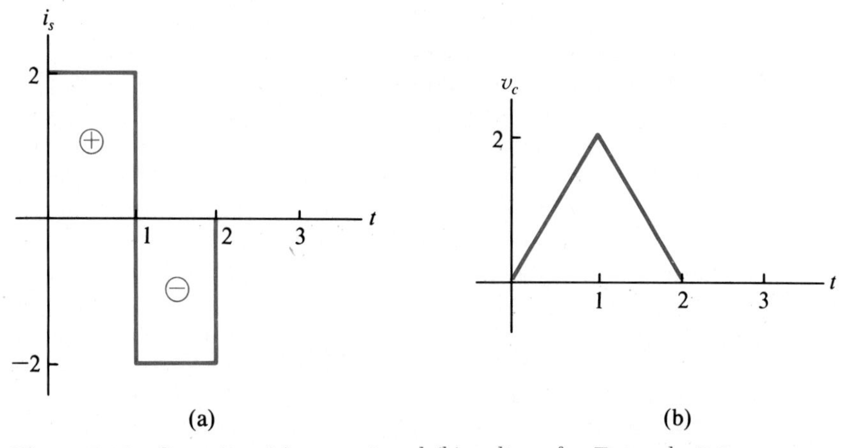

Figure 5.13 Capacitor (a) current and (b) voltage for Example 5.7.

Example 5.7

If a 1 F capacitor is charged by $i_s(t)$ in Fig. 5.13(a), given $v_C(0) = 0$ find $W_C(t)$ for $t = 0$, $t = 1$, and $t = 2$.

Solution: The current waveform is described by

$$i_s(t) = \begin{cases} 0 & -\infty < t < 0 \\ 2 & 0 \leq t \leq 1 \\ -2 & 1 \leq t \leq 2 \\ 0 & 2 < t \end{cases}$$

so

$$v_C(t) = \begin{cases} 0 & -\infty < t < 0 \\ \dfrac{1}{C}\displaystyle\int_0^t 2d\alpha = 2t & 0 \leq t \leq 1 \\ v_C(1) + \dfrac{1}{C}\displaystyle\int_1^t (-2)d\alpha = 2 - 2(t-1) & 1 \leq t \leq 2 \\ 0 & 2 \leq t \end{cases}$$

The stored energy is $W_C(0) = \dfrac{1}{2}Cv_C^2(0) = 0\text{ J}$

$$W_C(1) = \frac{1}{2}(1)(2)^2 = 2\text{ J}$$

$$W_C(2) = \frac{1}{2}(1)(0)^2 = 0\text{ J}$$

The energy stored in the capacitor is a maximum when the capacitor voltage is a maximum at $t = 1$. At $t = 2$ the stored energy is again zero. Figure 5.13(b) shows the voltage waveform that results from $i_s(t)$. For $0 \leq t \leq 1$ the source is supplying power to the capacitor and for $1 \leq t \leq 2$ the capacitor is supplying power to the source. At $t = 2$ all of the energy that was stored in the capacitor has been returned to the source, for a net dissipation of zero.

5.3 PARALLEL CIRCUITS AND CURRENT DIVISION

When capacitors are connected in parallel, as in Fig. 5.14, they all have a common voltage, so $i_1 = C_1 dv/dt$, $i_2 = C_2 dv/dt$, . . . , $i_k(t) = C_k dv/dt$, for $1 \leq k \leq n$, and the current into the parallel configuration can then be written as

$$i(t) = \left[C_1 + C_2 + \cdots + C_n\right]\frac{dv}{dt}$$

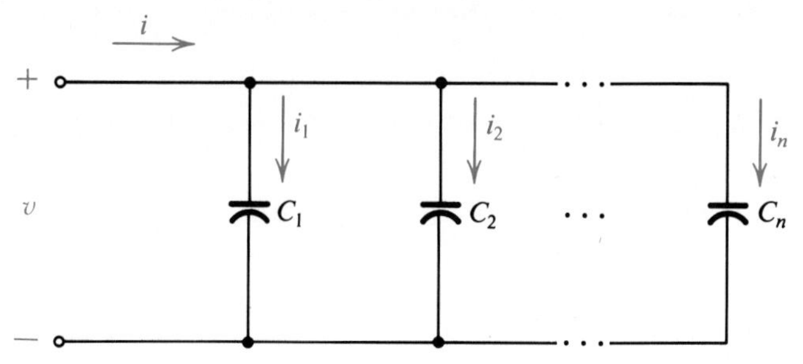

Figure 5.14 Capacitors in parallel.

Letting

$$C = C_1 + C_2 + \cdots + C_n$$

we have $i(t) = C dv/dt$. A parallel configuration of capacitors behaves like a capacitor whose value is equal to the sum of the individual parallel capacitors. This property is useful for making a needed value of C from available components.

Capacitors in parallel combine like conductances in parallel. On the other hand, inductors in parallel combine like resistors in parallel.

For the parallel configuration of inductors in Fig. 5.15:

$$i(t) = i_1(t) + i_2(t) + \cdots + i_n(t) = i_1(t_0) + i_2(t_0) + \cdots + i_n(t_0)$$
$$+ \frac{1}{L_1}\int_{t_0}^{t} v(\alpha)d\alpha + \frac{1}{L_2}\int_{t_0}^{t} v(\alpha)d\alpha + \cdots + \frac{1}{L_n}\int_{t_0}^{t} v(\alpha)d\alpha$$

After rearranging we get

$$i(t) = i_1(t_0) + i_2(t_0) + \cdots + i_n(t_0) + \left[\frac{1}{L_1} + \frac{1}{L_2} + \cdots + \frac{1}{L_n}\right]\int_{t_0}^{t} v(\alpha)d\alpha$$

Letting

$$i(t_0) = i_1(t_0) + i_2(t_0) + \cdots + i_n(t_0)$$

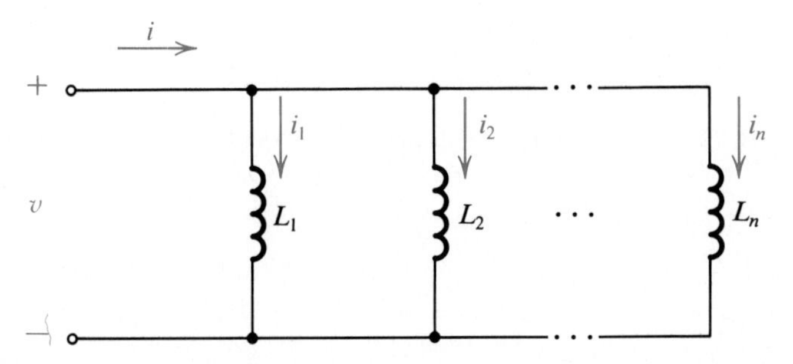

Figure 5.15 Inductors in parallel.

we have

$$\frac{1}{L} = \frac{1}{L_1} + \frac{1}{L_2} + \cdots + \frac{1}{L_n}$$

or

$$\boxed{L = \frac{1}{\dfrac{1}{L_1} + \dfrac{1}{L_2} + \cdots + \dfrac{1}{L_n}}}$$

This L is equivalent to the parallel configuration of inductors. Notice that the algorithm for L is the same as the algorithm for combining parallel resistors.

For current division, the current in each path of a parallel configuration of capacitors is obtained by using to express the total current. Therefore, the current in the kth path will be given by

$$\frac{i_k(t)}{i(t)} = \frac{C_k dv/dt}{[C_1 + \cdots + C_n]dv/dt}$$

and

$$\frac{i_k(t)}{i(t)} = \frac{C_k}{C_1 + \cdots + C_n}$$

Likewise, current division for n parallel inductors is governed by

$$\frac{i_k(t)}{i(t)} = \frac{\dfrac{1}{L_k}}{\dfrac{1}{L_1} + \cdots + \dfrac{1}{L_n}}$$

when each inductor is initially de-energized.

5.4 SERIES CIRCUITS AND VOLTAGE DIVISION

By now you might expect that capacitors in series combine like resistors in parallel. Referring to Fig. 5.16, we have

$$\begin{aligned} v(t) &= v_1(t) + v_2(t) + \cdots + v_n(t) \\ &= v_1(t_0) + \frac{1}{C}\int_{t_0}^{t} i(\alpha)d\alpha + v_2(t_0) + \frac{1}{C}\int_{t_0}^{t} i(\alpha)d\alpha \\ &\quad + \cdots + v_n(t_0) + \frac{1}{C_n}\int_{t_0}^{t} i(\alpha)d\alpha \\ &= v_1(t_0) + v_2(t_0) + \cdots + v_n(t_0) + \left[\frac{1}{C_1} + \frac{1}{C_1} + \cdots + \frac{1}{C_n}\right]\int_{t_0}^{t} i(\alpha)d\alpha \end{aligned}$$

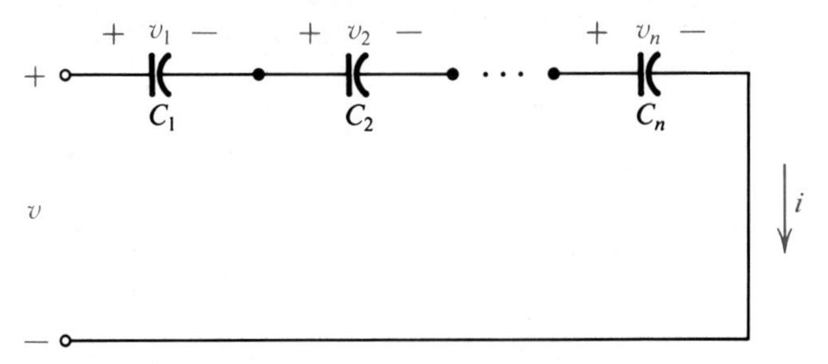

Figure 5.16 Capacitors in series.

Letting

$$v(t_0) = v_1(t_0) + \cdots + v_n(t_0)$$

and

$$\frac{1}{C} = \frac{1}{C_1} + \frac{1}{C_2} + \cdots + \frac{1}{C_n}$$

we conclude that the series circuit of capacitors behaves like a single capacitor whose value is given by:

$$C = \frac{1}{\frac{1}{C_1} + \frac{1}{C_2} + \cdots + \frac{1}{C_n}}$$

Likewise, the series combination of inductors in Fig. 5.17 can be replaced by an equivalent circuit

$$\begin{aligned} v(t) &= v_1(t) + v_2(t) + \cdots + v_n(t) \\ &= L_1\frac{di}{dt} + L_2\frac{di}{dt} + \cdots + L_n\frac{di}{dt} \\ &= \left[L_1 + L_2 + \cdots + L_n\right]\frac{di}{dt} \end{aligned}$$

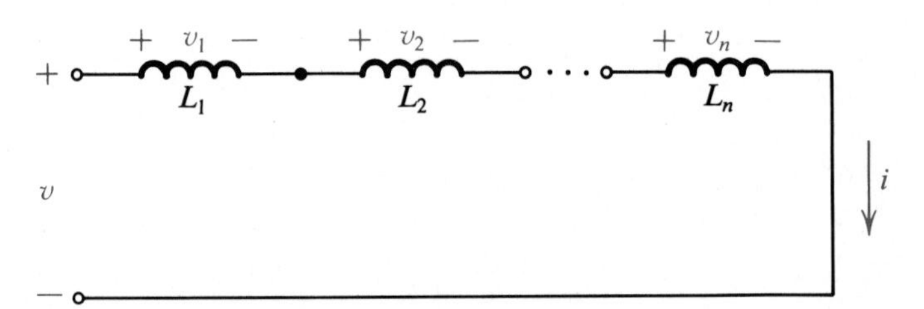

Figure 5.17 Inductors in series.

The equivalent inductor is

$$L = L_1 + L_2 + \cdots + L_n$$

condition

The voltage across initially uncharged capacitors connected in series divides according to

$$\frac{v_k(t)}{v(t)} = \frac{\dfrac{1}{C_k}\displaystyle\int_{t_0}^{t} i(\alpha)d\alpha}{\left[\dfrac{1}{C_1} + \cdots + \dfrac{1}{C_n}\right]\displaystyle\int_{t_0}^{t} i(\alpha)d\alpha} = \frac{\dfrac{1}{C_k}}{\dfrac{1}{C_1} + \cdots + \dfrac{1}{C_n}}$$

The voltage across inductors connected in series divides according to

$$\frac{v_k(t)}{v(t)} = \frac{L_k}{L_1 + \cdots + L_n}$$

5.5 MODELS AND INTUITION

The integral model for a capacitor lets us see capacitor voltage as an output and capacitor current as an input. In this model **input causes output, or capacitor current causes capacitor voltage.** Current causes charge to accumulate on the capacitor, or capacitor current *controls* capacitor voltage. It is essential to understand that charge accumulation requires current integration, i.e. current must be applied for a finite interval of time.

Skill Exercise 5.5

For the circuit shown in Fig. SE 5.5(a) find and sketch $i_R(t)$, $i_c(t)$, $v_c(t)$ and $v_o(t)$. Assume $v_c(0) = 0$ and that $v_s(t)$ is as shown.

Answer:

$$v_c(t) = \begin{cases} t/10 & 0 \le t \le 1 \\ (3 - t)/20 & 1 \le t \le 2 \\ 1/20 & t \ge 2 \end{cases}$$

$i_r(t) = i_c(t) = 0.5\, v_s(t)$

$v_o(t) = v_s(t) + v_c(t)$.

See Fig. SE 5.5(b).

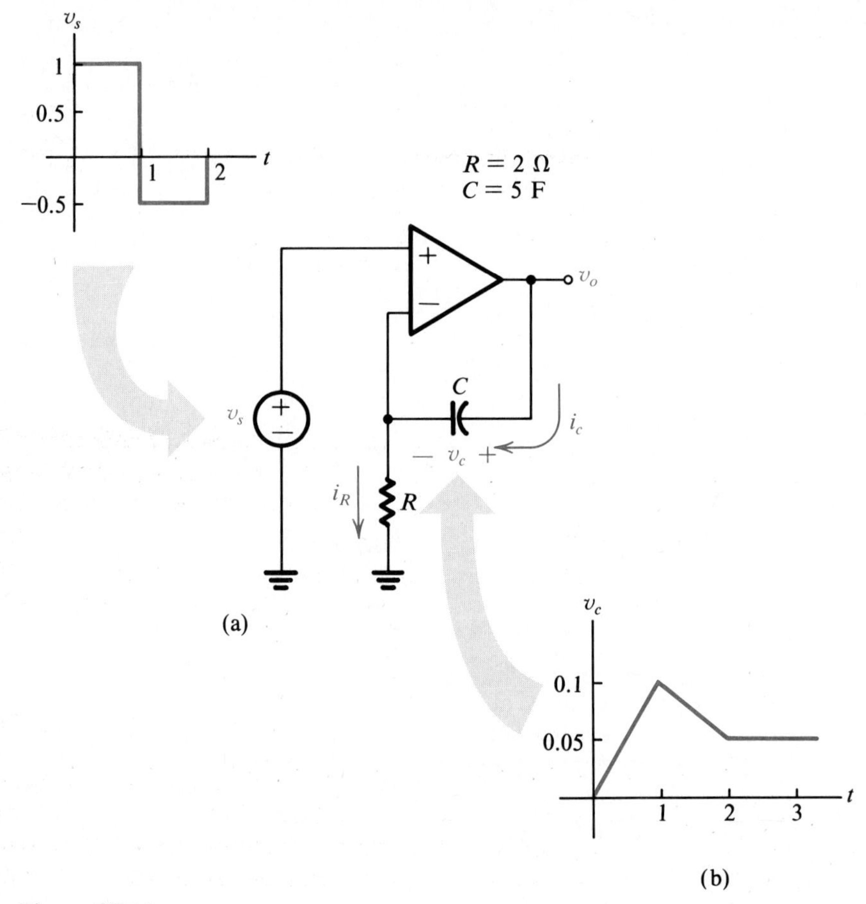

Figure SE5.5

5.6 CAPACITOR AND INDUCTOR BOUNDARY CONDITIONS

Physical sources deliver only finite power. This allows us to write two important boundary conditions that must be satisfied by capacitors and inductors:

> **If i_c is finite everywhere, v_c is continuous everywhere.**
> **If v_L is finite everywhere, i_L is continuous everywhere.**

These statements are known as the "continuity" conditions for capacitors and inductors. Mathematically speaking, if the curve for capacitor current is bounded (finite), the curve for capacitor voltage will be continuous.

Consequently, at any time t_0

$$v_c(t_0^+) = v_c(t_0^-)$$

and

$$i_L(t_0^+) = i_L(t_0^-)$$

where we use the notation

$$f(t_0^-) = \lim_{\substack{t \to t_0 \\ t < t_0}} f(t)$$

and

$$f(t_0^+) = \lim_{\substack{t \to t_0 \\ t > t_0}} f(t)$$

Example 5.8

Find $i_c(0^+)$ in Fig. 5.18 if $v_c(0^-) = 0$ and if $v_s(t) = 10$ for $t > 0$.

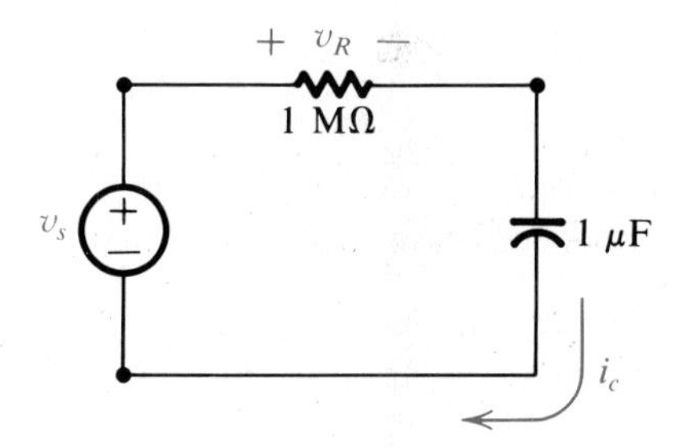

Figure 5.18

Solution: Since the source is finite, we have

$$v_c(0^+) = v_c(0^-) = 0 \text{ V}.$$

Therefore the initial capacitor current is

$$i_c(0^+) = \frac{1}{R} v_R(0^+) = \frac{1}{R} v_s(0^+) = 10\ \mu A.$$

Initially, the capacitor looks like a short circuit, since the initial current is determined only by the voltage drop across the resistor and the value of the resistor.

Example 5.9

Find $i(0^-)$, $v_L(0^+)$, and $v_c(0^+)$ if $i_L(0^-) = 0$, $v_c(0^-) = 0$ and if $v_s(t) = 10$ V for $t > 0$ in Fig. 5.19.

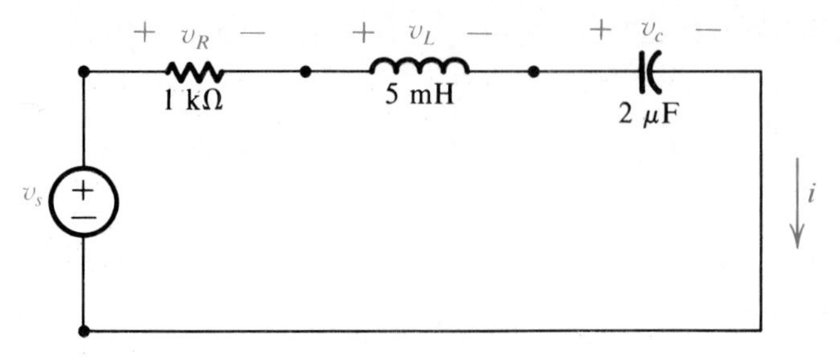

Figure 5.19

Solution: From the continuity condition: $v_c(0^+) = v_c(0^-) = 0$, and the initial inductor current must satisfy: $i_L(0^+) = i_L(0^-) = 0$. Therefore $v_R(0^+) = 0$, and from KVL we conclude that

$$v_R(0^+) + v_L(0^+) + v_c(0^+) = v_s(0^+)$$

and

$$v_L(0^+) = v_s(0^+) = 10 \text{ V}$$

or, the inductor looks like an open circuit because the current through it cannot change instantaneously. The initial current in the circuit is zero, due to the inductor, and the entire voltage of the source initially appears across the inductor.

Energy is required to produce a magnetic field, and a finite source cannot supply the energy instantaneously. The continuity condition for capacitors and inductors is analogous to the limitation that mechanical inertia imposes on the movement of a mass.

Example 5.10

Find $i_R(0^+)$, $i_L(0^+)$, $i_c(0^+)$ and $v_c(0^+)$ if $i_L(0^-) = 0$, $v_c(0^-) = 0$ and if $i_s(t) = 10$ A for $t > 0$. in Fig. 5.20.

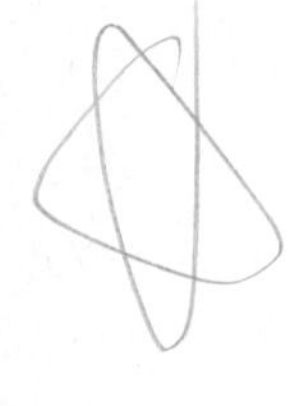

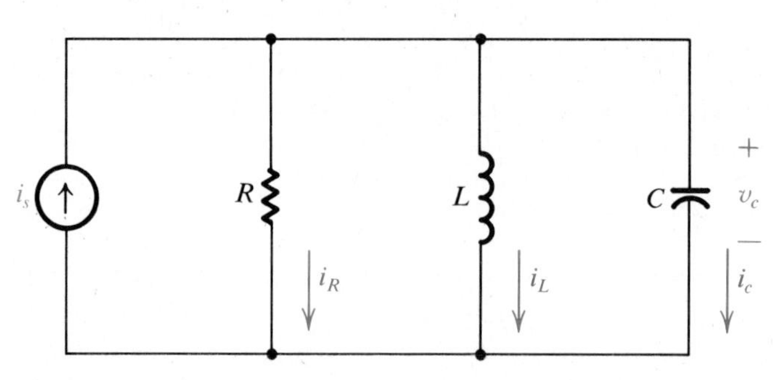

Figure 5.20

Solution: This circuit has R, L, and C connected in parallel with a current source, all having the same voltage, but different currents. Using the continuity property of the capacitor voltage and the inductor current we have $v_c(0^+) = v_c(0^-) = 0$, and $i_L(0^+) = i_L(0^-) = 0$, so we conclude that

$$i_R(0^+) = \frac{v_c(0^+)}{R} = 0$$

Using KCL at $t = 0^+$ gives

$$i_R(0^+) + i_L(0^+) + i_c(0^+) = i_s(0^+)$$

and since $i_R(0^+) = 0$ and $i_L(0^+) = 0$ the initial capacitor current is

$$i_c(0^+) = i_s(0^+) = 10 \text{ A}.$$

The entire source current is initially in the C-path. As charge accumulates on C to cause $v_c > 0$ there will be a current in the R-path, and the accumulation of flux linkages in L will cause i_L. Notice how the continuity condition for the capacitor voltage determines the behavior of the resistor current, and that the capacitor current is not restricted to be continuous, i.e. $i_c(0^-) = 0$ and $i_c(0^+) = 10$ A.

We will study the circuits of Examples 5.9 and 5.10 in greater detail later. For now we want to call attention to the difference between the initial values of the current and voltages and their DC values. In Example 5.9 the DC values of v_L and v_c, denoted by $v_L(\infty)$ and $v_c(\infty)$, are $v_L(\infty) = 0$ and $v_c(\infty) = 10$ V.

The capacitor voltage begins at a value of 0 V and increases, eventually reaching a value of 10 V. The inductor begins with a value of 10 V and decreases, reaching a value of 0 V.

Much of our focus in the following chapters will be on the nature of the transition that circuit variables make from their initial value to their DC or "steady state" value. You might wonder, for example, how long it takes for a circuit to reach its DC value from an initial condition, and how we can determine the actual curves for $v_c(t)$.

SUMMARY

This chapter presented models describing the circuit behavior of capacitors and inductors. If the plates of a capacitor are charged, a voltage drop will exist between them, and the charge can only change if the capacitor conducts a current. Similarly, if the coils of an inductor are surrounded by a magnetic flux, the inductor must be conducting a current, and this current can only change if a voltage is applied across the inductor's terminals. Now, the **integral model** specifies a capacitor's voltage in terms of the integration of the charging current, and specifies an inductor's current in terms of the accumulated magnetic flux, while the **derivative model** specifies the current that must exist if the voltage is changing (for capacitors), or the voltage that must exist if the current is changing (for inductors). We saw that the voltage waveform of capacitors cannot change instantaneously if the charging current is finite, and likewise, that the

current in an inductor cannot change instantaneously if the applied voltage is finite. We then noted that capacitors connected in parallel are equivalent to a capacitor whose value is the sum of their capacitances, while inductors connected in series are equivalent to an inductor whose value is the sum of their inductances. Finally, we presented current and voltage division rules for both inductors and capacitors.

Problems - Chapter 5

5.1 If $v_C(0^-) = 0$, find and draw $v_C(t)$ and $v_S(t)$ for $i_S(t)$ shown below.

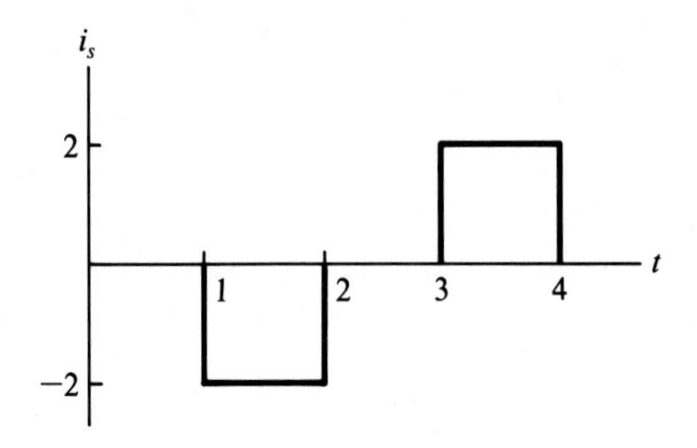

Figure P5.1

5.2 Find the dc values of v_c and v_R.

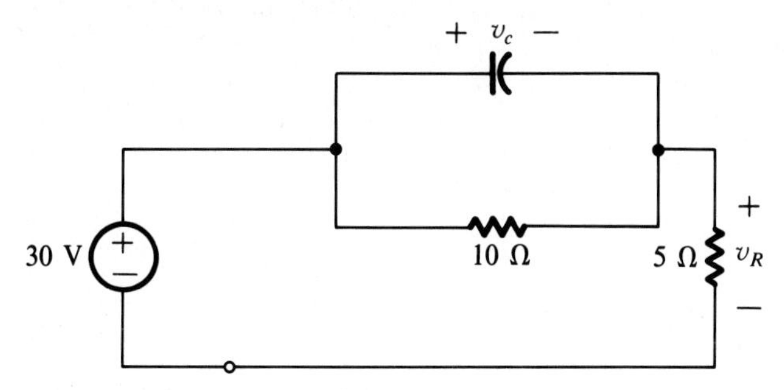

Figure P5.2

5.3 If $i(t)$ is the rectangular current pulse shown below, find $v(t)$, $p(t)$ and $w(t)$ for the capacitor.

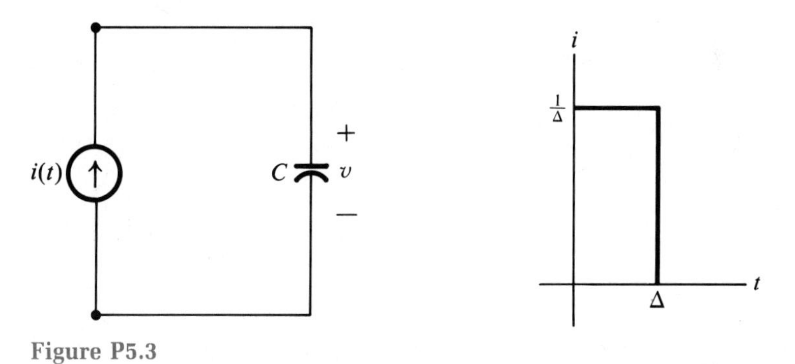

Figure P5.3

5.4 Find $i_R(t)$ and $v_o(t)$ for $v_s(t)$ given below, assuming that the capacitor is initially uncharged.

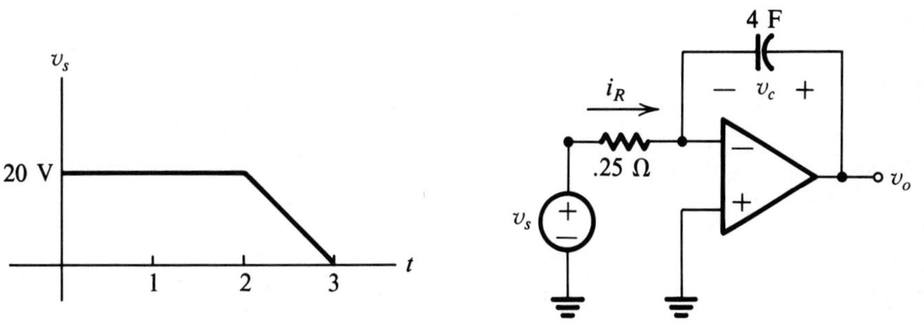

Figure P5.4

5.5 Repeat Problem 5.4 with $v_C(0^-) = 5$ V.

5.6 Find the dc capacitor voltage.

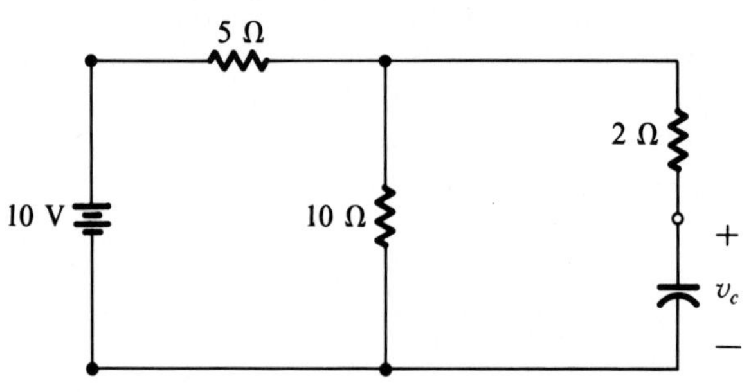

Figure P5.6

5.7 If $v_C(0^-) = 2$ V, find and draw $v_o(t)$ for $t > 0$.

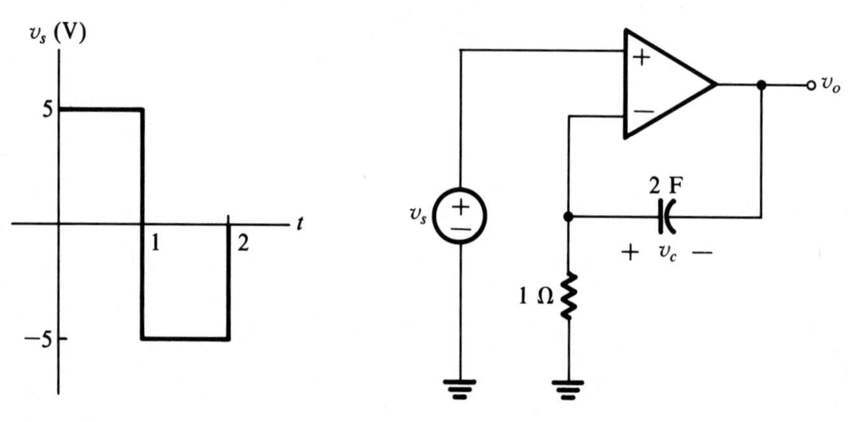

Figure P5.7

5.8 Repeat Problem 5.7 with C = 0.2F.

5.9 Find the dc values of i and v in Figure P5.9.

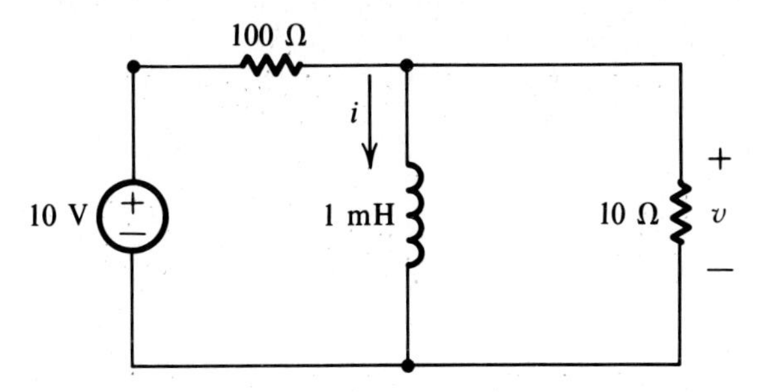

Figure P5.9

5.10 Insert a capacitor between the source and the 100 Ω resistor in Fig. P5.9 and find the dc values of i, v, and v_C.

5.11 Repeat Problem 5.10 if the resistor value is changed to 500 Ω.

5.12 If a 5 μF capacitor has $v(0^-) = 0$ and is charged by $i(t)$ having the shape shown below, find $v_C(t)$ for $t \geq 0$.

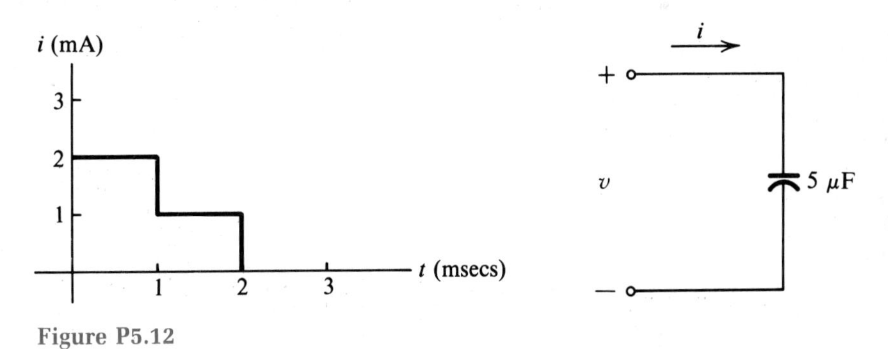

Figure P5.12

5.13 Repeat 5.12 with $v(0^-) = 10$ V.

5.14 If the voltage $v_c(t)$ across a 10 μF capacitor has the waveform shown, draw $i_c(t)$ and draw the graph of the energy stored in the capacitor, $W_c(t)$.

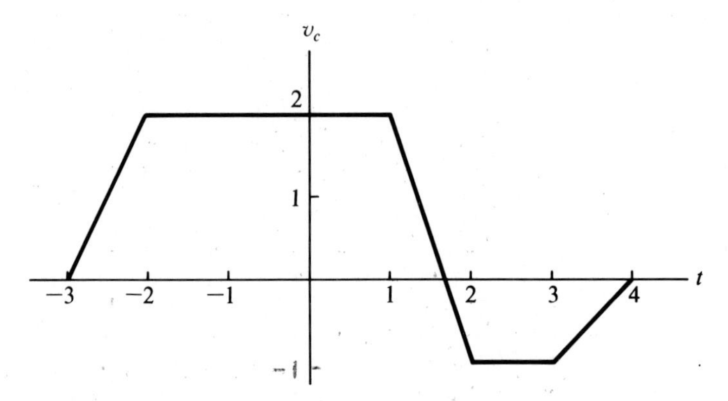

Figure P5.14

5.15 If $v(0^-) = 0$, find $i_1(t)$, $i_2(t)$ and $v(t)$.

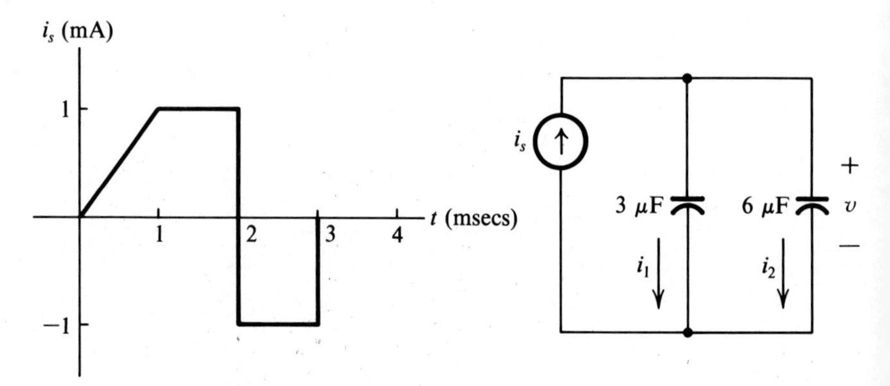

Figure P5.15

5.16 If the capacitors in Fig. P5.16 are initially uncharged and $v(t)$ has the waveform shown, find $v_{c3}(t)$.

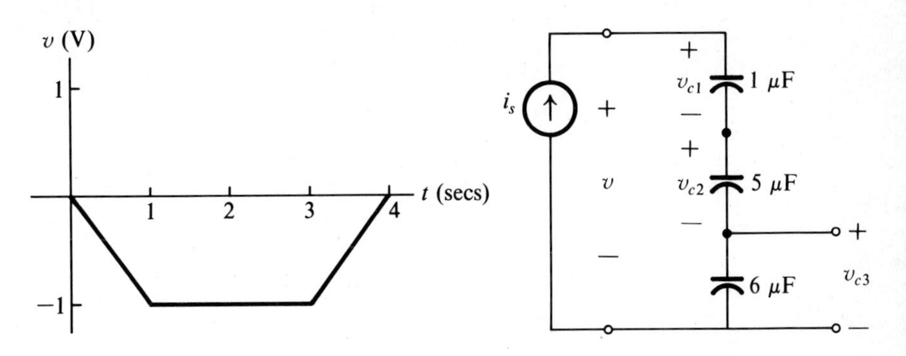

Figure P5.16

5.17 Repeat Problem 5.16 if $v_{c1}(0^-) = 5$ V and $v_{c2}(0^-) = -3$ V.

5.18 If $v_c(0^-) = 5$ V, find $v_c(0^+)$ and $i(0^+)$.

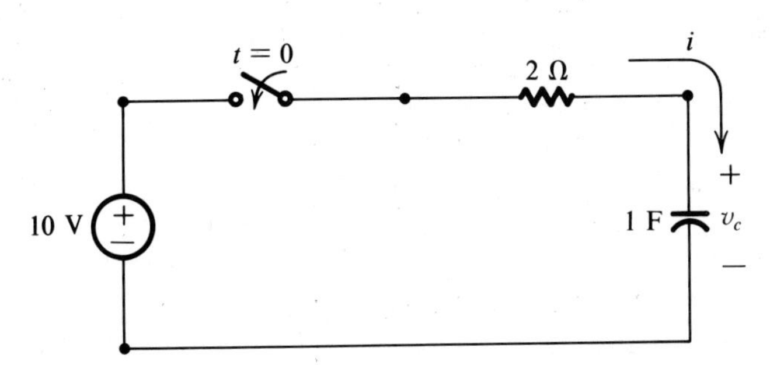

Figure P5.18

5.19 If $v_o(0) = 0$, $i_s = 5$ and the switch is initially closed, find $v_o(2^-)$, $v_o(2^+)$, and $v_o(4^+)$.

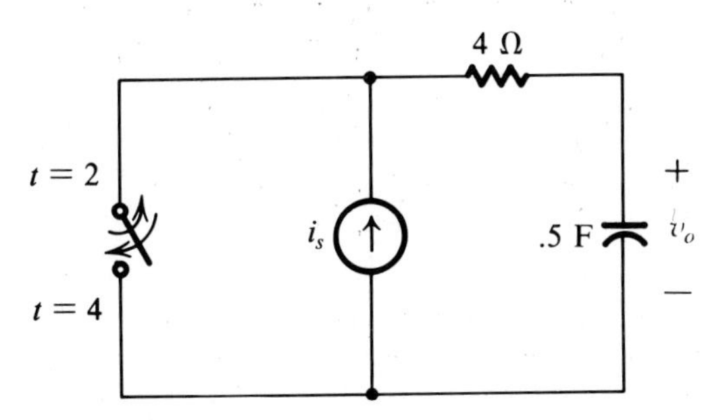

Figure P5.19

5.20 If $v_c(0^-) = 5$ V, find $v_c(0^+)$, $i_c(0^+)$ and $i_c(\infty)$.

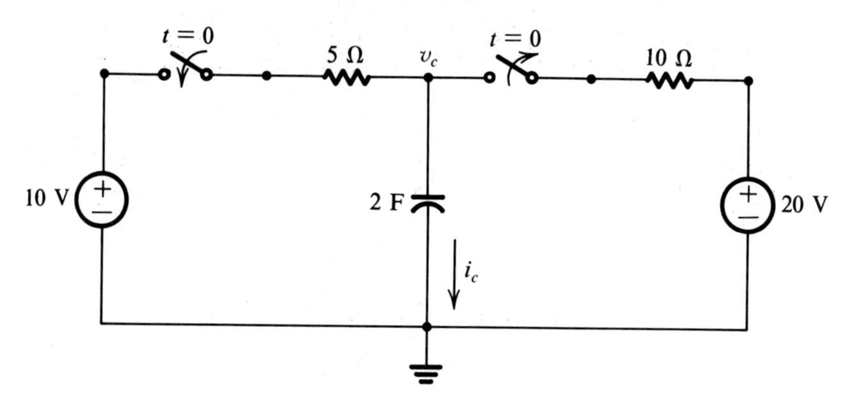

Figure P5.20

5.21 If $v_o(0^-) = 0$ and the switch is closed at $t = 0$, find $i_c(0^-)$, $i_c(0^+)$, $v_c(0^+)$ and $v_o(0^+)$. $V_c(0^-) = 5V$

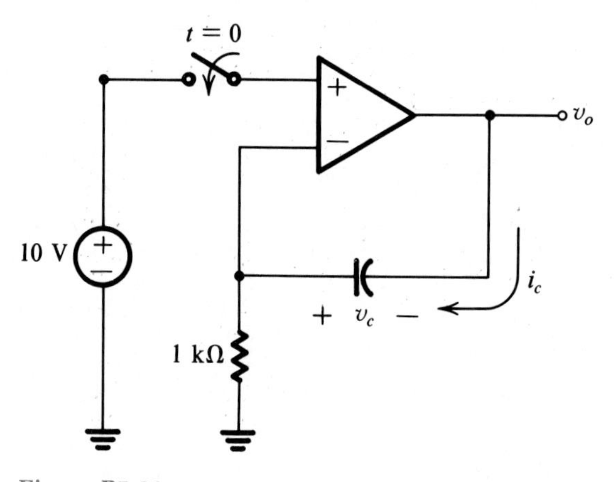

Figure P5.21

SIGNAL MODELS

INTRODUCTION

The voltages and currents in a circuit usually have values that vary with time. Although they are mathematical functions of time, engineers often call them signals rather than functions because of their importance in communications and other applications extending far beyond their mathematical name. We will use these terms interchangeably, calling the voltage or current applied to a circuit an input signal, and the circuit's measured response an output signal.

The widespread use of circuits as signal processors in electronic equipment means that we must understand both signal models and circuit models. By examining families of signals, this chapter will prepare us to understand circuit behavior for a broad, useful, and important set of input signals.

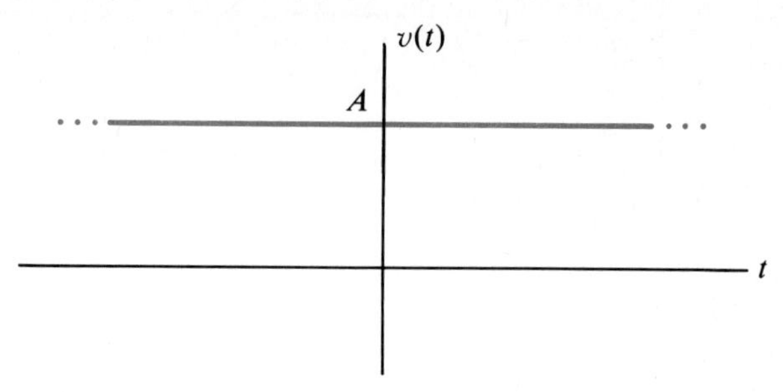

Figure 6.1 A constant signal.

6.1 CONSTANT SIGNALS

The constant signal shown in Fig. 6.1 is described by $v(t) = A$ for $-\infty < t < \infty$. We have already used it with DC circuits.

6.2 UNIT STEP

In the physical world we usually model signals and circuits in which the time reference begins at $t = 0$ rather than at $t = -\infty$. This accounts for situations in which a circuit is "turned on" or "switched on" at $t = 0$. The unit-step function shown in Fig. 6.2 is defined by

$$u(t) = \begin{cases} 0 & t < 0 \\ 1 & t > 0 \end{cases}$$

The value of $u(t)$ at $t = 0$ is left undefined. A related signal, the signum function shown in Fig. 6.3, is formed from the step function by

$$\text{Sgn}(t) = 2u(t) - 1$$

$$\text{Sgn}(t) = \begin{cases} -1 & t < 0 \\ 1 & t > 0 \end{cases}$$

The signal Sgn(t) is $+1$ if its argument is positive, and -1 if its argument is negative.

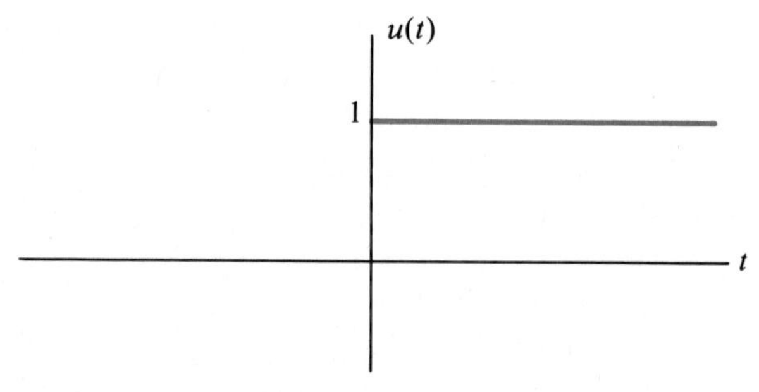

Figure 6.2 The unit-step function.

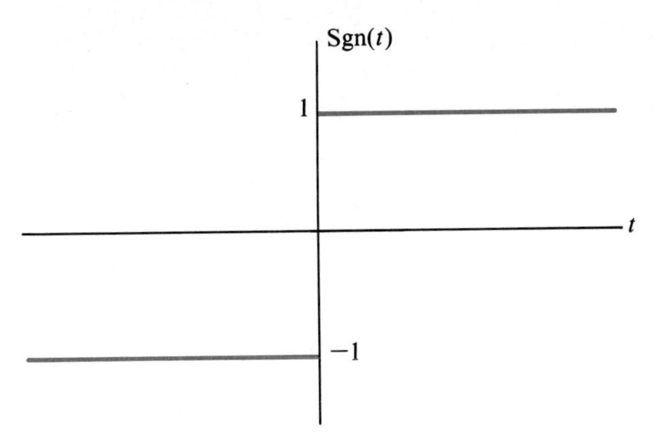

Figure 6.3 The signum function.

6.3 RAMPS AND POLYNOMIALS

Integration of a step function will generate a unit-ramp function r(t) shown in Fig. 6.4. Recall that integration and differentiation are inverse mathematical operations, and note that the slope (derivative) of the ramp function is a constant—the same as the height of the unit-step function whose integral produced the ramp.

We have

$$r(t) = \int_{-\infty}^{t} u(\alpha)d\alpha$$

$$= \int_{0}^{t} 1d\alpha = \begin{cases} 0 & t \leq 0 \\ t & t \geq 0 \end{cases}$$

The next member in the family is the parabola. We obtain it by integrating the ramp signal.

$$p(t) = \int_{-\infty}^{t} r(\alpha)d\alpha = \int_{0}^{t} \alpha d\alpha = \frac{\alpha^2}{2}\bigg|_0^t$$

$$= \begin{cases} 0 & t \leq 0 \\ \dfrac{t^2}{2} & t \geq 0 \end{cases}$$

The general polynomial signal obtained from the nth integration of the step function is given by

$$p_n(t) = \begin{cases} 0 & t \leq 0 \\ \dfrac{t^n}{n!} & t \geq 0 \end{cases} \qquad n = 0, 1, 2, \cdots$$

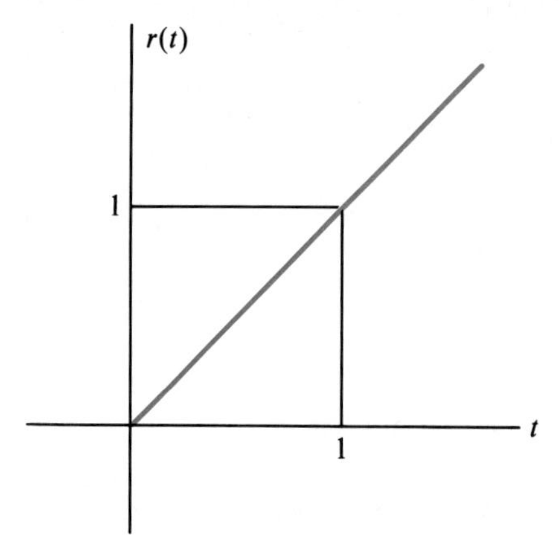

Figure 6.4 The unit-ramp signal.

The ramp, parabola, and other members of the family of functions obtained by integrating a step function are shown in Fig. 6.5.

If polynomial signals are generated by integrating the unit step, we might wonder what happens if we differentiate the unit step. The family of signals generated in this manner does not exist in the physical world because the derivative of a step function is not defined at the origin. We might be tempted to say that

$$\frac{d}{dt}\,u(t) = \begin{cases} 0 & t < 0 \\ \infty & t = 0 \\ 0 & t > 0 \end{cases}$$

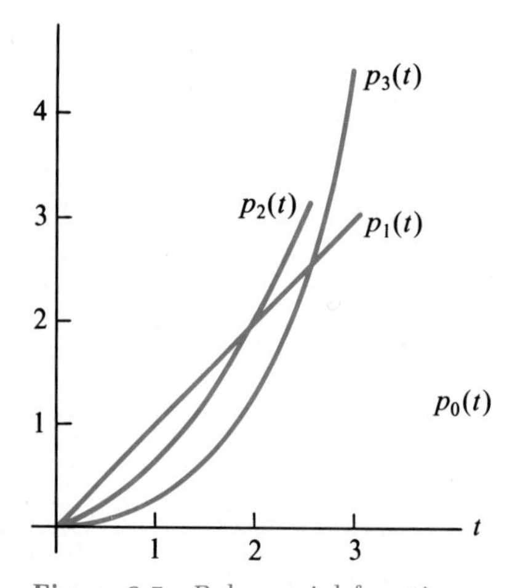

Figure 6.5 Polynomial functions.

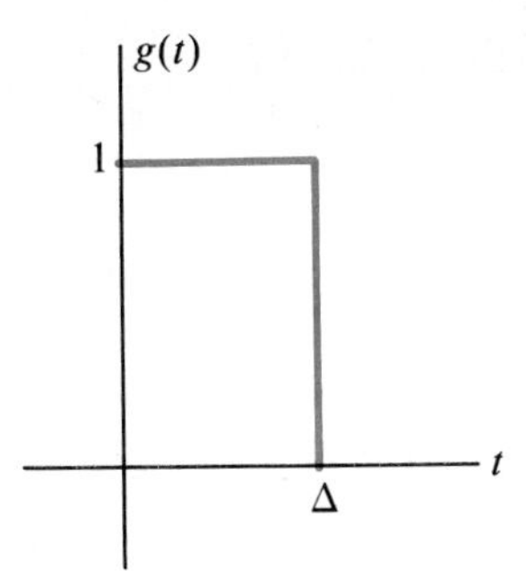

Figure 6.6 A rectangular pulse.

but remember that we did not assign a value to $u(0)$, so we cannot formally evaluate du/dt at $t = 0$! Somewhat confusing, isn't it? Treatment of derivatives of step functions requires that we next consider pulse functions.

6.4 PULSES AND IMPULSES

A unit pulse function, or gate function, is defined as

$$g(t) = \begin{cases} 0 & t < 0 \\ 1 & 0 < t < \Delta \\ 0 & \Delta < t \end{cases}$$

This signal is shown in Fig. 6.6, where Δ is the pulse width, or gate width.

Next, we note that changing the argument of a mathematical function $f(t)$ from t to $t - \Delta$ creates a new function $f(t - \Delta)$. This new function is called a **shifted copy** of $f(t)$ because its graph has the same shape as the graph of $f(t)$, but is shifted on the time axis by Δ units. If $\Delta > 0$ the shift is in the direction of increasing t, and vice versa, as shown in Fig. 6.7. When the shift is in the direction of increasing time, the signal is said to be *delayed*. Conversely, when the shift is in the direction of decreasing time, the function is said to be *advanced*, because its waveform features occur

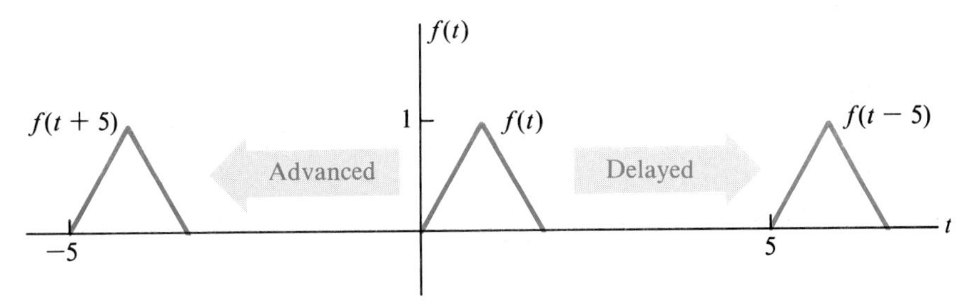

Figure 6.7 Delayed and advanced copies of a waveform. The delayed waveform always appears to the right of the original waveform; the advanced waveform always appears to the left.

earlier in time. The gate function can also be obtained by subtracting a delayed copy of a step function, $u(t-\Delta)$ from a step function

$$g(t) = u(t) - u(t-\Delta).$$

For $t > \Delta$, $u(t) = 1$ and $u(t-\Delta) = 1$, so $g(t) = 0$.

Now let's examine the behavior of a pulse, $g_\Delta(t)$, whose amplitude is scaled by the reciprocal of its pulse width, as shown in Fig. 6.8 for various choices of Δ. Each curve has unit area; pulses with short duration have high amplitude, and vice versa. Next, we form another signal by integrating the scaled pulse

$$u_\Delta(t) = \int_{-\infty}^{t} g_\Delta(\alpha)d\alpha$$

Each $u_\Delta(t)$ reaches and remains at a maximum value of 1 when $t = \Delta$. If $g_\Delta(t)$ is a current into a unit capacitor then $u_\Delta(t)$ is the capacitor voltage. It increases linearly for the duration of the constant current, and then the capacitor remains charged. After $t = \Delta$ the capacitor has 1 coulomb of charge. Note also that

$$g_\Delta(t) = \frac{d}{dt}u_\Delta(t)$$

Consider what happens when the value of Δ shrinks to zero. A shorter pulse width requires a higher amplitude of current to accumulate the same level of charge on the capacitor. The waveform of $u_\Delta(t)$ becomes the waveform of a unit step function, as shown in Fig. 6.9. This corresponds to a discontinuity, or step change, in the capacitor voltage. At first glance it might seem reasonable to have such a rapid change in capacitor voltage,

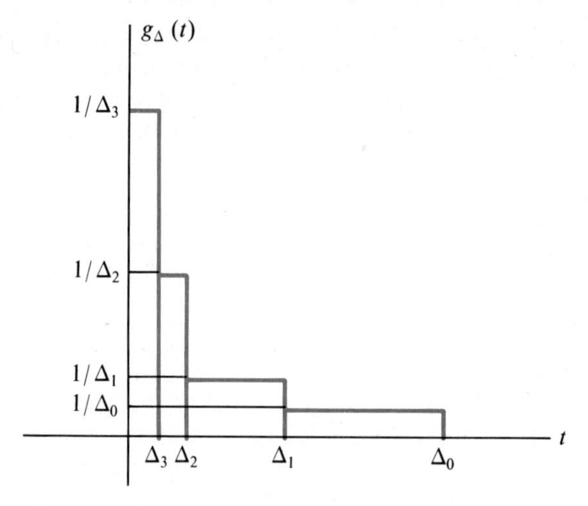

Figure 6.8 A family of rectangular pulses, each of which has unit area under its curve.

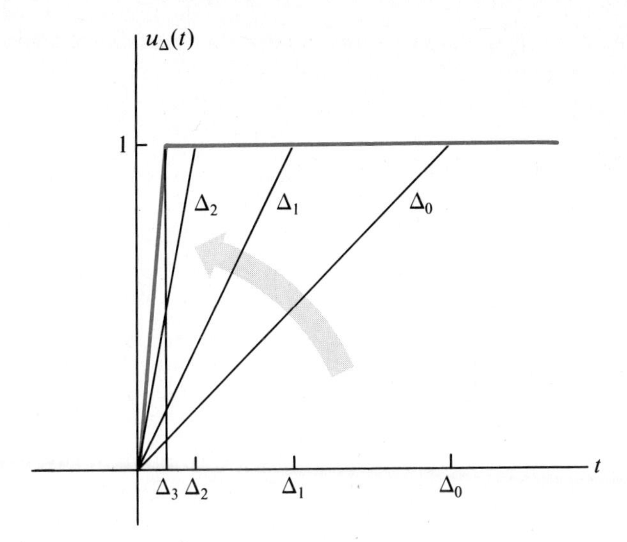

Figure 6.9 Curves for u_Δ as Δ shrinks to O. Notice how the leading edge of the waveform becomes steeper.

but remember that charge accumulation must accompany capacitor voltage, and that current must precede charge accumulation. As the pulse width shrinks, the amplitude of the pulse grows arbitrarily large. At the same time, each integral of the pulse still has a finite area under its curve! We have a situation in which $g_\Delta(t)$ does not approach a well-defined function, because its amplitude cannot be defined at $t = 0$, but its integral $u_\Delta(t)$ becomes the step function.

Without becoming ensnarled by the mathematical complexity of this example we simply state that we are ultimately interested in the *effect* caused by an input signal. In this case, the output signal converges to a step function as the input-pulse width shrinks, while the input signals themselves do not converge to an ordinary function. Nonetheless, with that motivation we define a **unit-impulse function** to be

$$\delta(t) = \lim_{\Delta \to 0} g_\Delta(t)$$

and

$$\delta(t) = \begin{cases} 0 & t < 0 \\ \infty & t = 0 \\ 0 & t > 0 \end{cases}$$

and we write

$$\delta(t) = \frac{d}{dt} u(t)$$

Note that $\delta(0) = \infty$ is a symbolic statement that $\delta(0)$ is *not defined*, because

$\delta(t)$ is formed from a sequence of functions that becomes arbitrarily large at $t = 0$ as Δ becomes small.

The unit-impulse function is also defined to have unit area

$$\int_{-\epsilon}^{\epsilon} \delta(t)\, dt = 1$$

Since $\delta(t)$ does not make sense as a mathematical function, we call it a **generalized function.** We use it in circuit analysis to summarize and symbolize the actual underlying problem of solving a sequence of circuit problems for physically realizable pulse functions. If $\delta(t)$ is the input to a circuit and $v(t)$ is the output, we say that $v(t)$ is the limit signal obtained by finding the response $v_\Delta(t)$ to the input $g_\Delta(t)$ and letting Δ shrink to 0. In practice, we don't bother to do this because finding the response to $\delta(t)$ directly leads to the same results. (By convention the height of the arrow used to symbolize $\delta(t)$ in Fig. 6.10 indicates the *area* of the pulse function g_Δ that is used to create the impulse, *not* the unbounded height of the underlying sequence of pulses.)

The impulse function has a very useful sampling property that describes how it affects other signals. If a function is integrated with an impulse, the result is the value of the function at the point where the impulse occurs

$$\int_{-\epsilon}^{\epsilon} f(t)\delta(t)\, dt = f(0) \tag{6.1}$$

and

$$\int_{\tau-\epsilon}^{\tau+\epsilon} f(t)\delta(t-\tau)\, dt = f(\tau) \tag{6.2}$$

To understand this property note the following Taylor and McLaurin series:

$$f(t) = f(0) + f'(0)t + \frac{1}{2}f''(0)t^2 + \cdots \tag{6.3}$$

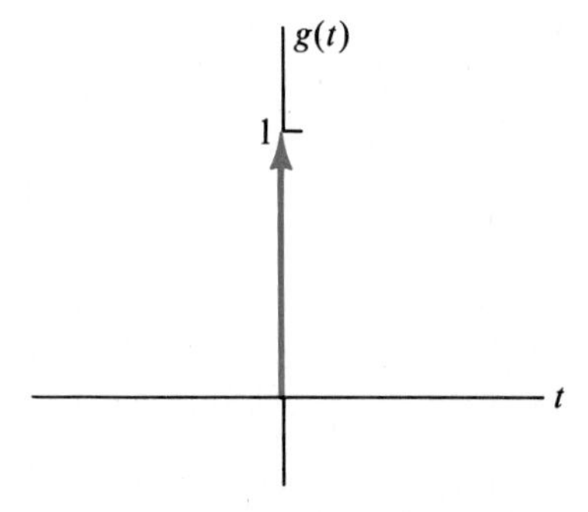

Figure 6.10 the "graph" of the unit impulse.

and

$$f(t) = f(\tau) + f'(\tau)(t - \tau) + \frac{1}{2}f''(\tau)(t - \tau)^2 + \cdots \quad (6.4)$$

Substituting (6.3) into (6.1) gives

$$\int_{-\epsilon}^{\epsilon} f(t)\delta(t)\,dt = \int_{-\epsilon}^{\epsilon} f(0)\delta(t)\,dt + \int_{-\epsilon}^{\epsilon} [f'(0)t + \frac{1}{2}f''(0)t^2 + \cdots +]\delta(t)dt$$
$$= f(0)$$

Noting that $f'(0)t + 1/2\, f''(0)t^2 + \cdots = 0$ for $t = 0$, and $\delta(t) = 0$ for $t \neq 0$ we have

$$\int_{-\epsilon}^{\epsilon} f(t)\delta(t) = f(0)\int_{-\epsilon}^{\epsilon} \delta(t)dt = f(0)$$

Equation (6.2) can be proved in similar manner [use (6.4)].

The impulse function is more than a mathematical anomaly. We use it to describe the physically impossible problem of charging a capacitor instantaneously, and in later chapters we'll see that it provides an important time-domain characterization of a circuit. Narrow pulses can be used to approximate impulses.

Example 6.1

Suppose a 2 F capacitor is charged by $i_{in}(t) = 10\,\delta(t - 1)$. Find and graph the capacitor voltage, $v_c(t)$.

Solution: $v_c(t) = 1/2 \int_0^t 10\delta(\alpha - 1)d\alpha = 5\,u(t - 1)$

The current source transfers all of the charge to the capacitor at $t = 1$ sec—instantaneously! Physically, this can't happen because it would take an infinite amount of power, but we'll take such liberties with the mathematical model. Note that the same result could be obtained by charging the capacitor with a pulse of current that turns on at $t = 1$, turns off at $t = 1 + \Delta$ and has a height of $1/\Delta$ amps, and then letting $\Delta \rightarrow 0$.

Example 6.2

Sketch $i(t)$ and $v_o(t)$ for the circuit in Fig. 6.11(a) and $v_{in}(t)$ as shown in Fig. 6.11(b).

Solution: The capacitor current shown in Fig. 6.11(c) is obtained by differentiating $v_{in}(t)$:

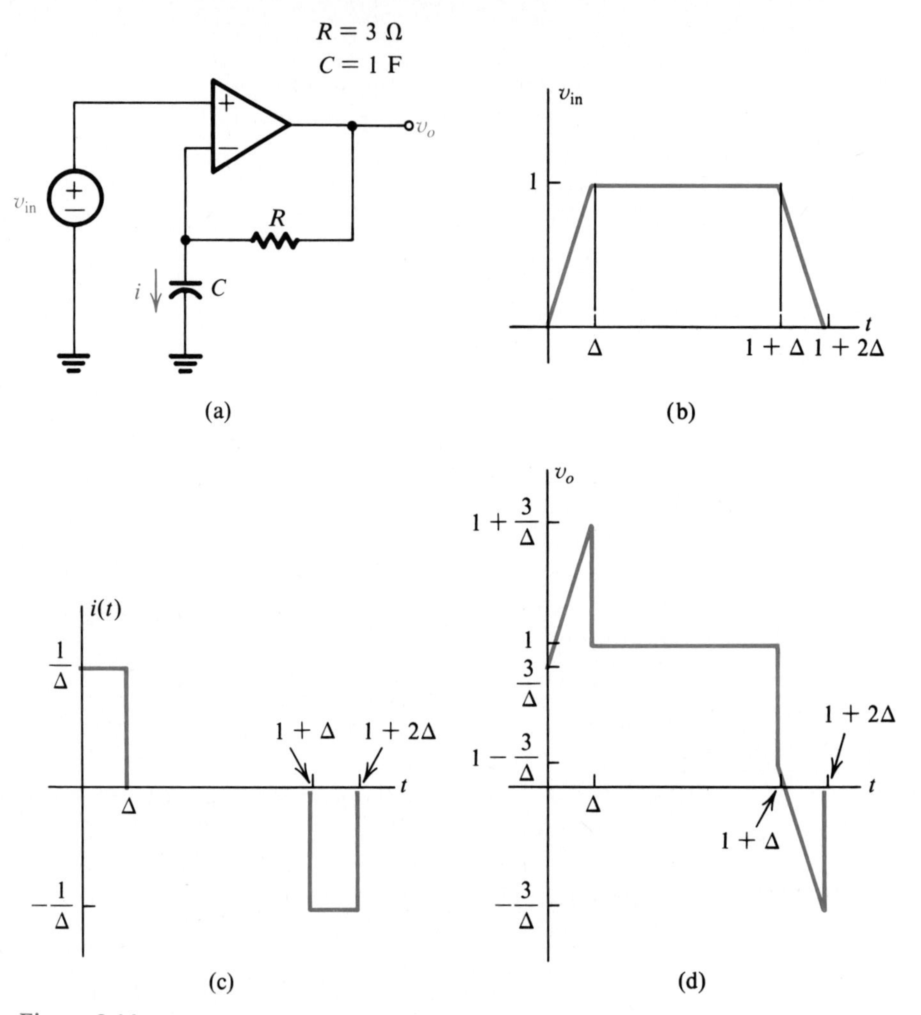

Figure 6.11

$$i(t) = C\frac{dv_{in}}{dt}$$

$$i(t) = \begin{cases} 0 & -\infty < t < 0 \\ 1/\Delta & 0 < t < \Delta \\ 0 & \Delta < t < 1 + \Delta \\ -1/\Delta & 1 + \Delta < t < 2\Delta + 1 \\ 0 & t > 2\Delta + 1 \end{cases}$$

The output voltage in Fig. 6.11(d) is formed by $v_o(t) = v_{in}(t) + 3\,i(t)$. For $\Delta > 0$ the pulse heights in $i(t)$ are bounded. As we try to make $v_{in}(t)$ a rectangular pulse [Fig. 6.12(a)] by letting $\Delta \rightarrow 0$ we see that

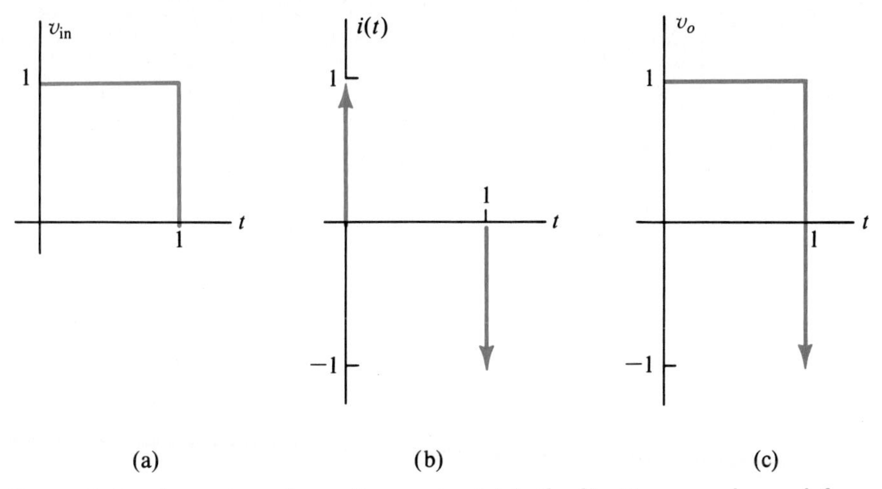

Figure 6.12 the rectangular voltage pulse (a) is the limiting waveform of the trapezoidal voltage pulse in Fig. 6.11(b), and the pair of impulses (b) represent the limiting behavior of the capacitor current in Fig.6.11(c). They result in the op amp output voltage (c).

$i(t)$ has unbounded pulse amplitude. This behavior is represented by the impulses shown in Fig. 6.12(b).

The output voltage shown in Fig. 6.12(c) consists of the rectangular pulse due to $v_{in}(t)$, added to the impulses due to the current "spikes" through the resistor. At $t = 0$ and $t = 1$ the source must deliver "infinite" power to instantaneously charge the capacitor. To see this, note that the power dissipated in the resistor will grow arbitrarily large as the voltage pulse of the source becomes rectangular. Physically, applying such a pulse could have the effect of damaging the op amp by exceeding the levels of current that it can conduct safely.

6.5 EXPONENTIAL SIGNALS

Exponential functions (signals) have a distinctive mathematical property—differentiating or integrating them produces another exponential. These signals play a major role in describing the behavior of electrical circuits. In fact, we'll see in later chapters that **every signal that has a physical significance can be described in terms of exponential signals.** A real exponential signal has the form

$$f(t) = Ke^{\sigma t} \tag{6.5}$$

where K and σ are fixed real numbers. Time-domain graphs of $f(t)$ are shown in Fig. 6.13 for $\sigma > 0$ and $\sigma < 0$. When $\sigma > 0$, $f(t)$ is said to be an

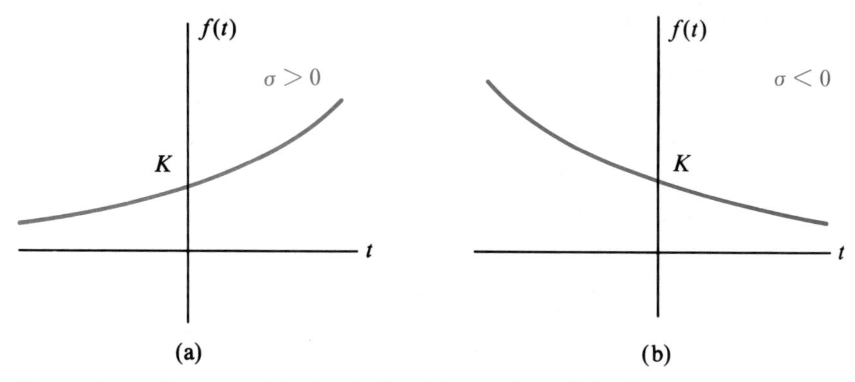

Figure 6.13 Exponential signals for (a) growth and (b) decay.

increasing, or growing, exponential, and when $\sigma < 0$, $f(t)$ is a decreasing, or decaying exponential. The time constant τ of an exponential signal is given by $\tau = 1/|\sigma|$ and when $\sigma < 0$ it can be used to rewrite (6.5) as $f(t) = Ke^{-t/\tau}$

The significance of the time constant can be understood from the graph of $e^{-t/\tau}u(t)$ shown in Fig. 6.14. An increment in time of one time constant causes the curve for $e^{-t/\tau}$ to drop to 36.8% of its initial value, and after 4 τ it has decreased by 98%. Thus τ provides a convenient measure of how quickly the curve of $f(t)$ changes.

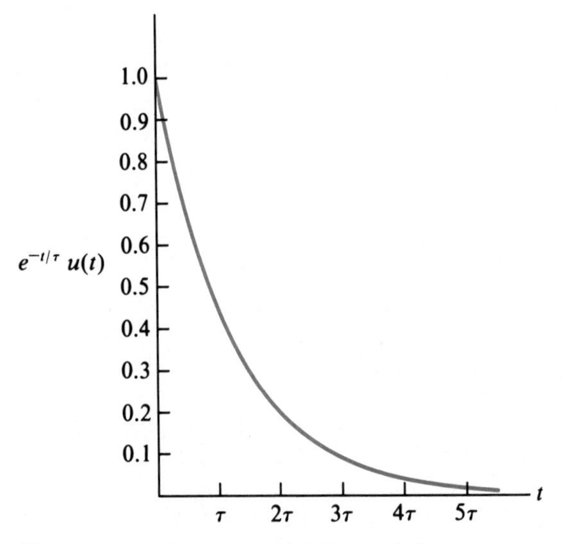

Figure 6.14 An exponential signal decays to approximately zero in 4 time constants.

6.6 SINUSOIDAL SIGNALS

A sinusoidal signal is described by

$$f(t) = A \cos(\omega_o t + \theta) \tag{6.6}$$

or by

$$f(t) = A \sin(\omega_o t + \theta) \tag{6.7}$$

where A is the signal amplitude, ω_o is its radian frequency, (rads/sec) and θ is the signal's phase angle. In general, the amplitude A can be positive or negative, but in either case

$$|A \cos(\omega_o t + \theta)| \leq |A|$$

and likewise

$$|A \sin(\omega_o t + \theta)| \leq |A|$$

Sinusoidal signals exhibit the property that their waveform, or graph, is periodic; that is, for some T^*

$$f(t + T^*) = f(t)$$

The smallest such value of T^* is called the **period** of the signal, and is denoted by T. The period of a sinusoidal signal is related to its radian frequency by

$$\boxed{T = \frac{2\pi}{\omega_o}}$$

and to its cyclic frequency f_o by

$$\boxed{T = \frac{1}{f_o}}$$

so that

$$\boxed{\omega_o = 2\pi f_o.}$$

The radian frequency ω_o has units of radians/sec (rads/sec), and the cyclic frequency f_o has units of Hertz (H_z). A signal having a high frequency has a short period, and conversely, a low-frequency signal has a long period. The periodicity of the sinusoidal signal is evident in the graph of the cosine function in Fig. 6.15. The period of the signal is the time measured between successive peaks of its waveform.

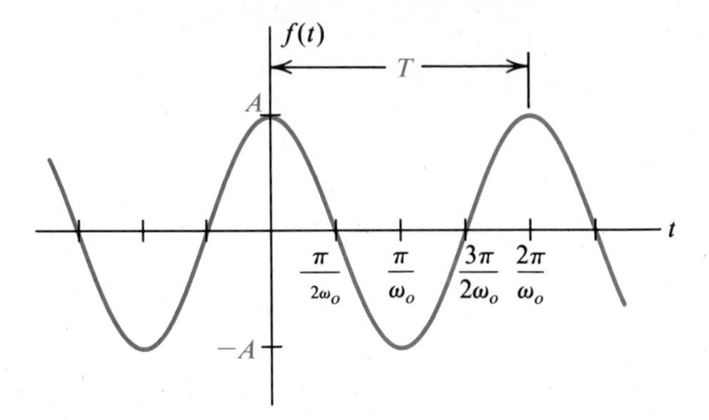

Figure 6.15 The parameters of a sinusoidal signal are its amplitude A, its radian frequency ω_o and its period T.

6.6.1 Phase Angle and Signal Delay

The phase angle θ of a sinusoid determines whether a signal is delayed or advanced along the time axis relative to its waveform when $\theta = 0$ (see Fig. 6.7). This effect can be seen by rewriting (6.6) as

$$f(t) = A \cos \left[\omega_o\left(t + \frac{\theta}{\omega_o}\right)\right]$$

and

$$f(t) = A \cos [\omega_o(t - \tau)]$$

with

$$\tau = \frac{-\theta}{\omega_o} \tag{6.8}$$

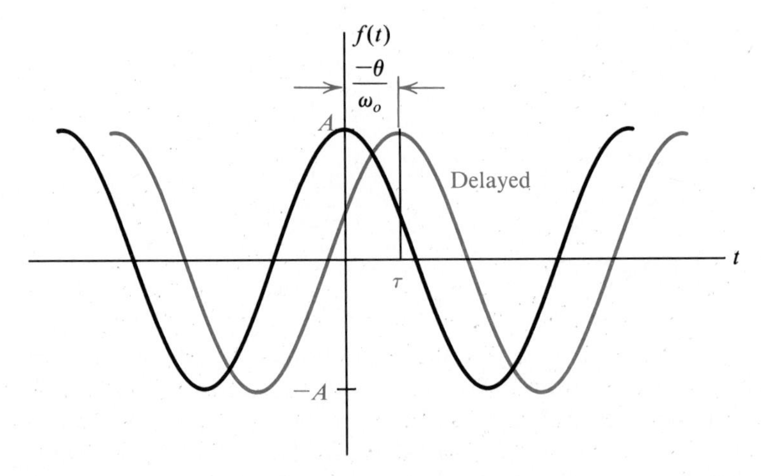

Figure 6.16 A sinusoidal signal's phase angle determines its translation on the time axis with $\tau = -\theta/\omega$.

Here we use τ to denote a time shift of $f(t)$ along the time axis. Note that τ is positive if and only if θ is negative, and vice versa.

A negative angle corresponds to signal delay, and a positive phase angle corresponds to signal advance. The amount of delay or advance is determined by (6.8). *A word of caution*—ω_o is customarily given in units of rads/sec, and θ is given in units of degrees, making it necessary to convert units when using (6.6), (6.7) and (6.8).

The time shift of a sinusoidal signal with a phase angle is also referred to as a phase shift. An example of a delayed sinusoidal signal is shown in Fig. 6.16.

Skill Exercise 6.1

If $f(t) = 10 \sin(2\pi t + \theta)$ determine the time shift of the signal with $\theta = -30°$ relative to the signal with $\theta = 0°$.

Answer: $\tau = .0833$ sec (delay).

6.6.2 Damped and Switched Sinusoids

The amplitude of a sinusoidal signal can be modified by including an exponential factor:

$$f(t) = Ae^{\sigma t} \sin(\omega_o t + \theta) \tag{6.9}$$

This results in a growing or decaying oscillatory signal, as shown in Fig. 6.17, depending on whether (a) $\sigma > 0$ or (b) $\sigma < 0$. The signal given in (6.7)

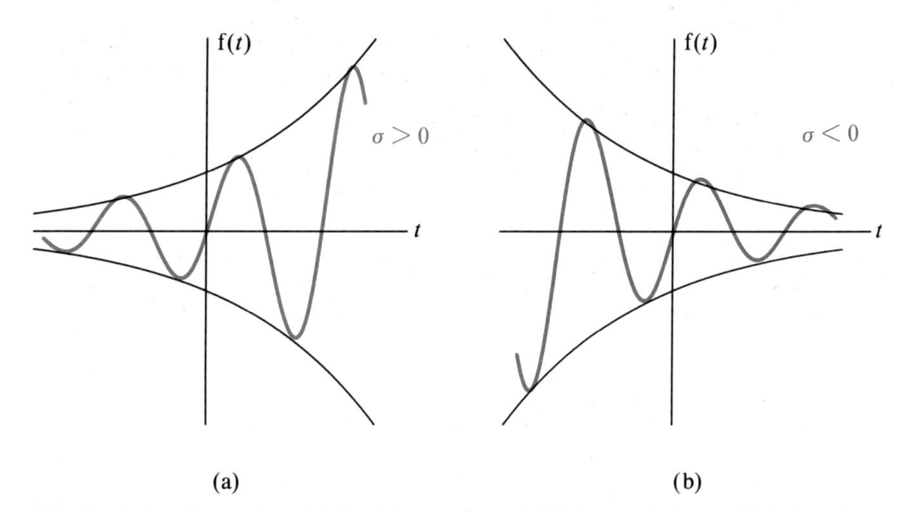

Figure 6.17 Examples of (a) undamped, and (b) damped sinusoidal signals.

with constant amplitude is the special case of (6.9) in which $\sigma = 0$. The exponential curves in Fig. 6.17 are referred to as the envelope of the signal. They correspond to the graph of the exponential factor, $Ae^{\sigma t}$, since

$$|f(t)| = |Ae^{\sigma t} \sin(\omega_o t + \theta) \leq |A|e^{\sigma t}| \sin(\omega_o t + \theta)| \leq |A|e^{\sigma t}$$

so that

$$-|A|e^{\sigma t} \leq f(t) \leq |A|e^{\sigma t}$$

switched, damped sinusoid is defined by

$$f(t) = Ae^{\sigma t} \sin(\omega_o t + \theta)u(t)$$

Note that the envelope of the signal decays to zero in approximately $4/\sigma$ sec and the zero crossing of the signal are spaced at an interval of $2\,\pi/\omega_o$ sec.

SUMMARY

Signals are mathematical expressions representing the voltages and currents in a circuit. This chapter defined the following important signals: the constant, unit-step, ramps, and polynomial signals; and pulses, impulses, exponentials, and damped and undamped sinusoids. Impulses were shown to be generalized signals obtained as the limit of suitably defined pulse signals. Sinusoidal signals were defined by their amplitude, radian frequency, and phase angle. Finally, it was shown that the phase angle of a sinusoid corresponds to either advancing or delaying the signal along the time axis.

Problems - Chapter 6

6.1 Using step functions, develop a model for $v_{in}(t)$.

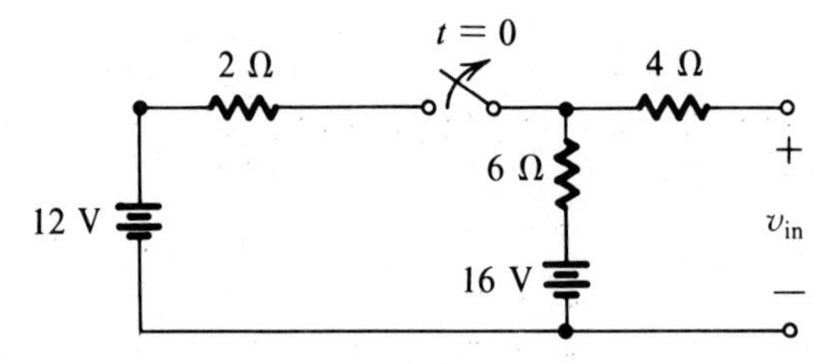

Figure P6.1

6.2 Estimate the time required for the signal $v_{in}(t) = 1000e^{-0.001t}u(t)$ to reach zero.

6.3 If $f(t) = 100 \sin 2\pi t$, and if $g(t) = f(t - 0.2)$,

a. Draw $f(t)$ and $g(t)$.

b. Find the phase angle of $g(t)$ relative to $f(t)$.

6.4 Draw the graph of $f(t) = e^{(-t/5)} \cos 2\pi t u(t)$.

6.5 If a sinusoidal signal is described by $g(t) = 100 \cos (400\pi t - \pi/6)$, find the signal's amplitude, phase angle, radian frequency, and cyclic frequency.

6.6 If $v_{in}(t) = 2\ u(t)$, find v_c and v_o, given that $v_c(0^-) = 0$.

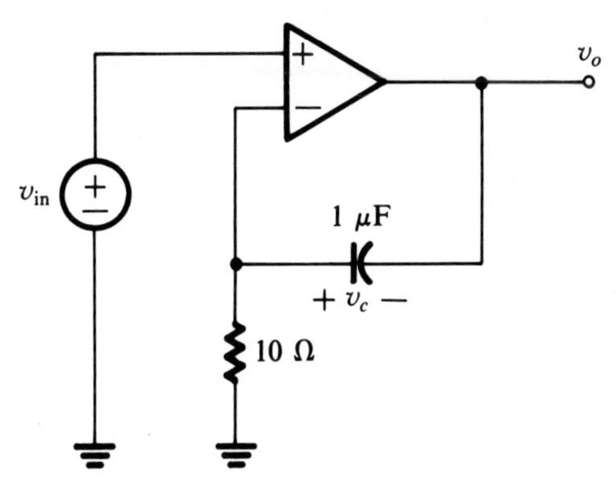

Figure P6.6

6.7 If $v(0) = 0$, find and draw $v_o(t)$.

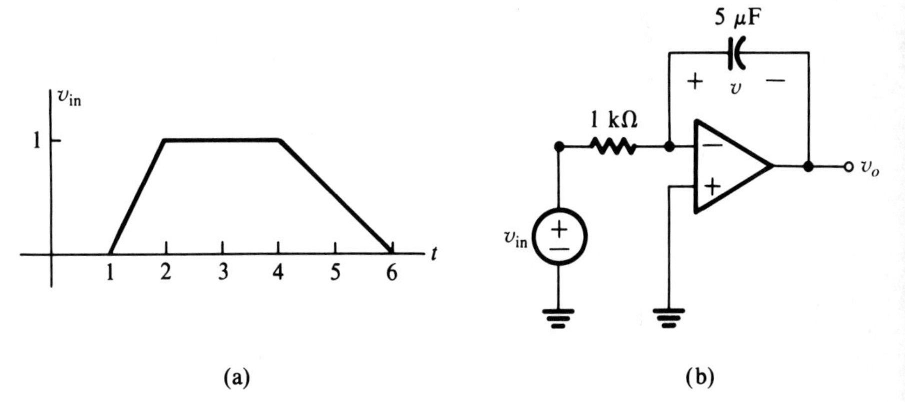

Figure P6.7

6.8 If the sawtooth-shaped voltage signal is observed across the capacitor, find and draw v_o, i and v_{in}, assuming that the capacitor is initially uncharged.

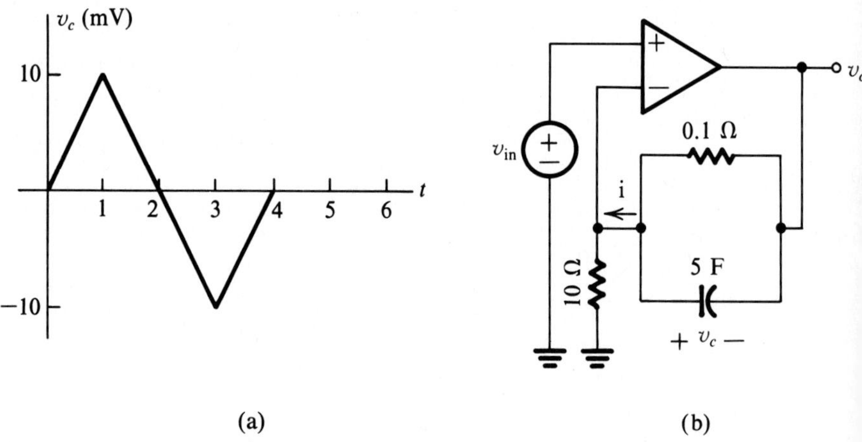

Figure P6.8

6.9 Describe $v_{in}(t)$ in terms of step functions and ramp functions.

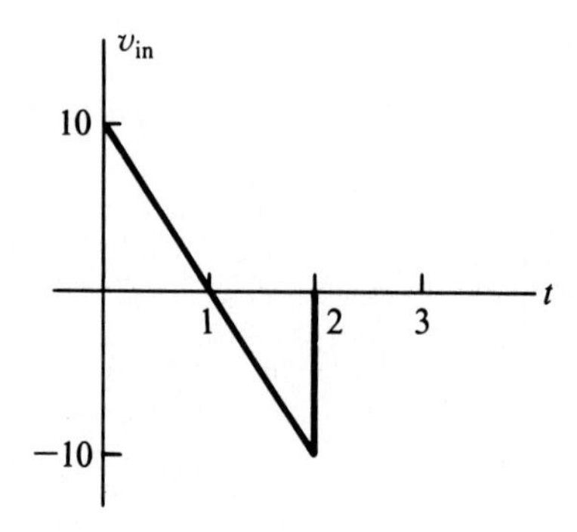

Figure P6.9

6.10 Draw $i(t)$ and $v_o(t)$.

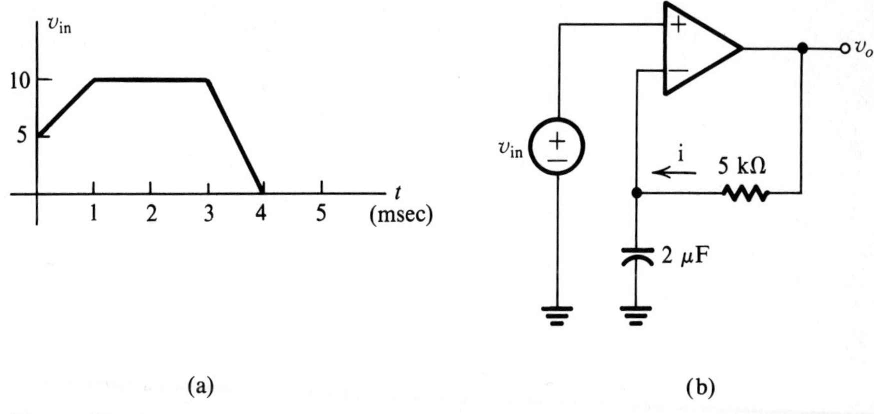

Figure P6.10

6.11 Find v_o in Fig. P6.6 if $v_{in}(t) = e^{-2000t}[u(t) - u(t - 0.005)]$.

6.12 If a sinusoidal signal is described by $v(t) = 15 \cos(400\pi t + 60°)$, find the generic parameters A, T, and τ.

6.13 Find an expression for $v(t)$ in Fig. P6.13.

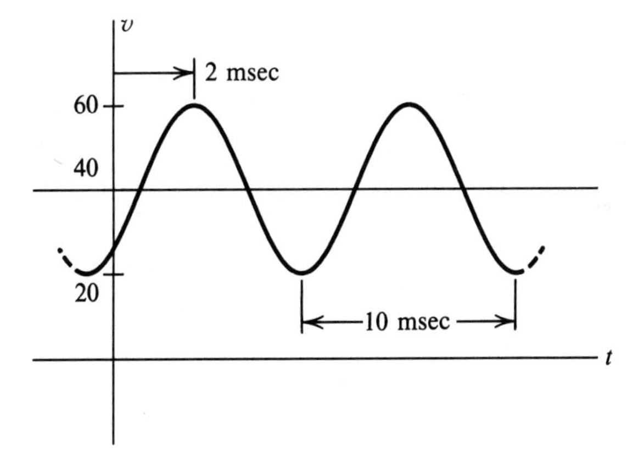

Figure P6.13

6.14 Estimate the time required for $v(t)$ to decay to zero if $v(t) = 100e^{-4000t} \cos(200\pi t + 45°)u(t)$.

6.15 Find an expression for $v(t)$ shown in Fig. P6.15.

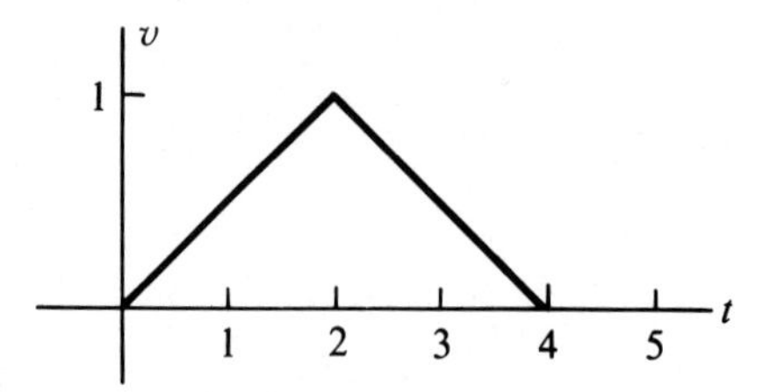

Figure P6.15

6.16 If v_{in} has the waveform shown, find and draw $i(t)$.

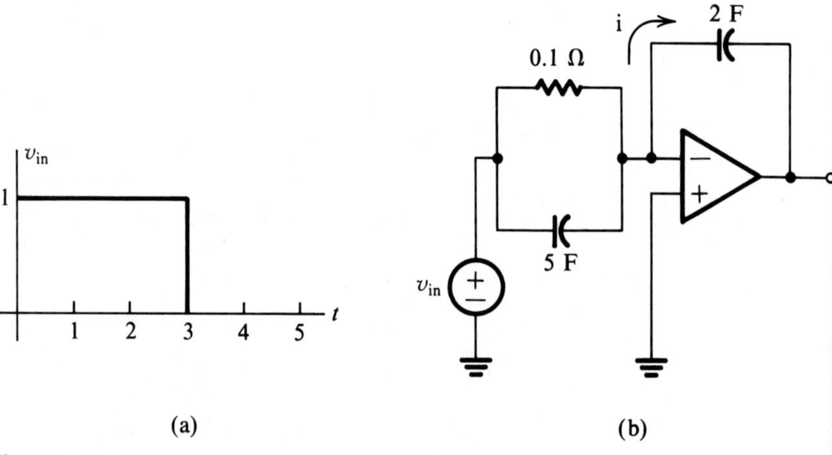

Figure P6.16

6.17 Find and draw v_o.

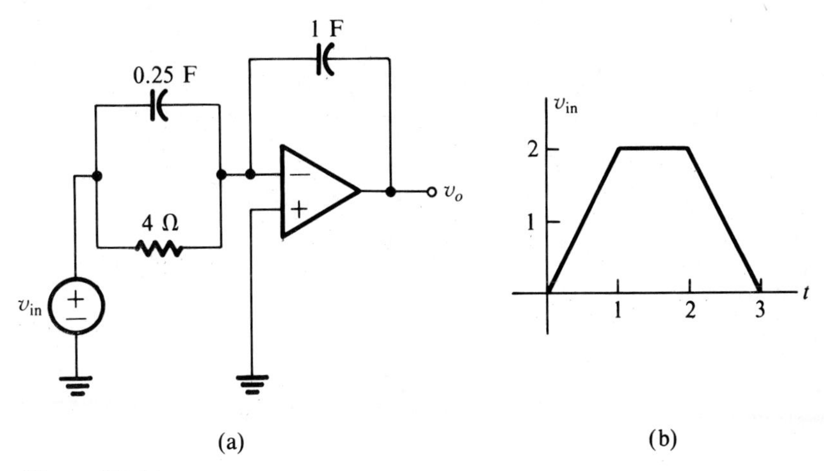

Figure P6.17

6.18 Repeat Problem 6.17 if $v_{in}(t)$ is a rectangular pulse of height 2 and width 3.

6.19 Find and draw $v_o(t)$.

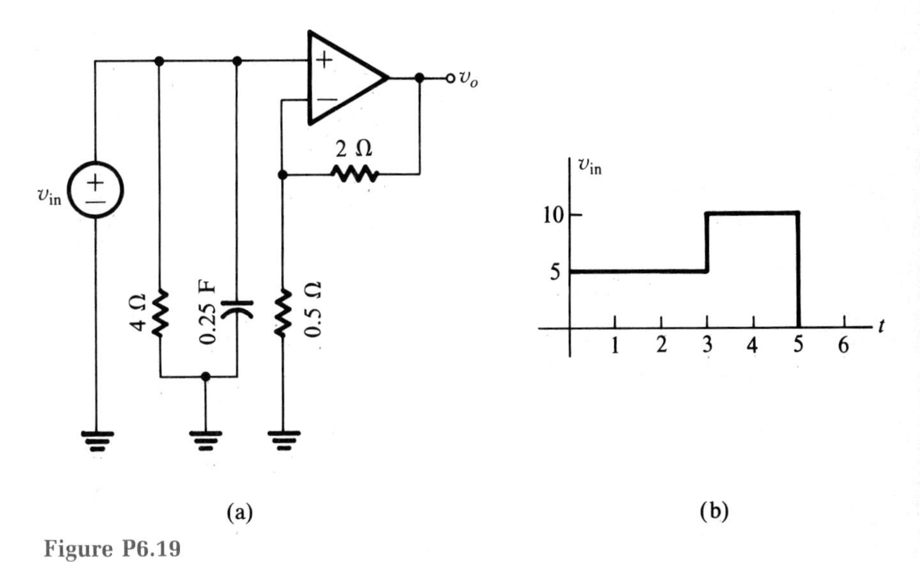

Figure P6.19

6.20 Find and draw i_c and v_o, assuming $v_c(0^-) = 1$ V.

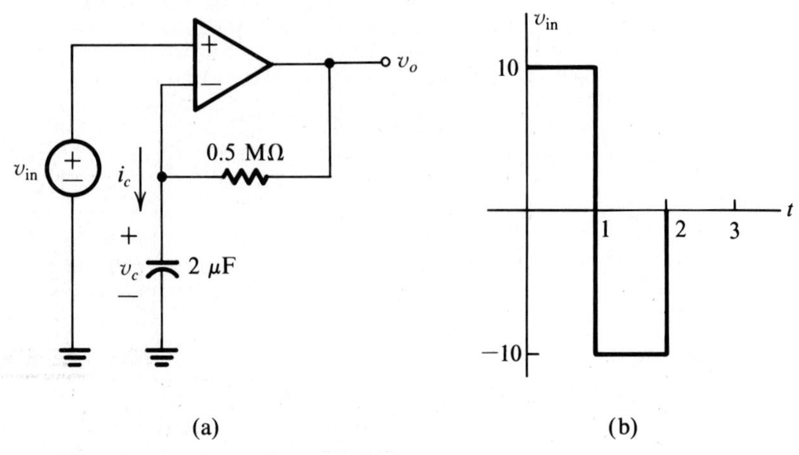

Figure P6.20

6.21 Find $v(t)$ if $v(0^-) = 0$.

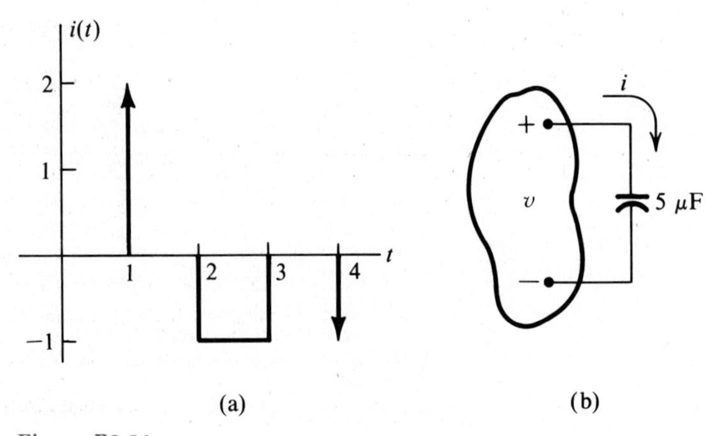

Figure P6.21

6.22 Write $v(t)$ in terms of the step and ramp functions.

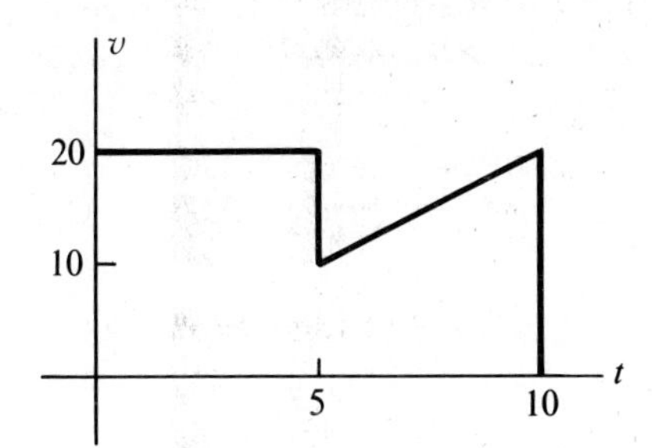

Figure P6.22

6.23 Find an expression for $i(t)$ in terms of step functions.

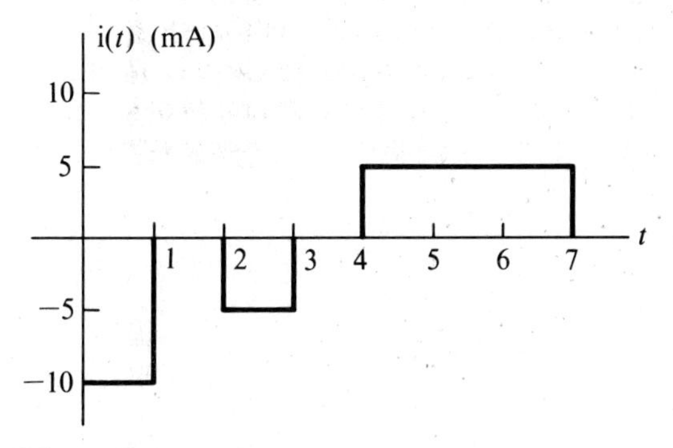

Figure P6.23

6.24 If $i(0^-) = 0$, find the current $i(t)$ through a 4 mH inductor for $t \geq 0$ when the voltage across it is $v_{in}(t) = u(t)$.

6.25 Repeat Problem 6.24 with $v_{in}(t) = 0.5t\ u(t)$.

6.26 If $v_o(0^-) = 0$ and $i(0^-) = 0$, find and draw $v_o(t)$ if $v_{in}(t) = r(t) - 2r(t - 2) + r(t - 3)$.

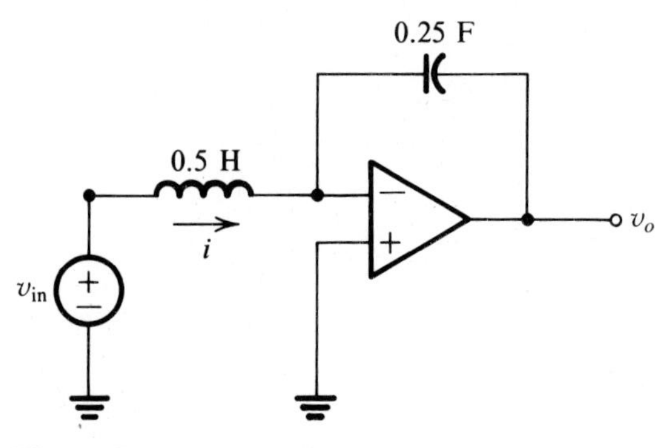

Figure P6.26

6.27 Given $V_{cc} = 15$ V, find the range of values for the signal amplitude X for which the op amp remains within its linear range of operation.

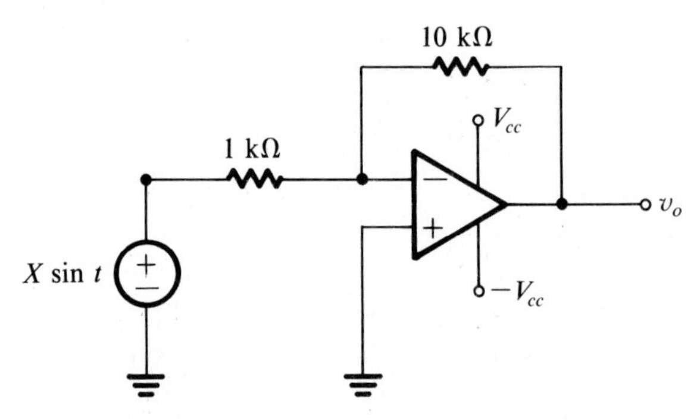

Figure P6.27

6.28 Given $V_{cc} = 15$ V, if $v_{in}(t) = 10 \sin 1000\pi t$, draw $v_o(t)$.

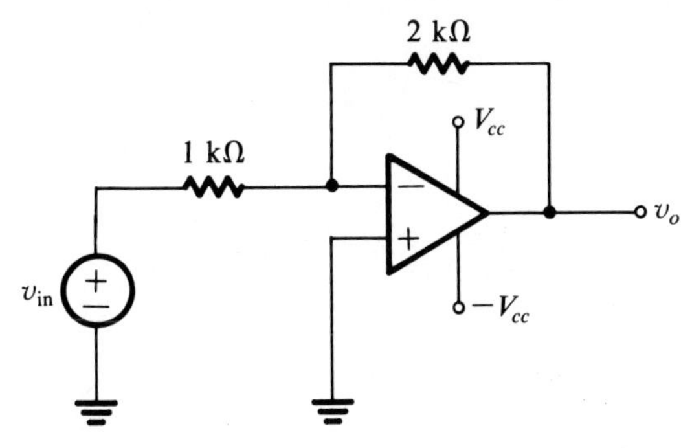

Figure P6.28

6.29 If $v_{in1}(t) = 5 \sin 1000\pi t$ and $v_{in2}(t) = 2 \sin 3000\pi t$, find $v_o(t)$.

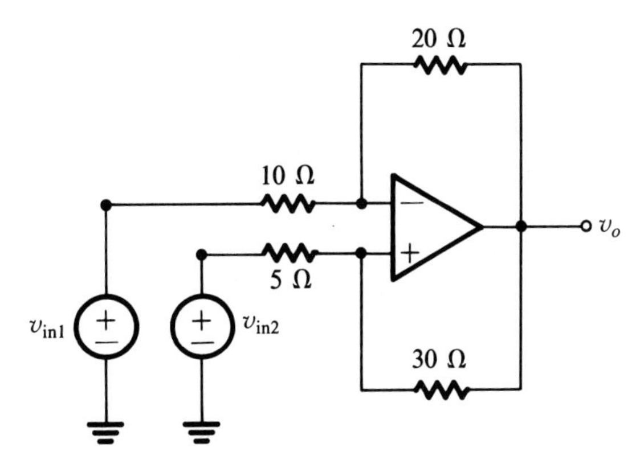

Figure P6.29

6.30 If $v_{in}(t) = 10 \cos 800\pi t$, find $i(t)$ and the power $P(t)$, dissipated in the 2 Ω resistor.

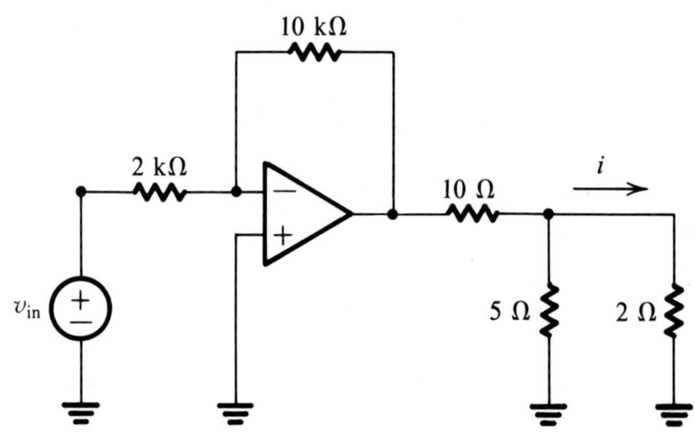

Figure P6.30

6.31 What is the power supplied by the source in Problem 6.30?

INPUT/OUTPUT MODELS FOR CIRCUIT RESPONSE

INTRODUCTION

Chapter 5 presented the input/output current–voltage relationships for capacitors and inductors. Now, just as individual resistors, op amps, capacitors, and inductors are described by models, circuits formed by interconnections of these components and sources have models that determine the values of all voltages and currents within the circuit. These models are usually more complex than individual device models, and require special methods for solution. This brief chapter will use KVL and KCL to develop differential-equation models for circuits containing capacitors and inductors, and then will show how to systematically construct the mathematical solution of the differential equation.

7.1 INPUT/OUTPUT MODELS

The first step in constructing the model of a circuit is to write a description of the circuit in terms of KVL and KCL. Applying KCL to the parallel RC circuit in Fig. 7.1 gives

$$i_c(t) + i_R(t) = i_{in}(t) \tag{7.1}$$

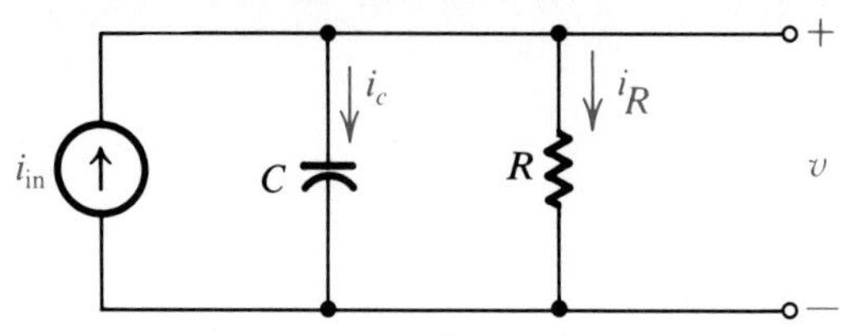

Figure 7.1 RC circuit driven by a current source.

Next, we use Ohm's law and the capacitor's derivative model in (7.1) to get

$$\frac{dv}{dt} + \frac{v}{RC} = \frac{1}{C} i_{in} \tag{7.2}$$

Equation (7.2) is an input/output (I/O) differential equation specifying an instantaneous relationship between the input source i_{in} and the output voltage $v(t)$.

In general, a separate model exists for each circuit variable's response to the source. The relationship between $i_{in}(t)$ and $i_c(t)$ in Fig. 7.1 is found by using the charge-storage model for the capacitor in (7.1)

$$i_c(t) + \frac{1}{R}\left[v(t_0) + \frac{1}{C}\int_{t_0}^{t} i_c(\alpha)d\alpha\right] = i_{in}(t)$$

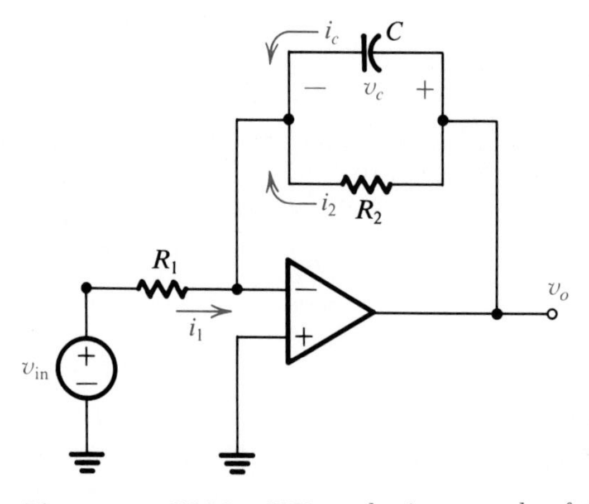

Figure 7.2 Writing KCL at the input node of the op amp leads to an input/output model for the circuit.

Taking $\frac{d}{dt}$ we get

$$\frac{d}{dt}i_c + \frac{1}{RC}i_c = \frac{d}{dt}i_{in} \tag{7.3}$$

The capacitor current $i_c(t)$ must satisfy (7.3) for a given $i_{in}(t)$. (Note that $i_c(t)$ can also be found by taking $i_c(t) = C\,dv_c/dt$.)

As another example, let's find the I/O model relating $v_o(t)$ to $v_{in}(t)$ in Fig. 7.2. We apply KCL at the op amp input node

$$i_c(t) + i_1(t) + i_2(t) = 0$$

and with $v_o(t) = v_c(t)$ we get

$$C\frac{dv_c}{dt} + \frac{v_{in}(t)}{R_1} + \frac{v_c(t)}{R_2} = 0$$

and

$$\frac{dv_o}{dt} + \frac{1}{R_2C}v_o = \frac{-1}{R_1C}v_{in} \tag{7.4}$$

Comparing (7.2) and (7.4), we conclude that $v_o(t)$ of the op amp circuit will behave like a parallel RC circuit being charged by a current source of value $-v_{in}(t)/R_1$. The next example will demonstrate that the task of finding the I/O model is sometimes cumbersome.

Example 7.1

Find the I/O model describing the relationship between $v_{in}(t)$ and $v(t)$ in Fig. 7.3.

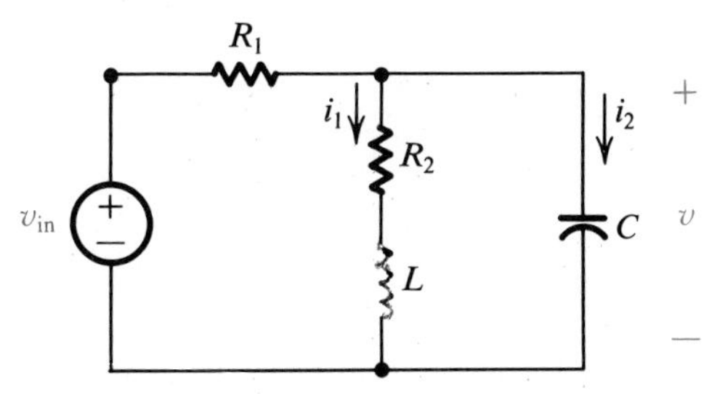

Figure 7.3 Writing the mesh equations for i_1 and i_2 leads to a second-order input/output equation relating $v(t)$ and $v_{in}(t)$.

Solution: One method is to first write an expression for i_2 in terms of v

$$C\frac{dv}{dt} = \frac{v_{in} - v}{R_1} - i_1$$

Taking the derivative of each term leads to a second-order differential equation describing $v(t)$ in terms of $v_{in}(t)$ and to eliminate di_1/dt

$$C\frac{d^2v}{dt^2} = \frac{dv_{in}}{R_1 dt} - \frac{1}{R_1}\frac{dv}{dt} - \frac{di_1}{dt} \tag{7.5}$$

To eliminate i_1 apply KVL to the branch containing the inductor

$$L\frac{di_1}{dt} = v - i_1 R_2$$

Substituting this expression into (7.5) and rearranging terms gives

$$\frac{d^2v}{dt^2} + \left[\frac{1}{R_1C} + \frac{R_2}{L}\right]\frac{dv}{dt} + \frac{R_1 + R_2}{R_1LC}v = \frac{1}{R_1C}\left[\frac{dv_{in}}{dt} + \frac{R_2}{L}v_{in}\right] \tag{7.6}$$

As an exercise, obtain (7.6) by writing mesh equations for i_1 and i_2. (You will need to write an integro-differential equation.)

The I/O model defined by (7.6) is a **second-order differential equation** because it involves the **second derivative** of $v(t)$. The steps required to obtain such models can be cumbersome and obscure. They usually require manipulation of integro-differential equations, a conceptually simple but tedious task. We'll provide an alternative in Chapter 13.

7.2 SOLVING THE CIRCUIT I/O MODEL

Circuit design and analysis requires that we find solutions to equations like (7.2), and (7.6) for specified input signals and initial conditions describing the energy-storage elements in the circuit. For example, we might want to know how the capacitor voltage responds to a step-input signal for current, or to a sinusoidal input, or how the circuit behaves if the source is removed and charge is present on the capacitor. Another important consideration is how fast the circuit responds to an input signal—if a step-input signal is applied, how long does it take for the output to reach a

DC value? How do the values of the circuit components affect the response of the circuit? How sensitive is the response of the circuit to changes in the value of the components? How can we design the circuit to have a desired response? To get a quantitative answer to these questions we must solve a differential equation. There are many different kinds of differential equations, and you will probably study them in later courses in mathematics. Here we will examine only differential equations that describe *RLC* op amp circuits.

In general, the circuit relationship between an input source and an output voltage or current is described by a differential equation of the form:

$$a_n \frac{d^n y}{dt^n} + a_{n-1} \frac{d^{n-1}}{dt^{n-1}} y + \cdots + a_1 \frac{dy}{dt} + a_0\, y(t) = b_m \frac{d^m}{dt^m} x(t)$$
$$+ b_{m-1} \frac{d^{m-1}}{dt^{m-1}} x(t) + \cdots + b_1 \frac{dx}{dt} + b_0 x(t) \qquad (7.7)$$

This is an *n*th order, constant-coefficient, ordinary linear differential equation expressing a mathematical relationship between a single output signal y and a single input signal x. The I/O equation embodies the physical constraints imposed by KVL, KCL, and the individual device models. Equation (7.6) is a second-order example of (7.7).

Next, we will show how to *construct* the solution for $y(t)$ in (7.7). The complete solution of the I/O model differential equation consists of two parts

$$y(t) = y_n(t) + y_p(t)$$

where y_n is called the **natural solution**, and y_p is called the **particular solution.** The natural solution satisfies the homogeneous equation obtained by setting the RHS (right-hand side) of 7.7 to zero

$$a_n \frac{d^n y}{dt^n} + a_{n-1} \frac{d^{n-1}}{dt^{n-1}} y + \cdots + a_1 \frac{dy}{dt} + a_0 y(t) = 0 \qquad (7.8)$$

Note that the natural solution satisfies the I/O model when the input source is turned off.

The **particular solution,** or **forced solution,** satisfies (7.7), and it must be added to the natural solution to account for the effect of a particular input (forcing function) x(*t*); note that $y_p(t)$ depends on x(*t*), but $y_n(t)$ does not.

7.2.1 The Natural Solution

The homogeneous differential equation (7.8) has a solution of the form

$$y(t) = Ke^{st}$$

where K and s are fixed parameters. To show this, we substitute $y(t)$ into (7.8), noting that

$$\frac{d^n}{dt^n}Ke^{st} = Ks^n e^{st}$$

The result is another equation:

$$a_n s^n Ke^{st} + a_{n-1}s^{n-1}Ke^{st} + \cdots + a_1 sKe^{st} + a_0 Ke^{st} = 0$$

or

$$[a_n s^n + a_{n-1}s^{n-1} + \cdots + a_1 s + a_0]Ke^{st} = 0$$

Since K must be chosen to be nonzero in order to have a nontrivial solution, it can be cancelled from the above expression. Likewise, e^{st} cannot be zero and we cancel it too, leaving the algebraic equation

$$a_n s^n + a_{n-1}s^{n-1} + \cdots + a_1 s + a_0 = 0 \tag{7.9}$$

This is called the **characteristic equation** of the circuit. If the value of s is chosen to satisfy the characteristic equation, the exponential $y(t) = Ke^{st}$ will solve the homogeneous equation.

The coefficients of the characteristic equation are the same as the coefficients of the homogeneous differential equation, and the power of s is the order of the corresponding derivative term,

$$\begin{array}{ccccccccc} a_n\dfrac{d^n y}{dt^n} & + & a_{n-1}\dfrac{d^{n-1}y}{dt^{n-1}} & + \cdots + & a_1\dfrac{dy}{dt} & + & a_0 y(t) = 0 \\ \updownarrow & & \updownarrow & & \updownarrow & & \updownarrow \\ a_n s^n & + & a_{n-1}s^{n-1} & + \cdots + & a_1 s & + & a_0 = 0 \end{array}$$

The characteristic equation can always be formed *by inspection* of the I/O equation, provided that we remember to put the response terms on the LHS (left-hand side) and the input terms on the RHS of the equation.

Example 7.2

Find the characteristic equation of the circuit I/O equation given by

$$\frac{d^2 v_o}{dt^2} + 5\frac{dv_o}{dt} + 6\,v_o(t) = \frac{7dv_{in}}{dt} + 2\,v_{in}(t)$$

Solution: $s^2 + 5\,s + 6 = 0.$

In general, the characteristic equation is an nth order algebraic equation. It will have n roots, denoted by $s_1, s_2, \cdots, s_n$. In factored form the characteristic equation is

$$(s - s_1)(s - s_2) \cdots (s - s_n) = 0$$

The roots of the characteristic equation are called the **natural frequencies** or the **characteristic frequencies** of the circuit because they determine the natural-solution component of its I/O differential-equation model. Note: use of the term "natural frequency" is not meant to imply that the natural solution is sinusoidal, or that the value of s will be the frequency of a sinusoid. It will be in some cases, but not always.

An nth order characteristic equation has n roots. If these roots are distinct their sum defines a parametric family of solutions given by

$$y_n(t) = K_1e^{s_1t} + K_2e^{s_2t} + \cdots + K_ne^{s_nt} \qquad (7.10)$$

where $K_1, K_2, \ldots, K_n$ are constants which can be arbitrarily chosen to create a variety of solutions to (7.8). Actually, each natural frequency s defines an exponential function that satisfies (7.8), but the **weighted sum** (or **superposition)** of the individual solutions given by (7.10) is the most general form of natural solution, since it includes the individual solutions. As an exercise, verify that (7.10) solves (7.8).

Example 7.3

Find the natural solution for the I/O DE (differential equation) of Example 7.2.

Solution: Note that the characteristic equation has roots $s = -3$ and $s = -2$.

First, we show that the homogeneous differential equation is satisfied by

$$v = K_1e^{s_1t} = K_1e^{-3t}$$

We do so by taking

$$\frac{d}{dt}v(t) = -3\ K_1e^{-3t}$$

and

$$\frac{d^2}{dt^2}v(t) = 9\ K_1e^{-3t}$$

Making substitutions in the homogeneous equation gives

$$\frac{d^2v}{dt^2} + 5\frac{dv}{dt} + 6\,v(t) \;=\; 9\,K_1e^{-3t} + 5\,(-3\,K_1e^{-3t}) + 6\,K_1e^{-3t}$$

$$= \;15\,K_1e^{-3t} - 15\,K_1e^{-3t} = 0$$

As an exercise, verify that K_2e^{-2t} is also a solution.

Yet another solution to Equation (7.8) can be formed by taking the sum of the individual exponentials. Let

$$v_n(t) \;=\; K_1e^{-3t} + K_2e^{-2t}$$

Then

$$\frac{dv_n}{dt}(t) \;=\; -3\,K_1e^{-3t} - 2\,K_2e^{-2t}$$

$$\frac{d^2v_n}{dt^2}(t) \;=\; 9\,K_1e^{-3t} + 4\,K_2e^{-2t}$$

and so

$$\frac{d^2v_n}{dt^2} + 5\frac{dv_n}{dt} + 6\,v_n(t) = 9\,K_1e^{-3t} + 4\,K_2e^{-2t}$$

$$-\,15\,K_1e^{-3t} - 10\,K_2e^{-2t} + 6\,K_1e^{-3t} + 6\,K_2e^{-2t} = 0$$

This example demonstrates that, in general, the roots of the characteristic equation define n different exponential solutions to (7.8). The sums of these solutions also define a solution.

Skill Exercise 7.1

A circuit I/O model has a characteristic equation given by $(s + 10)(s + 4)(s + 6) = 0$. Find the natural solution to the I/O equation in parametric form.

Answer: $y_n(t) = K_1e^{-10t} + K_2e^{-4t} + K_3e^{-6t}$

7.2.2 Transfer Functions and the Particular Solution

The **complete solution** of a circuit's I/O differential equation (7.7) is the sum of the natural solution and the particular solution. This section will show how to find the particular solution for a very important kind of input signal, the **exponential signal.** This won't be merely an academic exercise, because in Chapters 14, 15, and 16 we'll use superposition to represent *all* real-world signals in terms of exponential signals. There, we'll express an input signal as a sum of exponential signals and superimpose the responses of a circuit to those exponential signals to find the circuit's response to the given input signal.

First, we define a **transfer function, H(s)** as the ratio of the polynomials obtained from both sides of the I/O equation

$$\mathbf{H}(s) = \frac{b_m s^m + b_{m-1} s^{m-1} + \cdots + b_1 s + b_0}{a_n s^n + a_{n-1} s^{n-1} + \cdots + a_1 s + a_0} \tag{7.11}$$

The denominator polynomial is the same polynomial that defines the circuit's characteristic equation. It is formed from the left (*output*) side of (7.7), while the numerator polynomial is obtained from the terms on the right (*source*) side of the equation.

Example 7.4

Find the transfer function for the I/O equation of Ex. 7.2.

Solution: By inspection, the transfer function is

$$\mathbf{H}(s) = \frac{7\,s + 2}{s^2 + 5\,s + 6}$$

Next, suppose a circuit's input is an exponential signal having the generic form

$$x(t) = Xe^{st} \tag{7.12}$$

We call the fixed parameter s the **source exponential frequency,** and call the fixed parameter X the **source amplitude.** *When the source signal is exponential* [like (7.12)] *the particular solution will also be an exponential signal, and will have the same exponential frequency as the source exponential.* That is,

$$y_p(t) = Ye^{st} \tag{7.13}$$

where Y is called the amplitude of $y_p(t)$. Moreover, the input and output amplitudes, X and Y, are related by the transfer function according to

$$\boxed{Y = \mathbf{H}(s)X} \tag{7.14}$$

Therefore, the particular solution can be written

$$\boxed{y_p(t) = \mathbf{H}(s)Xe^{st}} \tag{7.15}$$

for a given value of s.

Before demonstrating that (7.13) defines the particular solution to (7.7), we **make note of the fact that** s**, used here without a subscript, is not to be confused with the natural frequencies** that we found when solving for the natural solution. The natural frequencies are properties of the circuit itself. The exponent s in (7.13) and (7.12) is just a fixed parameter of the applied input signal. In fact, we will see that in order for (7.15) to be valid, **the source parameter** s **must not have the same numerical value as a natural frequency of the circuit.** Another mathematical fact that we will not prove is that **the particular solution is unique.** Only one signal satisfies (7.7), and we've found it!

The transfer function is important because we need only evaluate it at the fixed source frequency, and then use (7.15) to find $y_p(t)$. $\mathbf{H}(s)$ determines the ratio of the amplitudes of the exponential-input and the (exponential) particular solution to the I/O differential equation.

The general proof that the particular solution has the exponential form defined by (7.13) and (7.15) when the source frequency is not a natural frequency is easy to construct. First, we insert $y_p(t)$ as defined by (7.13) into the LHS of (7.7), and insert $x(t)$ given by (7.12) into the RHS to get

$$a_n s^n Y e^{st} + a_{n-1} s^{n-1} Y e^{st} + \cdots + a_1 s Y e^{st} + a_0 Y e^{st} = b_m s^m X e^{st} + b_{m-1} s^{m-1} X e^{st} + \cdots + b_1 s X e^{st} + b_0 X e^{st}$$

which can be rewritten as

$$Y e^{st}[a_n s^n + a_{n-1} s^{n-1} + \cdots + a_1 s + a_0] = X e^{st}[b_m s^m + \cdots + b_1 s + b_0]$$

Since e^{st} is nonzero, it can be canceled on each side to obtain an algebraic relation between Y and X

$$Y[a_n s^n + \cdots + a_1 s + a_0] = X[b_m s^m + \cdots + b_1 s + b_0]$$

If the source frequency s is not a natural frequency the polynomial factor on the left side is nonzero because s is not a root of the characteristic equation. Therefore, the amplitude of the particular solution (for a given value of s) is

$$Y = \frac{b_m s^m + \cdots + b_1 s + b_0}{a_n s^n + \cdots + a_1 s + a_0} X$$

In summary, we have shown that the particular solution is an exponential having the same frequency as the source, and we have developed a simple algorithm that uses the circuit's transfer function to calculate the solution's amplitude. Thus, the particular-solution component is entirely specified.

The input/output relationship defined by $\mathbf{H}(s)$ is displayed in block diagram form in Fig. 7.4(a), where $\mathbf{H}(s)$ *represents* the circuit that is driven by Xe^{st}. We sometimes suppress the exponential factor and just show a

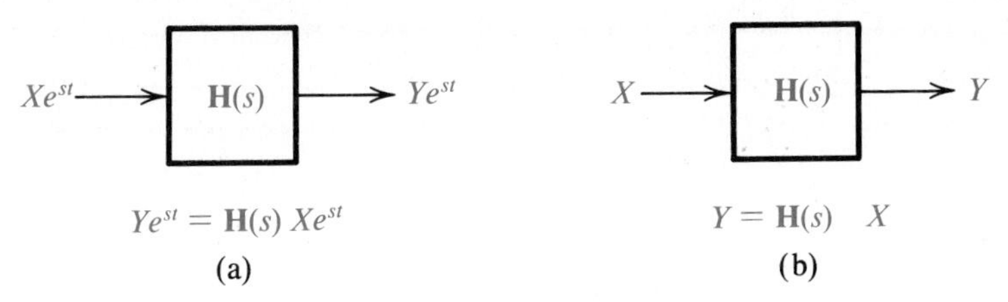

Figure 7.4 Symbolic representation of a circuit's input/output relationship. $\mathbf{H}(s)$ represents the circuit whose particular solution is Ye^{st} when the input is Xe^{st}.

block diagram that represents the relationship between the input and particular-solution amplitudes, as in Fig. 7.4(b).

Skill Exercise 7.2

A circuit has a transfer function given by

$$\mathbf{H}(s) = \frac{s + 5}{(s + 6)(s + 3)}$$

Find the particular solution to

a. $x(t) = 12e^{-2t}$
b. $x(t) = K$
c. $x(t) = 0$
d. $x(t) = 10e^{-5t}$
e. $x(t) = e^{-2.999t}$

Answer:
a. $y_p(t) = 9e^{-2t}$
b. $y_p(t) = 5K/18$
c. $y_p(t) = 0$
d. $y_p(t) = 0$
e. $y_p(t) = 666.8e^{-2.999t}$

7.2.3 The Complete Solution To an Exponential Input Signal

The **complete** solution to an exponential input of (7.7) is formed by adding the natural and particular solutions

$$y(t) = \sum_{i=1}^{n} K_i e^{s_i t} + \mathbf{H}(s)Xe^{st}$$

where e^{st} and $\mathbf{H}(s)$ are evaluated at the fixed frequency of the exponential source.

Skill Exercise 7.3

Find the complete solution to the I/O equation in Example 7.2 if $v_{in}(t) = 10e^{-t}$.

Answer: $y(t) = K_1e^{-3t} + K_2e^{-2t} - 25e^{-t}$

7.3 BOUNDARY CONDITIONS

The complete solution to the I/O differential equation is "generic" in the sense that the coefficients $K_1, K_2, \ldots, K_n$ have not been specified. These constants can only be specified if additional constraints, called **boundary conditions,** are imposed on the solution to (7.7). In fact, if n independent conditions are given, the values of the coefficients are uniquely determined (only one curve satisfies the differential equation and its boundary conditions).

Example 7.6

Find K_1 and K_2 so that the complete solution of the circuit model given in SE7.3 satisfies the boundary conditions given by $y(0) = 5$ and $dy/dt(0) = 1$.

Solution: The boundary conditions must be satisfied by the complete solution, so K_1 and K_2 must satisfy

$$y(0) = K_1 + K_2 - 25 = 5$$

and (after taking dy/dt):

$$\frac{dy}{dt}(0) = -3K_1 - 2K_2 + 25 = 1$$

Solving, we get $K_1 = -36$ and $K_2 = 66$. The unique solution satisfying both the differential equation *and* the boundary conditions is

$$y(t) = -36e^{-3t} + 66e^{-2t} - 25e^{-t}$$

Boundary conditions are used to model physical situations in which a circuit's I/O equation is valid only for a specific interval of time. For example, when a circuit is excited by a source for $t \geq 0$ the boundary conditions given at $t = 0$ summarize the history of the circuit for $t < 0$.

Example 7.7

Find the I/O model for $v(t)$ in the circuit in Fig. 7.5 for $t > 0$, and the boundary condition on the capacitor voltage at $t = 0^+$.

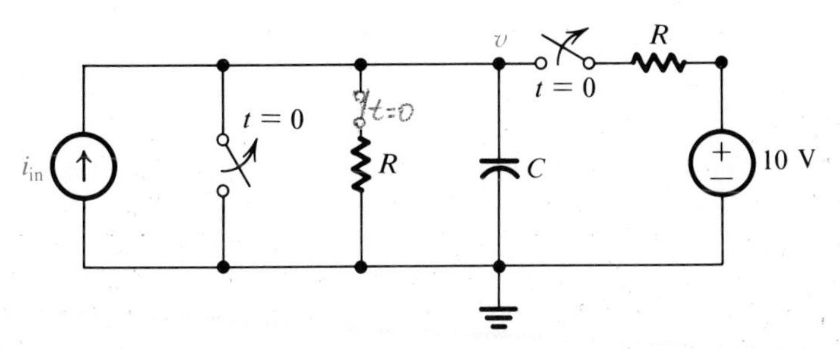

Figure 7.5 Circuit for Example 7.7

Solution: Instead of developing the I/O model of the circuit for $t < 0$, we *assume* that the capacitor is charged to a DC value by $t = 0$. Then, dividing the 10-volt source across the identical resistors, we get $v(0^-) = 5$. Since the capacitor voltage must be continuous at $t = 0$ we conclude that $v(0^+) = v(0^-) = 5$. For $t > 0$ the circuit has the configuration of a current source driving a parallel RC section. The I/O model for this parallel RC circuit was found in (7.2) to be

$$\frac{dv}{dt} + \frac{1}{RC} = \frac{1}{C}i_{in}$$

7.4 REPEATED NATURAL FREQUENCIES

The key step in constructing the natural solution of the I/O DE is to solve the characteristic equation for its roots, thereby determining the natural frequencies of the circuit. This can be done with a hand-held calculator, a personal computer, or a mainframe computer that executes a root-finding code. When the characteristic equation contains a repeated root, it will have a product factor of the form $(s - s_j)^r$, where $s - s_j$ is repeated r times. In this case, the natural solution will have exactly r exponential terms corresponding to s_j, but they are modified to include polynomial factors in t, so

$$y_n(t) = K_1 e^{s_1 t} + \cdots + K_{j1} e^{s_j t} + K_{j2} t e^{s_j t} + \cdots + K_{jr} t^{r-1} e^{s_j t} + \cdots$$

Skill Exercise 7.4

Find the natural solution of a circuit whose characteristic equation is $(s+4)^2(s+6)(s+10)^3 = 0$.

Answer: $y_n(t) = K_{11}e^{-4t} + K_{12}te^{-4t} + K_2e^{-6t} + K_{31}e^{-10t} + K_{32}te^{-10t} + K_{33}t^2e^{-10t}$

Whether the characteristic equation has distinct or repeated roots, the natural solution is parametric in exactly *n* coefficients. They can be uniquely determined if n independent boundary conditions are given.

7.5 TRANSFER FUNCTION (*S*-DOMAIN) MODELS OF CIRCUITS

Given an I/O equation we can form the transfer function, $\mathbf{H}(s)$ by inspection and conversely we can form the I/O equation from the transfer function.*

Example 7.8

Find the DE model of the circuit whose transfer equation is given by

$$\mathbf{H}(s) = \frac{s+2}{4s^3 + s + 1}$$

Solution: By inspection, the I/O DE is

$$4\frac{d^3y}{dt^3} + \frac{dy}{dt} + y(t) = \frac{dx}{dt} + 2x(t)$$

The transfer function is a circuit model equivalent to, but more compact than, its I/O differential equation. It contains all of the information needed to determine the natural frequencies, the natural solution, and the particular solution when the input signal is an exponential. Thus, it provides the information needed to construct the complete solution. The differential equation is called the **time-domain model** of the circuit, and the transfer function is called the ***s*-domain model.**

*We will not consider examples in which numerator and denominator factors of $\mathbf{H}(s)$ cancel each other.

Skill Exercise 7.5

Find the complete solution of the I/O model for the circuit with

$$\mathbf{H}(s) = \frac{10(s+2)}{(s+4)(s+10)(s+20)}$$

when the input is $x(t) = 5e^{-3t}$.

Answer: $y(t) = K_1e^{-4t} + K_2e^{-10t} + K_3e^{-20t} + \left(\frac{-50}{119}\right)e^{-3t}$

7.6 POLE-ZERO PATTERNS

The factored form of a transfer function's numerator and denominator polynomials displays their respective roots by

$$\mathbf{H}(s) = \frac{(s - z_1)(s - z_2) \cdots (s - z_m)}{(s - s_1)(s - s_2) \cdots (s - s_n)}$$

The numerator roots $z_1 \cdots z_m$ are called **zeros** of the transfer function because $\mathbf{H}(z_i) = 0$ and the particular solution for an exponential at that frequency will be zero. The denominator roots, the natural frequencies, are called the **poles** of $\mathbf{H}(s)$ because $\mathbf{H}(\cdot)$ is unbounded at these points, and the particular solution *cannot* be constructed from $\mathbf{H}(s)$.

In general, the poles and zeros will be complex numbers and the transfer function $\mathbf{H}(\cdot)$ will be defined on the complex plane. A graphic display of the poles and zeros of $\mathbf{H}(s)$ in the complex plane is called the **pole-zero pattern** of the circuit.

Example 7.9

Draw the pole-zero pattern of the circuit whose transfer function is $\mathbf{H}(s) = (s+1)/(s+2)$.

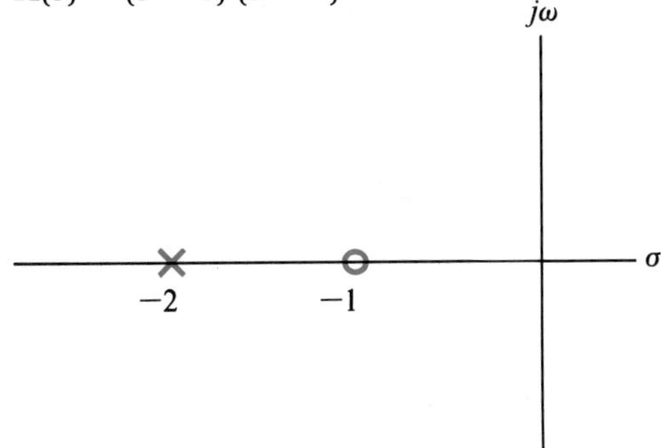

Figure 7.6 Pole-zero pattern for $\mathbf{H}(s) = \frac{s+1}{s+2}$

Solution: The transfer function has a zero at $s = -1$ and a pole at $s = -2$. The pole-zero pattern is shown in Fig. 7.6, with σ denoting the real axis, and ω denoting the imaginary axis.

The pole-zero pattern lets us predict a circuit's response. The location of the poles of **H**(s) corresponds to the location of the natural frequencies, and the location of the zeros corresponds to the frequencies at which the particular solution is zero.

Example 7.10

Find the input/output model, the transfer function, the natural frequency, and the pole-zero pattern of the circuit in Fig. 7.7

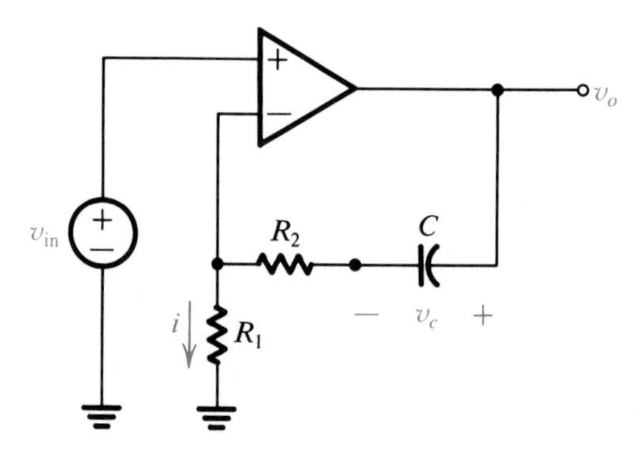

Figure 7.7

Solution: Writing and differentiating the KVL expression for $v_o(t)$ leads to

$$\frac{d}{dt}v_o = \frac{d}{dt}v_c + \left[1 + \frac{R_2}{R_1}\right]\frac{d}{dt}v_{in}$$

where

$$\frac{d}{dt}v_c = \frac{1}{R_1C}v_{in}(t)$$

so

$$\frac{dv_o}{dt} = \left[1 + \frac{R_2}{R_1}\right]\frac{dv_{in}}{dt} + \frac{1}{R_1C}v_{in}$$

Thus $v_o(t)$ is the sum of a scaled copy of $v_{in}(t)$ and the integral of $v_{in}(t)$. The first term is due to the input voltage causing a current in R_1 and R_2, and the sec-

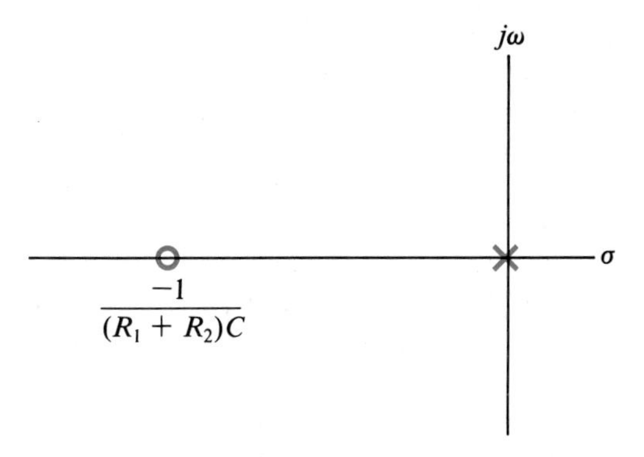

Figure 7.8 Pole-zero pattern for the op amp circuit in Fig. 7.7.

ond term is due to the integration of the capacitor current.

The transfer function

$$\mathbf{H}(s) = \frac{\left[1 + \frac{R_2}{R_1}\right]\left[s + \frac{1}{(R_1 + R_2)C}\right]}{s}$$

has a pole at $s = 0$, and a zero at $s = -1/(R_1 + R_2)C$. The pole at $s = 0$ in Fig. 7.8 has physical significance. If a constant input is applied to the circuit, the current through the capacitor will be constant and the output voltage will grow without bound (as a ramp signal). The pole of $\mathbf{H}(s)$ at the origin is a mathematical clue to this behavior.

Skill Exercise 7.6

Find the input/output transfer function for the circuit in Fig. SE7.6.

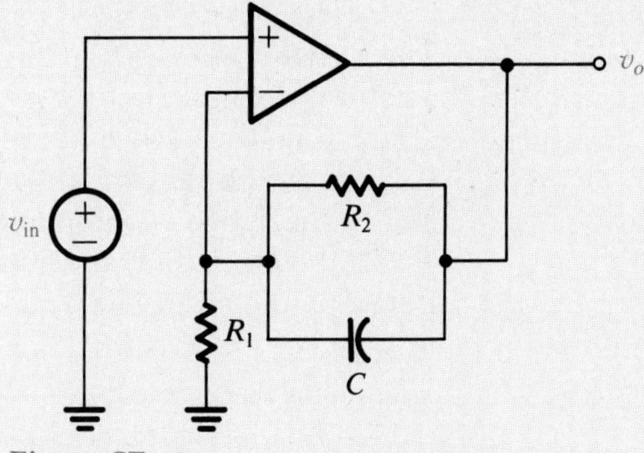

Figure SE7.6

Answer: $\mathbf{H}(s) = \dfrac{s + (R_1 + R_2)/(R_1R_2C)}{s + (1/R_2C)}$

7.7 SUPERPOSITION OF THE PARTICULAR SOLUTION

The particular solution of the input/output differential equation has the property of superposition. If the signal driving a circuit is a weighted sum of two or more exponential signals, the particular solution is the same weighted sum of the particular solutions due to each exponential. That is, if

$$x(t) = X_1 e^{s_{1p}t} + X_2 e^{s_{2p}t} + \cdots + X_k e^{s_{kp}t}$$

the particular solution to $x(t)$ is

$$y_p(t) = X_1 \mathbf{H}(s_{1p}) e^{s_{1p}t} + X_2 \mathbf{H}(s_{2p}) e^{s_{2p}t} + \cdots + X_k \mathbf{H}(s_{kp}) e^{s_{kp}t}$$

provided that none of the source exponential frequencies $s_{1p}, s_{2p}, \ldots, s_{kp}$ is a natural frequency of the circuit. (We will not prove the superposition property, but as an exercise, substitute $x(t)$ and $y_p(t)$ into 7.7 to verify that the property holds.) Note that $\mathbf{H}(s)$ is evaluated at each of the exponential source frequencies. Also note that $s_{1p}, s_{2p}, \ldots, s_{kp}$ are **fixed source frequencies.** Do not confuse them with the natural frequencies of the circuit.

Skill Exercise 7.7

If $\mathbf{H}(s) = 10/(s + 2)$, find the particular solution when $x(t) = 4e^{-t} - 5e^{-8t} - 3e^{-10t}$.

Answer: $y_p(t) = 40e^{-t} + 8.33e^{-8t} + 3.75e^{-10t}$

SUMMARY

The physical constraints imposed on a circuit by Kirchhoff's laws were combined with models for circuit components to develop an input/output differential equation (DE) describing the circuit. A method was presented for solving the I/O equation by finding its natural-solution and particular-solution (forced) components. The denominator polynomial of the circuit's transfer function determines the natural frequencies and the natural solution of the I/O equation. The transfer function also determines the particular solution of the I/O equation when the input signal is an exponential. Boundary conditions were used to specify the coefficients of the complete solution. Finally, we showed that the particular solution has the property of superposition; the I/O differential equation is a time-domain model; and the transfer function is an s-domain model.

Problems - Chapter 7

7.1 Find the I/O differential equation model describing v_o in terms of v_{in} for the series RL circuits in Fig. P7.1.

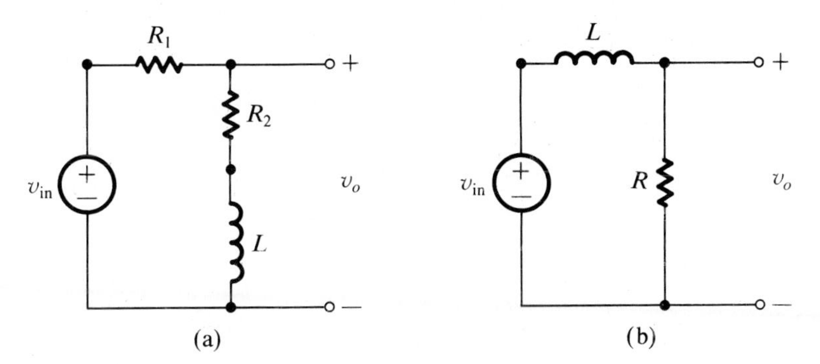

Figure P7.1

7.2 For a series RL circuit driven by a voltage source $v_{in}(t)$:

a. Find the I/O differential equation model relating the source voltage and the inductor current.

b. Find the characteristic equation.

c. Find the natural frequency.

d. Find the transfer function.

7.3 Find the differential equations describing $v(t)$ in Fig. P7.3 for $t < 0$ and for $t > 0$.

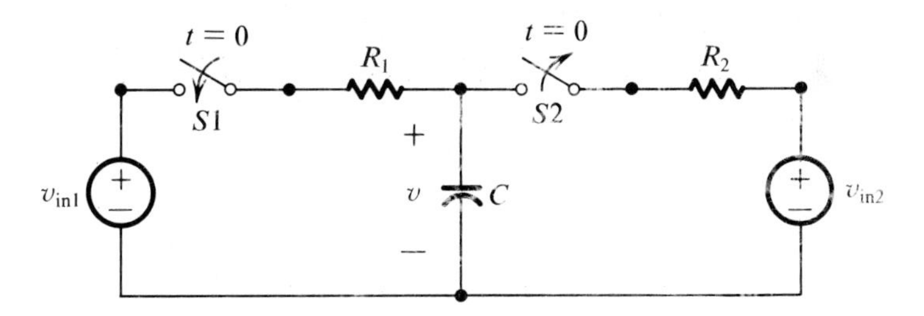

Figure P7.3

7.4 Write the I/O differential equations describing $i(t)$ for $t < 0$ and $t > 0$.

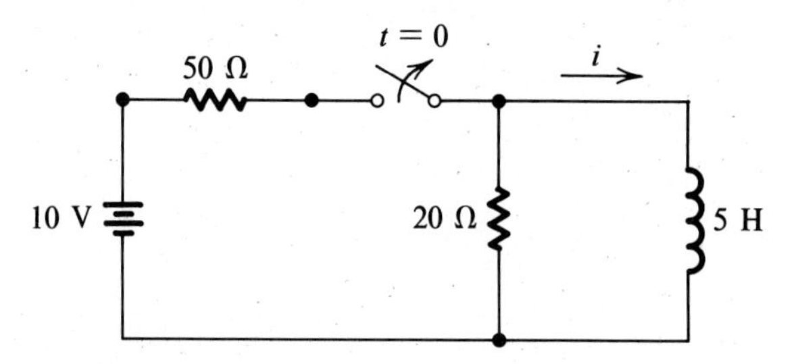

Figure P7.4

7.5 Find the I/O differential equation model relating v_{in} and v_o in Fig. P7.5.

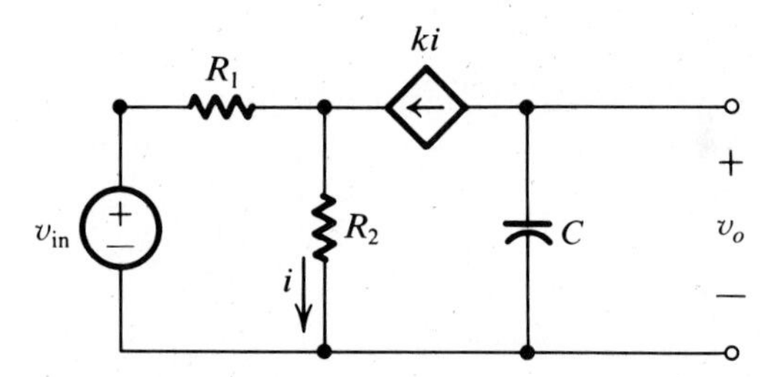

Figure P7.5

7.6 For the circuit shown in Fig. P7.6:

a. Find the I/O differential equation model.
b. Find the natural frequency.
c. Find the transfer function.

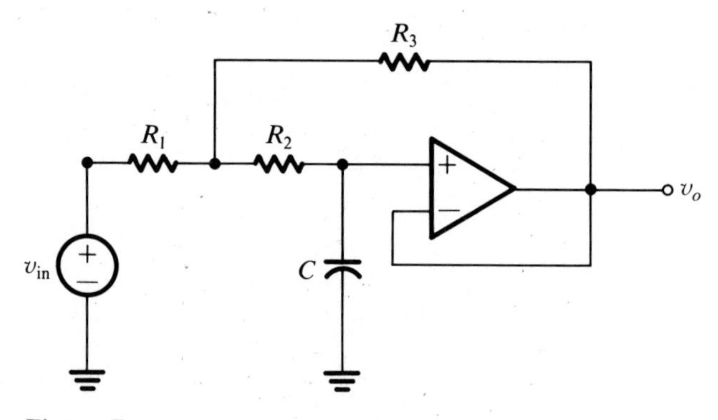

Figure P7.6

7.7 Find the differential equation describing $i(t)$ in terms of $v_{in}(t)$.

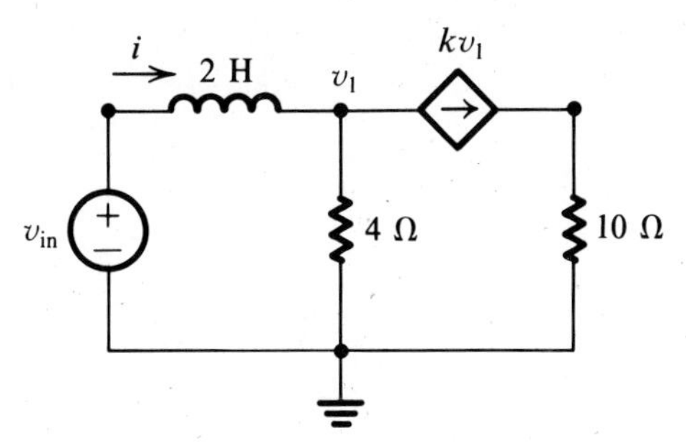

Figure P7.7

7.8 Find the differential equations describing $v(t)$ in the circuit in Fig. P7.8 for $t < 0$ and for $t > 0$.

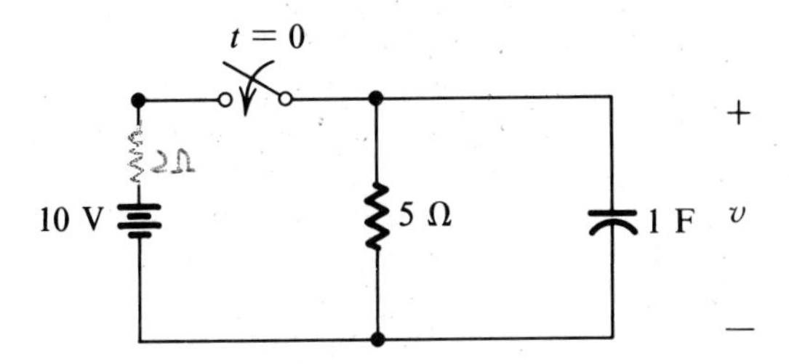

Figure P7.8

7.9 A circuit is described by $10\,dv/dt + 2v(t) = 5v_{in}(t)$.

a. Find the characteristic equation.
b. Find the natural frequency.
c. Find the natural solution for $v(t)$ in parametric form.

7.10 If a circuit's I/O model is given by $3di/dt + 12\ i(t) = 24i_{in}(t)$:

a. Find the circuit's transfer function relating $i(t)$ and $i_{in}(t)$.
b. Find the natural solution in parametric form.
c. Find the particular solution if $i_{in}(t) = 2e^{-6t}$.

7.11 Solve the following differential equations for $t > 0$, with the given boundary conditions.

a. $dv/dt + 4v(t) = 0, \quad v(0^-) = 100$
b. $di/dt + 0.5i = 100, \quad i(0^-) = 10$
c. $3\ dv/dt + 6v(t) = 12e^{-4t}, \quad v(0^-) = 8$
d. $3\ dv/dt + 6v(t) = 12e^{-2t}, \quad v(0^-) = 8$
e. $50\ di/dt + 5i(t) = 25\ dv_{in}/dt - 10v_{in}(t)$, with $v_{in}(t) = 2e^{-0.2t}$. $i(0^+) = 1$
f. Repeat e with $v_{in}(t) = 2e^{-0.101t}$. Hint: try $V_p(t) = Ate^{-2t}$

7.12 Find the boundary condition $v(0^-)$ in Fig. P7.12.

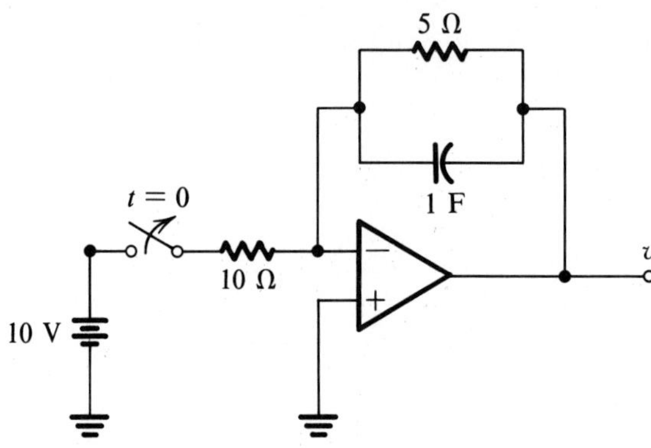

Figure P7.12

7.13 Draw the pole-zero patterns for the circuits whose I/O models are given below.

a. $\frac{dv}{dt} + 5v(t) = 10i_{in}(t)$

b. $100\frac{di}{dt} + 25i(t) = 5\frac{dv_{in}}{dt} + v_{in}(t)$

7.14 Find the transfer functions corresponding to the differential equations given in Problem 7.11e.

7.15 Draw the pole-zero patterns corresponding to the differential equations given in Problem 7.11.

7.16 Find the particular solution of $dy/dt + 8y = x(t)$ when

a. $x(t) = 5e^{-3t}$

b. $x(t) = 2e^{-6t} + 10e^{-4t}$

c. $x(t) = 5e^{-3t} + 2e^{-6t} + 10e^{-4t}$

7.17 Find the particular solution of the following differential equations when $x(t) = 100e^{-4t}$

a. $2\frac{dy}{dt} + 10y(t) = \frac{dx}{dt}$

b. $40\frac{dy}{dt} + 2y(t) = \frac{dx}{dt} + 2x(t)$

c. $\frac{dy}{dt} + 6y(t) = 2x(t)$

7.18 Solve $dy/dt + 6y(t) = x(t)$ for $t > 0$, subject to:

a. $y(0^+) = -4$ and $x(t) = 2e^{-3t}u(t)$

b. $y(0^+) = 2$ and $x(t) = [4e^{-t} + 10e^{-3t}]u(t)$

7.19 Find $y(t)$ for $t > 0$ if $y(0^+) = 5$ and

$$6\frac{dy}{dt} + 12y(t) = 3\frac{dx}{dt} + 24x(t)$$

a. $x(t) = 2e^{-t}u(t)$
b. $x(t) = 6e^{-4t}u(t)$
c. $x(t) = 10e^{-8t}u(t)$
d. $x(t) = [4e^{-t} + 10e^{-3t}]u(t)$

7.20 If a circuit has $\mathbf{H}(s) = (s + 10)/(s + 4)$, and if the particular solution to its I/O differential equation is $y_p(t) = 6e^{-2t}$, what must have been the input signal $x(t)$?

FIRST-ORDER CIRCUIT RESPONSE

INTRODUCTION

Chapter 7 developed differential-equation models for several circuits containing energy-storage elements—inductors and capacitors. It also presented the basic mathematical tools for systematically constructing solutions to these differential equations, subject to boundary conditions, when the circuits are driven by exponential input signals. In this chapter we move from the mathematical framework of differential equations to the physical world of circuits. We must therefore develop specific terminology that will help us understand circuit behavior and focus on circuits' physical response to applied signals, rather than on their abstract mathematical models.

When a differential equation describes either a voltage or a current in a circuit, its mathematical solution will be referred to as the response of the circuit. The I/O model describes the behavior of this voltage or current for a given source signal. Therefore, the response of the circuit is the solution to the I/O differential equation (subject to certain boundary conditions imposed by the physical circuit). Here we will restrict our attention to the response of first-order circuits to exponential signals.

This chapter examines *RC*-circuit response in depth because it plays such a key role in the behavior of modern integrated circuits. Grasping the behavior of this simple circuit is

the foundation for understanding more complex circuits, and for understanding the performance limits of integrated circuits that must respond quickly to pulses having very short duration. Our emphasis will be on obtaining a circuit's response to an exponential signal, including the step. Chapters 9 and 10 will examine the response of first-order circuits to other signals.

8.1 CAPACITOR RESPONSE

Chapter 7 showed that the generic form of the complete solution to a circuit's input/output (I/O) differential equation is the sum of the equation's natural solution and its particular solution, and that the solution parameters are specified by imposing boundary conditions representing the initial stored energy in the circuit. In general, we are interested in the "initial-state response" of a circuit, or how it responds to a given input signal *and* the initial state, or condition, of the circuit's stored energy. Because the response of a circuit is affected by both the boundary conditions and the input signal, it is useful to examine each of their effects on the circuit separately. If no source is applied (zero input), the response is due solely to the initial state, or condition of the energy-storage elements. If no energy is stored in the circuit (the circuit is in the so-called "zero state") the response will be due solely to the applied source. We'll begin by examining a capacitor's voltage response to a current source.

The solution to the I/O model relating the charging current to the terminal voltage of a capacitor (see Chapter 5) is described by

$$v_c(t) = v_c(t_0) + \frac{1}{C}\int_{t_0}^{t} i_c(\alpha)d\alpha \tag{8.1}$$

The **zero-input response (ZIR)**, is the response when no charging current (input) is applied to the capacitor. Under this condition $v_c(t) = v_c(t_0) = v_0$ for $t \geq t_0$. Similarly, if the capacitor is initially uncharged, or in the zero state, the output voltage caused by a given current is the **zero-state response** of the circuit

$$v_c(t) = \frac{1}{C}\int_{t_0}^{t} i_c(\alpha)d\alpha \qquad t \geq t_0 \tag{8.2}$$

The capacitor response when a current is input to an initially charged capacitor is the **initial-state response (ISR)** given in (8.1). We conclude that **the initial-state response is the sum of the zero-input response and the zero-state response.**

$$\boxed{\text{ISR} = \text{ZIR} + \text{ZSR}} \tag{8.3}$$

A circuit's response has one component due solely to the initial stored energy, and another component due solely to the applied source. The initial state response is the **superposition** of these two effects.

Example 8.1

Draw the ZIR, ZSR and ISR if $i_{in}(t)$ in Fig. 8.1 charges a 0.5 F capacitor, given $v_c(0) = 5$ V.

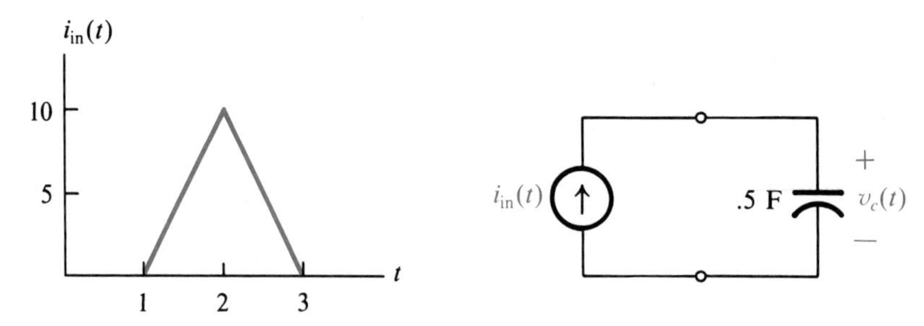

Figure 8.1 A capacitor driven by a triangular current pulse.

Solution: The ZIR is given by $v_{ZIR}(t) = 5$ for $t \geq 0$. The ZSR is obtained by integrating the charging current according to (8.2) to get

$$v_{ZSR}(t) = \begin{cases} 0 & 0 \leq t \leq 1 \\ 10t^2 - 20t + 10 & 1 \leq t \leq 2 \\ -10t^2 + 60t - 70 & 2 \leq t \leq 3 \\ 20 & t \geq 3 \end{cases}$$

Then the ISR is obtained by adding the ZIR and the ZSR to get

$$v_{ISR}(t) = \begin{cases} 5 & 0 \leq t \leq 1 \\ 15 + 10t^2 - 20t & 1 \leq t \leq 2 \\ -65 - 10t^2 + 60t & 2 \leq t \leq 3 \\ 25 & t \geq 3 \end{cases}$$

The graph of the ISR of $v_c(t)$ is shown in Fig. 8.2.

The important point to note about the response superposition in (8.3) is that the ZIR does not depend on the input signal, and it summarizes the "charging history" of the circuit before the input is applied. The ZSR, however, depends strongly on the input signal that is applied for $t \geq t_0$. The capacitor behaves like an ideal integrator in the sense that the capaci-

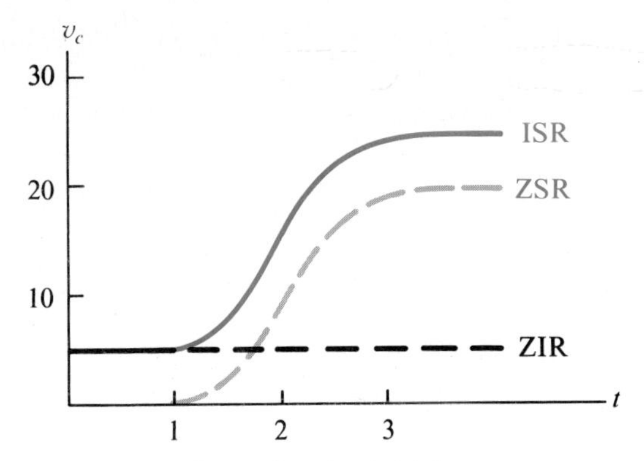

Figure 8.2 Curve for the initial-state response of the capacitor voltage in Fig. 8.1.

tor voltage is the scaled integral of the charging current, with the voltage being directly proportional to the accumulated charge. Thus, positive charging currents cause the capacitor voltage to increase, and vice versa. Response superposition (decomposition) applies to any circuit described by a linear differential equation.

8.2 SERIES-*RC*-CIRCUIT RESPONSE

The input/output model for the **parallel *RC* circuit** was presented in Chapter 7. Here we will develop the model for the **series *RC* circuit** and then show how to use the methods of Chapter 7 to find the response of the circuit. The concepts developed here also apply to other, more complex circuits.

The I/O model for the series RC circuit in Fig. 8.3 describes the relationship between the source voltage and the capacitor voltage, and the response of the circuit describes $v_c(t)$ whenever $v_{in}(t)$ is known for $t \geq t_0$, with t_0 given. In general, the capacitor voltage v_c will depend on the charging-current waveform for $t \geq t_0$ and on the voltage that was present when charging began (i.e., $v_c(t_0) = v_0$, a given voltage).

Instead of immediately plunging into the mathematical description of

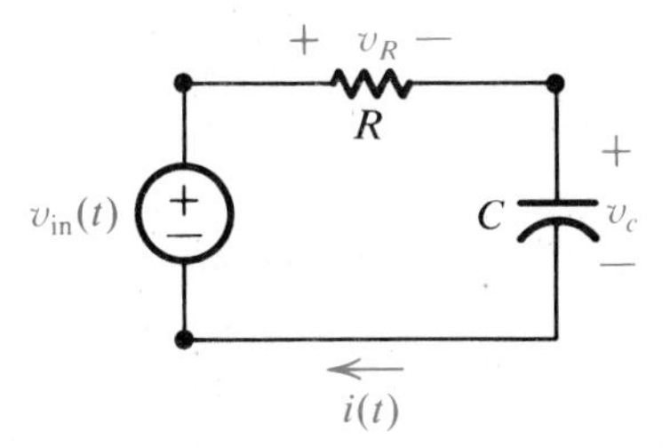

Figure 8.3 The source-driven series RC circuit.

KVL and KCL for the circuit, we want to dwell on the fact that the capacitor voltage is always *controlled* by the charging current. In the circuit we see that the charging current also passes through the series resistor. Therefore, by Ohm's Law

$$i(t) = \frac{1}{R}[v_{in}(t) - v_c(t)]$$

and the capacitor voltage caused by $i(t)$ is given by

$$v_c(t) = v_c(t_0) + \frac{1}{RC}\int_{t_0}^{t} [v_{in}(\alpha) - v_c(\alpha)]d\alpha \tag{8.4}$$

for $t \geq t_0$. The charging current will be positive if $v_{in} - v_c > 0$ and will be negative if $v_{in} - v_c < 0$. This allows us to conclude that $v_c(t)$ will increase as long as it is less than the source voltage, and vice versa. But what determines the shape of the curve for $v_c(t)$ for a given applied source?

Equation (8.4) describes $v_c(t)$ for a given $v_{in}(t)$ for $t \geq t_0$ but instead of solving this integral equation directly we differentiate both sides of (8.4) and rearrange terms to get the input/output differential equation

$$\boxed{\frac{d}{dt}v_c(t) + \frac{1}{RC}v_c(t) = \frac{1}{RC}v_{in}(t)} \tag{8.5}$$

For a given $v_{in}(t)$ the capacitor-voltage response must satisfy (8.5) for $t \geq t_0$ subject to the boundary condition that $v_c(t_0) = v_0$.

In general, each capacitor in a circuit will be controlled by its charging current, and the charging currents themselves will be determined by the overall arrangement of elements in the circuit, (it's topology). In this example the charging current is easily seen to be fixed (by Ohm's Law) for the series resistor, but this might be difficult to see in more complex circuits. Therefore, **it's usually advisable to approach circuits in a systematic way** by first writing KVL and/or KCL and then writing the individual device constraints. (As an exercise, obtain (8.5) by writing KVL and then using Ohm's Law and the derivative model for the capacitor.)

The series RC circuit's transfer function is

$$\mathbf{H}(s) = \frac{1/RC}{s + (1/RC)} \tag{8.6}$$

and its natural frequency is $s_1 = -1/RC$.

Now we are prepared to construct the particular and natural solutions of the differential equation and combine them with the boundary conditions to form the zero-input, zero-state, and initial-state responses of the circuit.

8.3 ZERO-INPUT RESPONSE (ZIR)

The zero-input response (ZIR) of the series RC circuit is the output voltage for $t \geq t_0$ when $v_c(t_0) = v_0$ and the source is turned off for $t \geq t_0$. Under these conditions the circuit diagram takes the simplified form shown in Fig. 8.4, and the RHS of (8.5) is set to zero. Many different charging histories could have led to the condition $v_c(t_0) = v_0$, but we are not concerned about the past. Knowing $v_c(t_0)$ is enough for us to determine the ZIR. The capacitor's initial stored energy, as represented by the boundary condition, causes a response for $v_c(t)$ and $i_c(t)$ to have a nonzero value for $t \geq t_0$ even though the source is turned off, because there is a path for current in the circuit to discharge the capacitor.

For simplicity we will use $t_0 = 0$ as the initial time, and after we have solved the problem with $t_0 = 0$ we will consider arbitrary initial times. Since the input signal is zero for $t \geq 0$, the particular solution is also zero i.e. $v_p(t) = 0$. Consequently, the complete solution of the I/O differential equation is the natural solution defined by the single natural frequency of the circuit. The zero-input response of the capacitor voltage is described generically by

$$v_c(t) = Ke^{-t/RC} \tag{8.7}$$

for $t \geq 0$, subject to the **ZIR boundary condition** that $v_c(0) = v_0$.

Equation (8.7) defines a parametric family of curves, each of which satisfies (8.5). However, there is one, and only one, value for K for which (8.7) also satisfies the ZIR boundary condition. This lets us specify the value of K by $v_c(0) = K = v_0$, so the ZIR is

$$v_c(t) = v_0 e^{-t/RC} \qquad t \geq 0 \tag{8.8}$$

These curves are decaying exponentials, whose natural rate of decay depends on the exponent factor RC. Letting $\tau = RC$ we can write (8.8) as

$$v_c(t) = v_0 e^{-t/\tau} \qquad t \geq 0$$

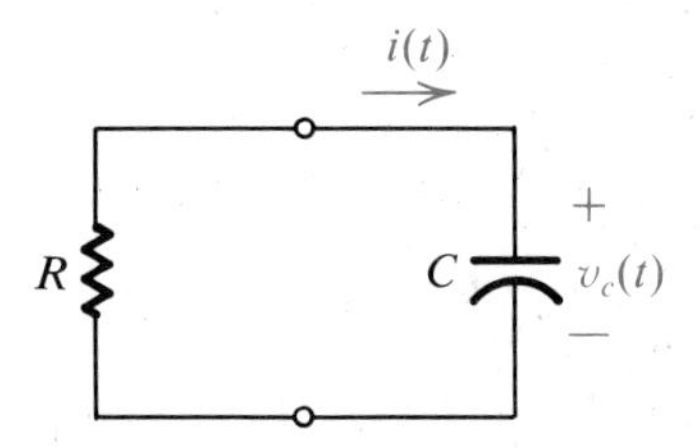

Figure 8.4 Circuit diagram for the zero-input response (ZIR).

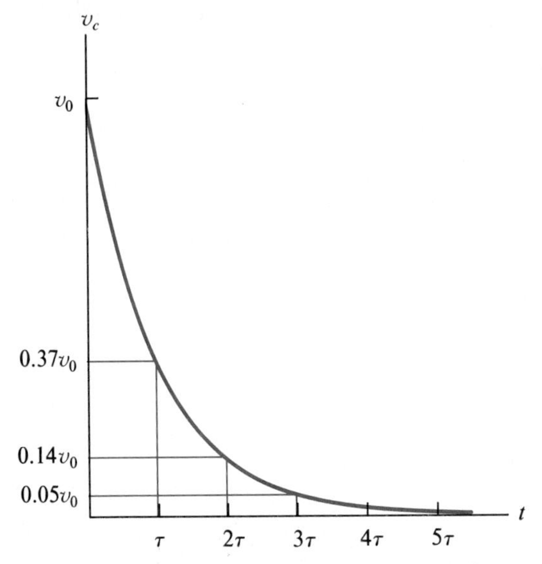

t	$e^{-t/\tau}$
τ	0.37
2τ	0.14
3τ	0.05
4τ	0.02
5τ	0.01

Figure 8.5 Series-*RC*-circuit capacitor voltage response with the time axis normalized by τ.

Figure 8.5 shows a graph of v_c with the time axis calibrated by integral multiples of τ, the time constant. As t becomes large, v_c becomes small, but never quite reaches 0.

The time constant τ of the exponential is an important reference quantity because it determines the length of time required for the exponential to decay to almost zero. In engineering work, we say that the exponential "decays to zero" in 4τ time units, even though it never quite reaches zero. If τ is relatively large, the value of t must also be large for the capacitor voltage to be small. Mathematically speaking, the exponent t/τ must be negative and have a magnitude of at least 4.

Example 8.2

If a series *RC* circuit has $R = 1000\ \Omega$ and $C = 1\ \mu F$ estimate the time required for an initial capacitor voltage to decay to zero.

Solution: The capacitor voltage never quite reaches zero, if you believe the mathematical model, but for engineering work we can consider it to be zero when only 2% of the initial voltage remains. This occurs after four time constants have elapsed, or

$$t = 4\,\tau = 4[10^3 \times 1 \times 10^{-6}] = 4 \text{ msec}$$

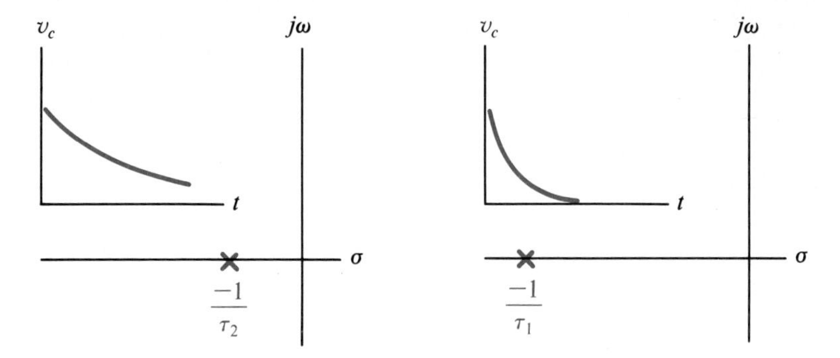

Figure 8.6 Comparison of pole-zero patterns for the series RC circuit with $\tau_2 > \tau_1$.

8.3.1 Time Constants and Pole-Zero Patterns

A circuit's time response is related to the location of its natural frequencies in the complex plane. Fig. 8.6 shows the pole-zero–pattern location of the natural frequencies corresponding to time constants τ_1 and τ_2 with $\tau_2 > \tau_1$.

The farther the location of the natural frequency from the origin, the faster the ZIR of the circuit.

If τ is large, the capacitor voltage will change slowly. In designing integrated circuits much effort goes into making C as small as possible so that signal levels can be reached quickly. If C is large, more current is needed to charge it to a given level, which requires more power. Thus, reducing C effectively reduces the power required by the circuit, a key factor when CMOS integrated circuits are used in wristwatches, calculators, memory cells, and portable instruments.

A circuit's time constant also depends directly on R. If R is large, the rate of decay will be small since

$$\frac{dv_C}{dt} = -\frac{1}{RC}v_C(t) = -\frac{1}{\tau}v_C(t)$$

Physically, a large value of R means i_R must be small (for a given v_c), so the capacitor will discharge more slowly. Note that if both R and C are doubled, the time constant is quadrupled, and that τ can be held fixed if increases in C are offset by decreases in R, or vice versa.

When the first-order I/O equation is written in the form

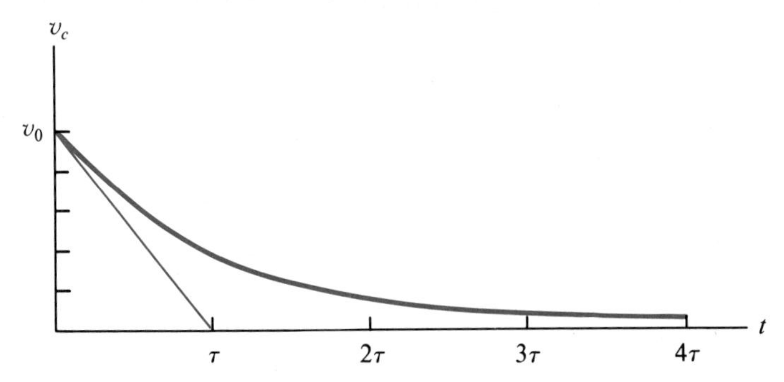

Figure 8.7 The circuit time constant determines the initial slope of the zero-input response of the capacitor voltage, $v_c(t)$.

$$\frac{dv_c}{dt} + \frac{1}{\tau}v_c = \frac{1}{\tau}v_{in}$$

it displays the natural frequency and the reciprocal of τ as the coefficient of v_c. The initial slope of the ZIR voltage is given by

$$\frac{dv_c}{dt}(0) = -\frac{1}{RC}v_c(0) = -\frac{1}{\tau}v_c(0) = -\frac{v_0}{\tau}$$

so the graph of $v_c(t)$ has a slope determined by τ, as indicated by the straight line in Fig. 8.7. This will aid us in choosing the time scale when we sketch the graph of $v_c(t)$, since the horizontal axis need only be 4 τ units in length.

The ZIR of the capacitor current for $t \geq 0$ is given by

$$i_c(t) = C\frac{dv_c}{dt} = -\frac{v_0}{R}e^{-t/RC}$$

or, since $i_c = -i_R$, we could have obtained i_c from $i_c(t) = -v_c(t)/R$. If $dv_c/dt \geq 0$ then $i_c < 0$ and vice versa; the current can also be written as

$$i_c(t) = i_0 e^{-t/RC}$$

with $i_c(0^+) = i_0 = -v_0/R$. This displays the fact that the initial current is determined (by Ohm's Law) together with the initial capacitor voltage across the resistor. It also allows us to write

$$\frac{di_c}{dt}(0^+) = -\frac{1}{RC}i_0 = -\frac{1}{\tau}i_0$$

so that the initial slope of the curve for $i(t)$ is easily found when drawing $i_c(t)$. A graph of i_c is shown in Fig. (8.8). The boundary conditions discussed for capacitors in Chapter 5 are apparent in this circuit. The initial capacitor current is limited by R; as t becomes large the capacitor looks like an open circuit with $i_c = 0$.

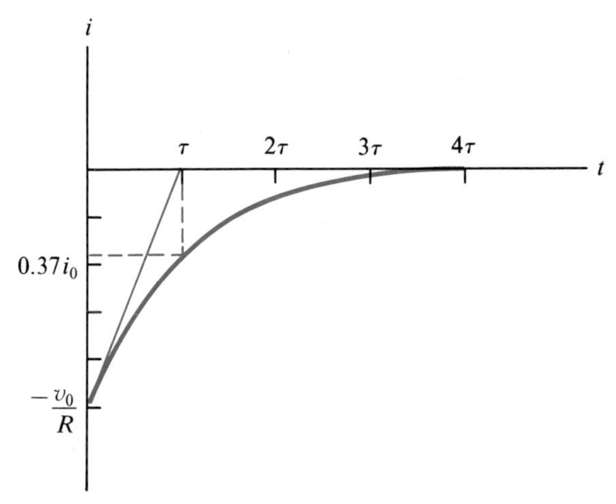

Figure 8.8 The circuit time constant determines the initial slope of the zero input response of the capacitor current, $i(t)$.

8.3.2 Steady-State Response

The steady-state component of the ZIR is obtained by letting t become arbitrarily large, or

$$v_{ss}(t) = \lim_{t\to\infty} v_c(t) = \lim_{t\to\infty} v_0 e^{-t/RC} = 0$$

In the absence of an applied source any charge on the capacitor decays to zero.

8.3.3 Transient Response

The transient-response component is the residual signal after the steady-state component is removed. So, for $t \geq 0$

$$v_{tr}(t) = v_0 e^{-t/RC}$$

The entire signal is the transient signal. The response reaches the steady state when the transient decays to zero (in approximately $4\,\tau$ seconds).

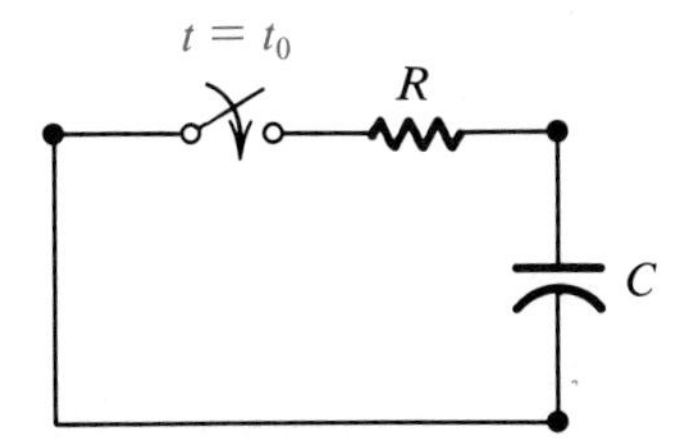

Figure 8.9 The capacitor cannot discharge until the switch is closed.

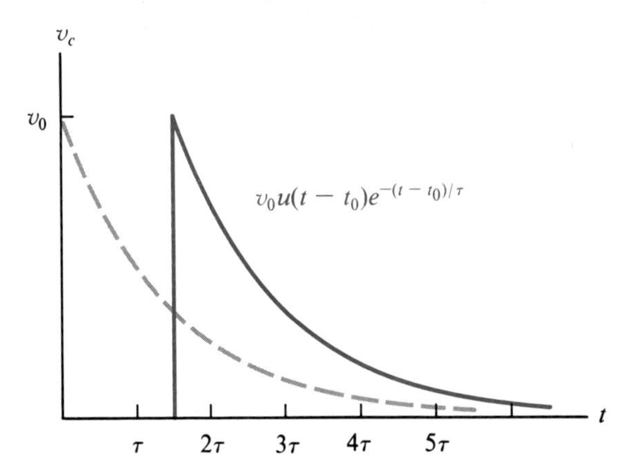

Figure 8.10 After the switch is closed in Fig. 8.9 the capacitor discharges. The curve followed by $v_c(t)$ has the same shape as it does when the switch is closed at $t = 0$, but the curve is translated on the time axis to begin at $t = t_0$.

8.3.4 ZIR Time Shift

If the switch in Fig. 8.9 is closed at $t = t_0$ instead of at $t = 0$ the zero-input response of the capacitor voltage is governed by the same homogeneous differential equation for $t > t_0$, but the boundary condition for the ZIR must be changed to $v_c(t_0) = v_0$. Applying this condition to the solution of (8.8) we have $Ke^{-t_0/RC} = v_0$ or $K = v_0e^{t_0/RC}$, and the ZIR is

$$v_c(t) = v_0e^{-(t-t_0)/RC}$$

for $t \geq t_0$. Changing the boundary condition for the ZIR merely translates the curve for the ZIR, putting its origin at $t = t_0$, as shown in Fig. 8.10.

Example 8.3

Suppose that the voltage source in Fig. 8.11 is such that $v_c(3) = 10$ V. Find $v_c(t)$ for $t \geq 3$.

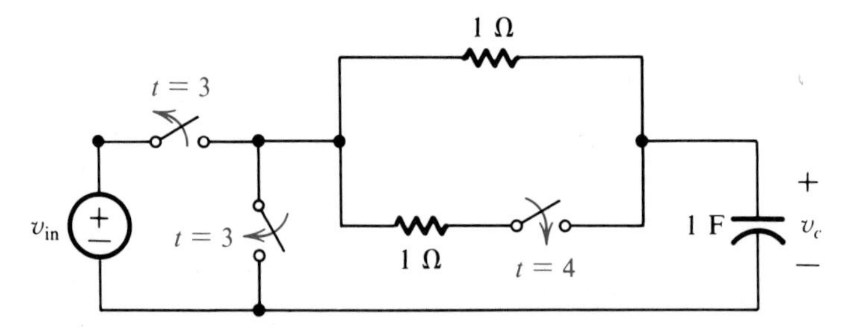

Figure 8.11 Response of the capacitor voltage in Fig. 8.11.

Solution: The ZIR for $3 \le t \le 4$ is not influenced by the change in the circuit resistance at $t = 4$. Once the resistance is changed the circuit displays a zero-input response for whatever voltage remained on the capacitor at $t = 4$.

For $3 \le t \le 4$ the ZIR will have the form

$$v_c(t) = Ke^{-(t-t_0)/RC}$$

where $t_0 = 3$, $K = 10$ and $RC = 1$, so

$$v_c(t) = 10e^{-(t-3)}$$

Next, consider $t \ge 4$. The ZIR will have the same form

$$v_c(t) = Ke^{-(t-t_0)/RC}$$

but with $K = v_c(4)$, $RC = 1/2$ and $t_0 = 4$. Therefore

$$v_c(t) = Ke^{-2(t-4)}$$

To find K we match this solution curve with the end point of the solution curve obtained for $0 \le t \le 3$:

$$v_c(4) = 10e^{-(4-3)} = 10e^{-1} = 3.68$$

so $K = 3.68$ and for $t \ge 4$

$$v_c(t) = 3.68e^{-2(t-4)}$$

The curve for $v_c(t)$ shown in Fig. 8.12 shows the effect at $t = 4$ of closing the switch and changing the circuit's time constant. The smaller τ leads to a faster

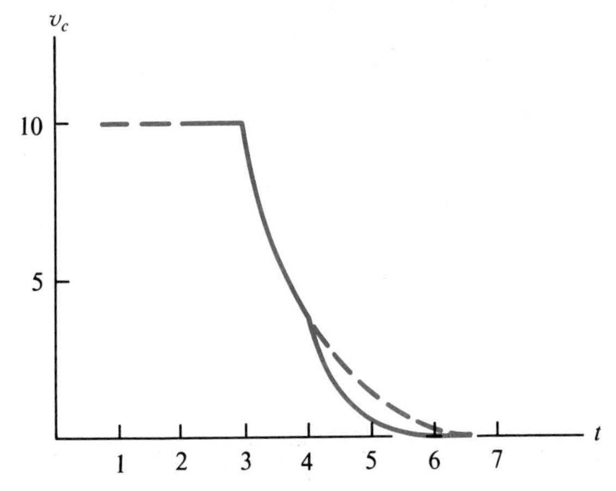

Figure 8.12 Capacitor-voltage response for the circuit in Fig. 8.11. Notice that the slope of v_c is discontinuous at $t = 3$ and $t = 4$.

decay than would have resulted without the additional resistive path.

The resistance in the RC circuit shapes the circuit response by determining how fast the capacitor voltage reaches its steady-state condition. Since the slope of the curve for v_C is proportional to the charging current, the charging current is discontinuous at $t = 4$, while v_C is continuous. Taking derivatives of v_C and scaling by $C = 1$ gives

$$i(t) = \begin{cases} -10e^{-(t-3)} & 3 \le t \le 4 \\ -7.36e^{-2(t-4)} & t \ge 4 \end{cases}$$

The discontinuity of $i(t)$ at $t = 4$ can be seen from

$$i(4^-) = -10e^{-(4-3)} = -3.68 \text{ A}$$

and

$$i(4^+) = -7.36e^{-2(4-4)} = -7.36 \text{ A}.$$

We see in Fig. 8.13 that a discontinuous but finite charging current leads to the continuous curve for voltage. In the absence of impulsive charging currents, capacitor voltage will always be continuous.

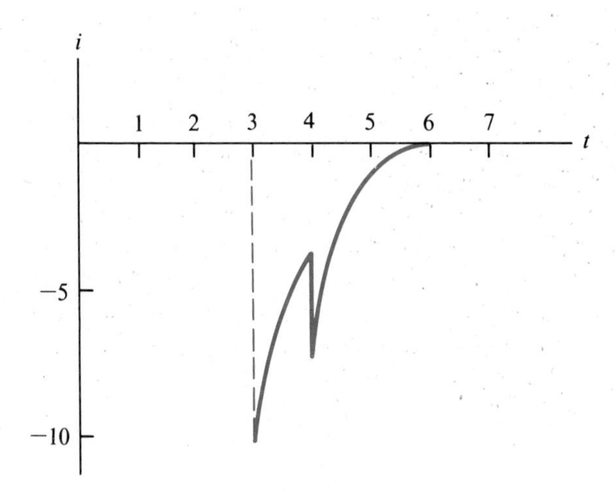

Figure 8.13 Capacitor charging current for the circuit in Fig. 8.11. Notice that $i(t)$ is discontinuous at $t = 3$, and $t = 4$.

8.4 ZERO-STATE RESPONSE (ZSR)

The previous section showed how to find the response of the series RC circuit when the source is turned off, and the capacitor is initially

charged. This section will show how to find the circuit's response when a source is applied and the capacitor has no initial charge.

A circuit is said to be in the zero state, or **at rest**, when all capacitor voltages and inductor currents are zero. A circuit that is in the zero state for $t \leq t_0$ is said to be **initially relaxed**, and the zero-state response of such a circuit to an input signal $x(t)$ that is applied for $t \geq t_0$ is the output signal for $t \geq t_0$. The ZSR of a circuit to an exponential source is found by forming the exponential particular solution (using the transfer function evaluated at the source signal's exponential frequency), adding it to the natural solution, and then applying the specific boundary conditions of the zero-state response.

8.4.1 ZSR with Step-Input Signal

The source-driven, series RC circuit in Fig. 8.3 will be in the zero state, or at rest, whenever $v_c = 0$. For convenience, suppose that $t_0 = 0$ and that $v_c(0) = 0$. If a step-input signal with height V is applied to the circuit, $v_{in}(t) = V$ for $t \geq 0$, or $v_{in}(t) = Vu(t)$. We must find $v_c(t)$ and $i(t)$ for $t \geq 0$. Then we must relate their behavior to the physical parameters of the circuit to be sure that our "intuitive feel" for the circuit's behavior is consistent with the mathematical results that we find.

The ZSR of the circuit is governed by the I/O differential-equation model of (8.5) and the ZSR boundary condition $v_c(0^-) = 0$. You may be tempted to begin immediately by writing down the solution to the mathematical model of the circuit. Instead, let's first try to gain some physical insight into its behavior. Notice that at time $t = 0^+$ we must have $v_c(0^+) = v_c(0^-) = 0$ since the capacitor voltage cannot change instantaneously in the absence of an impulsive current source. All the source voltage appears across R, and $v_R(0^+) = v_{in}(0^+) = V$. This causes an initial Ohm's-Law current given by

$$i(0^+) = \frac{1}{R} v_R(0^+) = \frac{V}{R}$$

This current is also in C, so C will begin to charge and increase in value. As v_c increases, the voltage v_R across R is reduced, since the source is fixed. So i must reduce. Also, i is initially positive, v_c will increase as long as $i > 0$, and i will remain positive as long as $v_R > 0$, $V - v_c > 0$ or $v_c < V$. This process continues until $v_c = V$. At that point, $i = 0$ and the capacitor is fully charged.

In summary, v_c begins at $v_c(0) = 0$ and charges until $v_c = V$. The current $i(t)$ begins at $i(0) = V/R$ and drops until $i = 0$. We know the initial and final values of our circuit variables, but still don't know how to *quantitatively* specify their intermediate values, or how to relate the characteristics of this process to the parameters of the circuit. We might expect that the charging process will be slower for large values of C, but does it take nanoseconds or weeks for C to charge? Similarly, if R is large, the current to charge C is relatively small, but how small? Does the circuit charge faster if V is made large? We've been able to answer some of these questions intuitively; others will require a mathematical solution. So let's begin!

The source is described by an exponential signal for $t \geq 0$ since $Vu(t) = Ve^{0t}$. Therefore, the particular solution of the I/O differential equation will be of the exponential form determined by the transfer function $v_p(t) = \mathbf{H}(s)\, Ve^{st}$ (where $\mathbf{H}(s)$ is given by (8.6) and s is to be evaluated at the source frequency, so $s = 0$, and $\mathbf{H}(0) = 1$). For $t \geq 0$ the particular solution is $v_p(t) = 1Ve^{0t} = V$. The natural solution is already known (from our work with the ZIR) to be $v_n(t) = Ke^{-t/RC}$.

The complete solution to the differential-equation model for the capacitor voltage is the sum of $v_n(t)$ and $v_p(t)$, or

$$v_c(t) = Ke^{-t/RC} + V \qquad t \geq 0 \tag{8.9}$$

(As an exercise, verify that $v_c(t)$ in (8.9) satisfies (8.5) with $v(t) = V$.) Next, we specify the ZSR by finding the value of K such that the solution in (8.9) satisfies both (8.5) and the ZSR boundary condition. We must have $v_c(0^+) = v_c(0^-) = 0$, so $K + V = 0$, and $K = -V$. Thus, the ZSR is

$$v_c(t) = -Ve^{-t/RC} + V = V[1 - e^{-t/RC}]u(t) \tag{8.10}$$

The graph of the zero-state response of v_c is shown in Fig. 8.14. The slope of v_c is given by

$$\frac{dv_c}{dt}(t) = V\left[\frac{1}{RC}e^{-t/RC}\right]u(t) + V[1 - e^{-t/RC}]\delta(t)$$

At $t = 0$

$$[1 - e^{-t/RC}] = 0$$

and for $t \neq 0$ $\delta(t) = 0$, so for all t

$$V[1 - e^{-t/RC}]\delta(t) = 0$$

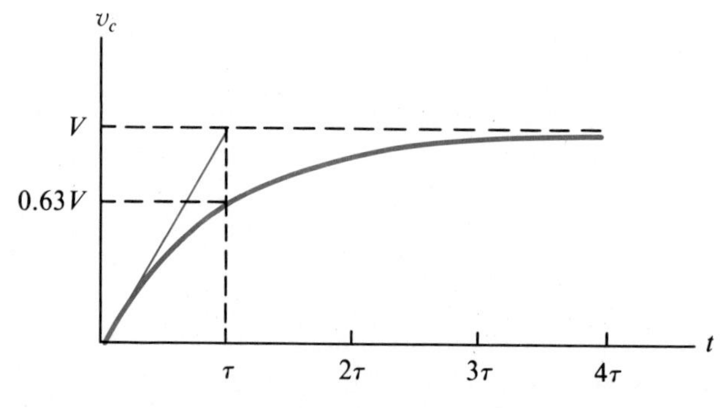

Figure 8.14 Zero-state response (ZSR) of the series RC circuit's capacitor voltage to a step input, $Vu(t)$.

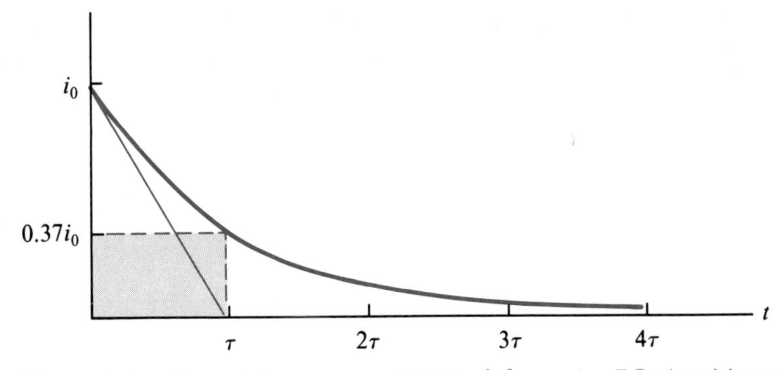

Figure 8.15 Zero-state response (ZSR) of the series RC circuit's capacitor current to a step input, Vu(t).

and

$$\frac{dv_C}{dt}(t) = \frac{V}{RC}e^{-t/RC}u(t)$$

The initial slope (i.e. when $t = 0$) is

$$\frac{dv_C}{dt}(0^+) = \frac{V}{RC} = \frac{V}{\tau}$$

This enables us to shape and calibrate the graph of v_C so that $v_c(\tau) = 0.63$ V and $v_c(4\,\tau) = V$.

The current for $t > 0$ is

$$i(t) = \frac{V - v_c(t)}{R} = \frac{V}{R}e^{-t/RC}u(t)$$

or

$$i(t) = i_0 e^{-t/\tau}u(t)$$

with $i_0 = V/R$. *The graph of* $i(t)$ is shown in Fig. 8.15 with initial slope given by

$$\frac{di}{dt}(0^+) = -\frac{i(0)}{\tau}$$

The mathematical description of the series RC circuit's behavior is consistent with our intuitive expectations. The capacitor voltage charges from 0 to V volts; the current discharges from V/R to 0 amps. Now, the initial slope of the response curve for $i(t)$ is inversely proportional to $\tau = RC$. So, if R or C is increased the value of τ increases and the initial slope decreases, corresponding to a slower process. These changes in circuit-parameter values can be related to changes in the circuit's pole-zero pattern, as depicted in Fig. 8.16. The farther the pole is located from the origin, the faster the circuit's ZSR to the step-input signal.

The steps we took to get the solution for $v_c(t)$ are summarized as

> 1. **Form an intuitive understanding of the circuit.**
> 2. **Develop the I/O model and transfer function.**
> 3. **Form the characteristic equation.**
> 4. **Find the natural frequency.**
> 5. **Find the parametric form of the natural solution.**
> 6. **Using the transfer function, find the particular solution.**
> 7. **Apply the boundary condition and obtain the complete solution.**
> 8. **Reconcile this solution with your intuitive expectations.**

With practice steps 2–7 can become somewhat routine, but steps 1 and 8 require much more effort. We purposely urge you to begin at step 1, not at step 2, because we think it is important to acquire as much insight as possible about the circuit before developing the model, as we often have a tendency to force our intuition to fit the model. Discrepancies between richly developed intuitive expectations for the circuit's behavior and its modeled behavior naturally suggest a need to discover the source of the discrepancy. Either the model has been developed and solved incorrectly, or our intuition is erroneous or incomplete. In either case, the challenge to truly understand the circuit can lead us on a rewarding path of discovery. This process is important in designing circuits, because we generally don't proceed in a random trial-and-error search for the "right" circuit. We begin intuitively and expansively from our basic understanding of fundamental circuits to develop new ones, and at each step we enrich our basic knowledge. However, if we view circuits merely as mathematical objects to be manipulated, we bring a needlessly restricted mindset to our design tasks.

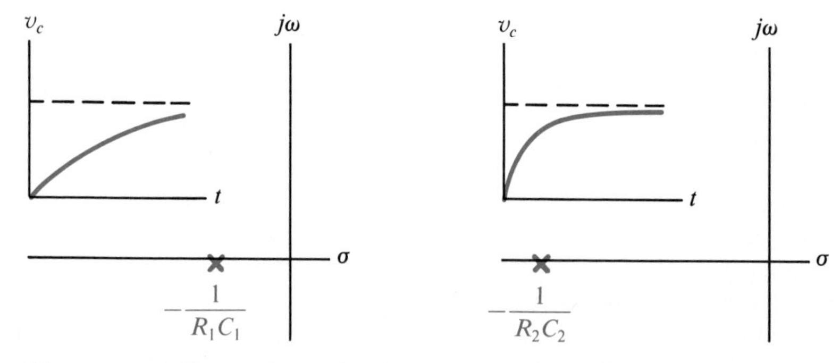

Figure 8.16 Comparison of pole-zero patterns and zero-state responses of the series RC circuit's capacitor voltage.

8.4.2 ZSR and Time Shift

The zero-state response of a circuit has the property that a time-domain translation of the input signal produces the same time-domain translation of the response. If $v_c(t)$ is the ZSR of the capacitor voltage when the input signal is $v_{in}(t)$, $v_c(t - t_0)$ will be the ZSR to $v_{in}(t - t_0)u(t - t_0)$. For example, recall that the ZSR of $v_c(t)$ to $Vu(t)$ in the series RC circuit is

$$v_c(t) = V[1 - e^{-t/RC}]u(t)$$

If the circuit is in the zero state at $t = t_0$ and if the input is $v_{in}(t) = V$ for $t \geq t_0$ the response will be

$$v_c(t) = V[1 - e^{-(t-t_0)/RC}]u(t - t_0)$$

In simplest terms, delaying the application of the source delays the zero-state response to the source. This delay is shown in Fig. 8.17. To obtain this result analytically, note that if the input is delayed until $t_0 > 0$, the circuit model is the same as before, so the ZSR will have the form

$$v_c(t) = Ke^{-t/RC} + V$$

for $t \geq t_0$. However, the boundary condition is now $v_c(t_0) = 0$, so

$$Ke^{-t_0/RC} + V = 0$$

and

$$K = -Ve^{t_0/RC}$$

so for $t \geq t_0$

$$v_c(t) = -Ve^{t_0/RC}e^{-t/RC} + V = V(1 - e^{-(t-t_0)/RC})$$

or

$$v_c(t) = V(1 - e^{-(t-t_0)/RC})u(t - t_0)$$

Likewise, delayed application of the voltage source causes a delay in the current, and

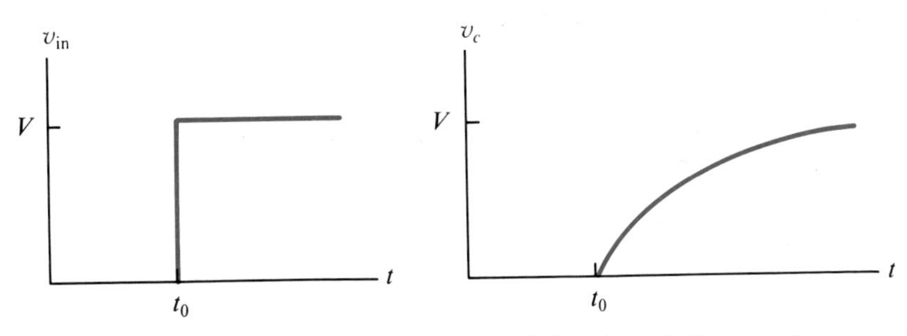

Figure 8.17 If the applied step input is delayed until time t_0, the zero-state response of the series RC circuit's capacitor voltage is delayed until time t_0.

$$i(t) = \frac{V}{R} e^{-(t-t_0)/RC} u(t - t_0)$$

As an exercise, obtain $i(t)$ by solving the current I/O model for $t \geq t_0$ with boundary condition $i(t_0) = V/R$.

8.4.3 ZSR and Steady-State Response

The steady-state value of a circuit's ZSR is obtained by observing the response after the source has been applied for a long time, e.g. $t > 4\tau$. For the series RC circuit we have

$$v_{ss} = \lim_{t\to\infty} v_c(t) = \lim_{t\to\infty} V[1 - e^{-t/RC}]u(t)$$

and because the exponential decreases as $t \to \infty$, we have $v_{ss} = V$.

The curve for $v_c(t)$ illustrates the fact that the capacitor charges to a steady-state voltage equal to the applied source voltage. Similarly

$$i_{ss} = \lim_{t\to\infty} i_c(t) = \lim_{t\to\infty} \frac{V}{R} e^{-t/RC} u(t) = 0$$

As $t \to \infty$, the current decays to a steady-state value of zero. In this steady-state condition, the capacitor behaves like an open circuit, i.e. $i_{ss} = 0$.

For exponential input signals, the steady-state response has a simple description in terms of the transfer function of the circuit. Recall that for $t \geq 0$ the ZSR is

$$v_c(t) = Ke^{-t/\tau} + \mathbf{H}(s)\, Ve^{st}$$

for the excitation value of s. Since its exponent is negative, the natural solution decays to zero regardless of the input and

$$v_{ss} = \lim_{t\to\infty} \mathbf{H}(s)\, Ve^{st}$$

For this reason, **the particular solution is sometimes called the steady-state response, but this is only the case when the natural solution decays to zero.** This happens when all of the natural frequencies are in the left half-plane. When such a circuit is driven by a step-input signal the steady-state response is determined by evaluating the transfer function at $s = 0$. For example, the step-input signal to the series RC circuit produces a steady-state capacitor voltage given by $v_{ss} = \mathbf{H}(0)V = V$ (because $\mathbf{H}(0) = 1$). In steady state, the capacitor voltage must equal the input voltage because the voltage drop across the resistor is zero.

Example 8.4

Obtain the steady-state current when a step-input voltage $vu(t)$ drives the series RC circuit.

Solution: The transfer function relating current to the applied voltage is

$$\mathbf{H}(s) = \frac{s/R}{s + (\ 1/RC)}$$

Since the natural frequency is in the left half-plane, we know that for the step-input signal the steady-state current is given by

$$i_{ss} = V\,\mathbf{H}(0) = 0$$

A DC input produces no steady-state current in the series *RC* circuit. This agrees with placing the zero at the origin in the transfer function's pole-zero pattern [Fig. 8.18]. The zero indicates that the particular solution to an exponential signal having frequency $s = 0$ must be zero. Also note that the location of the pole in the left half-plane means that the natural-solution component will also decay to zero. Thus, the steady-state value of the ZSR for current must be zero. In contrast, the pole-zero pattern of the voltage-transfer function shown in Fig. 8.18(b) does not have a zero at the origin, so the particular-solution component of $v_c(t)$ will be nonzero, as we have already found.

8.4.4 ZSR and Transient Response

The transient response of the capacitor voltage in the series *RC* circuit is given by

$$v_{tr}(t) = v_c(t) - v_{ss}(t)$$

The transient and steady-state components together form the entire signal by

$$v_c(t) = v_{tr}(t) + v_{ss}(t)$$

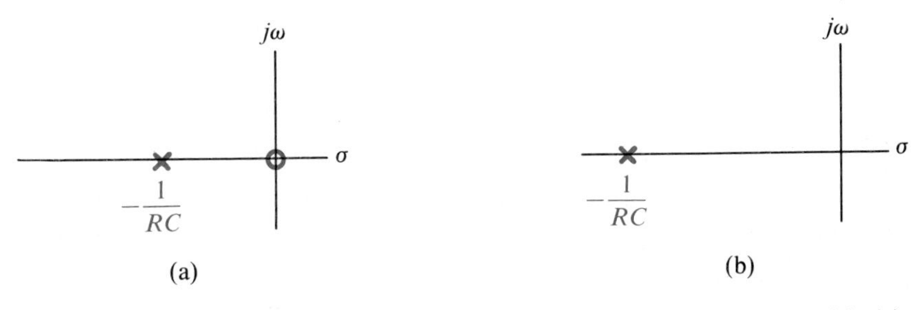

Figure 8.18 Comparison of pole-zero patterns for the series RC circuit's (a) current and (b) voltage transfer functions.

It follows from (8.10) that

$$v_{tr}(t) = V - Ve^{-t/RC} - V = -Ve^{-t/RC} \qquad t \geq 0$$

The transient and steady-state components of v_c are shown together with v_c in Fig. 8.19. The transient component of the response can be viewed as "bridging the gap" between the initial state and the steady state, given that the circuit cannot respond instantaneously.

It is *important* to note that:

i. The transient response decays approximately to zero in 4 τ secs.
ii. The duration of the transient response is inversely related to the distance of the transfer-function pole from the origin.
iii. The parameters of the circuit determine the distance of the pole from the origin by $s = -1/RC$.

8.5 THE INITIAL-STATE RESPONSE (ISR)

Given that a series RC circuit has a $v_c(t_0) = v_0$, the initial state response is $v_c(t)$ for $t \geq t_0$ when $v_{in}(t)$ is given for $t \geq t_0$. The initial state response obeys the same equation as the ZSR, except that the initial condition is nonzero. We don't care how $v_c(t_0)$ was established, even though it is implicit that charge had to be accumulated on the capacitor at some time in the past. For the moment, we'll consider $t_0 = 0$.

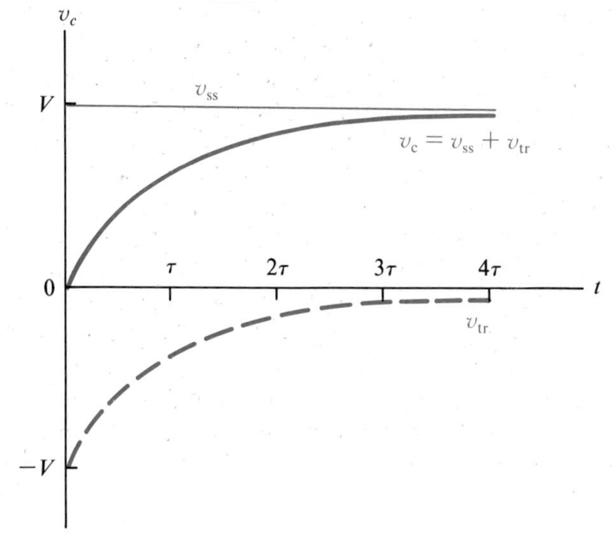

Figure 8.19 Transient and steady-state components of the capacitor voltage's zero-state response.

8.5.1 ISR and Step Input

The I/O model for the ISR and the ZSR have the same transfer function [see (8.6)]. Therefore, the complete solution to the I/O differential equation for the ISR is

$$v_c(t) = Ke^{-t/RC} + V$$

subject to the boundary condition that $v_c(t_0) = K + V = v_0$, so $K = v_0 - V$ and

$$v_c(t) = (v_0 - V)e^{-t/RC} + V$$

The ZSR is just a special case of the ISR in which $v_0 = 0$. The graph of $v_c(t)$ is shown in Fig. 8.20 for $(v_0 - V) < 0$ and for $(v_0 - V) > 0$. If $v_0 = V$ we have $v_c(t) = V$. The initial slope is

$$\frac{dv_c}{dt}(0) = \frac{-1}{RC}(v_0 - V) = -\frac{1}{\tau}(v_0 - V)$$

The initial slope is depicted by the straight line in Fig. 8.20. The ISR of the current is

$$i(t) = C\frac{dv_c}{dt} = -\frac{(v_0 - V)}{R}e^{-t/RC} = i_0 e^{-t/RC}$$

As an exercise, obtain this result using the I/O model for current. Notice that

$$i_0 = -\frac{v_0 - V}{R}$$

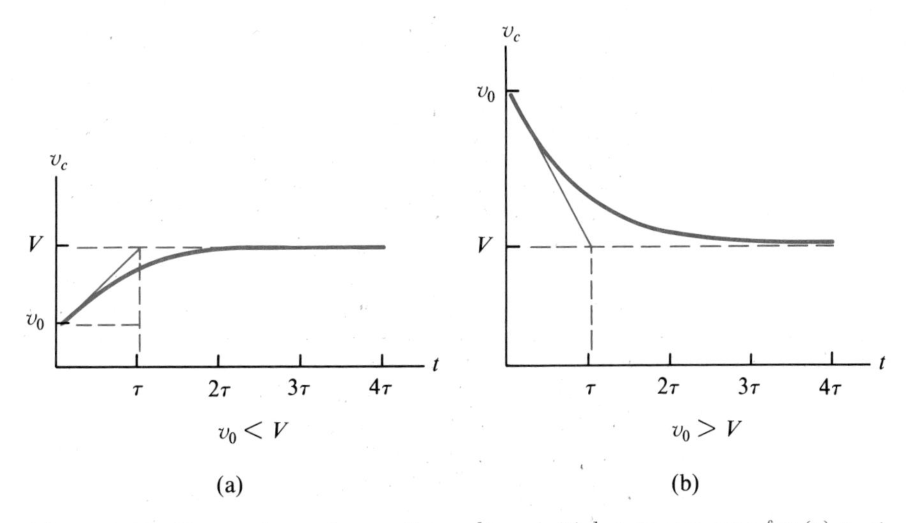

Figure 8.20 Comparison of capacitor-voltage initial-state response for (a) $v_0 < V$, and (b) $v_0 > V$.

The initial current will be negative if $(v_0 - V) > 0$, and vice versa.

A comparison of the solutions for ZSR and the ISR of the capacitor voltage leads to the following characterization of the general first-order response

$$y(t) = [\text{Initial Value} - \text{Final Value}]e^{-t/\tau} + \text{Final Value} \qquad t \geq 0$$

This form is valid for any first order I/O model. To use it, we need only know the initial value, the final value, and the time constant. If the initial value is given at an arbitrary t_0, the solution is

$$y(t) = [\text{Initial Value} - \text{Final Value}]e^{-(t-t_0)/\tau} + \text{Final Value} \qquad t \geq t_0$$

Example 8.5

Show that the general first-order response describes the ZIR of the series RC circuit.

Solution: The ZIR has an initial value of v_0, a final value of 0, and a time constant $\tau = RC$. Therefore

$$v_c(t) = [v_0 - 0]e^{-t/RC} + 0 = v_0 e^{-t/RC} \qquad t \geq 0$$

The general first-order response is helpful because we can frequently determine the initial and final values of the response by inspection.

Skill Exercise 8.1

Assuming that the right hand switch in Fig. SE 8.1 has been closed for a long time, and that both switches are switched at $t = 10$ find $v(t)$ for $t \geq 10$.

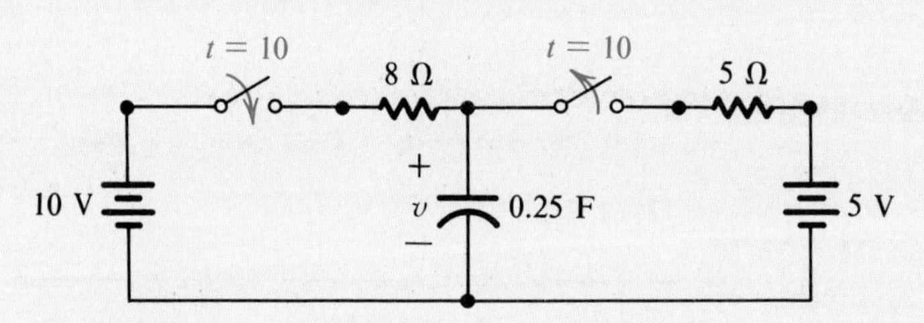

Figure SE8.1

Answer: $v(t) = [5 - 10]e^{-(t-10)/2} + 10 \qquad t \geq 10$

8.5.2 ISR and Steady-State Response

The steady-state value of the ISR to $Vu(t)$ is obtained from

$$v_{ss} = \lim_{t \to \infty} [(v_0 - V)e^{-t/RC} + V] = V$$

The steady-state voltage V is the same for the ISR as for the ZSR, but their curves for $v_c(t)$ are different because the ISR begins at v_0 while the ZSR begins at 0. Both reach the steady state in the same amount of time, approximately 4τ, *regardless of the magnitude of* v_0.

8.5.3 ISR and Transient Response

The ISR transient is obtained by removing the steady-state component of $v_c(t)$

$$v_{tr}(t) = v_c(t) - v_{ss} = (v_0 - V)e^{-t/RC} + V - V = (v_0 - V)e^{-t/RC} \qquad t \geq 0$$

The steady-state and transient components of $v_c(t)$ are shown in Fig. 8.21 with $v_c(t)$. The curve for $v_c(t)$ is obtained by adding the curve for v_{tr} to the curve for v_{ss}.

8.5.4 ISR and Superposition/Decomposition

The initial-state response simultaneously exhibits the effect of the initial stored energy and the source input signal. The initial-state response can be decomposed as follows:

$$v_c(t) = (v_0 - V)e^{-t/RC} + V = v_0 e^{-t/RC} + V(1 - e^{-t/RC}) \qquad t \geq 0$$

The initial-state response of the RC circuit is the sum of the zero-input response (ZIR) (see 8.8) and the zero-state response (ZSR). The superpo-

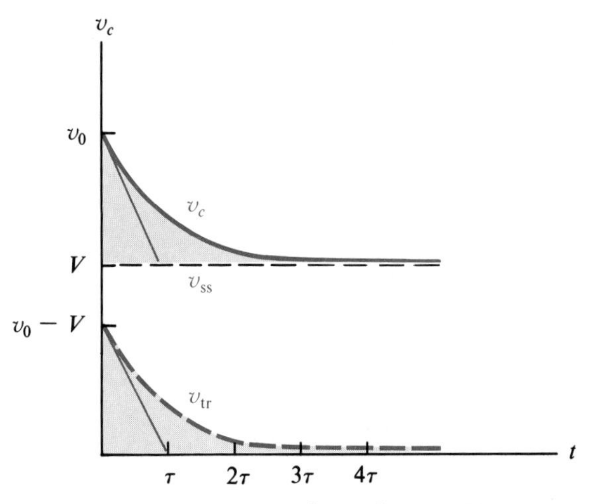

Figure 8.21 Transient and steady-state ISR components of v_c.

sition of the ZIR and the ZSR to form the ISR is shown in Fig. 8.22. The curve for the ISR shows the effect of the voltage source charging the capacitor to a steady-state value of V (the ZSR) superimposed on the effect of the capacitor discharging to zero from its initial voltage, the ZIR. The source looks like a short circuit to the ZIR, and the capacitor looks like it is initially uncharged for the ZSR component of the ISR.

It is important to note that **the transient for the ISR is not simply the ZIR.** Instead, it consists of the transient for the ZIR added to the transient for the ZSR.

8.6 SERIES *RC* CIRCUIT AND EXPONENTIAL INPUT*

Our understanding of circuits will be deepened if we examine how the RC circuit responds to input signals other than the familiar unit-step signal (an exponential with $s = 0$). For these other signals we need only find the zero-state–response component of the ISR because the zero-input–response component does not depend on the input signal. Because the exponential signal plays such an important role in our work, we will now examine the more general case in which the fixed exponent s is not zero.

Suppose that the series RC circuit has an exponential source described by $v_{in}(t) = Ve^{\hat{\sigma}t}$ $(t \geq 0)$ where $\hat{\sigma}$ must be real-valued and negative. If $\hat{\sigma}$ is not the same as the value of the circuit's natural frequency the transfer function can be used to find the particular-solution component of the circuit's I/O differential equation. Thus, the ZSR to the exponential is

$$v_c(t) = Ke^{-t/RC} + V\mathbf{H}(\hat{\sigma})e^{\hat{\sigma}t}$$

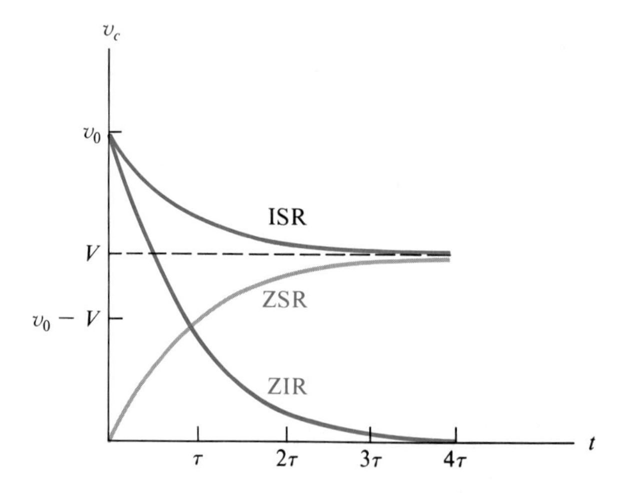

Figure 8.22 Zero-input–and zero-state–response components of the series RC circuit's initial-state response.

The value for K is obtained from the boundary condition $v_c(0^+) = v_c(0^-) = 0$, so $K + V\mathbf{H}(\hat{\sigma}) = 0$, and $K = -V\mathbf{H}(\hat{\sigma})$. The ZSR is

This is the same slope we found for the ZSR to a step input. Physically, the circuit responds to an exponential input as though it were a step for values of t that are small compared to the value of $1/|\hat{\sigma}|$. Once $t \geq 4/|\hat{\sigma}|$, the source will be effectively zero, but it initially looks like a step. So we might expect the circuit to exhibit a steplike response initially and then follow a decaying exponential curve. If the circuit time constant is large compared to the decay time of the source ($\tau \gg 4/|\hat{\sigma}|$), the response will follow the natural solution curve. If the circuit time constant is small compared to the decay time of the source ($\tau \ll 4/|\hat{\sigma}|$), the response will follow the source exponential. The exponential input "looks" like a step input over the 4τ units of time required for the natural solution to decay because it is changing very slowly compared to the natural solution.

The response of $v_c(t)$ is shown in Fig. 8.23(a) for a case in which $\hat{\sigma} > -1/RC$. The input "frequency," denoted by the dark circle, lies between the pole and the origin, and the response is characterized by the fact that the natural solution decays faster than the source. Once it has decayed, the solution is approximately the particular solution due to the exponential source. If we let $\hat{\sigma} \to 0$, the source becomes a step, and the response will look like the ZSR found earlier (see Fig. 8.14).

The response when $\hat{\sigma} < -1/RC$ is shown in Fig. 8.23(b). Here the source is short-lived compared to the duration of the natural solution. As a result, less charge accumulates on the capacitor before the source is effectively removed, so the voltage does not build to the level shown in Fig. 8.23(a) nor does it last as long.

8.7 *RL* CIRCUIT RESPONSE

The dynamic relationship between the input voltage and the inductor current in the series RL circuit shown in Fig. 8.24 follows directly from KVL

$$v_L + v_R = v_{\text{in}}(t)$$

and

$$L\frac{di}{dt} + iR = v_{\text{in}}(t)$$

and the transfer function of the circuit is

$$\mathbf{H}(s) = \frac{1}{Ls + R} = \frac{1/L}{s + (R/L)}$$

The circuit has a single natural frequency at $s = -R/L$. Since **the form of**

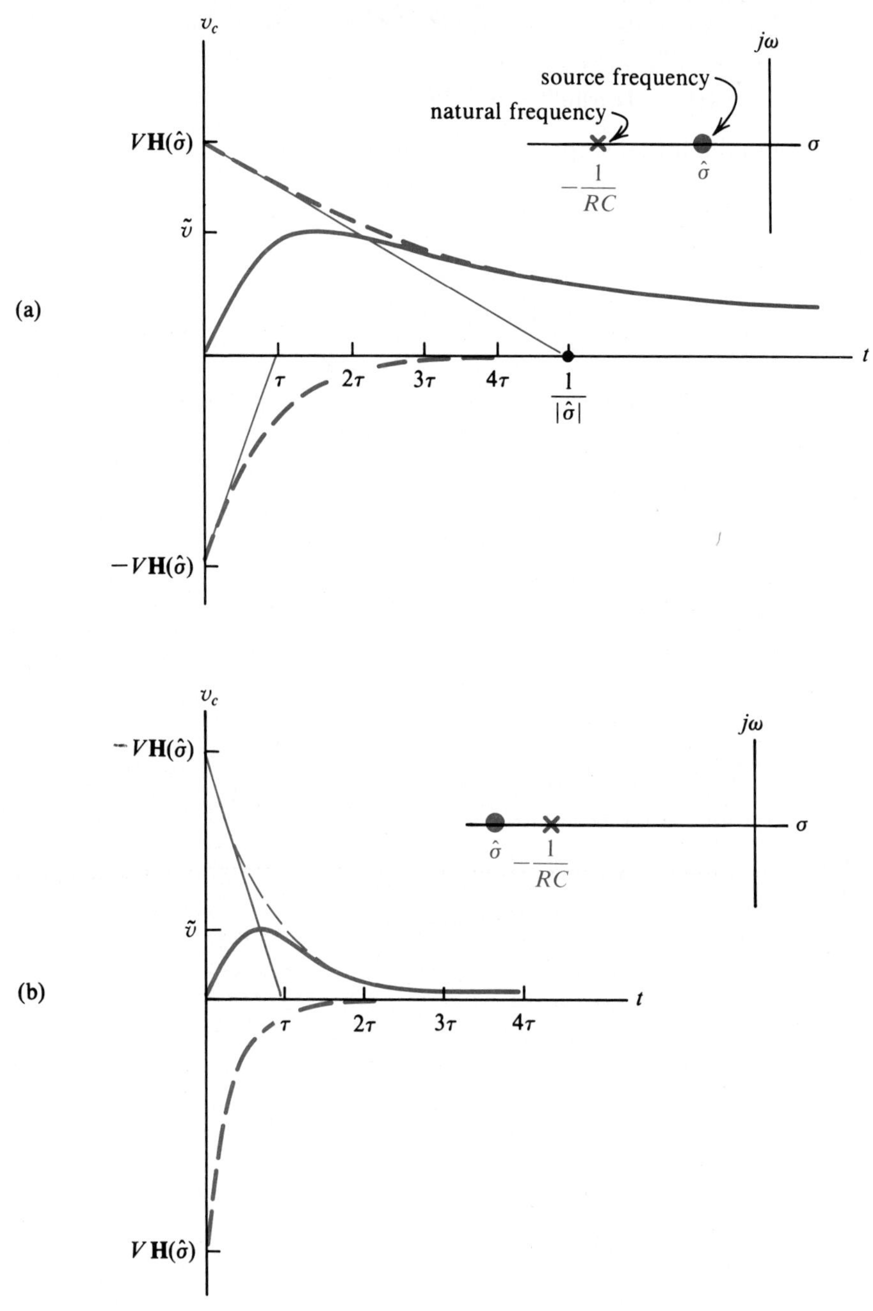

Figure 8.23 Zero-state response to an exponential input, (a) $v_{in}(t) = Ve^{\hat{\sigma}t}$, with (a) $\hat{\sigma}$ -1/RC, and (b) $v_{in}(t) = Ve^{\hat{\sigma}t}$, with $\hat{\sigma} < -1/RC$, and $H(\hat{\sigma}) < 0$.

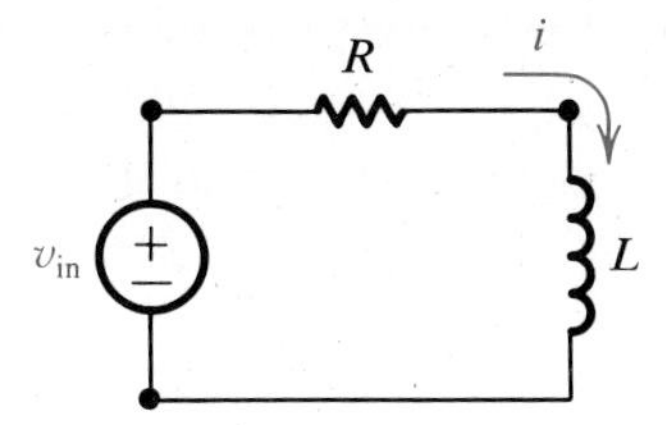

Figure 8.24 A series *RL* circuit.

H(s) for the series *RL* circuit's current response is the same as the form of the series *RC* circuit's voltage response, there is no need to repeat an extensive discussion of the *RL* circuit response. Instead, we will briefly discuss the ZIR and ZSR of the circuit.

8.7.1 *RL*–Circuit ZIR

The ZIR of the *RL* circuit is

$$i(t) = Ke^{-(R/L)t}$$

with K determined by the circuit's boundary condition. If $i(0^-) = i_0$ is given, the continuity property of inductor current requires that $i(0^+) = i(0^-) = i_0$, so $i(0^+) = K = i_0$, and

$$i(t) = i_0 e^{-(R/L)t}$$

In the absence of an applied voltage source the current in the *RL* circuit will decay to zero with a time constant of $\tau = L/R$. The ZIR is shown in Fig. 8.25.

Physically, the current in the resistor causes a voltage to appear across the inductor. This voltage has a polarity that reduces the magnetic flux (Lenz's Law), and thereby reduces the current in the inductor. As the magnetic field collapses to zero the current drops to zero.

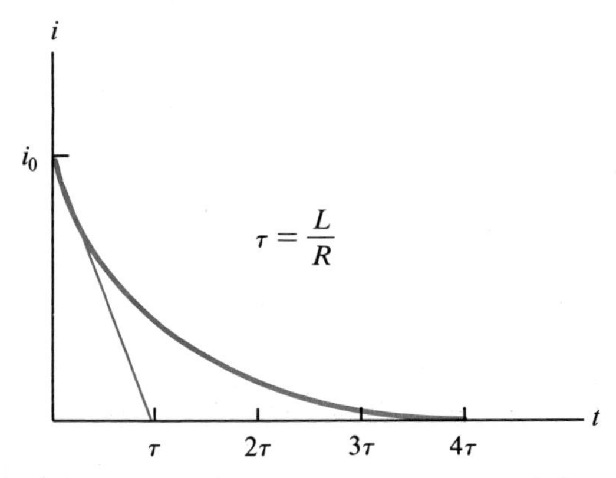

Figure 8.25 The zero-input response of the inductor current in the series RL circuit.

Skill Exercise 8.2

The switch in Fig. SE 8.2 has been closed long enough to establish steady state conditions in the circuit, and is opened at $t = 0$ sec. Find $i(t)$ for $t \geq 0$.

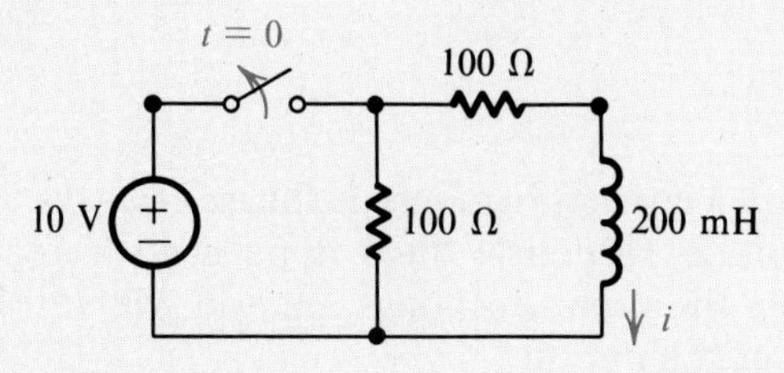

Figure SE8.2

Answer: $i(t) = 0.1e^{-1000t}$

8.7.2 *RL*–Circuit ZSR to Step Input

The ZSR to a step input with $v_{in}(t) = Vu(t)$ is $i(t) = Ke^{-(R/L)t} + \mathbf{H}(0)V = Ke^{-(R/L)t} + (V/R)$. The boundary condition for the ZSR is $i(0^+) = i(0^-) = 0$, so $K + V/R = 0$, $K = -V/R$, and

$$i(t) = \frac{V}{R}[1 - e^{-(R/L)t}]$$

The initial applied voltage is entirely across L, with $i(0^+) = 0$, and

$$v_L(0^+) = V = L\frac{di}{dt}(0^+)$$

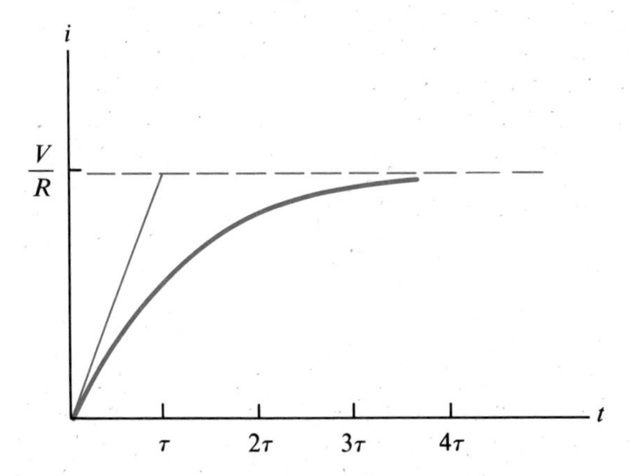

Figure 8.26 The zero-state response of the current in the series *RL* circuit.

so $di(0^+)/dt = V/L$. The initial voltage across the inductor will cause a magnetic field to develop, accompanied by current in the inductor. As the inductor develops current, the voltage across it is reduced by the amount of voltage that occurs across the resistor under Ohm's law. Eventually, the entire source voltage appears across the resistor, and inductor current is held to a constant value. The ZSR for $i(t)$ is shown in Fig. 8.26.

Example 8.6

If a series RL circuit has $R = 1000\ \Omega$, and $L = 1$ mH, find $i(t)$ when $v_{in}(t) = V$ for $t < 0$ and $v_{in}(t) = 2$ V for $t > 0$.

Solution: For $t \leq 0$ the response of the circuit must consist solely of the particular solution due to a step input, since any nonzero natural-solution component would have to vanish in a time that is negligible compared to the negative time axis. Alternately, let

$$i(t) = Ke^{-10^6 t} + \mathbf{H}(0)V = Ke^{-10^6 t} + \frac{1}{1000}V$$

Applying the boundary condition

$$i(0^-) = \frac{V}{R} = \frac{V}{1000} = K + \frac{V}{1000}$$

implies that $K = 0$. Therefore, $i(t) = V/1000$ for $t \leq 0$. For $t \geq 0$

$$i(t) = Ke^{-10^6 t} + 2V\mathbf{H}(0) = Ke^{-10^6 t} + \frac{2V}{1000}$$

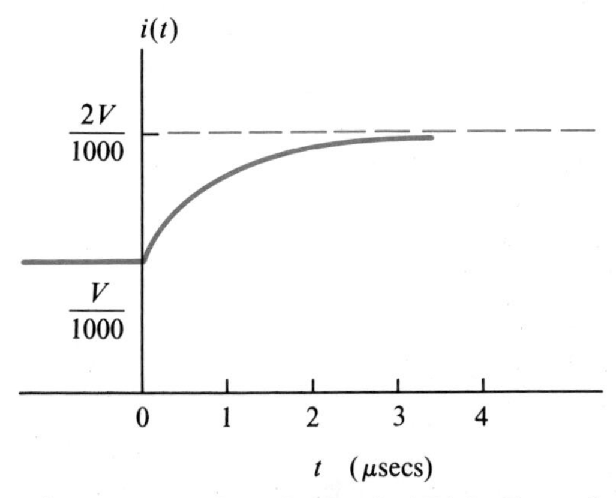

Figure 8.27 The solution for $i(t)$ in Example 8.6.

and

$$i(0^+) = K + \frac{2V}{1000} = i(0^-) = \frac{V}{1000}$$

so $K = -V/1000$ and $i(t) = (V/1000)(2 - e^{-10^6 t})$ for $t \geq 0$. The solution for $i(t)$ is shown in Fig. 8.27

SUMMARY

This chapter showed how to apply the mathematics of differential equations to the problem of finding the voltage and current responses of circuits containing capacitors and inductors. First-order *RC* and *RL* circuits were examined in detail to obtain the zero-input response (ZIR), zero-state response (ZSR) and initial-state response (ISR) of the circuit, and to expose the relationship of a circuit's physical parameters, pole-zero pattern, and time-domain response to an input signal. Finally, transient and steady-state components of each response were also considered.

Problems - Chapter 8

8.1 For the circuit in Fig. P8.1:

a. Find the transfer function relating v_o and v_{in}.

b. Find the ZIR of v_o if $v_c(0^-) = 5$ V.

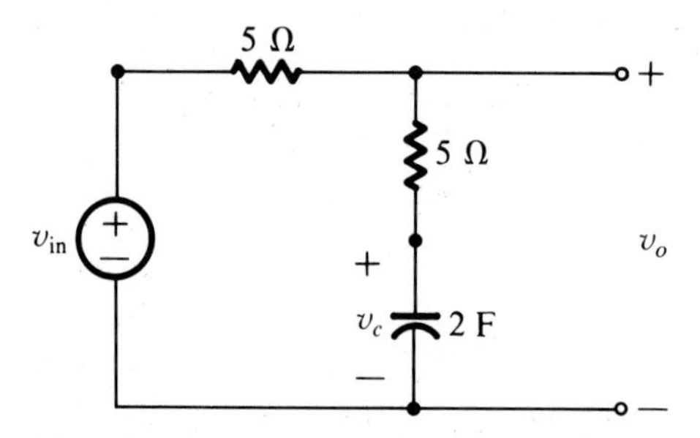

Figure P8.1

8.2 If $i(0^-) = 10$ A and $v_{in}(t) = 0$, find $v_o(t)$ for $t \geq 0$, with $R_1 = 2\ \Omega$, $R_2 = 8\ \Omega$ and $L = 2H$.

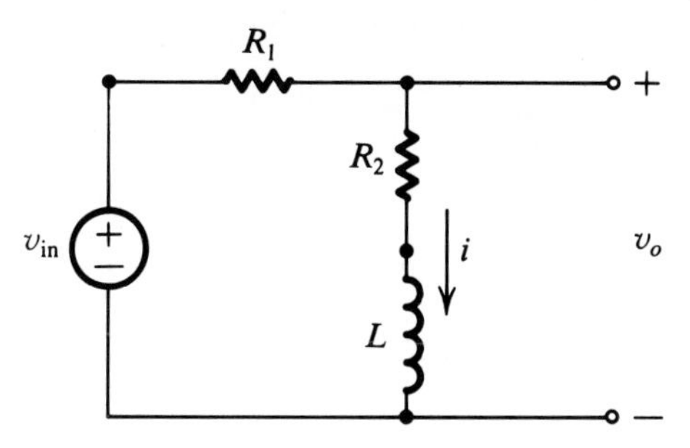

Figure P8.2

8.3 If switch S1 opens at $t = 0$ after being closed for a long time, and switch S2 closes, find the zero-input response of v_o for $t \geq 0$.

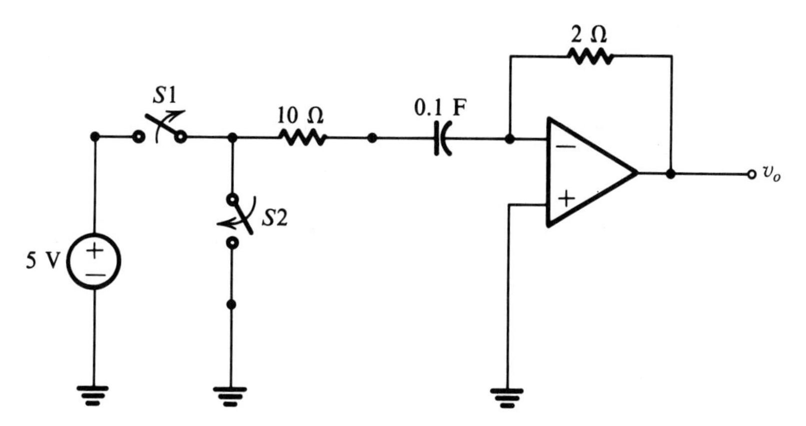

Figure P8.3

8.4 If $i_L(0^-) = 2$ A, find $v_o(t)$ for $t \geq 0$ when $v_{in}(t) = 0$.

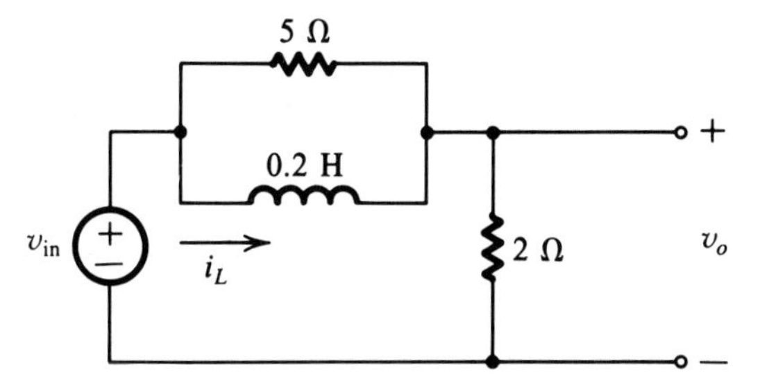

Figure P8.4

8.5 If $i(0^-) = 5$ A and $v_{in}(t) = 25\,u(t)$, find $v_o(t)$ for $t \geq 0$.

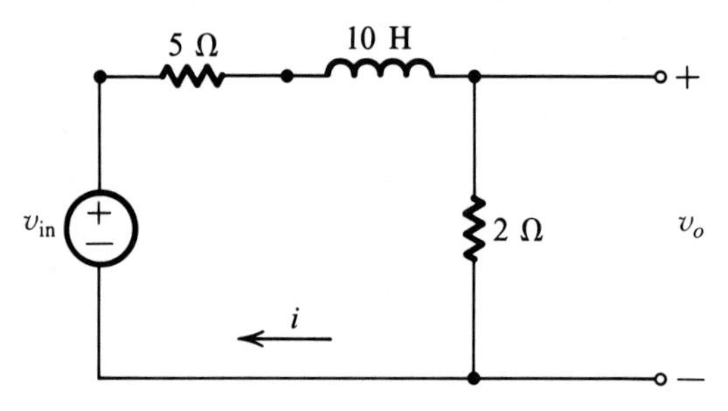

Figure P8.5

8.6 If $R_1 = 1\ \Omega$, $R_2 = 5\ \Omega$ and $C = 0.1$ F, what is the time constant for the ZIR of v_o?

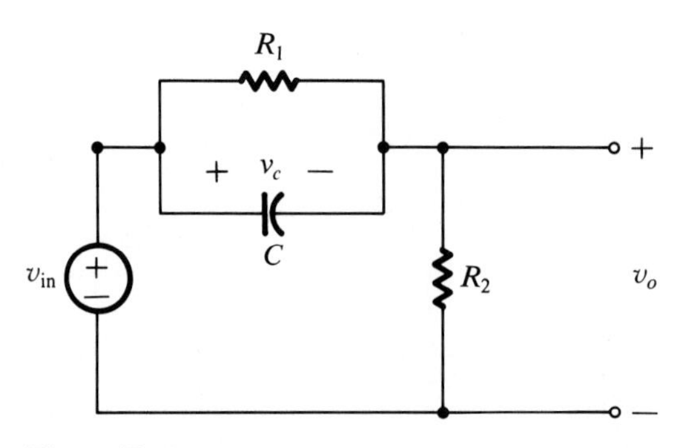

Figure P8.6

8.7 If a series RC circuit has

$$\frac{V_o}{V_{in}} = \mathbf{H}(s) = \frac{5s}{2s + 5}$$

find the ZIR of the circuit if $v_o(0^-) = v_o(0^+) = 50$ V.

8.8 If the circuit in Fig. P8.8 is initially relaxed, with $R_1 = 20\ \Omega$, $R_2 = 40\ \Omega$ and $C = 0.3$ F, find $i(0^+)$, $i(0^-)$, $v_c(0^+)$, $v_c(0^-)$, $v_o(0^-)$ and $v_o(0^+)$ in terms of v_{in}.

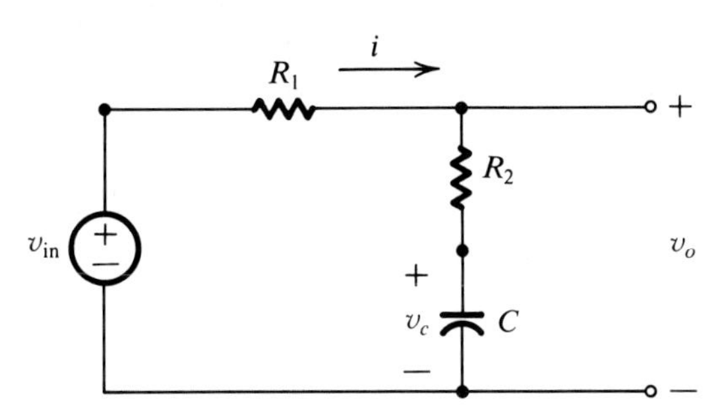

Figure P8.8

8.9 If $v_c(0^-) = 0$ in Fig. P8.9

a. Find $v_c(t)$ and $v_o(t)$ for $t \geq 0$ when $v_{in}(t) = 5u(t)$.

b. Draw a composite graph of v_c and v_o for $t \geq 0$.

c. Repeat a and b if $v_c(0^-) = 8$ V.

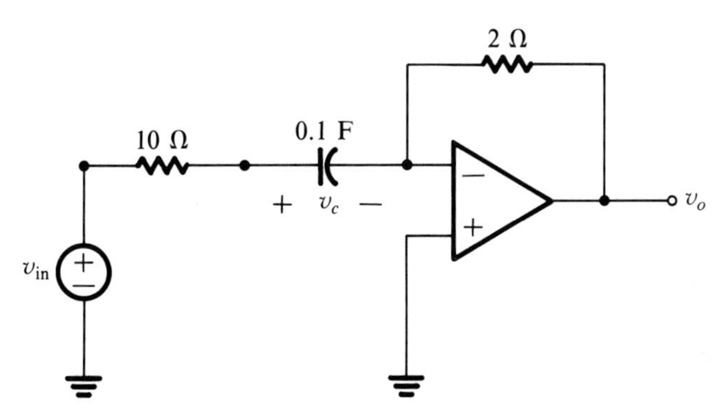

Figure P8.9

8.10 If the circuit in Fig. P8.8 has $R_1 = 5\ \Omega$, $R_2 = 5\ \Omega$ and $C = 2$ F,

a. Find the ZSR of v_o to $v_{in}(t) = 5u(t)$ V.

b. Find the ZIR of $v_o(t)$ if $v_c(0^-) = -5$ V.

c. Find the ISR of v_o to $v_{in}(t) = 5u(t)$ V if $v_c(0^-) = -5$ V, and show that ISR = ZIR + ZSR.

d. Draw a composite graph of the ZIR, ZSR, and ISR of the circuit when $v_{in}(t) = 5u(t)$ V and $v_c(0^-) = -5$ V.

8.11 If the circuit in Fig. P8.2 has $R_1 = 1\ \Omega$, $R_2 = 2\ \Omega$ and $L = 3H$:

a. Find the ZSR of v_o if $v_{in}(t) = u(t)$.

b. Repeat a for $v_{in}(t) = e^{-3t}u(t)$.

8.12 Find the ZSR of i to $v_{in}(t) = 10u(t)$.

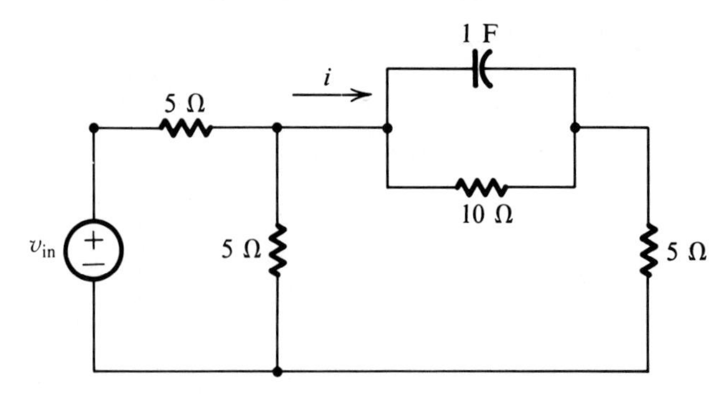

Figure P8.12

8.13 The transfer function relating v_o and v_{in} for the circuit in Fig. P8.13 has

$$\mathbf{H}(s) = \frac{R_2}{R_1 + R_2} \frac{1 + R_1C_1s}{sC_1R_1R_2/(R_1 + R_2) + 1}$$

a. Explain the physical significance of $\mathbf{H}(s)$ when $s = 0$.

b. If the capacitor is initially uncharged, find $v_c(t)$ and $v_o(t)$ for $t \geq 0$ when $v_{in}(t) = 10\, e^{-t/(R_1C_1)}u(t)$

c. Give a physical explanation for the fact that the forced response to the exponential input signal in b is zero.

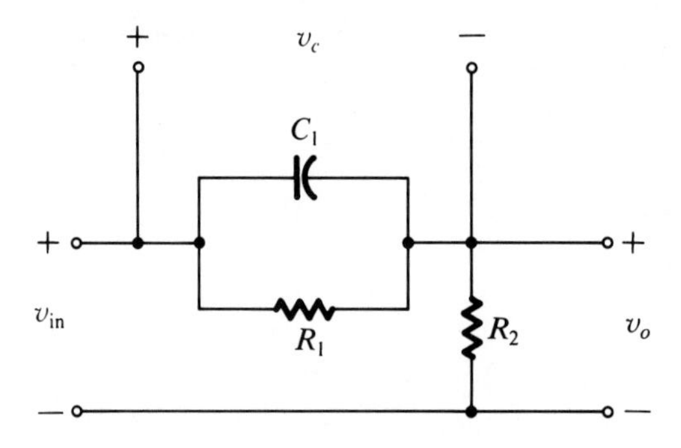

Figure P8.13

8.14 If both capacitors are initially uncharged in the op amp circuit in Fig. P8.14, find v_o when $v_{in}(t) = 5u(t)$.

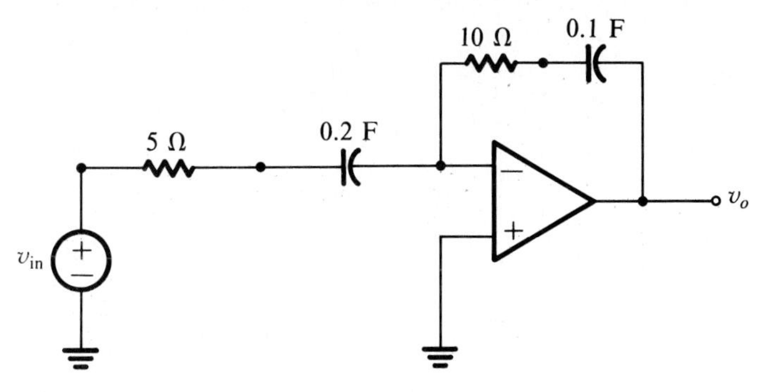

Figure P8.14

8.15 Repeat Problem 8.14 for the circuit in Fig. P8.15 with $i_{in}(t) = 2u(t)$.

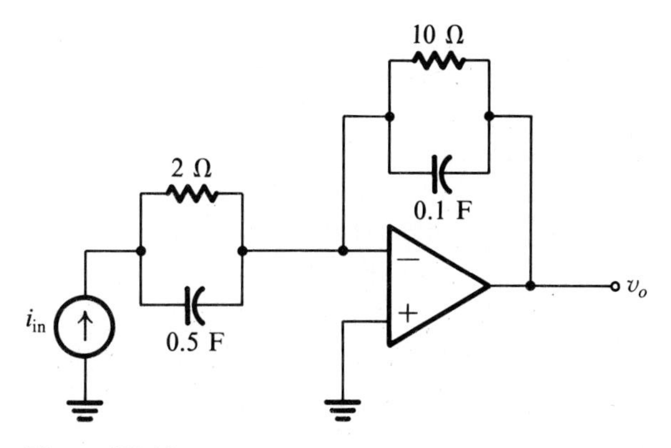

Figure P8.15

8.16 If both of the capacitors in the circuit in Fig. P8.16 are initially uncharged and $v_{in}(t) = 5u(t)$:

a. Find and draw $v_{o1}(t)$ and $v_{o2}(t)$ on the same coordinate system.

b. Find $\mathbf{H}_1(s)$ relating V_{in} and V_{o1}.

c. Find $\mathbf{H}_2(s)$ relating V_{o2} and V_{o1}.

d. Find $\mathbf{H}(s)$ relating V_{o2} and V_{in}. (Does $\mathbf{H}(s) = \mathbf{H}_1(s)\mathbf{H}_2(s)$?)

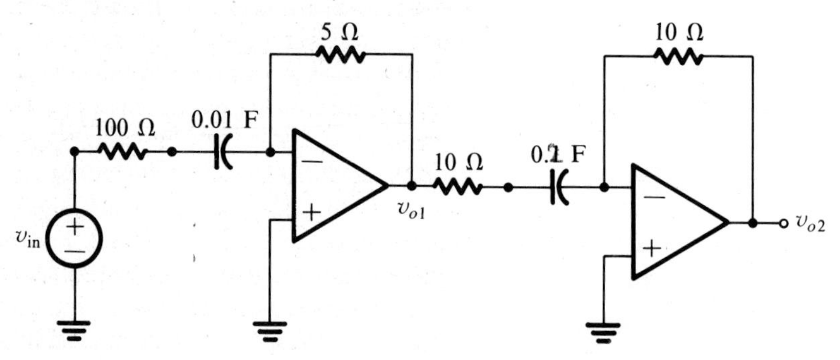

Figure P8.16

8.17 The circuit in Fig. P8.17 has unknown component values. However, it is known that when $v_c(0^-) = 20$ V the ZIR of the circuit is given by $v_o(t) = 2e^{-(1/200)t}u(t)$ for $t \geq 0$. Also, if the capacitor is initially uncharged and a source voltage given by $v_{in}(t) = 2e^{-(1/200)t}u(t)$ is applied, the response is $v_o(t) = -160e^{-(1/100)t} + 160e^{-(1/200)t}$ for $t \geq 0$. Find the response of the circuit for $t \geq 0$ when $v_c(0^-) = 100$ V and the source is given by $v_{in}(t) = 20e^{-(1/200)t}u(t)$.

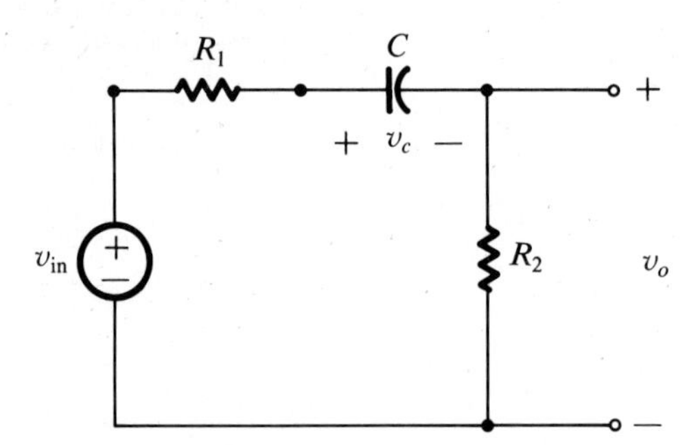

Figure P8.17

8.18 When the circuit in Fig. P8.17 has $R_1 = 10\ \Omega$, $R_2 = 20\ \Omega$, $C = 2$ F, and $v_c(0^-) = v_o$, its initial state response to the input signal $v_{in}(t) = 10e^{-(1/20)t}u(t)$ is $v_R(t) = [v_o - 10]e^{-(1/60)t}u(t) + 10e^{-(1/20)t}u(t)$.

a. Find an expression for the circuit's ZIR.

b. Find an expression for the circuit's ZSR to a unit step input signal.

c. Draw $v_R(t)$, the ZIR and the ZSR on the same coordinate system.

8.19 For a series RC circuit, draw a composite graph of the ZIR, the ZSR, and the ISR of v_R if $v_c(0^-) = 20$ V and $v_{in}(t) = 10u(t)$.

8.20 Explain why the transfer function relating v_o and v_{in} for the circuit in Fig. P8.17 should have a zero at $s = 0$.

8.21 If $v_{c1}(0^-) = 5$ V and $v_{c2}(0^-) = -2$ V, find v_o for $t \geq 0$ when the applied source has $v_{in}(t) = 5u(t)$.

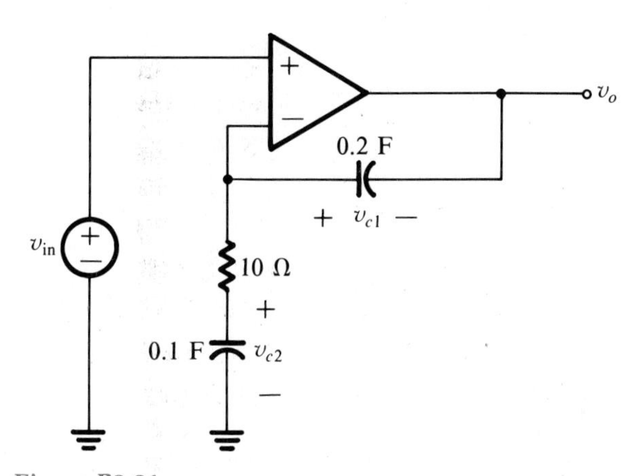

Figure P8.21

8.22 If $v_{c1}(0^-) = 10$ V and $v_{c2}(0^-) = 5$ V, find v for $t \geq 0$ when $v_{in}(t) = 10u(t)$.

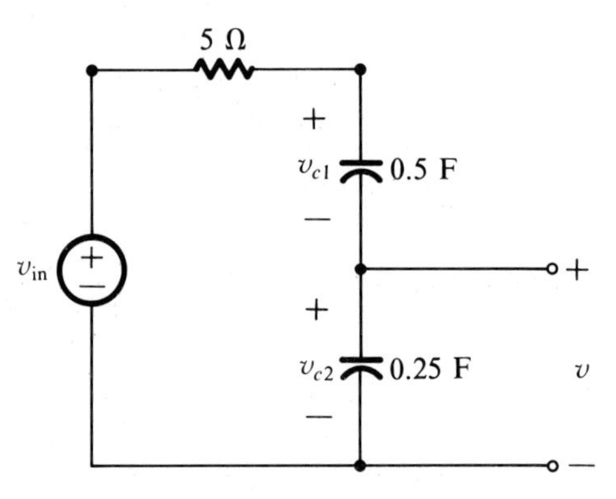

Figure P8.22

8.23 Find the ZSR of $v_o(t)$ when $v_{in}(t) = 5u(t)$.

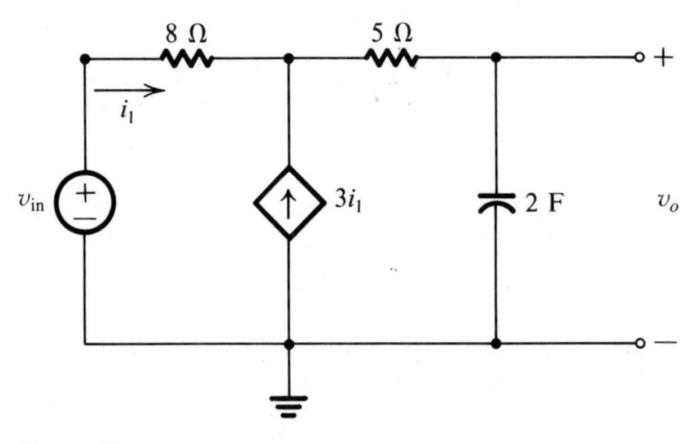

Figure P8.23

8.24 If $i(0^-) = 3$ A, and $v_{in}(t) = 5u(t)$,

a. Find $v_o(t)$ for $t \geq 0$.

b. Show that $v_o(t)$ has ISR = ZIR + ZSR.

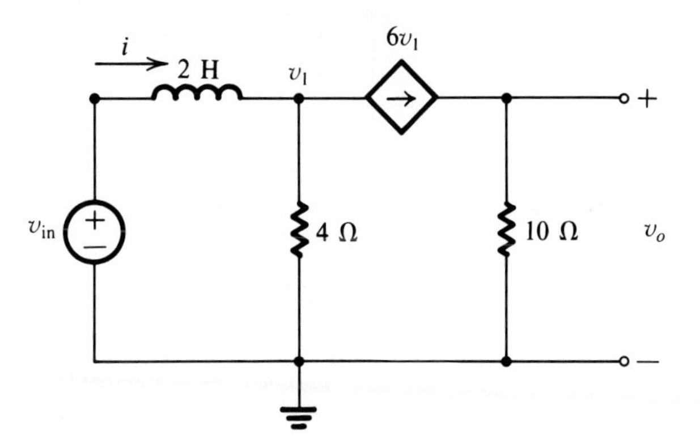

Figure P8.24

8.25 Given the circuit in Fig. P8.25 with $v_c(0^-) = 2$ V and $v_{in}(t) = 5u(t)$, find the ZIR, ZSR, and ISR of v_o. Show that the ISR = ZIR + ZSR.

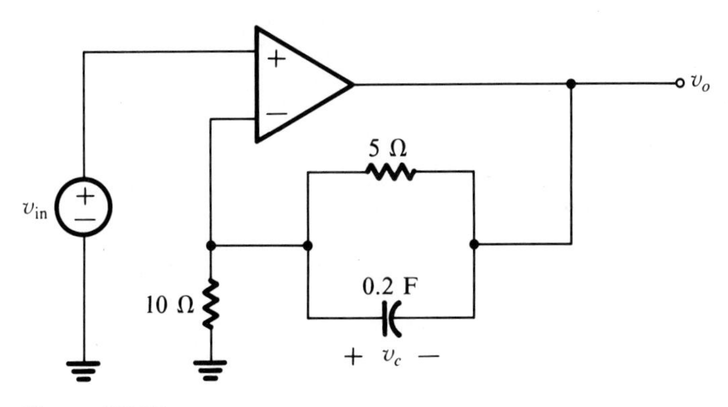

Figure P8.25

8.26 If the circuit shown has $v_c(0^-) = 1$ V and $v_{in}(t) = 5u(t)$, find v_o for $t \geq 0$.

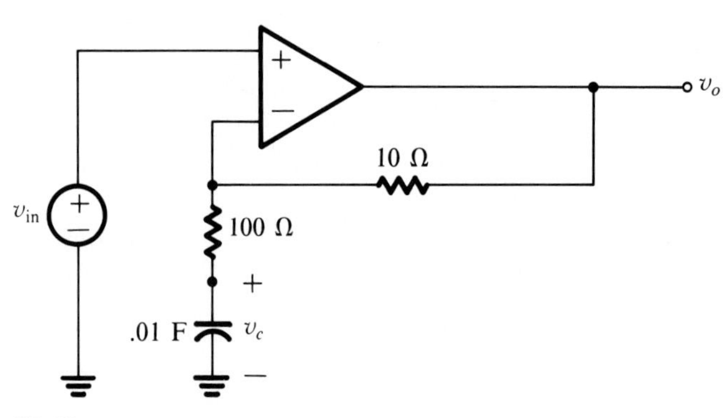

Figure P8.26

8.27 Find i and v_o for $t \geq 0$, given that $v_c(0^-) = 1$ V and $v_{in}(t) = Vu(t)$.

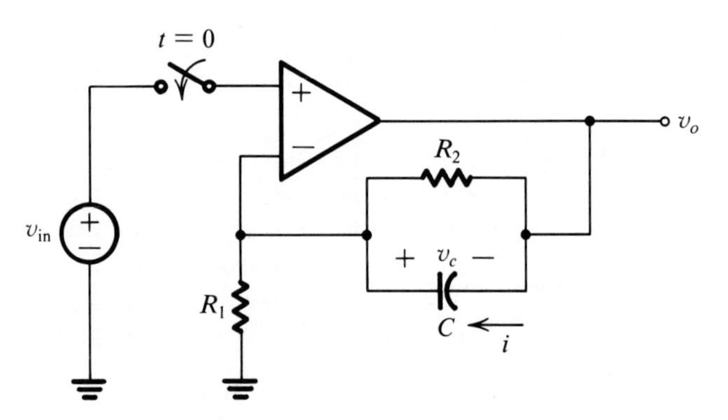

Figure P8.27

8.28 A load resistor R_L is added to the RC circuit in Fig. P8.28. Find an expression for the time constant of the transient component of the ZSR of v_o to a unit-step input signal, and draw a graph of τ vs R_L. Discuss the relationship between τ and R_L and explain how adding the load to the circuit affects the time required for v_o to reach a steady-state value for a step input signal.

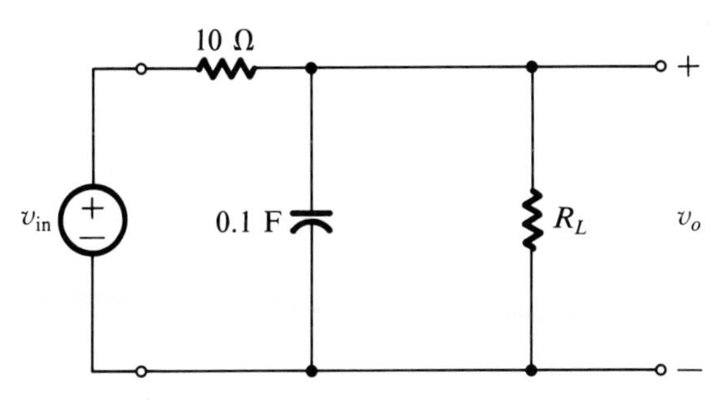

Figure P8.28

8.29 Physical capacitors differ from ideal capacitors because they cannot hold a charge indefinitely. This behavior can be modeled by placing a "leakage resistor" in parallel with the ideal capacitor. The value of this resistor determines how long it takes an initial charge to decay to zero. Suppose the switch in Fig. P8.29 was closed for a long time before opening at $t = 0$.

a. Specify the minimum leakage resistance of the capacitor R_L so that v_o will be at least 10% of its initial value two hours after the switch is opened.

b. Find the initial energy stored in C.

c. Find the total energy dissipated in R_L one day after the switch opens.

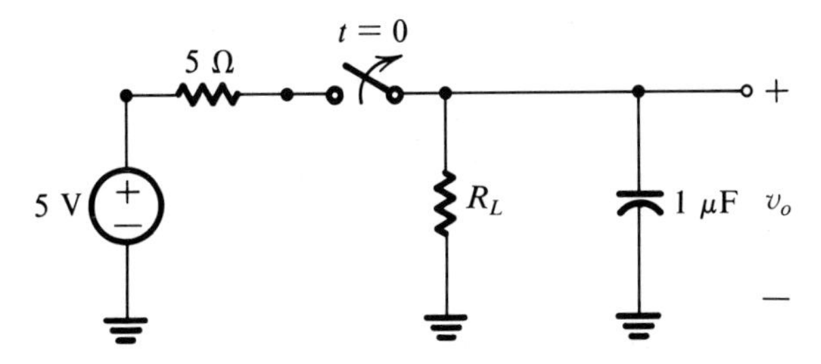

Figure P8.29

8.30 If the circuit shown in Fig. P8.30(a) has the output signal shown in Fig. P8.30(b) when $v_{in}(t) = 15u(t)$, find C.

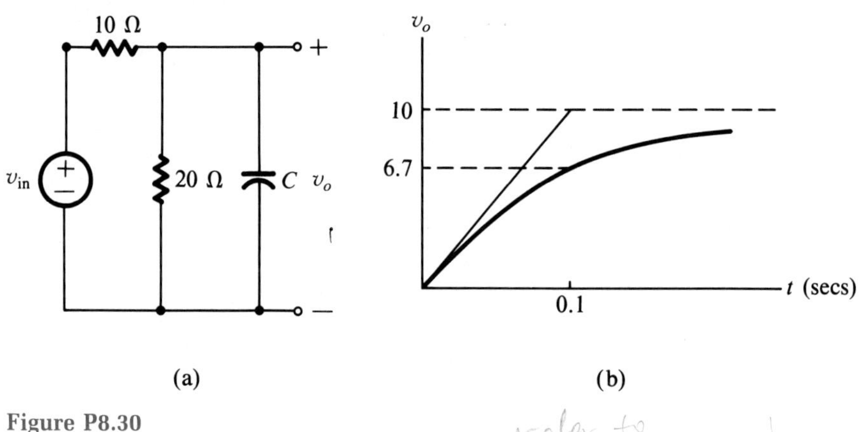

Figure P8.30

refer to correction manual

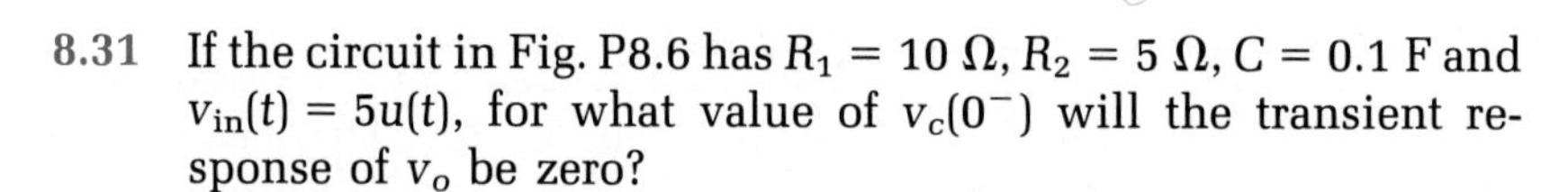

8.31 If the circuit in Fig. P8.6 has $R_1 = 10\ \Omega$, $R_2 = 5\ \Omega$, $C = 0.1$ F and $v_{in}(t) = 5u(t)$, for what value of $v_c(0^-)$ will the transient response of v_o be zero?

8.32 A series RL circuit has a time constant given by $\tau = L/R$. Discuss the behavior of τ with respect to changes in L or R, and provide a physical explanation for your conclusions.

8.33 Repeat Problem 8.31 with $C = 5$ F.

ADDITIONAL PROPERTIES OF FIRST-ORDER CIRCUITS

INTRODUCTION

The last two chapters showed how to solve the differential equation that describes a voltage or current in a circuit. They then introduced terminology to distinguish between the component of a circuit's response due to the circuit's initial stored energy, and that due to the input signal. This chapter will extend the concept of linearity introduced earlier for resistive circuits by showing that the zero-input and zero-state responses are linear. We will also demonstrate the use of superposition to find the ZSR when more than one source drives the circuit, or when the signal of a single source can be represented in terms of two or more other signals.

9.1 ZIR LINEARITY

The relationship between the ZIR of a circuit and the value of its initial inductor currents and capacitor voltages is linear. By this we mean that if a factor is used to scale (simultaneously) all of the initial capacitor voltages and inductor currents, the ZIR will be scaled by the same factor, and the ZIR due to a sum of initial conditions will be the sum of the ZIRs due to the individual conditions. We will not prove this for the general case, but instead will now demonstrate that it holds for the ZIR of the first-order series RC circuit.

Since the ZIR of the voltage in a series *RC* circuit depends on the initial capacitor voltage, two distinct initial voltages v_{01} and v_{02} will have the distinct ZIRs

$$v_{c1}(t) = v_{01}e^{-t/RC}$$
$$v_{c2}(t) = v_{02}e^{-t/RC}$$

respectively, for $t \geq 0$. If the initial capacitor voltage is a linear mix of these initial voltages ($v_0 = k_1v_{01} + k_2v_{02}$), then the ZIR will be

$$\begin{aligned} v_c(t) &= (k_1v_{01} + k_2v_{02})e^{-t/RC} \\ &= k_1v_{01}e^{-t/RC} + k_2v_{02}e^{-t/RC} \end{aligned}$$

If we let $k_1 = k_2 = 1$ we can conclude that the ZIR to a sum of the initial voltages is the sum of their individual ZIRs. If we let $k_2 = 0$, we can conclude that the ZIR to a scaled input voltage is the scaled value of its ZIR (i.e., doubling the initial capacitor voltage doubles the ZIR waveform).

9.2 ZSR SOURCE-SIGNAL ADDITIVITY

The zero-state response of a circuit is additive.

The zero-state response to the sum of *N* signals is the sum of their individual zero-state responses.

or:

$$\text{ZSR} = \text{ZSR}_1 + \text{ZSR}_2 + \cdots + \text{ZSR}_N$$

Example 9.1

Suppose that the series *RC* circuit in Fig. 9.1 is driven by a series connection of v_{in1} and v_{in2}. Find the ZSR of the capacitor voltage when

$$v_{in1}(t) = Vu(t)$$

and

$$v_{in2}(t) = Ve^{-(t-t_d)}u(t - t_d)$$

Solution: First, we note that $v_{in}(t)$ (shown in Fig. 9.2) is formed as

$$v_{in}(t) = Vu(t) + Ve^{-(t-t_d)}u(t - t_d)$$

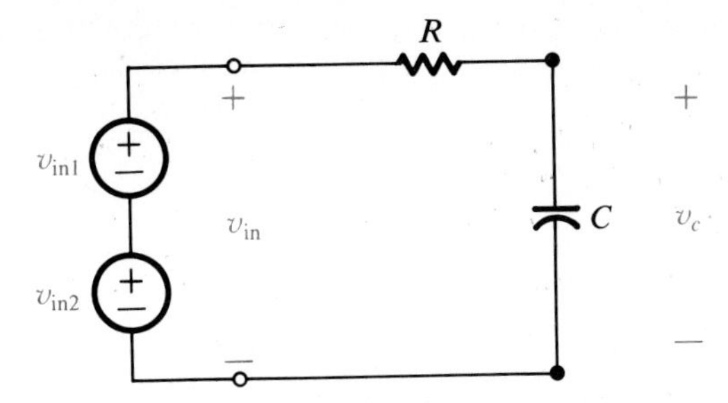

Figure 9.1 Superposition of source signals.

The ZSR to v_{in1} is given by

$$v_{c1}(t) = V(1 - e^{-t/RC})u(t)$$

Likewise, in Chapter 8 we showed that the ZSR to the real exponential source $Ve^{\hat{\sigma}t}$ is

$$v_c(t) = V\,\mathbf{H}(\hat{\sigma})[e^{\hat{\sigma}t} - e^{-t/RC}]u(t)$$

Therefore, the response when $\hat{\sigma} = -1$ and $RC \neq 1$ is given by

$$v_c(t) = \frac{V/RC}{-1 + (1/RC)}[e^{-t} - e^{-t/RC}]u(t)$$

If we apply the ZSR time-shifting property we get the response to $v_{in2}(t)$ as

$$v_c(t) = \frac{V/RC}{-1 + (1/RC)}[e^{-(t-t_d)} - e^{-(t-t_d)/RC}]u(t - t_d)$$

Then the composite input signal $v_{in}(t)$ has the zero-state response

$$v_c(t) = V[1 - e^{-t/RC}]u(t) + \frac{V/RC}{-1 + (1/RC)}[e^{-(t-t_d)} - e^{-(t-t_d)/RC}]u(t - t_d)$$

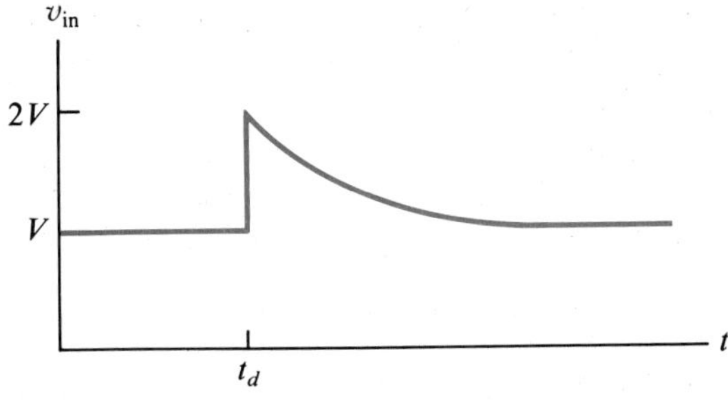

Figure 9.2 A composite source signal.

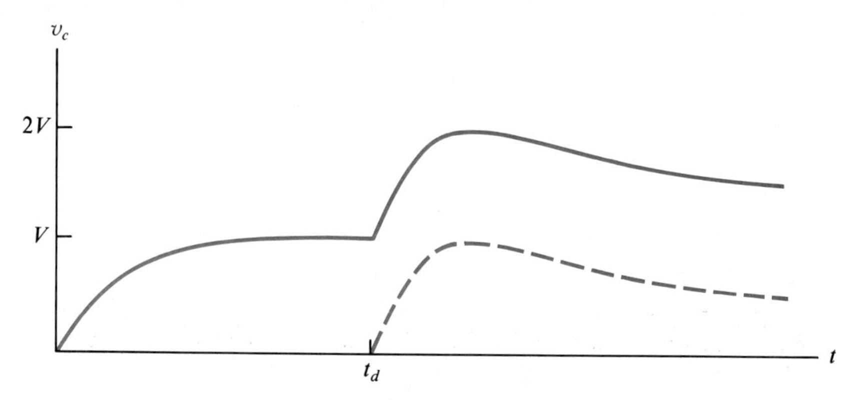

Figure 9.3 Superposition of the zero-state response.

The graph of $v_c(t)$ in Fig. 9.3 is the sum of the ZSR for the step-input signal component (Fig. 8.14) and the delayed, exponential input signal component [Fig. 8.23(a)]. The circuit initially responds to the step input by charging to V volts. When the exponential signal is applied, the increased current causes the capacitor to charge initially and then discharge (back to its steady-state value of V volts) as the exponential source decays to zero.

It should be noted that the principle of superposition that applies to the ZSR does not hold for the initial-state response.

Example 9.2

Find the ISR of the series RC circuit with $v_{in}(t)$ as given in Example 9.1.

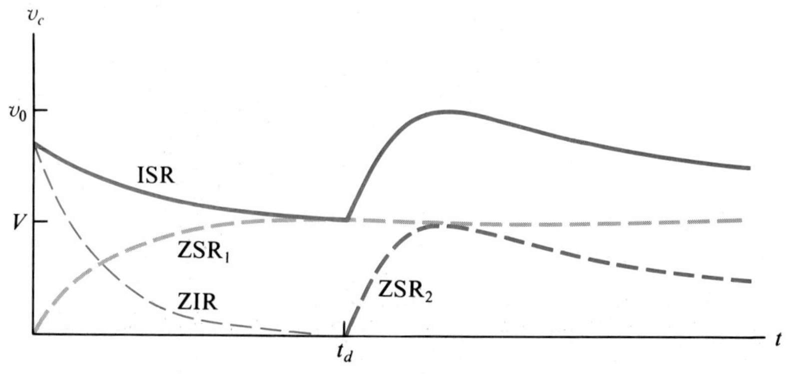

Figure 9.4 Initial-state response to a composite signal.

Solution: Adding the ZIR and the ZSR gives

$$v_c(t) = v_0 e^{-t/RC} u(t) + V(1 - e^{-t/RC}) u(t)$$

$$+ \frac{V/RC}{-1 + (1/RC)} [e^{-(t-t_d)} - e^{-(t-t_d)/RC}] u(t - t_d)$$

The graph of the ISR is given in Fig. 9.4 together with the ZIR and the ZSR components due to the exponential and step-input signals.

The source-signal additivity property does not apply to the ISR because the ZIR component of each individual ISR would be added N times. Instead

The initial-state response to the sum of N signals is the zero-input response of the circuit plus the sum of the N individual ZSR.

or

$$\text{ISR} = \text{ZIR} + \text{ZSR}_1 + \text{ZSR}_2 + \cdots + \text{ZSR}_N$$

9.3 ZSR SOURCE-SIGNAL SCALING (AMPLIFICATION)

Scaling an input signal changes the value of the signal uniformly by a factor of k, with the result that the geometric shape of the waveform is unchanged. In symbolic form, scaling a source $v_{in}(t)$ produces a new signal $\tilde{v}_{in}(t)$ given by $\tilde{v}_{in}(t) = kv_{in}(t)$. If $|k| > 1$, the new signal is an amplified copy of the old signal. We're familiar with this property on a day-to-day basis as we adjust the volume control of a stereo. This changes the loudness but not the shape of the signal that we hear. Wouldn't it be nice if we could economize by using our knowledge of the ZSR for the original signal to determine the ZSR for the scaled signal?

If the source signal is scaled by a factor of k, the ZSR of the circuit is scaled by the same factor.

Skill Exercise 9.1

Find the ZSR of the capacitor voltage in the series *RC* circuit when $v_{in}(t) = kVu(t)$.

Answer: $v_c(t) = k[1 - e^{-t/RC}[Vu(t)$.

Be careful to observe that scaling the source does *not* scale the ISR of a circuit because the ZIR component of the ISR is not affected by scaling the source.

The ISR to a scaled source is the ZIR plus the scaled ZSR of the original source.

So, if the source is scaled by k we have

$$\text{ISR} = \text{ZIR} + k\text{ZSR}$$

Skill Exercise 9.2

Find the ISR of the capacitor voltage in the series *RC* circuit when $v_{in}(t) = kVu(t)$.

Answer: $v_c(t) = v_0 e^{-t/RC} u(t) + k(1 - e^{-t/RC})Vu(t)$

9.4 ZSR LINEARITY AND SUPERPOSITION

The input/output relationship of a circuit is **linear** if it has both the additivity and scaling properties. That is, if $y_1(t)$ is the response to $v_{in1}(t)$ and $y_2(t)$ is the response to $v_{in2}(t)$, then the response to $k_1 v_{in1}(t) + k_2 v_{in2}(t)$ is $k_1 y_1(t) + k_2 y_2(t)$ for any scalars k_1 and k_2. When a circuit's input/output relationship is linear, the circuit is said to obey the principle of superposition because the net effect of several sources acting simultaneously is the sum of their individual effects. Although the linearity property is stated in terms of two signals, it can be extended in a pairwise fashion to apply to any number of signals.

9.5 ZSR AND ISR TO RAMP INPUT SIGNALS*

The ZSR of a circuit has the property that if a new input signal is formed by integrating or differentiating a given signal, the ZSR to the new signal is just the integral or derivative of the original input signal.

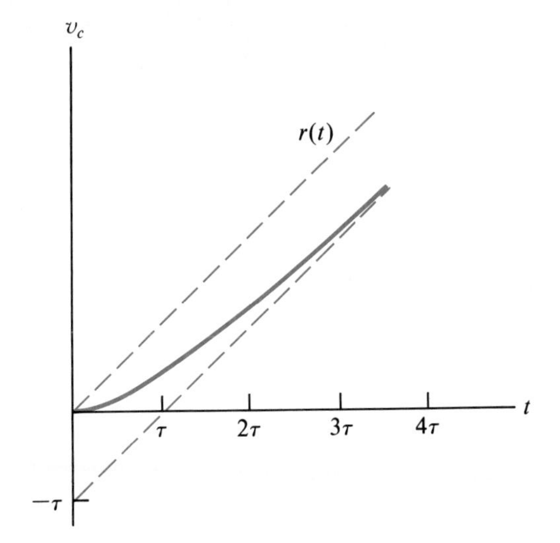

Figure 9.5 Zero-state response of a series RC circuit to a ramp signal source.

The ZSR of a circuit to a ramp input signal can be obtained by integrating the circuit's ZSR to a step input signal. Since the ZSR of a series RC circuit to a unit-step input is $v_c(t) = (1 - e^{-t/RC})u(t)$ the response to the ramp input $v_{in}(t) = tu(t)$ will be

$$v_c(t) = \int_0^t [1 - e^{-\alpha/RC}]d\alpha = [\alpha + RCe^{-\alpha/RC}]\Bigg|_{\alpha=0}^{\alpha=t}$$

$$= t + RCe^{-t/RC} - RC = t - \tau[1 - e^{-t/\tau}]$$

with $\tau = RC$. The ramp response of the series RC circuit is shown in Fig. 9.5.

The initial slope of the ramp response curve is 0, and the steady-state component of the response is $v_{ss}(t) = t - \tau$. As $t \to \infty$, the response approaches a ramp of the same slope as the input ramp, but offset by $-\tau$ volts. The response is said to **lag** the input. The transient response is $v_{tr}(t) = \tau e^{-t/\tau}$, and the initial slope of the transient component is

$$\frac{d}{dt}v_{tr}(0^+) = -1$$

Fig. 9.6 shows the ISR for a given v_0. The capacitor begins discharging from its initial voltage, but eventually the current due to the ramp source recharges the capacitor.

The ZSR and ISR for a ramp input illustrate a case where the steady-state component of the response signal is not a constant value, i.e. there is no "steady state" except in the sense that the transient solution has decayed to a negligible value.

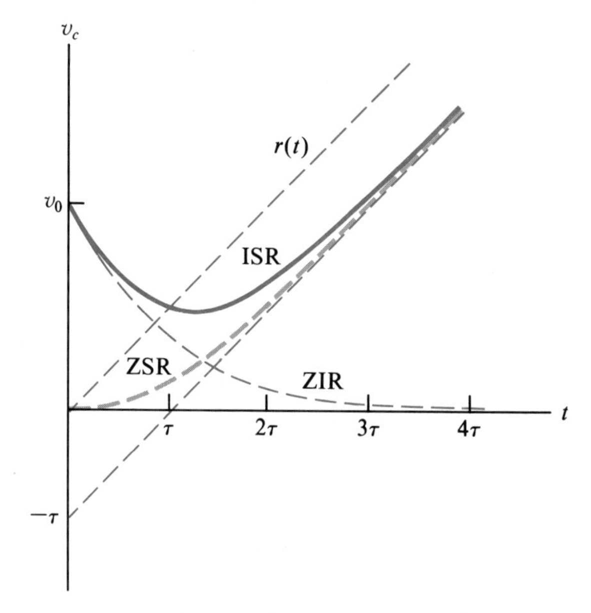

Figure 9.6 The initial-state response of a series *RC* circuit to a ramp input signal.

9.6 *RC*–CIRCUIT RESPONSE TO PULSE SIGNALS*

The response of an *RC* circuit to a rectangular pulse signal (see Chapter 6) is determined by the location of the circuit's natural frequency (pole) in the complex plane and by the width of the pulse Δ. The response will have one of the general shapes shown in Fig. 9.7, depending on the pulse width. Now, the input voltage pulse can be described in terms of step functions as

$$v_{in}(t) = V[u(t) - u(t - \Delta)]$$

Since we already know the ZIR of the circuit, we need only find its ZSR. For $0 \leq t \leq \Delta$ the circuit must behave as though a voltage step were applied. If the time constant of the circuit is small compared to Δ, $v_c(t)$ should grow exponentially up to V volts before $t = \Delta$. At $t = \Delta$ the source is turned off; then we expect to see the ZIR curve taking the output from V volts to zero, as shown in Fig. 9.7(a).

On the other hand, if $\tau \cong \Delta$ the current charging the capacitor does not have sufficient time to fully charge the capacitor to V volts before the source is removed. Therefore $v_0(\Delta)$ determines the ZIR curve for $t \geq \Delta$ as shown in Fig. 9.7(b). If $\tau \gg \Delta$, the output voltage is negligible for all t, and the circuit acts like a filter that blocks the input voltage pulse from appearing at the output terminal. [Fig. 9.7(c)]. These cases allow us to conclude that the **output pulse will have a relatively low amplitude compared to the input pulse when the input pulse width is short compared to the time constant of the filter.** When the circuit's pole is far from the origin, its time constant is small, and voltage pulses having narrow pulse

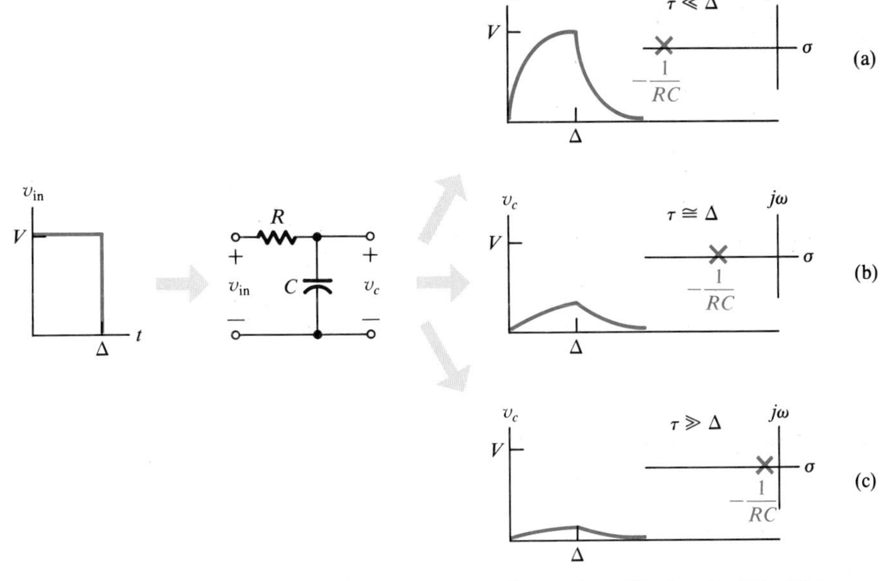

Figure 9.7 Comparison of pulse responses of a series RC circuit. "Best" transmission occurs when τ is much smaller than the pulse width.

widths can be passed by the circuit, in the sense that the input voltage causes a significant output voltage. The relatively short time constant of the circuit allows the capacitor to charge to the input voltage level before the pulse drops to zero.

The ZSR to a rectangular pulse is the sum of the ZSR to a step input and the ZSR to a delayed (by Δ) and scaled (by −1) copy of a step

$$v_c(t) = V(1 - e^{-t/RC})u(t) - V(1 - e^{-(t-\Delta)/RC})u(t - \Delta)$$

and

$$v_c(t) = \begin{cases} V(1 - e^{-t/RC}) & 0 \le t \le \Delta \\ (e^{\Delta/RC} - 1)Ve^{-t/RC} & t \ge \Delta \end{cases}$$

Note that $v_c(\Delta^-) = v_c(\Delta^+)$ and that $v_c(t) = v_c(\Delta)e^{-(t-\Delta)/RC}$ for $t \ge \Delta$.

This expression highlights the fact that the solution curve for $t \ge \Delta$ is the ZIR beginning at $t = \Delta$ with the initial voltage determined by the charge on the capacitor due to the step voltage that was applied for $0 \le t \le \Delta$. Removal of the source at $t = \Delta$ forces the capacitor to immediately begin discharging.

Skill Exercise 9.3

Show that for a pulse input, the ZSR of the current in the series RC circuit is given by

$$i(t) = \begin{cases} \dfrac{V}{R} e^{-t/RC} & 0 \le t \le \Delta \\ -\dfrac{V}{R}(1 - e^{-\Delta/RC})e^{-(t-\Delta)/RC} & t \ge \Delta \end{cases}$$

The current response given in Skill Exercise 9.3 has

$$i(\Delta^-) = \frac{V}{R} e^{-\Delta/RC}$$

$$i(\Delta^+) = -\frac{V}{R}(1 - e^{-\Delta/RC}) = \frac{V}{R} e^{-\Delta/RC} - \frac{V}{R} = i(\Delta^-) - \frac{V}{R}$$

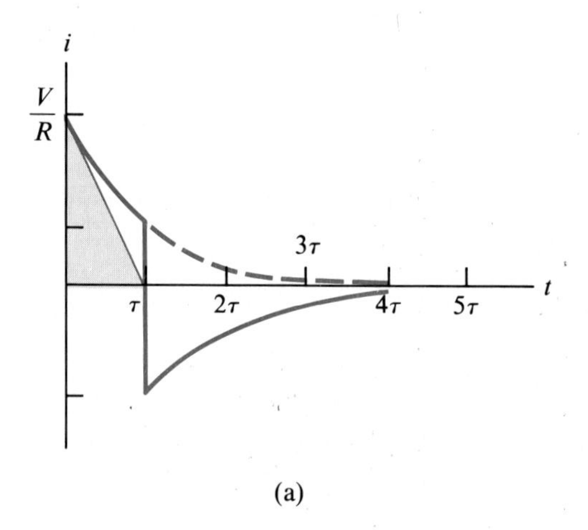

(a)

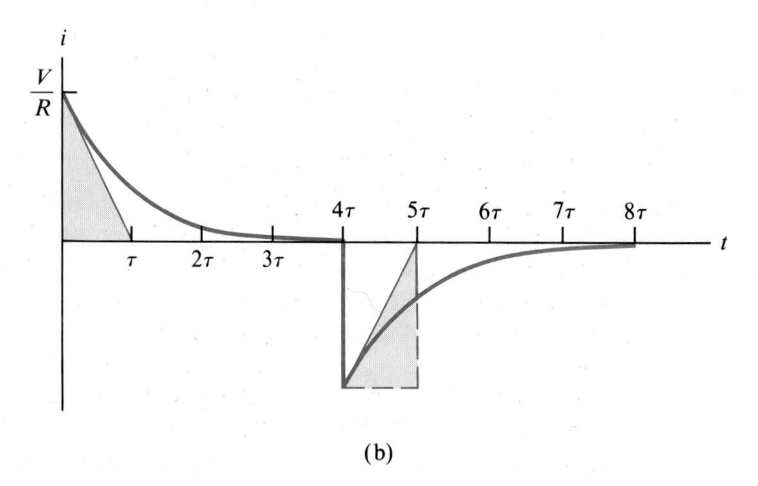

(b)

Figure 9.8 Zero-state response of current in a series RC circuit driven by a rectangular pulse with $\Delta = \tau$.

The current is discontinuous at $t = \Delta$. Although the capacitor voltage is continuous, the resistor voltage is discontinuous because the source is discontinuous. The source discontinuity causes the $-V/R$ drop in the ZSR current seen in Fig. 9.8(a) for $\Delta = \tau$ and in Fig. 9.8(b) for $\Delta = 4\tau$. Compare these curves to the voltage ZSR in Fig. 9.7 and note that the discontinuity in current causes a discontinuity in the *slope* of the capacitor voltage curve because $i(t) = C\, dv/dt$ for a capacitor.

9.7 *RC*–CIRCUIT IMPULSE RESPONSE

The impulse response of the series *RC* circuit is its response to $v_{in}(t) = V\delta(t)$. It can be obtained by first finding the ZSR to a properly chosen sequence of pulses, and then letting the pulse width shrink to zero (see Chapter 6).

The rectangular pulse of height V/Δ in Fig. 9.9 will have a ZSR given by

$$v_c(t) = \begin{cases} \dfrac{V}{\Delta}(1 - e^{-t/RC}) & 0 \le t \le \Delta \\ \dfrac{V}{\Delta}(1 - e^{-\Delta/RC})e^{-(t-\Delta)/RC} & t \ge \Delta \end{cases}$$

For $t \ge \Delta$ we have

$$v_c(t) = \frac{V}{\Delta}e^{-(t-\Delta/RC)} - \frac{V}{\Delta}e^{-t/RC} = \frac{V}{\Delta}e^{-t/RC}(e^{\Delta/RC} - 1)$$

Next, we take a Taylor series expansion of $e^{\Delta/RC}$ to get

$$e^{\Delta/RC} = 1 + \frac{\Delta}{RC} + \frac{1}{2!}\left(\frac{\Delta}{RC}\right)^2 + \cdots$$

Using this expansion gives

$$v_c(t) = \frac{V}{\Delta}e^{-t/RC}\left[\frac{\Delta}{RC} + \frac{1}{2!}\left(\frac{\Delta}{RC}\right)^2 + \cdots\right]$$

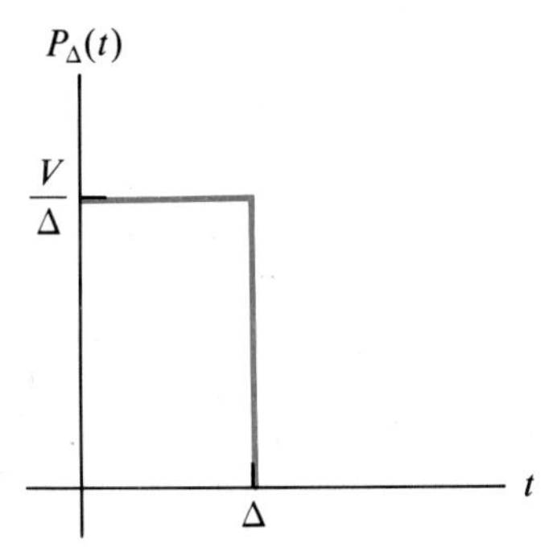

Figure 9.9 A rectangular pulse.

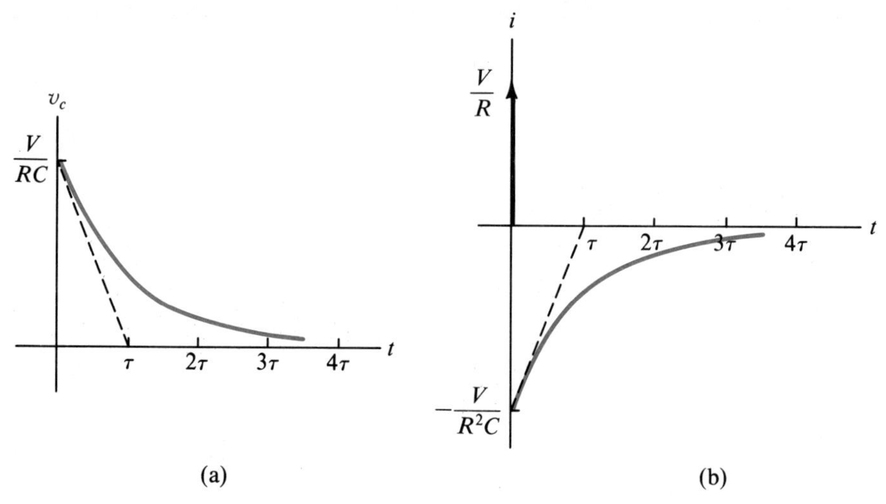

Figure 9.10 The impulse responses for (a) voltage and (b) current in a series RC circuit.

Cancelling Δ gives

$$v_c(t) = Ve^{-t/RC}\left[\frac{1}{RC} + \frac{1}{2!}\left(\frac{\Delta}{RC}\right)^2 + \cdots\right]$$

As $\Delta \to 0$ all but the first term will vanish, leaving **the impulse response**

$$v_c(t) = \frac{Ve^{-t/RC}}{RC}u(t)$$

The current that charges the capacitor can be obtained from $i(t) = C\,dv_c/dt$

$$i(t) = \frac{-CV}{(RC)^2}e^{-t/RC}u(t) + \frac{CV}{RC}e^{-t/RC}\delta(t) = -\frac{V}{R^2C}e^{-t/RC}u(t) + \frac{V}{R}\delta(t)$$

The current in the circuit consists of an impulse of area V/R and an exponential decay term for $t > 0$. The impulse of voltage applied by the source causes an impulse of current that instantaneously establishes a voltage on the capacitor, a voltage that subsequently decays to zero as a ZIR of the circuit. The impulse response of $v_c(t)$ is shown in Fig. 9.10(a), and the impulse response of $i(t)$ is shown in Fig. 9.10(b).

Skill Exercise 9.4

Obtain the impulse response of $v_c(t)$ in a series RC circuit by formally differentiating the circuit's ZSR to a unit-step input.

9.8 ZSR SUPERPOSITION OF MULTIPLE SOURCES

In Section 9.2 we applied the principle of superposition to obtain the zero-state response to a *signal* that is composed of a weighted sum of other signals. The next example uses the principle of superposition to obtain the zero-state response of a first-order circuit containing multiple *sources*.

Example 9.3

If $v_{in}(t) = 5u(t)$ and $i_{in}(t) = 2e^{-t}u(t)$ find the initial state response of $v(t)$ for $t \geq 0$ when $v(0^-) = 1$ V in Fig. 9.11.

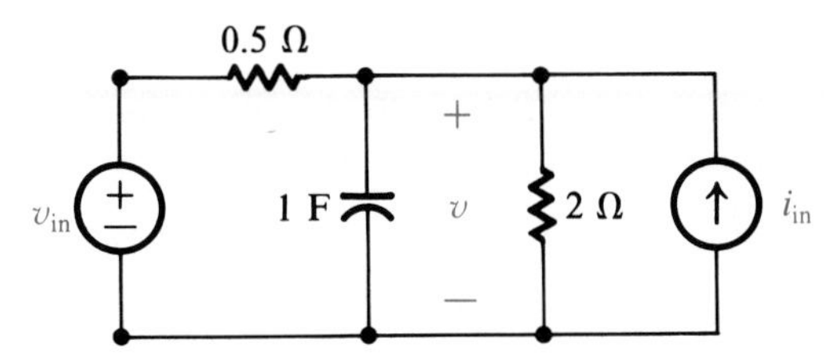

Figure 9.11 Source superposition.

Solution: Writing KCL at the capacitor node gives

$$\frac{v}{2} + \frac{dv}{dt} + 2(v - v_{in}) = i_{in}(t)$$

and so

$$\frac{dv}{dt} + \frac{5}{2}v(t) = 2v_{in}(t) + i_{in}(t)$$

With both sources removed the natural solution is

$$v(t) = Ke^{-5t/2}$$

If $i_{in}(t)$ is removed, the transfer function relating $v(t)$ and $v_{in}(t)$ is

$$\mathbf{H}_v(s) = \frac{2}{s + \frac{5}{2}}$$

and $\mathbf{H}_v(0) = \frac{4}{5}$.

Next, if $v_{in}(t)$ is removed, the transfer function between $i_{in}(t)$ and $v(t)$ is

$$\mathbf{H}_I(s) = \frac{1}{s + \frac{5}{2}}$$

and the particular solution to $i_{in}(t)$ is

$$v(t) = \frac{2e^{-t}}{-1 + \frac{5}{2}} = \frac{2e^{-t}}{\frac{3}{2}} = \frac{4}{3}e^{-t}$$

Therefore, the complete solution for the I/O differential equation is

$$v(t) = Ke^{-5t/2} + 4 + \frac{4}{3}e^{-t}$$

Invoking the boundary condition $v(0^+) = v(0^-)$ or $K + 4 + 4/3 = 1$ gives $K = -13/3$. The ISR of the circuit to *both sources* is

$$v(t) = -\frac{13}{3}e^{-5t/2} + 4 + \frac{4}{3}e^{-t}$$

for $t \geq 0$. (As a check, verify that $v_{ss} = 4$ V is consistent with the circuit and its sources).

As an alternative approach to solving the problem, we will add the circuit's ZIR and its ZSR to each source to show that the ZSR is the sum of the ZSRs to the individual sources, and that the ISR found above is the sum of its ZIR and its ZSR.

With both sources operative the ZSR of the circuit is

$$v(t) = Ke^{-5t/2} + 4 + \frac{4}{3}e^{-t}$$

with boundary condition $v(0^+) = K + 4 + 4/3 = 0$. So $K = -16/3$ and

$$v(t) = -\frac{16}{3}e^{-5t/2} + 4 + \frac{4}{3}e^{-t}$$

The ZIR is $v(t) = Ke^{-5t/2}$ with $v(0^+) = K = v(0^-) = 1$, so

$$v(t) = e^{-5t/2}$$

The ZSR due to v_{in} is given by

$$v(t) = Ke^{-5t/2} + 4$$

and $v(0^+) = K + 4 = 0$. So $K = -4$ and

$$v(t) = -4e^{-5t/2} + 4$$

The ZSR due to i_{in} is

$$v(t) = Ke^{-5t/2} + \frac{4}{3}e^{-t}$$

with $v(0^+) = K + 4/3 = 0$. So $K = -4/3$ and

$$v(t) = -\frac{4}{3}e^{-5t/2} + \frac{4}{3}e^{-t}$$

As claimed, the ZSR to both sources is the sum of their individual ZSRs or

$$v_{ZSR}(t) = \underbrace{-4e^{-5t/2} + 4}_{\text{Due to } v_s} \underbrace{- \frac{4}{3}e^{-5t/2} + \frac{4}{3}e^{-t}}_{\text{Due to } i_s}$$

$$= -\frac{16}{3}e^{-5t/2} + 4 + \frac{4}{3}e^{-t}$$

Also, the ISR is the sum of the ZIR and the ZSR, so

$$v_{ISR}(t) = \underbrace{e^{-5t/2}}_{\text{ZIR}} \underbrace{- \frac{16}{3}e^{-5t/2} + 4 + \frac{4}{3}e^{-t}}_{\text{ZSR}}$$

$$= -\frac{13}{3}e^{-5t/2} + 4 + \frac{4}{3}e^{-t}$$

This agrees with the result of the direct approach.

SUMMARY

The zero-input response of a circuit is linear with respect to the initial capacitor voltage, and the zero-state response is linear with respect to a sum of source signals and with respect to multiple driving sources. Each component of a source signal produces its own response component; likewise, each source produces its own zero-state–response component. But remember that the initial-state response does *not* satisfy source superposition because changing the sources does not change the initial stored energy in the circuit. The property of superposition will enable us to understand the response of higher-order circuits to signals that are not themselves exponentials, but can be represented as a discrete or continuous weighted sum of exponentials.

Problems - Chapter 9

9.1 Find $v_o(t)$ for $t \geq 0$ if $R_1 = 5\Omega$, $R_2 = 1\Omega$, $C = 0.2$ F, $v_c(0^-) = 0$ and $v_{in}(t) = 10[u(t-2) - 2u(t) - 4) + u(t-6)]$.

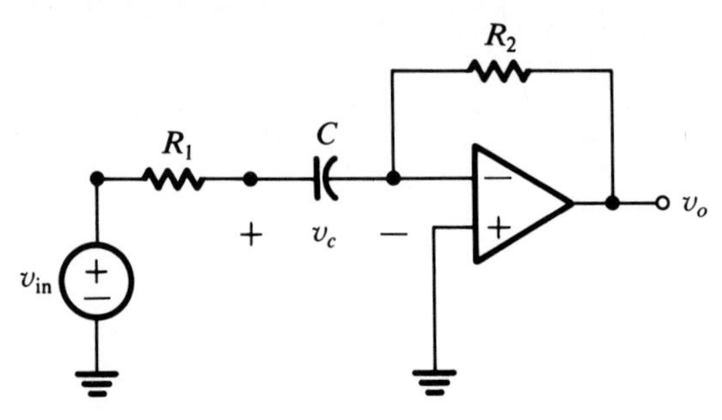

Figure P9.1

9.2 Find $v_o(t)$ for the circuit in Fig. P9.2(b) if $v_c(0^-) = 4$ V and the source is the signal shown in Fig. P9.2(a). Switch 1 closes at $t = 0$ and opens at $t = 6$ secs, while switch 2 closes at $t = 6$ secs.

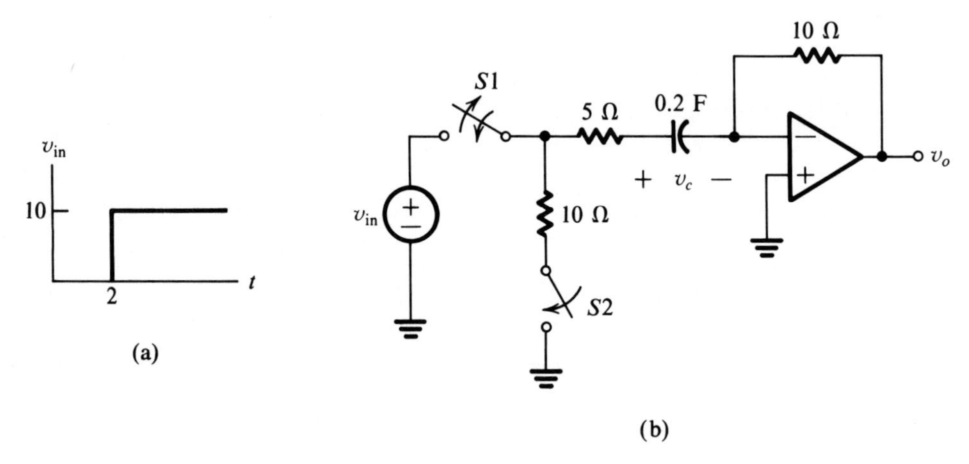

Figure P9.2

9.3 The input to the op amp circuit in Fig. P9.1 is the signal $v_{in}(t) = 5[u(t-1) - u(t-2)]$ and $R_1 = 20\ \Omega$, $R_2 = 5\ \Omega$ and $C = 0.05$ F.

a. Find and draw the ZSR of v_o for $t \geq 0$.

b. Repeat a with $C = 0.01$ F.

c. Compare the results of a and b, and give a physical explanation for the differences observed.

9.4 The circuit shown in Fig. P9.4 has $R_1 = 10\ \Omega$, $R_2 = 5\ \Omega$ and $C = 0.05$ F.

a. If $v_c(0^-) = 0$ V, find and draw the transient and steady-state responses of $v_o(t)$ when $v_{in}(t) = 10u(t)$.

b. Repeat a with $v_c(0^-) = 5$ V.

c. Repeat a with $v_c(0^-) = 0$ V and $v_{in}(t) = 4e^{-2t}u(t)$.

d. Repeat a with $v_c(0^-) = 5$ V and $v_{in}(t) = 4e^{-2t}u(t)$.

e. Repeat a with $v_c(0^-) = 0$ V and $v_{in}(t) = [10 + 4e^{-2t}]u(t)$.

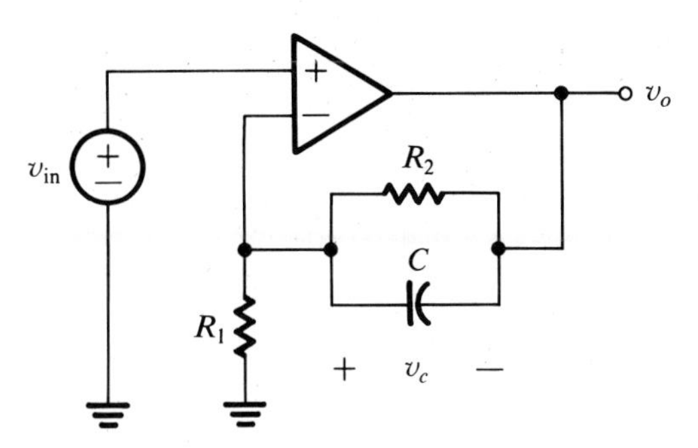

Figure P9.4

9.5 If the op amp circuit in Fig. P9.4 has $R_1 = 50$ kΩ, $R_2 = 100$ kΩ and $C = 1\ \mu$F, find and draw $v_o(t)$ when $v_{in}(t) = -5/2$ $(t - 2)[u(t) - u(t - 2)$.

9.6 Find and draw the ZSR of $i(t)$ to $v_{in}(t) = 5u(t)$ and $i_{in}(t) = 2u(t)$.

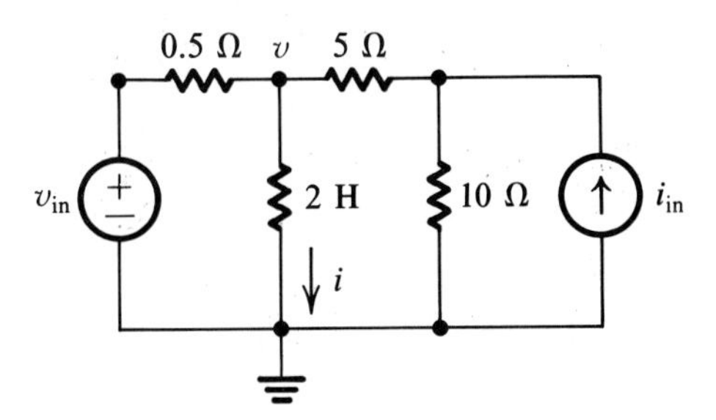

Figure P9.6

9.7 If $v(0^-) = 0$ and $v_{in}(t) = 10u(t) - 5u(t-1)$, find $v(t)$ for $t \geq 0$.

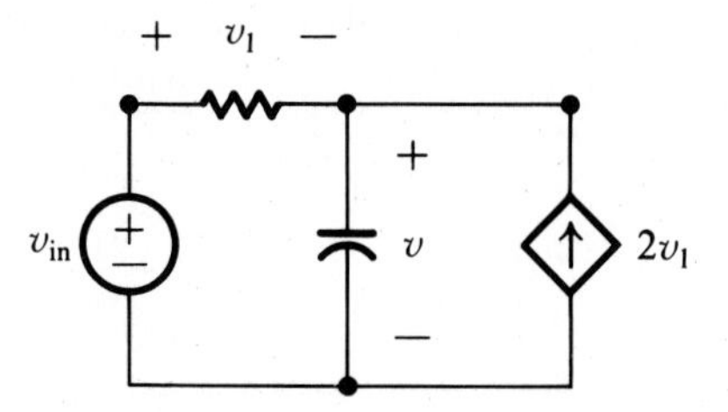

Figure P9.7

9.8 Draw the graph of $v_o(t)$ for the input signals given below, and discuss the apparent differences in their shape. Use $R_1 = 2\ \Omega$, $R_2 = 4\ \Omega$, $C = 1\ \mu F$ and $v_c(0^-) = 0$.

a. $v_{in}(t) = 10e^{-t/3.95}u(t)$.

b. $v_{in}(t) = 10e^{-t/4}u(t)$.

a. $v_{in}(t) = 10e^{-t/4.05}u(t)$.

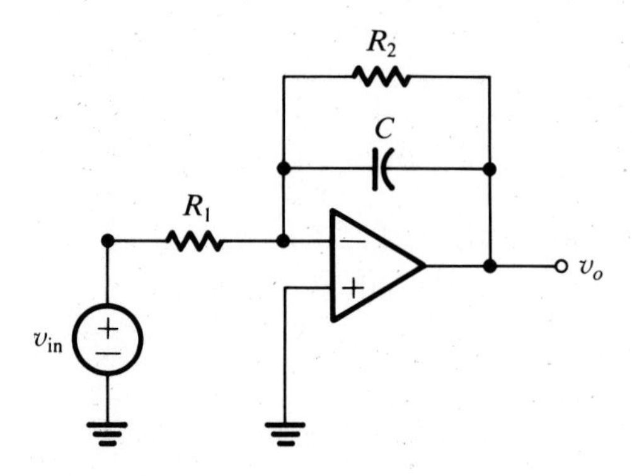

Figure P9.8

9.9 A claim has been made that if $i(t) = 10e^{-.5t}$ for $t > -10$ in Fig. P9.9, then $v_o(t) = 0$ for $t \geq 0$. Prove or disprove the claim, and provide a physical explanation supporting your answer.

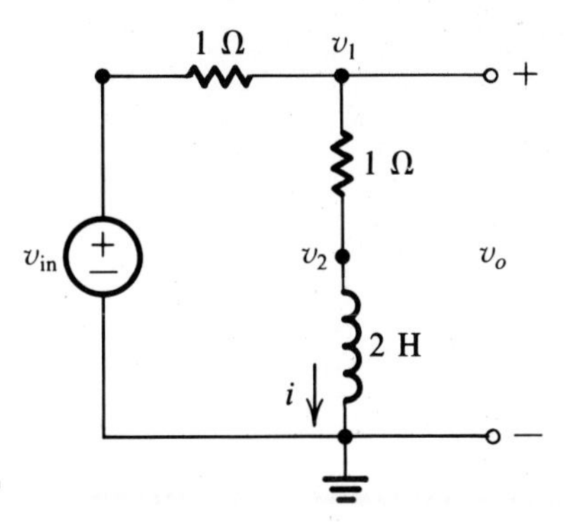

Figure P9.9

9.10 Estimate the time required for the circuit to reach steady state when both sources are unit-step input signals.

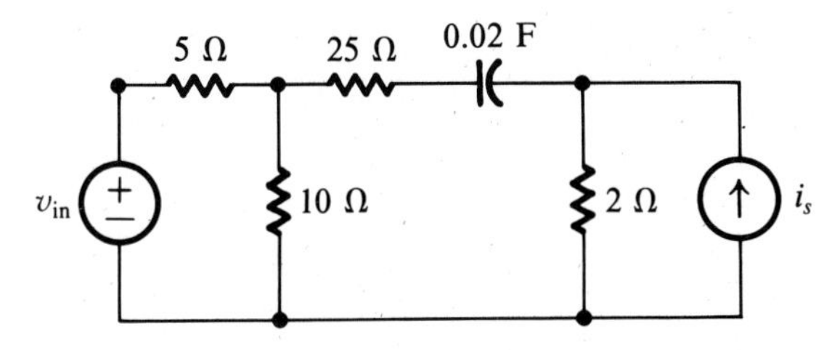

Figure P9.10

9.11 If a series *RL* circuit with R = 10 Ω and L = 5 H has $v_{in}(t) = 0.5t^2u(t)$, find the ZSR of $i(t)$.

9.12 If $R_1 = 100\ \Omega$, $R_2 = 25\ \Omega$ and $C = 0.02$ F, find $v_o(t)$ for all t when v_{in} has the waveform shown.

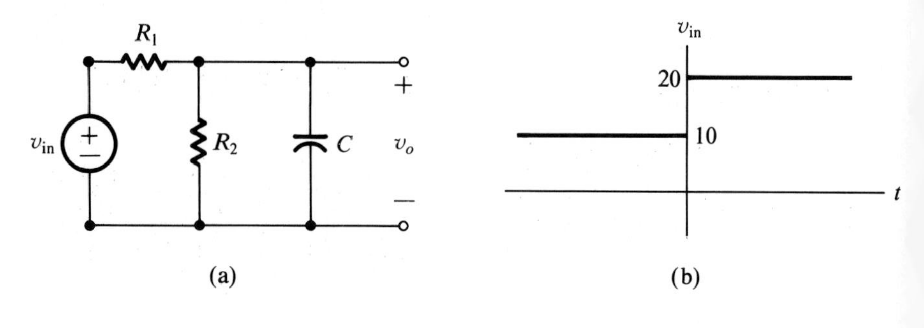

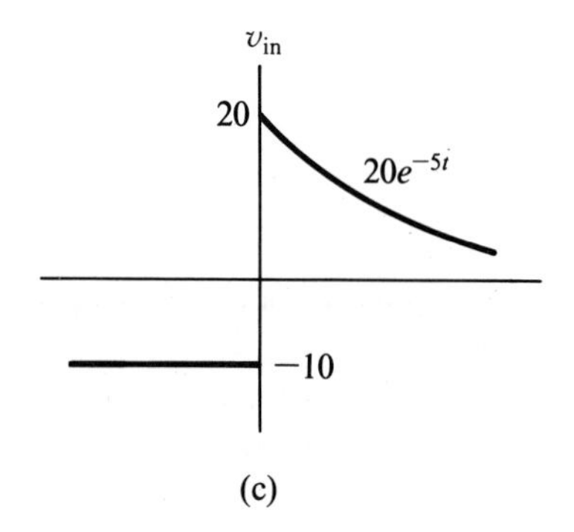

Figure P9.12

9.13 Repeat Problem 9.12 with $R_1 = 200\ \Omega$, $R_2 = 50\ \Omega$ and $C = 0.05$ F.

9.14 Find the ZSR of the current in a series *RL* circuit to the following input source voltages:

a. $v_{in}(t) = \delta(t)$
b. $v_{in}(t) = u(t)$
c. $v_{in}(t) = tu(t)$

9.15 Find the ISR of $i(t)$ in Problem 9.6 when $i(0^-) = 5$ A.

9.16 If $R_1 = 5\ \Omega$, $R_2 = 2\ \Omega$ and $C = 0.5$ F in Fig. P9.8, find the ZSR of $v_o(t)$ when $v_{in} = \delta(t)$.

FIRST-ORDER CIRCUITS WITH SINUSOIDAL INPUT

INTRODUCTION

Chapters 7–9 showed how to find a circuit's response by solving its input/output differential equation to get the natural- and particular-solution components, and how to find the complete solution to the I/O equation by using appropriate boundary conditions to obtain the circuit's zero-input, zero-state, and initial-state responses. We saw that the natural solution to the differential equation did not depend on the input signal, but that the particular solution did. Then we used the circuit's transfer function to construct the particular solution when the input signal was a real exponential. This chapter will expand the class of input signals for which solutions can be constructed by using *complex* exponential signals to represent damped and undamped sinusoidal signals. Knowing how to solve the *mathematical problem* of constructing the particular solution to the input/output differential equation when the input signal is a sinusoid will enable us to solve the *circuit problem* of finding a circuit's zero-state and initial-state responses to a sinusoid.

Sinusoidal signals play a major role in electrical engineering because many physical signals either are sinusoids or can be expressed as a sum of sinusoids. For example, laboratory instruments such as function generators provide a convenient source of sinusoidal voltages, and allow the operator to select

the amplitude and frequency of the sinusoidal voltage. Also, stereo systems and other electronic equipment are often specified in terms of their response to sinusoidal signals.

The objectives of this chapter are to:

1. **Develop familiarity with complex exponential signals.**
2. **Use complex signals to represent sinusoidal signals.**
3. **Learn how to construct the particular solution of the I/O differential equation when the circuit has a sinusoidal input signal (damped or undamped).**
4. **Develop a foundation for later work using Fourier series, Fourier transforms, and Laplace transforms to solve circuit problems with an even broader class of input signals.**

We will use complex exponential signals to represent sinusoids. We can thereby apply the analytical approach already developed for exponential signals directly to sinusoids and avoid having to develop a separate approach. Now might be a good time to review the material on complex numbers in Appendix B if your grasp of this material needs to be refreshed.

10.1 COMPLEX EXPONENTIAL SIGNALS

Complex exponential signals do not have a physical existence, but do provide a framework for representing real-world sinusoid signals. As an example of a complex signal (denoted by boldface symbols), consider

$$\mathbf{f}(t) = e^{j\omega t} \tag{10.1}$$

This signal is of the familiar exponential form $x(t) = Xe^{st}$ where we have let $X = 1$ and $s = j\omega$. Up to this point we have restricted s to real numbers. Now we will let s be a complex number on the j-axis; later we will let s be any complex number, or any value in the "s-domain."

Euler's law for complex numbers states that

$$e^{j\omega t} = \cos \omega t + j \sin \omega t$$

so $\mathbf{f}(t)$ can be written as

$$\mathbf{f}(t) = \cos \omega t + j \sin \omega t$$

Thus, at each time t a complex exponential [like $\mathbf{f}(t)$ in (10.1)] has both a real part and an imaginary part, and

$$\cos \omega t = \text{Re}\{e^{j\omega t}\}$$

$$\sin \omega t = \text{Im}\{e^{j\omega t}\}$$

Conversely, a real sinusoidal signal can always be written as either the real or the imaginary part of a complex signal. The signal $e^{j\omega t}$ can be thought of as being a rotating unit vector in the complex plane, as shown in Fig. 10.1. At each instant t the projection of this vector onto the real axis determines its real component. Likewise, its projection onto the imagi-

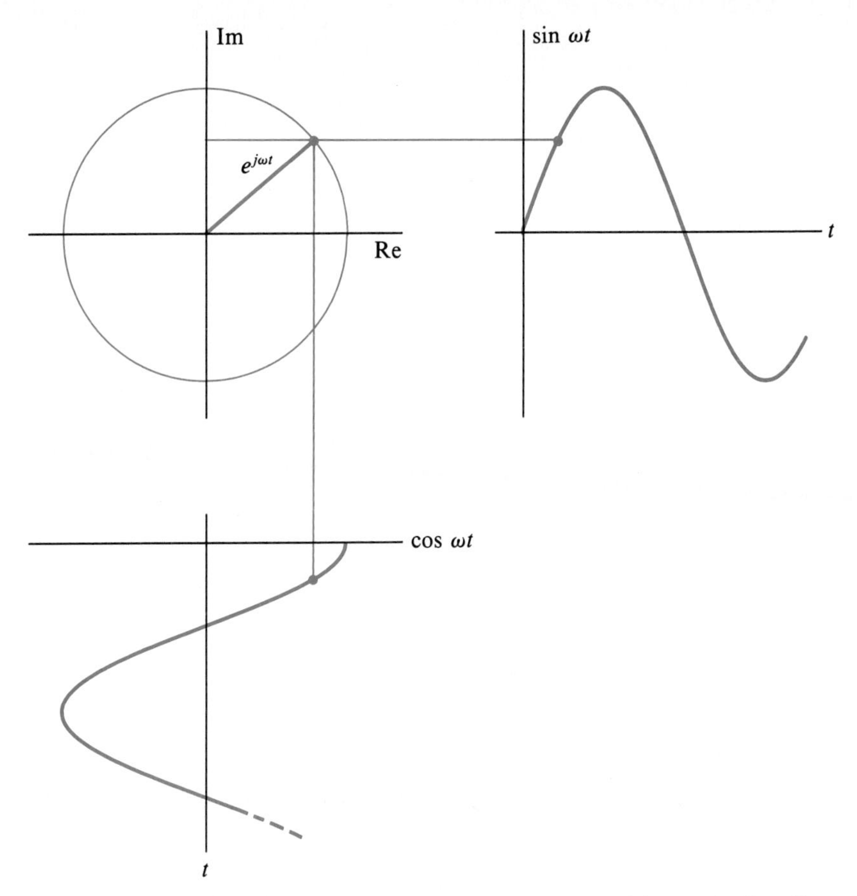

Figure 10.1 Real and imaginary components generated by projecting the rotating vector for $e^{j\omega t}$ onto the real and imaginary axis.

nary axis gives its imaginary component. These projections are drawn in Fig. 10.1. Note that the vector has a constant length, or magnitude, since at any time t

$$|e^{j\omega t}| = \sqrt{\cos^2 \omega t + \sin^2 \omega t} = 1$$

Another signal of general interest is the *real*, exponentially damped and phase-shifted signal given by

$$f(t) = Ve^{\sigma t} \cos(\omega t + \phi) \tag{10.2}$$

To express the real signal $f(t)$ in terms of a complex exponential signal, we use Euler's law and the scaling property of complex numbers to get

$$f(t) = Ve^{\sigma t}\text{Re}\{e^{j(\omega t + \phi)}\} \tag{10.3a}$$
$$= \text{Re}\{Ve^{\sigma t}e^{j(\omega t + \phi)}\} \tag{10.3b}$$
$$= \text{Re}\{Ve^{j\phi}e^{(\sigma + j\omega)t}\} \tag{10.3c}$$
$$= \text{Re}\{\mathbf{V}e^{st}\} = \text{Re}\{\mathbf{f}(t)\} \tag{10.3d}$$

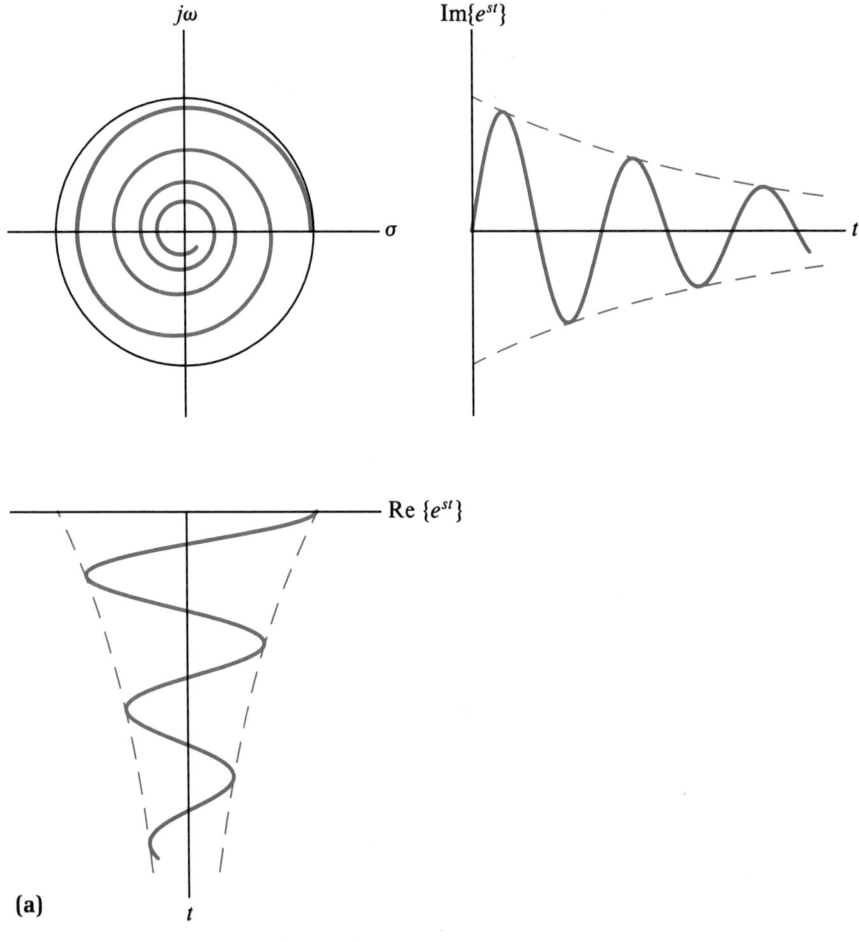

Figure 10.2 Exponential signal components: (a) damped and (b) undamped.

where

$$\mathbf{V} = Ve^{j\phi} = V\underline{/\phi}$$

and

$$\mathbf{f}(t) = \mathbf{V}e^{st}$$

V, σ, ω and ϕ are signal parameters. $\mathbf{V}$ is called the **complex signal phasor,** and $s = \sigma + j\omega$ is called the **complex frequency** with the **damping factor** σ, and the **radian frequency** ω. V is called the **phasor amplitude** of the complex signal, with $V = |\mathbf{V}|$. The complex phasor $\mathbf{V}$ determines the value of the signal at $t = 0$, with

$$\mathbf{f}(0) = Ve^{j\phi} = V \cos \phi + jV \sin \phi$$

This, too, is a vector in the complex plane; it specifies the initial value of the real and imaginary parts of $\mathbf{f}(0)$.

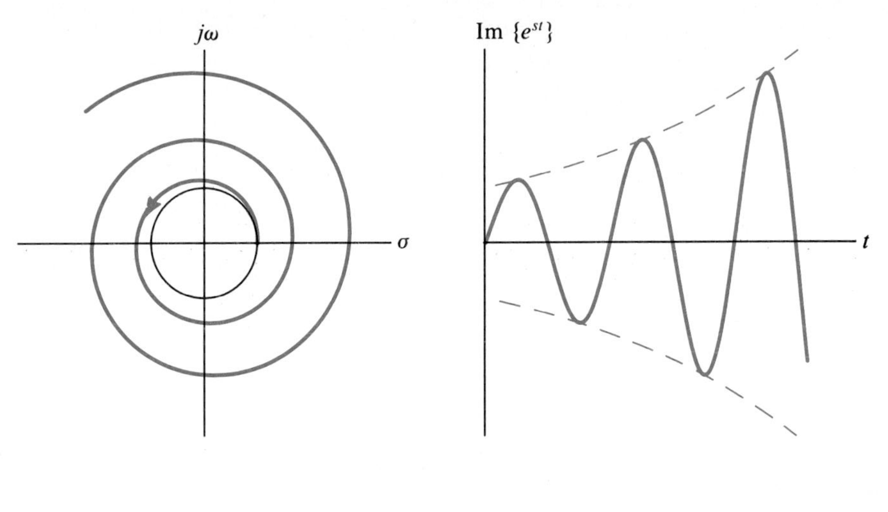

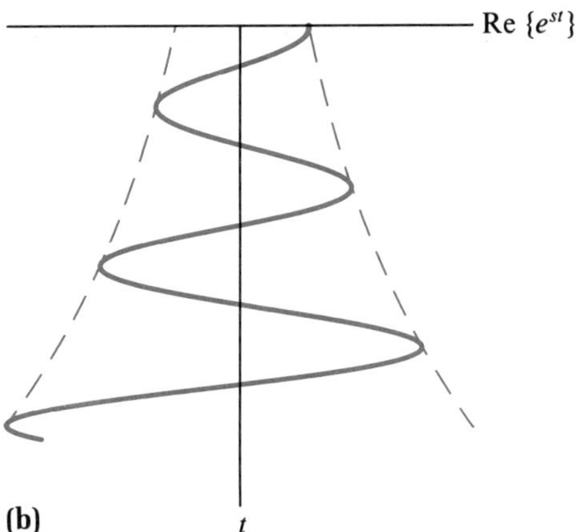

(b)

Figure 10.2 (continued)

The angle ϕ is called a **phasor angle** because it is the angle of the complex phasor $Ve^{j\phi}$. It is also the **phase angle** of the time-domain signals corresponding to the real and imaginary parts of the complex exponential signal $\mathbf{f}(t)$.

The time-domain portion of $\mathbf{f}(t)$ in (10.3d) can be written as

$$e^{(\sigma+j\omega)t} = e^{\sigma t}e^{j\omega t} \tag{10.4a}$$

$$= e^{\sigma t}(\cos \omega t + j \sin \omega t) \tag{10.4b}$$

$$= e^{\sigma t} \cos \omega t + je^{\sigma t} \sin \omega t \tag{10.4c}$$

Thus, $\mathbf{f}(t)$ has real and imaginary parts corresponding to damped sinusoids when $\sigma < 0$ (since $e^{\sigma t}$ decreases as t increases), and growing sinusoids when $\sigma > 0$. The complex signal defined by (10.4a) is shown in Fig. 10.2 as a vector rotating in the complex plane with (a) decaying or (b) growing length. The projected real and imaginary components are also shown.

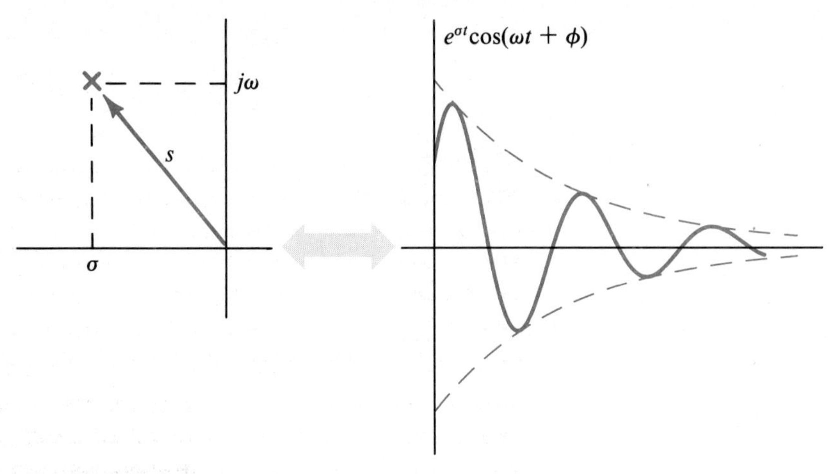

Figure 10.3 The location of s in the complex plane determines the time-domain parameters σ and ω of the damped sinusoidal signal.

A real, damped and phase-shifted sinusoid can also be written in terms of a complex exponential as

$$
\begin{aligned}
g(t) &= Ve^{\sigma t} \sin(\omega t + \phi) \\
&= Ve^{\sigma t}\mathrm{Im}\{e^{j(\omega t+\phi)}\} \\
&= \mathrm{Im}\{Ve^{\sigma t}e^{j(\omega t+\phi)}\} \\
&= \mathrm{Im}\{Ve^{j\phi}e^{(\sigma+j\omega)t}\} \\
&= \mathrm{Im}\{\mathbf{V}e^{\mathbf{s}t}\}
\end{aligned}
$$

Thus, a real, damped cosine or sine signal can be obtained as either the real or imaginary part, respectively, of a damped complex signal.

Notice how the expression of $\mathbf{f}(t)$ given by (10.3c) displays the amplitude, phase angle, damping factor, and frequency of the signal in (10.2). In fact, the location of s in the complex plane allows us to obtain σ and ω by inspection, as shown in Figure 10.3.

Example 10.1

Find the signal parameters of $f(t)$, where

$$f(t) = 10e^{-2t} \cos\left(3t - \frac{\pi}{4}\right)$$

Solution: $f(t) = 10e^{-2t}\mathrm{Re}\{e^{j(3t-\pi/4)}\}=\mathrm{Re}\{10e^{-j\pi/4}e^{(-2+j3)t}\}$

and so

$f(t) = \mathrm{Re}\{\mathbf{V}e^{\mathbf{s}t}\}$

where

$$\mathbf{V} = Ve^{j\phi} = 10e^{-j\pi/4}$$

and

$$s = \sigma + j\omega = -2 + j3$$

Thus, the signal parameters are $V = 10$, $\phi = -\frac{\pi}{4}$, $\sigma = -2$ and $\omega = 3$.

In summary, real-world signals can be expressed in terms of complex exponentials by using

$$Ve^{\sigma t}\cos(\omega t + \phi) = \text{Re}\{Ve^{j\phi}e^{(\sigma+j\omega)t}\} = \text{Re}\{\mathbf{V}e^{st}\}$$
$$Ve^{\sigma t}\sin(\omega t + \phi) = \text{Im}\{Ve^{j\phi}e^{(\sigma+j\omega)t}\} = \text{Im}\{\mathbf{V}e^{st}\}$$

with $s = \sigma + j\omega$ and $\mathbf{V} = |\mathbf{V}|e^{j\phi} = Ve^{j\phi} = V\angle\phi$.

Skill Exercise 10.1

Find the complex signal representation of $f(t)$ and $g(t)$, where

$$f(t) = 22e^{-3t}\cos\left(100t - \frac{\pi}{4}\right)$$

and

$$g(t) = -13e^{-10t}\sin\left(2t + \frac{\pi}{6}\right)$$

Answer: $f(t) = \text{Re}\{22e^{-j\pi/4}e^{(-3+j100)t}\}$
$g(t) = \text{Im}\{-13e^{j\pi/6}e^{(-10+j2)t}\}$

Real-world sinusoids can also be represented by a sum of complex exponentials if we use the trigonometric identities

$$\cos\alpha = \frac{e^{j\alpha} + e^{-j\alpha}}{2}$$

and

$$\sin\alpha = \frac{e^{j\alpha} - e^{-j\alpha}}{2j}$$

to write

$$Ve^{\sigma t}\cos(\omega t + \phi) = Ve^{\sigma t}\frac{e^{j(\omega t+\phi)} + e^{-j(\omega t+\phi)}}{2}$$

$$Ve^{\sigma t}\sin(\omega t + \phi) = Ve^{\sigma t}\frac{e^{j(\omega t+\phi)} - e^{-j(\omega t+\phi)}}{2j}$$

This forms damped sine and cosine signals as a superposition of damped complex exponential signals.

10.2 THE PARTICULAR SOLUTION FOR A SINUSOIDAL INPUT SIGNAL

Complex exponential signals are especially useful in circuit analysis because they allow us to include sinusoids in the same framework that we use to treat exponential signals. We will now show how to find the particular solution of a circuit's I/O differential equation for a complex exponential signal and will then construct the series *RC* circuit's zero-state response to a sinusoidal signal.

Only real signals can be generated in the physical world! Nevertheless, we can apply complex exponential signals to the abstract mathematical model of a circuit to obtain a complex response signal, as shown in Fig. 10.4, where both $\mathbf{V}$ and s are possibly complex.

Note: We show a time-domain signal entering a block containing a transfer function to *symbolize* the fact that the circuit represented by the transfer function operates on the time-domain input signal to produce the time-domain output signal. The signals are in the time domain, and the transfer-function model of the circuit is in the s, or complex-frequency, domain.

When the input to a circuit model (not the physical circuit!) is a complex exponential signal, the solution to the I/O differential equation will have a natural solution and a particular solution. The natural solution does not depend on the input signal. However, the particular-solution component will be complex and will have the exponential form presented in Chapters 7–9, but with *complex* signal parameters

$$\mathbf{y}_P(t) = \mathbf{V}\mathbf{H}(s)e^{st} = \mathbf{Y}e^{st} \tag{10.5}$$

provided that s is not a natural frequency of the circuit described by the transfer function $\mathbf{H}(s)$. Since $\mathbf{H}(s)$ may, in general, be a complex number, we have

$$\mathbf{H}(s) = |\mathbf{H}(s)|e^{j\theta(s)} \tag{10.6}$$

and so

$$\mathbf{Y} = \mathbf{H}(s)\mathbf{V} \tag{10.7}$$

with

$$\theta(s) = \measuredangle \mathbf{H}(s).$$

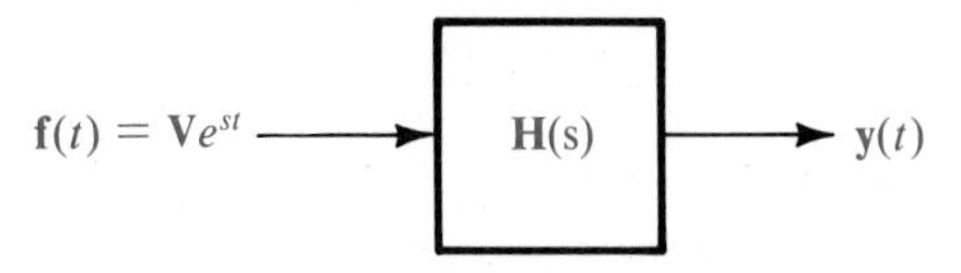

Figure 10.4 Complex input signal and complex output signal.

Therefore, **the algorithm for finding the complex output phasor is the same as the algorithm for finding the amplitude of the exponential particular solution to a real exponential input signal.** The complex output phasor is the product of the evaluated transfer function and the complex input phasor.

When the input signal is a complex exponential signal the particular solution to the I/O differential equation is complex, with

$$\begin{aligned}\mathbf{y}_p(t) &= Ve^{j\phi}|\mathbf{H}(s)|e^{j\theta(s)}e^{st}\\ &= V|\mathbf{H}(s)|e^{\sigma t}e^{j[\omega t + \phi + \theta(s)]}\\ &= V|\mathbf{H}(s)|e^{\sigma t}\{\cos[\omega t + \phi + \theta(s)] + j\sin[\omega t + \phi + \theta(s)]\}\end{aligned}$$

Equations (10.5) and (10.6) state that **the particular solution to a complex exponential is itself a complex exponential.** Using the **superposition** property of the particular solution, we associate the real part of this response with the real part of the complex input signal, and the imaginary part of the response with the imaginary part of the input signal. Thus, letting the notation $x \rightarrow y$ express the fact that y is the particular solution when x is the exponential input, we have

$$\text{Re}\{\mathbf{V}e^{st}\} \rightarrow \text{Re}\{\mathbf{V}\mathbf{H}(s)e^{st}\}$$

$$\text{Im}\{\mathbf{V}e^{st}\} \rightarrow \text{Im}\{\mathbf{V}\mathbf{H}(s)e^{st}\}$$

or

$$Ve^{\sigma t}\cos(\omega t + \phi) \rightarrow V|\mathbf{H}(s)|e^{\sigma t}\cos[\omega t + \phi + \theta(s)]$$

$$Ve^{\sigma t}\sin(\omega t + \phi) \rightarrow V|\mathbf{H}(s)|e^{\sigma t}\sin[\omega t + \phi + \theta(s)]$$

Thus, we conclude that if a circuit is linear, a damped sinusoidal (sine or cosine) input signal causes a sinusoidal particular solution to the I/O differential equation. This mathematical property of the solution to the differential equation allows us to quickly solve the circuit problem of finding the response to a damped sinusoidal source in the same manner as we found the response to a real exponential source.

Example 10.2

The series RC circuit with $\tau = 10$ has the voltage-transfer function

$$\mathbf{H}(s) = \frac{1/10}{s + (1/10)}$$

If the input *to the model* is the complex signal

$$\mathbf{V}_{in}(t) = 5e^{j45°}e^{(-1+j.5)t}$$

find the particular solution.

Solution: The input signal is described by $\mathbf{V} = 5e^{j45°}$. The *fixed* source frequency is $s = -1 + j.5$. Therefore, the particular solution is

$$\mathbf{v}_p(t) = 5e^{j45°}\,\mathbf{H}(-1 + j.5)e^{(-1+j.5)t}$$

$$= 5e^{j45°}\frac{\frac{1}{10}}{-1 + j.5 + \frac{1}{10}}e^{(-1+j.5)t}$$

$$= 5e^{j45°}\frac{0.1}{1.03e^{j150.95°}}e^{(-1+j.5)t}$$

$$= 0.49e^{-j105.95°}e^{(-1+j.5)t}$$

Note that the particular solution has a complex phasor of $\mathbf{V}_p = 0.49\underline{/-105.95°}$. With Euler's law the **complex particular solution** can be written as

$$\mathbf{V}_p(t) = 0.49e^{-t}e^{j(0.5t-105.95°)}$$

$$= 0.49e^{-t}[\cos(0.5t - 105.95°) + j\sin(0.5t - 105.95°)]$$

Likewise, the complex input signal can be written as

$$\mathbf{v}_{in}(t) = 5e^{j45°}e^{(-1+j.5)t}$$

$$= 5e^{-t}e^{j(0.5t+45°)}$$

$$= 5e^{-t}[\cos(0.5t + 45°) + j\sin(0.5t + 45°)]$$

Using superposition, we claim that

$$\text{Re}\{\mathbf{v}_{in}(t)\} \rightarrow \text{Re}\{\mathbf{v}_p(t)\}$$

and

$$\text{Im}\{\mathbf{v}_{in}(t)\} \rightarrow \text{Im}\{\mathbf{v}_p(t)\}$$

or that

$$5e^{-t}\cos(0.5t + 45°) \rightarrow 0.49e^{-t}\cos(0.5t - 105.95°)$$

and

$$5e^{-t}\sin(0.5t + 45°) \rightarrow .49e^{-t}\sin(0.5t - 105.95°)$$

This result can be verified by direct substitution in the I/O equation.

Before proceeding further, let us summarize the important concepts that have been discussed so far:

1. The particular solution to a damped cosine (sine) is a damped cosine (sine).
2. The s-domain frequency of the particular solution to a complex exponential input signal is the same as the s-domain frequency of the input signal (damping factor and radian frequency).

TABLE 10.1 INPUT/OUTPUT RELATIONSHIPS FOR COMPLEX EXPONENTIAL SIGNALS.

COMPLEX SIGNAL	COMPLEX FREQUENCY	COMPLEX AMPLITUDE	MAGNITUDE	ANGLE
INPUT: $\mathbf{V}e^{st}$	s	$\mathbf{V}$	$\lvert\mathbf{V}\rvert$	$\measuredangle\mathbf{V}$
OUTPUT: $\mathbf{VH}(s)e^{st}$	s	$\mathbf{VH}(s)$	$\lvert\mathbf{V}\rvert\lvert\mathbf{H}(s)\rvert$	$\measuredangle\mathbf{V} + \measuredangle\mathbf{H}(s)$

3. The complex amplitude of the particular solution is the complex amplitude of the input signal *scaled* by the magnitude of the value of the transfer function evaluated at the input frequency.
4. The phasor angle of the complex output phasor is the phasor angle of the input signal *plus* the angle of the transfer function evaluated at the input frequency.
5. The complex output phasor is the *product* of the complex input phasor and the value of the transfer function evaluated at the input frequency.

The transfer function now has greater significance because it *completely* summarizes the effect of the circuit on the complex exponential input signal, and thereby determines the particular solution to a sinusoidal signal. This important relationship is expressed in Fig. 10.5. Again, remember that $\mathbf{H}(s)$ represents the circuit that operates on the actual time-domain signals. The operation is defined by the relationship between the complex phasors and $\mathbf{H}(s)$.

The input/output relationship for complex exponential signals is summarized in Table 10.1 for a fixed value of s, s not a natural frequency.

Skill Exercise 10.2

If the model of a first-order circuit has

$$\mathbf{H}(s) = \frac{0.5s}{s + 10}$$

find the particular solution to

$$\mathbf{v}_{in}(t) = 100e^{-5t}e^{j(-10t+\pi/3)}$$

Answer: $\mathbf{v}_p(t) = 50e^{-5t}e^{j(-10t+6.87°)}$

Example 10.3

Find $v_o(t)$ when $v_{in}(t) = 5e^{-3t}\cos(2t + 30°)$, $t \geq 0$, and $\mathbf{H}(s) = 1/(s + 1)$.

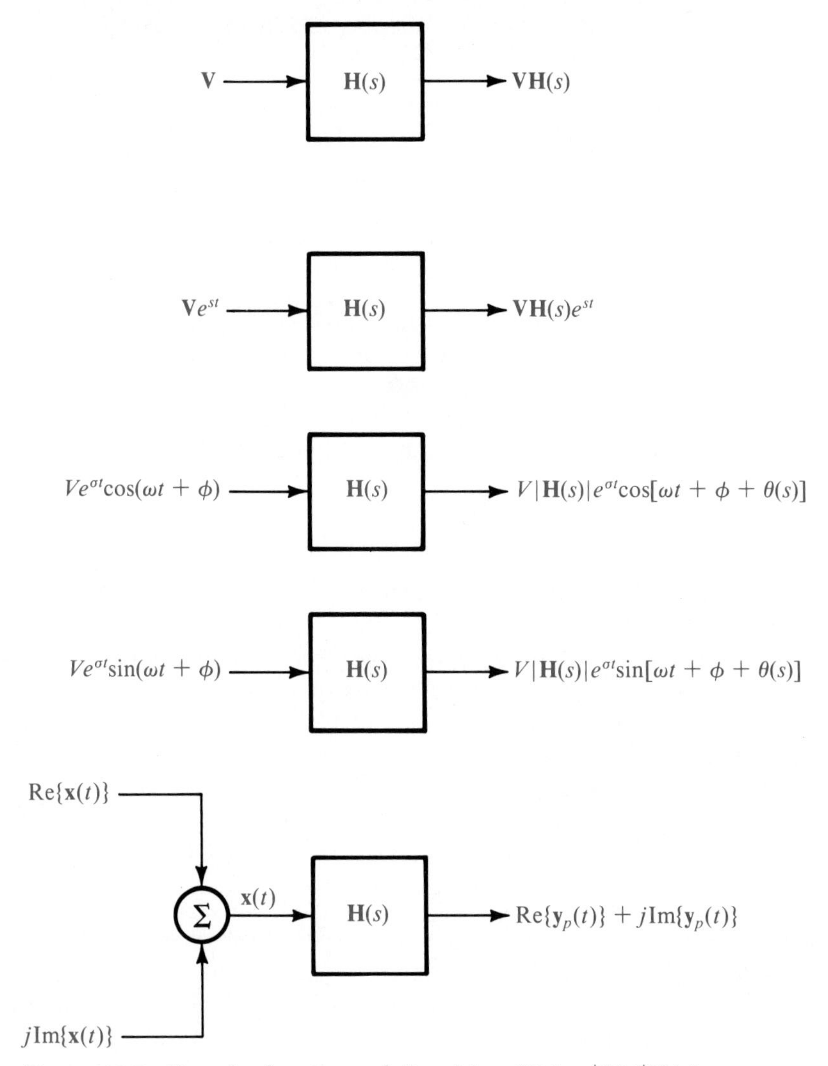

Figure 10.5 Transfer function relationships, $\mathbf{H}(s) = |\mathbf{H}(s)|\underline{/\Theta(s)}$.

Solution: First we describe $v_{in}(t)$ in terms of a complex signal

$$v_{in}(t) = \mathrm{Re}\{5e^{j30°}e^{(-3+j2)t}\} \qquad t \geq 0$$

then we identify the signal parameters $\mathbf{V} = 5\angle 30°$ and $s = -3 + j2$, with $\sigma = -3$ and $\omega = 2$. (With practice the signal parameters can be obtained by inspection of the input signal.) Next, we evaluate the transfer function at the source frequency

$$\mathbf{H}(s) = \mathbf{H}(-3 + j2) = 0.35\underline{/-135°}$$

so we know that v_{in} will have a particular solution given by

$$v_p(t) = \text{Re}\{\mathbf{H}(s)\,\mathbf{V}e^{st}\}$$
$$= \text{Re}\{0.35e^{-j135°}5e^{j30°}e^{(-3+j2)t}\}$$
$$= 1.75e^{-3t}\cos(2t - 105°)$$

The same result can be obtained more quickly by using (10.7) to get the effect of the change in magnitude and the change in phase angle. Thus, $\mathbf{H}(-3 + j2) = 0.35\angle -135°$ so $|\mathbf{H}(-3 + j2)| = 0.35$, $\sphericalangle\mathbf{H}(-3 + j2) = -135°$ and

$$5e^{-3t}\cos(2t + 30°) \rightarrow 5(0.35)e^{-3t}\cos(2t + 30° - 135°)$$

$$5e^{-3t}\cos(2t + 30°) \rightarrow 1.75e^{-3t}\cos(2t - 105°)$$

In terms of the complex phasors:

$$5\angle 30° \rightarrow 5(0.35)\angle(30° - 135°) = 1.75\angle -105°$$

This also leads to $v_p(t)$ by inspection.

Skill Exercise 10.3

Find the particular solution when $v_{in}(t) = 2e^{-5t}\cos(3t - 45°)$ and $\mathbf{H}(s) = s/(s + 1)$.

Answer: $v_p(t) = 2.33e^{-5t}\cos(3t - 39.09°)$

10.3 ZSR TO SINUSOIDAL INPUT

The ZSR of a circuit to a sinusoidal input signal is found by adding the natural and particular solutions to the circuit's I/O differential equation, and then applying the boundary conditions.

Example 10.4

The first-order op amp circuit in Fig. 10.6 has a transfer function given by

$$\mathbf{H}(s) = \frac{1/\tau}{s + (1/\tau)}$$

where

$$\tau = \frac{[R_1R_2 + R_2R_3 + R_1R_3]C}{R_3}$$

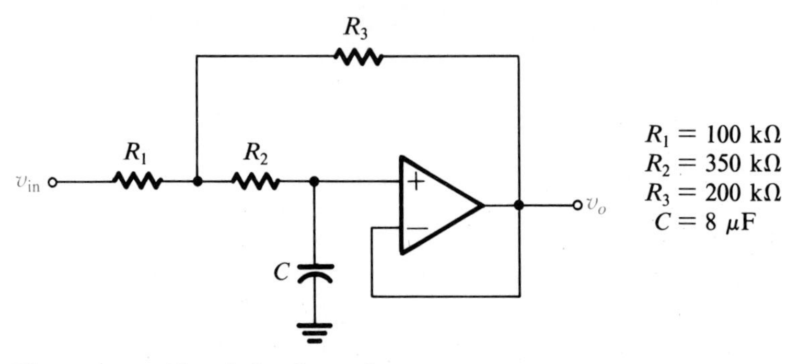

Figure 10.6 Circuit for Example 10.4

If $v_{in}(t) = 50 \sin(t + 30°)$, find the ZSR of $v_o(t)$, and its transient and steady-state components for the indicated circuit parameter values.

Solution: The ZSR will have the form

$$v(t) = Ke^{s_1 t} + I_m\{\mathbf{V}\mathbf{H}(s)e^{st}\}$$

where $Ke^{s_1 t}$ is the natural solution to the first-order differential equation model of the circuit, and $\mathbf{VH}(s)e^{st}$ is the particular solution. For the given component values, $\mathbf{H}(s) = 0.2/(s + 0.2)$. By inspection, $v_{in}(t) = \text{Im}\{50e^{j30°}e^{jt}\}$, so the complex source frequency is $s = j1$, and the source phase angle is $\phi = 30°$. Evaluating the transfer function, we get

$$\mathbf{H}(j1) = \frac{0.2}{j1 + 0.2} = 0.196 \angle -78.7°$$

so the particular solution is

$$\begin{aligned} v_p(t) &= 0.196(50) \sin(t + 30° - 78.7°) \\ &= 9.8 \sin(t - 48.7°) \end{aligned}$$

Next, the natural solution and the particular solution are combined to get the complete solution

$$v_o(t) = Ke^{-0.2t} + 9.8 \sin(t - 48.7°)$$

The last step is to use the boundary condition for the ZSR:

$$\begin{aligned} v_o(0^+) = 0 &= K + 9.8 \sin(-48.7°) \\ K &= -9.8(-0.75) = 7.35 \end{aligned}$$

and so

$$v_o(t) = 7.35e^{-2t} + 9.8 \sin(t - 48.7°) \qquad (9.34)$$

The graph of $v_o(t)$ is shown in Fig. 10.7

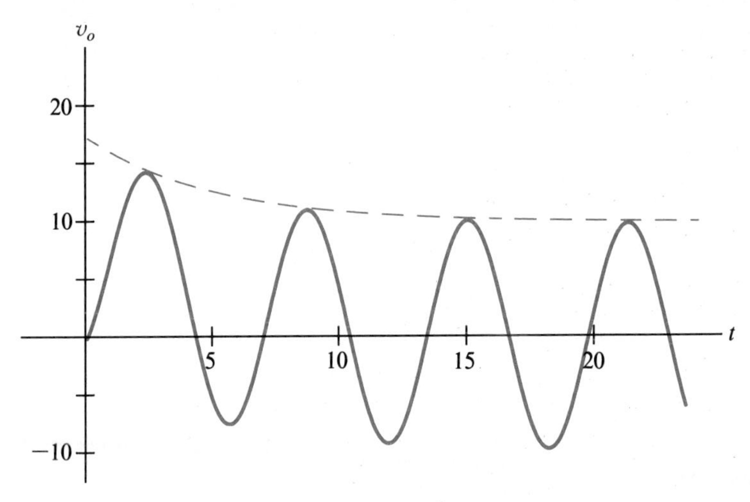

Figure 10.7 Graph of $v_o(t)$ for Example 10.4

The transient response decays to zero in approximately 20 sec (4 time constants).

10.4 SINUSOIDAL STEADY-STATE RESPONSE

If the transfer function of a circuit has all of its natural frequencies in the left half-plane, the natural solution will eventually decay to zero. Thus, an undamped sinusoidal input signal produces a steady-state sinusoidal signal equal to the particular solution. If

$$v_{in}(t) = V \cos(\omega t + \phi)$$

the steady-state response is

$$v_{ss}(t) = V|\mathbf{H}(j\omega)| \cos[\omega t + \phi + \measuredangle \mathbf{H}(j\omega)]$$

The sinusoidal steady-state response is determined completely by the transfer function, evaluated at $s = j\omega$ for a given ω, as depicted in Fig. 10.8 and summarized in Table 10.2. $\mathbf{H}(s)$ represents the circuit driven by $v_{in}(t)$.

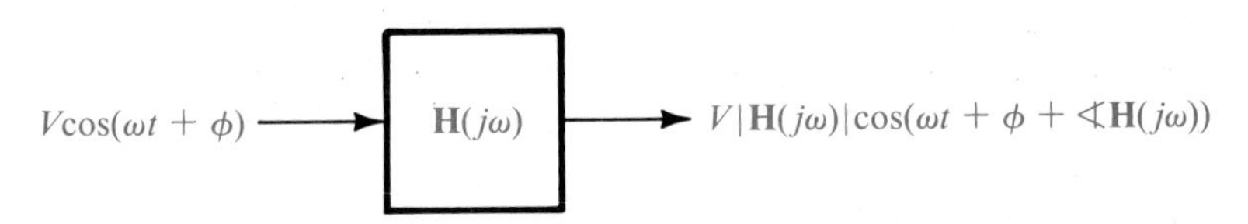

Figure 10.8 Sinusoidal steady-state input/output signal relationships.

TABLE 10.2 STEADY-STATE INPUT/OUTPUT RELATIONSHIPS

SIGNAL	FREQUENCY	AMPLITUDE	PHASE
INPUT	ω	V	ϕ
OUTPUT	ω	$V\|\mathbf{H}(j\omega)\|$	$\phi + \measuredangle\mathbf{H}(j\omega)$

The amplitude of the steady-state response to a sinusoidal input signal is equal to the product of the input amplitude and the magnitude of the transfer function, and the phase angle of the response is equal to the sum of the phase angle of the input signal and the angle of the transfer function. **The phase angles of the output signal and the input signal differ by the angle of the transfer function** $\measuredangle\mathbf{H}(j\omega)$. Therefore, the input signal can be written as

$$\cos(\omega t + \phi) = \cos\omega\left(t - \frac{-\phi}{\omega}\right) = \cos\omega(t - \tau_1)$$

and the output signal can be written as

$$\cos[\omega t + \phi + \measuredangle\mathbf{H}(j\omega)] = \cos\omega\left[t - -\left(\frac{\phi}{\omega} + \measuredangle\frac{\mathbf{H}(j\omega)}{\omega}\right)\right]$$

$$= \cos\omega(t - \tau_2)$$

with $\tau_1 = -\phi/\omega$, $\tau_2 = -\phi/\omega - \measuredangle\mathbf{H}(j\omega)/\omega$ and $\tau_2 - \tau_1 = -\measuredangle\mathbf{H}(j\omega)/\omega$. **The graph of the steady-state output signal will be delayed relative to the input signal by**

$$\tau = -\frac{\measuredangle\mathbf{H}(j\omega)}{\omega} = \frac{-\theta(j\omega)}{\omega}$$

on the time axis relative to the graph of the sinusoidal input signal, as shown in Figure 10.9. Since ω is the same, both signals have the same period.

Since $|\mathbf{H}(j\omega)|$ and $\theta(j\omega)$ play such an important role in determining the sinusoidal steady-state response of a circuit, their graphs are given special names. The graph of $|\mathbf{H}(j\omega)|$ versus ω is called the **magnitude response,** and the graph of $\theta(j\omega)$ is called the **phase response.** This engineering jargon is commonly used, but it can be a source of confusion because the graphs are not responses in the usual sense—they do not result from an applied input signal! Instead, they display the values of $|\mathbf{H}(j\omega)|$ and $\measuredangle\mathbf{H}(j\omega)$ that would determine the steady-state output signal for a given input signal whose frequency is ω. In calculating the response, only one point of data from each curve is used, corresponding to ω for the sinusoidal input. Sometimes $|\mathbf{H}(j\omega)|$ and $\measuredangle\mathbf{H}(j\omega)$ are called the **frequency-domain**

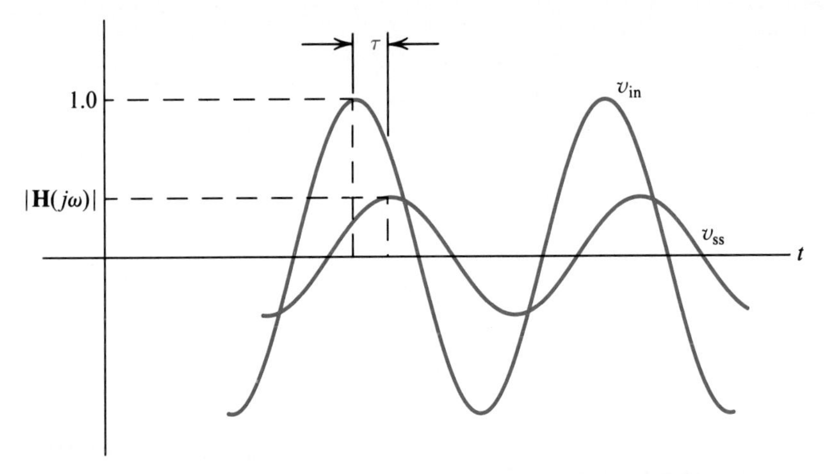

Figure 10.9 The sinusoidal steady-state output signal is delayed relative to the sinusoidal input signal.

response characteristics of the circuit, because they summarize the information needed to determine the response of the circuit to any sinusoidal input signal.

Example 10.5

Find the magnitude- and phase-response characteristics of the op amp circuit in Example 10.4.

Solution: The circuit has

$$\mathbf{H}(j\omega) = \frac{1}{1 + j\omega\tau}$$

so

$$|\mathbf{H}(j\omega)| = \frac{1}{\sqrt{1 + \omega^2\tau^2}}$$

and $\theta(j\omega) = \sphericalangle\mathbf{H}(j\omega) = -\tan^{-1}\omega\tau$. These graphs are shown in Figs. 10.10(a) and 10.10(b).

For this example, the sinusoidal steady-state output-voltage signal will always have an amplitude less than or equal to the input signal, and the output signal will be delayed on the time axis relative to the input signal. The circuit is said to exhibit a **lowpass frequency response** because $|\mathbf{H}(j\omega)| \rightarrow 0$ as $\omega \rightarrow \infty$, i.e. the output amplitude due to a given sinusoidal input signal will be smaller at high frequencies than at low frequencies.

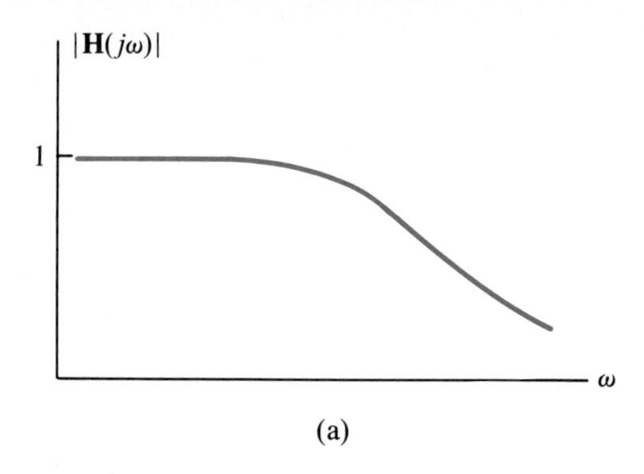

(a)

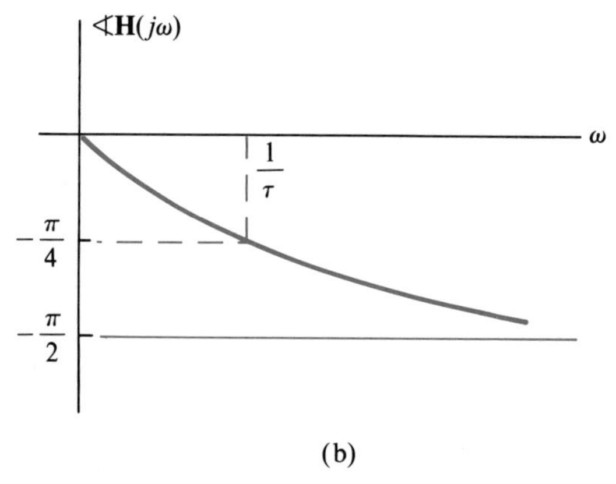

(b)

Figure 10.10 The (a) magnitude and (b) phase response of the op amp circuit in Fig. 10.6.

The physical explanation for the circuit's lowpass frequency response can be understood by considering the two input sinusoids shown in Fig. 10.11. Both signals have the same amplitude, but different frequency. The steady-state output voltage ultimately depends on the amount of charge that accumulates on the capacitor. With a high-frequency signal the capacitor has less time to accumulate charge before the current reverses and the capacitor starts discharging, so the high-frequency signal produces a smaller output voltage even though the input amplitude is fixed, and $|\mathbf{H}(j\omega)| \to 0$ as $\omega \to \infty$. This charge–voltage mechanism also explains the behavior of the phase characteristic as $\omega \to \infty$. Since the capacitor voltage is approximately zero, the charging current will be at the same phase angle as the source, and the capacitor itself introduces a voltage phase lag of 90° relative to the current.

We would also expect the output amplitude to nearly equal the input amplitude if the circuit's time constant τ is small relative to the duration of the charging cycle (if $\tau \ll 2\pi/\omega$). This can be seen by noting that

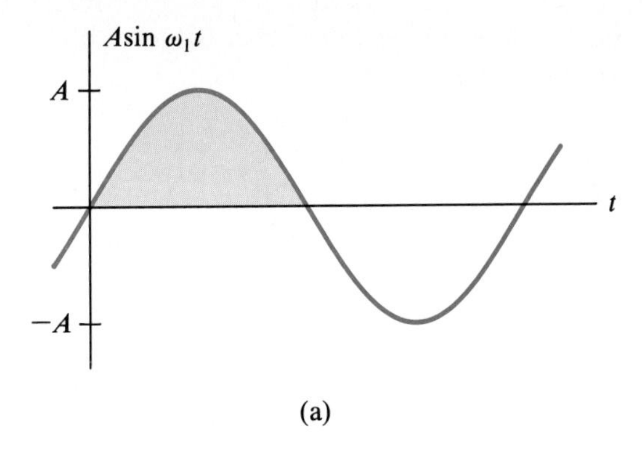

(a)

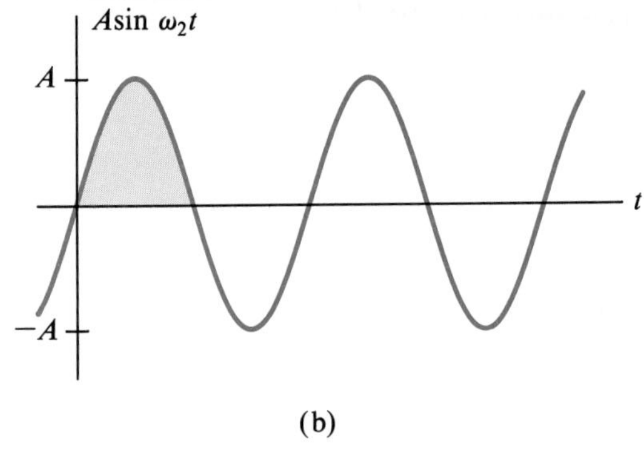

(b)

Figure 10.11 Comparison of sinusoidal signals having the same amplitude, but different radian frequencies, with $\omega_1 < \omega_2$.

$$|\mathbf{H}(j\omega)| \cong \frac{1}{\sqrt{1 + \omega^2\tau^2}} = 1$$

if $\omega^2\tau^2 \ll 1$, or if $\tau \ll 1/\omega$.

This relationship displays the tradeoff between the time domain and the frequency domain; if ω is large τ must be very small if the circuit is to replicate the input signal. On the other hand, for a given τ a sinusoid with period T must have

$$\boxed{\omega \ll \frac{1}{\tau}} \quad \text{or} \quad \boxed{\frac{T}{2\pi} \gg \tau}$$

to have the capacitor voltage closely follow the input voltage.

We'll examine frequency-domain response characteristics in greater detail in Chapter 16, and we'll use them to design signal filters whose graphs of $|\mathbf{H}(j\omega)|$ and $\measuredangle\mathbf{H}(j\omega)$ have desirable shapes.

10.5 SUPERPOSITION OF SINUSOIDAL SIGNALS

In many practical applications, the signal driving a circuit is actually the sum of two or more sinusoidal signals. In this case we construct the particular solution to the input/output differential equation by using the superposition property of the particular solution to the I/O DE (See Chapter 7.) First, we represent each sinusoid in terms of a complex exponential signal. Next, we find the complex exponential particular solution to each complex input exponential (taking care to evaluate the transfer function $\mathbf{H}(s)$ at the frequency corresponding to the term being evaluated). Then we superimpose the real and/or imaginary parts of these complex response exponentials to get the particular solution. If the complex input signal in Fig. 10.12 is

$$\mathbf{v}_{\text{in}}(t) = \mathbf{V}_1 e^{s_1 t} + \mathbf{V}_2 e^{s_2 t} + \cdots + \mathbf{V}_N e^{s_N t}$$

the complex particular solution is

$$\mathbf{y}_p(t) = \mathbf{V}_1\mathbf{H}(s_1)e^{s_1 t} + \mathbf{V}_2\mathbf{H}(s_2)e^{s_2 t} + \cdots + \mathbf{V}_N\mathbf{H}(s_N)e^{s_N t}$$

provided that the source frequencies $s_1, s_2, \ldots, s_N$ are distinct from any of the circuit's natural frequencies. Using superposition, we conclude that if the real input signal is

$$v_{\text{in}}(t) = V_1 \cos(\omega_1 t + \phi_1) + V_2 \cos(\omega_2 t + \phi_2) + \cdots + V_N \cos(\omega_N t + \phi_N)$$

the real particular solution is

$$y_p(t) = V_1|\mathbf{H}(j\omega_1)|\cos[\omega_1 t + \phi_1 + \theta(j\omega_1)] + V_2|\mathbf{H}(j\omega_2)|\cos[\omega_2 t + \phi_2 + \theta(j\omega_2)] + \cdots + V_N|\mathbf{H}(j\omega_N) \cos[\omega_N t + \phi_N + \theta(j\omega_N)]$$

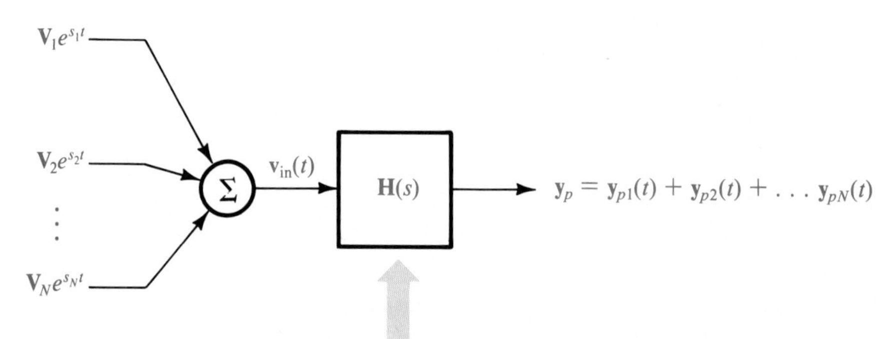

Figure 10.12 Superposition of particular solutions to complex exponential input signals.

Example 10.6

Suppose a circuit has $\mathbf{H}(s) = 1/(s + 0.2)$. If $v_{in}(t) = 10 \cos(t + 30°) - 5 \cos(2t + 45°)$ find the ZSR.

Solution: Using the superposition property of the particular solution we have

$$v_p(t) = 10|\mathbf{H}(j1)| \cos[t + 30° + \measuredangle\mathbf{H}(j1)]$$
$$-5|\mathbf{H}(j2)| \cos[2t + 45° + \measuredangle\mathbf{H}(j2)]$$

Since

$$\mathbf{H}(j1) = \frac{1}{j1 + 0.2} = \frac{1}{1.02\underline{/78.69°}} = 0.98\underline{/-78.69°}$$

and

$$\mathbf{H}(j2) = \frac{1}{j2 + 0.2} = \frac{1}{2.01\underline{/84.29°}} = 0.50\underline{/-84.29}$$

we have

$$v_p(t) = 10(.98) \cos(t + 30° - 78.69°)$$
$$- 5(0.50) \cos(2t + 45° - 84.29°)$$
$$= 9.8 \cos(t - 48.69°) - 2.5 \cos(2t - 39.29°)$$

The complete solution is

$$v_o(t) = Ke^{-0.2t} + 9.8 \cos(t - 48.69°)$$
$$- 2.5 \cos(2t - 39.29°)$$

For the ZSR

$$v(0^+) = K + 9.8(0.66) - 2.5(0.77) = K + 4.53 = 0$$

so $K = -4.53$ and the ZSR is

$$v_o(t) = -4.53e^{-0.2t} + 9.8 \cos(t - 48.69°)$$
$$- 2.5 \cos(2t - 39.29°)$$

Skill Exercise 10.4

Suppose the input signal in Example 10.5 is changed to $v_{in}(t) = 20 \cos(t + 30°) - 10 \sin(2t + 45°)$. Find the ZSR.

Answer: $v_o(t) = -9.06e^{-0.2t} + 19.6 \cos(t - 48.69°) - 5.0 \cos(2t - 39.29°)$

SUMMARY

This chapter showed how to construct the particular solution to the input/output differential equation of a circuit driven by a sinusoidal signal. Any real-world sinusoidal signal can be embedded within a complex exponential phasor signal whose real or imaginary part is the sinusoid. The particular solution to a complex exponential signal can be constructed in the same way as the particular solution to a real exponential signal—the circuit's transfer function specifies the complex amplitude of the particular solution, including its magnitude and phasor angle. The particular solution to a real sinusoidal signal is taken as the real or imaginary part of the corresponding complex particular solution, depending on whether the source was expressed as a cosine or a sine signal. The transfer function completely determines the circuit's steady-state response to a given sinusoidal signal if the circuit's natural frequencies are all in the left half-plane; and the steady-state response to a signal consisting of a sum of sinusoids is the sum of its response to each of them.

Problems - Chapter 10

10.1 Find the real and imaginary parts of the complex signals given below.

a. $f(t) = e^{-3000t - j45°}$

b. $f(t) = 100\underline{/30°}\, e^{-10t} e^{j(100t - 20°)}$

c. $f(t) = 20e^{-j\pi/3} e^{(-2+j4)t}$

d. $f(t) = 50\underline{/60°}\, e^{-10t} e^{j(100t - 20°)}$

10.2 Express the following signals in terms of complex exponential signals.

a. $f(t) = 10 \cos 50\pi t$

b. $f(t) = 1000 \cos (120\pi t - 30°)$

c. $f(t) = -500e^{-5t} \cos (60\pi t + \pi/4)$

d. $f(t) = -100 \sin (400\pi t + \pi/6)$

e. $f(t) = 1000e^{-10t} \sin (200\pi t - \pi/10)$

f. $f(t) = 10 + \cos 4\pi t$

10.3 Find the signal parameters of the following complex exponential signals.

a. $\mathbf{v}(t) = 100e^{-3t} e^{j30°}$

b. $\mathbf{v}(t) = 20e^{j\pi/4} e^{-10t} e^{j40t}$

c. $\mathbf{v}(t) = 4e^{-5t} e^{j(10t - \pi/3)}$

d. $\mathbf{v}(t) = e^{-2t}[5 \cos (10t + \pi/3) + 5j \sin (10t + \pi/3)]$

10.4 If the input/output relationship of a circuit is described by $\mathbf{H}(s) = 2s/(s + 100)$ find the particular solution of the I/O differential equation for the complex input signals given below.

a. $\mathbf{v}(t) = 10e^{j100t}$
b. $\mathbf{v}(t) = 20e^{(-100+j100)t}$
c. $\mathbf{v}(t) = 5e^{-50t}e^{(j75t-30°)}$
d. $\mathbf{v}(t) = 10e^{0t}$

10.5 Find the particular solution $v_p(t)$ of the input/output differential equation of the circuit described by the transfer function $\mathbf{H}(s) = 10/(2s + 100)$ for each of the real input signals given below.
a. $v(t) = \text{Re}\{10e^{j5t}\}$
b. $v(t) = \text{Im}\{20e^{j(50t-45°)}\}$
c. $v(t) = \text{Re}\{50e^{-25t}e^{j(100t-30°)}\}$
d. $v(t) = 5 \cos 20t$
e. $v(t) = 2 \cos (40t - 30°)$
f. $v(t) = -5 \cos (100t + \pi/4)$
g. $v(t) = 3 \sin (2\pi t + \pi/6)$
h. $v(t) = -100e^{-20t} \sin (12\pi t - \pi/2)$
i. $v(t) = 1000e^{-40t} \cos (50\pi t - 60°)$
j. $v(t) = \cos (0.0001t + 30°)$

10.6 A circuit is described by the transfer function $\mathbf{H}(s) = 10/(s + 5)$ and $v_{in}(t) = 10 + 2 \sin (2\pi t + 30°)$.
a. Find the natural solution of the I/O equation.
b. Find the particular solution of the I/O equation.
c. Find the complete solution.
d. Repeat a, b, and c when

$$v_{in}(t) = 5e^{-3t} \cos (4\pi t + 60°) + 1$$

10.7 Find the steady-state response $y_{ss}(t)$ of a circuit whose transfer function is $\mathbf{H}(s) = 10s/(s + 10)$ when the input signal is given be
a. $v(t) = 10u(t) + 5 \sin (2\pi t + 45°)$
b. $v(t) = [3 \cos (6\pi t + \pi/3) + 2e^{-t} \sin 2\pi t]u(t)$

10.8 The op amp in Fig. 5.7 has $R = 1$ MΩ and $C = 1$ μF.
a. If $v_o(0^-) = 0$ and if $v_{in}(t) = 5 \sin 2\pi t\ u(t)$, find and draw $v_o(t)$ for $t \geq 0$.
b. Repeat a for $v_{in}(t) = 5 \sin 20\pi t\ u(t)$.
c. Repeat a with $v_o(0^-) = 5$ V.
d. Repeat a with C changed to 0.1 μF.
e. Repeat a with C changed to 10 μF.
f. Observe and explain the difference between $v_o(t)$ for parts a, d, and e.

10.9 The switch is closed at $t = 0$ in the circuit shown in Fig. P10.9.
a. Find and draw $v_o(t)$ if $v_c(0^-) = 0$ and $v_{in}(t) = 10 \sin 10\pi t$.
b. Repeat a with $C = 0.5$ μF.
c. Repeat a with $v_{in}(t) = 10 \sin 100\pi t$.

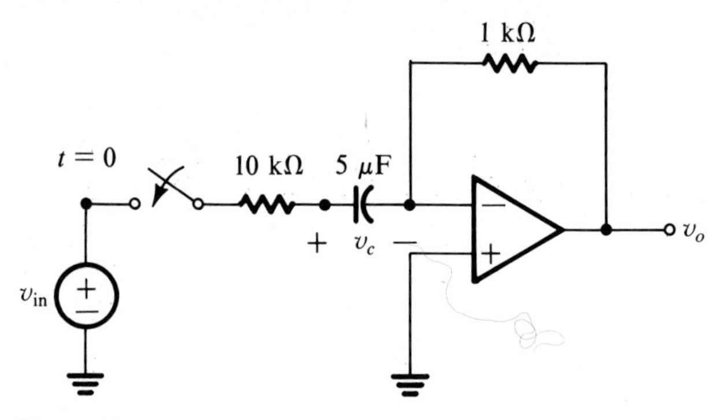

Figure P10.9

10.10 For the circuit of Fig. P10.10, $R_1 = 2000\ \Omega$, $R_2 = 100\ \text{k}\Omega$, $C = 5\ \mu\text{F}$, $v_c(0^-) = 0$ and the input signal is given by $v_{in}(t) = 10 \sin 0.4\pi t$. Find

a. $v_o(t)$
b. the transient component of $v_o(t)$.
c. the steady-state component of $v_o(t)$.

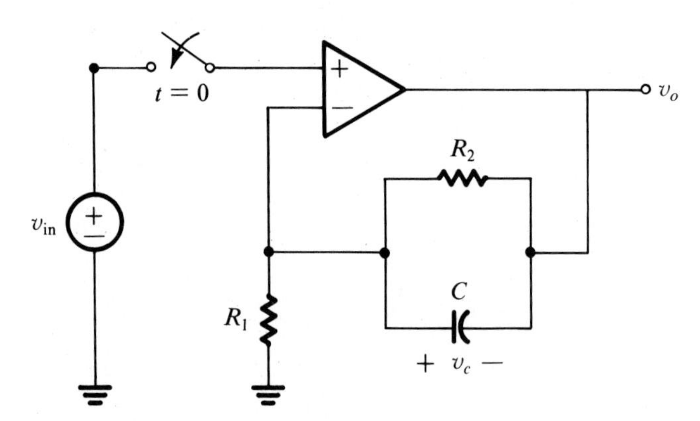

Figure P10.10

10.11 For the circuit of Fig. P10.11, $i(0^-) = 0$ and the switch closes at $t = 0$, with $v_{in}(t) = 100 \sin (20\pi t - 30°)$.

a. Find $i(t)$ for $t \geq 0$.
b. Find the transient component of $i(t)$.
c. Find the steady-state component of $i(t)$.
d. Display $i(t)$, $i_{tr}(t)$ and $i_{ss}(t)$ on the same graph.

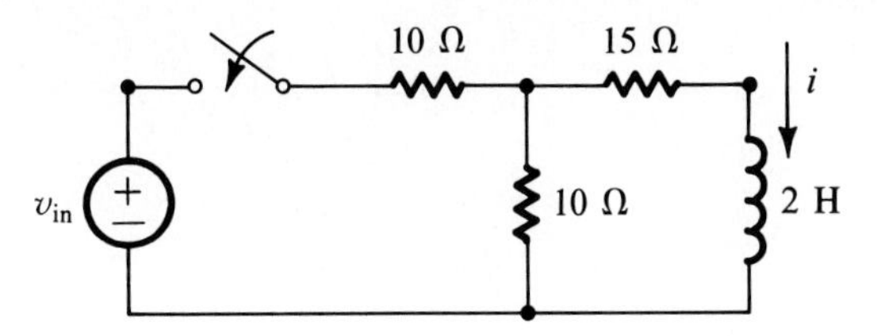

Figure P10.11

10.12 If $i(0^-) = 0$ and the switch closes at $t = 20$ msec, find $v(t)$ for $t \geq 0$ if $i_{in}(t) = 10 \sin 20\pi t \, u(t)$.

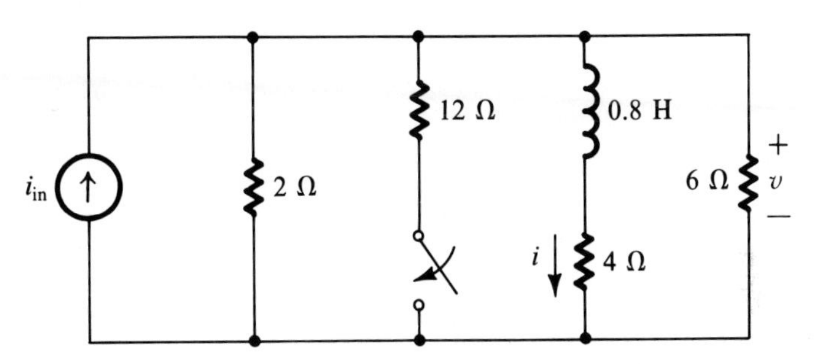

Figure P10.12

10.13 Find the inductor current $i(t)$ if $i(0^-) = 0$ and the voltage source is described by $v_{in}(t) = 5 \sin (100\pi t + 30°) \, u(t)$.

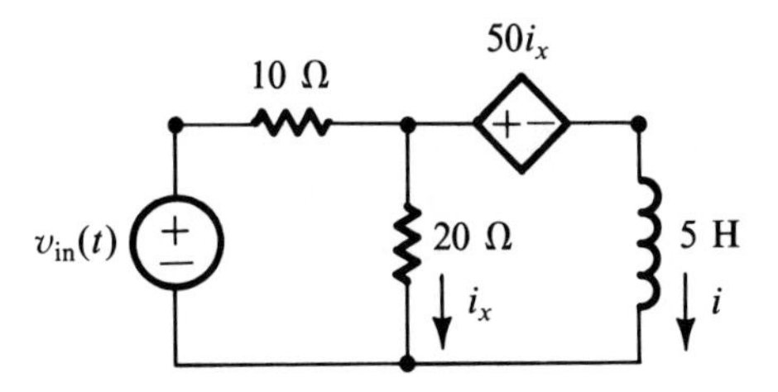

Figure P10.13

10.14 If the circuit of Fig. P10.10 has $R_1 = 1 \, \Omega$, $R_2 = 5 \, \Omega$, $C = 1$ F and $v_c(0^-) = 5$ V, find $v_o(t)$ for $t \geq 0$ when $v_{in}(t) = 5 \sin 2\pi t$.

10.15 If $i(0^-) = 2$ A, find $i(t)$ for $t \geq 0$ when the current source is described by $i_{in}(t) = 2 \cos (2\pi t + 40°)$.

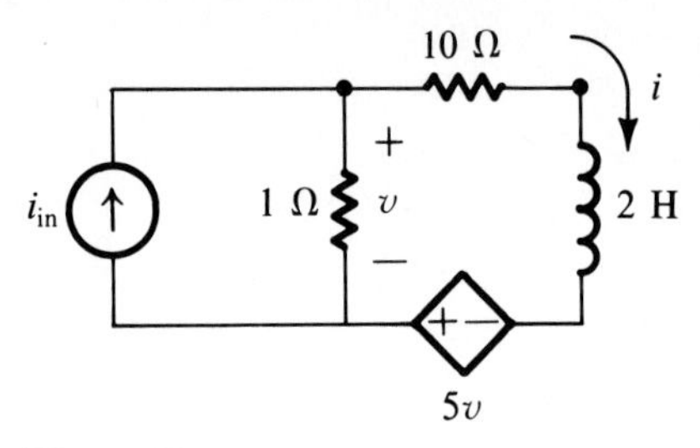

Figure P10.15

10.16 If $v_{in}(t) = 50 \cos 3\pi t$, find $v_o(t)$ for $t \geq 0$.

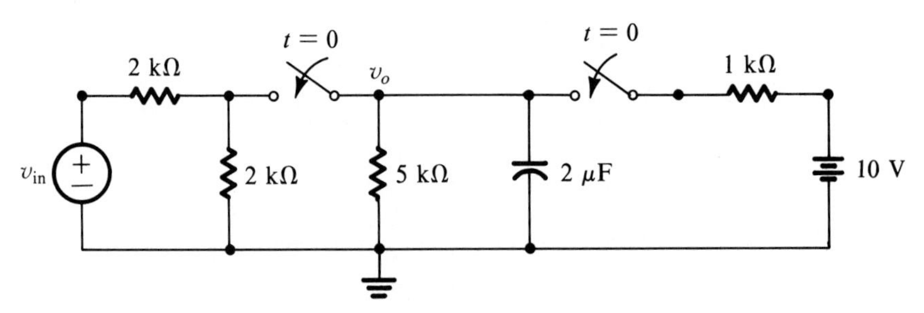

Figure P10.16

10.17 Suppose an integrator has $R = 10$ MΩ and $C = 1$ μF.

a. If $v_o(0^-) = 0$ V and $v_{in}(t) = 5 \sin \pi/10\, t$, find and draw $v_o(t)$ for $t \geq 0$.

b. Repeat a with $v_{in}(t) = 5 \sin \pi/2\, t$.

c. Repeat a with $v_o(0^-) = 10$ V.

d. Repeat a with $v_o(0^-) = 10$ V and $v_{in}(t)$-5 $\sin \pi/2\, t$.

10.18 Find $v_o(t)$ for $t \geq 0$ in Fig. P10.18

a. if $v_{in}(t) = 10 \sin 8\pi t$ and $v_o(0^-) = 0$ V.

b. Repeat a with $v_o(0^-) = 5$ V.

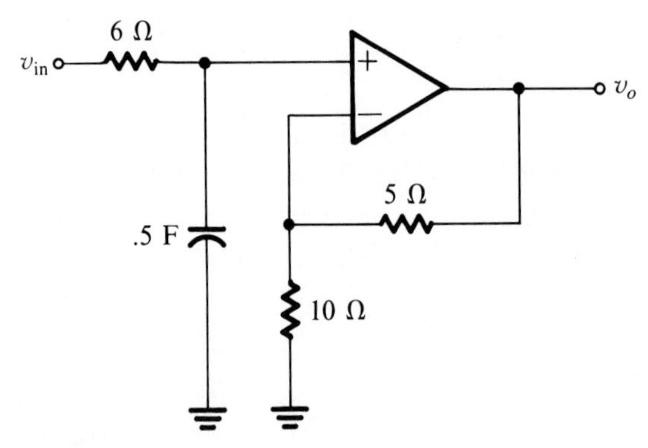

Figure P10.18

10.19 If $v_{in}(t) = 5 \sin 6\pi t\, u(t)$

a. Find $v(t)$ for $t \geq 0$ if $v(0^-) = 0$.

b. Find the ZIR of the circuit if $v(0^-) = 10$ V.

c. Repeat a if $v(0^-) = 20$ V.

d. Repeat a if $v_{in}(t) = 25 \sin 6\pi t$ and $v(0^-) = 40$ V.

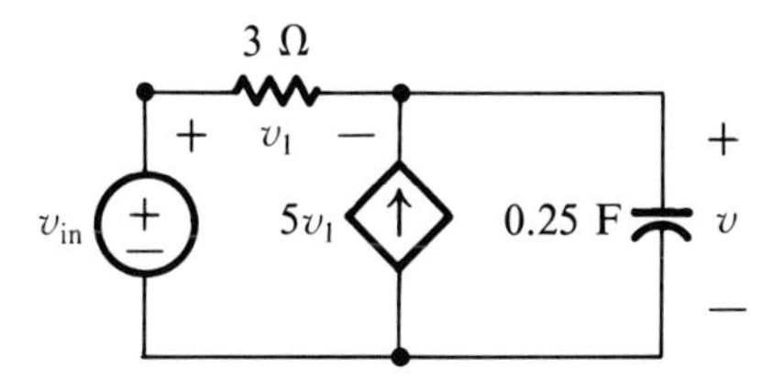

Figure P10.19

10.20 The circuit in Fig. 10.20 is driven by two current sources, with $i_{in1}(t) = 10 \sin 8\pi t\, u(t)$, and $i_{in2}(t) = 5 \cos 6\pi t\, u(t)$.

a. Find the ZSR of $v(t)$.

b. Find $v(t)$ for $t \geq 0$ if $i(0^-) = 2$ A.

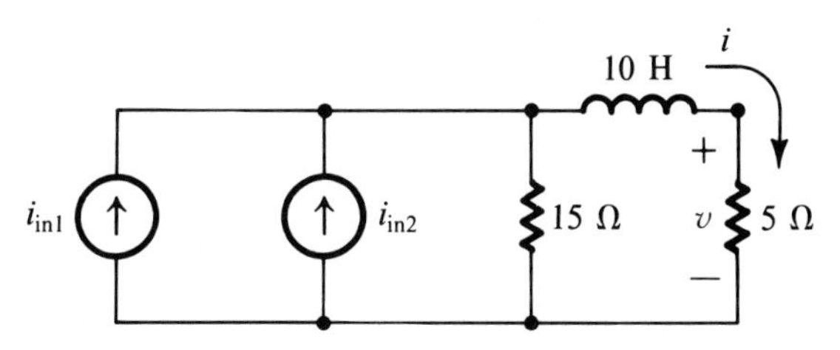

Figure P10.20

10.21 If two sources are connected in series with a series *RL* circuit having $R = 1$ kΩ, $L = 10$ H, $i(0^-) = 0$ and $v_{in1}(t) = 10u(t)$, $v_{in2}(t) = 5 \sin 400\pi t\, u(t)$, find $i(t)$ for $t \geq 0$.

10.22 If a source with $v_{in}(t) = 10 \sin 10\pi t$ drives a series *RC* circuit with $R = 50$ kΩ and $C = 1$ μf, draw a graph of $v_{in}(t)$ and the capacitor's steady-state output voltage on the same coordinate system.

10.23 Let $i_{in}(t) = 10u(t)$, $v_{in}(t) = 5 \cos 20\pi t\, u(t)$, $R_1 = 2500$ Ω, $R_2 = 2500$ Ω, $R_3 = 500$ Ω and $L = 1$ H.

a. Using source superposition, find $i(t)$ for $t \geq 0$ if $i(0^-) = 0$.

b. Repeat a with $i(0^-) = 2$ A.

c. Using source replacement, repeat a.

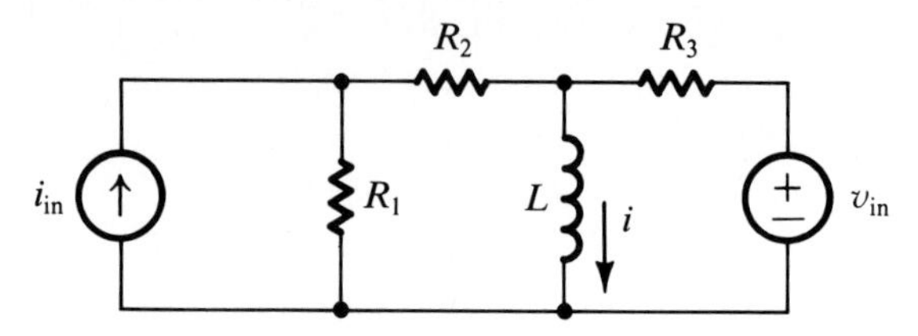

Figure P10.23

10.24 If a series RC circuit has $R = 1\ M\Omega$, $C = 2\ \mu F$, find the value of $v_o(0^-)$ for which $v_o(t) = v_{ss}(t)$ for $t \geq 0$ (i.e., the transient response is zero), if $v_{in} = 25u(t)$.

10.25 A series RC circuit with $R = 1\ \Omega$, $C = 1/3$ F is connected to a switch and a voltage source having $v_{in}(t) = 5 \sin 2t$. The switch closes at $t = 0$ and opens at $t = 2$ sec.

a. Find and draw $v_o(t)$ for $t \geq 0$ if $v_o(0^-) = 0$.

b. Repeat a if $v_o(0^-) = 10$ V.

10.26 If $\mathbf{H}_R(s)$ and $\mathbf{H}_c(s)$ are transfer functions relating $v_{in}(t)$ to $v_R(t)$ and $v_c(t)$ in a series **RC** circuit, draw a graph of $|\mathbf{H}_R(j\omega)|$, $|\mathbf{H}_c(j\omega)|$, $\measuredangle \mathbf{H}_R(j\omega)$ and $\measuredangle \mathbf{H}_c(j\omega)$ vs ω, and compare the characteristics of the sinusoidal steady-state responses.

10.27 A series RC circuit with a time constant of 1 sec has $v(0^-) = 0$.

a. Find $v(t)$ for $t \geq 0$ when $v_{in}(t) = 10 \sin \pi t/10\ [u(t) - u(t - 5)]$.

b. Repeat a when $v_{in}(t) = 10 \sin \pi t/10\ [u(t) - u(t - 0.5)]$.

c. Sketch the results of a and b.

10.28 If $v_o(0^-) = 0$ V and $v_{in}(t) = \sin 2\pi t$, find and draw $v_o(t)$ for $t \geq 0$.

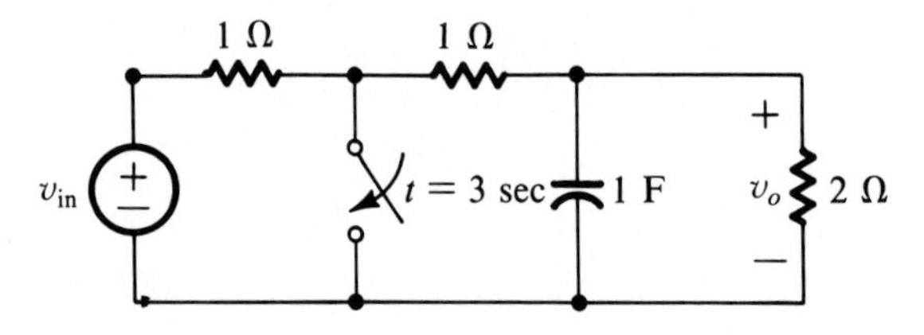

Figure P10.28

10.29 Find $i(t)$ for $t \geq 0$ if $i(0^-) = 10$ A and $v_{in}(t) = 5 \sin 4\pi(t - 1)\ u(t - 1)$.

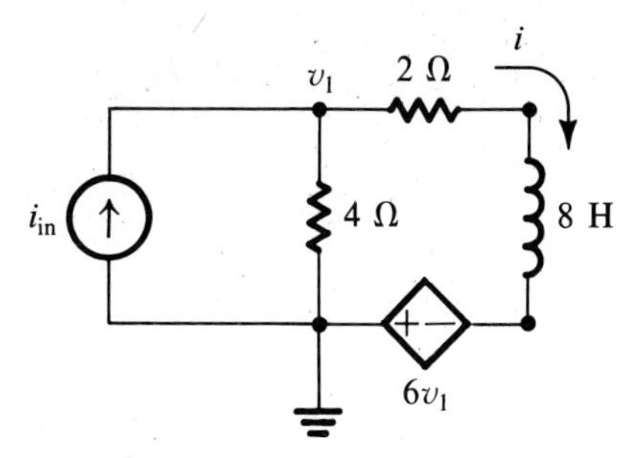

Figure P10.29

10.30 The circuit in Fig. P10.30 has $R_1 = 20\ k\Omega$, $R_2 = 80\ k\Omega$ and $C = 8\ \mu F$, with $v_{in}(t) = 10\cos(10\pi t + 45°)\ u(t)$ and $i_{in}(t) = 20\cos(40\pi t - 30°)\ u(t)$.

a. Using superposition, find $v(t)$ for $t \geq 0$ if $v(0^-) = 0$.

b. Using source replacement, find $v(t)$ for $t \geq 0$ if $v(0^-) = 0$.

c. Repeat a if $v(0^-) = 5$ V.

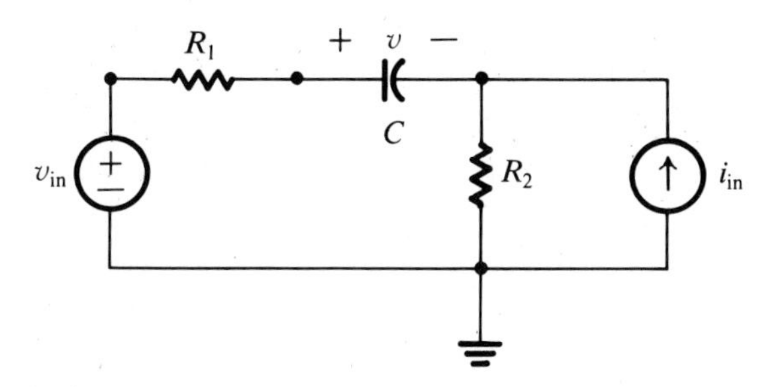

Figure P10.30

10.31 If $v_{in1}(t) = u(t)$ and $v_{in2}(t) = \sin 2\pi t\ u(t)$,

a. Find $v_o(t)$ for $t \geq 0$ and draw v_o for $0 \leq t \leq 5$ sec.

b. Estimate the time required for the op amp to saturate if $V_{cc} = 15$ V. (See Ch.4)

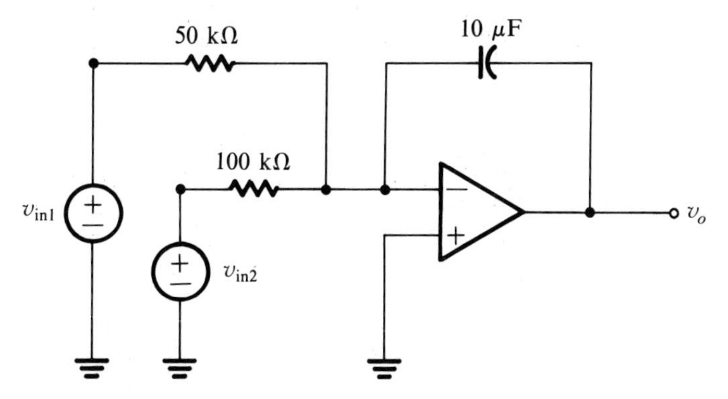

Figure P10.31

SECOND-ORDER CIRCUIT RESPONSE

INTRODUCTION

This chapter will show how to solve the I/O DE of a second-order circuit for its natural solution, and then how to combine the natural solution, particular solution, and boundary conditions to construct the zero-input, zero-state, and initial-state responses of the circuit.

11.1 SERIES *RLC* CIRCUIT

The series connection of a resistor, inductor, and capacitor with a voltage source shown in Fig. 11.1 has two independent energy-storage elements because it is possible to establish the initial conditions $i_L(0^-)$ and $v_c(0^-)$ independently of each other. We expect the capacitor will behave in a way that resists sudden changes in its voltage, and that the inductor will resist sudden changes in its current.

11.1.1 Physical Behavior

The key to understanding the behavior of many circuits is to understand what causes a voltage or a current to change. Consider the series *RLC* circuit redrawn in Fig. 11.2 to correspond to the ZIR (i.e., without the source). Since $i_c = i_L$ the capacitor voltage will increase if $i_L > 0$ and vice versa. Also, the inductor current will increase if $v_L > 0$ and vice versa. So **the inductor voltage effectively controls the inductor current, which con-**

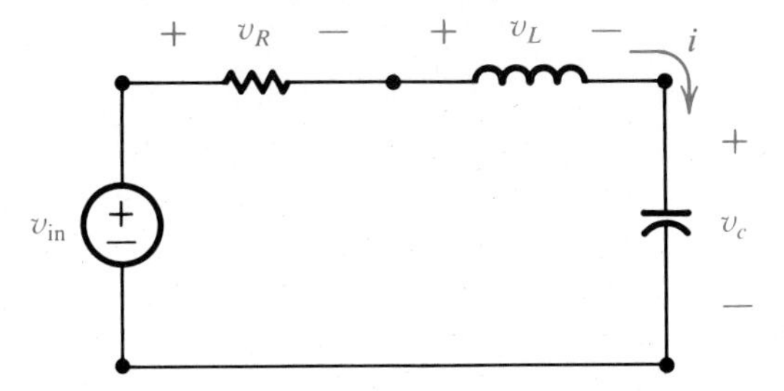

Figure 11.1 The series *RLC* circuit.

trols the capacitor voltage. What controls the inductor voltage? To answer this question, we use KVL to get $v_L = -(v_c + v_R)$. The inductor voltage will be positive if $v_c + v_R < 0$ and will be negative if $v_c + v_R > 0$.

Since $v_R = i_L R$ we can write

$$v_L > 0 \quad \text{if} \quad v_c < -i_L R \tag{11.1a}$$

$$v_L < 0 \quad \text{if} \quad v_c > -i_L R \tag{11.1b}$$

$$v_L = 0 \quad \text{if} \quad v_c = -i_L R \tag{11.1c}$$

The inductor current will be *increasing* if the capacitor voltage satisfies (11.1a), *decreasing* if the capacitor voltage satisfies (11.1b), and *constant* if it satisfies (11.1c). Thus, the values of v_c and i_L determine whether v_L is causing the inductor current to increase or decrease, which determines whether the capacitor voltage is increasing or decreasing. *This mechanism is the key to understanding the physical operation of the circuit.*

Skill Exercise 11.1

If a series *RLC* circuit has $v_c(0^-) = 5$ V and $i_L(0^-) = -2$ A, with $R = 15\ \Omega$, $L = 2$H and $C = 3$ F, determine $dv_c(0^+)/dt$ and $di_L(0^+)/dt$.

Answer: $dv_c(0^+)/dt = -2/3$ V/sec and $di_L(0^+)/dt = +12.5$ A/sec.

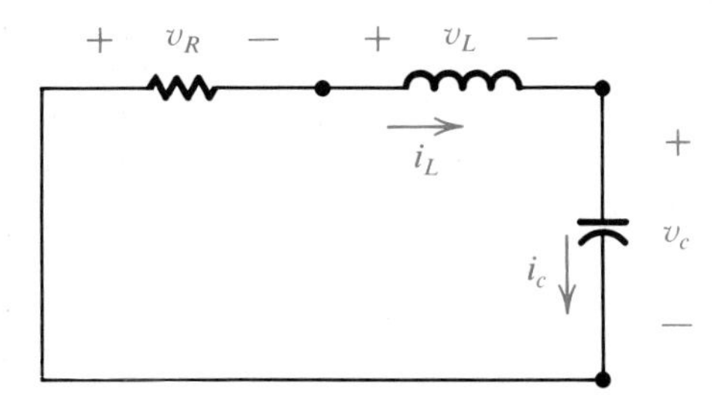

Figure 11.2 The series *RLC* circuit with the source removed for the zero-input response.

11.1.2 Series *RLC* I/O Models

The I/O model relating the applied source voltage to the capacitor voltage in the series RLC circuit in Fig. 11.1 is obtained by first writing the KVL equation

$$v_R(t) + v_L(t) + v_c(t) = v_{in}(t)$$

Since $i_R = i_L = i_c = i = Cdv_c/dt$, the inductor voltage is

$$v_L(t) = L\frac{di}{dt} = LC\frac{dv_c^2}{dt^2}$$

and the resistor voltage is

$$v_R(t) = Ri(t) = RC\frac{dv_c}{dt}$$

The KVL equation simplifies to

$$\frac{d^2v_c}{dt^2} + \frac{R}{L}\frac{dv_c}{dt} + \frac{1}{LC}v_c(t) = \frac{1}{LC}v_{in}(t) \qquad (11.2)$$

The I/O transfer function relating the input voltage to the capacitor voltage is obtained by inspection of the I/O DE

$$\mathbf{H}_v(s) = \frac{\dfrac{1}{LC}}{s^2 + \dfrac{R}{L}s + \dfrac{1}{LC}} \qquad (11.3)$$

The current in the circuit can also be chosen as an output signal. One way to find an expression for $i(t)$ is to solve (11.2) for $v_c(t)$, then differentiate and scale $v_c(t)$ by C. A second approach develops the I/O differential equation for $i(t)$ using the KVL equation and models for R, L and C behavior

$$Ri(t) + L\frac{di}{dt} + v_c(t_o) + \frac{1}{C}\int_{t_o}^{t} i(\alpha)d\alpha = v_{in}(t)$$

Taking the derivative of both sides of the equation eliminates the integral and leads to the differential-equation model of the relationship between $i(t)$ and $v_{in}(t)$

$$\frac{d^2i}{dt^2} + \frac{R}{L}\frac{di}{dt} + \frac{1}{LC}i(t) = \frac{1}{L}\frac{dv_{in}}{dt}$$

The transfer function for the current is

$$\mathbf{H}_I(s) = \frac{\dfrac{1}{L}s}{s^2 + \dfrac{R}{L}s + \dfrac{1}{LC}} \qquad (11.4)$$

Note that the voltage and current transfer functions given by (11.3) and (11.4) have the same characteristic equation. Therefore, the natural frequencies of the voltage and current will be the same. In general, all circuit variables will have the same natural frequencies.

11.2 I/O MODEL FOR A SECOND-ORDER OP AMP CIRCUIT

11.2.1 Physical Behavior

The op amp circuit in Fig. 11.3 is a second-order circuit because it contains two capacitors whose initial conditions can be established independently of each other. The voltage-follower configuration of the op amp causes v_o to appear across C_1, and causes v_{c2} to appear across R_2. The output voltage will increase if $i_{c1} > 0$ and vice versa. The current i_{c1} has the same polarity as v_{c2}. Therefore, the voltage on C_2 controls i_{C1}. The control of the current i_{C2} can be determined by first noting that $i_{C2} = i_{in} - i_{C1}$ and using Ohm's Law

$$i_{C2} = \frac{v_{in} - v_1}{R_1} + \frac{v_o - v_1}{R_2} = \frac{v_{in} - v_1}{R_1} - \frac{v_{c2}}{R_2}$$

Suppose a step-input voltage is applied to the circuit. In steady state the node voltages must satisfy $v_1 = v_o = v_{in}$. Otherwise i_{in} would not be zero and the voltages across C_1 and C_2 would be changing. When the circuit is not in steady state the current $i_{in}(t)$ has two components. One component drives v_{C1} in proportion to the difference in voltage between v_1 and v_o. The other component drives C_2, depending on the difference between v_{in} and v_1 and the difference between v_1 and v_o.

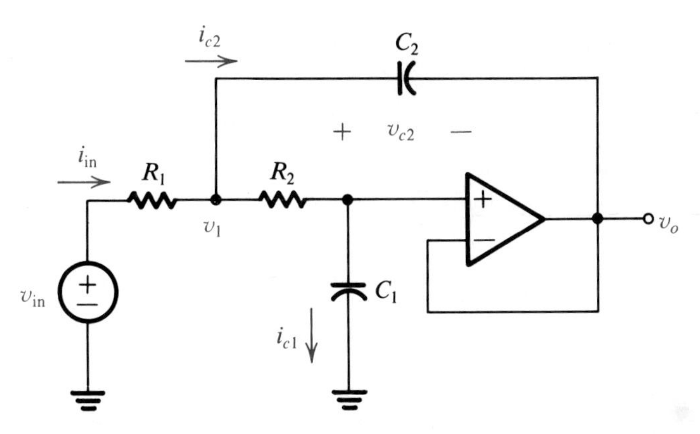

Figure 11.3 Second-order op amp circuit.

11.2.2 I/O Model

Applying KCL and KVL to the op amp circuit gives

$$\frac{v_{in} - v_1}{R_1} = C_2(\dot{v}_1 - \dot{v}_o) + C_1\dot{v}_o \tag{11.5}$$

and

$$v_1 = R_2C_1\dot{v}_o + v_o \tag{11.6}$$

where "dots" denote differentiation. Substituting (11.6) into (11.5) and rearranging leads to the second-order I/O model of the circuit

$$\frac{d^2v_o}{dt^2} + \frac{1}{C_2}\left(\frac{1}{R_1} + \frac{1}{R_2}\right)\frac{dv_o}{dt} + \frac{v_o}{R_1C_1R_2C_2} = \frac{v_{in}}{R_1C_1R_2C_2} \tag{11.7}$$

11.3 SECOND-ORDER NATURAL SOLUTIONS

The previous sections demonstrated how to develop the I/O model of second-order circuits. This section will introduce special terminology that applies to second-order circuits, and will show how to construct the natural solution to their I/O differential equation.

Recall (see Chapter 7) that the natural solution to the I/O differential equation depends on the roots of its characteristic equation. In the case of a second-order I/O model, we write the homogeneous differential equation in the standard form given by

$$\frac{d^2y}{dt^2} + 2\alpha\frac{dy}{dt} + \omega_n^2 y = 0 \tag{11.8}$$

A **second-order characteristic equation** will have the generic form

$$\boxed{s^2 + 2\alpha s + \omega_n^2 = 0} \tag{11.9}$$

with *real* parameters α and ω_n specified by the circuit components. The parameter α is called the **damping factor** of the circuit because, as we will see, it determines the exponential decay of the transient component of the circuit's underdamped response; ω_n is called the **undamped natural frequency** of the circuit because it determines the period of the circuit's oscillation *if* the circuit is undamped.

Skill Exercise 11.2

Find expressions for α and ω_n for the transfer function of the series *RLC* circuit.

Answer: $\alpha = R/2L$ and $\omega_n = 1/\sqrt{LC}$.

Skill Exercise 11.3

Find expressions for α and ω_n of the op amp circuit described by (11.7).

Answer: $\alpha = (R_1 + R_2)/(2C_2R_1R_2)$ and $\omega_n = 1/(R_1C_1R_2C_2)^{1/2}$.

The roots of the generic second-order characteristic equation are

$$s_1 = -\alpha + \sqrt{\alpha^2 - \omega_n^2} \tag{11.10a}$$

$$s_2 = -\alpha - \sqrt{\alpha^2 - \omega_n^2} \tag{11.10b}$$

Depending on α and ω_n, a second-order circuit will have either two distinct real natural frequencies (overdamped response), two identical real natural frequencies (critically damped response), or a complex pair of natural frequencies (underdamped response).

There are three possible pole patterns for a second-order circuit, since the natural solution of its input/output differential equation can have one of three generic forms. *Each leads to a distinct zero-input, zero-state, and initial-state response of the circuit* because the natural solution of the differential equation is used (with the particular solution) to construct each of these responses. The following sections will treat all three cases by first finding the natural frequencies and constructing the natural solution. Then the boundary conditions will be applied to the complete solution of the I/O DE to find the ZIR, ZSR, and ISR. It is essential that we know how to solve the second-order I/O differential equation for its natural solution because that is used to form the ZIR, ZSR and ISR of circuits.

Case I:
Overdamped (Distinct Real Natural Frequencies): $\alpha^2 > \omega_n^2$

When $\alpha^2 > \omega_n^2$ the radicals in (11.10) have positive arguments. In this case, the circuit is said to be overdamped, and the roots of the characteristic equation are the *real* numbers

$$s_1 = -\alpha + \sqrt{\alpha^2 - \omega_n^2} \tag{11.11a}$$

and

$$s_2 = -\alpha - \sqrt{\alpha^2 - \omega_n^2} \tag{11.11b}$$

The natural solution of an overdamped second-order circuit's I/O differential equation is given by

$$y_N(t) = K_1 e^{s_1 t} + K_2 e^{s_2 t}$$

Example 11.1

The series RLC circuit has two real natural frequencies if $\alpha^2 > \omega_n^2$, i.e. if $R^2/4L^2 > 1/LC$, or if $R > 2\sqrt{L/C}$. Its natural frequencies are defined by the physical component values according to

$$s_1 = \frac{-R}{2L} + \sqrt{\frac{R^2}{4L^2} - \frac{1}{LC}}$$

$$s_2 = \frac{-R}{2L} - \sqrt{\frac{R^2}{4L^2} - \frac{1}{LC}}$$

Skill Exercise 11.4

Calculate α, ω_n and the natural frequencies of the series RLC circuit, and draw the pole-zero patterns of $\mathbf{H}_v(s)$ and $\mathbf{H}_I(s)$ when $R = 10\ \Omega$, $L = 1$ H and $C = 1$ F.

Answer: $\alpha = 5$ and $\omega_n = 1$. The circuit is overdamped with natural frequencies $s_1 = -0.1$, and $s_2 = -9.9$. The pole-zero patterns for $\mathbf{H}_v(s)$ and $\mathbf{H}_I(s)$ are given in Fig. SE 11.4.

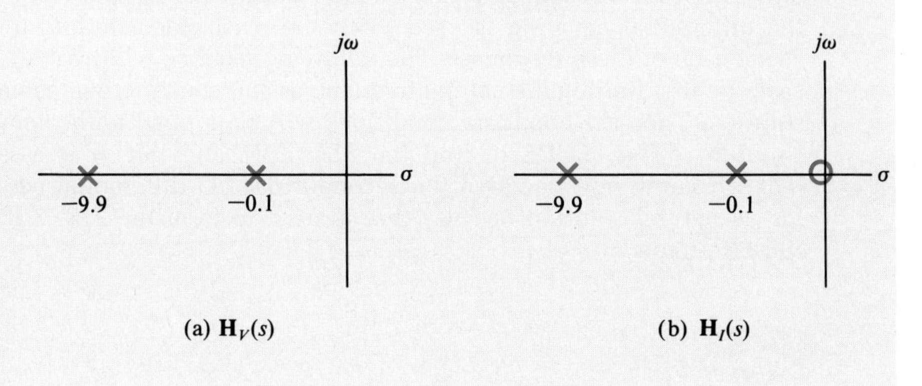

(a) $\mathbf{H}_V(s)$ (b) $\mathbf{H}_I(s)$

Example 11.2

Find the time constants and the natural solution for v_C and i_L in the series RLC circuit when $R = 4\ \Omega$, $L = 1$ H, and $C = 0.5$ F.

Solution: For this circuit $\alpha = 2$, $\omega_n = \sqrt{2}$, $s_1 = -0.59$ and $s_2 = -3.41$. The time constants are $\tau_1 = 1.69$ sec and $\tau_2 = 0.29$ sec, and the natural solution is

$$y_N(t) = K_1 e^{-0.59t} + K_2 e^{-3.41t}$$

The first response component will endure for about 6.76 sec, while the second will last for only 1.16 sec (i.e., four time constants). As an exercise, verify that the expression defined by $y_N(t)$ satisfies (11.2) if it is substituted for v_c with $v_{in} = 0$.

Skill Exercise 11.5

Find the natural solution $Y_N(t)$ of the differential equation

$$\frac{d^2y}{dt^2} + 7\frac{dy}{dt} + 10y(t) = x(t)$$

Answer: $y_N(t) = K_1e^{-2t} + K_2e^{-5t}$.

Case II: Critically Damped (Repeated Natural Frequencies): $\alpha^2 = \omega_n^2$

If $\alpha = \omega_n$ a second-order circuit's characteristic equation has two identical pole factors of the form $(s - s_1)$. The two roots to the characteristic equation will be identical, and the natural frequencies will be $s_1 = s_2 = -\alpha$. The pole-zero pattern will have a double, or repeated, pole as shown in Fig. 11.4 for $\mathbf{H}_V(s)$ and $\mathbf{H}_I(s)$ of the series *RLC* circuit. A second-order circuit with identical natural frequencies is said to be **critically damped.**

The natural solution of a critically damped second-order differential equation has the generic form

$$y_N(t) = K_1e^{s_1t} + K_2te^{s_2t} \tag{11.12}$$

This can be verified by substituting (11.12) into (11.8) with $\alpha = \omega_n$ and $s_1 = s_2 = -\alpha$. The graph of the factor $te^{-\alpha t}$ is shown in Figure 11.5 to illustrate how it reaches a maximum value of e^{-1}/α at $t = 1/\alpha$, one time constant, and then decays to zero. Thus, the maximum amplitude and the time of the maximum are inversely proportional to the time constant. The initial slope of the curve is always 1.

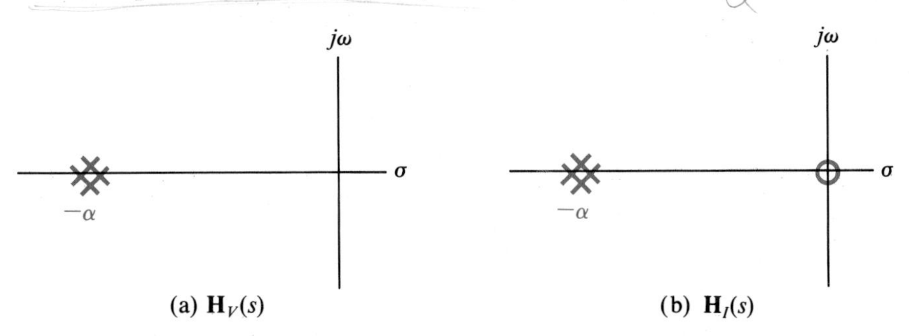

Figure 11.4 Pole-zero patterns of the critically damped series *RLC* circuit's (a) voltage and (b) current transfer functions.

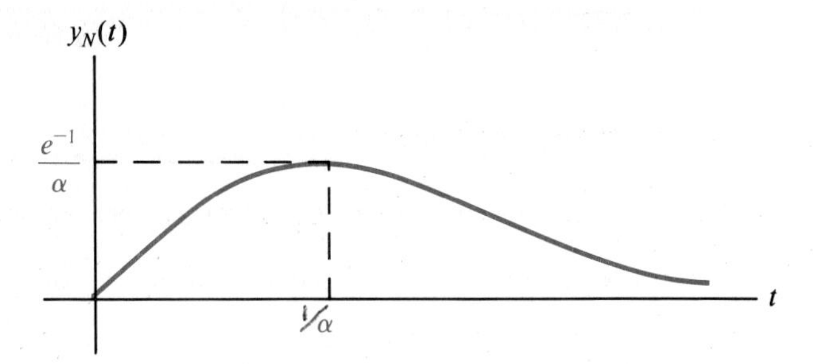

Figure 11.5 Natural solution of the critically damped second-order circuit's input/output equation.

Example 11.3

A series *RLC* circuit will be critically damped if $\alpha = R/2L = \sqrt{1/LC} = \omega_n$, or if $R = 2\sqrt{L/C}$.

Skill Exercise 11.6

Find the condition for which the op amp circuit described by (11.7) will be critically damped.

Answer: $(R_1 + R_2)^2/(R_1R_2) = 4C_2/C_1$.

Case III:
Underdamped (Complex Conjugate Natural Frequencies): $\alpha^2 < \omega_n^2$

When $\alpha^2 < \omega_n^2$ the radicals in (11.11) have negative argument, so the roots of the characteristic equation will be complex numbers. In this case we define a **damped frequency of oscillation** $\omega_d = \sqrt{\omega_n^2 - \alpha^2}$ and write the natural frequencies as

$$s_1 = -\alpha + j\omega_d$$
$$s_2 = -\alpha - j\omega_d$$

The natural frequencies s_1 and s_2 are a conjugate pair of complex numbers because $s_1^* = s_2$.

Example 11.4

The series *RLC* circuit has

$$\omega_d = \sqrt{\frac{1}{LC} - \frac{R^2}{4L^2}}$$

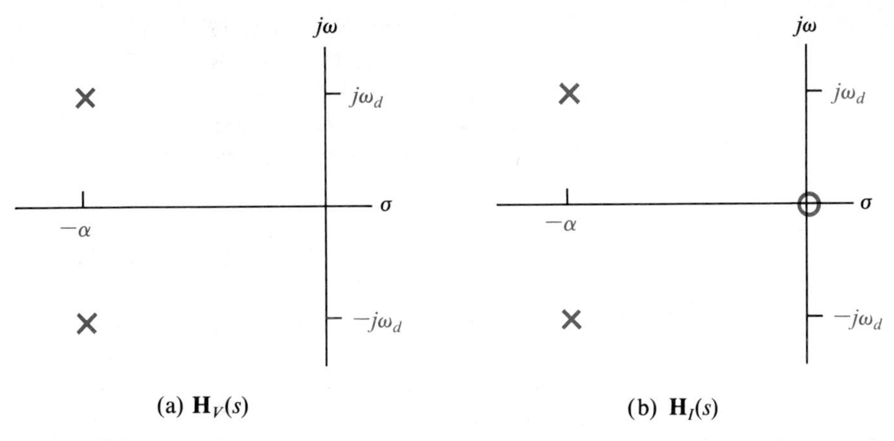

(a) $\mathbf{H}_V(s)$ (b) $\mathbf{H}_I(s)$

Figure 11.6 Pole-zero patterns of the underdamped series *RLC* (a) voltage and (b) current transfer functions.

and the circuit will be underdamped if $R < 2\sqrt{L/C}$. The pole-zero patterns for $\mathbf{H}_v(s)$ and $\mathbf{H}_I(s)$ of the underdamped circuit are shown in Fig. 11.6.

Now we will show how to find the natural solution of an underdamped circuit. If the coefficients of the I/O differential equation are real numbers its underdamped natural frequencies will *always* occur as a complex conjugate pair with $s_2 = s_1^*$ and because s_1 and s_2 are distinct, its natural solution will have the form

$$y_N(t) = K_1 e^{s_1 t} + K_2 e^{s_2 t}$$

This expression can be simplified by writing K_1 and K_2 in polar form.

$$\begin{aligned} y_N(t) &= |K_1| e^{j\psi_1} e^{-\alpha t} e^{j\omega_d t} + |K_2| e^{j\psi_2} e^{-\alpha t} e^{-j\omega_d t} \\ &= [\,|K_1| e^{j(\omega_d t + \psi_1)} + |K_2| e^{-j(\omega_d t - \psi_2)}] e^{-\alpha t} \end{aligned}$$

Since $y_N(t)$ is real-valued we must have

$$\mathrm{Im}\{y_N(t)\} = 0$$

or

$$|K_1| \sin(\omega_d t + \psi_1) - |K_2| \sin(\omega_d t - \psi_2) = 0$$

This equation can only be satisfied for arbitrary ω_d if $|K_1| = |K_2|$ and $\psi_1 = -\psi_2$. Thus, **the complex roots and their solution coefficients are both complex conjugate pairs.** The natural solution is

$$y_N(t) = |K_1| e^{-\alpha t} [e^{j(\omega_d t + \psi_1)} + e^{-j(\omega_d t + \psi_1)}]$$

Applying Euler's Law, we get

$$y_N(t) = 2|K_1| e^{-\alpha t} \cos(\omega_d t + \psi_1)$$

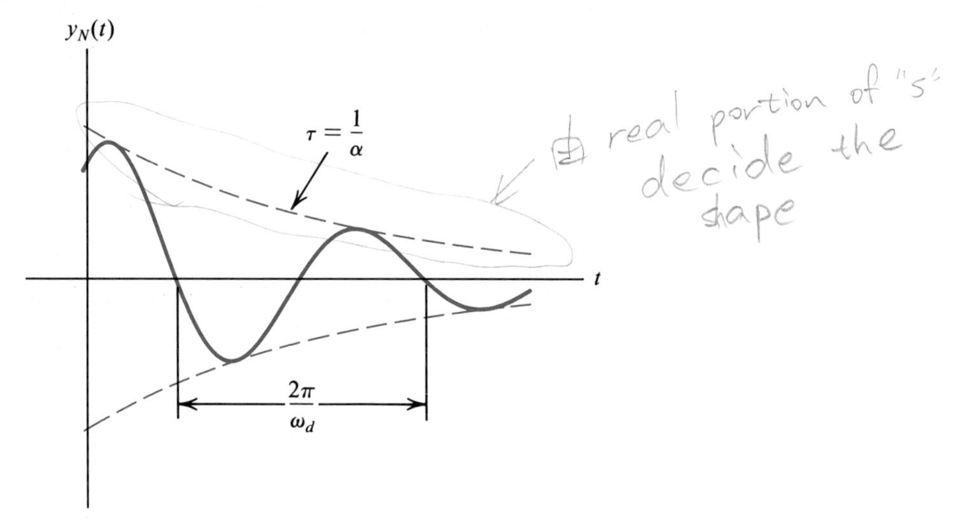

Figure 11.7 Natural solution of the underdamped second-order circuit's input/output equation.

Lastly, we let $|K| = |K_1|$ and $\phi = \psi_1 + \pi/2$ to obtain the standard form of the natural solution of an underdamped second-order differential equation:

$$y_N(t) = 2|K|e^{-\alpha t} \sin(\omega_d t + \phi) \tag{11.13}$$

The underdamped-natural-solution coefficients K and ϕ are determined by the boundary conditions of the circuit for either the ZIR, ZSR, or ISR.

The parameters α and ω_d play key roles in determining the shape of the damped oscillatory response of $y_N(t)$. The response decays to zero with a time constant of $\tau = 1/\alpha$ and the damped sinusoid has a period of $T = 2\pi/\omega_d$. (See Fig. 11.7) *A helpful observation* in graphing the underdamped response is that the number of oscillations that occur before the waveform decays to zero is obtained by dividing four time constants by the period of one oscillation

$$N = 4\tau/T = 2\omega_d/(\pi\alpha)$$

Sometimes $y_N(t)$ is written as

$$y_N(t) = e^{-\alpha t}(A \cos \omega_d t + B \sin \omega_d t)$$

This form is equivalent to (11.13), because it can be shown that

$$A \cos \omega_d t + B \sin \omega_d t = \sqrt{A^2 + B^2} \sin(\omega_d t - \phi)$$

with

$$\phi = \tan^{-1}(B/A - \pi/2)$$

Although these expressions are equivalent, (11.13) is preferred because it has a form that is easily graphed.

The natural frequencies of an underdamped second-order circuit have the property that

$$|s_1| = |s_2| = \sqrt{\alpha^2 + \omega_d^{\,2}} = \omega_n$$

That is, **the complex poles lie on a circle of radius ω_n.** It can also be shown that

$$\measuredangle s_1 = \pi - \tan^{-1} \omega_d/\alpha$$

and

$$\measuredangle s_2 = \pi + \tan^{-1} \omega_d/\alpha$$

These relationships between s_1, s_2, α, ω_n and ω_d are shown in Fig. 11.8. The damping factor α determines the time constant of decay in the natural solution.

It is sometimes convenient to describe the poles of an underdamped second-order circuit in terms of a **damping ratio** δ where δ is defined by $\delta = \alpha/\omega_n$. Then the parameters α and ω_d can be expressed by

$$\alpha = \delta\omega_n$$

and

$$\omega_d = \omega_n\sqrt{1 - \delta^2}$$

When $0 \le \delta \le 1$ the parameter δ is related to the angle ψ in Fig. 11.9 by $\delta = \cos \psi$. For a fixed ω_n the poles migrate in the pattern shown in Fig. 11.10 as δ varies from 0 to ∞.

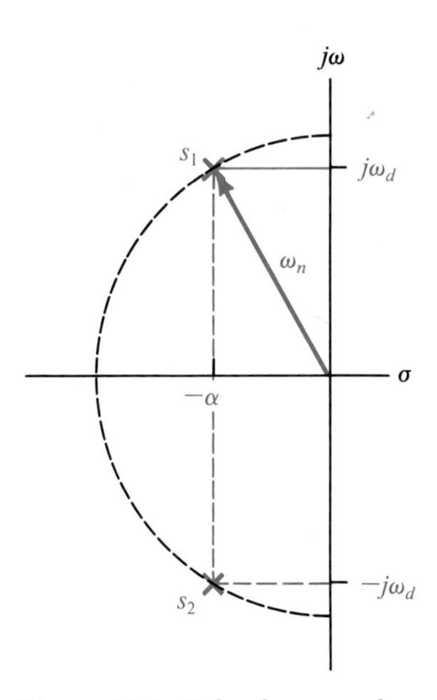

Figure 11.8 Poles for second-order transfer function with underdamped parameters.

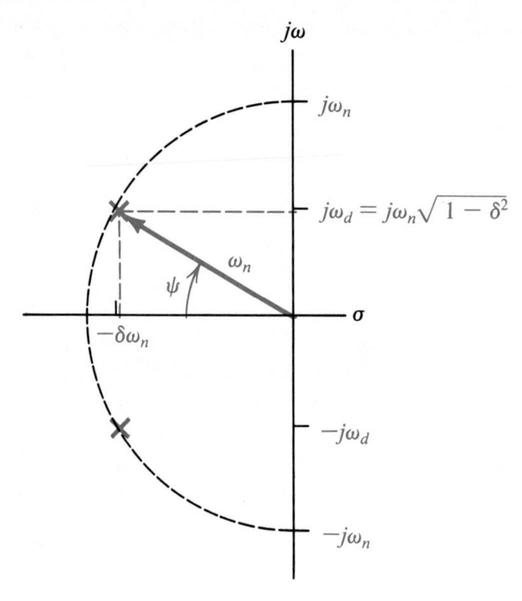

Figure 11.9 Relationship between the location of second-order poles and the damping ratio of the natural solution.

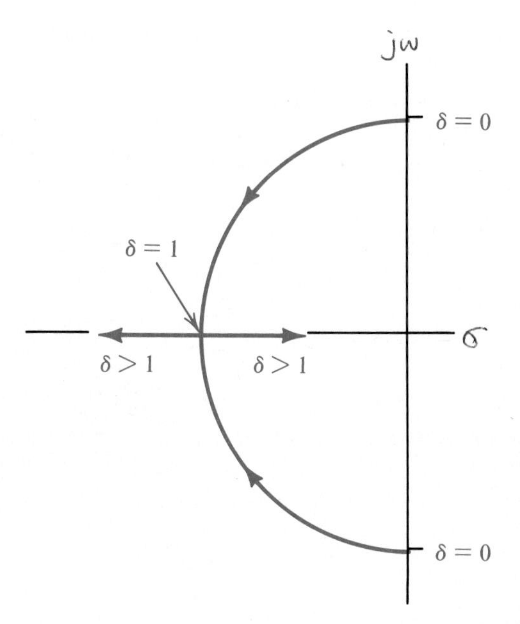

Figure 11.10 The poles of a second-order transfer function migrate in the complex plane as the damping ratio δ varies from 0 to ∞.

Lastly, the characteristic equation of a second-order circuit is sometimes written as

$$s^2 + 2\delta\omega_n s + \omega_n{}^2 = 0$$

to display δ and ω_n, and as

$$s^2 + 2\alpha s + (\alpha^2 + \omega_d{}^2) = 0$$

to reveal α and ω_d.

Table 11.1 summarizes the three generic forms of the natural solution of a second-order I/O differential equation and the parameters for a series *RLC* circuit. In each case, the solution contains two parameters that must be specified by the circuit's response boundary conditions.

TABLE 11.1 SECOND-ORDER NATURAL SOLUTION FOR SERIES *RLC* CIRCUIT.

SERIES *RLC* PARAMETERS	CASE	$y_N(t)$
$\alpha = \dfrac{R}{2L}$	Overdamped/Real	
$\omega_n = \sqrt{\dfrac{1}{LC}}$	$s_1 = -\alpha + \sqrt{\alpha^2 - \omega_n^2}$ $s_2 = -\alpha - \sqrt{\alpha^2 - \omega_n^2}$	$y_N(t) = K_1 e^{s_1 t} + K_2 e^{s_2 t}$
$\omega_d = \sqrt{\omega_n^2 - \alpha^2}$	Critically Damped/Real $s_1 = s_2 = -\alpha$	$y_N(t) = K_1 e^{-\alpha t} + K_2 t e^{-\alpha t}$
	Underdamped (Complex) $s_1 = -\alpha + j\omega_d$ $s_2 = -\alpha - j\omega_d$	$y_N(t) = 2\lvert K\rvert e^{-\alpha t} \sin(\omega_d t + \phi)$

11.4 BOUNDARY CONDITIONS

11.4.1 Series *RLC* Circuit

Solving a second-order I/O DE for the circuit's ZIR, ZSR, or ISR requires specifying two response boundary conditions. These depend on the structure of the circuit not on its damping, and follow directly from KVL, KCL, and the physical continuity conditions of v_c and i_L. For a given $v_o = v_c(0^-)$ and $i_o = i_L(0^-)$ the circuit has

$$v_c(0^+) = v_c(0^-) = v_o \tag{11.14a}$$

$$i_L(0^+) = i_L(0^-) = i_o \tag{11.14b}$$

Solving for the capacitor's voltage response also requires specifying of $dv_c(0^+)/dt$. This second boundary condition follows from

$$\dot{v}_c(0^+) = \frac{1}{C} i_L(0^+) = \frac{1}{C} i_L(0^-) = \frac{1}{C} i_o$$

Likewise, the boundary condition for the first derivative of the inductor current is given by

$$\frac{d i_L(0^+)}{dt} = \frac{1}{L} v_L(0^+) = \frac{1}{L}[v_{in}(0^+) - v_R(0^+) - v_c(0^+)]$$
$$= \frac{1}{L}[v_{in}(0^+) - i_L(0^+)R - v_c(0^+)]$$

From the continuity of v_C and i_L we get

$$\frac{d i_L(0^+)}{dt} = \frac{1}{L}[v_{in}(0^+) - i_L(0^-)R - v_c(0^-)]$$
$$= \frac{1}{L}[v_{in}(0^+) - i_o R - v_o]$$

It is important to note that the first derivatives of v_C and i_L are not necessarily continuous, and that the first derivative of i_L depends on the initial value of the applied voltage source—i.e., the applied source determines the initial slope of the curve for i_L for a given v_o and i_o.

The boundary condition equations for v_C and i_L and for their first derivatives are actually general, and by specific choices apply to the ZIR, ZSR, and ISR of the series *RLC* circuit. For example, for the ZSR boundary conditions of v_C we would let $v_o = 0$ and $i_o = 0$ in (11.14a) and (11.14b), and for the ZIR we would let $v_{in}(0^+) = 0$.

Table 11.2 summarizes the boundary conditions for the series *RLC* circuit for each of the three cases. They are a mathematical statement of the physical fact that the capacitor voltage and the inductor current cannot change instantaneously. They also confirm our physical insight, since the ZSR boundary condition states that the applied source causes a change in the *derivative* of the inductor current, which leads to subsequent changes in the other variables.

TABLE 11.2 SERIES *RLC* BOUNDARY CONDITIONS.

	$v_c(0^+)$	$\dot{v}_c(0^+)$	$i_L(0^+)$	$\dot{i}_L(0^+)$
ZIR	v_o	$\frac{1}{C} i_o$	i_o	$-\frac{1}{L}[i_o R + v_o]$
ZSR	0	0	0	$\frac{1}{L} v_{in}(0^+)$
ISR	v_o	$\frac{1}{C} i_o$	i_o	$\frac{1}{L}[v_{in}(0^+) - i_o R - v_o]$

11.4.2 Op Amp Circuit*

The op amp circuit (see Fig. 11.3) defined by (11.7) has boundary conditions

$$v_o(0^+) = v_{C1}(0^+) = v_{C1}(0^-) \tag{11.15a}$$

and

$$\dot{v}_o(0^+) = \dot{v}_{C1}(0^+) = \frac{1}{R_2C_1}v_{C2}(0^+) = \frac{1}{R_2C_1}v_{C2}(0^-) \tag{11.15b}$$

The voltage v_{C2} determines whether C_1 charges or discharges (how?), and with the time constant R_2C_1 determines the initial slope of the waveform for v_o *independently of the applied source*. Also

$$C_2\dot{v}_{C2}(0^+) = \frac{v_{in}(0^+)}{R_1} - v_{C2}(0^-)\left(\frac{1}{R_1} + \frac{1}{R_2}\right) - \frac{v_{C1}(0^-)}{R_1}$$

For given $v_{C1}(0^-)$ and $v_{C2}(0^-)$ the applied source voltage determines the initial slope of the curve for v_{C2} (and i_{C1}). This effect would be apparent in a computer simulation of the curve for $v_{C2}(t)$.

11.5 ZIR, ZSR, AND ISR

The previous sections showed how to derive the I/O model and the boundary conditions for a second-order circuit. This section will present a detailed discussion of the zero-input, zero-state, and initial-state responses of second-order circuits, with special emphasis on the series *RLC* example, for each of the three possible natural solutions—overdamped, critically damped, and underdamped. In each case **the circuit model's natural- and particular-solution components will be combined with the boundary conditions to solve for its response.**

11.5.1 Overdamped Circuit ZIR

The zero-input response of a circuit is obtained when the sources are removed. Therefore, the particular solution is zero, and the ZIR of v_c in the overdamped series *RLC* circuit is given by

$$v_c(t) = K_1e^{s_1t} + K_2e^{s_2t} \tag{11.16}$$

subject to the boundary conditions

$$v_c(0^+) = K_1 + K_2 = v_o$$

and

$$\dot{v}_c(0^+) = s_1K_1 + K_2s_2 = \frac{1}{C}i_o$$

These equations must be solved for K_1 and K_2 to specify the response.

Example 11.5

Find the ZIR of v_c and i_L in the series *RLC* circuit with $R = 4\ \Omega$, $L = 1$ H, and $C = 0.5$ F, when $v_o = 4$ V and $i_o = 2$ A.

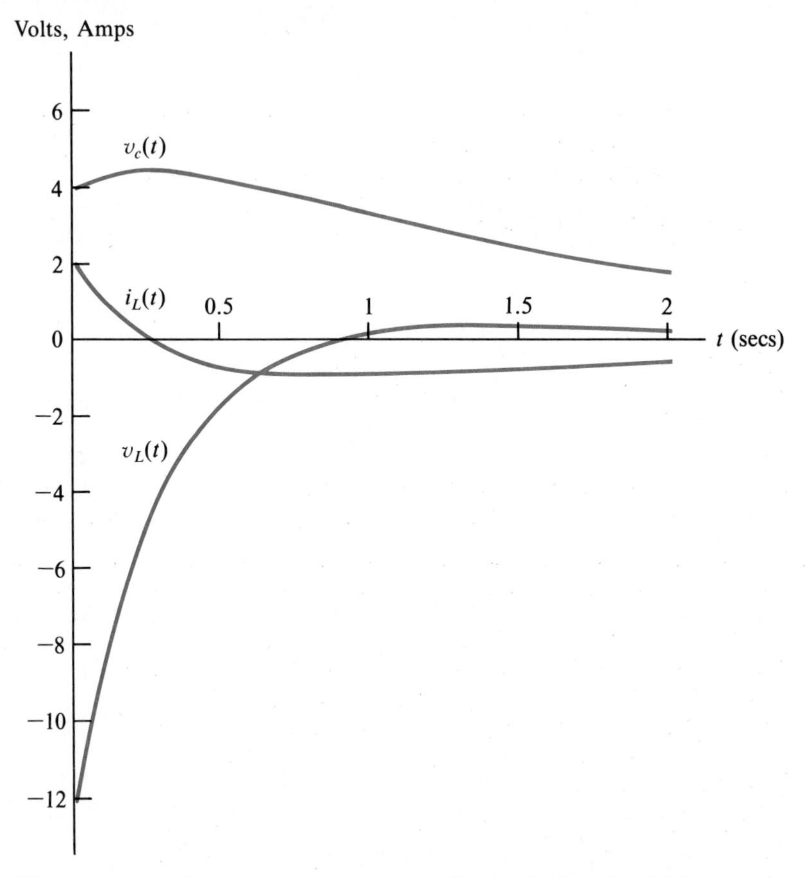

Figure 11.11 the zero-input response of $v_c(t)$, $i_L(t)$ and $v_L(t)$ for an over-damped second-order series *RLC* circuit.

Solution: The natural frequencies are $s_1 = -0.59$ and $s_2 = -3.41$. Solving $K_1 + K_2 = 4$ and $s_1K_1 + s_2K_2 = 4$ gives $K_1 = 6.26$ and $K_2 = -2.26$. For $t \geq 0$ the ZIR is $v_c(t) = 6.26e^{-0.59t} - 2.26e^{-3.41t}$. (Verify that $v_c(t)$ satisfies the boundary conditions.)

The graph of $v_c(t)$ is shown in Fig. 11.11 along with $i_L(t)$ and $v_L(t)$. The curve for $i_L(t)$ was obtained by noting that $i_L(t) = i_c(t)$ and that $i_c(t) = Cdv_c(t)/dt$, so

$$i_L(t) = -1.85e^{-0.59t} + 3.85e^{-3.41t}$$

The curve for $v_L(t)$ was found from KVL with $i_L(t)$ and $v_c(t)$ known

$$v_L(t) = -i_L(t)R - v_c(t)$$

$$= 1.14e^{-0.59t} - 13.14e^{-3.41t}$$

Note that $v_L(0^+) = -i_L(0^+)R - v_c(0^+) = -12$ V.

The initial conditions in Example 11.5 were such that the capacitor voltage initially increases because the initial inductor current i_L is positive. However, the voltage drops across v_R and v_c make the inductor voltage negative, so the current must drop, thereby making v_c charge less quickly, as indicated by the change in slope of $v_c(t)$. The inductor current eventually reaches zero, and because v_L is still negative, the current becomes negative. Thus v_c reaches a maximum value at the point where $i_L = 0$ and then starts to discharge. Meanwhile i_L has been dropping, with the result that v_R drops and v_L becomes less negative, and so i_L drops less quickly. Eventually $v_L = 0$ and the current reaches a minimum value, then turns upward as v_L goes positive. The capacitor voltage continues to decay to zero along with i_L and v_L. Because R is relatively large in this example v_L can become positive while v_c is positive (since the voltage across the resistor offsets the capacitor voltage for small currents). So v_c does not decay very rapidly. The large, negative, initial inductor voltage controls the current in the circuit which in turn controls the capacitor voltage. Note that the current does not drop indefinitely, because v_L eventually becomes positive (due to the polarity change of v_R as i_L changes polarity).

The response curves in Fig. 11.11 all decay to zero as $t \to \infty$. This can be explained by noting that the circuit has an initial amount of energy stored in its electric and magnetic fields. The resistor dissipates this energy in the form of heat, and in the ZIR case no applied source supplies additional energy.

Skill Exercise 11.7

Verify that the general solution for the ZIR coefficients K_1 and K_2 in (11.16) is given by

$$K_1 = \frac{s_2 v_o - \frac{1}{C} i_o}{s_2 - s_1} \qquad K_2 = \frac{-s_1 v_o + \frac{1}{C} i_o}{s_2 - s_1}$$

11.5.2 Overdamped Circuit ZSR

The zero-state response of an overdamped second-order circuit is obtained by combining the natural solution of its I/O differential equation with the equation's particular solution, and then applying the boundary conditions for the ZSR. If the input signal is an exponential ($x(t) = Xe^{st}$) the ZSR of the circuit is obtained from

$$y(t) = K_1e^{s_1t} + K_2e^{s_2t} + X\mathbf{H}(s)e^{st} \tag{11.17}$$

subject to the boundary conditions imposed by the circuit.

Example 11.6

The series RLC circuit with a step-input signal has $v_{in}(t) = Vu(t) = Ve^{st}$ for $t > 0$. The source exponential frequency is zero, so $\mathbf{H}_v(0) = 1$ and $\mathbf{H}_I(0) = 0$. The particular-solution component of the capacitor voltage is the source amplitude V and the particular-solution component of the current is zero. Since the poles of the circuit are in the left half-plane, the particular solutions also define the steady-state values of the voltage and current when the input is a step. The ZSR of $v_c(t)$ is given generically by

$$v_c(t) = K_1e^{s_1t} + K_2e^{s_2t} + V$$

The ZSR boundary conditions for v_c are $v_c(0^+) = 0$ and $dv_c(0^+)/dt = 0$. Solving for K_1 and K_2 leads to

$$v_c(t) = \left(\frac{s_2}{s_1 - s_2}e^{s_1t} - \frac{s_1}{s_1 - s_2}e^{s_2t} + 1\right)V$$

Note that $K_1 < 0$ and $K_2 > 0$ because $0 > s_1 > s_2$ [see (11.11)]. Since $|s_1| < |s_2|$, we conclude that $|K_1| > |K_2|$. The slower component of the natural-solution part of v_c is negative and has the larger initial magnitude. So we expect the response curve for v_c to begin at $t = 0$ with an initial slope of 0 and then to follow the slower component after $t = 4\tau_2$ (four time constants of the faster component).

The ZSR of the capacitor voltage can also be expressed in terms of the circuit time constants as

$$v_c(t) = \left(\frac{\tau_1}{\tau_2 - \tau_1}e^{-t/\tau_1} - \frac{\tau_2}{\tau_2 - \tau_1}e^{-t/\tau_2} + 1\right)V$$

These graphs of v_c and its "fast" and "slow" components are shown in Fig. 11.12. Notice that $dv_c(0^+)/dt = 0$ because the inductor initially acts like an open circuit.

Differentiating and scaling $v_c(t)$ gives the current in the series path:

$$i(t) = CV\frac{s_1s_2}{s_1 - s_2}[e^{s_1t} - e^{s_2t}]$$

Since $0 > s_1 > s_2$, $i(t)$ is positive for $t > 0$ if $V > 0$, meaning that a step input will cause v_c to charge from 0 to its steady-state value. The initial and final values of the current are zero. Its maximum value occurs at

$$t^* = \frac{\tau_1\tau_2\ln(\tau_2/\tau_1)}{\tau_2 - \tau_1}$$

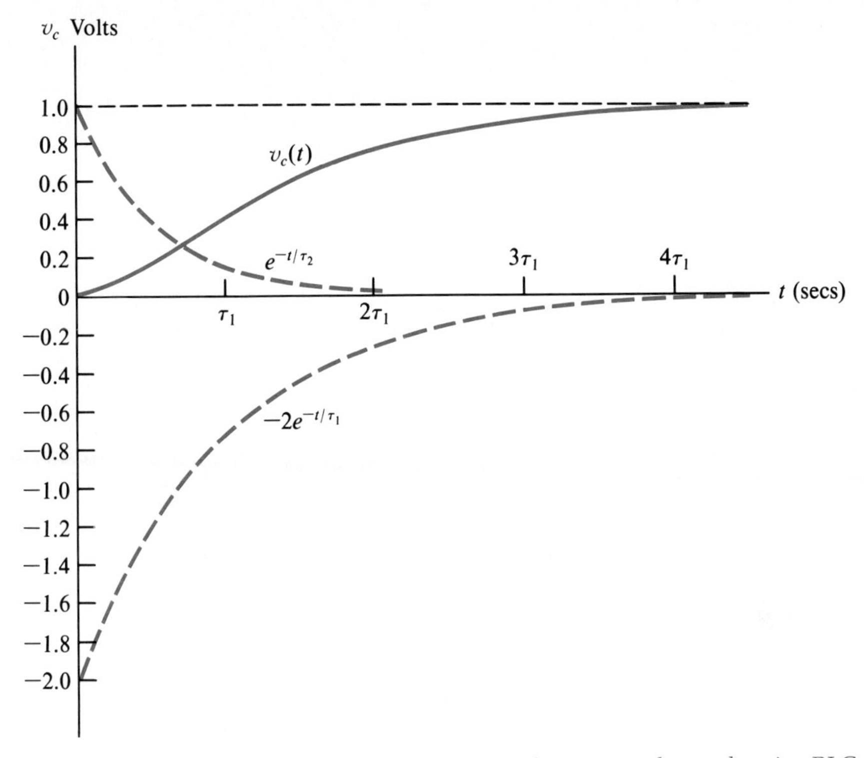

Figure 11.12 The zero-state response of $v_c(t)$ for an overdamped series RLC circuit driven by a unit-step input, with $\tau_2 = 0.5\tau_1$.

The location of t^* is shown on the graph of $i(t)$ in Fig. 11.13.

Lastly, the curve of the inductor voltage is obtained from $v_L(t) = L di_L(t)/dt$, and

$$v_L(t) = \frac{VLCs_1s_2}{s_1 - s_2}(s_1e^{s_1t} - s_2e^{s_2t})$$

Example 11.7

Find the ZSR of $v_c(t)$, $i(t)$ and $v_L(t)$ in the series RLC circuit with $v_{in}(t) = Vu(t)$, $R = 4\ \Omega$, $L = 1$ H and $C = 0.5$ F.

Solution: The circuit's natural frequencies and time constants are $s_1 = -0.59$, $s_2 = -3.41$, $\tau_1 = 1.69$ sec and $\tau_2 = 0.29$ sec. The coefficients of the response are given by $K_1 = -1.21$ V and $K_2 = 0.21$ V. Therefore

$$v_c(t) = (-1.21e^{-0.59t} + 0.21e^{-3.41t} + 1)Vu(t)$$
$$i(t) = (0.36e^{-0.59t} - 0.36e^{-3.41t})Vu(t)$$

and

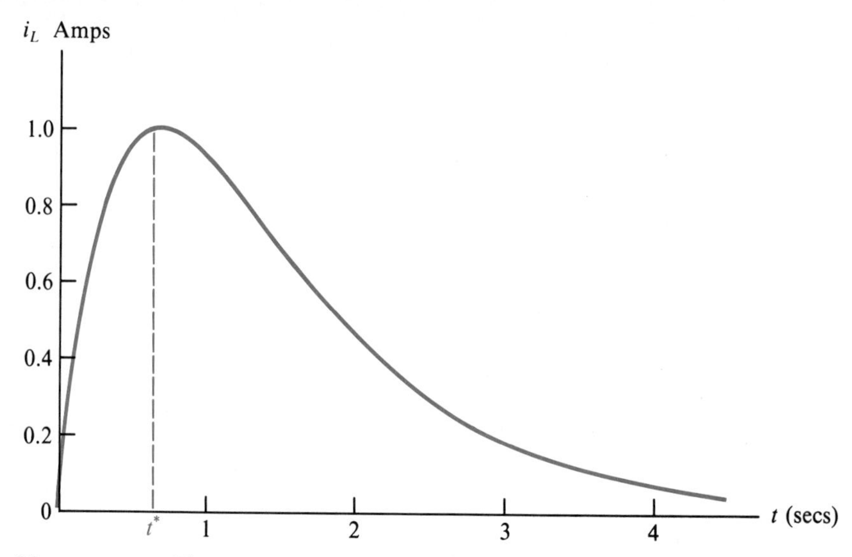

Figure 11.13 The zero-state response of $i_L(t)$ for an overdamped series *RLC* circuit driven by a step input signal $v_{in}(t) = Vu(t)$, with $V = \tau_1/C$ and $\tau_2 = 2\tau_1$.

$$v_L(t) = (-0.21e^{-0.59t} + 1.21e^{-3.41t})Vu(t)$$

As an exercise, verify that these solutions satisfy the zero-state-response boundary conditions.

The ZSR curves of v_c, i and v_L in Example 11.7 with $v_{in}(t) = u(t)$ are shown in Fig. 11.14. The entire applied voltage initially appears across the inductor. This causes the current to become positive, which causes the capacitor voltage to become positive. Current will increase as long as v_L is positive. When the inductor voltage becomes negative, the current will begin to decrease. This makes v_c increase at a slower rate. In the overdamped circuit, the current decreases to zero and the capacitor voltage reaches a steady-state value equal to the applied step-input voltage. The inductor causes the initial current to be zero, and the capacitor causes the steady-state current to be zero. (Recall that the pole-zero pattern for the current's transfer function has a zero at the orgin, indicating that the steady-state response to a step will be zero, as the capacitor behaves like an open circuit in the steady state. The current reaches a maximum at $t = 0.62$ sec, corresponding to the zero crossing of $v_L(t)$ and the change in slope of $v_c(t)$.

11.5.3 ZSRs for *RL*, *RC* and *RLC* Circuits

The effect of the capacitor and the inductor in the overdamped second-order circuit can be further appreciated by comparing the ZSR of

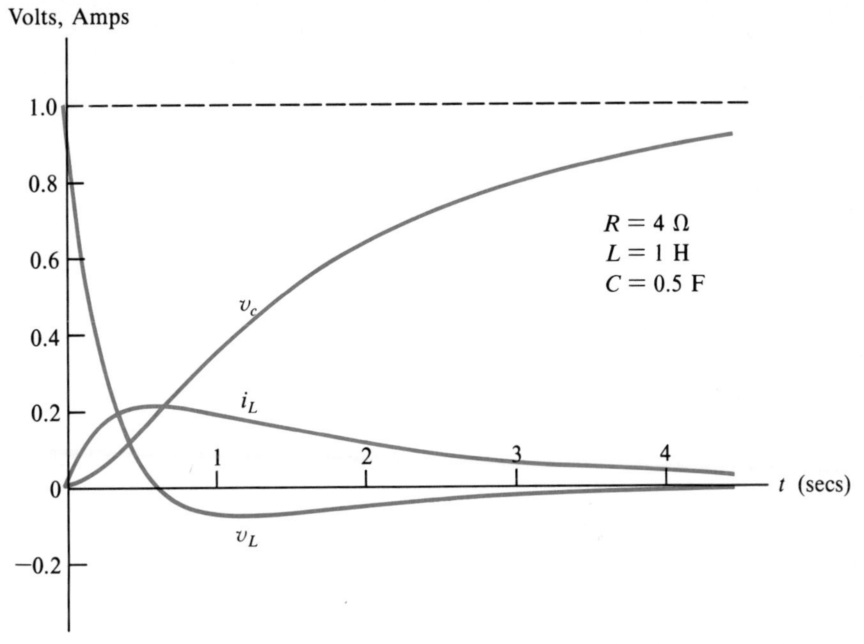

Figure 11.14 The zero-state responses of $v_c(t)$, $i_L(t)$ and $v_L(t)$ to a step input signal.

the *RLC* circuit to the ZSRs of the first-order *RL* and *RC* circuits having the same values of the circuit parameters. Table 11.3 shows a comparison of the time constants of the three circuits.

TABLE 11.3 COMPARISON OF TIME CONSTANTS FOR SERIES RL, RC AND RLC CIRCUITS.

CIRCUIT	TIME CONSTANT(S)	$R(\Omega)$	L(H)	C(F)
RL	.25 sec	4	1	0
RC	2.00 sec	4	0	0.5
RLC	.29 sec, 1.69 sec	4	1	0.5

Figure 11.15 shows the step response of the current in the three circuits. The *RC* circuit develops current instantaneously, the *RL* circuit develops current gradually, and the *RLC* circuit initially behaves like an *RL* circuit (with $v_c = 0$) but then makes a transition—behaving like an *RC* circuit as the buildup of capacitor voltage makes the inductor voltage drop to zero.

The *RLC* circuit has two response components and two time constants. The shorter time constant is characteristic of series *RL* behavior—current turns on very quickly in response to a source voltage applied across the inductor. The value of this time constant is slightly larger than

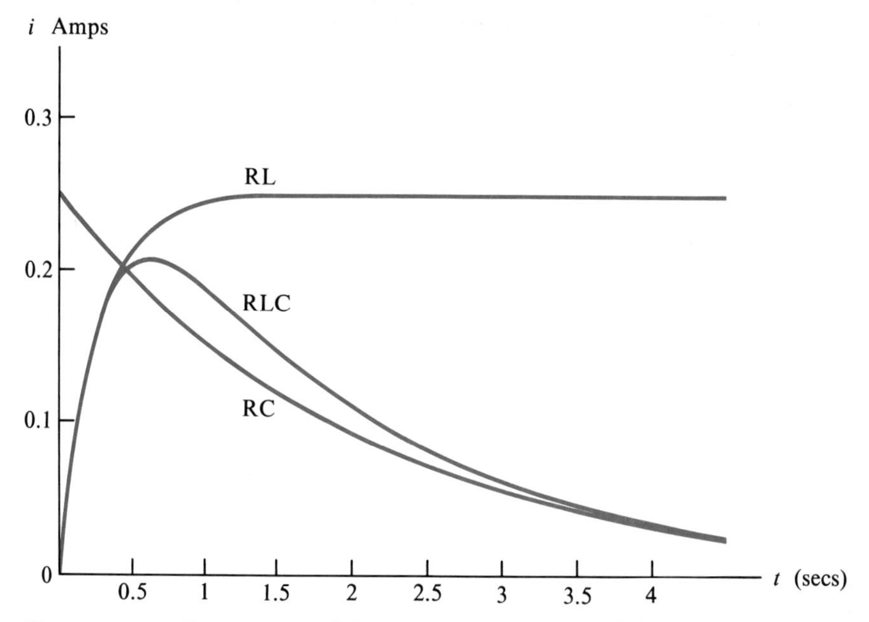

Figure 11.15 Comparison of the zero-state response of the current in series *RL*, *RC* and *RLC* circuits.

that of the pure *RL* time constant, and the response is slightly slower because the capacitor voltage effectively reduces the voltage across the inductor. The longer *RLC* time constant is characteristic of the *RC*–circuit charging process, with current decaying exponentially to zero. This time constant is shorter than that for a pure *RC* process because the inductor is contributing to the reduction of current.

11.5.4 Overdamped Circuit ISR

The ISR of the overdamped circuit has the same generic form as the ZSR

$$v_c(t) = K_1 e^{s_1 t} + K_2 e^{s_2 t} + V$$

but with boundary conditions $v_c(0^+) = v_o$ and $dv_c/dt(0^+) = i_o/C$ describing the initial state response. Solving for K_1 and K_2

$$K_1 = \frac{s_2(v_o - V) - i_o/C}{s_2 - s_1} \qquad K_2 = \frac{-s_1(v_o - V) + i_o/c}{s_2 - s_1} \tag{11.18}$$

These expressions show that v_c and i_L depend on the initial capacitor voltage and the initial inductor current. The response depends on *R*, *L* and *C* because these component values determine s_1 and s_2; changes in these parameters will effect changes in the response. (One way to examine these effects would be to produce and display the curves for v_c and i_L using a personal computer, and to compare solutions for various choices of *R*, *L* and *C*.)

Skill Exercise 11.8

Verify that $v_c(t)$ [with K_1 and K_2 defined by (11.18)] has $v_c(0^+) = v_o$ and $i_L(0^+) = i_o$.

Example 11.8

Find the ISR of the overdamped series RLC circuit to $v_{in}(t) = Vu(t)$ with $R = 4\ \Omega$, $L = 1$ H, $C = 0.5$ F, $v_o = 4$ V and $i_o = 2$ A.

Solution: The source and the circuit parameters are the same as in Example 11.7. Therefore, the ISR has the same generic expression as the ZSR

$$v_c(t) = K_1e^{-0.59t} + K_2e^{-3.41t} + V$$

but the boundary conditions are those of the ISR:

$$v_c(0^+) = K_1 + K_2 + V = 4$$

and

$$dv_c(0^+)/dt = -0.59K_1 - 3.41K_2 = 4$$

Solving for K_1 and K_2 gives the ISR

$$v_c(t) = (6.26 - 1.21\ V)e^{-0.59t} - (2.26 - 0.21\ V)e^{-3.41t} + V$$

for $t \geq 0$. (As an exercise, verify that the same result is obtained by adding the overdamped ZIR in Example 11.5 and the overdamped ZSR in Example 11.7.)

11.5.5 Critically Damped Circuit ZIR, ZSR, and ISR

The steps taken to find the zero-input, zero-state, and initial-state response of a critically damped circuit are identical to those that were just taken for an overdamped circuit. The important distinction is that **the natural solution of the circuit's I/O DE must be the one that corresponds to the circuit's repeated natural frequencies.**

Example 11.9

The capacitor voltage and inductor current in a critically damped, series RLC circuit driven by a step $v_{in}(t) = Vu(t)$ are described by

$$v_c(t) = (v_o - V)e^{-\alpha t} + [\alpha(v_o - V) + i_o/C]te^{-\alpha t} + V$$

and

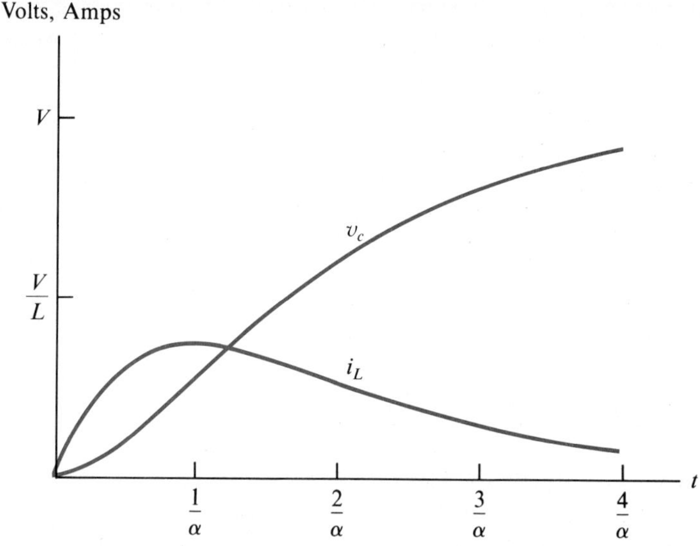

Figure 11.16 Zero-state response of $v_c(t)$ and $i_L(t)$ in a critically damped, series *RLC* circuit driven by a step input, and the pole-zero patterns of the circuit's voltage and current transfer functions.

$$i_L(t) = i_o e^{-\alpha t} + [\alpha i_o + (V - i_o R - v_o)/L]te^{-\alpha t}$$

As an exercise, use Table 11.2 to verify these expressions. Figure 11.16 shows the ZSR of v_c and i_L to a step input.

The critically damped, series *RLC* circuit will be treated in the problems at the end of the chapter.

11.5.6 Underdamped Circuit ZIR

A circuit is underdamped when its physical parameters have values that allow energy to "shuttle" between energy-storage elements before the circuit reaches the steady state. This condition will be evident in the alternating peaks of the inductor current and capacitor voltage. The process continues until the total energy stored in the circuit is compatible with the steady-state physical constraints imposed on the capacitor and inductor by the applied source (i.e., for a step-input signal $i_L(\infty) = 0$ and $v_c(\infty) = V$).

Example 11.10

The ZIR of an underdamped circuit has the generic form

$$y(t) = 2|K|e^{-\alpha t} \sin(\omega_d t + \phi) \tag{11.21}$$

subject to the ZIR boundary conditions of the circuit. Using this generic solution with the ZIR boundary conditions of the capacitor voltage in the series *RLC* circuit gives

$$v_c(0^+) = v_o$$
$$\frac{dv_c(0^+)}{dt} = \frac{i_o}{C}$$

and so

$$2|K| \sin \phi = v_c(0^+) = v_o$$

and

$$-2\alpha|K| \sin \phi + 2\omega_d |K| \cos \phi = \frac{dv_c(0^+)}{dt} = \frac{i_o}{C}$$

Solving the first equation for $|K|$ leads to

$$|K| = v_o/(2 \sin \phi)$$

which can be substituted into the second equation to get

$$\tan \phi = \frac{\omega_d v_o}{\frac{1}{C} i_o + \alpha v_o}$$

and

$$\phi = \tan^{-1} \frac{\omega_d v_o}{\frac{1}{C} i_o + \alpha v_o}$$

A word of caution—since two values of ϕ give the same value for their tangent, we must choose ϕ such that $|K| > 0$.

Example 11.11

Find the ZIR of v_c in the series *RLC* circuit with $R = 4\ \Omega$, $L = 1$ H and $C = 0.05$ F when $v_o(0^-) = 4$ volts, and $i_L(0^-) = 2$ amps.

Solution: The ZIR of $v_c(t)$ is

$$v_c(t) = 2|K|e^{-2t} \sin (4t + \phi)$$

with boundary conditions $v_c(0^+) = v_c(0^-) = 4$ and $dv_c(0^+)/dt = i_L(0^+)/C = i_L(0^-)/C = 40$. We could evaluate the expressions that were developed in Example 11.10 to obtain $|K|$ and ϕ; instead we'll solve for $|K|$ and ϕ to increase our skill at finding the response of underdamped circuits, so

$$v_c(0^+) = 2|K| \sin \phi = 4$$

and

$$\frac{dv_c(0^+)}{dt} = -4|K| \sin \phi + 8|K| \cos \phi = 40$$

The first equation gives $|K| = 2/\sin \phi$, so the second equation gives $\tan \phi = 1/3$. The arctangent function has two values that are consistent with $\tan \phi = 1/3$: $\phi_1 = 18.43°$ or $\phi_2 = 198.43°$. Only the value of $\phi = 18.43°$ is consistent with $|K| > 0$; with this choice we get $|K| = 6.32$. Then

$$v_c(t) = 12.64e^{-2t} \sin (4t + 18.34°)$$

The response curves for $v_c(t)$, $i_L(t)$ and $v_L(t)$ are shown in Fig. 11.17. Verify that the approximate number of oscillations in v_c before decay is $N = 1.27$. Notice how the curves for the ZIR display the following control mechanisms in the circuit: 1) the current decreases as long as $v_L < 0$ and vice versa, 2) the capacitor voltage increases as long as $i_L > 0$ and vice versa, and 3) the inductor current increases as long as $v_L < -v_c/R$.

Skill Exercise 11.9

Find the ZIR of v_c and i_L in the series RLC circuit when $R = 5\ \Omega$, $L = 2$ H, $C = 0.005$ F, $v_o = 5$ V and $i_o = -2$ A.

Answer: $v_c(t) = 40e^{-1.25t} \sin (9.92t + 172.82°)$

and

$$i_L(t) = -2e^{-1.25t} \cos (9.92t)$$

11.5.7 Underdamped Circuit ZSR

The ZSR of the underdamped series RLC circuit is its response to an applied source when both the capacitor and the inductor have no initial stored energy.

Example 11.12

If the underdamped series RLC circuit is driven by $v_{in}(t) = Vu(t)$ the ZSR of the capacitor voltage is given by

$$v_c(t) = 2|K|e^{-\alpha t} \sin (\omega_d t + \phi) + V$$

subject to $v_c(0^+) = 0$ and $dv_c(0^+)/dt = 0$. These boundary conditions re-

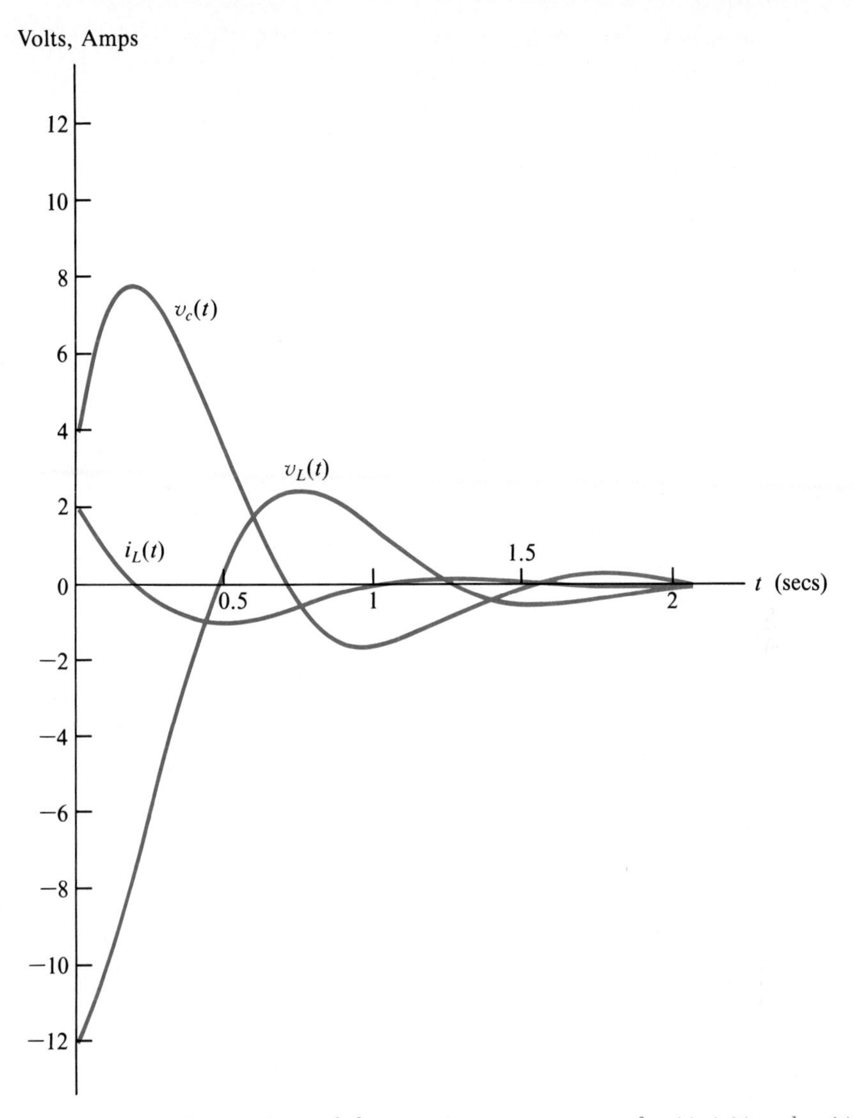

Figure 11.17 Comparison of the zero-input responses of $v_c(t)$, $i_L(t)$ and $v_L(t)$ in an underdamped series RLC circuit.

quire that $2|K| \sin \phi + V = 0$ and $-2\alpha|K| \sin \phi + 2\omega_d |K| \cos \phi = 0$. Solving the first equation gives $|K| = -V/(2 \sin \phi)$. Substituting this expression into the second boundary-condition equation gives $\tan \phi = \omega_d/\alpha$. As in the case of the underdamped ZIR, ϕ must be chosen so that $|K| > 0$.

Since the particular solution of the I/O DE describing the inductor current is zero when the input is a step signal, the ZSR of the inductor current is

$$i_L(t) = 2|K|e^{-\alpha t} \sin (\omega_d t + \phi)$$

subject to $i_L(0^+) = 0$ and $di_L(0^+)/dt = V/L$. (Note that some, but not all, of the boundary conditions of the ZSR are zero.) Applying the boundary conditions gives

$$2|K| \sin \phi = 0$$

and

$$-2\alpha|K| \sin \phi + 2\omega_d|K| \cos \phi = V/L$$

The first equation is solved by $\phi = 0$ and by $\phi = \pi$. With either choice, the second equation has

$$|K| = V/(2\omega_d L \cos \phi)$$

Therefore, if $V > 0$, choose $\phi = 0°$, and if $V < 0$, choose $\phi = \pi$ to ensure that $|K| > 0$. Consequently,

$$i_L(t) = \begin{cases} V/(\omega_d L)e^{-\alpha t} \sin \omega_d t & V > 0 \\ -V/(\omega_d L)e^{-\alpha t} \sin (\omega_d t + \pi) & V < 0 \end{cases}$$

which simplifies to

$$i_L(t) = V/(\omega_d L)e^{-\alpha t} \sin \omega_d t$$

The curves for v_c and i_L when $\alpha = 4$ and $T = 0.5$ are shown in Fig. 11.18, where α determines the decay envelope of the exponential response and ω_d determines the "period" of the damped oscillation. In each case the shape of the response curve is determined by the circuit's pole-zero pattern, and the steady-state values of the step responses are determined by $\mathbf{H}_V(0)$ and $\mathbf{H}_I(0)$.

Skill Exercise 11.10

Find the ZSR of v_c to $v_{in}(t) = Vu(t)$ in the series RLC circuit when $R = 4\ \Omega$, $L = 1$ H and $C = 0.05$ F.

Answer: $v_c(t) = 1.12\ Ve^{-2t} \sin (4t - 116.57°) + V$

11.5.8 Underdamped Circuit ISR

The last case to consider is the initial-state response of the underdamped circuit.

Example 11.13

In the series RLC circuit, the initial-state response of v_c to $v_{in}(t) = Vu(t)$ is described by

$$v_c(t) = 2|K|e^{-\alpha t} \sin (\omega_d t + \phi) + V$$

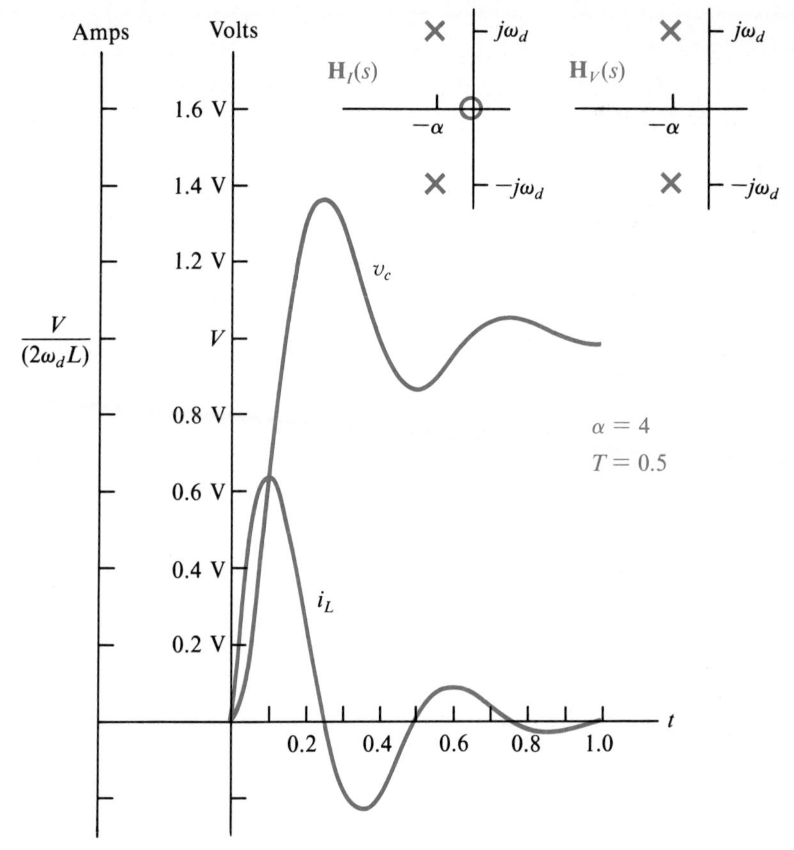

Figure 11.18 The zero-state responses of $v_c(t)$ and $i_L(t)$ to a step input signal, and the pole-zero patterns of the underdamped circuit's voltage and current transfer functions.

subject to the ISR boundary conditions

$$v_c(0^+) = 2|K| \sin \phi + V = v_o$$

and

$$dv_c(0^+)/dt = -2\alpha|K| \sin \phi + 2\omega_d|K| \cos \phi = i_o/C$$

Solving these equations gives

$$|K| = \frac{(v_o - V)}{2 \sin \phi}$$

and

$$\phi = \tan^{-1} [\omega_d(v_o - V)]/[i_o/C + \alpha(v_o - V)]$$

Once again, we note that the chosen value of ϕ must be consistent with $|K| > 0$.

Skill Exercise 11.11

Find the ISR of v_c and i_L to $v_{in}(t) = u(t)$ for the series *RLC* circuit when $R = 4\ \Omega$, $L = 1$ H, $C = 0.05$ F, $v_o = 4$ and $i_o = 2$.

Answer: $v_c(t) = 11.88e^{-2t} \sin(4t + 14.63°) + 1$

$i_L(t) = 0.05[-23.76e^{-2t} \sin(4t + 14.63°) + 47.52e^{-2t} \cos(4t + 14.63°)]$

The expression for $i_L(t)$ in the preceding Skill Exercise is obtained by differentiating the expression for $v_c(t)$. Doing so for an underdamped circuit gives a result that is not readily graphed. It then becomes necessary to use the identity

$$A \cos \alpha + B \sin \alpha = \sqrt{(A^2 + B^2)} \cos(\alpha - \psi)$$

where

$$\tan \psi = B/A$$

If only the ratio B/A is known the angle ψ is ambiguous, since $\tan \psi$ is the same for angles in the first and third quadrant, and for angles in the second and fourth quadrant. To resolve this dilemma, hand-held calculators evaluate $\eta = \tan^{-1} B/A$ and display angles with $-\pi/2 \leq \eta \leq \pi/2$. If A and B are known the quadrant containing the angle can be determined from the auxillary calculations: $\cos \psi = A/\sqrt{(A^2 + B^2)}$ and $\sin \psi = B/\sqrt{(A^2 + B^2)}$ so the actual angle is obtained from the value returned by the calculator according to

$$\psi = \tan^{-1} B/A \qquad \text{if } A \geq 0,\ B \geq 0$$
$$\psi = \tan^{-1} B/A + 180° \qquad \text{if } A \leq 0,\ B \geq 0$$
$$\psi = \tan^{-1} B/A + 180° \qquad \text{if } A \leq 0,\ B \leq 0$$
$$\psi = \tan^{-1} B/A \qquad \text{if } A \geq 0,\ B \leq 0$$

Example 11.14

The expression for $i_L(t)$ in SE 11.11 can be simplified by noting that $A = 47.52$, $B = -23.76$, $\psi = -26.57°$ and $\sqrt{(A^2 + B^2)} = 53.13$, giving

$$i_L(t) = 2.66e^{-2t} \cos(4t + 41.20°) = 2.66e^{-2t} \sin(4t + 131.20°)$$

11.6 ZIR AND ZSR*

The ZSR curves of v_c obtained in the previous examples are shown together in Fig. 11.19. For the same initial conditions the overdamped circuit takes the longest time to dissipate the initial stored energy. The critically damped case marks a boundary between the overdamped and underdamped cases. It represents the fastest possible response decay

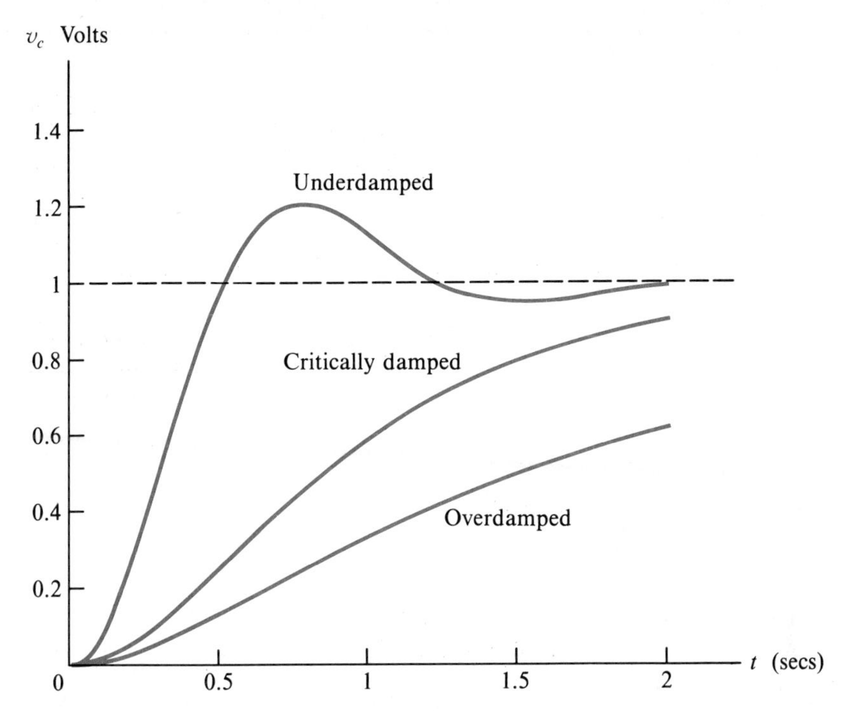

Figure 11.19 Comparison of zero-state responses of $v_c(t)$ for overdamped, critically damped, and underdamped series *RLC* circuits.

without oscillation. The energy dissipated in the transition to steady state is the same in each case, but the waveforms of the transition process differ radically.

11.7 SINUSOIDAL RESPONSE OF SECOND-ORDER CIRCUITS

The particular solution of the I/O DE for a second-order circuit driven by a sinusoidal source is obtained just as it was for a first-order circuit—the sinusoidal signal is modelled as the real or imaginary part of a complex exponential signal. Then the transfer function is evaluated at the complex frequency of the source to give a complex quantity whose magnitude and angle determine the magnitude and phase angle of the particular solution to a sinusoidal input. The only difference between the second- and first-order problems is that the algebraic form of the circuit's transfer function may have numerator and denominator polynomials of power two.

Example 11.15

Find the ISR of $v(t)$ in the series/parallel circuit shown in Fig. 11.20 when $v_{in}(t) = u(t) \sin 2t$, with $i(0^-) = 0.5$ A and $v(0^-) = 1.5$ V.

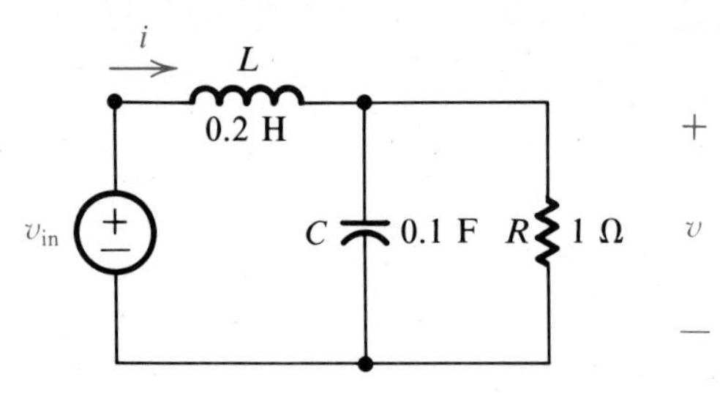

Figure 11.20 Series/parallel *RLC* circuit.

Solution: It can be shown that the transfer function relating v_{in} to v is given by

$$\mathbf{H}(s) = \frac{\dfrac{1}{LC}}{s^2 + s\dfrac{1}{RC} + \dfrac{1}{LC}}$$

For the given parameters the circuit has a pair of complex poles with $\alpha = 5$ and $\omega_d = 5$. Since $\sin 2t = \text{Im}\{e^{j2t}\}$ the particular solution of its I/O DE to the sinusoidal input signal is given by $v_p(t) = \text{Im}\{\mathbf{H}(j2)e^{j2t}\}$. Since $\mathbf{H}(j2) = 0.997\underline{/-23.499°}$, the particular solution is $v_p(t) = 0.997 \sin(2t - 23.499°)$. The generic form of the ISR is the sum of $v_p(t)$ and the undamped natural solution

$$v(t) = 2|K|e^{-5t} \sin(5t + \phi) + 0.997 \sin(2t - 23.499°)$$

where $|K|$ and ϕ are determined by the ISR boundary conditions

$$v(0^+) = 2|K| \sin \phi + 0.997 \sin(-23.499°) =$$

and

$$\begin{aligned} dv(0^+)/dt &= -10|K| \sin \phi + 10|K| \cos \phi \\ &\quad + (2)(.997) \cos(-23.499°) \\ &= i(0^+) - v(0^+)/R = i(0^-) - v(0^-)/R = - \end{aligned}$$

These equations are solved by $|K| =$ and $\phi =$ 1 ° to give

$$\begin{aligned} v(t) = &\ e^{-5t} \sin(5t + 1\ ) \\ &+ .997 \sin(2t - 23.499°) \end{aligned}$$

The graph of $v(t)$ in Fig. 11.21 has a transient response that decays to zero in approximately 0.8 sec, and an oscillation period of 0.4π sec superimposed on the steady-state oscillation period of π sec. After

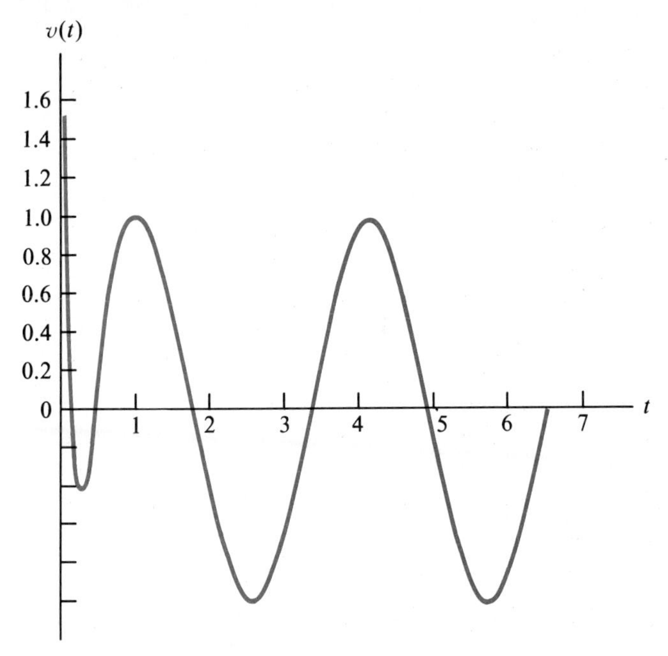

Figure 11.21 Solution for $v(t)$ in Example 11.24.

the transient decays to zero, the response is a scaled and shifted copy of the sinusoidal input signal.

Skill Exercise 11.12

If the circuit of Example 11.15 has $R = 2.236\ \Omega$, $L = 0.2$ H, and $C = 0.01$ F, find the ZSR of $v(t)$ when $v_{in}(t) = 10 \cos 5t\, u(t)$.

Answer: $v(t) = -(8.61 + 212.87t)e^{-22.36t} + 9.52 \cos (5t - 25.21°)$

11.8 PULSE RESPONSE*

The ZSR of a second-order circuit to a rectangular pulse signal is formed by taking the difference of its response to a step signal and to a delayed step signal. Of the three possible response types, the underdamped response is the easiest to visualize, especially when the duration of the pulse is long compared to the time constant of the circuit.

Example 11.16

Find the response of the capacitor voltage in the series *RLC* circuit of Skill Exercise 11.10 to a pulse input signal described by

$$v_{in}(t) = V[u(t) - u(t - \beta)]$$

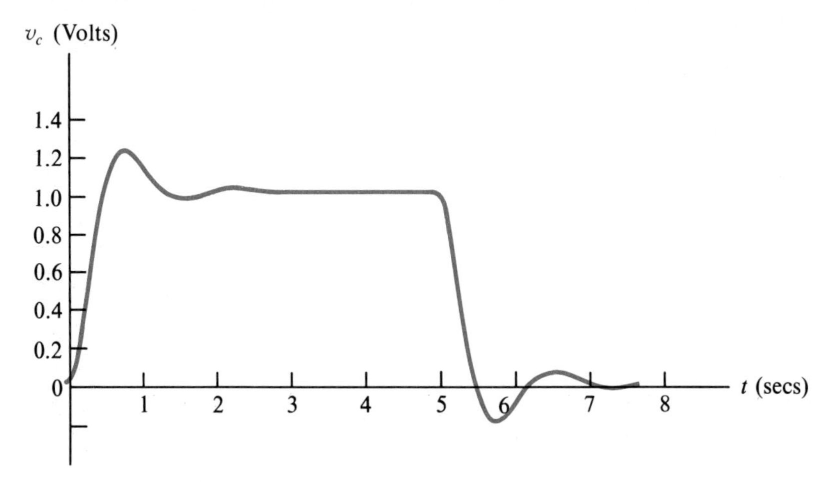

Figure 11.22 The response of the underdamped series *RLC* circuit to a pulse input signal.

Solution: In SE11.10 we showed that

$$Vu(t) \rightarrow 1.12[e^{-2t} \sin(4t - 116.57°) + 1]Vu(t)$$

By superposition

$$V[u(t) - u(t-\beta)] \rightarrow [1.12e^{-2t} \sin(4t - 116.57°) + 1]Vu(t) - [1.12e^{-2(t-\beta)} \sin(4(t-\beta) - 116.57°) + 1]Vu(t-\beta)$$

Figure 11.22 shows $v_c(t)$ when $\beta = 5$ sec. Notice how the circuit responds to the leading edge of the pulse, reaches a steady-state value to a step input, and then decays from that value at the trailing edge of the pulse.

SUMMARY

The characteristic equation of a second-order circuit has two natural frequencies. Depending on the value of the circuit parameters, these natural frequencies may be a pair of distinct real values (overdamped), identical real values (critically damped), or a complex-conjugate pair of values (underdamped). A distinct form of the natural solution to the circuit's I/O equation applies to each case of damping. In all cases, two distinct coefficients must be obtained by applying the boundary conditions of the ZIR, ZSR, or ISR of the circuit. These mathematical boundary conditions reflect physical constraints—an inductor's current must be continuous and a capacitor's voltage must be continuous.

The solution of a second-order I/O differential equation is a direct extension of the methods used in Chapters 8 and 9 for first-order circuits. The transfer function specifies the circuit's natural frequencies, the generic form of the natural-solution component to an exponential signal and the particular-solution component of the I/O equation for a complex exponential input signal.

Problems - Chapter 11

11.1 A second-order circuit's transfer function has poles at $s = -20 \pm j10$ and a zero at $s = -7.5$. Estimate the time required for the circuit to reach the steady state when it is driven by a step input.

11.2 For each of the transfer functions given below, determine whether the circuit is overdamped, critically damped, or underdamped.

a. $$\mathbf{H}(s) = \frac{s + 10}{s^2 + 12s + 35}$$

b. $$\mathbf{H}(s) = \frac{1}{s^2 + 8s + 17}$$

c. $$\mathbf{H}(s) = \frac{1}{s^2 + 12s + 36}$$

11.3 A circuit is described by

$$4\frac{d^2v}{dt^2} + 2\frac{dv}{dt} + v = 2\frac{dv_{in}}{dt} + v_{in}$$

a. Find the transfer function of the circuit.
b. Find the circuit's natural frequencies.
c. Draw the circuit's pole-zero pattern.

11.4 Develop a table summarizing the boundary conditions for i_L and v_c in a parallel *RLC* circuit driven by a current source $i_{in}(t)$. (See Table 11.2)

11.5 A second-order circuit is described by the differential equation $d^2i/dt^2 + 3\,di/dt + 2i = d^2v/dt^2 + 2\,dv/dt$. Find the particular solution to the differential equation when

a. $v(t) = 5 \cos t$.
b. $v(t) = e^{-1t} \cos t$.

11.6 For the op amp circuit in Fig. P11.6:

a. Find expressions for $v_o(0^+)$ and $dv_o(0^+)/dt$ in terms of the circuit parameters R_1, R_2, R_3, C_1 and C_2, the initial capacitor voltages, $v_{c1}(0^-)$ and $v_{c2}(0^-)$ and the source, $v_{in}(0^+)$.

b. If $v_{in}(t) = u(t)$, what is $v_o(\infty)$?

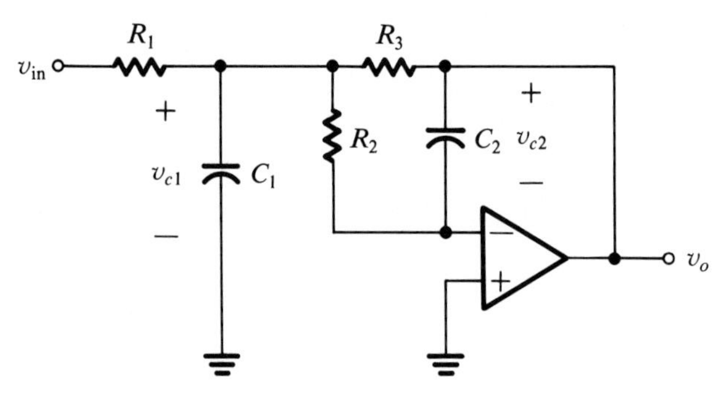

Figure P11.6

11.7 A second-order circuit has a zero-state response described by

$$v(t) = 10[1 - e^{-t} \cos(20\pi t)]u(t)$$

a. Sketch $v(t)$.

b. How many cycles of oscillation occur in $v(t)$ before the signal reaches the steady state?

c. Draw the pole pattern of the circuit.

d. If the circuit is a series *RLC* circuit with L = 0.5 H, find the values of *R* and *C*.

11.8 Repeat Problem 11.7 using:

a. $v(t) = 10[1 - e^{-40t} \cos(20\pi t)]u(t)$.

b. $v(t) = 10te^{-t}u(t)$.

c. $v(t) = 10[e^{-t} - e^{-0.5t}]u(t)$.

11.9 A parallel *RLC* circuit has $L = 1$ mH and $C = 0.3\ \mu$F.

a. Find the value of *R* for which the circuit is critically damped.

b. Find the value of *R* for which the circuit's damped frequency of oscillation is $\omega_d = 5(10^4)/\sqrt{3}$.

11.10 If a parallel *RLC* circuit has the transfer function:

$$\frac{\mathbf{V}_c}{\mathbf{I}_g} = \mathbf{H}(s) = \frac{100{,}000s}{0.5s^2 + 100s + 202{,}392.15}$$

a. What is the steady-state value of $v_c(t)$ if the input is a unit step?

b. What are the values of *R*, *L* and *C*?

c. If a step-input signal is applied, how long does it take for the inductor current to reach steady state?

11.11 If a parallel *RLC* circuit has $R = 10\ \Omega$, $L = 40$ H and $C = 0.025$ F, find the inductor current for $t \geq 0$ if $i_L(0^-) = -10$ A and $v_C(0^-) = 12$ V.

11.12 If a parallel *RLC* circuit has $R = 10\ \Omega$ and $L = 0.5$ H, for what range of values of C is the circuit underdamped?

11.13 If the op amp circuit in Fig. 11.3 has $R_1 = 10$ kΩ, $R_2 = 20$ kΩ and $C_1 = 1\ \mu$F, for what values of C_2 is the circuit (a) overdamped, (b) critically damped, and (c) underdamped?

11.14 Find the ZSR of the inductor current in a parallel *RLC* circuit when $i_{in}(t) = 5u(t)$, with $R = 10\ \Omega$, $L = 10$ H, and $C = 0.025$ F.

11.15 Repeat Problem 11.14 with $R = 10\ \Omega$, $L = 9.102$ H, and $C = 0.025$ F.

11.16 The circuit in Fig. P11.16 has a transfer function given by

$$\frac{\mathbf{V}}{\mathbf{V}_{in}} = \frac{\dfrac{1}{R_1C}\left[s + \dfrac{R_2}{L}\right]}{s^2 + s\left[\dfrac{1}{R_1C} + \dfrac{R_2}{L}\right] + \dfrac{R_1 + R_2}{R_1LC}}$$

a. Find and draw the zero-state response of $v(t)$ to $v_{in}(t) = 5u(t)$ when $R_1 = 2\ \Omega$, $R_2 = 4\ \Omega$, $L = 3$ H and $C = 1/3$ F.

b. If $v(0^-) = 8$ V and $i(0^-) = 2$ A, find and draw $v(t)$ for $t \geq 0$ when $v_{in}(t) = 5u(t)$.

c. Repeat b with $v(t) = 5e^{-t}u(t)$.

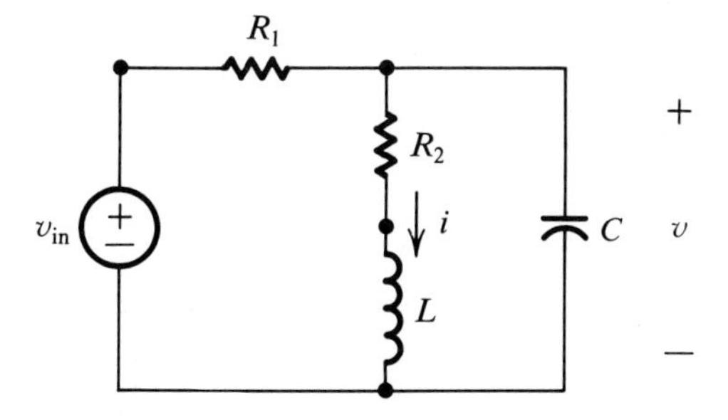

Figure P11.16

11.17 The switch in Fig. P11.17 is moved from position *a* to position *b* at time $t = 0$. Assume that the switch had been closed for at least 2 msec.

a. Find the I/O differential-equation model describing the capacitor voltage for $t > 0$.
b. Draw the pole-zero pattern of the circuit.
c. Find the circuit's transfer function (for $t > 0$).
d. Determine whether the circuit is underdamped, critically damped, or overdamped.
e. Find the zero-input response.
f. Find the zero-state response.
g. Find the initial-state response by solving the I/O equation directly, and compare your answer to the results obtained in e and f.

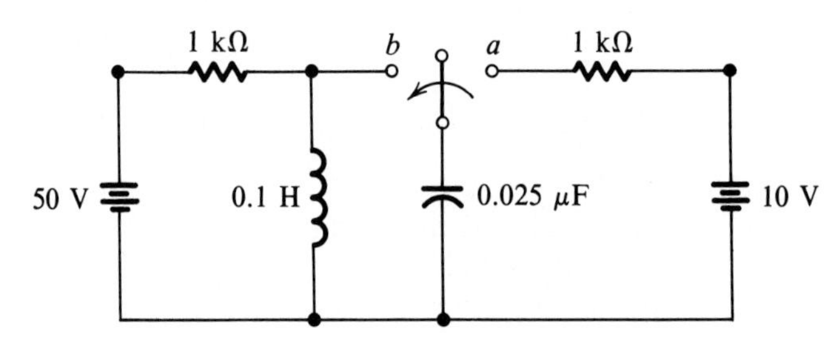

Figure P11.17

11.18 Find the ZIR and ISR of $i_2(t)$ when $v_{in}(t) = 5e^{-4t}u(t)$, $i_1(0^-) = 4$ A and $i_2(0^-) = -6$ A. Also find the steady-state values of i_1 and i_2.

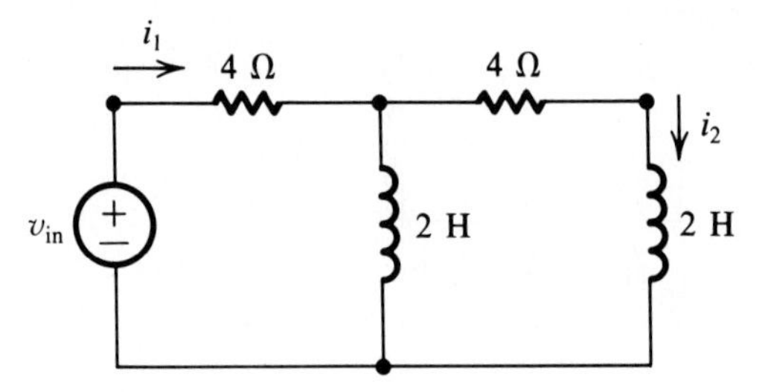

Figure P11.18

11.19 Find an expression for the ZSR of $v_o(t)$ when $v_{in}(t) = Vu(t)$. What is the steady-state value of v_o?

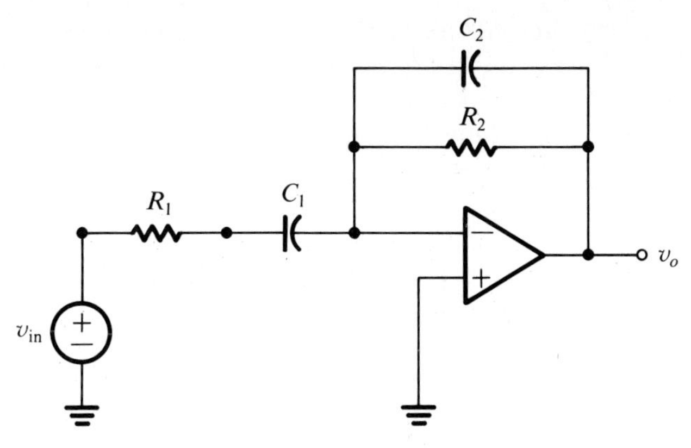

Figure P11.19

11.20 Repeat Problem 11.19 with $v_{in}(t) = Vu(t - 5)$.

11.21 Find the ZSR of $v_o(t)$ when $v_{in}(t) = 20 \sin(100t - 40°)$ for $t > 0$. What is the steady-state value of v_o?

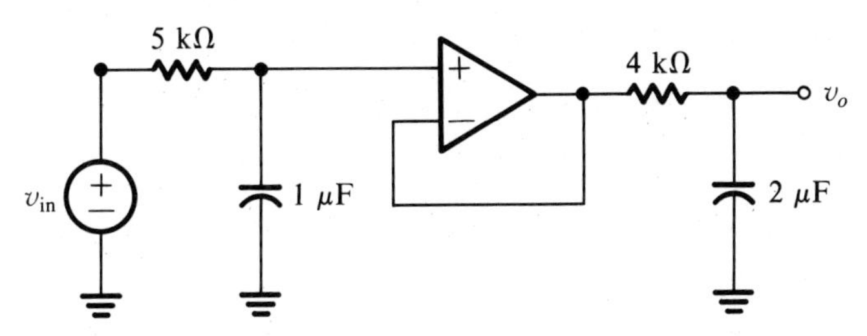

Figure P11.21

11.22 The circuit in Fig. 11.20 has no stored energy at $t = 0$.

a. Find $i(t)$ for $t > 0$ when a unit step of voltage is applied, with $L = 8$ H, $C = 200\ \mu$F and $R = 100\ \Omega$.

b. Repeat a with $L = 50$ mH.

c. Repeat a with $L = 12.5$ H.

11.23 The capacitor voltage and the input current of a circuit have the following I/O transfer function:

$$\mathbf{H}(s) = \frac{10}{s^2 + 2s + 1}$$

a. Find the ZSR when the input current is $i_{in}(t) = 10 \sin t$.

b. Find the ZIR of $v(t)$ if $v(0^-) = 10$ V and $dv(0^-)/dt = -2$ V/sec.

11.24 Let $v(0^-) = 20$ V and $i(0^-) = -5$ A.

a. Find $i(t)$ when $v_{in}(t) = 10u(t)$ and $L = 500$ H, $C = 500\ \mu$F, and $R = 500\ \Omega$.

b. Repeat a with $L = 889$ H.

c. Repeat a with $L = 2.01$.

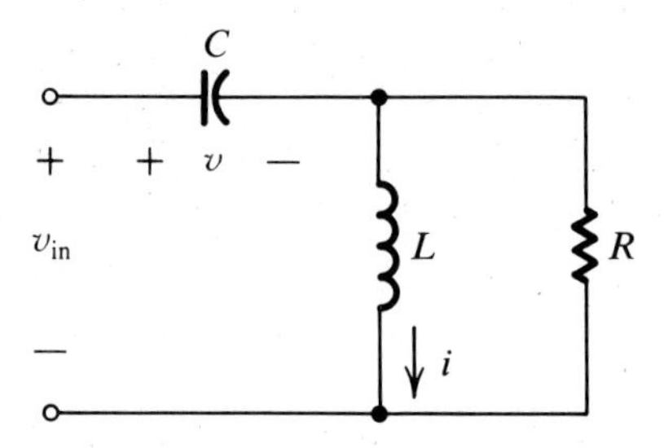

Figure P11.24

11.25 Consider the op amp circuit in Fig. P11.25.

a. Find the ZSR to $v_{in}(t) = 10u(t)$.

b. Repeat a for $v_{in}(t) = 20 \sin(2t)$.

c. Repeat a for $v_{in}(t) = e^{-2t}[u(t) - u(t - 0.5)]$.

d. Find the transient of v_o in each case.

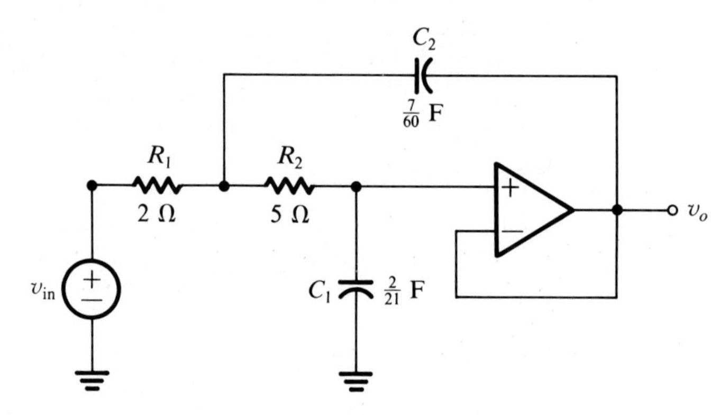

Figure P11.25

11.26 Find the ZSR of $i(t)$ if $L = 0.5$ H, $C = 0.00125$ F, $R = 20\ \Omega$ and

a. $i_{in}(t) = 5 \sin 2\pi t$.

b. $i_{in}(t) = e^{-4t} \sin 2\pi t$.

c. Find the transient and steady-state components of the ZSR in a and b.

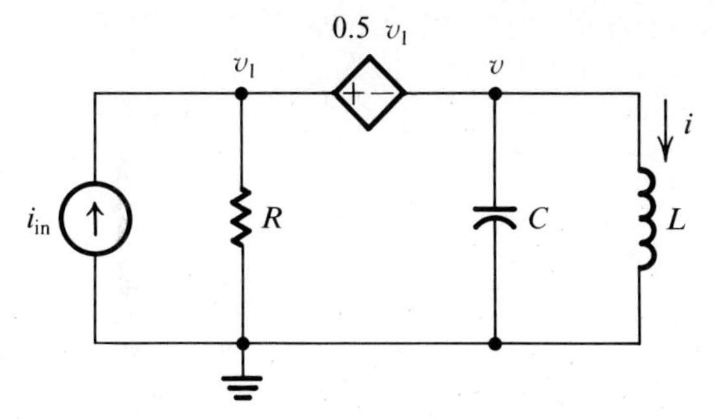

Figure P11.26

11.27 Repeat Problem 11.26 with $L = 1.28$ H.

11.28 Repeat Problem 11.26 with $L = 5.57$ mH.

11.29 Find the ZSR to a unit step if $R = 5\ \Omega$, $C = 0.02$ F, $k_1 = 2$ and $k_2 = 5$ with

a. $L = 2$ H.

b. $L = 12.5$ H.

c. $L = 12.58$ mH.

d. Find the transient and steady-state components of the ZSR in a, b and c.

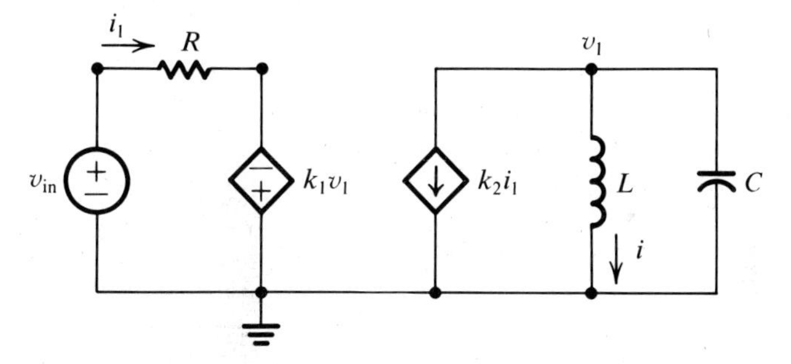

Figure P11.29

11.30 Find the ZIR of $v(t)$ and $i(t)$ if $v(0^-) = 10$ V and $i(0^-) = -5$ A, and find the ZSR to a unit step if $R = 20\ \Omega$, $L = 25$ mH, $k = 4$ and

a. $C = 10\ \mu$F.
b. $C = 20\ \mu$F.
c. $C = 2.5\ \mu$F.

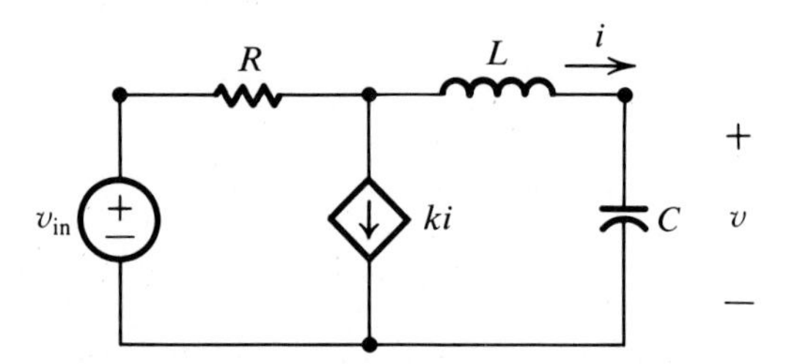

Figure P11.30

11.31 The circuit in Fig. P11.31 has $i(0^-) = 2$ mA and $v(0^-) = 5$ V.

a. Find $v(t)$ for $t \geq 0$ if $i_{in}(t) = 5\cos(3t + 45°)$.
b. Find $v(t)$ for $t \geq 0$ if $i_{in}(t) = 5u(t)$.
c. Estimate the time required for $v(t)$ to reach the steady state in a and b.

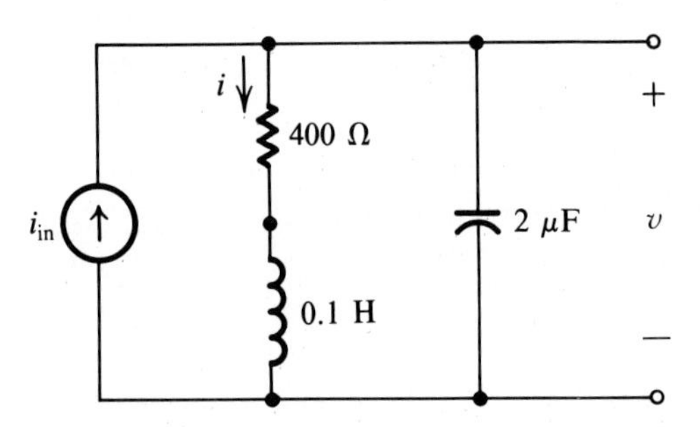

Figure P11.31

11.32 Consider the op amp circuit in Fig. P11.32.

a. Find the transfer function relating v_o to v_{in}.
b. If $v_1(0^-) = 5$ V and $v_o(0^-) = -10$ V, find $v_o(0^+)$ and $dv_o(0^+)/dt$ if $v_{in}(0^+) = 1$ V, assuming $R_1 = 1\ \Omega$, $R_2 = 4\Omega$, $R_3 = 2\Omega$, $C_1 = 4$ F and $C_2 = 2$F.
c. If $R_1 = 1\ \Omega$, $R_2 = 1\ \Omega$, $R_3 = 2\ \Omega$, $C_1 = 2$ F and $C_2 = 1$ F, find $v_o(t)$ for $t \geq 0$ if $v_o(0^+) = 0$ and $dv_o(0^+)/dt = -2$ when $v_{in}(t) = 4u(t)$.

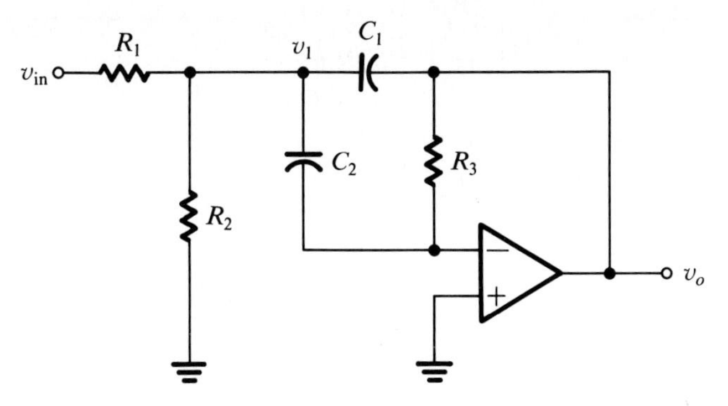

Figure P11.32

11.33 Let $i_{in}(t) = 5u(t)$ and $v_{in}(t) = 10e^{-0.5t}u(t)$.

a. Find the ZSR of $v(t)$ for $t \geq 0$ if $C = 0.08$ F, $L = 12.5$ H, and $R = 25\ \Omega$.

b. Repeat a with $C = 2$ F.

c. Repeat a with $C = 7.36$ mF.

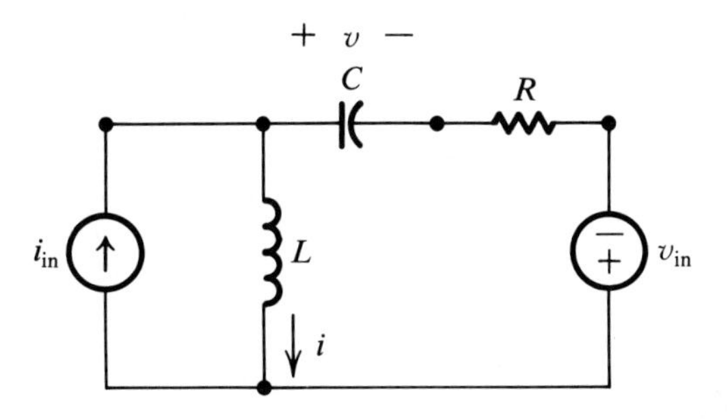

Figure P11.33

11.34 Let $i(0^-) = 4$ A, $v(0^-) = 2$ V, and $C = 2$ F in the circuit in Figure P11.33.

a. Find the ZIR of $v(t)$.

b. Find the ISR of $v(t)$ $i_{in}(t) = 5u(t)$ and $v_{in}(t) = 10e^{-0.5t}u(t)$.

11.35 If a series *RLC* circuit with $R = 1\ \Omega$, $L = 2/(1 + 4\pi^2)$ H and $C = 0.5$ F, find and draw the ZSR of v_c to $v_{in}(t) = 10e^{-0.4t}u(t)$.

GENERALIZED IMPEDANCE, ADMITTANCE, AND S-DOMAIN ANALYSIS

INTRODUCTION

Chapter 7 presented the powerful concept of the transfer function and used it to *construct* the particular solution to a circuit's input/output differential equation when the input signal had exponential form. Transfer functions are also called *s*-domain models of a circuit—signifying that the model depends on the complex frequency *s* and that the domain of *s* is the entire complex plane. In contrast, a circuit's differential equation is called a time-domain model.

Examples in the previous chapters found transfer functions indirectly by first finding the circuit's differential equation. This chapter presents a direct, algebraic method for finding the transfer function, and simple but powerful tools for analyzing circuits in the *s*-domain. The method is important because it is direct, and because it simplifies circuit analysis.

12.1 GENERALIZED IMPEDANCE AND ADMITTANCE

A transfer function exists between any independent signal source and any current or voltage in a circuit. It determines the forced response of a given circuit variable to a given exponential source (the particular solution). When the transfer function relates a circuit voltage to a current source it is called a **generalized impedance** and is denoted by $\mathbf{Z}(s)$ instead of $\mathbf{H}(s)$. If

source is current source

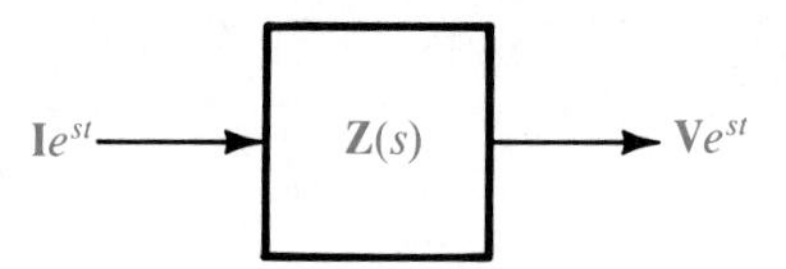

Figure 12.1 The s-domain impedance model of a circuit relates the output exponential $\mathbf{V}e^{st}$ to the input exponential $\mathbf{I}_e{}^{st}$.

the current source $i(t)$ is an exponential signal, the particular-solution component of a given $v(t)$ will also be an exponential. The relationship between these exponential signals is represented by the input/output block diagram in Fig. 12.1, where $\mathbf{Z}(s)$ *symbolizes* the circuit whose impedance is $\mathbf{Z}(s)$ and whose time-domain forced response to $\mathbf{I}e^{st}$ will be $\mathbf{V}e^{st}$. Moreover, since $\mathbf{Z}(s)$ is actually a transfer function, we know that the forced response component of $v(t)$ is defined by

$$v_p(t) = \mathbf{V}e^{st} = \mathbf{Z}(s)\mathbf{I}e^{st}$$

for fixed s and the impedance is given by

$$\boxed{\mathbf{Z}(s) = \frac{\mathbf{V}}{\mathbf{I}}}$$

For a given s the generalized impedance of a circuit and the complex amplitudes $\mathbf{V}$ and $\mathbf{I}$ have the same relationship as in Ohm's Law for resistors:

$$\boxed{\mathbf{V} = \mathbf{I}\mathbf{Z}(s)}$$

This relationship forms the complex amplitude of the forced response from the complex amplitude of the input signal for a given value of s. We must be careful to remember that $v(t) \neq i(t)\mathbf{Z}(s)$ and in general $\mathbf{Z}(s) \neq v(t)/i(t)$ because $v(t)$ includes the natural as well as the particular solution of the circuit's I/O equation.

The generalized impedance of an inductor can be derived by noting that its current and voltage are related by $v(t) = L\,di/dt$. If the current is a complex exponential $i(t) = \mathbf{I}e^{st}$ the forced response of the voltage will be the complex exponential $v(t) = \mathbf{V}e^{st}$. We conclude that $\mathbf{V}e^{st} = sL\mathbf{I}e^{st}$ and $\mathbf{Z}(s) = \mathbf{V}/\mathbf{I} = sL$.

The transfer function of a circuit is also referred to as a **network function.** When the input and output signals are both voltages or both currents, the transfer function is called the **gain function** of the circuit, or the **I/O gain.**

When an exponential voltage source is applied to a circuit and a current is measured, the transfer-function relationship between the signals is called a **generalized admittance.** When the input signal is the complex exponential $\mathbf{V}e^{st}$ the forced response is the exponential $\mathbf{I}e^{st}$. Then the **admittance function** of the circuit defines the ratio

$$\mathbf{Y}(s) = \frac{\mathbf{I}}{\mathbf{V}}$$

and the complex amplitude of the current is

$$\mathbf{I} = \mathbf{V}\mathbf{Y}(s)$$

The generalized admittance for a capacitor can be derived by noting that $i(t) = C\,dv/dt$. When the source is $v(t) = \mathbf{V}e^{st}$ the particular solution for $i(t)$ will be $\mathbf{V}\mathbf{Y}(s)e^{st}$, where $\mathbf{Y}(s) = sC$. The generalized impedance and admittance functions are related in the same way as resistance and conductance—they are reciprocals.

$$\mathbf{Z}(s) = \frac{1}{\mathbf{Y}(s)}$$

The s-domain generalized impedance functions for the basic R, L and C circuit components are given in Table 12.1

TABLE 12.1 S-DOMAIN MODELS FOR *R, L* AND *C.*

	R	*L*	*C*
$\mathbf{Z}(s)$	R	sL	$\frac{1}{sC}$
$\mathbf{Y}(s)$	$\frac{1}{R}$	$\frac{1}{sL}$	sC

Skill Exercise 12.1

Verify the generalized impedance and admittance functions given for the resistor in Table 12.1.

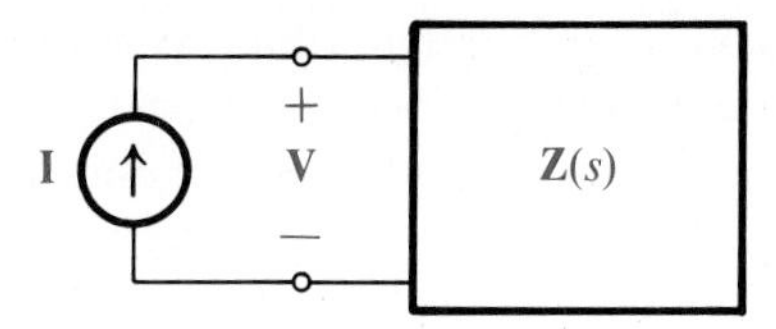

Figure 12.2 The driving-point impedance is calculated from **V** and **I** at the input terminals of the circuit.

In general, an impedance **Z**(s) is called the **driving-point or input impedance** of a circuit when the response voltage is measured at the pair of circuit nodes that connect the stimulus current source to the circuit, as shown in Fig. 12.2. The driving-point admittance of a circuit is the reciprocal of its driving-point impedance. All other admittances and impedances are called **transfer admittances or transfer impedances.**

In the remainder of the book we will sometimes take the liberty of drawing circuit diagrams with the complex amplitudes **V** and **I** (corresponding to the particular-solution component of the time-domain signals $v(t)$ and $i(t)$ when the circuit is driven by an exponential source). When complex amplitude labels are used, remember that they are associated with a time-domain signal; when reference is made to the transfer function between two time-domain signals, remember that the transfer function defines the particular-solution component of one when the other is an exponential signal source. It is also common practice to refer to **V** and **I** as though they were physical quantities rather than mathematical abstractions.

Example 12.1

In Chapter 11 we saw that a series *RLC* circuit has a transfer function between v_{in} and i given by

$$\mathbf{H}_I(s) = \frac{s/L}{s^2 + s\dfrac{R}{L} + \dfrac{1}{LC}}$$

$\mathbf{H}_I(s)$ is the driving-point admittance between the applied voltage and the current into the circuit at the source nodes.

12.2 GENERALIZED KVL AND KCL

Kirchhoff's voltage and current laws apply to the time-domain voltages and currents in a circuit as well as to the s-domain complex amplitudes **V** and **I.** When an exponential source is applied to a circuit the forced response of each node voltage and each branch current has an exponential form. This can be made apparent by noting that we could construct an I/O transfer function between the source and each node voltage and between the source and each branch current. By evaluating these transfer functions at the source frequency we can evaluate the complex amplitudes of the forced responses, and produce transformed copies of the source signal at each node and branch. Each node voltage will have the form $\mathbf{V}_m e^{st}$ and each branch current will have the form $\mathbf{I}_k e^{st}$. (The branch voltages are also exponential because each is a difference of exponential node voltages having the same complex frequency.) Around any mesh of the circuit $\Sigma v_m(t) = 0$ where the v_m are measured between nodes. Writing each $v_m(t)$ in terms of its natural- and particular-solution components gives

$$\Sigma[v_n(t) + v_p(t)]_m = 0$$

If all of the natural frequencies are in the LHP (left half-plane) the natural-solution component vanishes and the node voltage becomes its particular solution. We conclude that in general the particular solution components of the node voltages satisfy KVL around any mesh:

$$\Sigma v_{pm}(t) = 0$$

If the source has exponential form the particular solution of the branch voltages can be written as $v_{pm}(t) = \mathbf{V}_m e^{st}$, so

$$\Sigma \mathbf{V}_m e^{st} = 0$$

Cancelling e^{st} from each term in the sum gives a generalized form of KVL (GKVL):

$$\text{GKVL: } \Sigma \mathbf{V}_m = 0$$

When a circuit is driven by an exponential source, Kirchhoff's voltage law is satisfied by the complex amplitudes of the forced-response (particular-solution) branch voltages around any mesh or loop of a circuit.

Example 12.2

A series *RL* circuit is shown in Fig. 12.3 with each component labeled to show its voltage forced response to the exponential source voltage. Apply GKVL to obtain a relationship between $\mathbf{V}$, $\mathbf{V}_R$ and $\mathbf{V}_L$.

Solution: Applying KVL to the forced response of each voltage gives

$$\mathbf{V}e^{st} = \mathbf{V}_R e^{st} + \mathbf{V}_L e^{st}$$

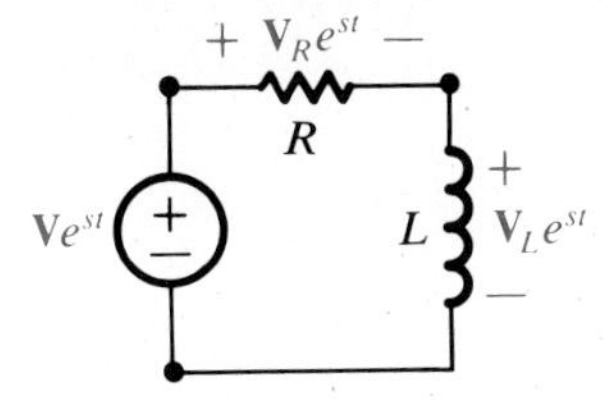

Figure 12.3 Exponential signal source and forced-response components for inductor and resistor voltages in a series *RL* circuit.

or

$$\mathbf{V} = \mathbf{V}_R + \mathbf{V}_L.$$

Thus, the complex amplitudes of the source, resistor, and inductor voltages satisfy KVL. (Food for thought: How can we determine $\mathbf{V}_R$ and $\mathbf{V}_L$ without writing a differential equation?)

Example 12.3

Find the driving-point impedance and the transfer functions between the voltage source and the resistor voltage, and between the voltage source and the capacitor voltage in Fig. 12.4.

Solution: Kirchhoff's voltage law for the series *RC* circuit has

$$v(t) = v_R(t) + v_C(t)$$

If $v(t) = \mathbf{V}e^{st}$ the GKVL states that $\mathbf{V} = \mathbf{V}_R + \mathbf{V}_C$. Also, since the forced response of the current in the circuit has the form $i(t) = \mathbf{I}e^{st}$ we can use the impedance models of the resistor and the capacitor to write $\mathbf{V}_R = \mathbf{I}R$ and $\mathbf{V}_C = \mathbf{I}(1/sc)$.

With these substitutions for $\mathbf{V}_R$ and $\mathbf{V}_C$ GKVL becomes

$$\mathbf{V} = \mathbf{I}(R + 1/sC)$$

and

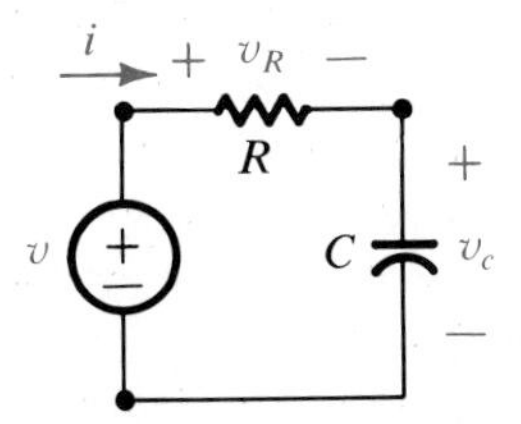

Figure 12.4 Circuit for Example 12.3.

$$\mathbf{I} = \frac{\mathbf{V}}{R + \dfrac{1}{sC}}$$

$$\mathbf{V}_R = \frac{\mathbf{V}R}{R + \dfrac{1}{sC}}$$

$$\mathbf{V}_C = \frac{\mathbf{V}\dfrac{1}{sC}}{R + \dfrac{1}{sC}}$$

(Note that the expressions for $\mathbf{V}_R$ and $\mathbf{V}_C$ are voltage dividers—whose division algorithms are determined by impedances.) The driving-point impedance between the source voltage and the input current is

$$\mathbf{Z}(s) = \frac{\mathbf{V}}{\mathbf{I}} = R + \frac{1}{sC}$$

which is the sum of the impedances connected in series with the source. The voltage gains between the source and the component voltages are

$$\mathbf{H}_R(s) = \frac{\mathbf{V}_R}{\mathbf{V}} = \frac{R}{R + \dfrac{1}{sC}} = \frac{sRC}{sRC + 1}$$

and

$$\mathbf{H}_C(s) = \frac{\mathbf{V}_C}{\mathbf{V}} = \frac{\dfrac{1}{sC}}{R + \dfrac{1}{sC}} = \frac{1}{sRC + 1}$$

The preceding example showed how s-domain transfer functions can be derived without using time-domain differential equations. Once the transfer functions are found, we can draw the pole-zero patterns; define the roots of the characteristic equation (the natural frequencies); and construct the generic form of the natural solution, the particular solution to an exponential input, and the complete solution to the circuit's I/O DE. All that remains is to apply appropriate boundary conditions to find the time-domain zero-input, zero-state, and initial-state responses. In short, s-domain models exploit your background in solving a circuit's I/O DE, but bypass the step of actually deriving the differential equation.

Next, consider KCL at a node of a circuit when the circuit's input signal is an exponential. If i_k denotes the current in the kth branch entering the node, KCL at the node implies that $\Sigma i_k(t) = 0$. Expanding $i_k(t)$ to show its natural and forced-response components gives

$$\Sigma[i_n(t) + i_p(t)]_k = 0$$

for any t. Expressing the forced-response components as exponentials gives

$$\Sigma(i_n(t) + \mathbf{I}e^{st})_k = 0$$

and letting $t \rightarrow \infty$ eliminates the natural solutions, leaving

$$\Sigma \mathbf{I}_k e^{st} = 0$$

Cancelling the exponential gives the generalized form of KCL (GKCL):

$$\text{GKCL: } \Sigma \mathbf{I}_k = 0$$

When a circuit is driven by an exponential signal source, Kirchhoff's current law is satisfied by the complex amplitudes of the forced-response components of the branch currents entering any node of the circuit.

Example 12.4

It is common practice to refer to the driving-point impedance as the impedance "seen" at its driving terminals. Find the input impedance "seen" by the current source at the terminals (a, b) in Fig. 12.5.

Solution: We begin by re-drawing the circuit with the labels shown in Fig. 12.6 (With practice, this step can be eliminated.) Then we write GKCL in terms of the complex amplitudes of the circuit's forced responses to an exponential signal.

$$\text{GKCL: } \mathbf{I} = \mathbf{I}_R + \mathbf{I}_L + \mathbf{I}_C$$

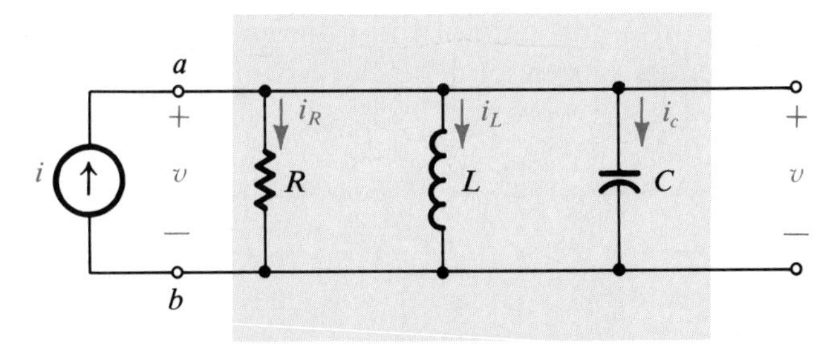

Figure 12.5 Circuit for Example 12.4.

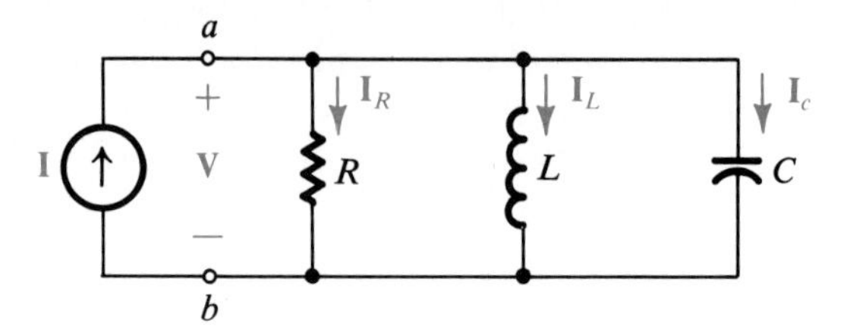

Figure 12.6 Circuit for Example 12.4 redrawn to show the complex amplitudes of the forced responses.

The component admittance models from Table 12.1 can be substituted into this equation to get

$$\mathbf{I} = \left[\frac{1}{R} + \frac{1}{sL} + sC\right]\mathbf{V}$$

The driving point impedance is

$$\mathbf{Z}(s) = \frac{\mathbf{V}}{\mathbf{I}} = \frac{1}{\dfrac{1}{R} + \dfrac{1}{sL} + sC}$$

or

$$\mathbf{Z}(s) = \frac{\dfrac{1}{C}s}{s^2 + s\dfrac{1}{RC} + \dfrac{1}{LC}}$$

The previous example used GKCL to derive the input impedance of the parallel *RLC* circuit without finding a differential equation. Even when an I/O differential equation is sought, it is usually easier to recover it from a transfer function than to derive it directly. The recovery process is straightforward because the numerator polynomial of the transfer function corresponds to the input side of the differential equation, and the denominator polynomial corresponds to the output variable.

Example 12.5

Find the I/O DE relating the capacitor voltage to the source current in the parallel RLC circuit of Example 12.4.

Solution: By inspection we have

$$\mathbf{Z}(s) = \frac{\frac{1}{C}s}{s^2 + s\frac{1}{RC} + \frac{1}{LC}}$$

$$\frac{d^2v}{dt^2} + \frac{1}{RC}\frac{dv}{dt} + \frac{1}{LC}v(t) = \frac{1}{C}\frac{di}{dt}$$

Skill Exercise 12.2

Obtain $\mathbf{Z}(s)$ for the parallel *RLC* by deriving the differential equation directly.

Skill Exercise 12.3

Find the generalized input admittance seen by the current source driving a parallel *RL* circuit.

Answer: $\mathbf{Y}(s) = \frac{1}{R} + \frac{1}{sL}$

12.3 SERIES AND PARALLEL CIRCUITS

Impedances in series combine like resistors in series. Referring to Fig. 12.7(a) and using GKVL, we can write

$$\mathbf{V} = \mathbf{I}\mathbf{Z}_1(s) + \mathbf{I}\mathbf{Z}_2(s) + \cdots + \mathbf{I}\mathbf{Z}_n(s)$$

so the input impedance is given by

$$\mathbf{Z}(s) = \frac{\mathbf{V}}{\mathbf{I}} = \mathbf{Z}_1(s) + \mathbf{Z}_2(s) + \cdots + \mathbf{Z}_n(s)$$

Impedances in series combine by addition.

Example 12.6

Find the driving-point impedance at (a, b) for the series RLC circuit in Fig. 12.8 and the I/O transfer function relating $v_c(t)$ to $v(t)$ (through $\mathbf{V}_c$ and $\mathbf{V}$).

Solution: Adding the series impedances gives

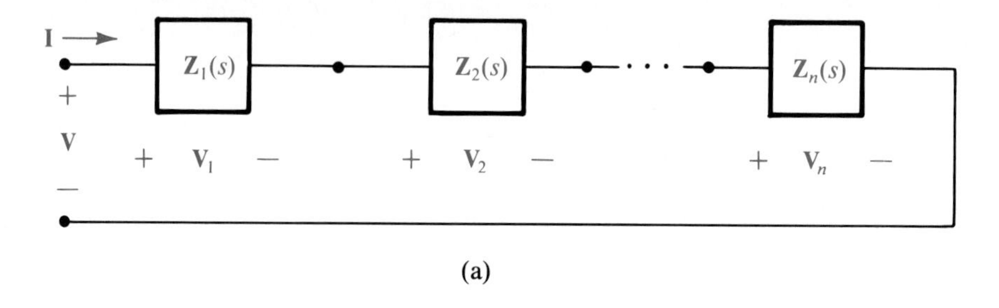

(a)

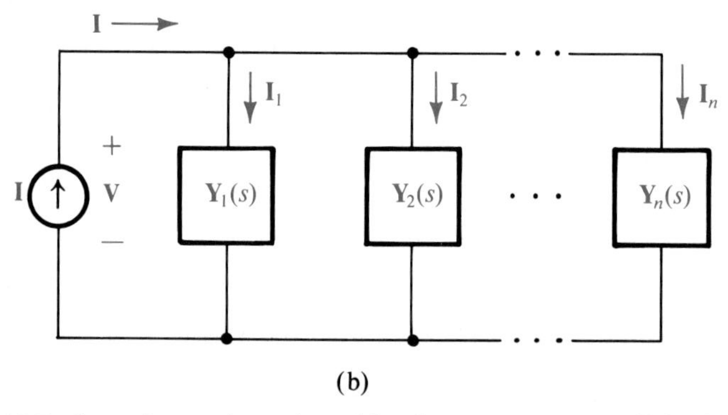

(b)

Figure 12.7 Impedances in series add; admittances in parallel add.

$$\mathbf{Z}(s) = R + sL + \frac{1}{sC} = \frac{s^2LC + sRC + 1}{sC}$$

Next, since $\mathbf{Z}(s) = \mathbf{V}/\mathbf{I}$

$$\mathbf{I} = \frac{sC}{s^2LC + sRC + 1}\mathbf{V}$$

The above expression determines the complex current amplitude, for given source amplitude and frequency. To find the transfer function between the source and $\mathbf{V}_c$ note that

$$\mathbf{V}_c = \mathbf{I}\mathbf{Z}_c(s) = \frac{sC\mathbf{V}}{s^2LC + sRC + 1}\frac{1}{sC}$$

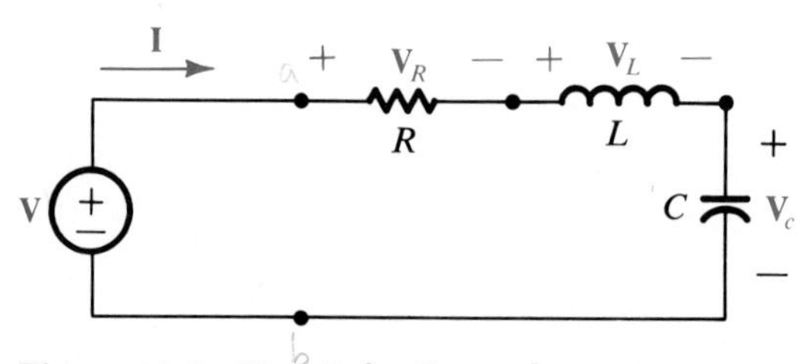

Figure 12.8 Circuit for Example 12.6.

or

$$\frac{\mathbf{V}_c}{\mathbf{V}} = \frac{1/LC}{s^2 + s\frac{R}{L} + \frac{1}{LC}}$$

Recall that the same result was found in Chapter 11.

The voltage-gain relationship in this example illustrates the fact that multiplying the complex amplitude of an exponential signal by s corresponds to differentiating the time-domain signal, and vice versa. In the example the complex amplitudes and the corresponding time-domain signals are related by

$$\mathbf{V}_c = \frac{\frac{1}{LC}\mathbf{V}}{s^2 + s\frac{R}{L} + \frac{1}{LC}} \longleftrightarrow v_c(t)$$

$$s\mathbf{V}_c = \frac{\frac{s}{LC}\mathbf{V}}{s^2 + s\frac{R}{L} + \frac{1}{LC}} \longleftrightarrow \frac{dv_c}{dt}$$

and

$$\mathbf{I} = Cs\mathbf{V}_c = \frac{\frac{s}{L}\mathbf{V}}{s^2 + s\frac{R}{L} + \frac{1}{LC}} \longleftrightarrow \frac{Cdv_c}{dt} = i(t)$$

The complex amplitude for $i(t)$ is the same expression implied by $\mathbf{H}_I(s)$ in Chapter 11 and in Example 11.1.

Parallel admittances combine like parallel conductances. The input admittance for the combination of admittances connected in parallel in Fig. 12.7(b) can be found by using GKCL:

$$\mathbf{I} = \mathbf{I}_1 + \mathbf{I}_2 + \cdots + \mathbf{I}_n$$

With $\mathbf{I}_1 = \mathbf{Y}_1(s)\mathbf{V}$, $\mathbf{I}_2 = \mathbf{Y}_2(s)\mathbf{V}$, . . . , $\mathbf{I}_n = \mathbf{Y}_n(s)\mathbf{V}$ the input current amplitude is

$$\mathbf{I} = \mathbf{Y}_1(s)\mathbf{V} + \mathbf{Y}_2(s)\mathbf{V} + \cdots + \mathbf{Y}_n(s)\mathbf{V}$$

and

$$\boxed{\mathbf{Y}(s) = \mathbf{Y}_1(s) + \mathbf{Y}_2(s) + \cdots + \mathbf{Y}_n(s)}$$

Admittances in parallel combine by addition.

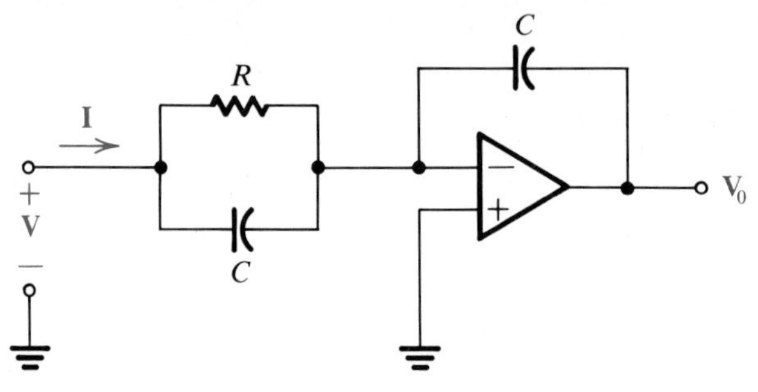

Figure 12.9 Circuit for Example 12.7.

Example 12.7

Find the generalized input admittance of the op amp circuit in Fig. 12.9.

Solution: The input admittance of the circuit is given by the ratio **I/V** at the input terminals, under the assumption that a source is connected to the parallel *RC* section. Whatever type the source is (current or voltage), the complex amplitudes of **V** and **I** at the input must satisfy GKCL and GKVL, so

$$\mathbf{I} = \frac{\mathbf{V}}{R} + sC\mathbf{V}$$

and

$$\mathbf{Y}(s) = \frac{1}{R} + sC$$

This is just the admittance of the parallel combination of *R* and *C*.

12.4 GENERALIZED VOLTAGE AND CURRENT DIVISION

In Chapter 1 we developed voltage- and current-division rules for resistive circuits. Now we'll develop voltage- and current-division rules for impedances and admittances. The input impedance for the series connection of impedances in Fig. 12.7(a) was found to be

$$\mathbf{Z}(s) = \mathbf{Z}_1(s) + \mathbf{Z}_2(s) + \cdots + \mathbf{Z}_n(s)$$

Using GKVL we write

$$\mathbf{V} = \mathbf{V}_1 + \mathbf{V}_2 + \cdots + \mathbf{V}_n = \mathbf{I}\mathbf{Z}_1(s) + \mathbf{I}\mathbf{Z}_2(s) + \cdots + \mathbf{I}\mathbf{Z}_n(s)$$

The voltage across any of the individual impedances is

$$\mathbf{V}_k = \mathbf{I}\mathbf{Z}_k(s)$$

and since

$$\mathbf{I} = \frac{\mathbf{V}}{\mathbf{Z}(s)}$$

we have

$$\mathbf{V}_k = \mathbf{V}\frac{\mathbf{Z}_k(s)}{\mathbf{Z}(s)}$$

The (complex) voltage amplitudes divide according to

$$\frac{\mathbf{V}_k}{\mathbf{V}} = \frac{\mathbf{Z}_k(s)}{\mathbf{Z}_1(s) + \mathbf{Z}_2(s) + \cdots + \mathbf{Z}_n(s)}$$

The voltage drop across an individual impedance is

$$\mathbf{V}_k = \frac{\mathbf{Z}_k(s)\mathbf{V}}{\mathbf{Z}_1(s) + \mathbf{Z}_2(s) + \cdots + \mathbf{Z}_n(s)}$$

This is the same voltage division algorithm that would apply if all the elements were resistors.

Example 12.8

This example demonstrates how voltage division provides a shortcut to deriving the transfer function. Find the voltage amplitude across the inductor in Fig. 12.8 for a given source amplitude **V**.

Solution: By generalized voltage division:

$$\frac{\mathbf{V}_L}{\mathbf{V}} = \frac{\mathbf{Z}_L}{\mathbf{Z}_R + \mathbf{Z}_L + \mathbf{Z}_C}$$

and

$$\mathbf{V}_L = \frac{s^2\mathbf{V}}{s^2 + s\dfrac{R}{L} + \dfrac{1}{LC}}$$

Of course, the same result can be obtained by other, slower methods.

Skill Exercise 12.4

Find the voltage amplitude across the capacitor in a series *RLC* circuit for a given exponential source amplitude **V**.

Answer: $$\mathbf{V}_c = \frac{\dfrac{1}{LC}\mathbf{V}}{s^2 + s\dfrac{R}{L} + \dfrac{1}{LC}}$$

Next, we develop a generalized current-division algorithm for the parallel admittances shown in Fig. 12.7(b). The circuit's admittance is

$$\frac{\mathbf{I}}{\mathbf{V}} = \mathbf{Y}(s) = \mathbf{Y}_1(s) + \mathbf{Y}_2(s) + \cdots + \mathbf{Y}_n(s)$$

and the common voltage can be written as

$$\mathbf{V} = \frac{1}{\mathbf{Y}(s)}\mathbf{I}$$

An individual branch current is

$$\mathbf{I}_k = \mathbf{Y}_k(s)\mathbf{V} = \frac{\mathbf{Y}_k(s)}{\mathbf{Y}(s)}\mathbf{I}$$

which can be expressed as a generalized current-division algorithm

$$\boxed{\frac{\mathbf{I}_k(s)}{\mathbf{I}} = \frac{\mathbf{Y}_k}{\mathbf{Y}(s)}}$$

The current amplitude in the kth path is

$$\boxed{\mathbf{I}_k = \frac{\mathbf{Y}_k(s)\mathbf{I}}{\mathbf{Y}(s)}}$$

Example 12.9

Find the current amplitude in the inductor path in Fig. 12.6.

Solution: Using the generalized current-division algorithm we get:

$$\mathbf{I}_L = \frac{\mathbf{Y}_L}{\mathbf{Y}_R + \mathbf{Y}_L + \mathbf{Y}_C}\mathbf{I}$$

$$= \frac{(1/sL)\mathbf{I}}{\frac{1}{R} + \frac{1}{\mathbf{s}L} + \mathbf{s}C}$$

$$= \frac{(1/LC)\mathbf{I}}{s^2 + s\frac{1}{RC} + \frac{1}{LC}}$$

12.5 CIRCUIT MODELING AND ANALYSIS—AN EASIER WAY

The concepts of generalized impedance and generalized admittance provide a direct, algebraic method for modeling circuits: We can analyze circuits by analyzing complex signal amplitudes in the s-domain and then transforming them back into corresponding time-domain signals. That is, we work with the circuit properties that shape complex-signal amplitudes rather than with the entire time-domain signal. This approach emphasizes constructing and interpreting the circuit response, rather than manipulating differential equations. It simplifies circuit analysis because transfer functions are easier to derive than differential equations.

Example 12.10

Find the gain $\mathbf{H}(s) = \mathbf{V}_o/\mathbf{V}_{in}$ of the voltage-follower op amp circuit in Fig. 12.10.

Solution: Since the voltage source is effectively applied across R_1, the voltage-division relationship

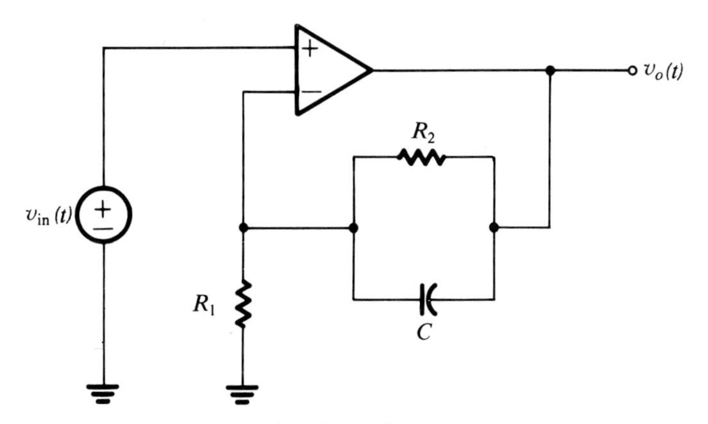

Figure 12.10 Circuit for Example 12.10.

$$\frac{\mathbf{V}_{in}}{\mathbf{V}_o} = \frac{R_1}{R_1 + \mathbf{Z}_2}$$

must be satisfied by the complex amplitudes of v_{in} and v_o, where $\mathbf{Z}_2$ is the impedance created by R_2 and C in parallel. So

$$\frac{\mathbf{V}_o}{\mathbf{V}_{in}} = 1 + \frac{\mathbf{Z}_2}{R_1}$$

$$= 1 + \frac{(1/sC)R_2}{R_2 + (1/sC)} \frac{1}{R_1}$$

and

$$\mathbf{H}(s) = \frac{s + (1/R_2C) + (1/R_1C)}{s + (1/R_2C)}$$

Without a differential equation!

Example 12.11

Find the input/output voltage-gain function $\mathbf{H}(s) = \mathbf{V}_o/\mathbf{V}_{in}$ and the ZSR of the op amp circuit in Fig. 12.11.

Solution: Writing GKCL at the node for v_1 gives

$$\frac{\mathbf{V}_{in} - \mathbf{V}_1}{R_1} - \frac{\mathbf{V}_1}{R_2} - sC_1\mathbf{V}_1 + sC_2(\mathbf{V}_o - \mathbf{V}_1) = 0$$

which can be arranged to

$$\mathbf{V}_1\left[-\frac{1}{R_1} - \frac{1}{R_2} - sC_1 - sC_2\right] + sC_2\mathbf{V}_o = -\frac{\mathbf{V}_{in}}{R_1}$$

At the op amp input node we also apply GKCL. There

$$sC_1\mathbf{V}_1 = -\frac{1}{R_3}\mathbf{V}_o$$

so

$$\mathbf{V}_1 = \frac{-1}{sR_3C_1}\mathbf{V}_o$$

Substituting this expression into the first GKCL equation produces

$$\frac{-1}{sR_3C_1}\left[-\frac{1}{R_1} - \frac{1}{R_2} - sC_1 - sC_2\right]\mathbf{V}_o + sC_2\mathbf{V}_o = -\frac{\mathbf{V}_{in}}{R_1}$$

Multiplying both sides by sR_3C_1 gives

$$\left[\frac{R_1 + R_2}{R_1R_2} + s(C_1 + C_2)\right]\mathbf{V}_o + s^2R_3C_1C_2\mathbf{V}_o = -s\frac{R_3}{R_1}C_1\mathbf{V}_{in}$$

Grouping terms leads to the voltage-gain function

$$\mathbf{H}(s) = \frac{\mathbf{V}_o}{\mathbf{V}_{in}} = \frac{\dfrac{-1}{R_1C_2}s}{s^2 + s\dfrac{C_1 + C_2}{R_3C_1C_2} + \dfrac{R_1 + R_2}{R_1R_2R_3C_1C_2}}$$

Notice that the pole-zero pattern of $\mathbf{H}(s)$ will have a zero at the origin (corresponding to the physical fact that a step input will produce zero steady-state output voltage). Physically, C_1 passes no DC current, thereby shorting the output to ground through R_3.

Skill Exercise 12.5

Explore the physical behavior of the op amp circuit in Fig. 12.11.

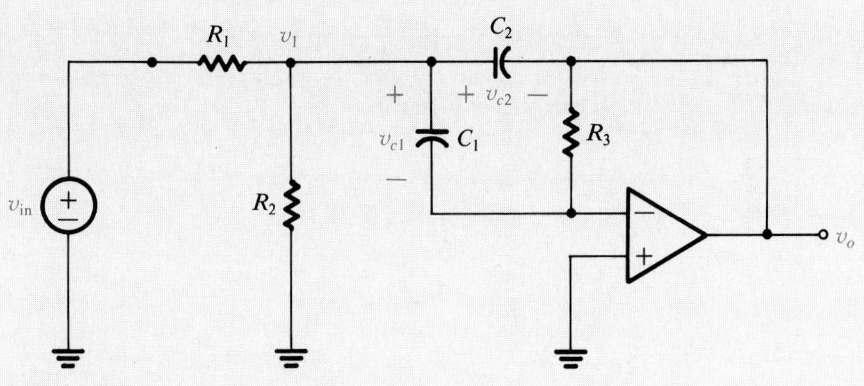

Figure 12.11 Circuit for Example 12.11.

12.6 SOURCE SUPERPOSITION

When a circuit contains more than one source a given circuit variable may be influenced by all of the sources, so we must derive the transfer function between each source and the voltage or current of interest. Once the transfer functions are found, the zero-input response of the circuit is found by simultaneously removing *all* of the sources (all voltage and current sources are set to zero). The roots of the characteristic polynomial in the transfer functions determine the natural solution and the generic form of the ZIR. The zero-state response can be found by either of two methods. The first is to find and superimpose the ZSR due to the individual sources acting with the others "turned off"; the second is to superimpose the forced-response components of the sources. The first method requires solving as many ZSR–boundary-condition problems as there are sources. The second requires only one boundary-condition solution. Likewise, the ISR can be found by adding the ZIR to the ZSR of the circuit, or by direct evaluation using the superimposed forced responses due to simultaneous application of the sources. In both cases, care must be taken to evaluate each transfer function at the complex frequency corresponding to the exponential source for which the function was derived. A first-order example of source superposition was given in Chapter 9. Here we treat a second-order example.

Example 12.12

The *RLC* circuit shown in Fig. 12.12 has transfer functions between i_{in1} and v, and i_{in2} and v, that are given by

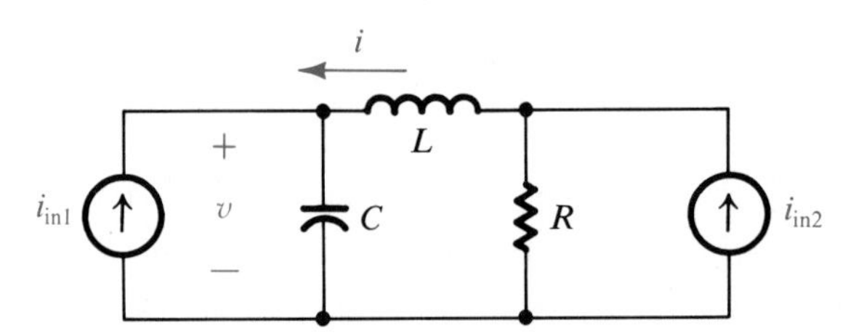

Figure 12.12 Circuit for Example 12.12.

$$\mathbf{H}_1(s) = \frac{\dfrac{1}{C}\left(s + \dfrac{R}{L}\right)}{s^2 + \dfrac{R}{L}s + \dfrac{1}{LC}}$$

and

$$\mathbf{H}_2(s) = \frac{\dfrac{R}{LC}}{s^2 + \dfrac{R}{L}s + \dfrac{1}{LC}}$$

Find the ISR of $v(t)$ when $i_{in1}(t) = 2u(t)$, $i_{in2}(t) = u(t)\sin(\pi/8)t$, $v(0^-) = 1$ V and $i(0^-) = 1$ A, with $R = 1\ \Omega$, $L = 1$ H and $C = 1$ F.

Solution: For the given component values we have

$$\mathbf{H}_1(s) = \frac{s+1}{s^2+s+1} \qquad \mathbf{H}_2(s) = \frac{1}{s^2+s+1}$$

The underdamped circuit parameters are $\alpha = 0.5$, $\omega_n{}^2 = 1$ and $\omega_d = \sqrt{3}/2$. To construct the ISR we add the natural response to the forced responses due to the sources:

$$v(t) = 2|K|e^{-0.5t} \sin(\sqrt{3}t/2 + \phi) + 2\mathbf{H}_1(0)e^{0t} + \text{Im}\{\mathbf{H}_2(j\pi/8)e^{j\pi t/8}\}$$

where $\mathbf{H}_1(0) = 1$ and $\mathbf{H}_2(j\pi/8) = 1.07 \angle -24.91°$. Therefore,

$$v(t) = 2|K|e^{-0.5t} \sin(\sqrt{3}t/2 + \phi) + 2 + 1.07 \sin\left(\frac{\pi t}{8} - 24.91°\right)$$

with boundary conditions given by

$$v(0^+) = v(0^-) = 1$$

$$dv(0^+)/dt = 1/C[i_{in1}(0^+) + i(0^+)] = 3$$

or

$$2|K| \sin\phi + 2 + 10.7 \sin(-24.91°) = 1$$

and

$$-|K| \sin\phi + (\sqrt{3})|K| \cos\phi + 1.07(\pi/8) \cos(-24.91°) = 3$$

The solution has $|K| = -0.2747/\sin\phi$. and $\tan\phi = -0.2030$. Choosing $\phi = -11.47°$ and $|K| = 1.38$ gives the ISR

$$v(t) = 2.76e^{-0.5t} \sin\left(\frac{\sqrt{3}}{2}t - 11.47°\right) + 1.07 \sin\left(\frac{\pi}{8}t - 24.91°\right) + 2$$

Figure 12.13 shows the graph of $v(t)$. Notice that after the transient vanishes $v(t)$ reaches a steady-state value in response to the step input at i_{in1} and the sinusoidal input at i_{in2}.

12.7 POLE-ZERO GEOMETRY*

The pole-zero patterns of a circuit's impedance, admittance, and other transfer functions provide insight into its physical behavior. We saw in

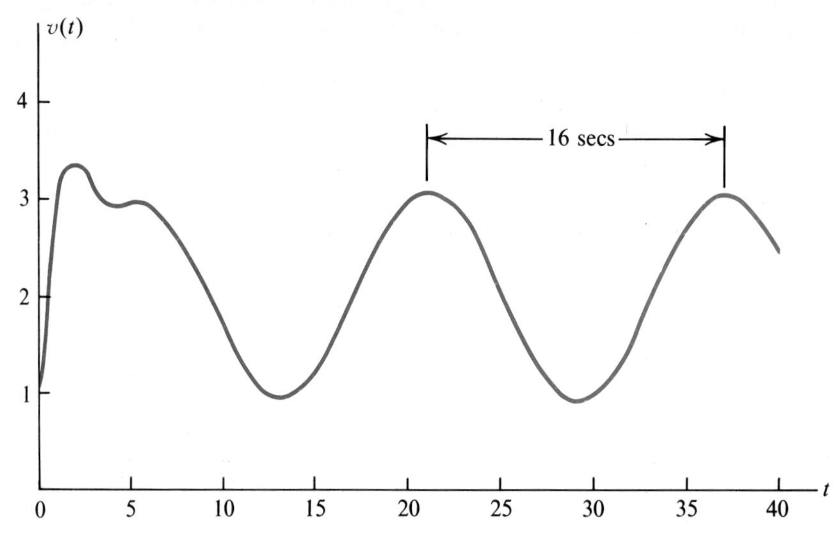

Figure 12.13 Graph of $v(t)$ in the circuit of Fig. 12.12.

Example 12.12 that the transfer function between the first current source and the capacitor voltage has a zero at $s = -R/L$. Thus, a source of the form $\mathbf{I}e^{-(R/L)t}$ will not produce a forced-response component of voltage across the capacitor. The physical basis for this can be understood in terms of the series RL circuit shown in Fig. 12.14. For the given source the *forced* responses are

$$v_R(t) = \mathbf{I}Re^{-(R/L)t}$$

and

$$v_L(t) = L\, d/dt(\mathbf{I}e^{-(R/L)t}) = -\mathbf{I}Re^{-(R/L)t} = -v_R(t).$$

Therefore, the forced response for v_c is zero.

Physically, the series RL circuit "looks" like a short-circuit to the exponential source, since v_R and v_L cancel each other.

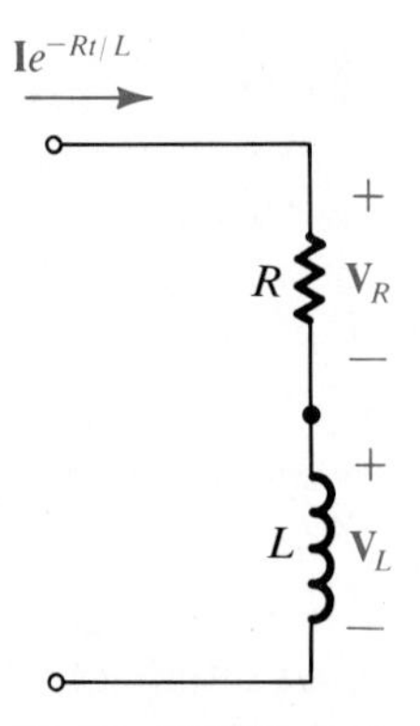

Figure 12.14 Series RL circuit.

Skill Exercise 12.6

Show that if $v(t) = \mathbf{V}e^{-t/RC}$ in Fig. SE12.6 the parallel admittance "looks" like an open circuit (that is, $i(t) = \mathbf{I}e^{-t/RC}$ and $\mathbf{I} = 0$). Is it true that $\mathbf{I}_R = \mathbf{I}_c = 0$?

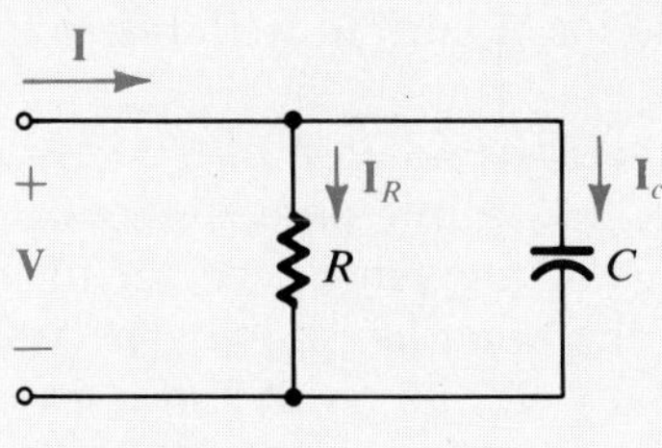

Figure SE12.6

The pole-zero patterns of capacitor and inductor admittance functions can be physically interpreted in terms of short and open circuits. Fig. 12.15 shows the pole-zero patterns of $\mathbf{Z}(s)$ and $\mathbf{Y}(s)$ for a capacitor and inductor. The zero at the origin of the inductor impedance corresponds to the fact that an inductor behaves like a short circuit to a constant current source ($v_L = 0$ whatever the level of the current). On the other hand, the pole at the origin of the circuit's admittance function is a mathematical expression of the fact that an inductor behaves like an integrator with

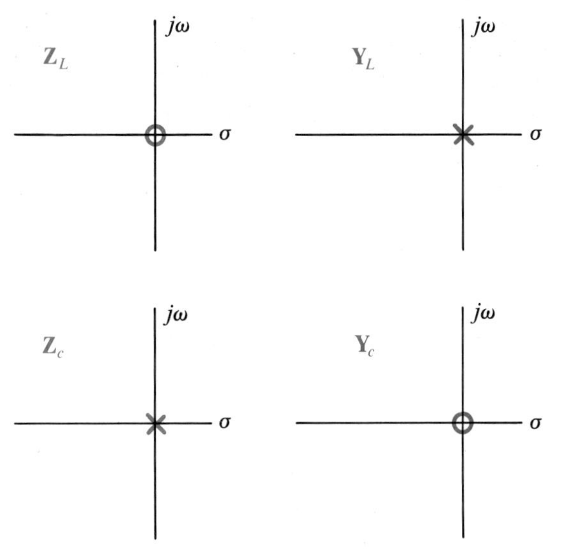

Figure 12.15 Pole-zero patterns of $\mathbf{Z}(s)$ and $\mathbf{Y}(s)$ for an inductor and a capacitor.

respect to applied voltage. Integration of a constant voltage will produce a ramp of current. Thus, we cannot use **H**(s) to evaluate the amplitude of its forced response, because it grows without bound!

Z(s) for a capacitor has a pole at the origin, corresponding to the behavior of an integrator. A step of current will cause a ramp of capacitor voltage. Similarly, the zero of **Y**(s) at the origin represents the infinite impedance/open-circuit–admittance behavior of current with respect to a constant applied voltage.

Pole-zero patterns provide a geometric interpretation of the complex value of a transfer function. For example, the simple pole shown in Fig. 12.16(a) corresponds to $\mathbf{H}(s) = K/(s - s_1)$. For a given s a vector may be drawn from the origin to the point s in the complex plane. The vector $s - s_1$ is the vector drawn from the pole at s_1 to the point s. If a transfer function has two poles, as shown in Fig. 12.16(b), vectors can be drawn from both poles to s. For the first example

$$|\mathbf{H}(s)| = \frac{|K|}{|s - s_1|}$$

and for the second

$$|\mathbf{H}(s)| = \frac{|K|}{|s - s_1|\,|s - s_2|}$$

Since $|s - s_k|$ if the length of the vector $s - s_k$, we conclude that the magnitude of **H**(s) is inversely proportional to the product of the lengths of the vectors drawn from its poles to the point s.

Next, we consider the pole-zero pattern (PZP) shown in Fig. 12.17, which corresponds to

$$\mathbf{H}(s) = \frac{K(s - z_1)(s - z_2)}{(s - s_1)(s - s_2)}$$

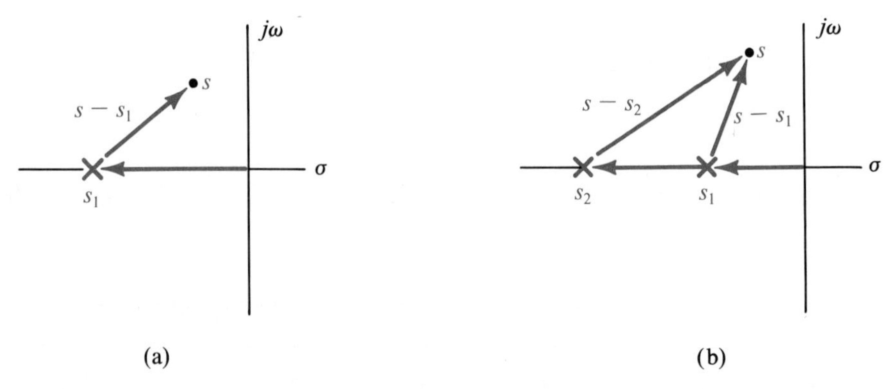

Figure 12.16 (a) A simple pole, and (b) a vector drawn from the pole to the point s.

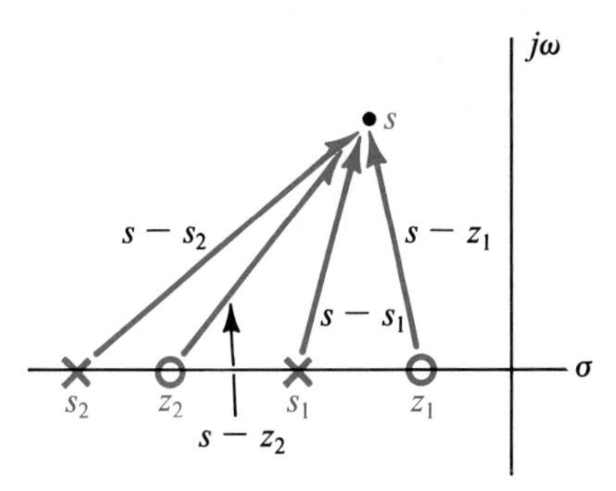

Figure 12.17 Vectors drawn from poles and zeros to the point s.

With vectors drawn from the poles and the zeros to the point s the magnitude of **H**(s) is given by

$$\mathbf{H}(s) = \frac{|K|\,|s - z_1|\,|s - z_2|}{|s - s_1|\,|s - s_2|}$$

The magnitude of **H**(s) is directly proportional to the product of the lengths of the vectors drawn from its zeros to the point s and inversely proportional to the product of the lengths of the vectors drawn from its poles to s. The scale factor K is not represented on the pole-zero diagram, so many different transfer functions can have the same PZP. If **H**(s) is written in factored form as

$$\mathbf{H}(s) = \frac{K \prod_{i=1}^{m} (s - z_i)}{\prod_{j=1}^{n} (s - s_j)}$$

Then

$$|\mathbf{H}(s)| = \frac{|K| \prod_{i=1}^{m} |s - z_i|}{\prod_{j=1}^{n} |s - s_j|}$$

Consequently, $|\mathbf{H}(s)|/|K|$ can be calculated directly from the measurements of the lengths of vectors on the pole-zero patterns.

Skill Exercise 12.7

If $K = -2$ evaluate $\mathbf{H}(j2)$ for $\mathbf{H}(s)$ represented by the PZP in Fig. SE12.7.

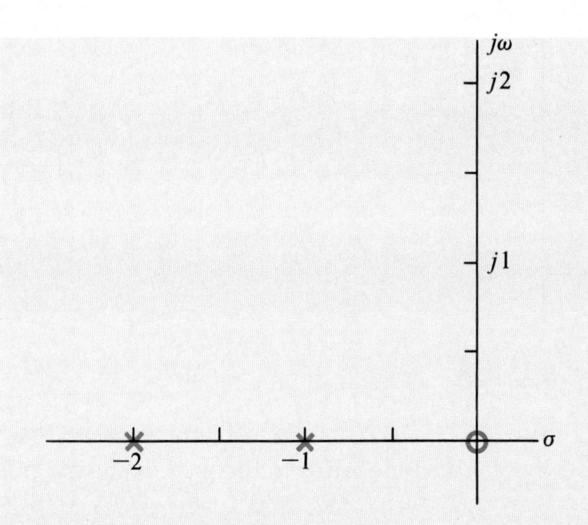

Figure SE12.7

Answer: $|\mathbf{H}(j2)| = |-2|(l_3)/(l_1 l_2) = 2\sqrt{10}$ where l_1, l_2 and l_3 are the vector lengths.

The geometric interpretation of $|\mathbf{H}(s)|$ enables us to visualize the behavior of $|\mathbf{H}(s)|$ at various locations of s in the complex plane. When s is near the pole s_j the vector $s - s_j$ is short, so its reciprocal is large. Likewise, when s is near the zero z_k the length of $s - z_k$ is small and $|\mathbf{H}(s)|$ is likewise small.

The behavior of ang $\mathbf{H}(s)$ [$\measuredangle\mathbf{H}(s)$] has a geometric interpretation too. Note in Fig. 12.18 that

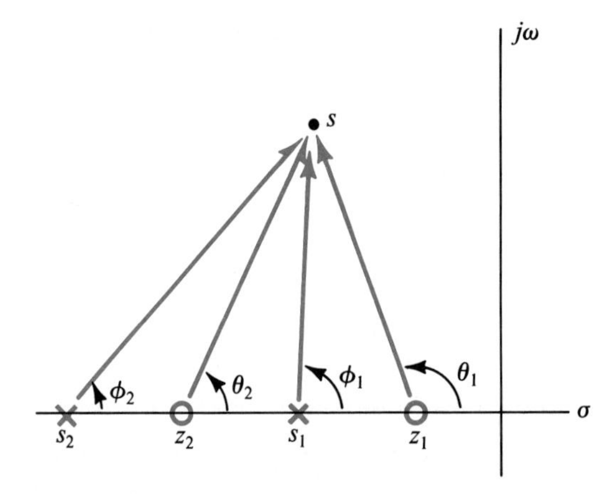

Figure 12.18 The angles of the vectors drawn from poles and zeros to the point s determine $\measuredangle\mathbf{H}(s)$.

$$\measuredangle\, \mathbf{H}(s) = \measuredangle\, (s - z_1) + \measuredangle\, (s - z_2) - \measuredangle\, (s - s_1) - \measuredangle\, (s - s_2)$$

where $\phi_i = \measuredangle\, (s - z_1)$ and $\phi_j = \measuredangle\, (s - s_j)$. The angle of $\mathbf{H}(s)$ is equal to the sum of the angles of the vectors drawn from the zeros to s minus the sum of the angles drawn from the poles to s, or

$$\measuredangle\, \mathbf{H}(s) = \sum_{j=1}^{m} \phi_j - \sum_{i=1}^{n} \phi_i$$

12.8 GENERALIZED NODAL ANALYSIS*

Nodal equations can be written to describe KCL for a circuit in terms of the complex amplitudes of the node-voltage-vector response to exponential sources. Since the GKCL equations that describe a circuit with admittances/impedances between its nodes have the same algebraic form as the KCL equations for a resistive network, the generalized nodal-equation model will have the form (see Chapter 3) $\mathbf{GV} = \mathbf{BU}$ where the matrix $\mathbf{G}$ embodies the coupling admittances between nodes and the matrix $\mathbf{B}$ determines the coupling of the sources to the nodes. The vector $\mathbf{V}$ contains the exponential node-voltage amplitudes, and $\mathbf{U}$ is a vector of exponential current- and voltage-source amplitudes, corresponding to the exponential signals $\mathbf{v}(t) = \mathbf{V}e^{st}$ and $\mathbf{u}(t) = \mathbf{U}e^{st}$. The system of generalized nodal equations has the solution $\mathbf{V} = \mathbf{G}^{-1}\mathbf{BU}$

Nodal equation models for a circuit were covered in detail in Chapters 2 and 3 for a variety of situations, and they are easily extended to treat the generalized case presented here. Rather than derive those results again, we'll solve some examples.

Example 12.13

Find the generalized nodal equations for the circuit in Fig. 12.19 with $R_1 = 1\ \Omega$, $R_2 = 0.5\ \Omega$, $L = 0.25$ H and $C = 5$ F.

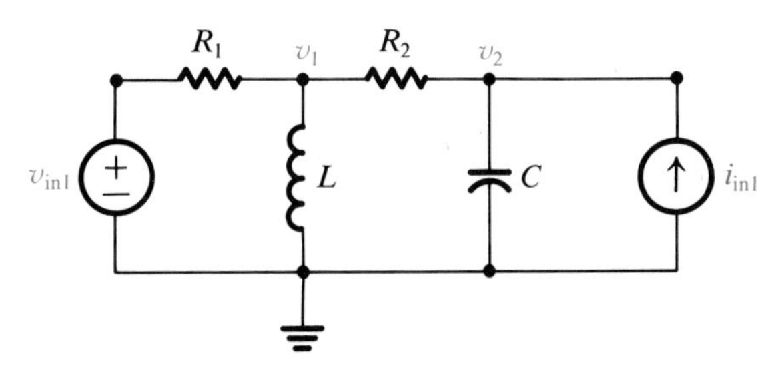

Figure 12.19 Circuit for Example 12.13.

Solution: The generalized nodal model of the circuit has source-vector components $v_{in1}(t) = \mathbf{V}_{in1}e^{st}$ and

$i_{in1}(t) = \mathbf{I}_{in1}e^{st}$. We begin by applying GKCL at the nodes

$$\mathbf{V}_1\left[\frac{1}{R_1} + \frac{1}{sL} + \frac{1}{R_2}\right] - \frac{1}{R_1}\mathbf{V}_{in1} - \frac{1}{R_2}\mathbf{V}_2 = 0$$

and

$$-\frac{1}{R_2}\mathbf{V}_1 + \left[\frac{1}{R_2} + sC\right]\mathbf{V}_2 = \mathbf{I}_{in1}$$

In matrix form

$$\begin{bmatrix} \frac{1}{R_1} + \frac{1}{R_2} + \frac{1}{sL} & -\frac{1}{R_2} \\ -\frac{1}{R_2} & \frac{1}{R_2} + sC \end{bmatrix} \begin{bmatrix} \mathbf{V}_1 \\ \mathbf{V}_2 \end{bmatrix} = \begin{bmatrix} \frac{1}{R_1} & 0 \\ 0 & 1 \end{bmatrix} \begin{bmatrix} \mathbf{V}_{in1} \\ \mathbf{I}_{in1} \end{bmatrix}$$

Inverting the coefficient matrix gives

$$\begin{bmatrix} \mathbf{V}_1 \\ \mathbf{V}_2 \end{bmatrix} = \frac{\begin{bmatrix} \left(\frac{1}{R_2} + sC\right) & \frac{1}{R_2} \\ \frac{1}{R_2} & \left(\frac{1}{R_1} + \frac{1}{R_2} + sL\right) \end{bmatrix}}{\left(\frac{1}{R_1} + \frac{1}{R_2} + \frac{1}{sL}\right)\left(\frac{1}{R_2} + sC\right) - \frac{1}{R_2^2}} \begin{bmatrix} \frac{1}{R_1} & 0 \\ 0 & 1 \end{bmatrix} \begin{bmatrix} \mathbf{V}_{in1} \\ \mathbf{I}_{in1} \end{bmatrix}$$

$$= \frac{\begin{bmatrix} \left(\frac{1}{R_2} + sC\right)\frac{1}{R_1} & \frac{1}{R_2} \\ \frac{1}{R_1R_2} & \left(\frac{1}{R_1} + \frac{1}{R_2} + sL\right) \end{bmatrix}}{\left(\frac{1}{R_1} + \frac{1}{R_2} + \frac{1}{sL}\right)\left(\frac{1}{R_2} + sC\right) - \frac{1}{R_2^2}} \begin{bmatrix} \mathbf{V}_{in1} \\ \mathbf{I}_{in1} \end{bmatrix}$$

At this point the form of the corresponding homogeneous differential equations for $v_1(t)$ and $v_2(t)$ can be recognized, along with the characteristic equation. For the given component values

$$\begin{bmatrix} \mathbf{V}_1 \\ \mathbf{V}_2 \end{bmatrix} = \frac{\begin{bmatrix} (2 + 5s) & 2 \\ 2 & \left(3 + \frac{4}{s}\right) \end{bmatrix}}{\left(3 + \frac{4}{s}\right)(2 + 5s) - 4} \begin{bmatrix} \mathbf{V}_{in1} \\ \mathbf{I}_{in1} \end{bmatrix}$$

Therefore, the particular solutions of v_1 and v_2 to exponential inputs at $v_{in1}(t)$ and $i_{in1}(t)$ are determined by the transfer functions

$$\mathbf{H}_{11}(s) = \frac{\mathbf{V}_1}{\mathbf{V}_{in1}} = \frac{5s^2 + 2s}{15s^2 + 22s + 8}$$

$$\mathbf{H}_{12}(s) = \frac{\mathbf{V}_1}{\mathbf{I}_{in1}} = \frac{2s}{15s^2 + 22s + 8}$$

$$\mathbf{H}_{21}(s) = \frac{\mathbf{V}_2}{\mathbf{V}_{in1}} = \frac{2s}{15s^2 + 22s + 8}$$

$$\mathbf{H}_{22}(s) = \frac{\mathbf{V}_2}{\mathbf{I}_{in1}} = \frac{3s + 4}{5s^2 + 22s + 8}$$

Skill Exercise 12.8

Obtain $\mathbf{H}_{11}(s)$ and $\mathbf{H}_{21}(s)$ in Example 12.13 by inspecting the circuit with i_{in1} set to zero. Obtain $\mathbf{H}_{12}(s)$ and $\mathbf{H}_{22}(s)$ from the circuit with v_{in1} set to zero.

12.9 GENERALIZED THEVENIN AND NORTON EQUIVALENT CIRCUITS*

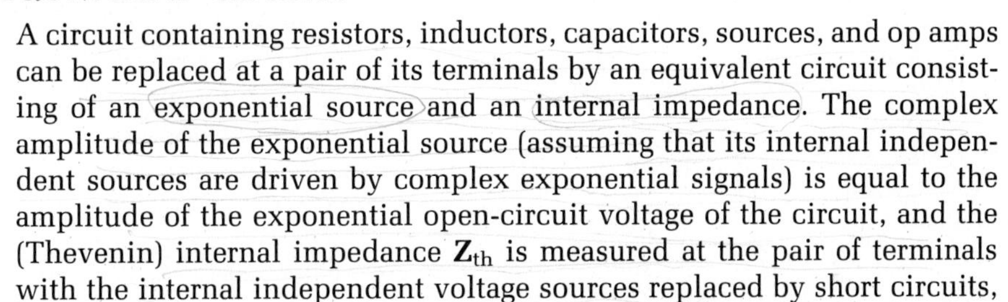

A circuit containing resistors, inductors, capacitors, sources, and op amps can be replaced at a pair of its terminals by an equivalent circuit consisting of an exponential source and an internal impedance. The complex amplitude of the exponential source (assuming that its internal independent sources are driven by complex exponential signals) is equal to the amplitude of the exponential open-circuit voltage of the circuit, and the (Thevenin) internal impedance $\mathbf{Z}_{th}$ is measured at the pair of terminals with the internal independent voltage sources replaced by short circuits, and the internal independent current sources replaced by open circuits.

Example 12.14

Find the generalized Thevenin equivalent of the circuit shown in Fig. 12.20(a). Use the equivalent circuit to obtain the ZSR of $v_o(t)$ to $v_{in}(t) = 5u(t)$.

Solution: The open-circuit voltage is

$$\mathbf{V}_{oc} = \frac{sL}{R + sL}\mathbf{V}_{in}$$

and the equivalent impedance is

$$\mathbf{Z}_{th} = \frac{sRL}{sL + R} = \frac{sR}{s + R/L}$$

The equivalent circuit is shown in Fig. 12.20(b). To find the ZSR, note that

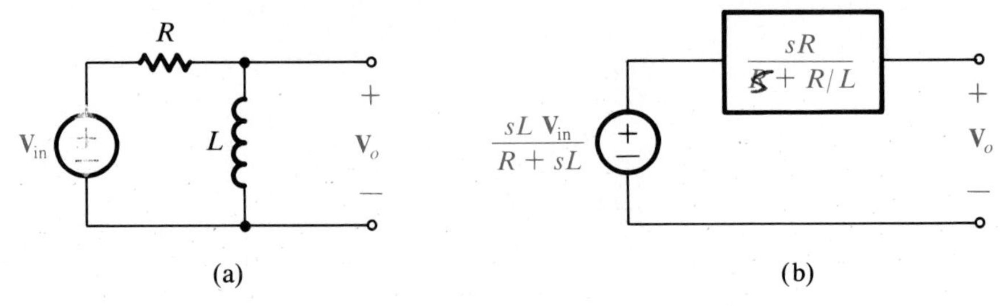

Figure 12.20(a) Circuit for Example 12.14.
Figure 12.20(b) Generalized Thevenin equivalent circuit.

$$\mathbf{V}_o = \frac{sL\,\mathbf{V}_{in}}{R + sL}$$

with $s = 0$ and $\mathbf{V}_{in} = 5$. So $\mathbf{V}_o = 0$ and the boundary conditions for the ZSR lead to $v_o(t) = 5e^{-(R/L)t}$.

The generalized Norton equivalent of a circuit has a current source $\mathbf{I}_{sc}$ in parallel with $\mathbf{Y}_N(s)$ (the Norton admittance of the circuit). $\mathbf{I}_{sc}$ is the complex amplitude of the short-circuit forced response of current to the internal exponential input sources. $\mathbf{Y}_N(s)$ is the internal admittance measured with all internal independent sources set to zero. The important relationship between the Thevenin and Norton circuits that holds for the resistive case also holds for circuits modeled by impedances

$$\mathbf{Z}_{th}(s) = \frac{1}{\mathbf{Y}_N(s)}$$

$$\mathbf{Z}_{th}(s) = \frac{\mathbf{V}_{oc}}{\mathbf{I}_{sc}}$$

Combining the Thevenin and Norton models leads to the source conversion illustrated in Fig. 12.21. Replacing a voltage source by a current source or vice versa can be a useful tool to simplify a circuit.

Example 12.15

Find the Thevenin circuit seen by the capacitor in Fig. 12.22.

Solution: The open-circuit voltage at (a, b) is found from

$$\mathbf{I} = \frac{\mathbf{V}_{in} - \mathbf{V}_{oc}}{R_1} = \frac{-k\mathbf{V}_{oc} + \mathbf{V}_{oc}}{R_2}$$

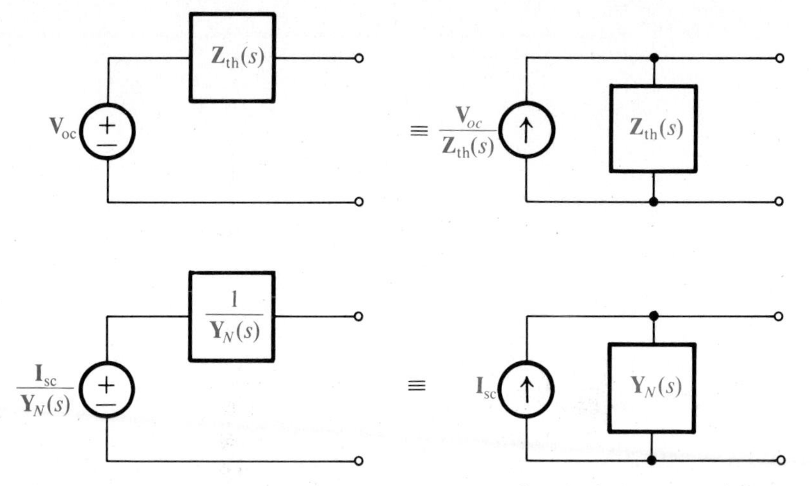

Figure 12.21 Source conversion using generalized Thevenin and Norton equivalent circuits.

so the Thevenin voltage source is

$$\mathbf{V}_{oc} = \frac{R_2\mathbf{V}_{in}}{(1-k)R_1 + R_2}$$

The short-circuit current from a to b is

$$\mathbf{I}_{sc} = \mathbf{V}_{in}/R_1$$

Therefore, the Thevenin impedance is

$$\mathbf{Z}_{th}(s) = \frac{\mathbf{V}_{oc}}{\mathbf{I}_{sc}} = \frac{R_1R_2}{(1-k)R_1 + R_2}$$

When $k = 1$ the impedance seen by C is just R_1 because no current may flow in R_2.

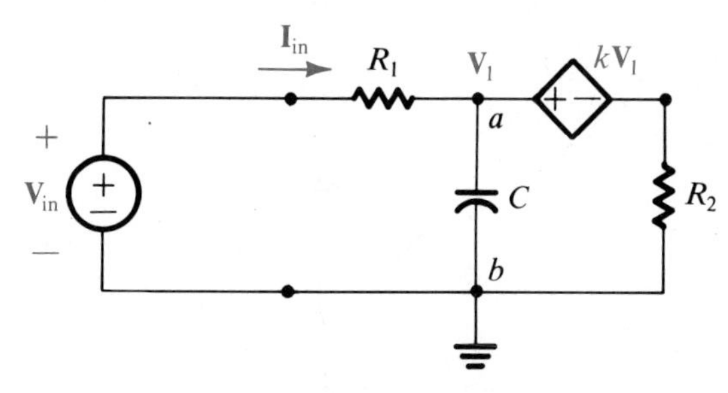

Figure 12.22 Circuit for Example 12.15.

SUMMARY

This chapter showed that the transfer-function model of a circuit can be found without having to derive the time-domain differential equation from KVL, KCL, and the element models for resistors, inductors, and capacitors. Impedance and admittance models of resistors, inductors, and capacitors specify a relationship between the complex amplitudes of (exponential) voltage and current for the device. These device transfer functions are called their generalized impedance and admittance functions.

The complex amplitudes of the forced responses of voltages and currents in a circuit obey generalized versions of KVL and KCL, which can be written and solved in terms of complex signal amplitudes and the element impedances and admittances, rather than in terms of the time-domain signals. The transfer-function model is an alternative to the time-domain differential-equation model. It happens to be a lot easier, in most cases, to find a transfer function by using impedances or admittances and KVL and KCL than to write and manipulate differential equations.

The generalized impedance of a circuit determines the ratio of the complex amplitude of an exponential voltage signal to the complex amplitude of an exponential current signal. Impedances obey the same algebraic rules as are obeyed by resistors; both may be added when they are connected in series. Admittances and conductances in parallel combine by addition. Voltage- and current-division algorithms for complex signal amplitudes have the same form as they have in the resistive case.

Problems - Chapter 12

12.1 For the circuits shown below, find

a. the impedance function at the input terminals of the circuit.

b. the I/O differential equation that would describe the behavior of the voltage at the input terminals if a current source was attached.

c. the I/O differential equation that would describe the behavior of the current into the circuit if a voltage source was attached to its input terminals.

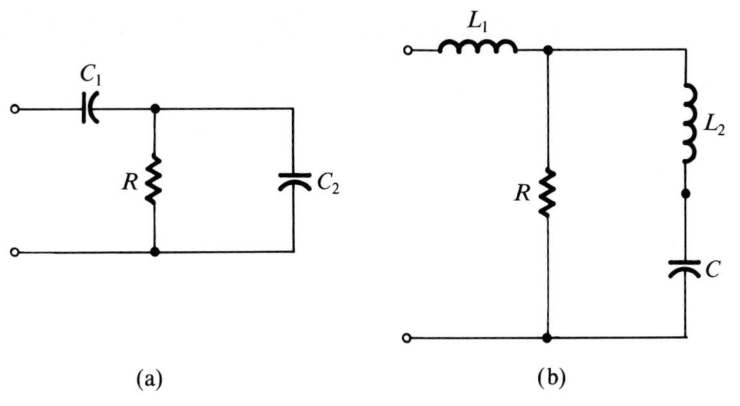

Figure P12.1

12.2 Find the transfer function for $\mathbf{H}(s) = \mathbf{V}_o/\mathbf{V}_{\text{in}}$.

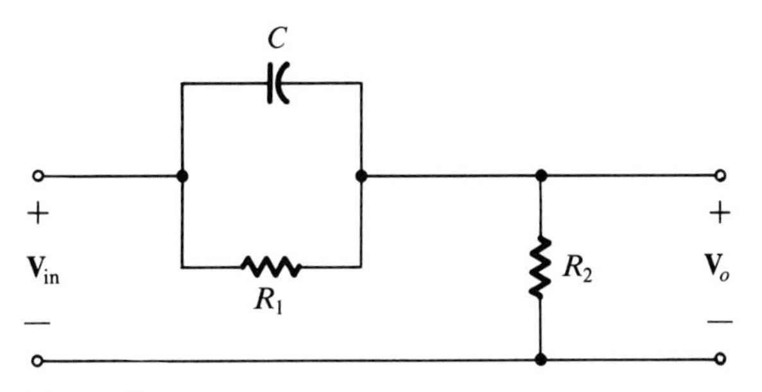

Figure P12.2

12.3 A circuit has an input impedance given by

$$\mathbf{Z}(s) = \frac{s^2 + 2s + 17}{s^2 + 8s + 17}$$

a. Estimate the time required for the input current to reach its steady-state value if a step-input voltage source is applied.

b. Estimate the time required for the voltage at the input terminals to reach steady state if a step current source is applied.

12.4 Using impedance methods, find the I/O differential equation for the circuits described in Problem 7.1.

12.5 Obtain the transfer functions relating $\mathbf{V}_o$ to $\mathbf{V}_{\text{in}}$ in the circuit given in Problem 7.5, and find the locus of the circuit's natural frequency in the complex plane if the gain parameter k of the controlled source is allowed to vary.

12.6 Use impedance methods and transfer functions to solve these problems from Chapter 8.
a. 8.5 **b.** 8.9(a) **c.** 8.10(a) **d.** 8.12 **e.** 8.14

12.7 Find the transfer functions for $\mathbf{V}_R/\mathbf{V}_{in}$ and $\mathbf{V}_c/\mathbf{V}_{in}$ for a series *RLC* circuit driven by a voltage source.

12.8 Find the transfer functions for $\mathbf{I}_c/\mathbf{I}_{in}$ and $\mathbf{I}_L/\mathbf{I}_{in}$ for a parallel *RLC* circuit driven by a current source.

12.9 Find the gain $\mathbf{H}(s) = \mathbf{V}_o/\mathbf{V}_{in}$ in Fig. P12.9.

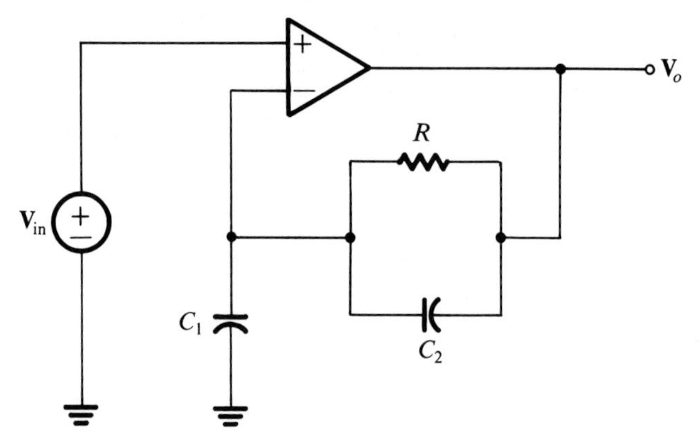

Figure P12.9

12.10 Find the impedance functions seen by the source at (c, d) and by the source at (e, f) in Figure P12.10 before and after the switch is moved from *a* to *b*.

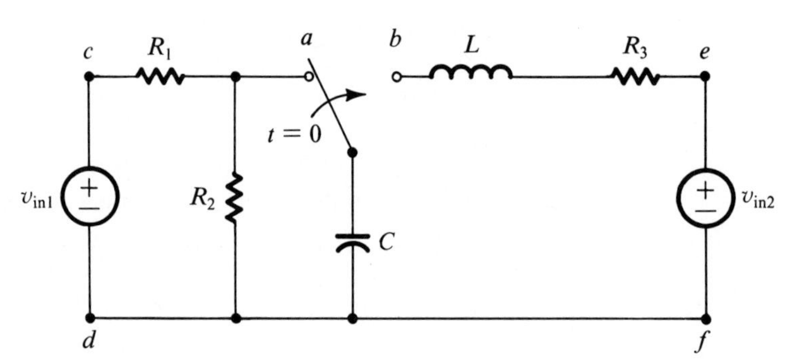

Figure P12.10

12.11 Find the transfer function between an applied voltage source and the inductor current in Fig. P12.11.

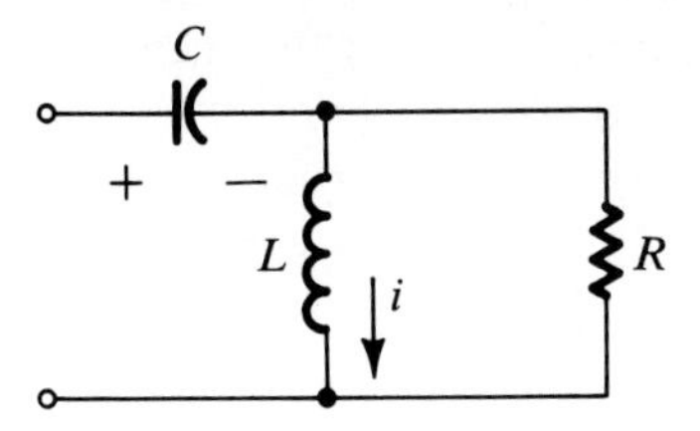

Figure P12.11

12.12 The natural frequencies of the circuit in Fig. P12.12 depend on the values of the circuit's components and on the gain parameter of the controlled voltage source. If R = 9/8 Ω, L = 36 H and C = 1/9 F, draw the locus of the natural frequencies of the current response as k varies.

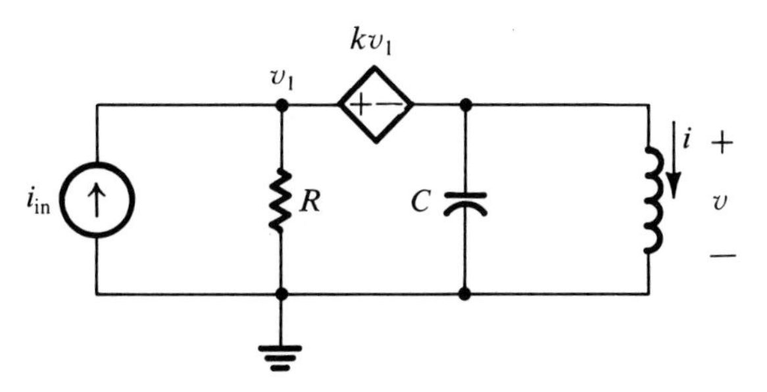

Figure P12.12

12.13 Find an expression for the voltage gain $\mathbf{V}_o/\mathbf{V}_{in}$.

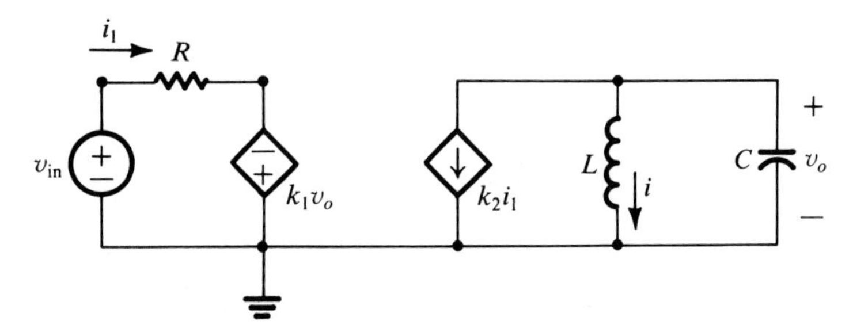

Figure P12.13

12.14 Find the Thevenin equivalent impedance seen (a) by the voltage source and (b) by the capacitor in Fig. P12.13.

12.15 Find the transfer admittance $\mathbf{H}(s) = \mathbf{I}/\mathbf{V}_{in}$ in Fig. P12.13.

12.16 Find the locus of the poles of the input impedance in the circuit in Fig. P12.16 if

a. $R_1 = 5\ \Omega$, $R_2 = 20\ \Omega$, $L_1 = 2$ H and L_2 varies.

b. $R_1 = 5\ \Omega$, $R_2 = 20\ \Omega$, $L_2 = 2$ H and L_1 varies.

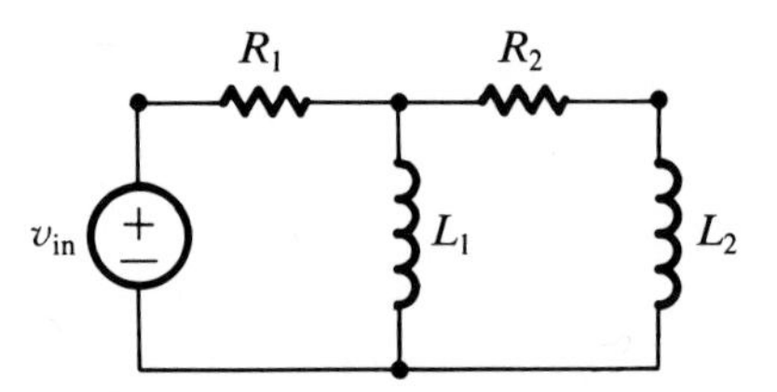

Figure P12.16

12.17 Let the op amp circuit shown in Fig. P11.25 have generic components.

a. Find the voltage gain $\mathbf{H}(s) = \mathbf{V}_o/\mathbf{V}_{in}$.

b. Determine the locus of the poles of $\mathbf{H}(s)$ as C_2 varies, with $0 < C_2 < \infty$.

12.18 Find $\mathbf{I}_1/\mathbf{V}_{in}$ and $\mathbf{I}_2/\mathbf{V}_{in}$ for the op amp circuit in Figure P12.18.

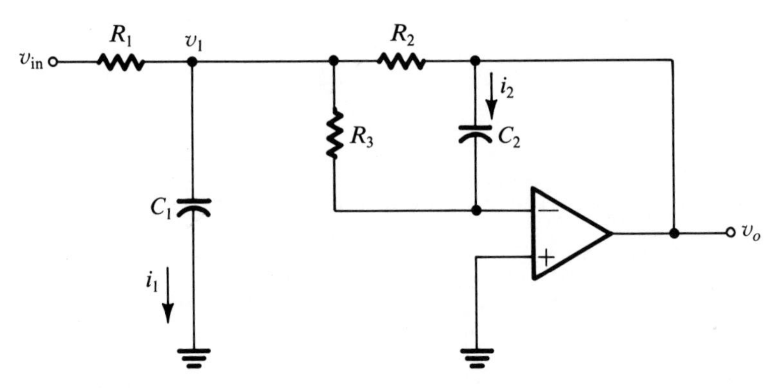

Figure P12.18

12.19 Find the Thevenin equivalent of the circuit shown in Fig. P12.19.

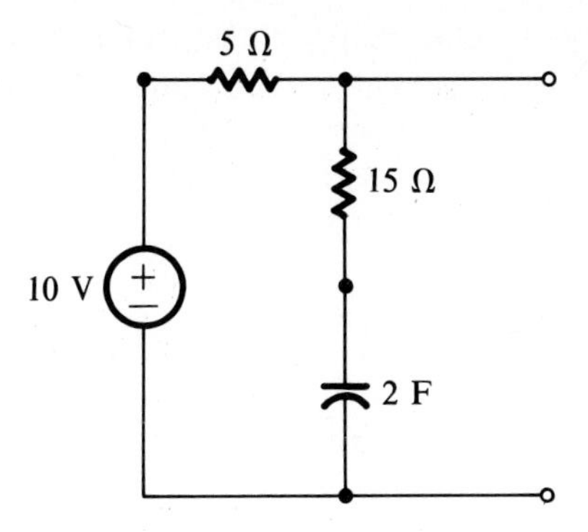

Figure P12.19

12.20 Use mesh analysis to find $\mathbf{I}_1/\mathbf{V}_{in}$ and $\mathbf{I}_2/\mathbf{V}_{in}$.

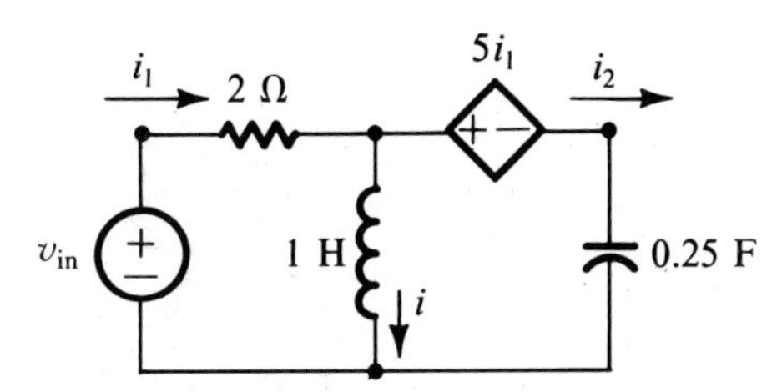

Figure P12.20

12.21 Find $\mathbf{V}/\mathbf{V}_{in}$, and the Thevenin equivalent at (a, b) in the circuit shown below.

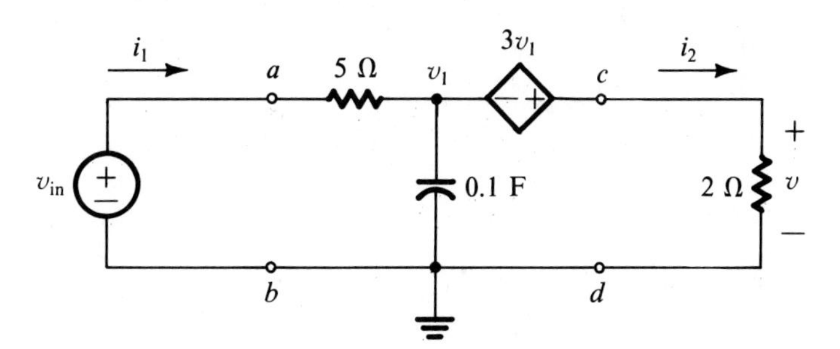

Figure P12.21

How to find I_{sc} = ? in this case

12.22 Using matrix nodal analysis, find the transfer functions relating each source to each of the node voltages.

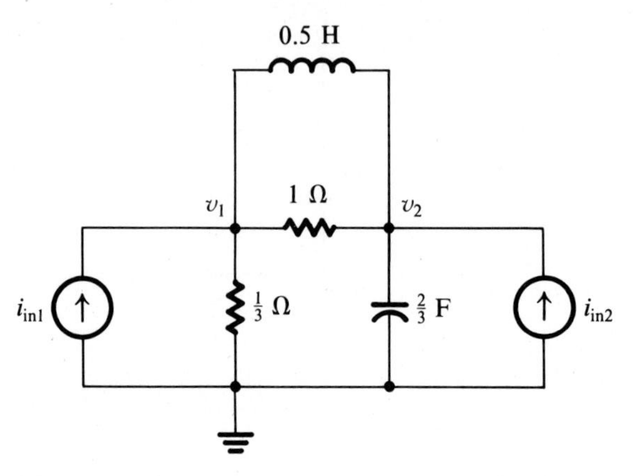

Figure P12.22

12.23 Evaluate $\mathbf{H}(-1 + j)$ for the transfer functions whose pole-zero pattern is shown.

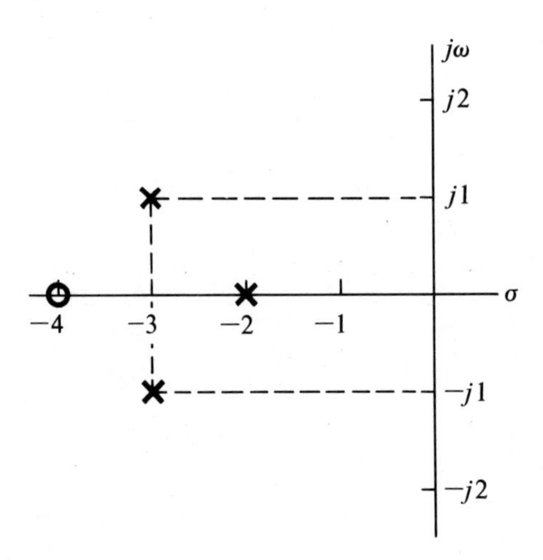

Figure P12.23

FREQUENCY-DOMAIN MODELS AND SINUSOIDAL STEADY-STATE ANALYSIS

INTRODUCTION

This chapter will develop frequency-domain models to simplify the analysis of circuits operating in the sinusoidal steady state. Sinusoidal signals are important because many physical circuits are driven either by sinusoidal signals or by signals that can be represented in terms of sinusoids. The electrical distribution system in the United States, for example, as well as that in many other countries, generates electricity in a sinusoidal form because it is convenient and efficient to do so. Even signals that are not sinusoidal can be defined in terms of sinusoids, as we will see in Chapter 14. Consequently, manufacturers usually specify the electrical characteristics of an appliance or a motor in terms of its frequency-domain response to sinusoidal signals.

When a circuit is driven by a sinusoid its time-domain response consists of the sum of its transient and steady-state response components. If only the steady-state response is of interest and if the source is sinusoidal, the *s*-domain models of Chapter 12 can be used with the restriction that $s = j\omega$. In this case these models are called frequency-domain models.

This chapter exploits two very important consequences of superposition for *RLC*/op amp circuits:

(1) When a circuit is excited by a sinusoidal source all of the steady-state currents and voltages in the circuit are either sinusoidal signals or sums of sinusoidal signals.
(2) When a circuit is driven by more than one sinusoidal source having the same frequency, the steady-state response of any voltage or current in the circuit will be a sinusoid at the same frequency as the sources.

Sinusoidal steady-state circuit analysis is greatly simplified because there is no need to use boundary conditions to solve for the coefficients of the natural component of the time response. Thus, it is feasible to solve for all of the response variables, rather than just one.

Now might be a good time to review the material on sinusoidal signals in Chapters 6 and 10.

13.1 IMPEDANCE, ADMITTANCE, AND PHASORS

A circuit's s-domain model determines its complex exponential forced response to a complex exponential input signal. We saw in Chapter 10 that s-domain models can be used to define the time-domain response of a circuit to a sinusoidal signal because a sinusoid can be represented by a complex exponential signal. When $s = j\omega$ we refer to these models as frequency-domain models because s is associated with the frequency of a sinusoid. When a circuit driven by a sinusoidal current source is in the steady state, the phasor corresponding to the sinusoidal steady-state voltage at the driving point is obtained by evaluating the generalized input impedance at $s = j\omega$ to form

$$\mathbf{V} = \mathbf{I}\mathbf{Z}(j\omega) \tag{13.1a}$$

where

$$v(t) = \text{Re}\{\mathbf{V}e^{j\omega t}\} \tag{13.1b}$$

$$i(t) = \text{Re}\{\mathbf{I}e^{j\omega t}\} \tag{13.1c}$$

for a given, fixed frequency ω. In this case, we refer to $\mathbf{Z}(j\omega)$ as the input impedance function of the circuit, and $\mathbf{Y}(j\omega)$ as the input admittance, and we reserve the descriptor "generalized" for these functions when they are evaluated at an arbitrary but fixed complex frequency s (as in Chapter 11). It is customary to write $v(t)$ and $i(t)$ as in (13.1) even though the signals are established at some point in time.

The complex phasors of the steady-state voltage and current at the input nodes of the circuits in Fig. 13.1 are related by the driving-point impedance (i.e. the impedance at the pair of nodes that are connected to the source) according to

$$\frac{\mathbf{V}}{\mathbf{I}} = \mathbf{Z}(j\omega) \tag{13.2a}$$

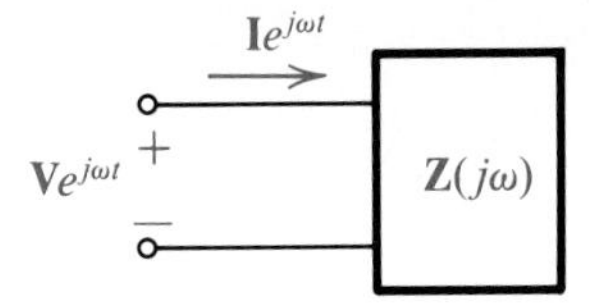

Figure 13.1 The complex exponential signals for voltage and current at the input terminals of a circuit are related by the impedance function $\mathbf{Z}(j\omega)$.

Likewise, the ratio of the current phasor to the voltage phasor determines the input admittance of the circuit

$$\frac{\mathbf{I}}{\mathbf{V}} = \mathbf{Y}(j\omega) \tag{13.2b}$$

and

$$\boxed{\mathbf{Z}(j\omega) = \frac{1}{\mathbf{Y}(j\omega)}} \tag{13.3}$$

Note that (13.2a) and (13.2b) are in the familiar form of Ohm's Law, and that (13.3) describes a reciprocal relationship like the one for the resistance and conductance of a resistor.

It makes no difference whether the circuit is driven by a voltage source or by a current source—the phasor relationship for the ratio $\mathbf{V}/\mathbf{I}$ is governed by the input impedance $\mathbf{Z}(j\omega)$. $\mathbf{Z}(j\omega)$ and $\mathbf{Y}(j\omega)$ determine ratios of phasors, *not* ratios of time-domain signals, so in general $\mathbf{Z}(j\omega)$ does not equal $v(t)/i(t)$. Impedance, like resistance, is a property of a circuit, not of its signal source. It determines *how* the circuit will respond *if* it is excited by a complex exponential signal.

At the input port the magnitude of the input-voltage phasor is

$$|\mathbf{V}| = |\mathbf{I}||\mathbf{Z}(j\omega)| \tag{13.4}$$

so that

$$\frac{|\mathbf{V}|}{|\mathbf{Z}(j\omega)|} = |\mathbf{I}|$$

and

$$|\mathbf{I}| = |\mathbf{V}||\mathbf{Y}(j\omega)|$$

The angles of the phasors of the input voltage and current satisfy

$$\measuredangle\mathbf{Z}(j\omega) = \measuredangle\mathbf{V} - \measuredangle\mathbf{I} \tag{13.5}$$

so that

$$\measuredangle\mathbf{V} = \measuredangle\mathbf{I} + \measuredangle\mathbf{Z}(j\omega)$$

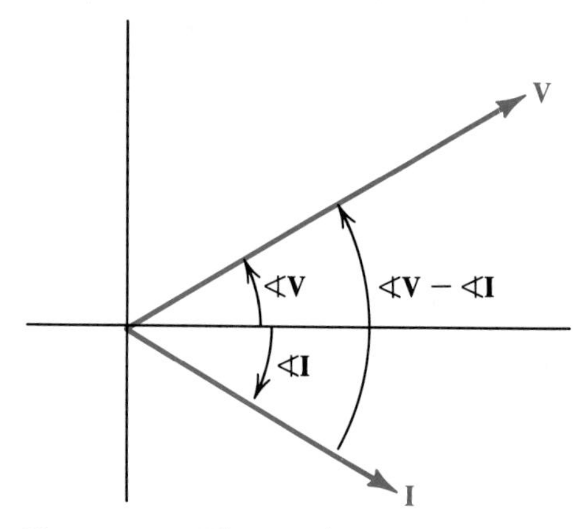

Figure 13.2 Phasor relationships for **V** and **I**.

and

$$\sphericalangle \mathbf{I} = \sphericalangle \mathbf{V} + \sphericalangle \mathbf{Y}(j\omega)$$

Fig. 13.2 shows the relationship between the phasor angles of **V** and **I.**

$\mathbf{Z}(j\omega)$ and $\mathbf{Y}(j\omega)$ are called **frequency-domain models** because they are functions of frequency, and because they determine the relationship between the phasors of current and voltage when the source is a complex exponential whose real and imaginary parts are (undamped) sinusoids with frequency ω. **V** and **I** are the frequency-domain **phasors** corresponding to the time-domain, steady-state *waveforms* of the input voltage and current. If the time-domain signals are expressed in terms of the cos $(\cdot)$ function, they are obtained by taking the real part of the phasor signal and writing

$$v(t) = |\mathbf{V}| \cos (\omega t + \phi_v)$$

and

$$i(t) = |\mathbf{I}| \cos (\omega t + \phi_i)$$

where the phase angles ϕ_v and ϕ_i are obtained from the phasor angles as

$$\phi_v = \text{ang } \mathbf{V} = \sphericalangle \mathbf{V}$$

and

$$\phi_i = \text{ang } \mathbf{I} = \sphericalangle \mathbf{I}$$

The expressions developed in (13.4) and (13.5) for the phasors of the current and voltage steady-state waveforms can also be used to express the voltage and current waveforms in terms of the impedance and admittance as

$$\begin{aligned} v(t) &= |\mathbf{I}||\mathbf{Z}(j\omega)| \cos (\omega t + \phi_v) \\ &= |\mathbf{I}||\mathbf{Z}(j\omega)| \cos [\omega t + \phi_i + \sphericalangle \mathbf{Z}(j\omega)] \end{aligned} \quad (13.6a)$$

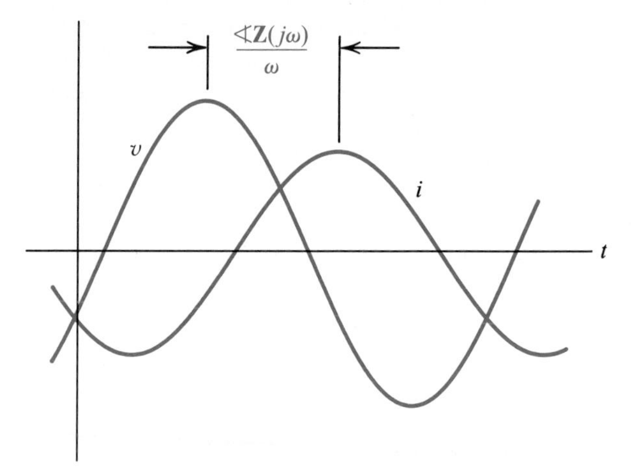

Figure 13.3 Steady-state, time-domain relationship between $v(t)$ and $i(t)$.

and

$$i(t) = |\mathbf{V}||\mathbf{Y}(j\omega)|\cos[\omega t + \phi_V - \sphericalangle \mathbf{Z}(j\omega)] \quad (13.6b)$$

or, alternately,

$$i(t) = |\mathbf{V}||\mathbf{Y}(j\omega)|\cos[\omega t + \phi_V + \sphericalangle \mathbf{Y}(j\omega)]$$

since ang $\mathbf{Z}(j\omega) = -$ang $\mathbf{Y}(j\omega)$.

The frequency-domain properties of a circuit determine the time-domain relationship of its steady-state voltage and current waveforms. The angle of the complex impedance is the phase-angle difference between the time-domain voltage and current signals, and so determines the amount of time displacement between their waveforms. Fig. 13.3 shows a graph of $v(t)$ and $i(t)$ in the steady-state and demonstrates how the impedance angle affects the phase-angle difference of $v(t)$ and $i(t)$. When $\sphericalangle \mathbf{Z}(j\omega)$ is positive, the waveform of $v(t)$ is *advanced* (on the time axis) relative to the waveform of $i(t)$ and vice versa.

Example 13.1

If a circuit has $\mathbf{Z}(j\omega) = 10 - j15$, find $v(t)$ in the steady state when $i(t) = 4\cos(\omega t + 20°)$.

Solution: In polar form $\mathbf{Z}(j\omega) = 18.028 \underline{/-56.31°}$, so $v(t) = |4||\mathbf{Z}(j\omega)|\cos[\omega t + 20° + \sphericalangle \mathbf{Z}(j\omega)] = 72.111\cos(\omega t - 36.31°)$.

The study of circuits operating in the steady state is often called **steady-state analysis** or **AC analysis** to signify the special behavior of circuits when they are driven by sinusoidal signals and reach steady state. In steady-state analysis the phasors **V** and **I** are referred to as though they were the actual time-domain signals. This can be a source of confusion, so when misunderstanding seems likely we will use the explicit terminology and refer to the voltage phasor and current phasor by name.

The impedance and admittance of resistors, inductors, and capacitors are obtained from direct evaluation of the expressions for their generalized impedance and admittance as they were given in Chapter 12, with $s = j\omega$, so here we only summarize their values in Table 13.1. Notice that they describe the behavior of $\mathbf{Z}(j\omega)$ and $\mathbf{Y}(j\omega)$ along the entire j axis (i.e., for any given value of ω).

TABLE 13.1 IMPEDANCE AND ADMITTANCE VALUES OF *R, L* AND *C*

	R	*L*	*C*
$\mathbf{Z}(j\omega)$	R	$j\omega L$	$1/j\omega C$
$\mathbf{Y}(j\omega)$	$1/R$	$1/j\omega L$	$j\omega C$

The waveforms of sinusoidal steady-state v and i for a resistor are shown in Fig. 13.4. The voltage and current of a resistor are said to be **in phase,** meaning that their phase-angle difference, and hence their time displacement, is zero. A graph of $v(t)$ and $i(t)$ for a resistor will have maximum and minimum values of v and i occurring simultaneously on the time axis. In the time domain the steady-state values of v and i for a resistor are given by:

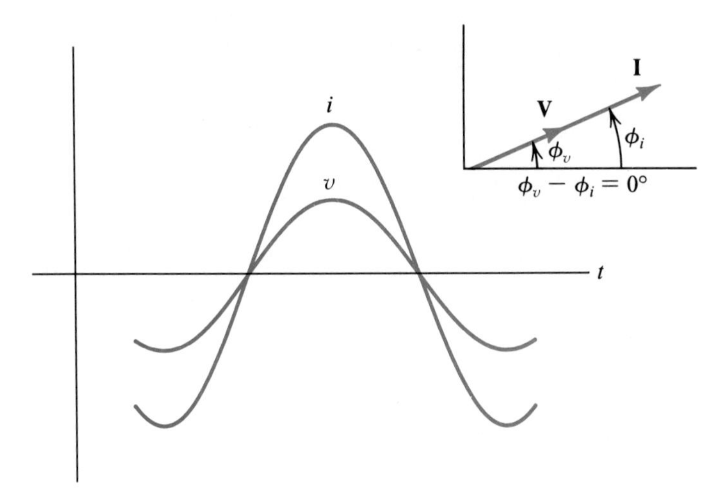

Figure 13.4 Phasor and phase-angle relationships for the steady-state waveforms of voltage and current for a resistor. Voltage and current are "in phase."

$$i(t) = |\mathbf{I}| \cos (\omega t + \phi_i)$$

and

$$v(t) = |\mathbf{V}| \cos (\omega t + \phi_v)$$

with $\phi_i = \phi_v$. From Table 13.1 we have:

$$|\mathbf{V}| = |\mathbf{I}|R$$

so the expression for $v(t)$ can also be written as

$$v(t) = |\mathbf{I}|R \cos (\omega t + \phi_v)$$

The information for an inductor in Table 13.1 has $\measuredangle\mathbf{V} = \measuredangle\mathbf{I} + 90°$, so the steady-state waveforms of v and i are *always* out of phase by 90°, as shown in Fig. 13.5, and the *current waveform is said to* ***lag*** *the voltage waveform* because its peak value always occurs one-quarter cycle *after* the peak value of current (on an interval of measurement not exceeding one period). Recall that in Chapter 5 we saw that the applied voltage must cause the accumulation of flux linkages in the inductor *before* the inductor will conduct current. This accounts for the time displacement of the steady-state v and i waveforms that is evident in Fig. 13.5. The time-domain waveforms for v and i of an inductor are

$$i(t) = |\mathbf{I}| \cos (\omega t + \phi_v - 90°)$$

and

$$v(t) = |\mathbf{V}| \cos (\omega t + \phi_v)$$

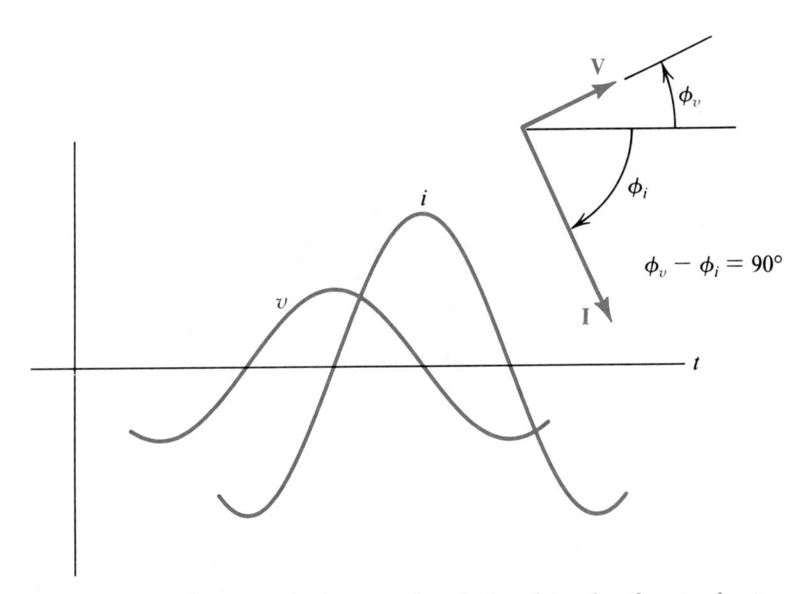

Figure 13.5 Phasor and phase-angle relationships for the steady-state waveforms of voltage and current for an inductor. Voltage leads current by 90°.

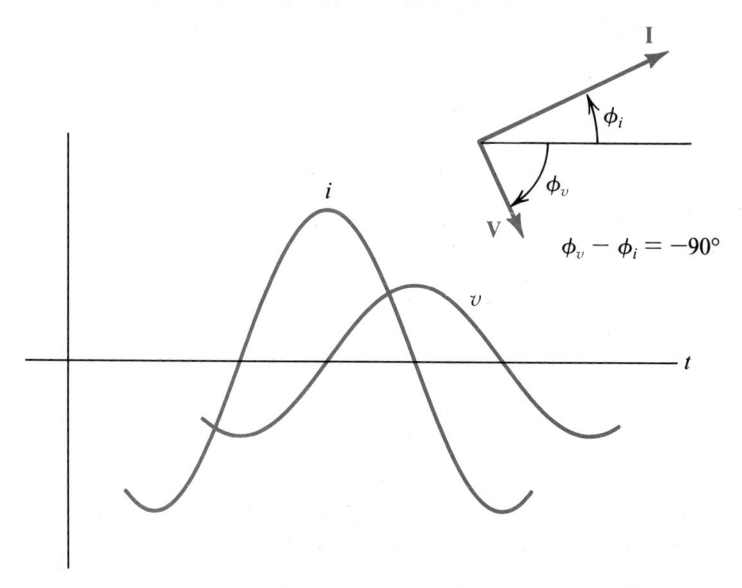

Figure 13.6 Phasor and phase-angle relationships for the steady-state waveforms of voltage and current for a capacitor. Current leads voltage by 90°.

For a capacitor $\measuredangle \mathbf{V} = \measuredangle \mathbf{I} - 90°$. If the sinusoidal source is expressed in terms of the sin (·) function rather than the cos (·) function the expressions for v and i of a capacitor become

$$i(t) = |\mathbf{I}| \sin(\omega t + \phi_i)$$
$$v(t) = |\mathbf{V}| \sin(\omega t + \phi_i - 90°)$$

Figure 13.6 shows that the sinusoidal steady-state waveforms of v and i for a capacitor are always out of phase by 90°, and the current waveform is said to **lead** the voltage waveform (by 90°) because its maximum value occurs before the peak value of the voltage. A charging current must provide charge to the capacitor before voltage can appear between its terminals.

When the source is expressed in terms of the cos (·) function, the steady-state time-domain waveforms for v and i are given by

$$i(t) = |\mathbf{I}| \cos(\omega t + \phi_i)$$

and

$$v(t) = |\mathbf{V}| \cos(\omega t + \phi_i - 90°)$$

If the source is expressed in terms of the sin (·) function, we take the imaginary part of the phasor signal and replace cos (·) with sin (·) in the above expressions for $v(t)$ and $i(t)$.

Skill Exercise 13.1

(a) If a capacitor has $i(t) = 5 \sin(\omega t - 30°)$, find an expression for the steady-state capacitor voltage, $v(t)$. (b) If an inductor has $v(t) = 10 \cos(\omega t - 45°)$, find an expression for the steady-state inductor current, $i(t)$.

Answer: (a) $v(t) = 5/\omega C \sin(\omega t - 120°)$, (b) $i(t) = 10/\omega L \cos(\omega t - 135°)$

The phase-angle difference for the current and voltage of individual resistors, capacitors, and inductors can also be compared by drawing their phasors and waveforms together in the complex plane for a common **I**, as shown in Fig. 13.7. The phasors **V** and **I** are aligned in the case of a resistor, meaning that the corresponding steady-state time-domain signals are such that peaks in voltage and current occur simultaneously. For the inductor, the voltage phasor *leads* the current phasor by 90° (current lags voltage), or the time-domain waveforms have voltage peaks occurring one-quarter cycle before the peaks of current. The current phasor for a capacitor *leads* its voltage phasor by 90° (voltage lags current), and the waveform of current leads the waveform of voltage by one-quarter cycle.

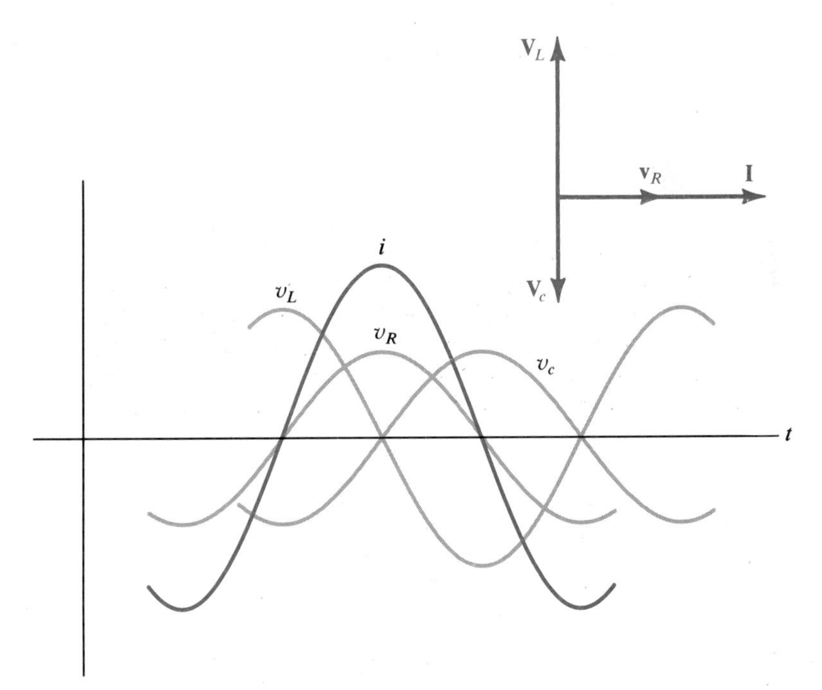

Figure 13.7 Comparison of the phasors and waveforms of $v(t)$ and $i(t)$ for a resistor, inductor, and capacitor.

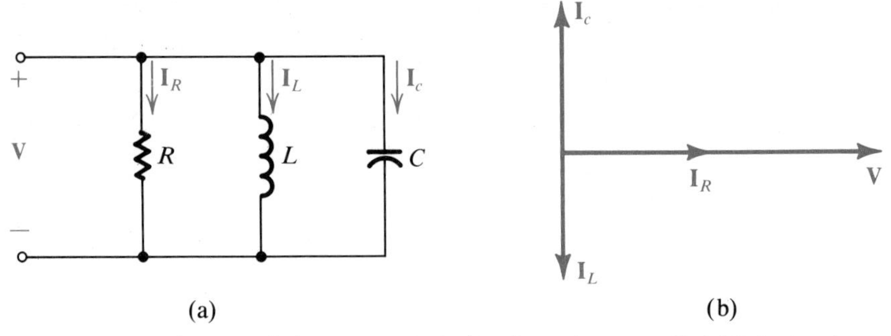

Figure 13.8 Phasors of the currents and voltage in a parallel *RLC* circuit.

Note too, that for the same current the inductor voltage leads the capacitor voltage by 180°, meaning that in the steady state v_L and v_c will always have opposite polarity when connected in series.

It is common practice to describe the steady-state behavior of circuits by the phasors of their voltages and currents, and to show phasors on circuit diagrams instead of time-domain signals. The components of a parallel *RLC* circuit have a common applied voltage. The capacitor and inductor current phasors in Fig. 13.8 differ by 180°, so their corresponding time-domain signals will always have opposite polarity in the steady state.

Just as a circuit has a time-domain response to input signals, it is also said to have a **frequency response** to sinusoidal sources. The frequency response of an individual component consists of a description of how the magnitudes and angles of its voltage and current phasors are affected by changes in the frequency of the source.

A graph of $|\mathbf{Z}(j\omega)|$ is called the **magnitude response** of $\mathbf{Z}(j\omega)$, and a graph of ang $\mathbf{Z}(j\omega)$ is called its **phase response.** Table 13.2 can be used to develop the graphs of $|\mathbf{Z}(j\omega)|$ and ang $\mathbf{Z}(j\omega)$ shown in Fig. 13.9. The magnitude of a resistor's impedance is *fixed;* for an inductor impedance *increases linearly* with ω; for a capacitor it *varies inversely* with ω.

TABLE 13.2 IMPEDANCE AND ADMITTANCE FOR *R, L* AND *C.*

	R	*L*	*C*
$\lvert\mathbf{Z}(j\omega)\rvert$	R	ωL	1/ωC
$\measuredangle\mathbf{Z}(j\omega)$	0°	90°	−90°
$\lvert\mathbf{Y}(j\omega)\rvert$	1/R	1/ωL	ωC
$\measuredangle\mathbf{Y}(j\omega)$	0°	−90°	90°

At $\omega = 0$ an inductor behaves like a short circuit to current, in the sense that the steady-state voltage across the inductor will be zero what-

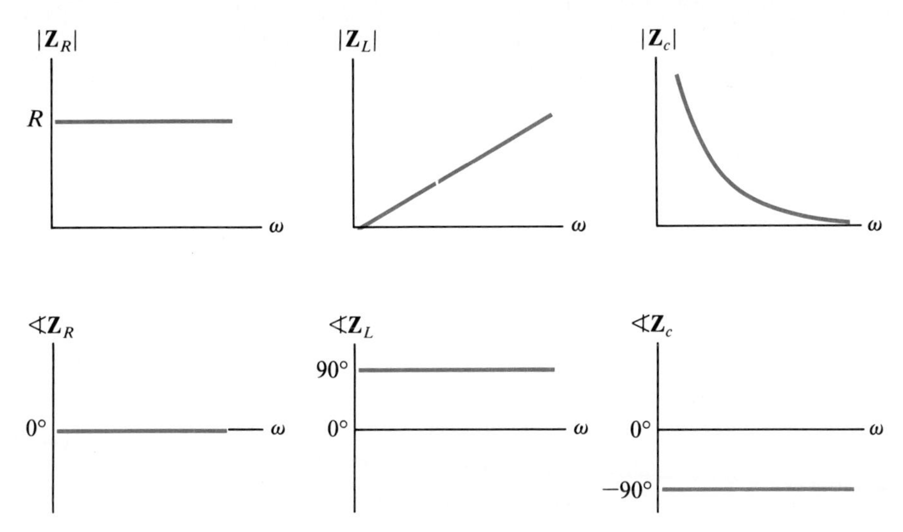

Figure 13.9 Magnitude and angle of $\mathbf{Z}_R(j\omega)$,$\mathbf{Z}_L(j\omega)$ and $\mathbf{Z}_c(j\omega)$ vs ω.

ever the level of the current. Recall that in Chapter 8 we examined the zero-state time response (ZSR) of a series *RL* circuit to a step voltage source, and found that in the steady state a step function can be considered to be a cosine function with its frequency set to 0. There we saw that the response of the current in the *RL* circuit grew exponentially from its initial value of 0 to a steady-state value determined by the resistor and the height of the applied voltage step, and that the voltage across the inductor reached a steady-state value of zero, for arbitrary step height. In contrast, when the sinusoidal frequency ω is arbitrarily large, an inductor behaves like an open circuit, in the sense that a given steady-state voltage will produce a relatively small current (the amplitude of the sinusoid will be small). An attempt to force high-frequency current through an inductor requires a relatively high level of voltage across its terminals.

At $\omega = 0$ the impedance of a capacitor is "infinite," a mathematical clue that a capacitor behaves like an open circuit at that frequency, since the steady-state component of current will be zero. At high frequency the impedance of a capacitor is relatively small—it behaves like a short circuit to current, allowing sinusoidal current to flow with only a small voltage drop across the capacitor. This means that relatively high levels of current are needed to provide enough charge to sustain a level of capacitor voltage at high frequency.

Example 13.2

If $i_{in}(t) = 10 \sin(1200t + 30°)u(t)$, find the frequency-domain phasors and the steady-state time-domain signals of $v_L(t)$ and $v_c(t)$ in Fig. 13.10.

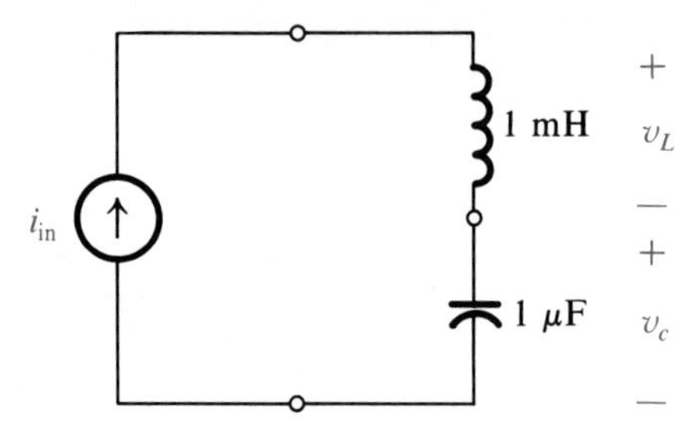

Figure 13.10 Circuit for Example 13.2.

Solution: $\mathbf{V}_L = \mathbf{I}_{in}\mathbf{Z}_L(j\omega) = 10e^{j30°}j1200(10^{-3}) = 12e^{j120°}$

and

$\mathbf{V}_c = \mathbf{I}_{in}\mathbf{Z}_c(j\omega) = 10e^{j30°}1/[j1200(10^{-6})] = 8333.3e^{-j60°}$

Next, we use the voltage phasors to form the steady-state time-domain signals that they represent. Since the current source is given in terms of a sin (·) function, we take

$v_L(t) = \text{Im}\{\mathbf{V}_L e^{j\omega t}\} = \text{Im}\{12e^{j120°}e^{j1200t}\} = 12 \sin(1200t + 120°)$

Likewise,

$v_c(t) = |\mathbf{V}_c| \sin(\omega t + \phi_v) = 8333.3 \sin(1200t - 60°)$

Skill Exercise 13.2

If $v(t) = 10 \sin(\omega t - 45°)u(t)$, find the steady-state time waveforms for i_L and i_c in Fig. SE13.2 when (a) $\omega = 0.1$ rad/sec and (b) $\omega = 10$ rad/sec. For the same $v(t)$, find (c) $i_L(t)$, $i_C(t)$ and $i(t)$ when $\omega = 1/\sqrt{10}$ rad/sec.

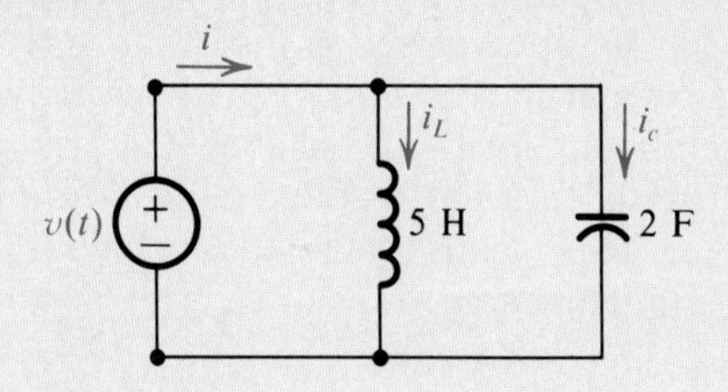

Figure SE13.2

Answer: (a) $i_L(t) = 20 \sin(0.1t - 135°)$ and $i_C(t) = 2 \sin(0.1t + 45°)$
(b) $i_L(t) = 0.2 \sin(10t - 135°)$ and $i_C(t) = 200 \sin(10t + 45°)$
(c) $i_L(t) = -i_C(t) = 6.325 \sin(1/\sqrt{10}\, t - 135°)$ and $i(t) = 0$

13.2 STEADY-STATE CIRCUIT ANALYSIS WITH PHASORS

When a circuit is driven by a sinusoidal source all of the steady-state voltages and currents within the circuit are sinusoidal signals. The voltage and current at the pair of nodes connected at the source are sinusoidal, and all of the internal steady-state signals are also sinusoidal. A transfer function can be constructed between the input source and any voltage or current in the circuit. If the input signal is a sinusoid, that voltage or current is also a sinusoid, having the same frequency as the source, with magnitude and phase angle specified by the value of the transfer function. Because the same steps can be repeated to construct a transfer function between the source and any node voltage or branch current, we conclude that the node voltages and branch currents are all sinusoidal and must all have the same frequency as the source.

Kirchhoff's current and voltage laws apply to the frequency-domain voltage and current *phasors* within a circuit, not just to the time-domain signals they represent! This is just a special case (with $s = j\omega$) of the generalized form of KVL and KCL presented in Chapter 12, so we will not give a proof.

$$\Sigma \mathbf{V}_k = 0 \tag{13.7a}$$

$$\Sigma \mathbf{I}_k = 0 \tag{13.7b}$$

The voltage and current phasors of all of the sinusoidal steady-state signals in a circuit must satisfy KVL and KCL. **The physical laws that govern the instantaneous values of current and voltage in a circuit likewise govern their phasors.** As a result, the steady-state behavior of a circuit can be modeled using frequency-domain phasors of signals rather than the time-domain signals themselves. The frequency-domain phasors completely characterize the steady-state time-domain signals in the circuit. Algebraic solution of (13.7) for the phasors in a circuit simultaneously provides the information about the amplitude and phase angle of the time-domain sinusoidal signals in the circuit—with no need to write or solve a differential equation!

Applying KCL at the nodes of a circuit in frequency-domain analysis leads to a matrix equation having the same form as the nodal analysis model that was developed in Chapter 2 for resistor circuits

$$\mathbf{YV} = \mathrm{B}\mathbf{U} \tag{13.8}$$

where $\mathbf{Y}$ is an admittance matrix, $\mathbf{V}$ is a vector of node-voltage phasors (relative to a ground node), B is a coupling matrix, and $\mathbf{U}$ is a vector of independent current- and voltage-source phasors, each of which has the same excitation frequency. (The details are left to the reader.) Premultiplying both sides of (13.8) by the inverse of the admittance matrix gives the solution

$$\mathbf{V} = \mathbf{Y}^{-1}\mathbf{B}\mathbf{U}$$

This expression has the same form as the solution of the matrix model for a resistive circuit.

Example 13.3

Write KCL for the node-voltage phasors of the circuit in Fig. 13.11 and solve for the steady-state values of $v_1(t)$ and $v_2(t)$ when $i(t) = 5\cos 120\pi t u(t)$, $R = 1/50\ \Omega$, $L = 1/320$ H and $C = 0.1$ F.

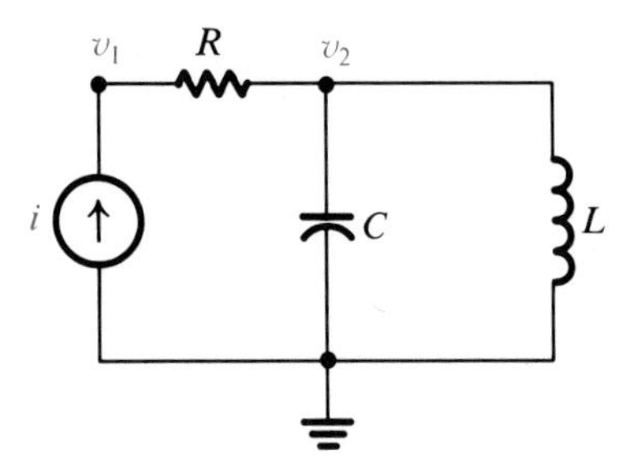

Figure 13.11 Circuit for Example 13.3.

Solution: The nodal equations have the form

$$\begin{bmatrix} 1/R & -1/R \\ -1/R & (1/R + j\omega C + 1/j\omega L) \end{bmatrix} \begin{bmatrix} \mathbf{V}_1 \\ \mathbf{V}_2 \end{bmatrix} = \begin{bmatrix} \mathbf{I} \\ 0 \end{bmatrix}$$

where **I** is the phasor of the current source. Inverting the **Y** matrix leads to

$$\begin{bmatrix} \mathbf{V}_1 \\ \mathbf{V}_2 \end{bmatrix} = \frac{\begin{bmatrix} (1/R + j\omega C + 1/j\omega L) & 1/R \\ 1/R & 1/R \end{bmatrix} \begin{bmatrix} \mathbf{I} \\ 0 \end{bmatrix}}{(1/R + j\omega C + 1/j\omega L)/R - 1/R^2}$$

$$= \frac{\begin{bmatrix} (50 + j(12\pi - 2.667/\pi))5 \\ 5(50) \end{bmatrix}}{[50 + j(12\pi - 2.667/\pi)]50 - 2500}$$

$$= \begin{bmatrix} 0.169 \angle -53.6^\circ \\ 0.135 \angle -90.0^\circ \end{bmatrix}$$

In the steady state

$$v_1(t) = 0.169 \cos (120\pi t - 53.6^\circ)$$
$$v_2(t) = 0.135 \cos (120\pi t - 90.0^\circ)$$

Skill Exercise 13.3

Write the nodal equations for the node-voltage phasors of the circuit in Fig. SE13.3.

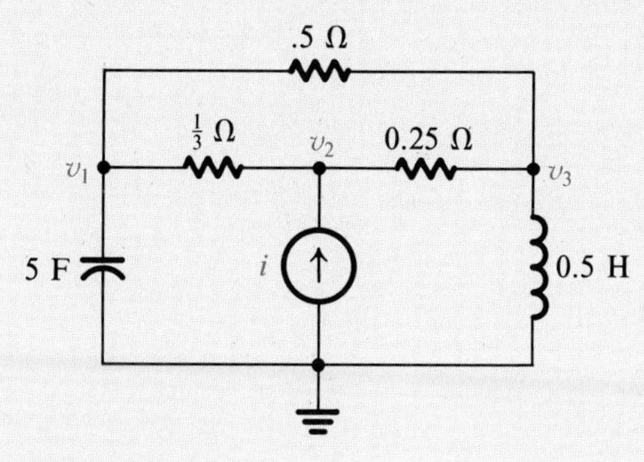

Figure SE13.3

Answer:
$$\begin{bmatrix} (5+j5\omega) & -3 & -2 \\ -3 & 7 & -4 \\ -2 & -4 & (6-j2/\omega) \end{bmatrix} \begin{bmatrix} \mathbf{V}_1 \\ \mathbf{V}_2 \\ \mathbf{V}_3 \end{bmatrix} = \begin{bmatrix} 0 \\ \mathbf{I} \\ 0 \end{bmatrix}$$

13.3 PHASOR MODELS OF CIRCUITS

Because the voltage and current phasors of a circuit completely determine its sinusoidal steady-state behavior, it is sometimes convenient to describe a circuit in a way that makes it easier to calculate its phasors. The **phasor model of a circuit** is formed by replacing its component labels with the value of the component's (sinusoidal) impedance for a *given* source frequency. The values of the circuit's voltage and current phasors can be obtained by writing and solving KVL and KCL using the element impedances and admittances given by its phasor model.

Example 13.4

The op amp circuit in Fig. 13.12 is labeled with *impedance* values and voltage phasors for $\omega = 2000\pi$ rad/sec. Find the phasor for $\mathbf{V}_o$ when $v_g(t) = 10\cos(2000\pi t - 30°)$ in steady state.

Solution: Since $\mathbf{V}_2$ is at ground, the nodal equations for the $\mathbf{V}_1$ node and the input node of the op amp reduce to

$$\begin{bmatrix} (5+10+j2+j2.5) & -j2 \\ j2.5 & 2 \end{bmatrix} \begin{bmatrix} \mathbf{V}_1 \\ \mathbf{V}_o \end{bmatrix} = \begin{bmatrix} 5\mathbf{V}_g \\ 0 \end{bmatrix}$$

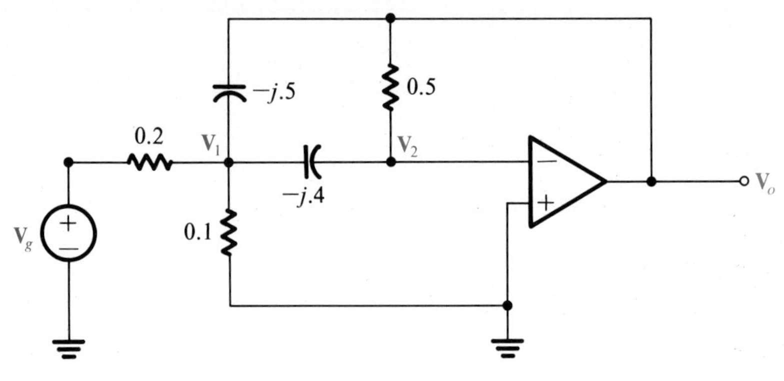

Figure 13.12 Circuit for Example 13.4.

so

$$\begin{bmatrix} \mathbf{V}_1 \\ \mathbf{V}_o \end{bmatrix} = \frac{\begin{bmatrix} 2 & j2 \\ -j2.5 & (15 + j4.5) \end{bmatrix}}{2(15 + j4.5) - j2.5(-j2)} \begin{bmatrix} 50 \angle -30^\circ \\ 0 \end{bmatrix}$$

$$= \frac{\begin{bmatrix} 100 \angle -30^\circ \\ -125 \angle 60^\circ \end{bmatrix}}{26.571 \angle 19.799^\circ} = \begin{bmatrix} 3.76 \angle -49.80^\circ \\ 4.70 \angle -139.80^\circ \end{bmatrix}$$

and

$$v_1(t) = 3.76 \cos (2000\pi t - 49.80^\circ)$$
$$v_o(t) = 4.70 \cos (2000\pi t - 139.80^\circ)$$

Warning: The phasor model of a circuit is valid only for the source frequency that was used to calculate the component impedances. If the source frequency is changed to a different value, all of the impedances must be recalculated.

Example 13.5

Write the nodal equations for the circuit shown in Fig. 13.13.

Solution: Because the circuit contains a controlled voltage source it will be necessary to write a constraint equation and a supernode equation

$$\mathbf{V}_1 - \mathbf{V}_2 = 10\mathbf{V}_1$$

and

$$3\mathbf{V}_1 + j5\mathbf{V}_2 - j2\mathbf{V}_2 = \mathbf{I}_g$$

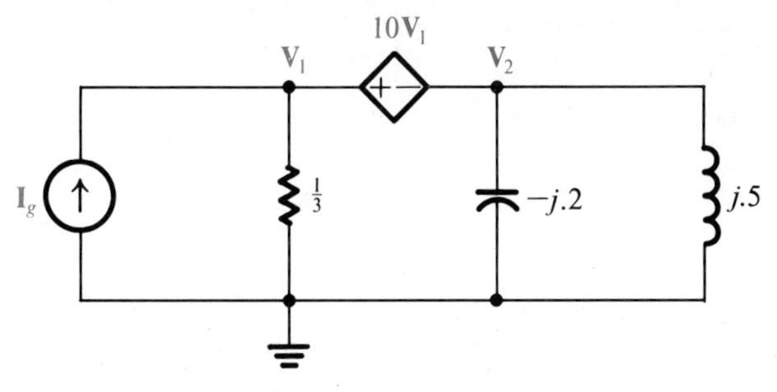

Figure 13.13 Circuit for Example 13.5.

In matrix form

$$\begin{bmatrix} -9 & -1 \\ 3 & j3 \end{bmatrix} \begin{bmatrix} \mathbf{V_1} \\ \mathbf{V_2} \end{bmatrix} = \begin{bmatrix} 0 \\ \mathbf{I_g} \end{bmatrix}$$

Chapter 12 presented voltage- and current-division results for generalized impedances and admittances. Using these results with $s = j\omega$ gives voltage- and current-division rules for the phasors of sinusoidal steady-state signals. For example, if $\mathbf{Z}_1(j\omega)$ and $\mathbf{Z}_2(j\omega)$ are connected in series, the voltage phasors divide according to

$$\frac{\mathbf{V}_1}{\mathbf{V}} = \frac{\mathbf{Z}_1(j\omega)}{\mathbf{Z}_1(j\omega) + \mathbf{Z}_2(j\omega)}$$

where $\mathbf{V}$ is the phasor or the voltage across the series combination.

In the sinusoidal steady state, the phasor of the voltage across impedances in series divides according to the ratio of the individual impedances to the total series impedance. As an exercise, show that the phasor of current into a parallel connection of admittances divides according to the ratio of the branch admittances to the total admittance of the parallel combination.

Example 13.6

In the circuit shown in Fig. 13.14 the voltage and current phasors are obtained as follows:

$$\mathbf{V}_1 = \frac{3 - j5}{3 - j5 + 4 + j10}\mathbf{V}_g = \frac{5.831\,\underline{/-59.036^\circ}}{8.602\,\underline{/35.538^\circ}}\mathbf{V}_g$$

$$= 0.678\,\underline{/-94.574^\circ}\,\mathbf{V}_g$$

$$\mathbf{V}_2 = \frac{4 + j10}{8.602\,\underline{/35.538^\circ}}\mathbf{V}_g = 1.252\,\underline{/32.661^\circ}\,\mathbf{V}_g$$

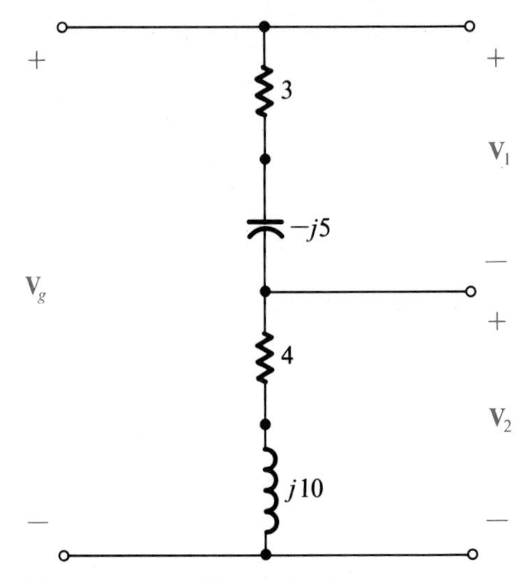

Figure 13.14 Circuit for Example 13.6.

$$\mathbf{I} = \frac{1}{8.602 \angle 35.538^\circ} \mathbf{V}_g = 0.116 \angle -35.538^\circ \, \mathbf{V}_g$$

Phasor diagrams are sometimes used to demonstrate KCL and KVL for a circuit, since the vector sum of the voltage phasors around any closed path must be zero, and the vector sum of the current phasors at a node must be zero. A phasor diagram is formed by drawing the circuit's voltage and/or current phasors to scale in the complex plane.

Example 13.7

If $\mathbf{V}_g = 10 \angle 0^\circ$ in Example 13.6, the voltage phasor diagram in Fig. 13.15 displays KVL. Note that $\mathbf{V}_1 + \mathbf{V}_2 = \mathbf{V}_g$.

Example 13.8

Find the current phasors, the phasor diagram, and the steady-state signals of the circuit in Fig. 13.16 if $\omega = 1$.

Solution:

$$\mathbf{I}_1 = \frac{5\,\mathbf{I}}{5 - j2 + j3} = 0.981\,\mathbf{I} \angle -11.31^\circ$$

$$\mathbf{I}_2 = \frac{-j2\,\mathbf{I}}{5 + j} = 0.392\,\mathbf{I} \angle -101.31^\circ$$

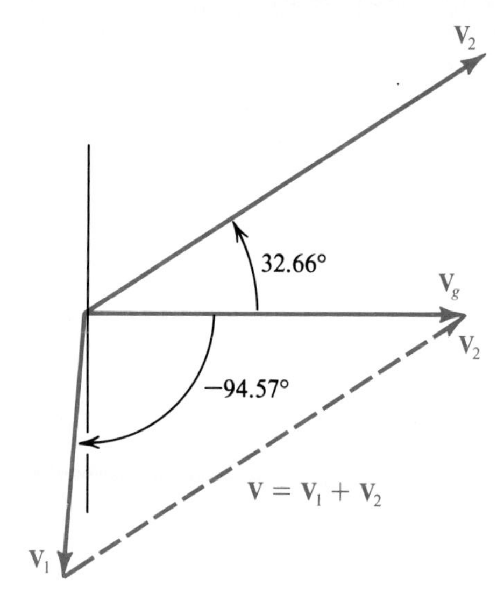

Figure 13.15 ~~Circuit~~ Phasor diagram for Example 13.7.

$$\mathbf{I}_3 = \frac{j3\ \mathbf{I}}{5 + j} = 0.588\ \mathbf{I}\ \underline{/\ 78.69^\circ}$$

and

$$i_1(t) = .981\ |\mathbf{I}| \cos(\omega t - 11.31^\circ)$$

$$i_2(t) = .392\ |\mathbf{I}| \cos(\omega t - 101.31^\circ)$$

$$i_3(t) = .588\ |\mathbf{I}| \cos(\omega t + 78.69^\circ)$$

The current phasor diagram in Fig. 13.17 has $\mathbf{I} = \mathbf{I}_1 + \mathbf{I}_2 + \mathbf{I}_3$

13.4 SOURCE SUPERPOSITION

The matrix nodal model for the sinusoidal steady-state behavior of a circuit [see (13.8)] can be used to demonstrate the principle of superposition

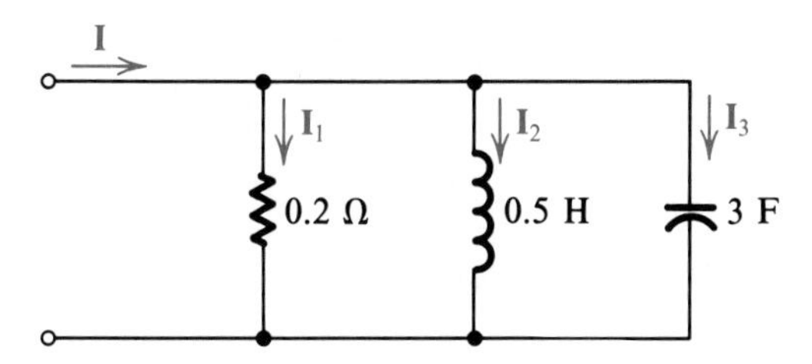

Figure 13.16 Circuit for Example 13.8.

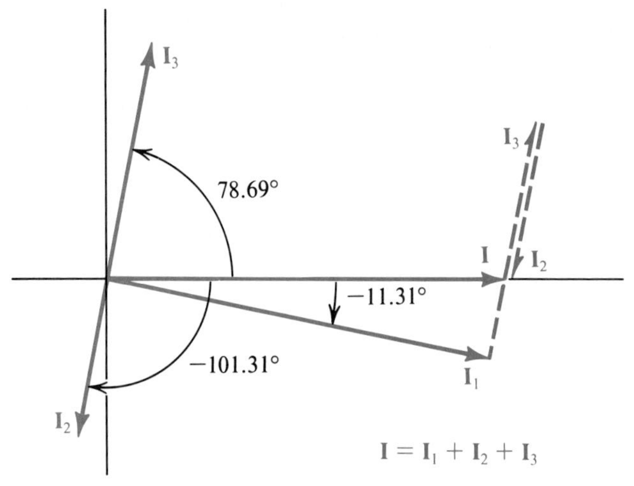

Figure 13.17 Phasors for Example 13.8.

for circuits that are driven by more than one independent sinusoidal source. If the sources have the same frequency a vector of source phasors can be formed as the sum of vectors in which all but one source is turned off

$$\mathbf{V} = [\mathbf{Y}^{-1}\mathrm{B}]\begin{bmatrix}\mathbf{U}_1\\0\\\vdots\\0\end{bmatrix} + [\mathbf{Y}^{-1}\mathrm{B}]\begin{bmatrix}0\\\mathbf{U}_2\\\vdots\\0\end{bmatrix} + \cdots + [\mathbf{Y}^{-1}\mathrm{B}]\begin{bmatrix}0\\0\\\vdots\\\mathbf{U}_m\end{bmatrix}$$

The effect of the multiplication is that the single source term in the column vector is multiplied by the entries in the corresponding column of $[\mathbf{Y}^{-1}\mathrm{B}]$, and all other terms are zero. Therefore

$$\mathbf{V} = [\mathbf{Y}^{-1}\mathrm{B}]_1\mathbf{U}_1 + [\mathbf{Y}^{-1}\mathrm{B}]_2\mathbf{U}_2 + \cdots + [\mathbf{Y}^{-1}\mathrm{B}]_m\mathbf{U}_m$$

where $\mathbf{U}_k$ is the phasor of the kth source (either current or voltage), and $[\mathbf{Y}^{-1}\mathrm{B}]_k$ denotes the kth column of $[\mathbf{Y}^{-1}\mathrm{B}]$. This displays the fact that **the sinusoidal steady-state node-voltage phasors consist of the sum of the node-voltage phasors that would be caused by each of the independent sources acting separately.** That is, the node-voltage phasors obey the principle of superposition with respect to the applied independent sources. *Caution:* power does *not* obey the principle of superposition because it is a nonlinear function of current and voltage.

Example 13.9

Find the phasor of $v(t)$ in Fig. 13.18 when $v_{\text{in1}}(t) = 5\cos(4t + 30°)$ and $v_{\text{in2}}(t) = 3\cos(4t - 20°)$.

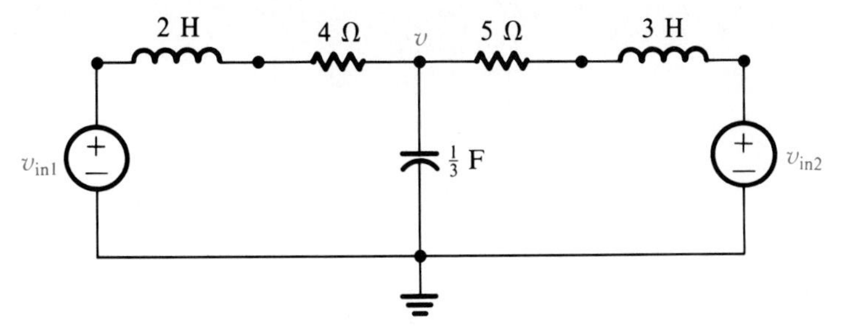

Figure 13.18 Circuit for Example 13.9.

Solution: The circuit has the generic form shown in Fig. 13.19, where the components of the original circuit have been replaced by symbolic impedances. The nodal model can be formed by taking

$$\frac{\mathbf{V}_{in1} - \mathbf{V}}{\mathbf{Z}_1} + \frac{\mathbf{V}_{in2} - \mathbf{V}}{\mathbf{Z}_2} = \frac{\mathbf{V}}{\mathbf{Z}_3}$$

so

$$\frac{\mathbf{V}}{\mathbf{Z}_1} + \frac{\mathbf{V}}{\mathbf{Z}_2} + \frac{\mathbf{V}}{\mathbf{Z}_3} = \frac{\mathbf{V}_{in1}}{\mathbf{Z}_1} + \frac{\mathbf{V}_{in2}}{\mathbf{Z}_2}$$

or

$$\mathbf{V}(\mathbf{Y}_1 + \mathbf{Y}_2 + \mathbf{Y}_3) = \mathbf{Y}_1\mathbf{V}_{in1} + \mathbf{Y}_2\mathbf{V}_{in2}$$

and

$$\mathbf{V} = \frac{\mathbf{Y}_1\mathbf{V}_{in1} + \mathbf{Y}_2\mathbf{V}_{in2}}{\mathbf{Y}_1 + \mathbf{Y}_2 + \mathbf{Y}_3}$$

For given values of $\mathbf{V}_{in1}$ and $\mathbf{V}_{in2}$ this expression determines the value of the phasor $\mathbf{V}$ at a given frequency ω. Here we have $\omega = 4$ rad/sec for both sources, so we must evaluate $\mathbf{Y}_1(j4)$, $\mathbf{Y}_2(j4)$ and $\mathbf{Y}_3(j4)$. We find

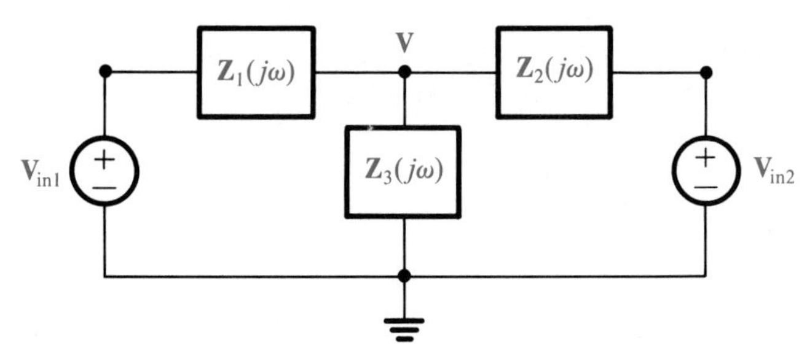

Figure 13.19 Generic form of the circuit in Fig. 13.18.

$\mathbf{Z}_1(j4) = 4 + j8 = 8.944 \underline{/63.435°}$
$\mathbf{Z}_2(j4) = 5 + j12 = 13.000 \underline{/67.380°}$
$\mathbf{Z}_3(j4) = -j3/4 = 0.750 \underline{/-90°}$

and

$\mathbf{Y}_1(j4) = 0.112 \underline{/-63.435°} = 0.050 - j.100$
$\mathbf{Y}_2(j4) = 0.077 \underline{/-67.380°} = 0.030 - j.071$
$\mathbf{Y}_3(j4) = 1.333 \underline{/90°} = j1.333$

Using $\mathbf{V}_{in1} = 5 \underline{/30°}$ and $\mathbf{V}_{in2} = 3 \underline{/-20°}$ we get

$$\mathbf{V} = \frac{(0.112 \underline{/-63.435°})(5 \underline{/30°}) + (0.077 \underline{/-67.380°})(3 \underline{/-20°})}{(0.050 + 0.030) + j(-.100 - 0.071 + 1.333)}$$

$$= \frac{0.560 \underline{/-33.435°} + 0.231 \underline{/-87.380°}}{0.080 + j1.162}$$

$$= .619 \underline{/-134.535°}$$

In the time domain $v(t) = .619 \cos(4t - 134.535°)$.

When sinusoidal sources of different frequencies are applied to a circuit the admittance matrix in the nodal model (13.8) must be calculated separately for each source frequency, and then used to determine the voltages due to each separate source. These phasors cannot be added, since their sum does not correspond to a sum of their corresponding sinusoids. Instead, the individual node-voltage phasors must be used to calculate their individual contribution to the steady-state signal.

Example 13.10

If the sources in Example 13.9 are changed so that $v_{in2}(t) = 3 \cos(10t - 20°)$ find $v(t)$.

Solution: The impedance and admittance values of the circuit must be recalculated with $j\omega = j10$:

$\mathbf{Z}_1(j10) = 4 + j20 = 20.396 \underline{/78.690°}$
$\mathbf{Z}_2(j10) = 5 + j30 = 30.414 \underline{/80.538°}$
$\mathbf{Z}_3(j10) = -0.3\mathbf{j} = 0.3 \underline{/-90°}$

and

$\mathbf{Y}_1(j10) = 0.049 \underline{/-78.690°} = 0.010 - j.048$
$\mathbf{Y}_2(j10) = 0.033 \underline{/-80.538°} = 0.005 - j.033$
$\mathbf{Y}_3(j10) = 3.333 \underline{/90°} = j3.333$

The contribution to $v(t)$ due to $v_{in1}(t)$ is obtained from Example 13.9:

$$\mathbf{V} = \frac{(0.112 \angle -63.435°)(5 \angle 30°)}{0.080 + j1.162}$$
$$= 0.481 \angle -119.497°$$

and

$$v(t) = 0.481 \cos (4t - 119.497°)$$

The contribution to $v(t)$ due to $v_{in2}(t)$ is obtained from the values that were calculated above using voltage division according to the ratio $\mathbf{Y}_2/(\mathbf{Y}_1 + \mathbf{Y}_2 + \mathbf{Y}_3)$:

$$\mathbf{V} = \frac{(0.033 \angle -80.538°)(3 \angle -20°)}{(0.010 + 0.005) + j(-0.048 - 0.033 + 3.333)}$$
$$= 0.030 \angle 169.726$$

and

$$v(t) = 0.030 \cos (10t + 169.726°)$$

The sum of the responses to the separate sources is

$$v(t) = 0.481 \cos (4t - 119.497°) + 0.030 \cos (10t + 169.726°)$$

Note that $v(t)$ cannot be written as a single sinusoid.

13.5 THEVENIN AND NORTON EQUIVALENT CIRCUITS

The Thevenin model of a circuit in the steady state consists of a voltage-source phasor in series with an impedance. The value of the source phasor corresponds to the value of the circuit's steady-state open-circuit voltage; the impedance is calculated at the circuit's terminals with all of the internal independent sources turned off. This model requires that *all* internal sources have the same frequency. Since the Thevenin and Norton equivalent circuits for the sinusoidal steady state are just special cases of the generalized Thevenin and Norton equivalent circuits that were derived in Chapter 12 with $s = j\omega$, we will not derive the models again.

Example 13.11

Find the sinusoidal steady-state Thevenin equivalent at (a, b) for the phasor circuit model in Fig. 13.20.

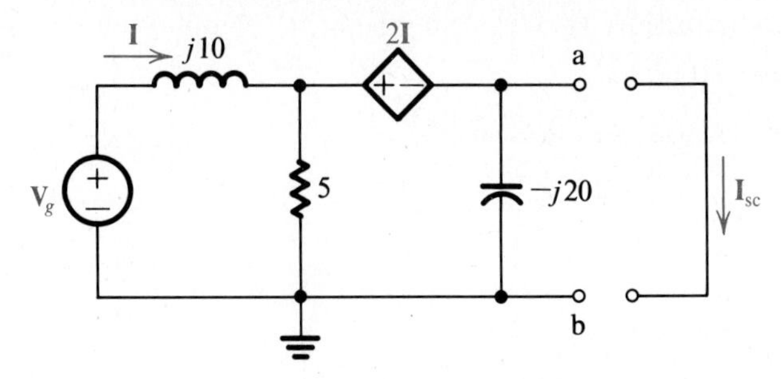

Figure 13.20 Circuit for Example 13.11.

Solution: Our approach will be to calculate the short-circuit current and the open-circuit voltage at the terminals of the circuit. If the terminals are shorted together the current in the dependent source is $\mathbf{I}_{sc}$ so KVL implies that

$$(\mathbf{I} - \mathbf{I}_{sc})5 = 2\,\mathbf{I}$$

or

$$\mathbf{I}_{sc} = .6\,\mathbf{I}$$

and that

$$\mathbf{V}_g = \mathbf{I}(j10) + 2\mathbf{I} = \mathbf{I}(2 + j10)$$

so

$$\mathbf{I} = \mathbf{V}_g/(10.198 \angle 78.690^\circ)$$

and

$$\mathbf{I}_{sc} = 0.6\mathbf{V}_g/(10.198 \angle 78.690^\circ) = (0.059 \angle -78.690^\circ)\,\mathbf{V}_g$$

On the other hand, if the terminals are open, KVL for the outer loop gives

$$\mathbf{V}_{oc} + 2\mathbf{I} + j10\mathbf{I} = \mathbf{V}_g$$

By using Ohm's Law and current division for the path containing the capacitor we find

$$\mathbf{V}_{oc} = -j20[\mathbf{I} - 1/5\,(2\mathbf{I} + \mathbf{V}_{oc})]$$

or

$$\mathbf{V}_{oc} = 2.910 \angle -14.036^\circ\,\mathbf{I}$$

and

$$\mathbf{I} = 0.344 \angle 14.036^\circ\,\mathbf{V}_{oc}$$

Combining this expression with the KVL equation gives

$\mathbf{V}_{oc} + (2 + j10)(0.344 \angle 14.036^\circ)\,\mathbf{V}_{oc} = \mathbf{V}_g$

which can be solved for

$\mathbf{V}_{oc} = 0.278 \angle -76.627^\circ\,\mathbf{V}_g$

Lastly, the Thevenin impedance is

$\mathbf{Z}_{th} = \mathbf{V}_{oc}/\mathbf{I}_{sc} = 4.712 \angle 2.063^\circ$

The Norton equivalent of a circuit has a current-source phasor in parallel with an admittance (or an impedance). The phasor is equal to the phasor of the short-circuit current at the terminals of the circuit; the admittance is the admittance that would be measured at the input terminals when all of the internal sources are disabled. It is also the reciprocal of the Thevenin impedance.

Skill Exercise 13.4

Find the Norton equivalent of the circuit in Example 13.11.

Answer: $\mathbf{I}_N = 0.059 \angle -78.690^\circ$ and $\mathbf{Y}_N = 0.212 \angle -2.063^\circ$

Skill Exercise 13.5

Use source transformations to find the Thevenin equivalent of the circuit in Fig. SE13.5.

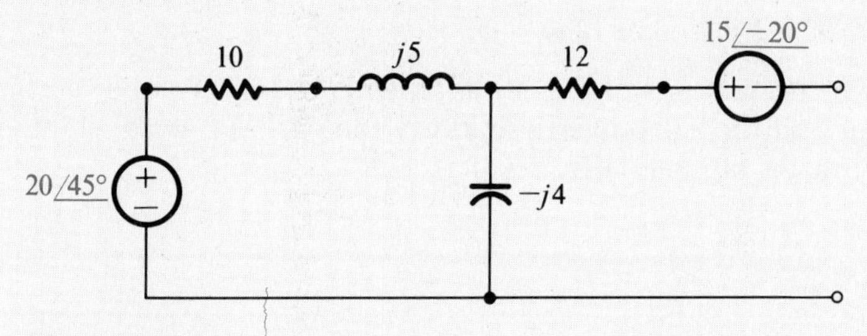

Figure SE13.5

Answer: $\mathbf{V}_{oc} = \sout{22.76 \angle -32.71^\circ}$ and $\mathbf{Z}_{th} = \sout{6.19 \angle -42.24^\circ}$
9.112 ∠−173.504° 14.026∠−17.020°

13.6 SINUSOIDAL STEADY-STATE POWER

When the two-terminal circuit in Fig. 13.1 is driven by a sinusoidal source the voltage and current at its input terminals **deliver an instantaneous power to the load,** according to

$$p(t) = v(t)\ i(t) \tag{13.9}$$

After the circuit reaches the steady state the voltage and current waveforms are

$$v(t) = V_m \cos(\omega t + \phi_v) \tag{13.10a}$$

and

$$i(t) = I_m \cos(\omega t + \phi_i) \tag{13.10b}$$

with $V_m = |\mathbf{V}|$, $I_m = |\mathbf{I}|$, $\mathbf{V} = V_m e^{j\phi_v}$ and $\mathbf{I} = I_m e^{j\phi_i}$.

Substituting (13.10) into (13.9) leads to an expression $p(t) = V_m I_m \cos(\omega t + \phi_v) \cos(\omega t + \phi_i)$ for the instantaneous steady-state power. But since

$$\cos A \cos B = 1/2\ [\cos(A + B) + \cos(A - B)]$$

the expression for $p(t)$ becomes

$$p(t) = 1/2\ V_m I_m [\cos(2\omega t + \phi_v + \phi_i) + \cos(\phi_v - \phi_i)]$$

Writing $p(t)$ in this form reveals the fact that the instantaneous power absorbed by the circuit consists of the sum of a constant term and a sinusoid whose frequency is *twice* the frequency of the source. Thus the power can be written as

$$p(t) = p_{DC} + p_{AC}(t)$$

with

$$p_{DC} = 1/2\ V_m I_m \cos(\phi_v - \phi_i)$$

and

$$p_{AC}(t) = 1/2\ V_m I_m \cos(2\omega t + \phi_v + \phi_i)$$

These two components of the instantaneous power are referred to as the DC power and the AC power to identify the part that is constant and the part that varies sinusoidally.

Example 13.12

If the circuit in Fig. 13.1 has $\mathbf{V} = 5\ \underline{/30^\circ}$ volts and $\mathbf{I} = 3\ \underline{/-45^\circ}$ amps, find expressions for p_{DC}, $p_{AC}(t)$ and $p(t)$.

Solution: Because the steady-state waveforms of $v(t)$ and $i(t)$ are $v(t) = 5 \cos(\omega t + 30^\circ)$ and $i(t) = 3 \cos(\omega t - 45^\circ)$ we get

$$p_{DC} = 1/2\ (5)(3) \cos(75^\circ) = 1.94 \text{ W}$$

$$p_{AC}(t) = 1/2\ (5)(3) \cos(2\omega t - 15^\circ) = 7.5 \cos(2\omega t - 15^\circ) \text{ W}$$

and

$$p(t) = 1.94 + 7.5 \cos(2\omega t - 15°) \text{ W}$$

13.7 AVERAGE POWER

In many applications it is important to have a measure of the power absorbed by a circuit over a specific interval of time. For convenience, the *average power* absorbed by the circuit is defined as

$$P_{av} = \frac{1}{T}\int_0^T p(t)\,dt \tag{13.11}$$

The average power is the energy absorbed by the circuit over one cycle of the input signal divided by the length of the signal period. It can be used in circuit design to specify cooling requirements if the power dissipation is at a level that might damage the circuit components.

When the instantaneous power defined by (13.9) is used in (13.11) the average power consists of the sum of the averages of the DC power and the AC power. The average of a cosine function over one or more periods is zero, so the average of the AC power is zero, and the only contribution to the average power is due to the DC component of the instantaneous power, or

$$P_{av} = \frac{1}{T}\int_0^T \frac{V_m I_m}{2}\cos(\phi_v - \phi_i)\,dt$$

and so

$$P_{av} = \frac{V_m I_m}{2}\cos(\phi_v - \phi_i) = \frac{V_m I_m}{2}\cos\phi \tag{13.12}$$

with $\phi = \phi_v - \phi_i$. The average power is proportional to the cosine of the phasor-angle difference between the voltage and current phasors. The **power factor** of the circuit is defined to be

$$PF = \cos(\phi_v - \phi_i) = \cos\phi$$

where ϕ is called the power-factor angle. Because the cosine function is bounded by 1 it follows that $|PF| \leq 1$. To understand the relationship between a circuit's average-power, power-factor, and time-domain signal waveforms, consider the graphs of hypothetical $v(t)$, $i(t)$ and $p(t)$ shown in Fig. 13.21. The waveform for the instantaneous power is positive whenever $v(t)$ and $i(t)$ have the same polarity, and negative when they have opposite polarity. The average power will be positive when the sub-interval over which the curve for $p(t)$ is positive is greater than the sub-interval over which it is negative. Of the two cases illustrated, the one with a phase-angle difference of 45° has a positive average power; the one with a phase-angle difference of 135° is negative.

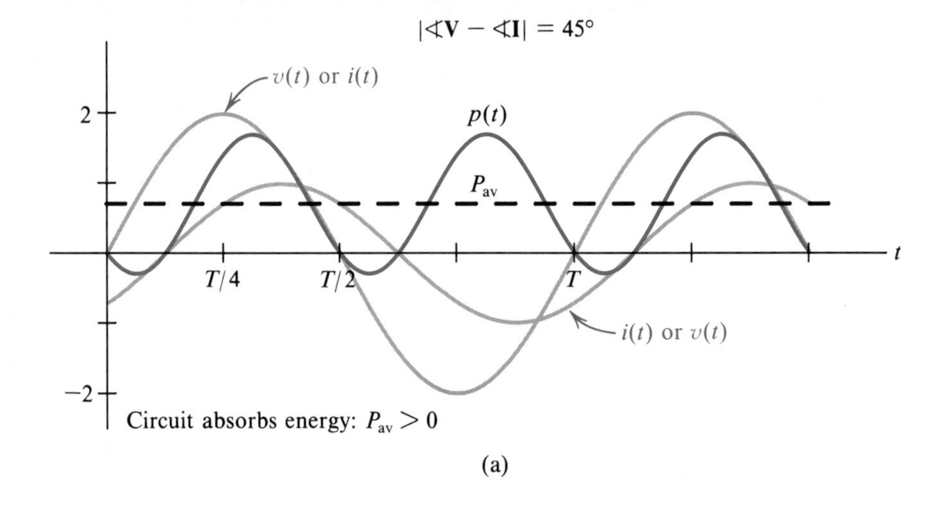

(a)

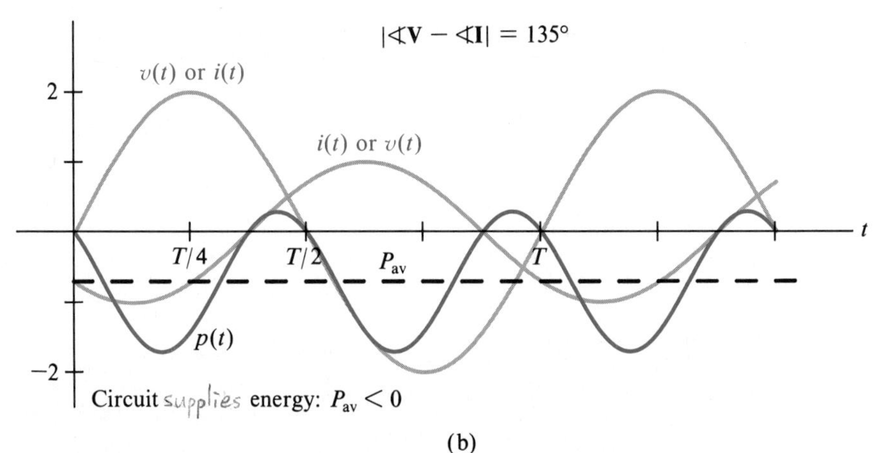

(b)

Figure 13.21 Time-domain phase relationships of **v(t)**, **i(t)** and **p(t)**, with (a) $P_{av} > 0$ and (b) $P_{av} < 0$.

By convention, a circuit is said to have a **lagging load-power factor** whenever the phasor of the current is between 0° and 90° *behind* the voltage phasor, and is said to have a **leading load-power factor** when the current phasor is between 0° and 90° *ahead* of the voltage phasor. In practical situations in which the load always absorbs power the power factor is nonnegative. The average power dissipated by the circuit will be positive when $-90° < \phi < 90°$ and negative when $90° < \phi < 270°$.

The angle of the driving-point impedance of a circuit determines its power factor because

$$\mathbf{Z}(j\omega) = \sphericalangle\mathbf{V} - \sphericalangle\mathbf{I}$$

and so

$$PF = \cos [\measuredangle \mathbf{Z}(j\omega)]$$

When **V** and **I** are 90° out of phase the average power is zero, and when the voltage and current phasors are aligned it is at a maximum. The frequency-domain driving-point impedance determines the time-domain waveform relationships between $i(t)$, $v(t)$ and $p(t)$.

Skill Exercise 13.6

Find the power factor for (a) an inductor, (b) a capacitor.

Answer: (a) $PF = 0$ and (b) $PF = 0$

13.8 POWER FACTOR AND TRANSMISSION LOSSES

A source delivers energy to a load efficiently whenever most, if not all, of the energy supplied by the source is absorbed by the load. Physical energy sources often must deliver power over a transmission system (wires!) to a load that is remotely located, as shown in Fig. 13.22, where V_m and I_m denote the amplitudes of the sinusoidal signals. When the circuit is operating in the sinusoidal steady state the amplitude of the current in the transmission path can be obtained from (13.12) as $I_m = 2\, P_{av}/(V_m PF)$, where P_{av} is the average power delivered to the load. If the power factor of the load is small, a greater I_m will be required in order for the source to deliver P_{av} to the load for a fixed source voltage V_m (thereby increasing the cost of the generation equipment). Another, often more serious consequence, is that a low power factor causes energy to be wasted in the operation of the transmission system. Suppose that the source is not an ideal source; that its internal resistance and the resistance of the transmission wires can be lumped together, as denoted by R_g in Fig. 13.23; and that the load can be represented by its impedance $\mathbf{Z}_L$. If the load is to be supplied by V_m and I_m the average power dissipation in the resistor R_g is given by

$$P_d = \frac{1}{2} I_m^2 R_g \tag{13.13}$$

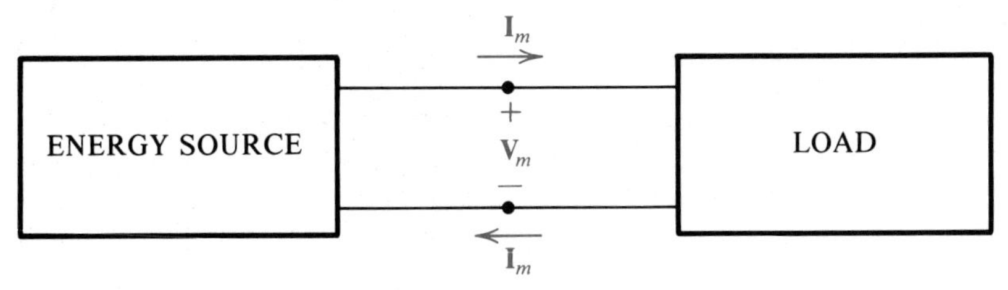

Figure 13.22 Energy source supplying power to a load.

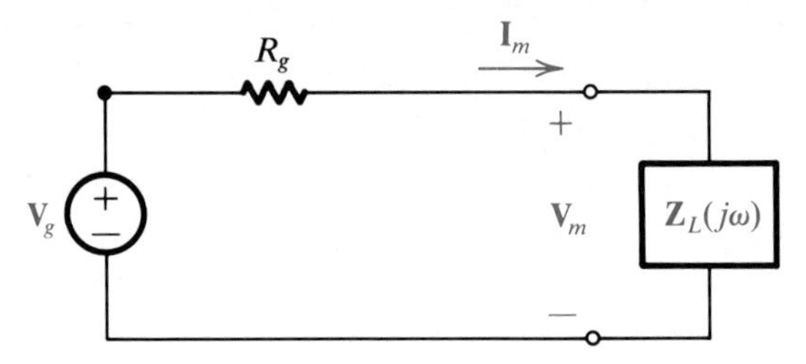

Figure 13.23 The average power supplied to $\mathbf{Z}_L(j\omega)$ by $\mathbf{V}g$ depends on R_g and $\mathbf{Z}_L(j\omega)$.

Thus, more power will be dissipated in R_g if I_m is large because of a small load-power factor. Another way to view this is to substitute (13.12) into (13.13) to display the effect of the power factor on P_d

$$P_d = \frac{2\ P_{av}^2 R_g}{V_m^2\ (PF)^2}$$

The power dissipated in the source/transmission resistance is inversely proportional to the *square* of the load-power factor for a fixed load requirement of P_{av} and V_m. If *PF* is smaller P_d will be larger.

The impedance of large electric motors (used in many industrial processes) is inherently inductive because the wire windings of electric motors effectively create large inductors. This causes a lagging load-power factor. Since a utility company would rather not waste energy by heating up the atmosphere, corrective steps are usually taken to change the power factor seen by the transmission system, thereby reducing the transmission-line current while simultaneously satisfying the customer's requirements. Placing a capacitor across an inductive load can correct the power factor.

Example 13.13

The average power that will be absorbed by the load in Fig. 13.24 is 5.75 kW, at a lagging power factor of 0.707 and a sinusoidal line voltage of $v_g(t) = 115\sqrt{2}\sin 120\pi t$.

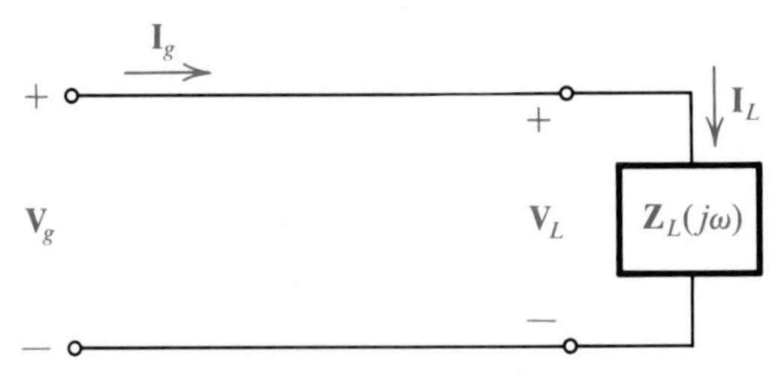

Figure 13.24 Circuit for Example 13.16.

a. Find the line-current phasor $\mathbf{I}_g$ and the steady-state line current $i_g(t)$.
b. Connect a 0.00163 F capacitor in parallel with the load and calculate the capacitor-current phasor, the line-current phasor, the line voltage, and the power factor seen by the transmission line.

Solution: a. Using V_m and I_m to denote the magnitude of the line voltage and current phasors, and noting that $\mathbf{I}_L = \mathbf{I}_g$ and $\mathbf{V}_L = \mathbf{V}_g$, we have $V_m = 115\sqrt{2}$. Next, we find the magnitude of the line-current phasor for the rated load conditions:

$$P_{av} = 1/2\, V_m I_m \cos(\phi_v - \phi_i) = 1/2(115\sqrt{2})I_m(0.707)$$
$$= 57.50\, I_m$$

Solving for I_m gives $I_m = 100$ A. The phasor of the given sinusoidal line and load voltages will be $\mathbf{V}_g = 115\sqrt{2}\,\underline{/0^\circ}$ and for a *lagging* power factor of 0.707 the line and load current phasor will be $\mathbf{I}_g = 100\,\underline{/-45^\circ}$. Therefore, the sinusoidal steady-state line and load current will be $i_g(t) = 100 \sin(120\pi t - 45^\circ)$.

(b) Placing a 0.00163 F capacitor across the load in Fig. 13.25 creates an additional component of line current for the capacitor path. Current *leads* voltage in a capacitor, so this current will cause the line current to have a smaller lagging phase angle and a larger power factor than it would without the capacitor. First we find the current phasor for the capacitor:

$$\mathbf{I}_c = \mathbf{V}_g\mathbf{Y}_c = (115\sqrt{2}\,\underline{/0^\circ})[120\pi(0.00163)\,\underline{/90^\circ}]$$
$$= 99.94\,\underline{/90^\circ}$$

The line current is the sum of the capacitor current and the load current, and will have a phasor given by $\mathbf{I}_g = \mathbf{I}_c + \mathbf{I}_L$. Since the load and the capacitor are in parallel, the load current will be the same as in part (a), so the line-current phasor is $\mathbf{I}_g = 76.51\,\underline{/22.46^\circ}$.

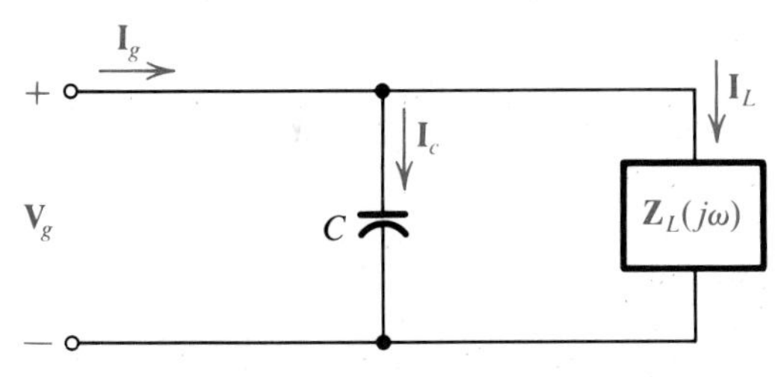

Figure 13.25 Adding the shunt capacitor improves the power factor at the source.

The new sinusoidal steady-state line current is $i_g(t) = 76.51 \sin(120\pi t + 22.46°)$. The power factor without the capacitor was 0.707 lagging. With the capacitor, the power factor is 0.924 leading.

Increasing the power factor from 0.707 to 0.924 in the previous example reduced the magnitude of the phasor-angle difference between the line-voltage and current phasors, reduced the phase angle between the line-voltage and current steady-state waveforms, and reduced the magnitude of the line current by nearly 25% while still meeting the power, voltage, and current requirements of the load. The capacitor acts as a reservoir of charge for the load by discharging during portions of the cycle when the inductive load requires high current levels, and storing charge when the load current drops. The net effect is that the transmission line will carry less current and will waste less energy.

Phasor diagrams are used to find the value of the shunt capacitor required to increase the power factor of a load. If a circuit with a lagging power factor has the load-phasor diagram shown in Fig. 13.26(a), the task is to find a shunt capacitor value such that placing it across the load causes a component of current to be added to $\mathbf{I}_L$ and results in a line current whose phasor is aligned with the load-voltage phasor, as shown in Fig. 13.26(b).

The current phasor of the parallel capacitor leads the load-voltage phasor by 90°. When the power factor is leading or lagging, the possible values of the sum of the capacitor- and load-current phasors lie on the locus shown in Fig. 13.27. We can choose the capacitor value c according to

$$|\mathbf{I}_c| = |\mathbf{I}_L \sin \phi| = |\mathbf{V}_L||j\omega C|$$

or

$$C = \frac{|\mathbf{I}_L \sin \phi|}{\omega|\mathbf{V}_L|}$$

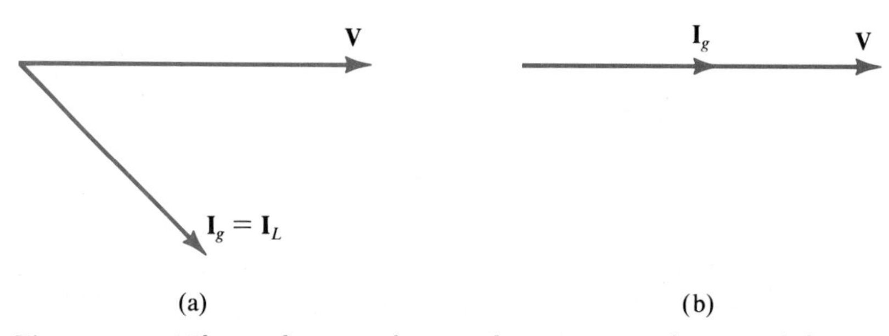

Figure 13.26 Phasor diagrams for (a) a lagging power factor and (b) a zero power factor.

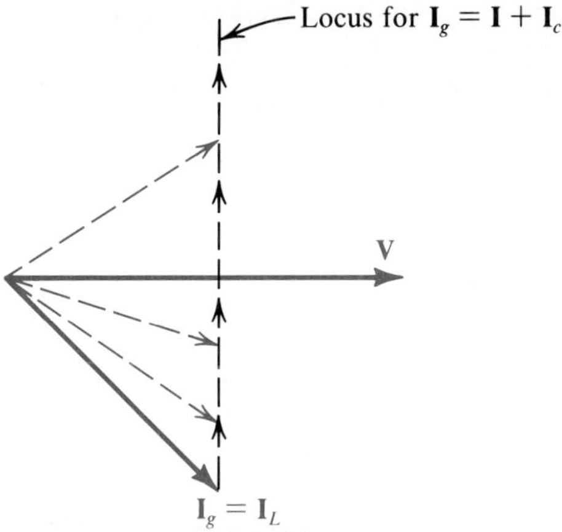

Figure 13.27 The locus of the source-current phasors for $I_g = I_L + I_c$

to give a capacitor-current phasor that *cancels* the imaginary component of the load-current phasor and results in a line current aligned with the load-voltage phasor. For this value of C the power factor of the load will be a minimum.

A smaller value of C will cause the line current to have an improved, but not optimum, lagging power factor; a larger value of C will create a leading power factor.

Example 13.14

For the circuit in Example 13.13 find the value of C that results in a unity power factor, and calculate the values of the line-current phasor and the steady-state line current with the capacitor inserted.

Solution: In Example 13.13 the value of $\mathbf{I}_L$ was found to be $\mathbf{I}_g = 100 \angle -45°$ so for a load voltage phasor of $\mathbf{V}_L = 115\sqrt{2}\angle 0°$ the value of C is

$$C = |100 \sin(-45°)|/[(120\pi)\,115\sqrt{2}] = 1153\ \mu\text{F}$$

For this choice of C the capacitor current is given by

$$\mathbf{I}_c = j\omega C\mathbf{V}_L = j120\pi(0.001153)(115\sqrt{2}) = j70.69$$

and the line current is

$$\mathbf{I}_g = \mathbf{I}_c + \mathbf{I}_L = j70.69 + 100\angle -45° = 70.7\angle 0°\ \text{A}$$

Using $C = 1153\ \mu$F cancels the imaginary component of the load current, leaving a line current that is in phase with the load-voltage phasor, and giving a load power factor of one. The resulting line current has the smallest possible magnitude that can deliver the rated power to the load at the given load voltage.

13.9 RMS VOLTAGE AND CURRENT

The **effective value** of a signal is defined by

$$V_{rms} = 1/T \left[\int_0^T v^2(t)\, dt \right]^{1/2}$$

The effective value of a signal is also called its **rms value.** It is formed by squaring, taking the mean value, and taking the square root of the signal waveform. If $v(t)$ is a sinusoid

$$V_{rms} = 1/T \int_0^T V_m^2 \cos^2 (\omega t + \phi_v) dt$$

$$= 1/T \int_0^T V_m^2 (1 - \cos [2(\omega t + \phi_v)])/2\, dt$$

$$= \sqrt{1/2\, V_m^2}$$

or

$$V_{rms} = \sqrt{2}/2\, V_m \quad (13.14a)$$

$$= 0.707\, V_m \quad (13.14b)$$

The same steps can be followed to obtain the rms value of the sinusoidal current:

$$I_{rms} = \sqrt{2}/2\, I_m \quad (13.15a)$$

$$= 0.707\, I_m \quad (13.15b)$$

In terms of rms values, the average power is

$$P_{av} = V_{rms} I_{rms} \cos (\phi_v - \phi_i)$$

The average power is the product of the rms voltage and current phasors and the power factor. We call V_{rms} and I_{rms} **effective values** of v and i because when the power factor is zero they deliver the same average power to a resistor as would be delivered by DC signals having the same values.

The electrical-supply system in the United States is based on generating electricity at a frequency of 60 Hz, and the distribution system provides sinusoidal signals characterized by either their peak or their rms value. Because calculation of average power with effective values is done in the same way as the calculation of power for DC signals, industrial equipment and household appliances are rated in terms of their average power and the effective values of their operating current and voltage (I_{rms}, V_{rms}) rather than their peak values (I_m and V_m).

Skill Exercise 13.7

If household voltage is rated at 115 rms volts at 60 Hz and if a toaster dissipates 900 watts, find the rms and peak currents supplied to the toaster. (Assume a resistive load.)

Answer: $I_{rms} = 7.8$ rms A and $I_m = 11.068$ A.

13.10 COMPLEX POWER

The calculation of average power in a phasor circuit model can be simplified by introducing the notion of **complex power.** Complex power (**S**) is defined in terms of the phasors of voltage (**V**) and current (**I**) as the quantity

$$\mathbf{S} = 1/2\ \mathbf{V}\mathbf{I}^*$$

Complex power has the property that the average power can be taken as its real part, because

$$\begin{aligned}\mathrm{Re}\{\mathbf{S}\} &= 1/2\ \mathrm{Re}\{|\mathbf{V}|e^{j\phi_v}|\mathbf{I}|e^{-j\phi_i}\}\\ &= 1/2\ |\mathbf{V}||\mathbf{I}|\cos(\phi_v - \phi_i) = P_{av}\end{aligned}$$

Note that the complex conjugate of the current phasor is used in the definition of **S** to provide $-\phi_i$ in the above expression for P_{av}. If "rms phasors" are defined as $\mathbf{V}_{rms} = 0.707\ \mathbf{V}$ and $\mathbf{I}_{rms} = 0.707\ \mathbf{I}$, then $\mathbf{S} = \mathbf{V}_{rms}\ \mathbf{I^*}_{rms}$. (Note that we do not call **S** a phasor since it does not correspond to a time-domain complex exponential signal).

The real part of **S** is the average power delivered to the circuit, and is called the **real power;** the imaginary part is called the **reactive power.** Complex power is given units of VA (volt-amps), real power is given units of watts, and imaginary power is given units of Vars (volt-amps reactive). Complex power is also called the **apparent power** because it is the power that would be calculated from measurements of the sinusoidal voltage and current amplitudes that ignored the power factor of the circuit. These amplitudes alone do not determine the power delivered to the circuit because they don't account for the phase relationship of their waveforms.

Reactive power is associated with the component of power due to current and voltage being 90° out of phase. It must not be ignored because it affects the line current required to service the load, even though, on the average, it is not dissipated at the load. Reactive power is due to current and voltage that alternately store and release energy in the electric and magnetic field of the circuit's capacitors and inductors. A relatively high reactive power causes high energy losses in transmission.

Since $\mathbf{V} = \mathbf{IZ}$, complex power can also be written as

$$\mathbf{S} = 1/2\ \mathbf{II}^*\mathbf{Z}(j\omega) \tag{13.16a}$$

$$= 1/2\ |\mathbf{I}|^2\mathbf{Z}(j\omega) \tag{13.16b}$$

$$= |\mathbf{I}^2{}_{\mathrm{rms}}|\mathbf{Z}(j\omega) \tag{13.16c}$$

We conclude that the angle of $\mathbf{S}$ is the angle of $\mathbf{Z}$: $\angle\mathbf{S} = \angle\mathbf{Z}(j\omega)$ and that

$$P_{\mathrm{av}} = 1/2\ |\mathbf{I}|^2\ |\mathbf{Z}|\cos(\phi_v - \phi_i)$$

and

$$P_{\mathrm{av}} = |\mathbf{I}_{\mathrm{rms}}|^2\ |\mathbf{Z}|\cos(\phi_v - \phi_i)$$

Now let's write the impedance $\mathbf{Z}$ in terms of its real and imaginary parts:

$$\mathbf{Z}(j\omega) = R(\omega) + jX(\omega) \tag{13.17}$$

The imaginary part X is called the **reactive component** of $\mathbf{Z}$, or the **reactance.** Substituting (13.17) in (13.16b) gives

$$\mathbf{S} = 1/2\ |\mathbf{I}|^2\ [R(\omega) + jX(\omega)]$$

Therefore, the real and imaginary parts of the complex power are

$$P_{\mathrm{av}} = 1/2\ |\mathbf{I}|^2 R(\omega) \tag{13.18a}$$

and

$$Q = 1/2\ |\mathbf{I}|^2 X(\omega) \tag{13.18b}$$

Because $\mathbf{I} = \mathbf{V}/\mathbf{Z}$ both components of $\mathbf{S}$ can be written in terms of the voltage phasor:

$$P_{\mathrm{av}} = 1/2\ |\mathbf{V}|^2 R(\omega)/|\mathbf{Z}(j\omega)|^2 = \frac{1/2\ |\mathbf{V}|^2 R(\omega)}{[\sqrt{R^2(\omega) + X^2(\omega)}]^2}$$

and

$$Q = 1/2\ |\mathbf{V}|^2\ X(\omega)/|\mathbf{Z}(j\omega)|^2 = \frac{1/2\ |\mathbf{V}| X(\omega)}{[\sqrt{R^2(\omega) + X^2(\omega)}]^2}$$

Skill Exercise 13.8

A circuit with $\mathbf{Z} = 200\,\underline{/\,30^\circ}$ absorbs 1000 volt-amps of power from a source. Find P_{av} and Q.

Answer: $P = 1000 \cos(30°) = 866$ watts and $Q = 1000 \sin(30°) = 500$ vars.

13.11 MAXIMUM POWER TRANSFER

A source delivers maximum power to a resistive circuit when the Thevenin resistance of the circuit is equal to the internal resistance of the source. If the load in Fig. 13.28 contains inductors and capacitors we write the source impedance $\mathbf{Z}_g$ and the load impedance $\mathbf{Z}_L$ in terms of their real and imaginary components as

$$\mathbf{Z}_g(j\omega) = R_g(j\omega) + jX_g(j\omega)$$
$$\mathbf{Z}_L(j\omega) = R_L(j\omega) + jX_L(j\omega)$$

For this circuit configuration the current phasor is

$$\mathbf{I}_g = \frac{\mathbf{V}_g}{\mathbf{Z}_g(j\omega) + \mathbf{Z}_L(j\omega)}$$
$$= \frac{\mathbf{V}_g}{R_g(j\omega) + R_L(j\omega) + j[X_g(j\omega) + X_L(j\omega)]}$$

Noting that the magnitude of the source voltage phasor is V_m and taking the magnitude of the current phasor we get

$$I_m = \frac{V_m}{\{[R_g(j\omega) + R_L(j\omega)]^2 + [X_g(j\omega) + X_L(j\omega)]^2\}^{\frac{1}{2}}}$$

When this expression is substituted into (13.18a) the average power becomes

$$P_{av} = \frac{1/2\ V_m^2 R_L(j\omega)}{[R_g(j\omega) + R_L(j\omega)]^2 + [X_g(j\omega) + X_L(j\omega)]^2}$$

The average power will be a maximum for a given ω, R_L and R_g if

$$X_g = -X_L$$

The remaining expression for P_{av} can be shown to be minimized with respect to R_L when $R_L = R_g$. When these conditions are satisfied the source and load impedances are a conjugate pair with

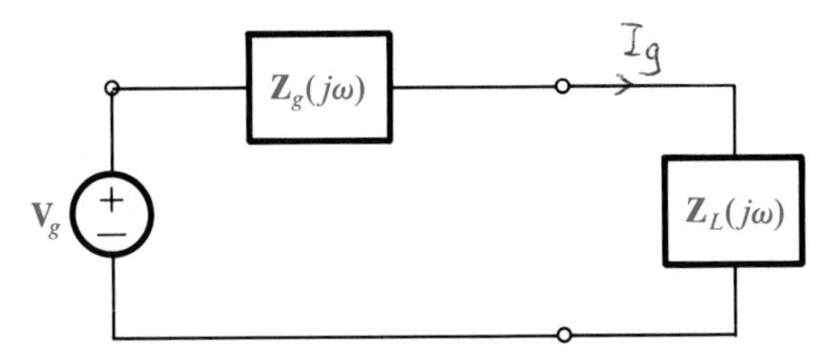

Figure 13.28 The maximum average power is delivered to $\mathbf{Z}_L$ when $\mathbf{Z}_L(j\omega) = \mathbf{Z}^*_g(j\omega)$.

$$\mathbf{Z}_g(j\omega) = \mathbf{Z}_L^*(j\omega)$$

The source and load are said to be matched when their impedances are a conjugate pair. Under this condition, the reactive components of their impedances are in series and exactly cancel each other, so that the current (phasor) in the series path can be written as

$$\mathbf{I} = \frac{\mathbf{V}_g}{2R_L(j\omega)}$$

When the source and load are matched the average power transferred to the load from the source is a maximum value and is given by

$$P_{\text{av max}} = \frac{V_m{}^2}{8\,R_L(j\omega)}$$

The condition for maximum power transfer specifies the load that will absorb the maximum average power from a given source, and, alternately, specifies the source that must be used to deliver the maximum average power to a fixed load.

13.12 THREE-PHASE CIRCUITS*

The electric generation system in the United States operates as a **three-phase system** in which the generators in a power plant produce three electrical signals simultaneously. When such generators are **balanced** the instantaneous power delivered to the external loads is a constant, which has the effect of reducing mechanical wear and stress on the generating equipment. This feature of three-phase power, coupled with the inherent efficiency of its transmission compared to single-phase (less wire for the same delivered power), provides a compelling reason for its use.

Suppose that three ideal voltage sources are arranged in the Y (wye) configuration shown in Fig. 13.29. Here one end of each source is connected to the reference node n and each source's remaining node is available for connecting a load. Next, suppose that the sources are defined to have the same sinusoidal frequency ω but have their phase angles adjusted so that their phasors are defined by

$$\mathbf{V}_{an} = V_m \angle 0^\circ \qquad \mathbf{V}_{bn} = V_m \angle -120^\circ \qquad \mathbf{V}_{cn} = V_m \angle -240^\circ \tag{13.19}$$

where V_m is a scaler. As a result, the three phasors form the phasor diagram shown in Fig. 13.30 (where they have the same magnitude and are pairwise separated by 120°). Such a three-phase source is said to be **balanced.** Its phasors sum to zero. (As an exercise, verify that $\mathbf{V}_{an} + \mathbf{V}_{bn} + \mathbf{V}_{cn} = \mathbf{0}$)

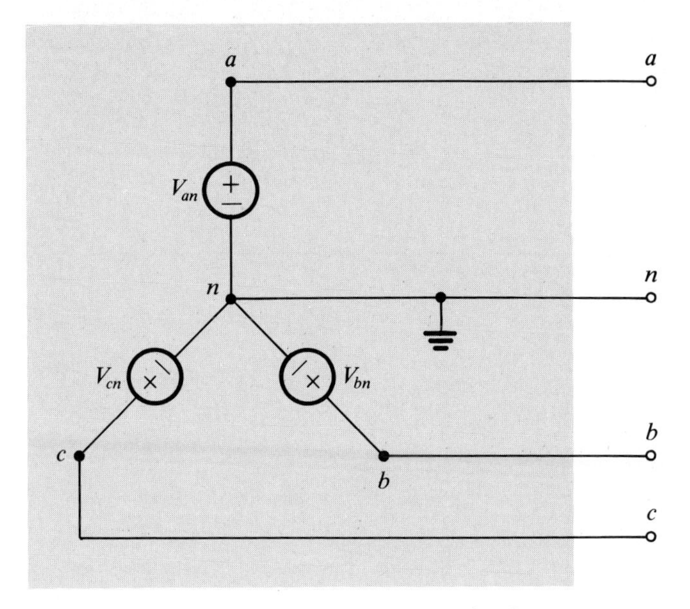

Figure 13.29 Three-phase connection of voltage sources.

When $\mathbf{V}_{bn}$ and $\mathbf{V}_{cn}$ have the phasor angles in relationship to $\mathbf{V}_{an}$ as shown in Fig. 13.30 the set of phasors is said to have a **positive phase sequence.** If $\mathbf{V}_{bn}$ and $\mathbf{V}_{cn}$ are interchanged the phase sequence is said to be negative. (We will develop results only for the case of a positive sequence of balanced sources.)

The steady-state waveforms corresponding to a positive phase sequence are shown in Fig. 13.31. The key feature to note is that the signals

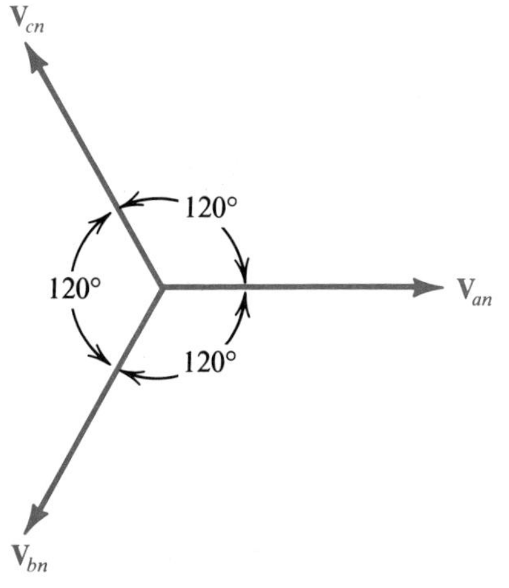

Figure 13.30 Phasors for a balanced set of Y-connected three-phase voltages.

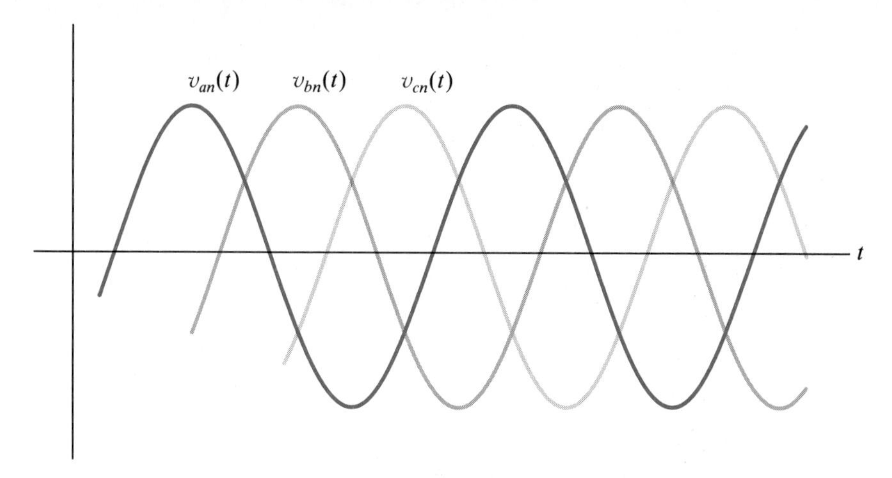

Figure 13.31 Waveforms of voltages in a balanced three-phase system.

have phase angles of 0°, −120° and +120°. The symmetry of this phase-angle set is due to the relationship between the rotating electromagnet of the generator's rotor circuit and the physical symmetry of the three coils of its stator circuits, which are mounted on the stator frame at angles of 120° from each other. The steady-state time-domain signals that result from such a configuration are described by

$$v_{an}(t) = \cos(\omega t)$$
$$v_{bn}(t) = \cos(\omega t - 120°)$$
$$v_{cn}(t) = \cos(\omega t - 240°)$$

These voltages are called the **phase voltages** of the three-phase system.

Before connecting a load to the circuit, let's use the phasors of the sources to determine the phasors of the voltages between the external terminals of the source (we want to find $\mathbf{V}_{ab}$, $\mathbf{V}_{bc}$ and $\mathbf{V}_{cd}$ in Fig. 13.29). These voltages are called the **line voltages** because they exist between the transmission lines connecting a three-phase source and its load. The line voltages are related to the phase voltages according to

$$\mathbf{V}_{ab} = \mathbf{V}_{an} - \mathbf{V}_{bn} = V_m \underline{/0°} - V_m \underline{/-120°}$$
$$\mathbf{V}_{bc} = \mathbf{V}_{bn} - \mathbf{V}_{cn} = V_m \underline{/-120°} - V_m \underline{/-240°}$$
$$\mathbf{V}_{ca} = \mathbf{V}_{cn} - \mathbf{V}_{an} = V_m \underline{/-240°} - V_m \underline{/0°}$$

Consider the first expression, with

$$\mathbf{V}_{ab} = V_m[1 \underline{/0°} - 1 \underline{/-120°}] = V_m[1 - (-1/2 - j\sqrt{3}/2)]$$
$$= V_m[3/2 + j\sqrt{3}/2] = V_m\sqrt{3}[\sqrt{3}/2 + j1/2] = \sqrt{3}\, V_m \underline{/30°}$$

Taking similar steps with $\mathbf{V}_{bc}$ and $\mathbf{V}_{ca}$ would show that

$$\mathbf{V}_{ab} = \sqrt{3}\, V_m \underline{/30°} \qquad \mathbf{V}_{bc} = \sqrt{3}\, V_m \underline{/-90°} \qquad \mathbf{V}_{ca} = \sqrt{3}\, V_m \underline{/150°}$$

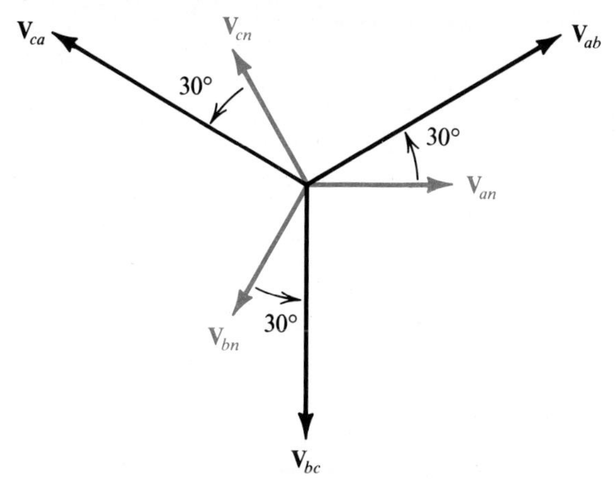

Figure 13.32 Phase- and line-voltage phasors in a balanced three-phase system.

Note that the line voltages corresponding to these phasors have sinusoidal steady-state waveforms, a magnitude that is equal to $\sqrt{3}$ times the phase voltages, and phasors that are still pairwise separated by 120°, but are rotated by 30° ahead of the phasor set that generates the phase voltages of the system. Thus, **if the phase voltages are balanced the line voltages will be balanced.** Fig. 13.32 shows the phasors of the phase voltage and the line voltages. The time-domain line voltages will lead the corresponding phase voltages by 30°.

Skill Exercise 13.9

Write the time-domain expressions for the line voltages of a balanced three-phase Y-connected source.

Answer:
$$v_{ab}(t) = \sqrt{3}\, V_m \cos(\omega t + 30°)$$
$$v_{bc}(t) = \sqrt{3}\, V_m \cos(\omega t - 90°)$$
$$v_{ca}(t) = \sqrt{3}\, V_m \cos(\omega t - 210°)$$

There are two common ways to connect a load to a Y-connected three-phase source: the "wye" (Y) and "delta" (Δ) load connections shown in Fig. 13.33. In the Y connection a fourth wire connects the reference node of the sources to the common node of the load branches. Suppose for the moment that the sources drive three identical Y-connected load impedances, denoted by **Z** in Fig. 13.34. Such a load is said to be **balanced.** The neutral wire carries no current in a balanced system. To see this, replace the sources by their Norton equivalents, as in Fig. 13.35, which shows

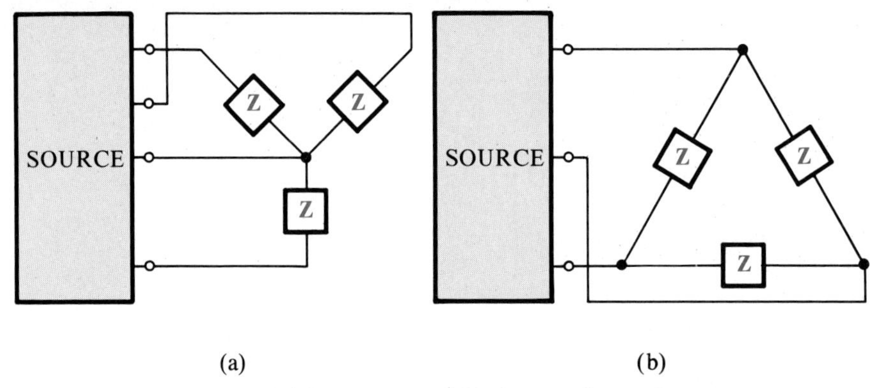

Figure 13.33 Wye- and delta-connected loads in a three-phase system.

three parallel current sources and an impedance equivalent to the three identical load impedances connected in parallel. If the three sources are added together

$$\frac{\mathbf{V}_{an}}{\mathbf{Z}} + \frac{\mathbf{V}_{bn}}{\mathbf{Z}} + \frac{\mathbf{V}_{cn}}{\mathbf{Z}} = \frac{1}{\mathbf{Z}}(\mathbf{V}_{an} + \mathbf{V}_{bn} + \mathbf{V}_{cn}) = 0.$$

This reduces the equivalent circuit to a parallel impedance. Consequently, the current in the fourth wire $\mathbf{I}_n$ must be zero. This path is called the **neutral path** because it connects two nodes at the same potential and carries no current—whatever the size of the impedance Z_n. In practice, the fourth wire is accomplished by grounding the transmission system and the load to earth, so that a path for current exists in the event that the system is slightly unbalanced.

When a balanced three-phase source is connected to a balanced load the circuit looks like three independent, single-phase circuits that are

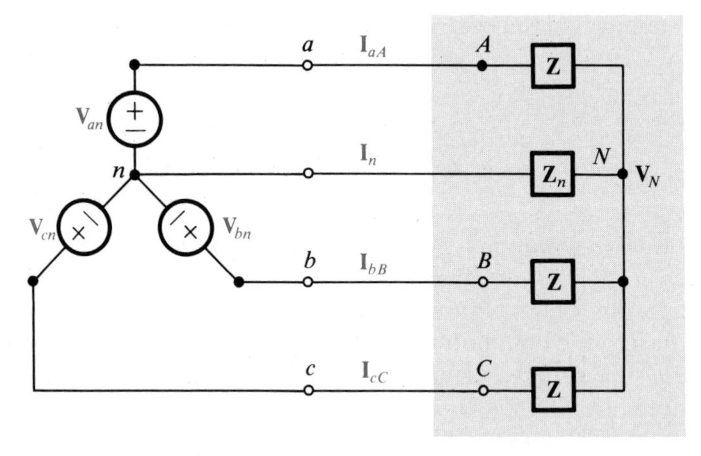

Figure 13.34 Wye- connected load with balanced impedances.

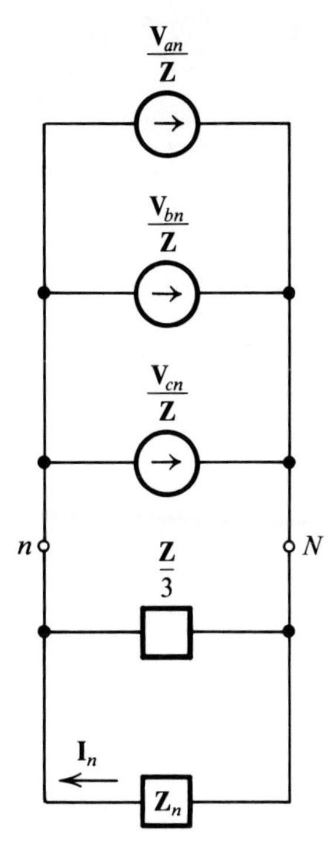

Figure 13.35 Norton equivalent circuit for a balanced three-phase system.

connected by a common node but have no effect on each other (because nodes n and N are at the same potential, though not necessarily physically connected). This is attractive because single-phase circuits are simpler to analyze. The line currents are given by

$$\mathbf{I}_{aA} = \mathbf{V}_{an}/\mathbf{Z} \qquad \mathbf{I}_{bB} = \mathbf{V}_{bn}/\mathbf{Z} \qquad \mathbf{I}_{cC} = \mathbf{V}_{cn}/\mathbf{Z}$$

Since $|\mathbf{I}_{aA}| = |\mathbf{I}_{bB}| = |\mathbf{I}_{cC}|$ and $\sphericalangle\mathbf{I}_{aA} - \sphericalangle\mathbf{I}_{bB} = +120°$ and $\sphericalangle\mathbf{I}_{aA} - \sphericalangle\mathbf{I}_{cC} = -120°$ we conclude that **the line currents are also balanced.** The line-current phasors lag the phasors of their respective phase voltages by the angle of the load impedance.

Example 13.15

If the loads in Fig. 13.34 have $\mathbf{Z} = 100 + j50$, find the average power dissipated when the sources have $V_m = 120$ volts.

Solution: With $\mathbf{V} = 120\,\underline{/0°}$ and $\mathbf{Z} = 111.8\,\underline{/26.57°}$ the line currents are given by

$$\mathbf{I}_{aA} = 1.07\,\underline{/-26.57°} \qquad \mathbf{I}_{bB} = 1.07\,\underline{/-146.57°}$$

$$\mathbf{I}_{cC} = 1.07\,\underline{/93.43°}$$

Since the average power is $P_{av} = 1/2\text{Re}\ \{\mathbf{VI}^*\}$ we have

$$P_a = 1/2\text{Re}\{(120\ \underline{/0°})\ (1.07\ \underline{/+26.57°})\} = 57.420\ \text{W}$$

$$P_b = 1/2\text{Re}\{(120\ \underline{/-120°})(1.07\ \underline{/146.57°})\} = 57.420\ \text{W}$$

$$P_b = 1/2\text{Re}\{(120\ \underline{/120°})(1.07\ \underline{/-93.43°})\} = 57.420\ \text{W}$$

In the preceding problem each phase dissipates the same average power. This is also true, in general, when a 3-ϕ (three-phase) system is balanced. It can further be shown that the total instantaneous power delivered by the sources is a constant.

Impedances connected to a three-phase source in the arrangement shown in Fig. 13.36 are said to be delta-connected (Δ-connected). The whole circuit is a Y–Δ three-phase circuit because the source is in a Y and the loads are in a Δ. Let's examine the relationship between the phase currents and the line currents at the load.

First, note that the phase voltages are the same as those for the Y-connected three-phase source

$$\mathbf{V}_{AB} = \sqrt{3}\ V_m\ \underline{/30°} \qquad \mathbf{V}_{BC} = \sqrt{3}\ V_m\ \underline{/-90°} \qquad \mathbf{V}_{CA} = \sqrt{3}\ V_m\ \underline{/210°}$$

The phase currents follow from Ohm's law:

$$\mathbf{I}_{AB} = \mathbf{V}_{AB}/\mathbf{Z} \qquad \mathbf{I}_{BC} = \mathbf{V}_{BC}/\mathbf{Z} \qquad \mathbf{I}_{CA} = \mathbf{V}_{CA}/\mathbf{Z}$$

Since the line voltages have equal magnitude, the phase currents will also have equal magnitude if the load impedances are identical. Thus, **if the line voltages are a balanced set the phase currents will be a balanced set.** From KCL the line currents are

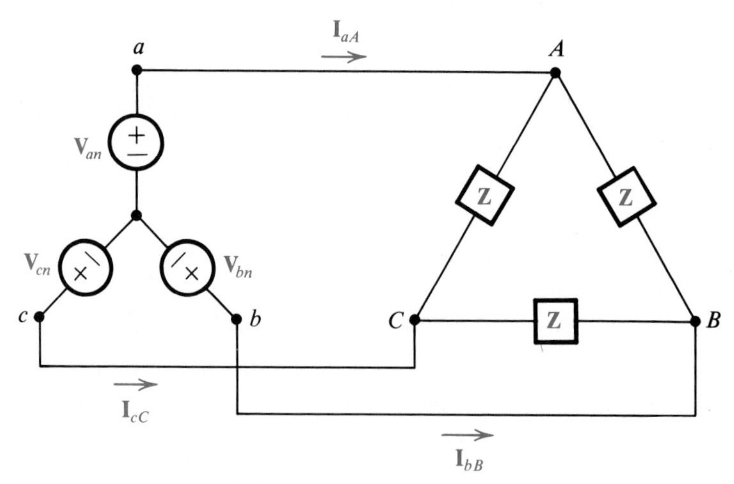

Figure 13.36 Y-to-Δ three-phase system.

$$\mathbf{I}_{aA} = \mathbf{I}_{AB} - \mathbf{I}_{CA} \qquad \mathbf{I}_{bB} = \mathbf{I}_{BC} - \mathbf{I}_{AB} \qquad \mathbf{I}_{cC} = \mathbf{I}_{CA} - \mathbf{I}_{BC}$$

Now consider

$$\begin{aligned} \mathbf{I}_{aA} &= 1/\mathbf{Z}[\mathbf{V}_{AB} - \mathbf{V}_{CA}] \\ &= 1/\mathbf{Z}(\sqrt{3}\, V_m \,\underline{/\,30^\circ} - \sqrt{3}\, V_m \,\underline{/\,-210^\circ}) \\ &= \sqrt{3}\, V_m/\mathbf{Z}(1 \,\underline{/\,30^\circ} - 1 \,\underline{/\,-210^\circ}) \\ &= 3\, V_m \,\underline{/\,0^\circ}\, /\mathbf{Z} \end{aligned}$$

Likewise, $\mathbf{I}_{bB} = \quad V_m \,\underline{/\,-120^\circ}\, /\mathbf{Z}$ and $\mathbf{I}_{cC} = \quad V_m \,\underline{/\,-240^\circ}\, /\mathbf{Z}$.

The phasors of the line currents have equal magnitude and are pairwise separated by 120°. Therefore, the line currents are a balanced set. Also note that

$$|\mathbf{I}_{aA}| = |\mathbf{I}_{bB}| = |\mathbf{I}_{cC}| = \quad V_m/|\mathbf{Z}|$$

while

$$|\mathbf{I}_{AB}| = |\mathbf{I}_{BC}| = |\mathbf{I}_{CA}| = \sqrt{3}\, V_m/|\mathbf{Z}|$$

So the magnitude of the line currents is $\sqrt{3}$ times the magnitude of the phase currents in a balanced 3-ϕ load. The line currents have

$$\measuredangle \mathbf{I}_{aA} = -\measuredangle \mathbf{Z} \qquad \measuredangle \mathbf{I}_{bB} = -120^\circ - \measuredangle \mathbf{Z} \qquad \measuredangle \mathbf{I}_{cC} = -240^\circ - \measuredangle \mathbf{Z}$$

and the phase currents have

$$\measuredangle \mathbf{I}_{AB} = 30^\circ - \measuredangle \mathbf{Z} \qquad \measuredangle \mathbf{I}_{BC} = -90^\circ - \measuredangle \mathbf{Z} \qquad \measuredangle \mathbf{I}_{CA} = -210^\circ - \measuredangle \mathbf{Z}$$

Consequently, the set of phasors of the phase currents in Fig. 13.37 *leads* the phasors of the line currents by 30°. **If balanced line voltages drive a balanced Δ load, the line currents will be balanced.**

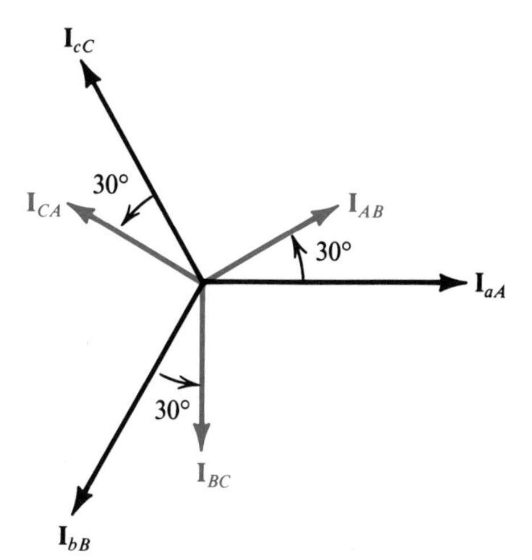

Figure 13.37 Phasors of line currents and phase currents for a balanced three-phase Δ-connected load.

Example 13.16

If each phase of the Δ-connected balanced load in Fig. 13.36 dissipates 25 kW at a lagging power factor of 0.9, find the line currents if the magnitude of each phase voltage is 220 V_{rms}.

Solution: The average power is given by $P_{av} = I_{rms}V_{rms}\cos\phi$, so the load current must have a magnitude given by

$I_{rms} = 2.5 \times 10^3/(220 \times 0.9) = 12.63\ A_{rms}$

Since the load is balanced, the phase currents will also be balanced, and

$\mathbf{I}_{ab} = 12.63\ \measuredangle\mathbf{Z} \qquad \mathbf{I}_{bc} = 12.63\ \measuredangle\mathbf{Z} - 120°$

$\mathbf{I}_{ca} = 12.63\ \measuredangle\mathbf{Z} - 240°$

The line current lags the phase current by 30°, and $\phi = \cos^{-1}(0.9) = 25.84°$, so $\measuredangle\mathbf{I}_{ab} = 25.84°$, $\measuredangle\mathbf{I}_{bc} = -94.16°$ and $\measuredangle\mathbf{I}_{ca} = -214.16°$

In practice, Δ connections of three-phase sources are not as common as Y connections because high currents can circulate in the Δ source loop when the loads are not balanced, thereby wasting energy. It is possible to show that a wye load has an equivalent delta load, and vice versa. Thus, a Y–Y source-load connection can be used to analyze a Y–Δ connection. So it is important to know how to convert from one configuration to the other.

13.12.1 Δ–Y Impedance Conversion

First, we'll show how to find a Y connection that is equivalent to a given Δ connection. If the two circuits in Fig. 13.38 are equivalent when they are connected to a circuit, they must also have the same impedances between corresponding pairs of terminals when the remaining terminal is not connected to an external circuit. Therefore

$$\mathbf{Z}_A + \mathbf{Z}_B = \frac{\mathbf{Z}_{AB}(\mathbf{Z}_{BC} + \mathbf{Z}_{CA})}{\mathbf{Z}_{AB} + \mathbf{Z}_{BC} + \mathbf{Z}_{CA}}$$

$$\mathbf{Z}_B + \mathbf{Z}_C = \frac{\mathbf{Z}_{BC}(\mathbf{Z}_{AB} + \mathbf{Z}_{CA})}{\mathbf{Z}_{AB} + \mathbf{Z}_{BC} + \mathbf{Z}_{CA}}$$

$$\mathbf{Z}_C + \mathbf{Z}_A = \frac{\mathbf{Z}_{CA}(\mathbf{Z}_{AB} + \mathbf{Z}_{BC})}{\mathbf{Z}_{AB} + \mathbf{Z}_{BC} + \mathbf{Z}_{CA}}$$

If the second equation is subtracted from the sum of the other two we get

$$\mathbf{Z}_A = \frac{\mathbf{Z}_{AB}\mathbf{Z}_{CA}}{\mathbf{Z}_{AB} + \mathbf{Z}_{BC} + \mathbf{Z}_{CA}}$$

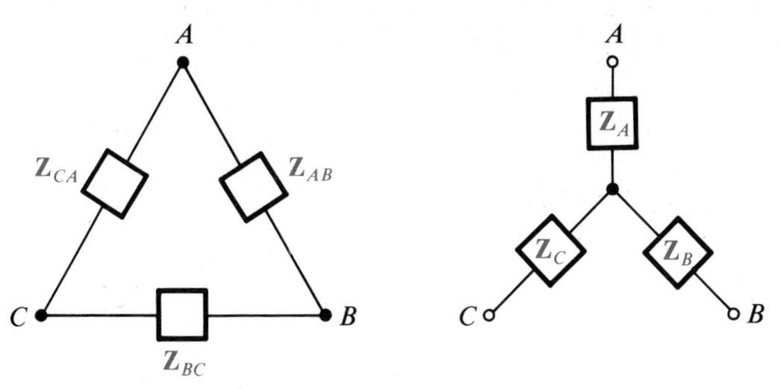

Figure 13.38 Δ-to-Y impedance conversion.

In a similar way we can find

$$\mathbf{Z}_B = \frac{\mathbf{Z}_{BC}\mathbf{Z}_{AB}}{\mathbf{Z}_{AB} + \mathbf{Z}_{BC} + \mathbf{Z}_{CA}} \qquad \mathbf{Z}_C = \frac{\mathbf{Z}_{CA}\mathbf{Z}_{BC}}{\mathbf{Z}_A + \mathbf{Z}_B + \mathbf{Z}_C}$$

13.12.2 Y–Δ Admittance Conversion

To convert a Y circuit to an equivalent Δ circuit, we argue that if the two circuits in Fig. 13.39 are equivalent, they must have the same admittance between any pair of terminals when the remaining terminal is shorted to one member of the pair. If we short node c to node a and measure the admittance between a and b we get

$$\mathbf{Y}_{AB} + \mathbf{Y}_{BC} = \frac{\mathbf{Y}_B(\mathbf{Y}_C + \mathbf{Y}_A)}{\mathbf{Y}_A + \mathbf{Y}_B + \mathbf{Y}_C}$$

Likewise, shorting node a to node b and measuring the admittance between b and c gives

$$\mathbf{Y}_{BC} + \mathbf{Y}_{CA} = \frac{\mathbf{Y}_C(\mathbf{Y}_A + \mathbf{Y}_B)}{\mathbf{Y}_A + \mathbf{Y}_B + \mathbf{Y}_C}$$

Then we short b to c and measure the admittance between c and a

$$\mathbf{Y}_{CA} + \mathbf{Y}_{AB} = \frac{\mathbf{Y}_A(\mathbf{Y}_B + \mathbf{Y}_C)}{\mathbf{Y}_A + \mathbf{Y}_B + \mathbf{Y}_C}$$

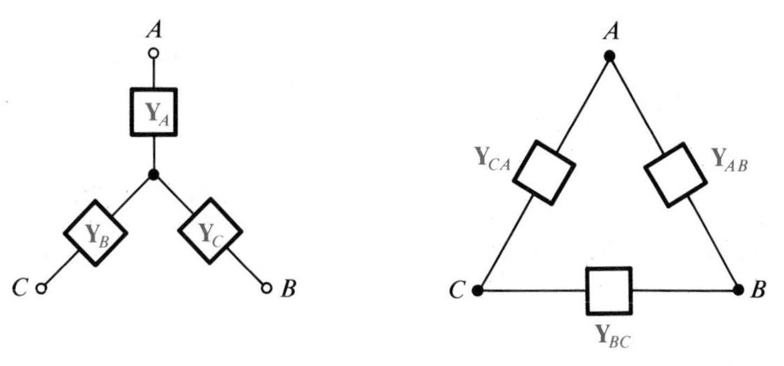

Figure 13.39 Y-to-Δ admittance conversion.

If the second equation is subtracted from the sum of the other two, for example, we get

$$\mathbf{Y}_{AB} = \frac{\mathbf{Y}_A\mathbf{Y}_B}{\mathbf{Y}_A + \mathbf{Y}_B + \mathbf{Y}_C}$$

$$\mathbf{Y}_{BC} = \frac{\mathbf{Y}_B\mathbf{Y}_C}{\mathbf{Y}_A + \mathbf{Y}_B + \mathbf{Y}_C}$$

$$\mathbf{Y}_{CA} = \frac{\mathbf{Y}_C\mathbf{Y}_A}{\mathbf{Y}_A + \mathbf{Y}_B + \mathbf{Y}_C}$$

Lastly, we take reciprocals of these expressions to obtain the equivalent impedances for Y–Δ conversion

$$\mathbf{Z}_{AB} = \frac{\mathbf{Z}_A\mathbf{Z}_B + \mathbf{Z}_B\mathbf{Z}_C + \mathbf{Z}_C\mathbf{Z}_A}{\mathbf{Z}_C}$$

$$\mathbf{Z}_{BC} = \frac{\mathbf{Z}_A\mathbf{Z}_B + \mathbf{Z}_B\mathbf{Z}_C + \mathbf{Z}_C\mathbf{Z}_A}{\mathbf{Z}_A}$$

$$\mathbf{Z}_{CA} = \frac{\mathbf{Z}_A\mathbf{Z}_B + \mathbf{Z}_B\mathbf{Z}_C + \mathbf{Z}_C\mathbf{Z}_A}{\mathbf{Z}_B}$$

So, given a Y-connected load, its equivalent Δ-connected load can be determined with these equations.

13.13 TRANSFORMERS

Efficient power-distribution and transmission systems usually operate at voltages that greatly exceed the voltages needed to operate loads. For example, a city's electrical utility might distribute voltages at a level that is a hundred times greater than needed to operate domestic appliances, but only ten times the level needed to operate industrial motors. Transformers provide an interface between the available source voltage or current and that which is required by a load. They make it possible for a common distribution system to efficiently service a variety of loads.

A **transformer** is formed when two coils of wire are sufficiently close to each other to share common flux linkages. Under these conditions the electrical characteristics of the two coils are mutually dependent on the common magnetic flux that links them. Such a configuration is represented by the circuit symbol shown in Fig. 13.40, where by convention the left coil is called the **primary,** and the right coil is called the **secondary.**

The labels on the transformer symbol have been adopted as a convention to indicate how the orientation of the mutual flux depends on the direction of their coil windings. The dots indicate which terminals the current must enter on each side of the transformer in order to have the mutual flux aligned in the same direction as the self-induced flux in each

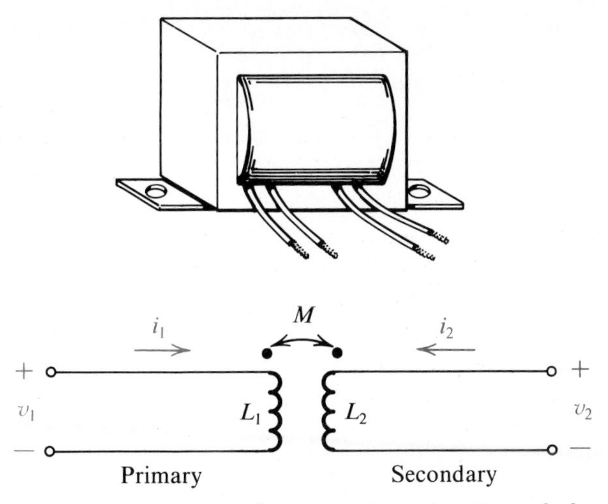

Figure 13.40 A transformer and its circuit symbol.

side of the coil. If both currents enter dotted terminals, or if neither enters a dotted terminal, the common-flux linkages are in the same direction as the flux linkages on each side of the coil. In this case the time-domain model of the transformer is given by a pair of coupled differential equations:

$$v_1(t) = L_1\frac{di_1}{dt} + M\frac{di_2}{dt}$$

$$v_2(t) = L_2\frac{di_2}{dt} + M\frac{di_1}{dt}$$

On the other hand, if one of the currents enters a dotted terminal and the other leaves a dotted terminal the direction of the common-flux linkages is opposite to the direction of the self-induced flux linkages, and the model equations must be written as

$$v_1(t) = L_1\frac{di_1}{dt} - M\frac{di_2}{dt}$$

$$v_2(t) = L_2\frac{di_2}{dt} - M\frac{di_1}{dt}$$

Physically, applying a voltage $v_1(t)$ will cause i_1 to change and will thereby cause flux linkages that induce a voltage v_2 in the secondary circuit. Likewise, applying v_2 will ultimately induce a voltage in the primary circuit. The mutual inductance M determines the electrical effect caused by the magnetic coupling between the primary and secondary coils of the transformer. The key point (which can be demonstrated from basic physics) is that ***M* appears with the same value in both equations.**

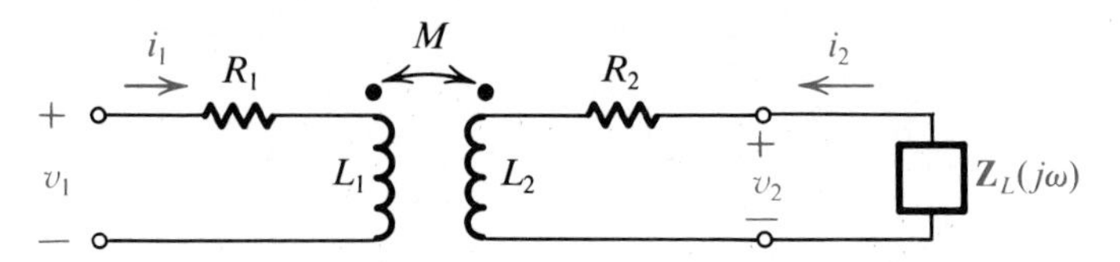

Figure 13.41 Transformer with a load impedance.

A more complete model of a transformer accounts for the resistance in the wire of the primary and secondary coils. Such a model is shown in Fig. 13.41, where R_1 and R_2 are the winding resistances of the primary and secondary, respectively.

The frequency-domain model for the sinusoidal steady-state behavior of the loaded transformer is obtained from

$$\mathbf{V}_1 = \mathbf{I}_1(R_1 + j\omega L_1) + j\omega M\mathbf{I}_2 \tag{13.20a}$$

and

$$\mathbf{V}_2 = \mathbf{I}_2(R_2 + j\omega L_2) + j\omega M\mathbf{I}_1 \tag{13.20b}$$

The load imposes the constraint that $\mathbf{V}_2 = -\mathbf{I}_2\mathbf{Z}_L$ so solving for $\mathbf{I}_2$ gives

$$\mathbf{I}_2 = \frac{-j\omega M\mathbf{I}_1}{R_2 + j\omega L_2 + \mathbf{Z}_L} \tag{13.21}$$

Substituting this expression into (13.20a) gives

$$\mathbf{V}_1 = \mathbf{I}_1(R_1 + j\omega L_1) + \frac{\omega^2 M^2}{R_2 + j\omega L_2 + \mathbf{Z}_L}\mathbf{I}_1 \tag{13.22}$$

Equation (13.21) specifies the secondary current (phasor) for a given source current (phasor), and (13.22) specifies the primary voltage as a function of the primary current (or vice versa). Equation (13.21) also yields an expression

$$\mathbf{Z}_{\text{in}} = R_1 + j\omega L_1 + \frac{\omega^2 M^2}{R_2 + j\omega L_2 + \mathbf{Z}_L} \tag{13.23}$$

for the input impedance of the transformer.

Writing the load impedance as $Z_L = R_L + jX_L$ and multiplying the numerator and denominator of the third term in (13.23) by the complex conjugate of its denominator gives

$$\mathbf{Z}_{\text{in}} = R_1 + j\omega L_1 + \mathbf{Z}_r$$

where the impedance $\mathbf{Z}_r$ is "reflected" from the secondary

$$\mathbf{Z}_r = \frac{\omega^2 M^2[(R_2 + R_L) - j(\omega L_2 + X_L)]}{(R_2 + R_L)^2 + (\omega L_2 + X_L)^2}$$

The input impedance of the loaded transformer is equal to the sum of the "self-impedance" of its primary circuit plus a reflected secondary impedance. The reflected impedance is equal to the conjugate of the loaded secondary circuit's self-impedance $(R_2 + R_L + j\omega L_2 + jX_L)$ scaled by the factor $\omega^2 M^2/[(R_2 + R_L)^2 + (\omega L_2 + X_L)^2]$. As far as the source is concerned, the electrical behavior of the primary side of the transformer is governed by $\mathbf{Z}_{in}$.

Example 13.17

If $v_1(t) = 240 \cos(377t + 20°)\, u(t)$ find the steady-state value of $i_1(t)$ in Fig. 13.42.

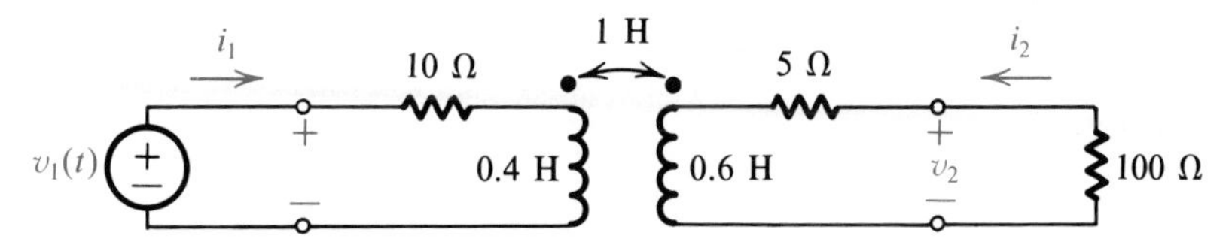

Figure 13.42 Circuit for Example 13.22.

Solution: Using (13.23) we get $\mathbf{Z}_{in} = 443.33 \angle -55.68°$ and so $i_1(t) = 0.54 \cos(377t + 75.68°)$.

13.14 IDEAL TRANSFORMERS

Physical transformers can be fabricated so that all but a negligible amount of the flux linkages entirely link both of the coils, and the magnetic path has a very low reluctance. Under these conditions the transformer model can be simplified to represent the fact that

$$\frac{v_1(t)}{v_2(t)} = -\frac{i_2(t)}{i_1(t)} = \frac{N_1}{N_2} = n$$

with N_1 and N_2 being the number of turns of the primary and secondary coils respectively, and n being the primary-to-secondary turns ratio (sometimes denoted as n:1). This relationship only applies when the currents, voltages, and dots are as in Fig. 13.43. Otherwise, the minus sign must be changed to a plus sign.

Ideal transformers are used in circuits that require a voltage or a current other than what is available from given source. By proper choice of n, a voltage gain can be developed between the primary and secondary sides of the transformer, and likewise for current. One application is to step down a household supply voltage to produce the relatively low voltage needed by the doorbell circuit. In contrast, a microwave oven needs a voltage source that is much greater than the domestic service. It uses a step-up transformer to create the high voltage needed by the microwave

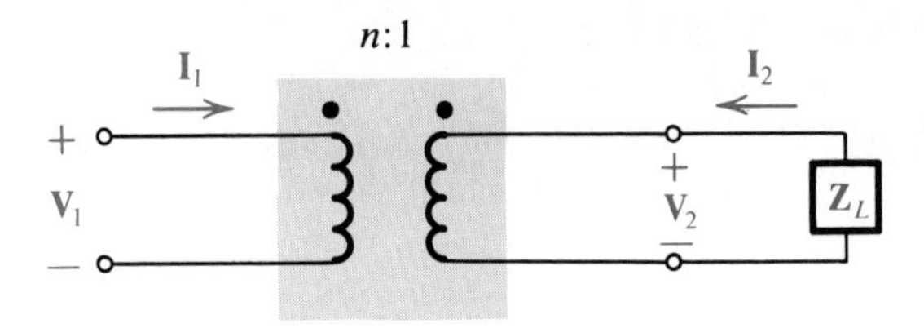

Figure 13.43 Ideal transformer with a load impedance.

tube. A step-down transformer will have $n > 1$, and a step-up transformer will have $n < 1$. When voltage is stepped up, the current is stepped down, and vice versa.

Skill Exercise 13.10

Find the turns ratio n to convert $115\mathbf{V}_{\text{rms}}$ to $10\mathbf{V}_{\text{rms}}$ for a household doorbell circuit.

Answer: $n = 11.5$

The instantaneous power delivered to the primary side of an ideal transformer is

$$P_{\text{in}}(t) = v_1(t)i_1(t)$$
$$= -\frac{N_1}{N_2}v_2(t)\,\frac{N_2}{N_1}\,i_2(t) = -v_2(t)\,i_2(t)$$

All the power delivered to the primary circuit is supplied to whatever load is connected to the secondary circuit. The sign change accounts for the defined polarity of v_2 relative to i_2.

If a load is connected to the secondary, we have

$$\mathbf{Z}_{\text{in}} = \frac{\mathbf{V}_1}{\mathbf{I}_1} = \frac{n\mathbf{V}_2}{(-1/n)\mathbf{I}_2} = -n^2\frac{\mathbf{V}_2}{\mathbf{I}_2} = -n^2(-\mathbf{Z}_L) = n^2\mathbf{Z}_L$$

An ideal transformer "reflects" its load impedance back to the source with a scale factor of n^2. Transformers are often used to match a resistive load to the resistance of the source to achieve maximum power transfer. If the source has a resistance given by R_g, then the turns ratio should be chosen so that $R_g = n^2R_L$, or

$$n^2 = R_g/R_L$$

SUMMARY

When a circuit driven by a sinusoidal source reaches the steady state, every voltage and current in the circuit is a sinusoidal signal. The amplitude and phase angle of each signal are determined by its transfer function and the sinusoidal source. The frequency of each sinusoid is the same as the frequency of the source.

If a circuit contains more than one source (having different frequencies), the signals in the circuit consist of the sum of the sinusoidal responses caused by each of the sources. The transfer function between a given source signal and the output (measured) signal determines the component due to that source signal.

A circuit in the steady state can be analyzed by forming the phasor model of the circuit for a given source frequency and writing KVL and KCL for the sinusoidal steady-state voltage and current phasors. Phasor models describe the circuit at a single frequency.

The average power dissipated by a circuit depends on the phase-angle difference between its time-domain voltage and current waveforms at the input terminals. The difference between these phase angles is also the angle of the input impedance, and the difference in angle between the input-voltage phasor and the input-current phasor. A load absorbs maximum average power from a source when its impedance is the complex conjugate of the source's impedance.

Electric utilities usually generate balanced three-phase signals because (a) such signals can be generated more efficiently; (b) they can deliver constant, total, instantaneous power to a balanced load; and (c) their distribution requires a smaller investment in wire to deliver the same level of power to a load. The line-to-line voltages in a Y configuration are $\sqrt{3}$ times the phase voltages of a Y source. Conversely, the line currents to a Δ load are $\sqrt{3}$ times the load's phase currents.

Transformers use magnetic coupling between two adjacent coils of wire to create a device whose primary/secondary voltage ratio, and secondary/primary current ratio is related to the secondary/primary turns ratio. Ideal transformers are used to convert voltage or current from one level to another without a loss in power.

Problems - Chapter 13

13.1 The phasor model of a parallel *RLC* circuit is shown in Fig. P13.1.

a. Find the phasors $\mathbf{I}_1$, $\mathbf{I}_2$ and $\mathbf{I}_3$.

b. If the circuit's indicated impedance values were calculated for a sinusoidal current source having $i_g(t) = 10\cos(120\pi t)\,u(t)$, find expressions for the steady state currents $i_1(t)$, $i_2(t)$ and $i_3(t)$.

c. Repeat a and b with the capacitor value doubled.

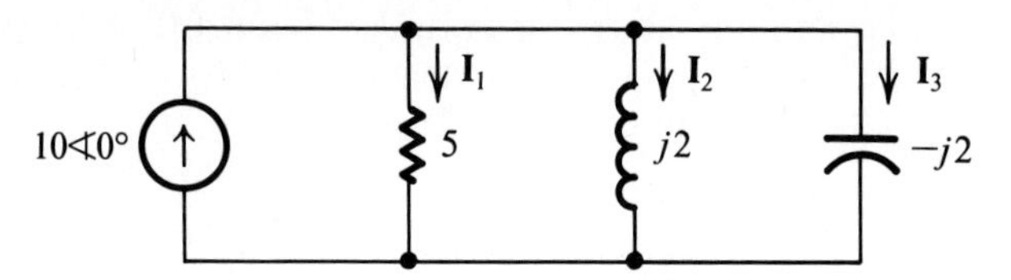

Figure P13.1

13.2 Draw the phasors found in parts a and c of Problem 13.1 and verify that they satisfy Kirchhoff's current law with the source phasor. Make a sketch to indicate how you would expect the magnitude and angle of each phasor to change if the source frequency were increased, for the same circuit parameters used in a.

13.3 Find the phasors $\mathbf{V}_R$, $\mathbf{V}_L$ and $\mathbf{V}_c$, and verify that they satisfy Kirchhoff's voltage law with the source.

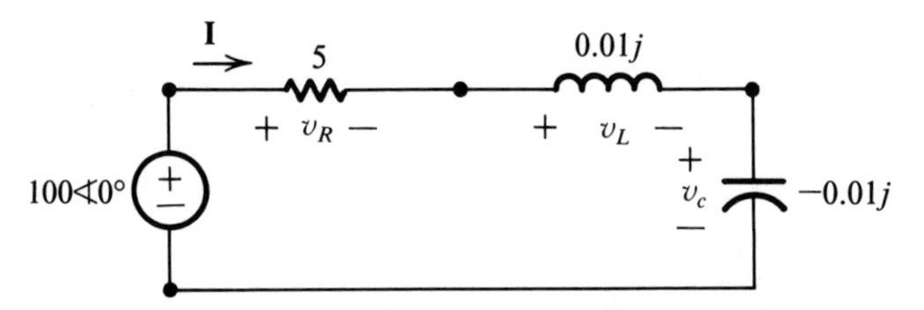

Figure P13.3

13.4 If the voltage source in Problem 13.3 is given by $v_{in}(t) = 100 \sin(400t + 30°)\, u(t)$, find $v_R(t)$, $v_L(t)$, and $v_c(t)$ in the steady state.

13.5 Calculate the impedance seen by the current source in Problem 13.1.

13.6 If the op amp circuit in Fig. P11.25b has $R_1 = 2\ \Omega$, $R_2 = 4\ \Omega$, $C_1 = 0.125$ F and $C_2 = 0.5$ F, find $v_o(t)$ in the steady state.

13.7 Find $i(t)$, $i_1(t)$, $i_2(t)$, $v_1(t)$ and $v_2(t)$ in the steady state if the source frequency is 60Hz.

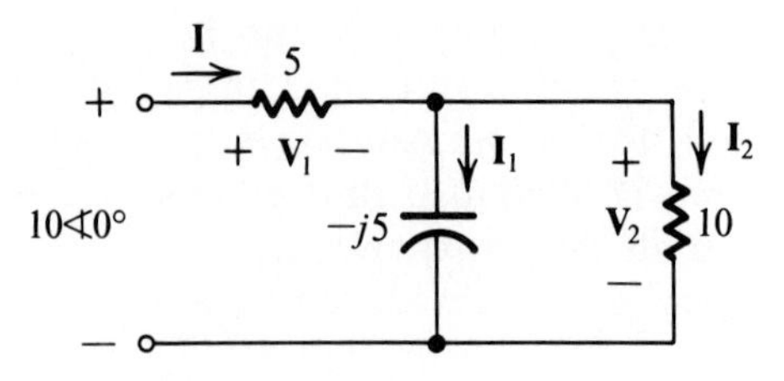

Figure P13.7

13.8 Find the phasors and the steady-state signals they represent in Fig. P13.8.

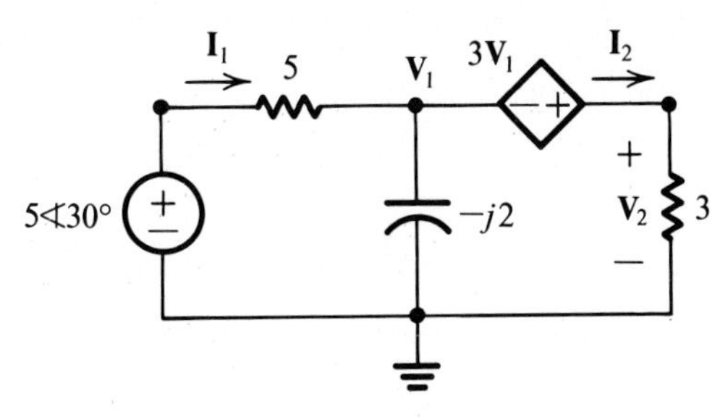

Figure P13.8

13.9 If the source frequency is 10 kHz in Fig. P13.9, find the steady-state value of $v_o(t)$.

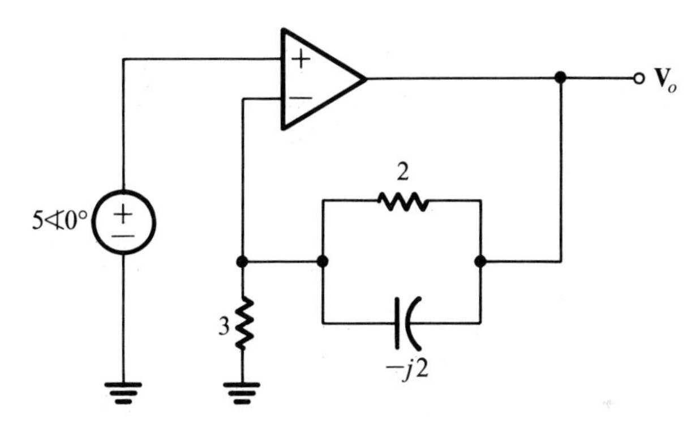

Figure P13.9

13.10 Find the average power delivered to the circuit by the current source in Problem 13.1.

13.11 Find the average power delivered to the circuit by the voltage source in Problem 13.3.

13.12 Find the Thevenin equivalent circuit seen by the source in Fig. P13.7.

13.13 Find the average power delivered by the source in Fig. P13.7.

13.14 If $R_1 = 2\ \Omega$, $R_2 = 4\ \Omega$, $R_3 = 5\ \Omega$, $C_1 = 1/3$ F and $C_2 = 0.5$ F, find the steady state value of $v_o(t)$ when $v_{in}(t) = 10\cos(6t + 30°)\, u(t)$ in Fig. 12.11.

13.15 Find the Thevenin equivalent circuit at (a, b) in Fig. P13.15. If a source is applied with $v_{in}(t) = 120 \cos 20t\, u(t)$, find $i_{in}(t)$ in the steady state.

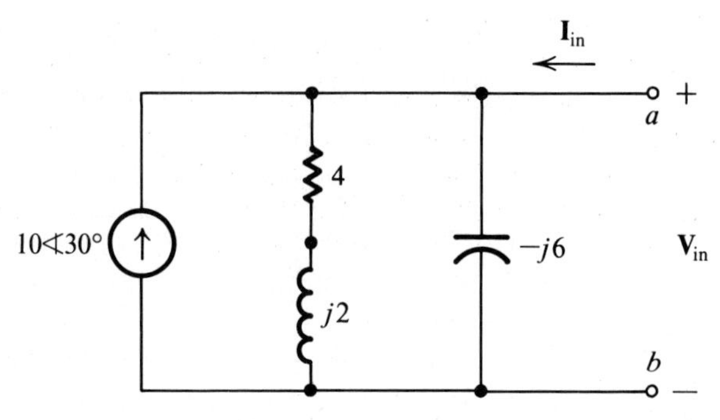

Figure P13.15

13.16 Find the Thevenin equivalent of the circuit "seen" by the 3 Ω resistor in Fig. P13.8.

13.17 Use the matrix nodal-analysis model to find $v_1(t)$ and $v_2(t)$ in the steady state when $i_{in1}(t) = 10 \cos 2t\, u(t)$ and $i_{in2}(t) = 4 \sin 2t\, u(t)$.

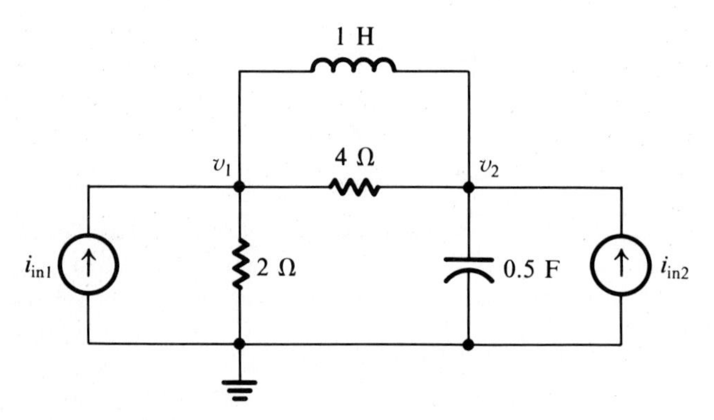

Figure P13.17

13.18 Repeat Problem 13.17 with $i_{in2}(t)$ changed to $i_{in2}(t) = 4 \sin 4t\, u(t)$.

13.19 If $v_{in}(t) = 10 \cos 10t\, u(t)$ find $v_1(t)$, $i(t)$ and $v(t)$ in the steady state. Find the average power delivered by the voltage source, and the average power dissipated by each of the circuit elements.

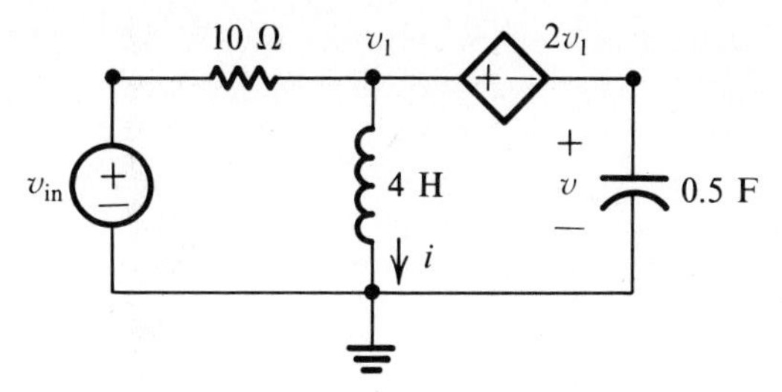

Figure P13.19

13.20 Calculate the power factor seen by a voltage source operating at a frequency of 60 Hz when it is connected across the terminals of the circuit in Fig. P13.19. What reactance should be connected in parallel with the source to change the power factor to unity?

13.21 If the source in Fig. P13.21 is operating at a frequency of 60 Hz, find $i(t)$ and $v(t)$ in the steady state, and the average power dissipated by the resistor.

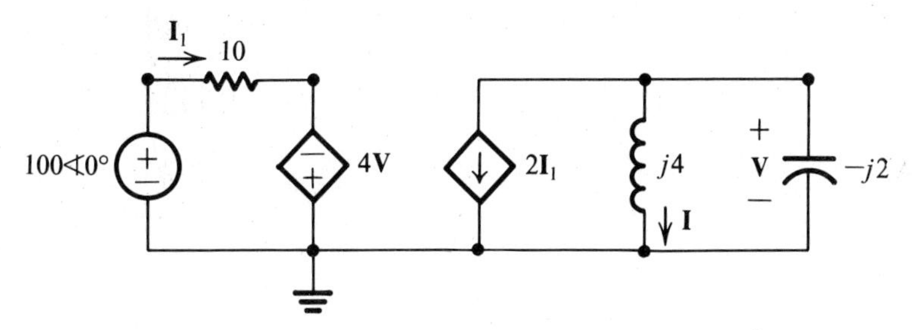

Figure P13.21

13.22 If a series configuration of a resistor, a capacitor, and a load are connected to a voltage source, what value of load impedance will dissipate the maximum average power? At the frequency of the source the resistor has $R = 100\ \Omega$ and the capacitor has a reactance of $-j40$.

13.23 A load is rated to dissipate 2500 watts at a lagging power factor of 0.8 with a line voltage of 220 V_{rms}.

a. Find the current supplied to the load.

b. Find the value of the capacitor that will create a unity power factor for the parallel impedance created by the load and the capacitor.

13.24 If the transformer shown in Fig. P13.24 is connected to a 500 Ω load, find the impedance seen by the source.

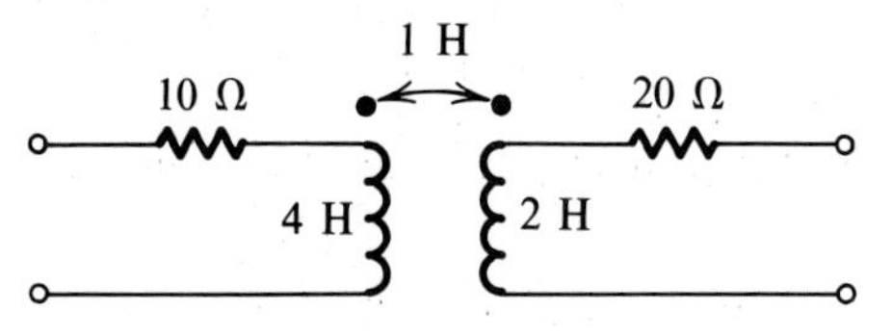

Figure P13.24

13.25 Find the average power dissipated by the load in Fig. P13.25.

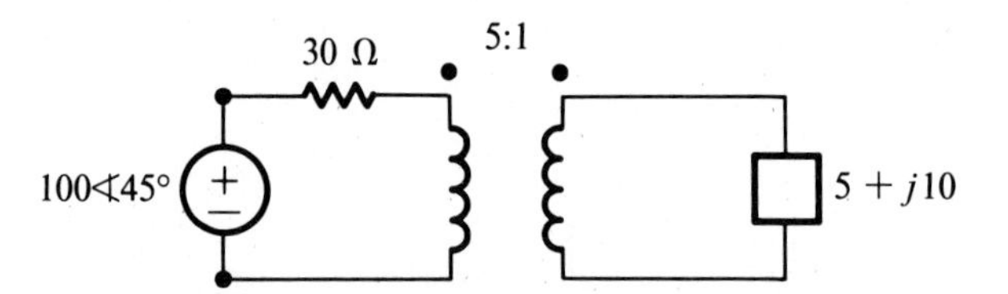

Figure P13.25

13.26 What is the impedance seen by the voltage source in Fig. P13.25?

13.27 If a 1000 V_{rms} source is to supply a load that requires a voltage of 25 V_{rms}, specify a transformer that will allow the source to be used.

13.28 Find the effective turns ratio between v_1 and v_2 in Fig. P13.28.

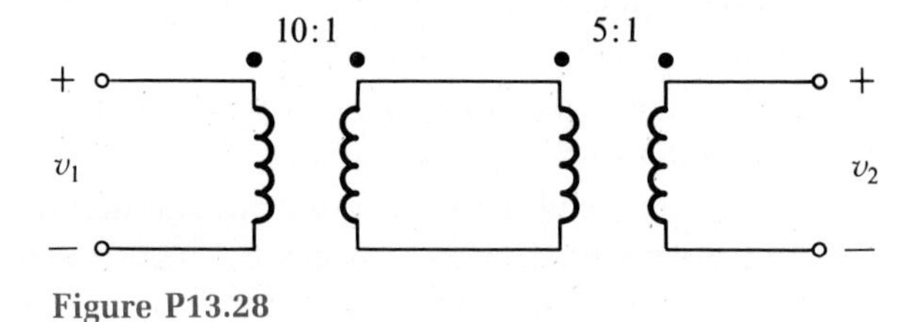

Figure P13.28

13.29 Find the impedance seen by $\mathbf{V}_{in}$ in Fig. P13.29.

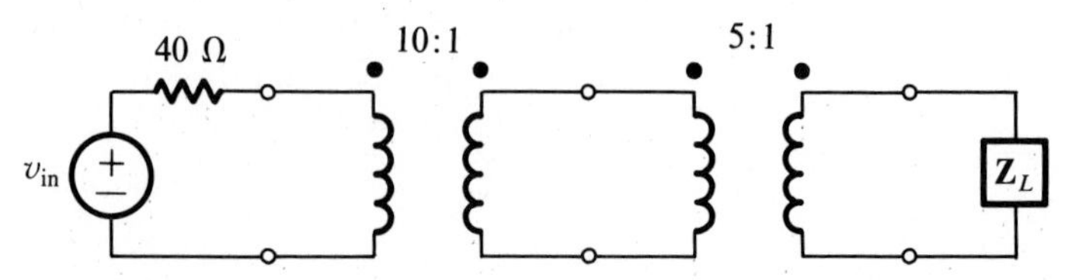

Figure P13.29

13.30 For the ideal transformer circuit in Fig. P13.30, find

a. the value of "a" to maximize the power delivered to the 200 Ω load.

b. **v** when the power to the 200 Ω load resistor is a maximum.

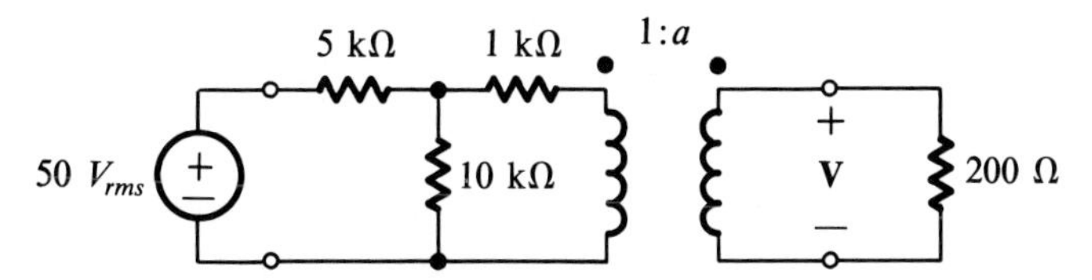

Figure P13.30

13.31 A balanced three-phase voltage source is connected in a Y configuration, with steady state phase voltages given by

$$v_{an}(t) = 170 \sin(377t + 40°)$$
$$v_{bn}(t) = 170 \sin(377t - 80°)$$
$$v_{cn}(t) = 170 \sin(377t - 200°)$$

Draw a phasor diagram for the phase and line voltages.

13.32 If the source in Problem 13.31 has an impedance $\mathbf{Z} = 2 + j$ in each phase, and is connected to a Δ load with each phase having $\mathbf{Z}_L = 2 + j4$, find and draw the phasors of the line and phase current at the load if the line impedance is 5 Ω.

13.33 A balanced Y-connected three-phase source delivers a total average power of 2800 watts to a Δ-connected load at a lagging power factor of 0.75. If the line voltage is 220 V_{rms}, find the phase-voltage phasors of the source, the line-current phasors, and the load-current phasors.

13.34 Find the Y equivalent of a Δ-connected balanced load having $\mathbf{Z}_L = 10 + j5$ in each branch.

13.35 The phase voltages of a balanced three-phase Δ-connected load are 220 V, and the line currents are 10 A. Find the impedance in each phase of the load, and find the power factor at the load.

THE FOURIER SERIES

INTRODUCTION

The previous chapters developed methods to analyze circuits whose inputs are exponential signals or signals that can be modeled by complex exponentials, such as sinusoids and steps. When a signal is not one of those already discussed, how do we determine the circuit's response? Do radically different methods have to be used or can the methods that have served us so far be extended beyond their present applications?

This chapter will show that *any* periodic signal (not just sines and cosines) is equivalent to a weighted discrete sum of complex exponentials. Then we'll use superposition and transfer functions to find a circuit's response to periodic signals—all because we know how a circuit responds to an exponential input signal!

14.1 PERIODIC SIGNALS

A signal $f(t)$ is said to be **periodic** if there is some value of time T such that $f(t) = f(t + nT)$ for all t. When a signal is periodic it cannot be distinguished from its shifted replica $f(t + T)$ or from $f(t + nT)$ when n is an integer. The smallest positive value of T for which a signal is periodic is called its period T.

Skill Exercise 14.1

Find the period of the signal shown in Figure SE14.1.

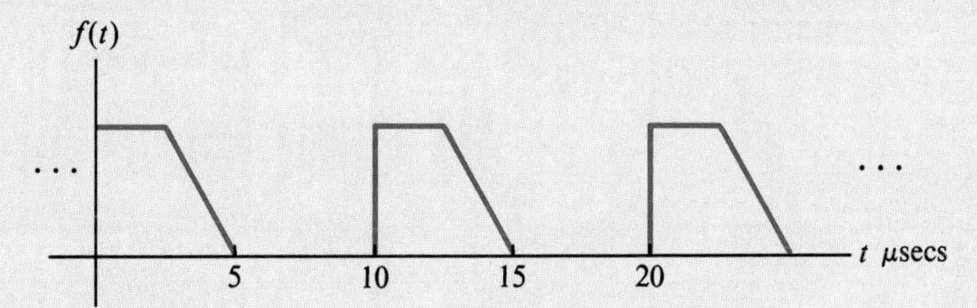

Figure SE14.1

Answer: $T = 10\ \mu\text{sec}$

14.2 THE PHASOR SIGNAL SET

In Chapter 10 we saw that a sinusoidal signal is periodic, and that it is exactly equal to the weighted sum of two phasor signals (complex exponential signals):

$$\sin \omega_o t = \frac{1}{2j} e^{j\omega_o t} - \frac{1}{2j} e^{-j\omega_o t}$$

A sinusoidal signal can also be written as a finite series, or discrete sum, of phasor signals as

$$\sin \omega_o t = \sum_{\substack{n=-1 \\ n\neq 0}}^{n=+1} \frac{1}{2jn} e^{jn\omega_o t}$$

The finite sum is formed by evaluating the complex exponentials and their weights with $n = \pm 1$ and adding the results. If we expand the evaluation limits of the sum to include all integers, the same signal can also be described by an infinite series

$$\sin \omega_o t = \sum_{n=-\infty}^{\infty} \mathbf{F}_n e^{jn\omega_o t}$$

provided that the complex weighting coefficient $\mathbf{F}_n$ is chosen exactly according to

$$\mathbf{F}_n = \begin{cases} \dfrac{1}{2jn} & n = \pm 1 \\ 0 & n \neq \pm 1 \end{cases}$$

Adding phasor signals with this choice of coefficients will create a sinusoid. By analogy, the flavor of a salad depends on the kind and amounts of ingredients that are mixed together to create it, and various mixtures create distinct flavors. Here, the ingredients of the function $f(t) = \sin \omega_o t$ are the two phasor signals $g_1(t) = e^{j\omega_o t}$ and $g_{-1}(t) = e^{-j\omega_o t}$ and their relative amounts are $\mathbf{F}_1 = \frac{1}{2j}$ and $\mathbf{F}_{-1} = \frac{-1}{2j}$.

Skill Exercise 14.2

Find coefficients $\mathbf{F}_n$ to express $\cos \omega_o t$ as a sum of phasor signals.

Answer: $\mathbf{F}_n = \frac{1}{2}$ for $n = \pm 1$ and $\mathbf{F}_n = 0$ for $n \neq \pm 1$.

The exponentials used to form $\sin \omega_o t$ are members of a family of signals, called the **phasor signal set** having the form $g_n(t) = e^{jn\omega_o t}$ for $n = 0, \pm 1, \pm 2, \pm 3 \ldots$. The phasor signal set has several important properties:

1. Each member of the phasor signal set is a complex periodic signal with $|e^{jn\omega_o t}| = 1$ and $\angle e^{jn\omega_o t} = n\omega_o t$. This conclusion results from Euler's law* and the fact that both $\cos n\omega_o t$ and $\sin n\omega_o t$ are periodic. The magnitude of $e^{j\omega_o t}$ was shown to be 1 in Chapter 10, and the angle of the signal is just
$$\angle e^{j\omega_o t} = \tan^{-1} \frac{\sin \omega_o t}{\cos \omega_o t} = \omega_o t$$
If we define $T = 2\pi/\omega_o$, then the radian frequency ω_n and period T_n of the nth phasor signal are given by $\omega_n = n\omega_o$ and $T_n = 2\pi/n\omega_o = T/n$.
2. The signals in the phasor signal set are said to be **harmonically related** because their frequencies are integer multiples of the fundamental frequency ω_o and their period is an integer fraction of the fundamental period (i.e., the fundamental period is n times the period of the nth signal). The signals $e^{jn\omega_o t}$ with $n \neq 1$ are called the **harmonics** of $e^{j\omega_o t}$.
3. The real parts ($\cos n\omega_o t$) of the phasor signals are harmonically related. In the same way, the imaginary parts of the phasor signals are harmonically related to the fundamental signal $\sin \omega_o t$ (as shown in Figure 14.1).
4. The phasor signals are orthogonal over any period T. (The significance of orthogonality is treated in Sections 14.3 and 14.5.4)

---

*$e^{jn\omega_o t} = \cos n\omega_o t + j \sin n\omega_o t$

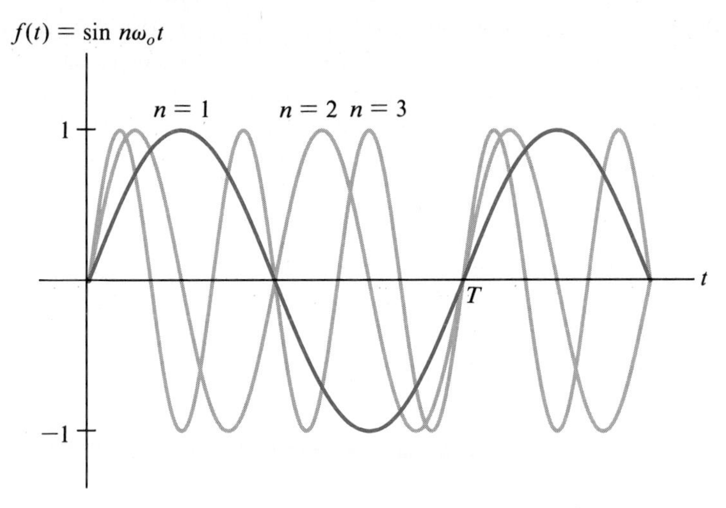

Figure 14.1 Harmonically related sinusoidal signals.

14.3 COMPLEX FOURIER SERIES

The Fourier theorem is a mathematical property of periodic signals with widespread application in engineering. It states that a periodic signal can be exactly represented by an infinite sum of phasor signals (provided it satisfies the Dirchlet conditions—which we will discuss shortly). If $f(t)$ is periodic, there are numbers $\mathbf{F}_n$ such that

$$f(t) = \sum_{n=-\infty}^{\infty} \mathbf{F}_n e^{jn\omega_o t} \tag{14.1}$$

The infinite sum is called the **complex Fourier series** of $f(t)$ and the phasor amplitudes $\mathbf{F}_n$ are called the complex Fourier coefficients of the signal.

The Fourier coefficients are defined by

$$\mathbf{F}_n = \frac{1}{T}\int_T f(t)e^{-jn\omega_o t}dt \qquad n = 0, \pm 1, \pm 2, \ldots \tag{14.2}$$

where the integration is over any interval of width T. The complex Fourier coefficients are phasors, and the sum of the phasor signals according to (14.1) produces the given periodic signal. To see that (14.2) properly defines the value of $\mathbf{F}_n$ used in (14.1), suppose that (14.1) is true for some choice of $\mathbf{F}_n$. Then, for the purpose of discussion, take an interval of evaluation of (14.1) as (0, T) and form the integral

$$\int_0^T f(t)e^{-jn\omega_o t}\,dt = \int_0^T \left[\sum_{k=-\infty}^{\infty} \mathbf{F}_k e^{jk\omega_o t}\right] e^{-jn\omega_o t}\,dt$$

$$= \int_0^T \sum_{k=-\infty}^{\infty} \mathbf{F}_k e^{j(k-n)\omega_o t}\,dt$$

$$= \sum_{k=-\infty}^{\infty} \int_0^T \mathbf{F}_k e^{j(k-n)\omega_o t}\,dt$$

If $k \neq n$

$$\int_0^T f(t)e^{-jn\omega_o t}\,dt = \sum_{\substack{-\infty \\ k\neq n}}^{\infty} \frac{\mathbf{F}_k e^{j(k-n)\omega_o t}}{(k-n)\omega_o}\Bigg|_0^T$$

$$= \frac{\mathbf{F}_k}{(k-n)\omega_o}[\cos(k-n)\omega_o T - 1] = 0$$

On the other hand, if $k = n$

$$\int_0^T f(t)e^{-jn\omega_o t}\,dt = \int_0^T \mathbf{F}_n e^{j(n-n)\omega_o t}\,dt$$

$$= \int_0^T \mathbf{F}_n dt = \mathbf{F}_n T$$

Solving for $\mathbf{F}_n$ shows that (14.2) defines the value of $\mathbf{F}_n$ that satisfies (14.1).

The Fourier series is said to "represent" $f(t)$ in the sense that $f(t)$ can be constructed from the Fourier coefficients $\mathbf{F}_n$. Although the mathematical proof of (14.1) is beyond the level of this text, we should consider two questions: (1) What guarantees that the Fourier coefficients exist? and (2) Given that the Fourier coefficients have been calculated, what guarantees that the series (14.1) converges to the actual value of the signal?

The assumption that the periodic signal $f(t)$ is **bounded** is sufficient to guarantee that the Fourier coefficients defined by (14.2) exist. To see this, observe that

$$|\mathbf{F}_n| = \left|\frac{1}{T}\int_T f(t)e^{-jn\omega_o t}dt\right|$$

$$\leq \frac{1}{T}\int_T |f(t)e^{-jn\omega_o t}|\,dt$$

$$\leq \frac{1}{T}\int_T |f(t)||e^{-jn\omega_o t}|\,dt$$

But since $|e^{-jn\omega_o t}| = 1$ for any values of n and t we conclude that

$$|\mathbf{F}_n| \leq \frac{1}{T}\int_T |f(t)|\,dt$$

The area under the curve of the absolute value of the signal provides a numerical bound for the magnitude of the Fourier coefficient. This area will be bounded on a finite interval if and only if $f(t)$ itself is bounded, i.e. if $|f(t)| < \infty$ for all $0 \leq t \leq T$. But existence of the Fourier coefficients does not in itself guarantee that the Fourier series converges to $f(t)$. Additional conditions, the so-called Dirchlet conditions, are needed.

Given the Fourier coefficients of the signal $f(t)$ the Fourier series formed by (14.1) will converge to $f(t)$ at all points where $f(t)$ is continuous, provided that the signal meets the Dirchlet conditions. The Dirchlet conditions require that $f(t)$ be bounded, have at most a finite number of finite discontinuities, and have a finite number of local maxima and minima within the interval $(0, T)$. The signal shown in SE14.1 meets the Dirchlet conditions. In fact, it is safe to say that all physical signals do. At points $\hat{t}$ where $f(t)$ is discontinuous, the Fourier series converges to

$$f(t) = \frac{1}{2}[f(\hat{t}^-) + f(\hat{t}^+)]$$

which is the midpoint of the left and right limits of $f(t)$ at $t = \hat{t}$.

Example 14.1

The periodic signal shown in Fig. 14.2 is called a pulse train because it consists of repeated rectangular pulses. The period of the signal determines its pulse repetition frequency, or its **pulse rate,** $1/T$; the ratio Δ/T of the pulse width to the period is called its **duty cycle.**

The Fourier series coefficients of the pulse train are obtained by evaluating (14.3) using the signal

$$f(t) = \begin{cases} A & |t| \leq \dfrac{\Delta}{2} \\ 0 & \dfrac{\Delta}{2} < |t| < \dfrac{T}{2} \end{cases}$$

Thus

$$\mathbf{F}_n = \frac{1}{T}\int_{-\Delta/2}^{\Delta/2} Ae^{-jn\omega_o t}\,dt$$

$$= \frac{-A}{jn\omega_o T}[e^{-jn\omega_o\Delta/2} - e^{jn\omega_o\Delta/2}]$$

Since $\sin x = (e^{jx} - e^{-jx})/2j$ we can reverse the order of the terms to get

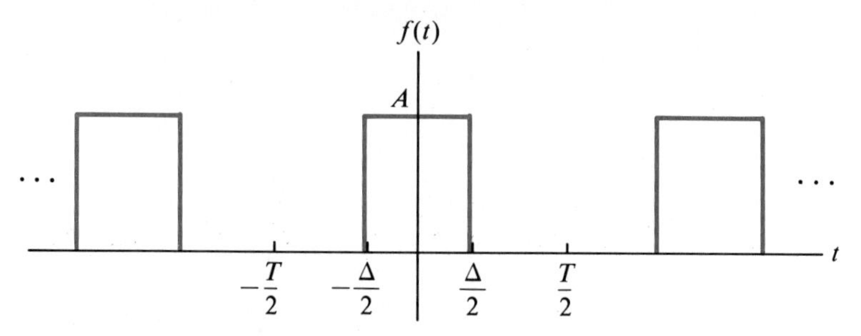

Figure 14.2 A rectangular periodic pulse train.

$$\mathbf{F}_n = \frac{2jA}{jn\omega_o T} \sin n\omega_o\Delta/2 = \frac{2A}{n\omega_o T} \sin n\omega_o\Delta/2$$

Since $\omega_o T = 2\pi$ this expression reduces to

$$\mathbf{F}_n = \frac{A}{n\pi} \sin n\omega_o\Delta/2 = \frac{A}{n\pi} \sin n\pi\frac{\Delta}{T}$$

The expression for $\mathbf{F}_n$ *cannot be used* to evaluate $\mathbf{F}_0$ since direct substitution with $n = 0$ leads to an attempt to evaluate 0/0. Instead, we can either calculate $\mathbf{F}_0$ from (14.2) with $n = 0$, or use L'Hospital's rule to get

$$\mathbf{F}_0 = \lim_{n\to 0} \frac{A}{\pi}\left[\omega_o \frac{\Delta}{2} \cos n\omega_o\Delta/2\right] = A\frac{\Delta}{T}$$

Thus, $\mathbf{F}_0$ is the product of the pulse height and the pulse duty cycle. In practice the series (14.2) is truncated and Fig. 14.3 shows how the pulse train with $\Delta/T = 1/2$ is approximated by the truncated series

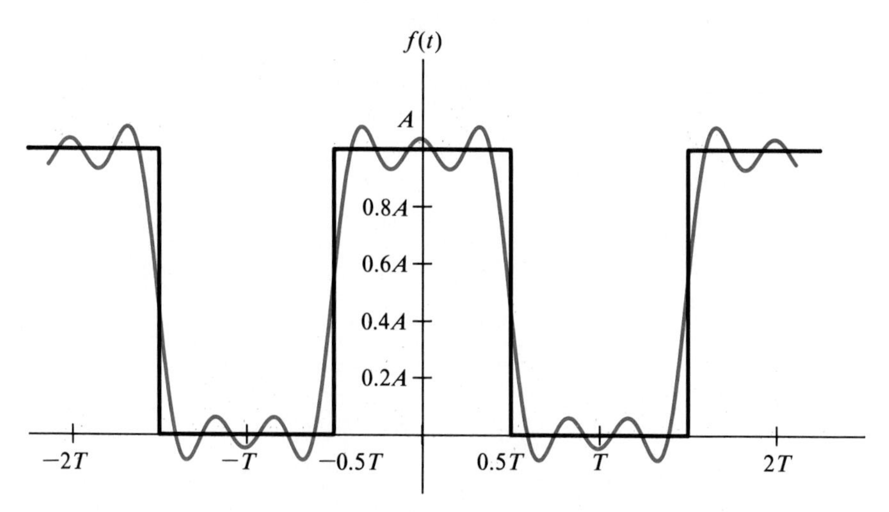

Figure 14.3 A five-term truncated Fourier series approximation to a rectangular pulse train, with $\Delta = T/2$.

$$f(t) \cong \sum_{n=-5}^{5} \mathbf{F}_n e^{jn\omega_o t}$$

Additional terms would provide a better representation of the signal.

14.4 FOURIER SERIES PROPERTIES*

The complex Fourier series of a periodic signal has several properties that simplify computation of the Fourier coefficients and check the accuracy of our work.

14.4.1 Uniqueness

A signal $f(t)$ and its Fourier coefficients have the property that one and only one signal corresponds to a given set of Fourier coefficients for a given ω_o and vice versa. The signal and its coefficients are said to form a **Fourier series pair,** which we symbolize by writing $f(t) \longleftrightarrow \mathbf{F}_n$. This property has practical significance because it creates a unique signal from its Fourier series coefficients.

Example 14.2

The impulse function (see Chapter 6) was shown to be a generalized function that engineers find helpful in modeling phenomena that occur over very short intervals of time. The signal $f(t)$ in Fig. 14.4 is a **periodic impulse train.** Its Fourier coefficients are given by

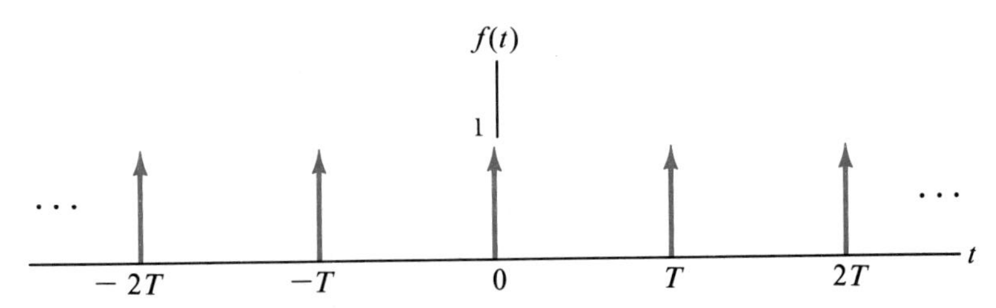

Figure 14.4 A periodic impulse train.

$$\mathbf{F}_n = \frac{1}{T}\int_{-T/2}^{T/2} \delta(t)e^{-jn\omega_o t}\, dt$$

Using the sampling property of the impulse, we get

$$\mathbf{F}_n = \frac{1}{T} \qquad n = 0, \pm 1, \pm 2$$

Thus, the behavior of the impulse train can be represented by the behavior of the Fourier series as

$$\delta(t) = \frac{1}{T} \sum_{n=-\infty}^{\infty} e^{+jn\omega_o t}$$

14.4.2 DC Value

The DC or average value f_{DC} of a periodic signal over one cycle is defined by

$$f_{DC} = \frac{1}{T} \int_0^T f(t)\, dt = \mathbf{F}_0$$

The DC value of a periodic signal is given by the $\mathbf{F}_0$ term of its complex Fourier series.

Example 14.3 (Halfwave Rectifier)

A rectifier is a device that converts a signal into one having only positive polarity. A rectifier's output signal is related to its input signal according to

$$v_o(t) = \begin{cases} v_{in}(t) & v_{in}(t) \geq 0 \\ 0 & v_{in}(t) \leq 0 \end{cases}$$

Rectifiers play an important role in electrical equipment because many electronic devices, such as transistors, require DC signals for their operation. If the equipment is not battery-operated, a power supply must provide the required DC signal. Rectifiers are needed because the power supply itself is supplied by an alternating signal, which must be converted to DC. Here we will be concerned with finding the Fourier series of the output of a rectifier.

A common way to describe a rectifier is by its input/output voltage-transfer characteristic, which is simply a graph of its instantaneous output voltage vs its instantaneous input voltage. A rectifier having the input/output transfer-characteristic shown in Figure 14.5 is called a halfwave rectifier because it effectively removes the negative half-cycle of a sinusoidal signal. If the input signal shown is described by $v_{in}(t) = V_{in} \sin \hat{\omega} t$ the output signal for one cycle will be

$$v_o(t) = \begin{cases} V_{in} \sin \hat{\omega} t & 0 \leq t \leq \frac{T}{2} \\ 0 & \frac{T}{2} \leq t \leq T \end{cases}$$

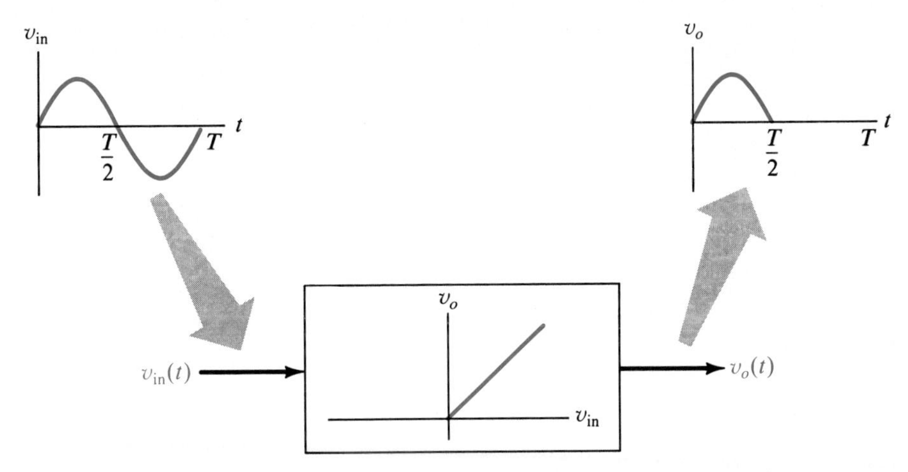

Figure 14.5 A halfwave rectifier with a simusoidal input signal and its rectified output signal.

where the period of both the input and the output signals is $T = 2\pi/\hat{\omega}$. The fundamental frequency of the output signal is ω_o and $\omega_o = \hat{\omega} = 2\pi/T$.

The Fourier coefficients of $v_o(t)$ are found by taking

$$\mathbf{F}_n = \frac{1}{T}\int_0^{T/2} V_{in} \sin \hat{\omega}t\, e^{-jn\omega_o t}\, dt$$

Noting that $\hat{\omega} = \omega_o$ and replacing $\sin \omega_o t$ by its exponential equivalent gives

$$\mathbf{F}_n = \frac{1}{T}\int_0^{T/2} V_{in}\left[\frac{e^{j\omega_o t} - e^{-j\omega_o t}}{2j}\right] e^{-jn\omega_o t}\, dt$$

$$= \frac{1}{T}\frac{V_{in}}{2j}\left[\frac{-e^{-j(n-1)\omega_o t}}{j(n-1)\omega_o} + \frac{e^{-j(n+1)\omega_o t}}{j(n+1)\omega_o}\right]_{t=0}^{t=T/2}$$

Factoring $j\omega_o$ from the denominator of each term and evaluating the limits with $\omega_o T = 2\pi$ leads to

$$\mathbf{F}_n = \frac{V_{in}}{4\pi}\left[\frac{-(n+1)e^{-j(n-1)\pi} - (n-1)e^{-j(n+1)\pi} - 2}{n^2 - 1}\right]$$

Using Euler's law

$$\mathbf{F}_n = \frac{V_{in}}{4\pi}\left[\frac{2j\, ne^{-jn\pi} \sin \pi + 2e^{-jn\pi} \cos \pi - 2}{n^2 - 1}\right]$$

And since $\sin \pi = 0$, and $\cos \pi = -1$, we get

$$\mathbf{F}_n = \frac{V_{in}}{4\pi}\left[\frac{-2e^{-jn\pi} - 2}{n^2 - 1}\right]$$

This expression can be further reduced, since $e^{-jn\pi} = \cos n\pi - j \overset{\nearrow 0}{\sin n\pi} = (-1)^n$ and therefore

$$\mathbf{F}_n = \frac{V_{in}}{2\pi}\left[\frac{(-1)^{n+1} - 1}{n^2 - 1}\right]$$

This expression for $\mathbf{F}_n$ is only valid for $n \neq \pm 1$ (otherwise the denominator is zero). $\mathbf{F}_1$ and $\mathbf{F}_{-1}$ must be calculated separately:

$$\mathbf{F}_1 = \frac{1}{T}\int_0^{T/2} V_{in} \sin \omega_o t \, e^{-j\omega_o t} \, dt$$

$$= \frac{V_{in}}{T}\int_0^{T/2} \frac{e^{j\omega_o t} - e^{-j\omega_o t}}{2j} e^{-j\omega_o t} \, dt$$

$$= \frac{V_{in}}{2jT}\left[\frac{T}{2} + \frac{e^{-j\omega_o T} - 1}{2j\omega_o}\right] = \frac{V_{in}}{4j}$$

Likewise, it can be shown that $\mathbf{F}_{-1} = -V_{in}/4j$.

In summary

$$\mathbf{F}_n = \begin{cases} \dfrac{V_{in}}{2\pi} \dfrac{(-1)^{n+1} - 1}{n^2 - 1} & n \neq \pm 1 \\ \dfrac{V_{in}}{4j} & n = 1 \\ -\dfrac{V_{in}}{4j} & n = -1 \end{cases}$$

The DC value of the signal is given by $\mathbf{F}_0 = V_{in}/\pi$.

Table 14.1 contains the magnitude and angle of the first five coefficients in the series for the halfwave rectified signal.

TABLE 14.1 FOURIER COEFFICIENTS OF THE HALFWAVE RECTIFIER OUTPUT SIGNAL

ω	n	$\lvert \mathbf{F}_n \rvert$	$\angle \mathbf{F}_n$
0	0	$\frac{V_{in}}{\pi}$	–
$\hat{\omega}$	± 1	$\frac{V_{in}}{4}$	$\pm\frac{\pi}{2}$
$2\hat{\omega}$	± 2	$\frac{V_{in}}{3\pi}$	$\pm\pi$
$3\hat{\omega}$	± 3	0	–
$4\hat{\omega}$	± 4	$\frac{V_{in}}{15\pi}$	$\pm\pi$
$5\hat{\omega}$	± 5	0	–

The DC and fundamental components of the signal have approximately the same value, and the remaining terms attenuate rapidly.

The halfwave rectifier has limited use as a DC signal source because it has significant phasor-signal content at frequencies other than DC. There are two ways to correct the situation. The first is to design a filter to suppress the unwanted signal content, the other is to build a better rectifier. We'll return to the matter of designing a filter in Chapter 17. Here, we'll next consider a fullwave rectifier (one that inverts the negative half-cycle of a sine wave).

Example 14.4 (Fullwave Rectifier)

The transfer characteristic of a fullwave rectifier is shown in Fig. 14.6. When the input signal is $v_{in}(t) = V_{in} \sin \hat{\omega}t$ the output signal has the waveform shown.

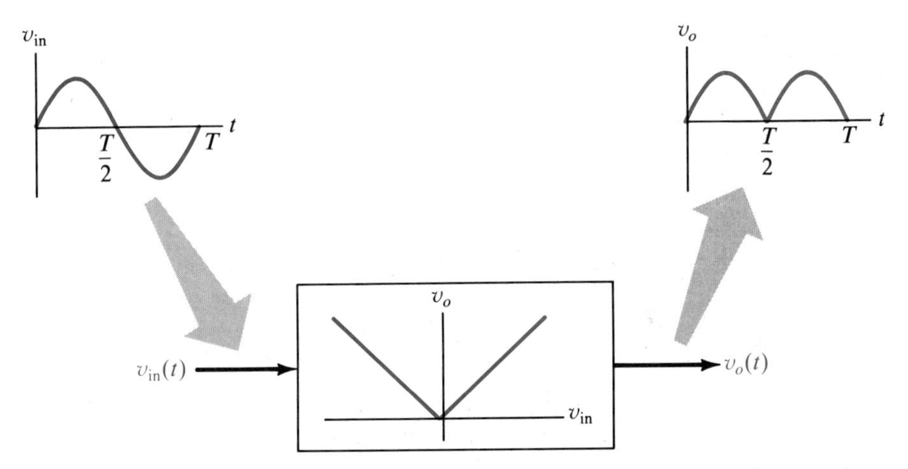

Figure 14.6 A fullwave rectifier with a sinusoidal input signal and its rectified output signal.

The fullwave rectifier has the distinctive feature of reversing the polarity of a negative input signal (i.e., its algebraic sign) without affecting its magnitude, so $v_o(t) = |v_{in}(t)|$. When the input signal is sinusoidal with a period $\hat{T} = 2\pi/\hat{\omega}$ the output signal is also periodic. But its period T is one-half the period of the input signal, $(T = \hat{T}/2)$. The fundamental frequency of a periodic signal is always defined in terms of its period, and **the output signal's fundamental frequency** is $\omega_o = 2\pi/T$. The waveform of the output signal is described by

$$v_o(t) = V_{in}|\sin \hat{\omega} t| \qquad 0 \le t \le T$$

Its Fourier coefficients are obtained from

$$\mathbf{F}_n = \frac{1}{T}\int_0^T V_{in}|\sin \hat{\omega} t| e^{-jn\omega_o t}\, dt$$

Since $T = \hat{T}/2$ it follows that $\omega_o = 2\hat{\omega}$ and

$$\mathbf{F}_n = \frac{2}{\hat{T}}\int_0^{\hat{T}/2} V_{in} \sin \hat{\omega} e^{-jn2\hat{\omega} t}\, dt$$

$$= \frac{1}{T}\int_0^T V_{in} \sin \hat{\omega} e^{-jn2\hat{\omega} t}\, dt$$

First we calculate $\mathbf{F}_0$ as

$$\mathbf{F}_0 = \frac{1}{T}\int_0^T V_{in} \sin \hat{\omega} t\, dt$$

$$= \frac{-V_{in}}{\hat{\omega} T} \cos \hat{\omega} t \Big|_0^T$$

$$= \frac{-V_{in}}{\hat{\omega} T} [\cos \hat{\omega} T - 1]$$

Noting that $\hat{\omega} T = \pi$ we write $\mathbf{F}_0 = 2V_{in}/\pi$.

For $n = \pm 1, \pm 2, \ldots$, calculating $\mathbf{F}_n$ for the fullwave rectifier is the same as calculating $2\mathbf{F}_{2n}$ for the coefficients of the halfwave rectifier, because the period of the halfwave signal is twice the period of the fullwave signal. Therefore, we can construct the coefficients of the fullwave rectifier from Table 14.1 without additional computation. If we let $\mathbf{F}_{Hk}$ denote the kth Fourier coefficient for the halfwave rectifier, then the fullwave Fourier coefficients are given by $\mathbf{F}_n = 2\mathbf{F}_{H(2n)}$. For example $\mathbf{F}_1 = 2(\mathbf{F}_{H2}) = 2(-1/3\pi)V_{in} = -(2/3\pi)V_{in}$ or, the halfwave rectifier signal's Fourier coefficients for $n = 2, 4, 6, 8, \ldots$ define the fullwave rectifier coefficients for $n = 1, 2, 3, 4, \ldots$.

In general, for the fullwave rectifier output signal,

$$\mathbf{F}_n = \begin{cases} \dfrac{2}{\pi} V_{in} & n = 0 \\ \dfrac{V_{in}}{\pi} \dfrac{(-1)^{2n+1} - 1}{(2n)^2 - 1} & n = \pm 1, \pm 2, \ldots \end{cases}$$

which reduces to

$$\mathbf{F}_n = \frac{2V_{in}}{\pi(1 - 4n^2)}$$

and

$$|\mathbf{F}_n| = \frac{2V_{in}}{\pi|4n^2 - 1|} \qquad n = 0, \pm 1, \ldots$$

The coefficient's magnitudes decrease according to $1/n^2$.

The magnitude and angle of $\mathbf{F}_n$ of v_o for the fullwave rectifier are summarized in Table 14.2. At this point, note that the DC component of the output signal fullwave rectifier has twice the magnitude of the DC component of the halfwave rectifier output signal. We'll make additional comparisons in Section 14.6.

TABLE 14.2 FOURIER COEFFICIENTS OF THE FULLWAVE RECTIFIER OUTPUT SIGNAL

ω	n	$\|\mathbf{F}_n\|$	$\angle\mathbf{F}_n$
0	0	$\frac{2}{\pi} V_{in}$	0
$2\hat{\omega}$	1	$\frac{2}{3\pi} V_{in}$	π
$4\hat{\omega}$	2	$\frac{2}{15\pi} V_{in}$	π
$6\hat{\omega}$	3	$\frac{2}{35\pi} V_{in}$	π
$8\hat{\omega}$	4	$\frac{2}{63\pi} V_{in}$	π

14.4.3 Linearity

If signals $f(t)$ and $g(t)$ have the same period T then the Fourier series of a signal formed by a linear combination of $f(t)$ and $g(t)$ is equal to the same linear combination of their individual Fourier series. Symbolically, if $f(t)$ and $g(t)$ have Fourier coefficients $\mathbf{F}_n$ and $\mathbf{G}_n$ respectively, the signal $a_1 f(t) + a_2 g(t)$ has Fourier coefficients $a_1\mathbf{F}_n + a_2\mathbf{G}_n$. (Proof is left as an exercise.)

14.4.4 Time Shift

If $f(t)$ is periodic with Fourier coefficients $\mathbf{F}_n$ the shifted signal $g(t) = f(t - \tau)$ is also periodic and has Fourier coefficients given by

$$\mathbf{G}_n = \mathbf{F}_n e^{-jn\omega_o\tau}$$

To prove this, we note that the signal $g(t)$ is a shifted copy of $f(t)$ so $g(t) = f(t - \tau)$ and

$$g(t) = \sum_{n=-\infty}^{\infty} \mathbf{F}_n e^{-jn\omega_o\tau} e^{jn\omega_o t}$$

But $g(t)$ must have a Fourier series given by

$$g(t) = \sum_{n=-\infty}^{\infty} \mathbf{G}_n e^{jn\omega_o t}$$

Because the Fourier series coefficients are unique, we conclude that $\mathbf{G}_n = \mathbf{F}_n e^{-jn\omega_o t}$.

The magnitudes and angles of $\mathbf{F}_n$ and $\mathbf{G}_n$ are related by $|\mathbf{G}_n| = |\mathbf{F}_n|$ and $\measuredangle \mathbf{G}_n = \measuredangle \mathbf{F}_n - n\omega_o\tau$. The nth phasor signal in the Fourier series of the shifted signal has a phasor angle equal to the phasor angle of the original signal minus a term that varies linearly with n. Recall that subtracting a phase angle θ from the argument of a sinusoid corresponds to delaying the signal by θ/ω (where ω is the frequency of the sinusoid). Alternately, to delay a signal by τ the angle $\omega\tau$ must be subtracted from the argument of the sinusoid. If a periodic signal is to be delayed by τ each of its phasor signal components must also be delayed by τ. Thus the nth phasor signal (which has frequency $\omega = n\omega_o$) must have the angle $n\omega_o\tau$ subtracted from its argument.

Example 14.5

The pulse-train signal g(t) shown in Fig. 14.7 is obtained by delaying the symmetrical pulse-train signal $f(t)$ (see Example 14.1) by $\tau = \Delta/2$.

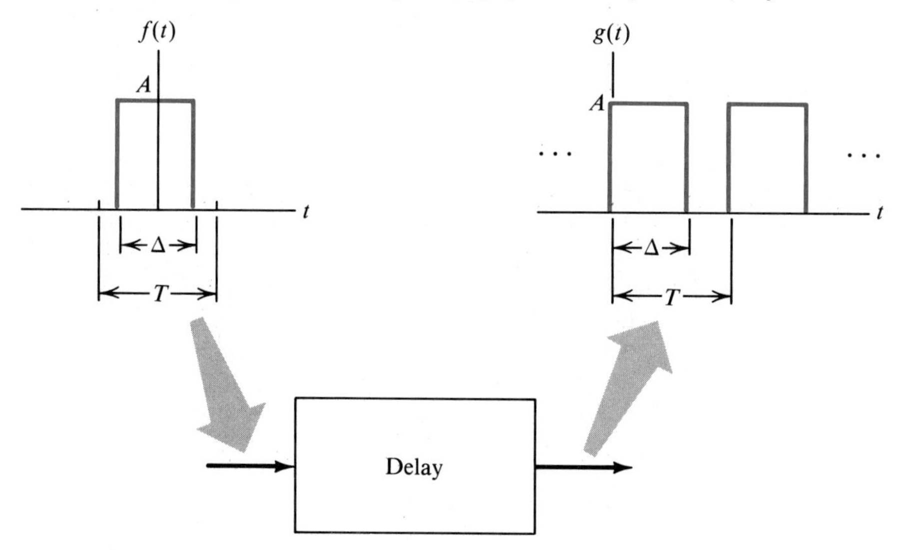

Figure 14.7 Delay of a symmetrical pulse train.

Therefore g(t) has Fourier coefficients

$$\mathbf{G}_o = A\frac{\Delta}{T} \text{ and } \mathbf{G}_n = \frac{A}{n\pi}\left(\sin n\omega_o\frac{\Delta}{2}\right)e^{-jn\omega_o\Delta/2}.$$

14.4.5 Fourier Derivative

When a periodic signal consists of segments of polynomials or linear segments it is usually much easier to exploit the derivative property of the

Fourier series than to derive the Fourier coefficients directly. If $f(t)$ has Fourier series coefficients $\mathbf{F}_n$ the signal $g(t) = f'(t)$ is a periodic signal, and if it has a Fourier series that series is defined by

$$g(t) = \sum_{n=-\infty}^{\infty} \mathbf{G}_n e^{jn\omega_o t}$$

where

$$\mathbf{G}_n = jn\omega_o \mathbf{F}_n \qquad n \neq 0, \tag{14.3}$$

provided that $f(t)$ is continuous at $t = \pm T/2$. We prove this by letting

$$f(t) = \mathbf{F}_o + \sum_{\substack{n=-\infty \\ n\neq 0}}^{\infty} \mathbf{F}_n e^{jn\omega_o t}$$

Then

$$f'(t) = 0 + \sum_{\substack{n=-\infty \\ n\neq 0}}^{\infty} jn\omega_o \mathbf{F}_n e^{jn\omega_o t} \tag{14.4}$$

Since the phasor signals are periodic $g(t) = f'(t)$ must also be periodic. If $g(t)$ has a Fourier series, it must be given by

$$g(t) = \sum_{n=-\infty}^{\infty} \mathbf{G}_n e^{jn\omega_o t} \tag{14.5}$$

Comparing (14.4) and (14.5) we conclude that $\mathbf{G}_o = 0$ and

$$\mathbf{G}_n = jn\omega_o \mathbf{F}_n \qquad n \neq 0.$$

The derivative property does not guarantee that $f'(t)$ has a Fourier series, only that *if* it does, the coefficients are defined by (14.3). The coefficients $\mathbf{F}_n$ can be found if $\mathbf{G}_n$ are known and vice versa. Usually $\mathbf{G}_n$ are easier to obtain, as the following example shows.

Example 14.6

A sawtooth signal $f(t)$ and its derivative $f'(t)$ are shown in Fig. 14.8. Derive the Fourier coefficients of $f(t)$ using the derivative property.

Solution: The DC component of $f(t)$ is zero, so $\mathbf{F}_o = 0$. Next, observe that

$$g(t) = f'(t) = \frac{A}{T} - A\,\delta(t) \qquad -\frac{T}{2} < t < \frac{T}{2}$$
$$= g_1(t) + g_2(t)$$

where $g_1(t) = A/T$ and $g_2(t) = -A\delta(t)$. The signal $g_1(t)$ is just a constant, with Fourier series coefficients

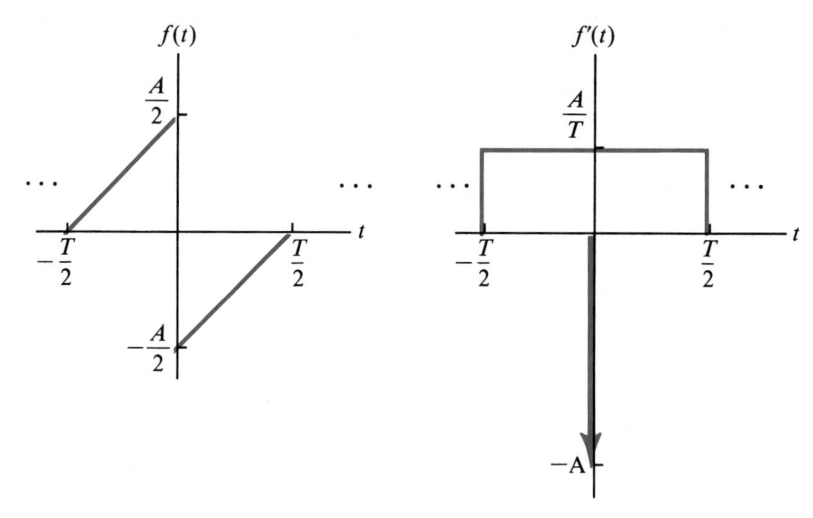

Figure 14.8 A sawtooth signal and its derivative.

$$\mathbf{G}_{n1} = \begin{cases} \dfrac{A}{T} & n = 0 \\ 0 & n \neq 0 \end{cases}$$

The function $g_2(t)$ is the impulse function, with Fourier coefficients

$$\mathbf{G}_{n2} = -\frac{A}{T} \qquad \text{all n.}$$

Using the linearity property of the Fourier series gives $\mathbf{G}_o = 0$ and $\mathbf{G}_n = -A/T$ ($n = \pm 1, \pm 2, \ldots$).

The derivative property implies that $\mathbf{G}_n$ and $\mathbf{F}_n$ are related by

$$\mathbf{F}_n = \frac{1}{jn\omega_o}\mathbf{G}_n \qquad n \neq 0$$

so

$$\mathbf{F}_n = \frac{1}{jn\omega_o}\left(-\frac{A}{T}\right) = \frac{jA}{2\pi n} \qquad n \neq 0$$

and

$$\mathbf{F}_o = 0$$

As an exercise obtain this result using (14.2).

A word of caution: When applying the derivative property, the DC value of $f(t)$ must be calculated separately, and cannot be inferred from the

DC component of $f'(t)$, since this implicitly involves dividing by zero in (14.3). Meaningless results can be obtained if this restriction is ignored.

If $f(t)$ is not continuous at $t = \pm T/2$ the derivative property must be modified to assure correct results. In this case $f(t)$ and $g(t) = f'(t)$ are defined over the interval $[(-T/2 - \epsilon), (T/2)]$ or over any $T + \epsilon$ interval. Then if $g(t)$ has a Fourier series, its coefficients are given by

$$\mathbf{G}_n = \lim_{\epsilon \to 0} jn\omega_o \mathbf{F}_n$$

This modification guarantees that an impulse function will be included in $g(t)$ when $f(t)$ is discontinuous at the end points of the interval chosen for evaluation. The next example demonstrates why this is important.

Example 14.7

Suppose a biased sawtooth signal has the shape shown in Fig. 14.9.

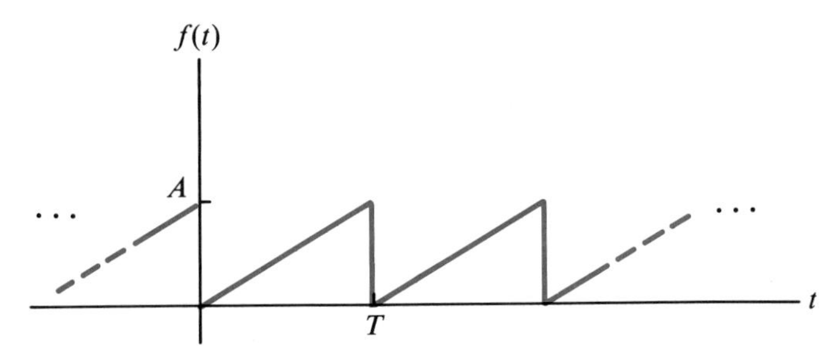

Figure 14.9 A biased sawtooth signal.

If we take $g(t) = f'(t) = A/T$ for $0 < t < T$ we are led to the *false* conclusion that the Fourier series of $g(t)$ has a single term (DC) with $G_n = A/T$ for $n = 0$ and $G_n = 0$ elsewhere. It would (erroneously) follow that

$$\mathbf{F}_n = \frac{1}{jn\omega_o}(0) = 0 \qquad n \neq 0$$

On the other hand, if we consider the derivative of $f(t)$ for $-\epsilon < t < T$ we get

$$g(t) = f'(t) = -\frac{A}{T}\delta(t) + \frac{A}{T}$$

so that $\mathbf{G}_o = 0$ and $\mathbf{G}_n = -A/T$. Using the derivative property to find $\mathbf{F}_n$ leads correctly to the same result $\mathbf{F}_n = jA/2\pi n$ as would be obtained by direct evaluation. Now $\mathbf{F}_o$ cannot be obtained from $\mathbf{G}_o$ but by inspection we find $\mathbf{F}_o = \frac{1}{2}\frac{A}{T}$.

14.4.6 Symmetry

If $f(t)$ is real valued, the complex Fourier series coefficients for $\pm n$ are a conjugate pair, that is $\mathbf{F}_n = \mathbf{F}_{-n}^*$. (Remember that * denotes the complex conjugate.)

This property follows directly from the algorithm for calculating the Fourier series coefficient (show it), and has two direct consequences. First, if two numbers are a conjugate pair they have the same magnitude and second, they have angles that are equal in magnitude but opposite in polarity, so

$$|\mathbf{F}_n| = |\mathbf{F}_{-n}|$$

and

$$\measuredangle\mathbf{F}_n = -\measuredangle\mathbf{F}_{-n}.$$

This property reduces computational effort by a factor of two because $\mathbf{F}_{-n}$ can be determined directly from $\mathbf{F}_n$.

14.4.7 Cosine Fourier Series

The symmetry property leads directly to the conclusion that the complex Fourier series is equivalent to the **cosine Fourier series**

$$f(t) = \sum_{-\infty}^{\infty} \mathbf{F}_n e^{jn\omega_o t} = \mathbf{F}_o + 2\sum_{n=1}^{\infty} |\mathbf{F}_n| \cos(n\omega_o t + \measuredangle\mathbf{F}_n)$$

This property is useful because it is easier to visualize the graph of the individual cosine terms than it is to visualize the combination of the phasor signals.

A pair of phasor signals corresponds to the pair of conjugate phasors $\mathbf{F}_n$ and $\mathbf{F}_{-n}$ shown in Fig. 14.10. The phasors rotate clockwise and counterclockwise, respectively, at a frequency $\omega = n\omega_o$. When the two phasor signals are added together in the Fourier series the result is the signal

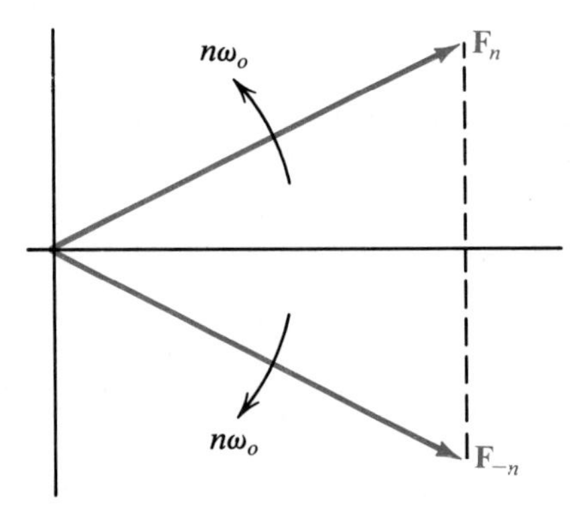

Figure 14.10 A pair of conjugate phasors.

$$g_n(t) = \mathbf{F}_n e^{jn\omega_o t} + \mathbf{F}_{-n} e^{-jn\omega_o t}$$
$$= |\mathbf{F}_n| e^{j(n\omega_o t + \measuredangle \mathbf{F}_n)} + |\mathbf{F}_{-n}| e^{-j(n\omega_o t - \measuredangle \mathbf{F}_{-n})}$$

and by symmetry, $\mathbf{F}_{-n} = \mathbf{F}_n{}^*$ so

$$g_n(t) = |\mathbf{F}_n| e^{j(n\omega_o t + \measuredangle \mathbf{F}_n)} + |\mathbf{F}_n| e^{-j(n\omega_o t + \measuredangle \mathbf{F}_n)}$$
$$= 2|\mathbf{F}_n| \cos(n\omega_o t + \measuredangle \mathbf{F}_n)$$

Thus $g_n(t)$ is a real signal when the pair of complex phasor signals are added in the complex Fourier series. Their sum creates a real signal because their imaginary components cancel each other at every instant of time.

The cosine form of the Fourier series may also be written as

$$f(t) = C_o + \sum_{n=1}^{\infty} C_n \cos(n\omega_o t + \phi_n)$$

with $C_o = \mathbf{F}_o$, $C_n = 2|\mathbf{F}_n|$ for $n \neq 0$ and $\phi_n = \measuredangle \mathbf{F}_n$ for $n \neq 0$.

Skill Exercise 14.3

Find the cosine series coefficients for the sawtooth signal in Example 14.7.

Answer: $C_o = \dfrac{A}{2T}$ $\quad C_n = \dfrac{A}{\pi n}$ $\quad \phi_n = \dfrac{\pi}{2}$

14.5 SIGNAL SPECTRA

The process of finding the Fourier coefficients of a signal is sometimes called spectral decomposition. The spectrum of a periodic signal consists of two parts forming graphic representations of the magnitude and phasor angle of its Fourier series coefficients.

14.5.1 Magnitude

The **magnitude spectrum** of a periodic signal is a graph with vertical lines corresponding to the magnitude of the signal's Fourier coefficient. The lines are located on the ω-axis at the points corresponding to $n\omega_o$. Since the Fourier coefficients obey the symmetry property ($|\mathbf{F}_n| = |\mathbf{F}_{-n}|$) **the magnitude spectrum of a signal is always symmetric.**

Example 14.8

The pulse-train signal in Example 14.1 has Fourier coefficients given by

$$\mathbf{F}_n = 2\frac{A}{T}\frac{\sin n\omega_o\Delta/2}{n\omega_o}$$

Multiplying both numerator and denominator by $\Delta/2$ gives

$$\mathbf{F}_n = 2\frac{A}{T}\frac{\Delta}{2}\frac{\sin n\omega_o\Delta/2}{n\omega_o\Delta/2}$$

The function $\frac{\sin x}{x}$ appear frequently in the analysis of communication circuits, and is called the **sample function.** It is given its own symbol

$$\text{Sa}(x) = \frac{\sin x}{x}$$

and is shown in Fig. 14.11, where L'Hospital's rule provides the value of Sa(0). Using this function, the Fourier coefficient of the pulse train is

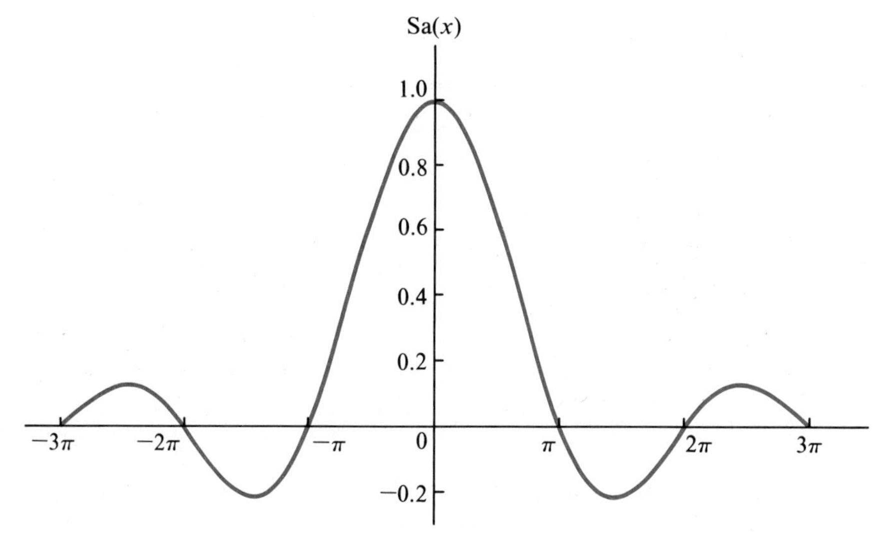

Figure 14.11 A graph of the sample function Sa (x).

$$\mathbf{F}_n = \frac{A\Delta}{T}\,\text{Sa}\left(n\omega_o\frac{\Delta}{2}\right)$$

and the magnitude spectrum of the signal will be given by

$$|\mathbf{F}_n| = \frac{A\Delta}{T}\left|\text{Sa}\left(n\omega_o\frac{\Delta}{2}\right)\right|$$

The symmetric magnitude spectrum in Fig. 14.12 has spectral lines spaced by ω_o and located at frequencies that are integer multiples of ω_o (at $\omega = n\omega_o$ for $n = 0, \pm 1, \pm 2, \ldots$). The lines are within an envelope specified by the function $\text{Sa}(\omega\Delta/2)$. Since the sample function has a value of 1

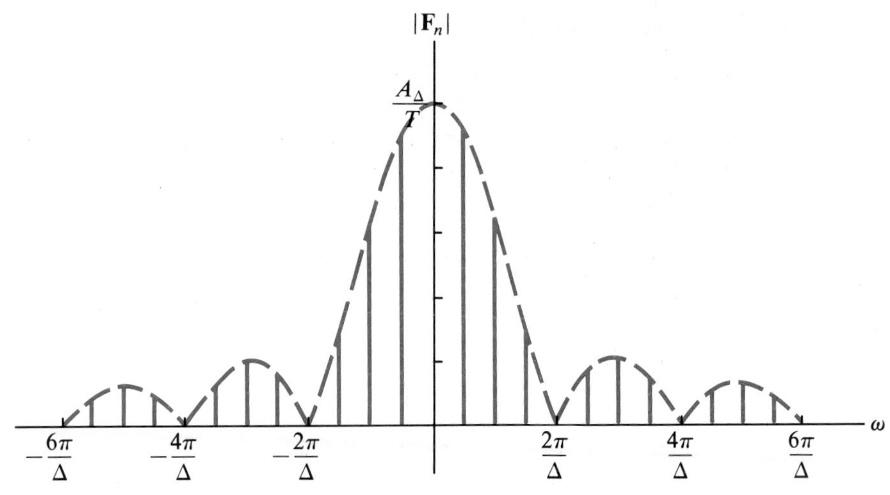

Figure 14.12 The magnitude spectrum of the rectangular pulse train.

at x = 0, the pulse train's spectrum has a value of $A\Delta/T$ at $\omega = 0$—or the DC signal content is equal to A times the duty cycle Δ/T. The zero crossings of the envelope of the spectrum are located at the points where

$$\mathrm{Sa}(\omega\Delta/2) = 0$$

or where the argument of $\mathrm{Sa}(\cdot)$ is an integer multiple of π. We must have

$$\omega\frac{\Delta}{2} = m\pi$$

so the zero crossing occurs at ω^* with

$$\omega^* = m\frac{2\pi}{\Delta}.$$

The zero crossings occur at values of ω that are integer multiples of 2π times the *reciprocal* of the pulse width of the signal. The location of the zero crossing on the ω-axis does not depend on the period T but only on the pulse width Δ.

The magnitude spectrum provides a visual display of the relative importance of the phasor signals that synthesize the pulse train signal. For example, if ω_o is small, the spectral lines will be relatively close together, corresponding to more low-frequency content in the signal—i.e., the signal is changing more slowly. Conversely, increasing ω_o has the effect of decreasing T, which causes the height of the spectral lines that are close to the origin to increase (for a fixed pulse width Δ). That is, increasing the duty cycle increases the relative DC spectral content of the signal.

The significant spectral content of a pulse train is arbitrarily taken to be the lines that lie within the first zero crossing of the spectral envelope on each side of the origin, because they account for a significant part of the power in the signal. The number of lines between the origin and the first zero crossing is the largest integer smaller than the number found by dividing the frequency interval to the zero crossing by the line spacing ω_o. Thus

$$N = \left[\frac{2\pi}{\Delta}\frac{1}{\omega_o}\right] = \left[\frac{T}{\Delta}\right] = \left[\frac{1}{\text{Duty Cycle}}\right]$$

Where [x] is the greatest integer smaller than x. The reciprocal of the duty cycle determines the significant lines in the spectrum. For a given pulse width a pulse train with a low duty-cycle value will have more significant lines than one with a high duty-cycle value. Figure 14.13 illustrates this relationship for two cases of a pulse-train duty cycle, with the same Δ in each case. Notice the changes in the relative height and zero crossing of the lines.

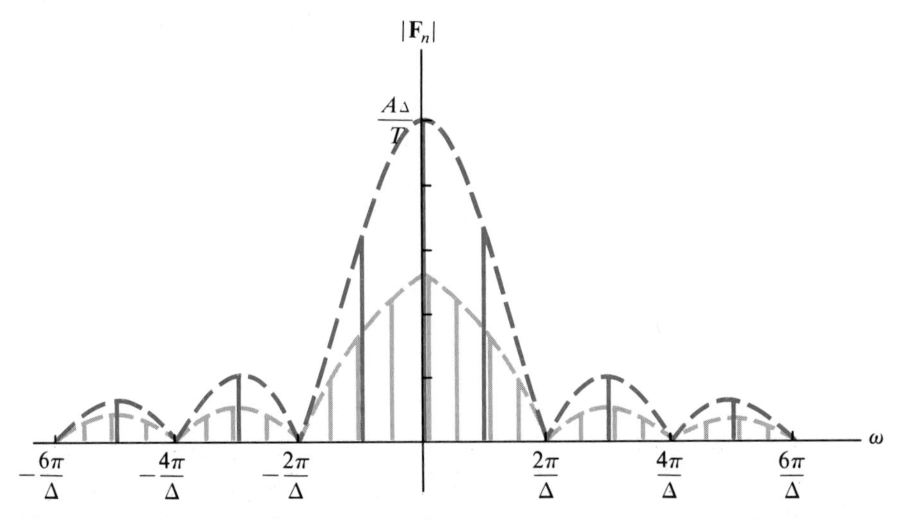

Figure 14.13 Magnitude spectra of the rectangular pulse train, with $\omega_{o2} = 2\omega_{o1}$.

If the pulse width Δ is made narrow, the location of the zero crossing moves away from the origin, so the signal must have more high-frequency content in order to synthesize the time-domain behavior of a narrow pulse.

If frequency is measured in Hz, rather than radians the first zero crossing occurs at $f^* = 1/\Delta$.

The reciprocal of the pulse width determines the significant spectral content of the signal. This suggests that if a circuit is to be affected by a narrow pulse train, it must be capable of responding to high-frequency sinusoidal signals.

14.5.2 Phase

The phase or angle spectrum of a periodic signal displays symbols representing the angles of its complex Fourier coefficients. The phase spectrum is always an odd function. (Why?)

Example 14.9

Figure 14.14 shows the phase spectrum of the rectangular pulse train signal. The angle of the coefficients is either 0 or $\pm\pi$ since $\mathbf{F}_n$ is real and therefore either positive or negative.

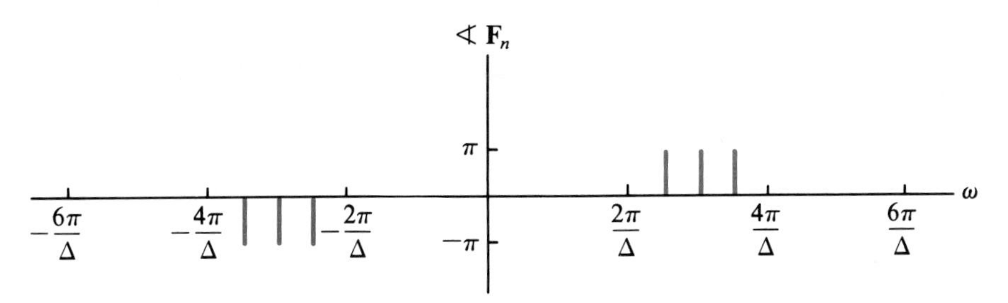

Figure 14.14 The phase spectrum of the rectangular pulse train.

14.5.3 Power

A signal's average power is calculated over its period according to

$$P_{av} = \frac{1}{T}\int_{-T/2}^{T/2} f^2(t)\, dt$$

In general a signal is considered by analogy to have power associated with it even when it is not an electrical voltage or current. The Fourier coefficients of a signal are related to its average power. First we note that since $f(t)$ is periodic, its Fourier series representation can be used to compute

$$P_{av} = \frac{1}{T}\int_{-T/2}^{T/2} f(t) \sum_{n=-\infty}^{\infty} \mathbf{F}_n e^{jn\omega_0 t}\, dt$$

By interchanging the order of integration and summation we get

$$P_{av} = \sum_{n=-\infty}^{\infty} \mathbf{F}_n \frac{1}{T}\int_{-T/2}^{T/2} f(t) e^{jn\omega_0 t}\, dt$$

Recognizing that

$$\mathbf{F}_{-n} = \frac{1}{T}\int_{-T/2}^{T/2} f(t) e^{jn\omega_0 t}\, dt$$

we arrive at

$$P_{av} = \sum_{n=-\infty}^{\infty} \mathbf{F}_n\mathbf{F}_{-n} = \sum_{n=-\infty}^{\infty} \mathbf{F}_n\mathbf{F}_n^*$$

The properties of complex numbers allow us to replace $\mathbf{F}_n\mathbf{F}_n^*$ with $|\mathbf{F}_n|^2 = \mathbf{F}_n\mathbf{F}_n^*$ so

$$P_{av} = \sum_{n=-\infty}^{\infty} |\mathbf{F}_n|^2$$

This can be simplified by using the symmetry property of the Fourier coefficients

$$\boxed{P_{av} = |\mathbf{F}_o|^2 + 2\sum_{n=1}^{\infty} |\mathbf{F}_n|^2}$$

This highlights the fact that the phasors $\mathbf{F}_n$ and $\mathbf{F}_{-n}$ must be treated as a pair, because both are used to synthesize a real-world signal from the pair of phasor signals.

The average power in a periodic signal is equal to the sum of the squares of the magnitudes of its Fourier coefficients. The **power spectrum** of the signal is a graphic display of lines corresponding to $|\mathbf{F}_n|^2$.

Example 14.10

The power spectrum of the pulse-train signal has

$$|\mathbf{F}_n|^2 = \left(A\frac{\Delta}{T}\right)^2 \mathrm{Sa}^2\left(n\omega_o\frac{\Delta}{2}\right)$$

The envelope of the power spectrum is governed by the behavior of $\mathrm{Sa}^2(x)$, which is shown by the dashed line in Fig. 14.15. The envelope of the magnitude spectrum decreases according to a factor of $1/n$ and the envelope of the power spectrum decreases by a factor of $1/n^2$. Relatively small signal components will contribute relatively less power content.

The expression for calculating the average power of a periodic signal associates the power content of a signal with the power content of its phasor signals. This has implications for filter design because the power contributions of the signal components are independent of each other, in the sense that the contribution due to the nth phasor signal pair is $|\mathbf{F}_n|^2$, regardless of the value of the other phasor coefficients. We'll see that fil-

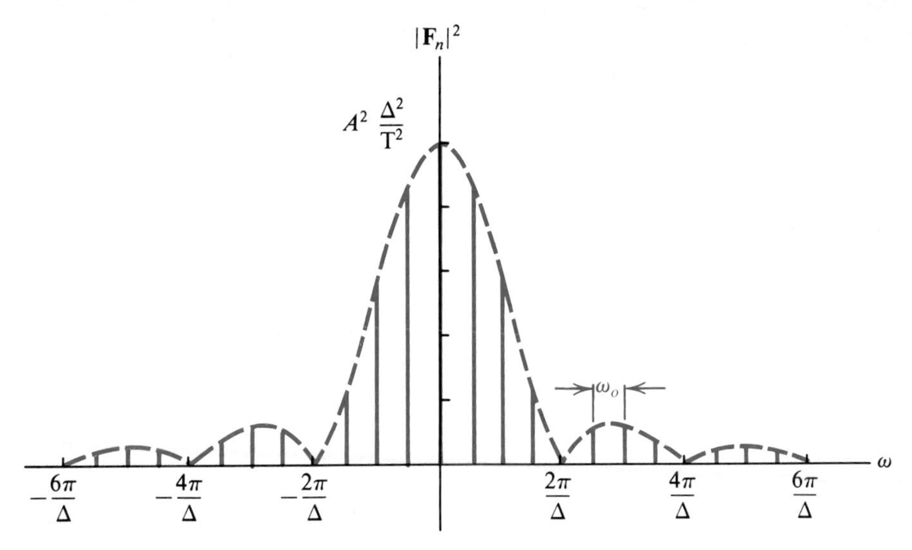

Figure 14.15 The power spectrum of the rectangular pulse train with generic parameters A, Δ and ω_o (or T).

ters can act to selectively change the relative heights of the Fourier coefficients of a signal.

14.5.4 Signal Orthogonality

Two signals g_1 and g_2 are said to be **orthogonal** on an interval of width T if

$$\int_T g_1(t)g_2{}^*(t)dt = \int_T g_1{}^*(t)g_2(t)\,dt = 0$$

By analogy with geometric vectors, when g_1 and g_2 are orthogonal to each other neither has a component in the direction of the other. The members of the phasor-signal family are mutually orthogonal. This can be demonstrated by taking

$$\int_0^T e^{jn\omega_o t}(e^{jm\omega_o t})^*\,dt = \int_0^T e^{j(n-m)\omega_o t}\,dt$$

$$= \frac{1}{j(n-m)\omega_o}\,e^{j(n-m)\omega_o t}\Big|_0^T$$

when $n \neq m$. Evaluating the limits with $\omega_o T = 2\pi$ gives

$$\int_0^T e^{jn\omega_o t}(e^{jm\omega_o t})^*\,dt = \begin{cases} 0 & n \neq m \\ T & n = m \end{cases}$$

Signal orthogonality plays an important role in the calculation of the average power in a signal, since

$$P_{av} = \frac{1}{T}\int_0^T f^2(t)\,dt$$

$$= \frac{1}{T}\int_0^T \sum_{n=-\infty}^{\infty} \mathbf{F}_n e^{jn\omega_o t} \sum_{m=-\infty}^{\infty} \mathbf{F}_m e^{jm\omega_o t}\,dt$$

$$= \frac{1}{T}\sum_{n=-\infty}^{\infty}\sum_{m=-\infty}^{\infty} \mathbf{F}_n\mathbf{F}_m \int_0^T e^{j(n+m)\omega_o t}\,dt$$

Orthogonality assures us that the integral will vanish for all indices n and m such that $n + m \neq 0$. Conversely, the integral will have the value T when $n + m = 0$ $(m = -n)$. Thus,

$$P_{av} = \frac{1}{T}\sum_{n=-\infty}^{\infty} \mathbf{F}_n\mathbf{F}_{-n}T = \sum_{n=-\infty}^{\infty} |\mathbf{F}_n|^2$$

which agrees with the result obtained earlier. **The power spectrum of a periodic signal depends only upon its magnitude spectrum.** The phase spectrum has no effect. The phase spectrum only affects the **relative location** of the phasor signal components on the time axis; their placement does not affect their orthogonality over the interval defined by the period of the signal. For a given magnitude spectrum the phase spectrum determines the waveform of the signal. Thus, many signals may have the same magnitude spectrum and power spectrum but dramatically different waveforms if their phase spectra differ.

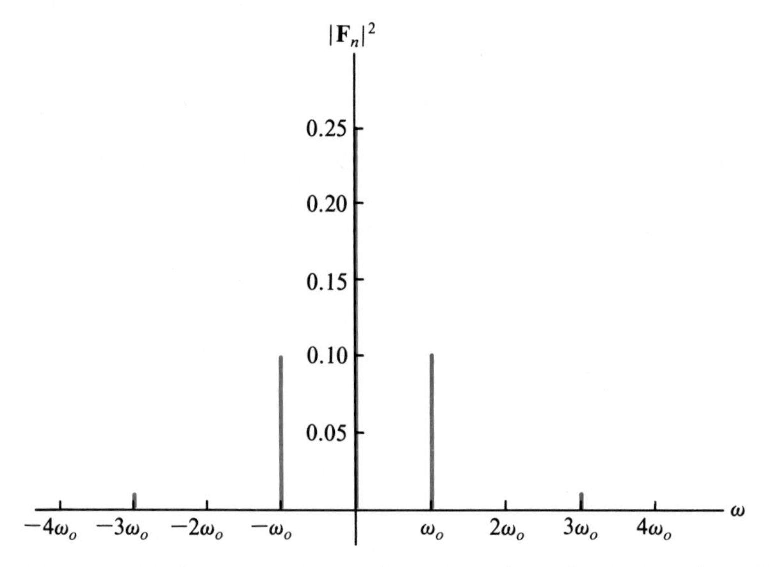

Figure 14.16 Power spectrum of a rectangular pulse train with a 50% duty cycle.

Example 14.11

The power spectrum of the pulse-train signal can be graphed from the data in Table 14.3 (with the results shown in Fig. 14.16). Table 14.3 also shows the percent of contribution of each pair of coefficients (for index n) to the total average power. The DC and first harmonics ($n = \pm 1$) account for over 90% of the power in the pulse-train signal (their spectral lines are within the envelope's first zero crossing).

TABLE 14.3 PULSE-TRAIN POWER CONTENT

n	$\lvert F_n \rvert$	$2\lvert F_n \rvert^2$	% of P_{av}
0*	$\frac{1}{2}$	$\frac{1}{4}$	50%
1	$\frac{1}{\pi}$	$\frac{2}{\pi^2}$	41%
2	0	0	0
3	$\frac{1}{3}\pi$	$\frac{2}{9\pi^2}$	5%
4	0	0	0
5	$\frac{1}{5\pi}$	$\frac{2}{25\pi^2}$	2%

*Table entry for $n = 0$ contains $\lvert \mathbf{F}_o \rvert^2$ instead of $2\lvert \mathbf{F}_o \rvert^2$.
Total Average Power = 0.5, $\Delta/T = 0.5$ and $A = 1$.

Example 14.12

The Fourier coefficients of the halfwave rectifier were evaluated in Example 14.3. Here we will calculate the rectifier's power spectrum and determine the relative power content in the output-signal components. Using Table 14.1 we construct Table 14.4 below.

TABLE 14.4 HALFWAVE RECTIFIER POWER OUTPUT

n	$\lvert F_n \rvert$	$2\lvert F_n \rvert^2$	% of P_{av}†
0*	$\frac{1}{\pi} V_{in}$	$\frac{1}{\pi^2} V_{in}^2$	40.53
1	$\frac{1}{4} V_{in}$	$\frac{1}{8} V_{in}^2$	50.00
2	$\frac{1}{3\pi} V_{in}$	$\frac{2}{9\pi^2} V_{in}^2$	9.01

TABLE 14.4 CONTINUED

n	$\|F_n\|$	$2\|F_n\|^2$	% of P_{av}†
3	0	0	0
4	$\frac{1}{15\pi} V_{in}$	$\frac{2}{225\pi^2} V_{in}^2$	.04
5	0	0	0

*Table entry for $n = 0$ contains $|\mathbf{F}_o|^2$
†$P_{av} = V^2{}_{in}/4$.

The first harmonic actually contributes more power to the output signal than does the DC component. The average power of the rectifier output is half the average power of the input sinusoid. We conclude that the halfwave rectifier is not very efficient because less than half of the input-signal power is available as output-signal power, and of that, less than one-half is power content at DC.

Example 14.13

Table 14.5 displays the Fourier coefficients and their relative contributions to the power in the output signal of a fullwave rectifier driven by a sinusoidal input signal. (See Example 14.4)

TABLE 14.5 FULLWAVE RECTIFIER POWER OUTPUT

ω	n	$\|F_n\|$	$2\|F_n\|^2$	% of P_{av}†
0*	0	$\frac{2}{\pi} V_{in}$	$\frac{4}{\pi^2} V_{in}^2$	81.10%
$2\omega_o$	1	$\frac{2}{3\pi} V_{in}$	$\frac{8}{9\pi^2} V_{in}^2$	18.01%
$4\omega_o$	2	$\frac{2}{15\pi} V_{in}$	$\frac{2}{225\pi^2} V_{in}^2$	.72%
$6\omega_o$	3	$\frac{2}{35\pi} V_{in}$	$\frac{8}{1225\pi^2} V_{in}^2$	.13%
$8\omega_o$	4	$\frac{2}{63\pi} V_{in}$	$\frac{8}{3969\pi^2} V_{in}$	.04%

*Table entry for $n = 0$ shows $|\mathbf{F}_o|^2$.
†$P_{av} = V_{in}^2/2$.

The average power of the output signal is the same as the average power of the input signal, so the fullwave rectifier is superior to the halfwave rectifier on that point. The spectrum of the fullwave rectifier has 81.1% of the signal's power at DC (the fullwave rectifier converts over 81% of the average sinusoidal power into DC power). Figure 14.17 shows a composite of the power spectra of the halfwave and fullwave rectifiers.

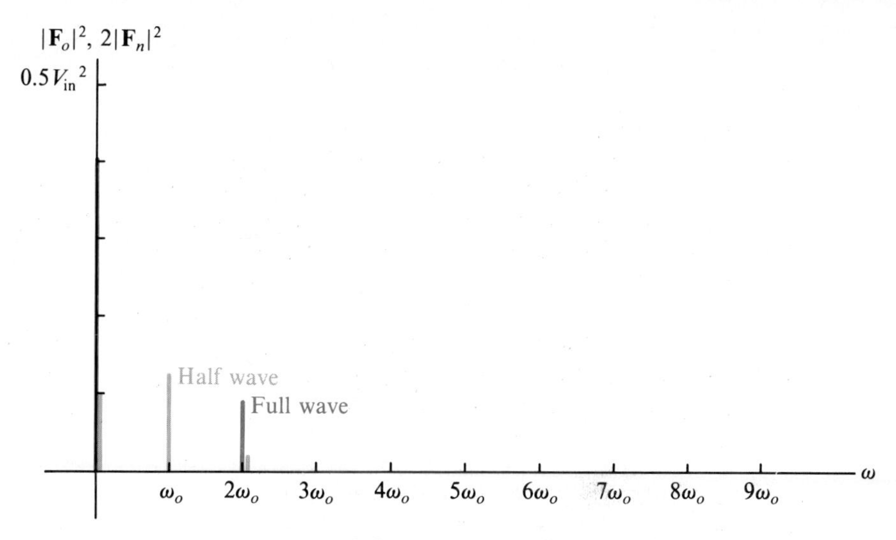

Figure 14.17 A comparison of the output-signal magnitude spectra of halfwave and fullwave rectifiers when the input signal is a sinusoid.

For purpose of comparison, the lines are shown relative to the fundamental frequency ω_o of the halfwave rectifier which is the same as the frequency of the input sinusoid. The fullwave rectifier does not have a spectral component at $\omega = \omega_o$ and its lines are spaced by $2\omega_o$; (this has implications for the design of any filter that might be used to eliminate the AC signal content, as will be shown in Chapter 17). The spacing between spectral lines is an indicator of how difficult it is to filter, or separate two sinusoidal signal components from each other by passing them through a circuit. Thus, it will be easier to build a filter to isolate the DC content for the fullwave rectifier than for a halfwave rectifier.

14.5.5 RMS Signal Value

The rms or effective value of a sinusoidal signal was defined in Chapter 13. When a signal $f(t)$ is periodic but not necessarily sinusoidal, its rms value is similarly defined as

$$f_{\text{rms}} = \left(\frac{1}{T}\int_0^T f^2(t)\, dt\right)^{1/2}$$

The rms value of the signal is the square root of the average power, so

$$f_{\text{rms}} = \sqrt{\sum_{-\infty}^{\infty} |\mathrm{F_n}|^2}$$

The rms value of a periodic signal is the square root of the sum of the squares of the magnitudes of its complex Fourier coefficients. This is equivalent to adding the average power due to the spectral components, and taking the square root of the sum. Because of the orthogonality property of the Fourier series, a given component's contribution to the rms value does not depend on the other components. A DC signal having the value of the periodic signal's rms value would deliver the same power as the periodic signal to a 1 Ω resistor.

14.6 CIRCUIT ANALYSIS—PERIODIC SIGNALS

In practice, many circuits are driven by signals of the form $x_{in}(t) = x(t)u(t)$, where $x(t)$ is periodic. We call $x_{in}(t)$ a **switched periodic signal.** The key observation is that the forced response to $x_{in}(t)$ for $t \geq 0$ is just the forced response to $x(t)$. Now note that $x(t)$ can be written in terms of a Fourier series, and recall that the forced response to a complex phasor signal is also a complex phasor signal

$$\mathbf{X}_n e^{jn\omega_o t} \longrightarrow \mathbf{X}_n \mathbf{H}(jn\omega_o) e^{jn\omega_o t}$$

where $\mathbf{X}_n$ denotes a Fourier coefficient of $x(t)$. Consequently, superposition can be used to write the *forced* response of the circuit to $x_{in}(t)$ as

$$y_f(t) = \sum \mathbf{X}_n \mathbf{H}(jn\omega_o) e^{jn\omega_o t} = \sum \mathbf{Y}_n e^{jn\omega_o t}$$

where the *output* Fourier coefficients are given by

$$\boxed{\mathbf{Y}_n = \mathbf{X}_n \mathbf{H}(jn\omega_o)}$$

The zero-state and initial-state responses of the circuit to $x_{in}(t)$ are described by

$$y(t) = y_n(t) + \sum \mathbf{X}_n \mathbf{H}(jn\omega_o) e^{jn\omega_o t}$$

where $y_n(t)$ denotes the natural solution. Appropriate boundary conditions must be used with this expression to determine the solution coefficients of $y_n(t)$.

If the natural solution decays to zero, and we'll assume that it does, the **steady-state output to a periodic input signal** is the superposition of the forced responses due to the individual phasor-signal components of the periodic signal $x(t)$. This creates a periodic steady-state output signal

$$y_{ss}(t) = \sum \mathbf{X}_n \mathbf{H}(jn\omega_o) e^{jn\omega_o t}$$

having the same period as the input signal.

Each phasor signal in the Fourier series for the input signal produces an output phasor signal whose complex amplitude is uniquely determined by the amplitude of the input signal and the evaluated transfer function of the circuit. This relationship is shown in Fig. 14.18, where $\mathbf{Y}_n = \mathbf{X}_n \mathbf{H}(jn\omega_o)$.

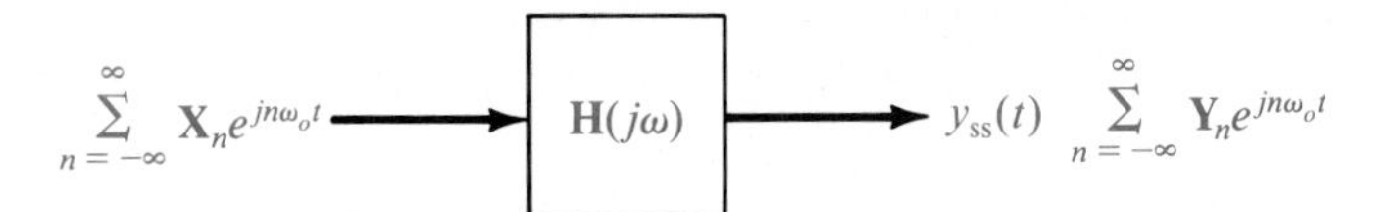

Figure 14.18 The steady-state response of a circuit to a periodic signal is determined by the circuit's transfer function, with $\mathbf{Y}_n = \mathbf{X}_n\ \mathbf{H}(j\omega)$.

Example 14.14

Find the steady-state output signal's Fourier coefficients for $|n| \leq 5$ when the circuit shown in Fig. 14.19 has an input signal described by the periodic biased sawtooth signal shown in Fig. 14.9. Draw the approximation to $v_o(t)$ using $|n| \leq 5$.

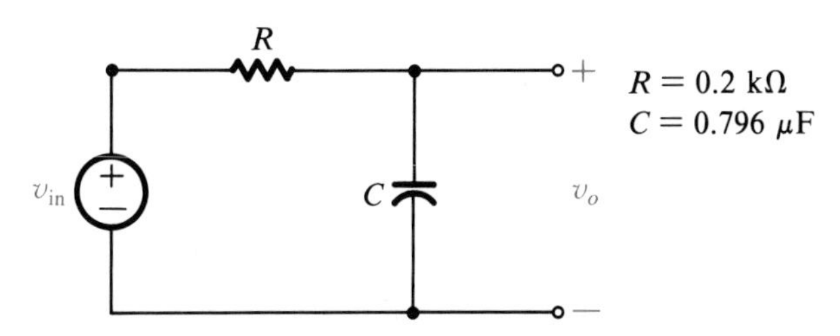

Figure 14.19 Circuit for Example 14.14.

Solution: In example 14.7 we saw that $v_{in}(t)$ has Fourier coefficients given by $\mathbf{F}_0 = A/2$ and $\mathbf{F}_n = jA/2\pi n$. The I/O transfer function is

$$\mathbf{H}(s) = \frac{\mathbf{V}_o}{\mathbf{V}_{in}} = \frac{1}{1 + sRC} = \frac{1}{1 + j\omega(0.159 \times 10^{-3})}$$

Next, we construct Table 14.6, with $\mathbf{X}_n$ as the Fourier coefficient of the input signal and $\mathbf{Y}_n$ as the coefficient of the corresponding output signal.

TABLE 14.6 INPUT/OUTPUT FOURIER COEFFICIENTS

n	$n\omega_o$	$\mathbf{X}_n$	$\mathbf{H}(jn\omega_o)$	$\mathbf{Y}_n = \mathbf{X}_n\mathbf{H}(jn\omega_o)$
0	0	$\frac{A}{2}$	1	$0.5\ A\ \underline{/0^\circ}$
1	$2\pi \times 10^3$	$j\frac{A}{2\pi}$	$0.707\ \underline{/-45^\circ}$	$0.1125\ A\ \underline{/45^\circ}$
2	$4\pi \times 10^3$	$j\frac{A}{4\pi}$	$0.447\ \underline{/-63.44^\circ}$	$0.0356\ A\ \underline{/26.56^\circ}$

TABLE 14.6 CONTINUED

n	$n\omega_o$	X_n	$H(jn\omega_o)$	$Y_n = X_n H(jn\omega_o)$
3	$6\pi \times 10^3$	$j\frac{A}{6\pi}$	$0.316 \angle -71.57°$	$0.0168\,A \angle 18.43°$
4	$8\pi \times 10^3$	$j\frac{A}{8\pi}$	$0.243 \angle -75.96°$	$0.0097\,A \angle 14.04°$
5	$10\pi \times 10^3$	$j\frac{A}{10\pi}$	$0.196 \angle -78.69°$	$0.0062\,A \angle 11.31°$

Since $\mathbf{Y}_{-n} = \mathbf{Y}_n{}^*$ there is no need to compute $\mathbf{Y}_{-n}$ separately. The graph of $v_o(t)$ for $|n| \leq 5$ is shown in Fig. 14.20. The computation of the values of the approximation to $v_o(t)$ was done using a personal computer. The waveform of v_o resembles the input ramp signal but cannot replicate it because the output series is truncated and because the spectrum of the output is not identical to the spectrum of the input signal.

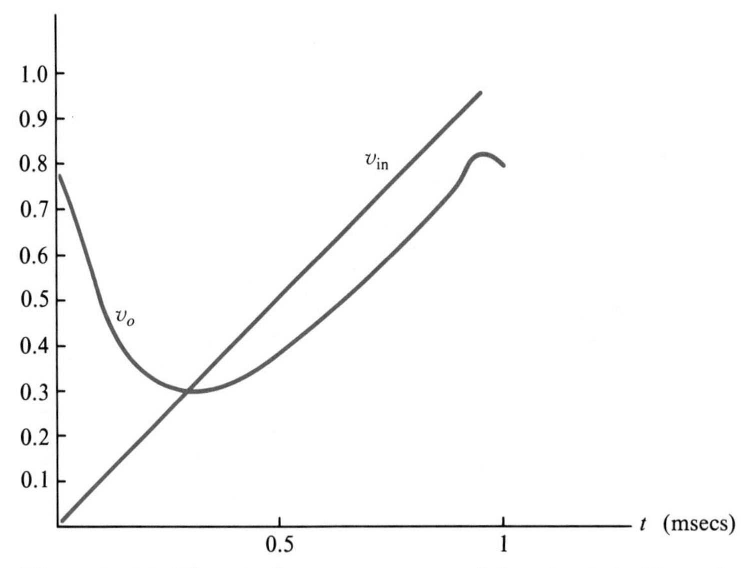

Figure 14.20 The steady-state response of the circuit in Example 14.14 to the biased sawtooth signal.

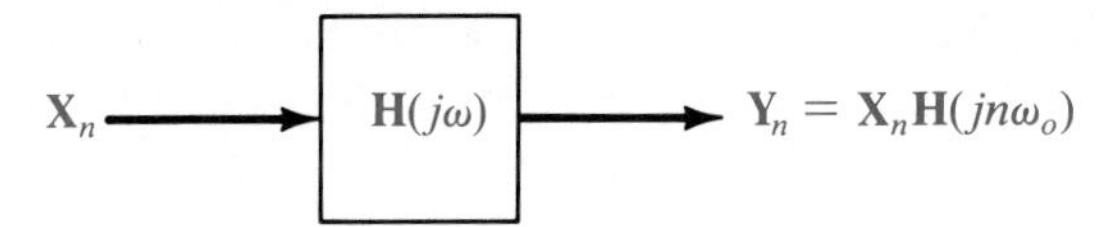

Figure 14.21 Input/output phasor relationship.

14.6.1 Input/Output Amplitude Spectra

The steady-state operation of a circuit can be viewed in terms of its periodic **time-domain** response to a periodic input signal, or in terms of the circuit's **frequency response.** The frequency response of a circuit defines the relationship between the phasor amplitudes of its input signal and the phasor amplitudes of any voltage or current in the circuit. This relationship is depicted in Fig. 14.21 where $\mathbf{X}_n$ denotes either a voltage- or a current-source phasor and $\mathbf{Y}_n$ denotes the phasor of a voltage or current in the circuit.

Superposition states that the steady-state output-signal spectrum is the term-by-term product of the input spectrum and the circuit's transfer function. Once again, the transfer function provides all of the information needed to characterize the circuit's response. There are two important consequences of the phasor I/O relationship. First, the output signal's magnitude spectrum is determined by the *product* of the input signal's magnitude spectrum and the circuit's magnitude response. Second, the output phase spectrum is given by the *sum* of the input phase spectrum and the phase response

$$|\mathbf{Y}_n| = |\mathbf{X}_n||\mathbf{H}(jn\omega_o)|$$
$$\sphericalangle\mathbf{Y}_n = \sphericalangle\mathbf{X}_n + \sphericalangle\mathbf{H}(jn\omega_o)$$

14.6.2 Input/Output Power Spectra

The power spectrum of a circuit's steady-state output signal is obtained from

$$|\mathbf{Y}_n|^2 = |\mathbf{X}_n|^2|\mathbf{H}(jn\omega_o)|^2$$

and the output signal's average power is

$$P_{\text{av}} = \sum_{-\infty}^{\infty} |\mathbf{Y}_n|^2 = \sum_{-\infty}^{\infty} |\mathbf{X}_n|^2|\mathbf{H}(jn\omega_o)|^2$$

and so

$$P_{\text{av}} = |\mathbf{X}_0|^2|\mathbf{H}(0)|^2 + 2\sum_{n=1}^{\infty} |\mathbf{X}_n|^2|\mathbf{H}(jn\omega_o)|^2$$

Unfortunately, the average power in a circuit's output signal usually cannot be found by inspection of the waveform. In fact, constructing the waveform itself requires a computer to calculate the Fourier coefficients. Furthermore P_{av} as given above is not defined in closed form. Thus, we must rely on approximations of the average power and the waveform. The economics of computation dictate that we choose an integer N and take

$$y_{ss}(t) \cong \sum_{n=-N}^{N} \mathbf{Y}_n e^{jn\omega_o t}$$

and

$$P_{av} \cong |\mathbf{X}_0|^2|\mathbf{H}(0)|^2 + 2\sum_{1}^{N} |\mathbf{X}_n|^2|\mathbf{H}(jn\omega_o)|^2 \qquad (14.27)$$

14.6.3 Approximation Error*

When a periodic signal $f(t)$ is approximated by a truncated Fourier series the error signal of the approximation is defined as

$$\xi(t) = f(t) - \sum_{-N}^{N} \mathbf{F}_n e^{jn\omega_o t}$$

The mean-square error, or the average power in the error signal, is

$$\overline{\xi}^2 = \frac{1}{T}\int_0^T \xi^2(t)\,dt$$

Using the expression for $\xi(t)$ in $\overline{\xi}^2$ gives (after exploiting the orthogonality of the phasor signals),

$$\overline{\xi}^2 = P_{av} - \sum_{n=-N}^{N} |\mathbf{F}_n|^2$$

This result demonstrates that adding terms to the series of a signal reduces the power in the error signal and improves the approximation. It can also be shown that this expression actually defines the minimum mean-square value of the error signal. No other choice of coefficients to form a series approximating $f(t)$ can produce a smaller value of $\overline{\xi}^2$. So the truncated series that uses the Fourier coefficients gives the *best* approximation to $f(t)$. That is, if $\xi_a(t) = f(t) - \sum a_n e^{jn\omega_o t}$ we are assured that $\overline{\xi_a}^2 \geq \overline{\xi}^2$.

14.7 TRIGONOMETRIC FOURIER SERIES

If Euler's law is used in (14.1) a trigonometric Fourier series can be defined by

$$f(t) = a_0 + \sum_{n=1}^{\infty} (a_n \cos n\omega_o t + b_n \sin n\omega_o t)$$

where

$$a_0 = \frac{1}{T}\int_0^T f(t)\,dt = \mathbf{F}_o.$$

and for $n \geq 1$

$$a_n = \frac{2}{T}\int_0^T f(t)\cos n\omega_o t\,dt = (\mathbf{F}_n + \mathbf{F}_n{}^*) = 2\mathrm{Re}\{\mathbf{F}_n\}$$

and

$$b_n = \frac{2}{T}\int_0^T f(t)\sin n\omega_o t\,dt = j(\mathbf{F}_n - \mathbf{F}_n{}^*) = 2j\mathrm{Im}\{\mathbf{F}_n\}$$

(The proof is left as an exercise.)

The complex Fourier series can be obtained from the trigonometric Fourier series by using $\mathbf{F}_n = (a_n - jb_n)/2$ for $n \neq 0$ and $\mathbf{F}_0 = a_o$.

The cosine form of the Fourier series can be obtained from the trigonometric series according to

$$C_n = 2|\mathbf{F}_n| = 2\frac{|a_n - jb_n|}{2}$$

with

$$C_n = \sqrt{a_n{}^2 + b_n{}^2}$$

Likewise $\phi_n = \sphericalangle\mathbf{F}_n = \sphericalangle(a_n - jb_n)/2$ and $\phi_n = -\tan^{-1}(b_n/a_n)$.

Table 14.7 summarizes the forms of the complex exponential, cosine, and trigonometric Fourier series. We prefer to work with the complex exponential and cosine forms almost exclusively because they fit in the framework we have developed—using transfer functions to determine a circuit's response to exponential signals.

TABLE 14.7 COMPLEX EXPONENTIAL, COSINE, AND TRIGONOMETRIC FOURIER SERIES COEFFICIENTS

$f(t) =$	$\Sigma \mathbf{F}_n e^{jn\omega_o t}$	$\Sigma C_n \cos(n\omega_o t + \phi_n)$	$\Sigma(a_n \cos n\omega_o t + j \sin n\omega_o t)$
	$\mathbf{F}_n = \frac{1}{T}\int_0^T f(t)e^{-jn\omega_o t}\,dt$	$C_0 = \mathbf{F}_0$	$a_0 = \frac{1}{T}\int_0^T f(t)\,dt$
	$\mathbf{F}_o = \frac{1}{T}\int_0^T f(t)\,dt$	$C_n = 2\lvert\mathbf{F}_n\rvert$	$a_n = \frac{2}{T}\int_0^T f(t)\cos n\omega_o t\,dt$
		$\phi_n = \sphericalangle\mathbf{F}_n$	$b_n = \frac{2}{T}\int_0^T f(t)\sin n\omega_o t\,dt$
	$\mathbf{F}_n = \frac{a_n - jb_n}{2}$	$C_n = \sqrt{a_n{}^2 + b_n{}^2}$	$a_0 = \mathbf{F}_0$
		$\phi_n = -\tan^{-1}\frac{b_n}{a_n}$	$a_n = 2\mathrm{Re}\{\mathbf{F}_n\}$
			$b_n = 2j\mathrm{Im}\{\mathbf{F}_n\}$

SUMMARY

This chapter demonstrated that periodic signals can be represented by a weighted sum of complex exponential signals or by a trigonometric series. The Fourier coefficients of a signal are the weights that determine the relative amount of the exponential in the periodic signal. They are obtained by direct evaluation.

The forced or particular-solution component of the solution of a linear differential equation to a periodic forcing function is, by superposition, the sum of its forced solutions to the individual exponentials that comprise the Fourier series of the periodic forcing function.

The family of complex exponentials used to represent a periodic signal is a mutually orthogonal set of signals. This allows the average power content of a periodic signal to be associated with the discrete frequencies that are associated with its Fourier series.

The magnitude and phase spectra of a signal consist of graphs of the values of the magnitude and angle of its complex Fourier series coefficients. The spectra provide a visual tool for appreciating the unique frequency-domain identity of a periodic signal. The average power spectrum of a periodic signal is the square of its magnitude spectrum. The magnitude spectrum of the steady-state response of a circuit to a periodic or switched-periodic signal is obtained as the product of the magnitude spectrum of the input signal and the magnitude of the circuit's input/output transfer function, with the transfer function evaluated at the harmonic frequencies of the input signal. The phase spectrum of the output signal is the sum of the phase spectrum of the input signal and the angle of the transfer function.

Problems - Chapter 14

14.1 Find the complex Fourier series coefficients of the periodic signals shown below, where T is the signal period.

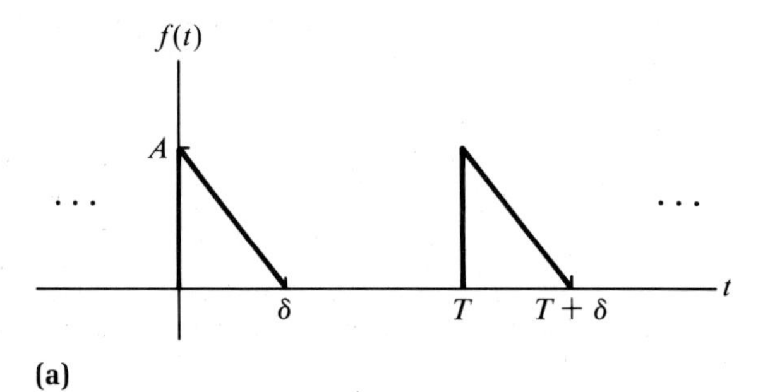

(a)

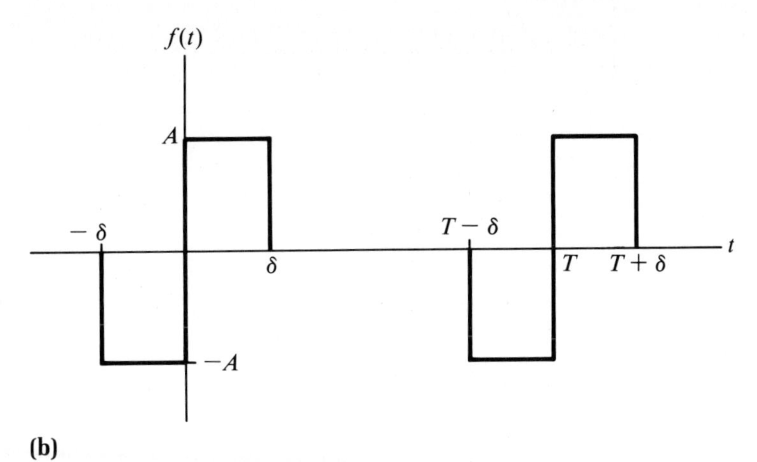

(b)

Figure P14.1

14.2 The signal $f_c(t)$ shown in Fig. P14.2(b) is a "clipped" version of the signal shown in Fig. P14.2(a) with $|K| < A$. Obtain the complex Fourier series coefficients for $f_c(t)$.

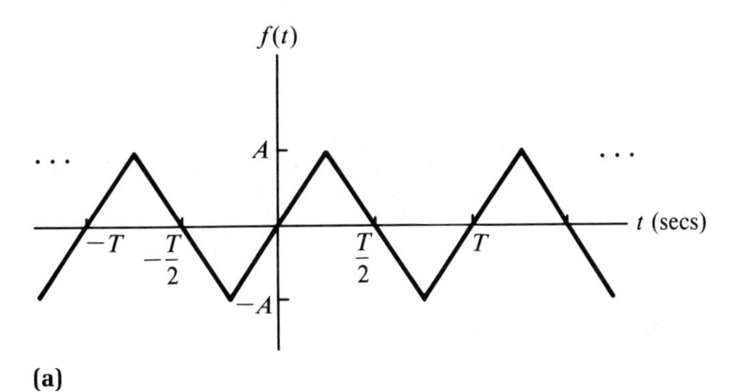

(a)

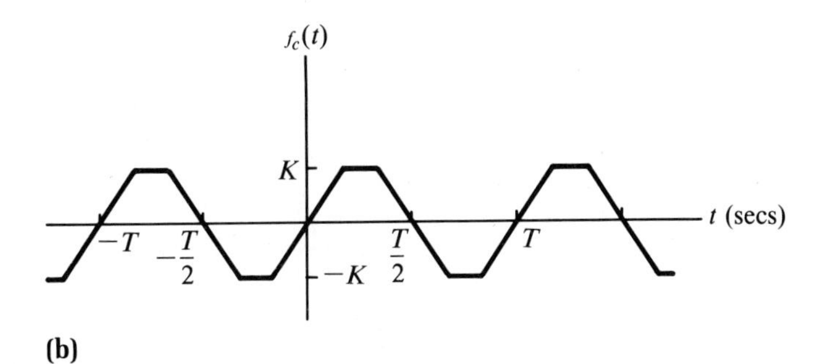

(b)

Figure P14.2

14.3 Obtain the complex Fourier series coefficients of the signal in Fig. P14.3.

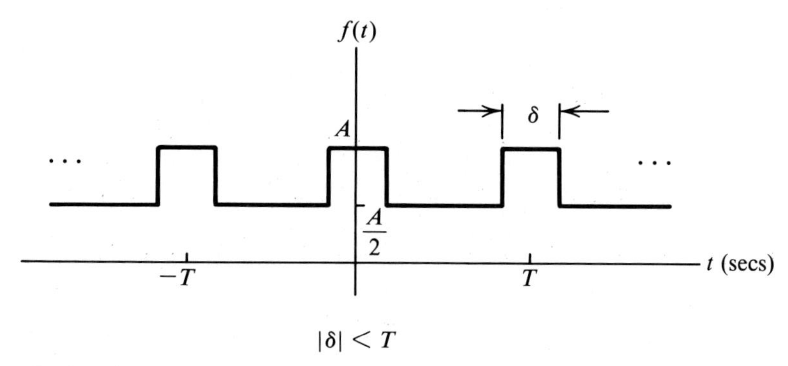

Figure P14.3

14.4 Obtain the Fourier series coefficients of the rectangular pulse train shown below.

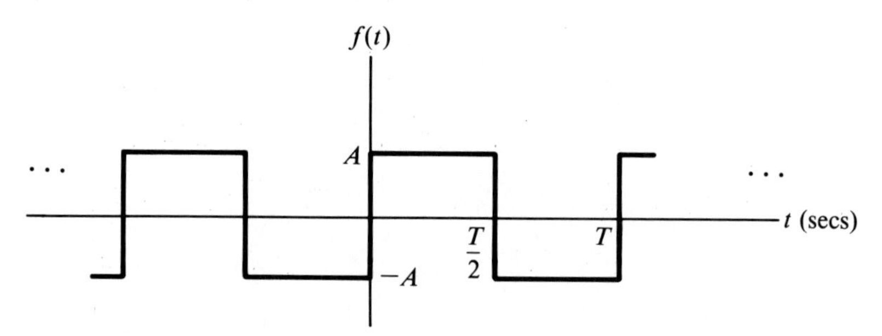

Figure P14.4

14.5 Obtain the Fourier series coefficients of the signal shown in Fig. P14.5.

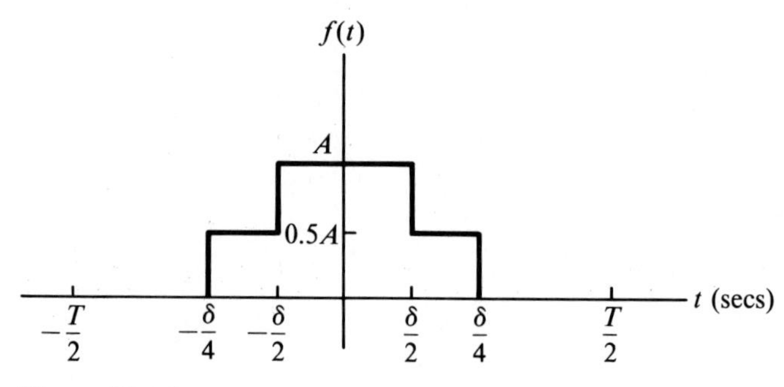

Figure P14.5

14.6 Find the DC content of the periodic signal in Fig. P14.5.

14.7 Draw the magnitude, phase, and power spectra of the signals in problems 14.2, 14.3, 14.4, and 14.5.

14.8 Find the average power and the power associated with the DC content of the signal in Fig. P14.5.

14.9 If the waveform of a current signal has been found to have the complex Fourier coefficients given below. Estimate the average power that the signal would deliver to a 10 Ω resistor.

TABLE P14.9

$n\omega_o$	$\|F_n\|$	$\angle F_n$
0	1.00	0°
ω_o	0.75	30°
$2\omega_o$	0.40	45°
$3\omega_o$	0.20	60°
$4\omega_o$	0.05	75°

14.10 If the signal in Fig. P14.3 is an input voltage to the circuit shown below, find the value of C for which the average power in the output signal is at least 70% of the average power in the input signal.

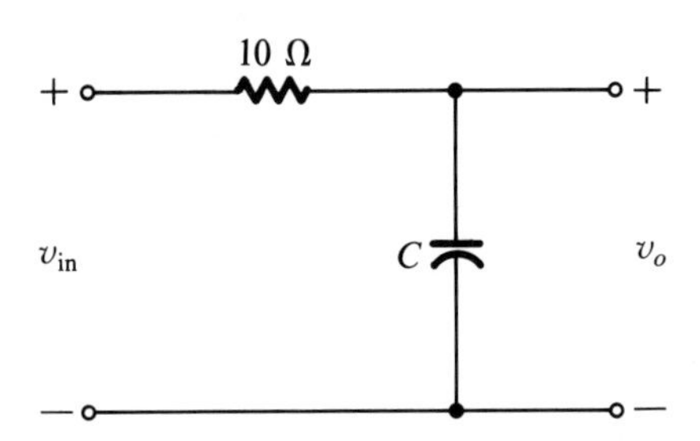

Figure P14.10

14.11 Repeat Problem 14.10 using the signal described in Problem 14.9.

14.12 Find the rms value of the steady-state input and output signals if the signal in Fig. 14.2(a) is the input to the circuit shown below, with $T = 1$ msec.

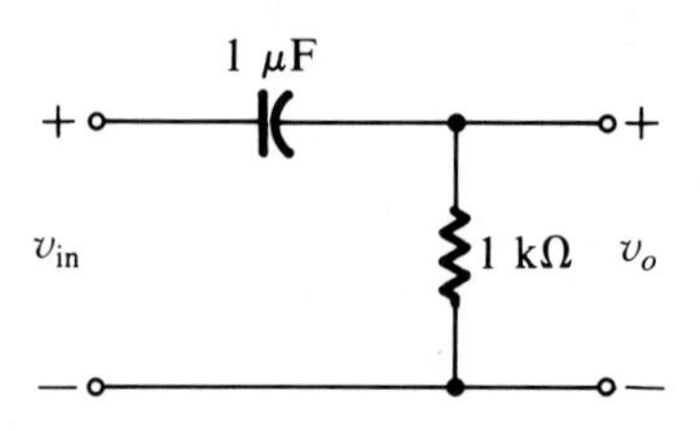

Figure P14.12

14.13 Construct a table showing the magnitude and power spectral components and the percentage of the total average power for the first five components of each of the signals in Fig. P14.2.

14.14 For each of the filters shown below, sketch the steady-state waveform for $v_o(t)$ when the filter is driven by the sawtooth signal shown in Fig. P14.14(c). Hint: consider the shape of $|\mathbf{H}(j\omega)|$ in relation to the spectral components of $v_{in}(t)$.

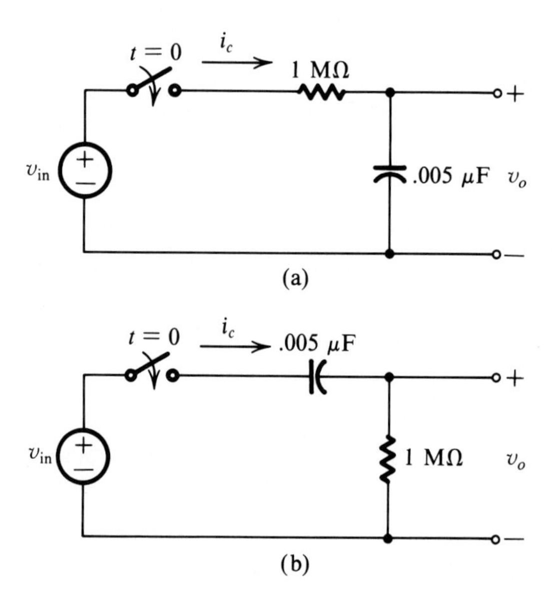

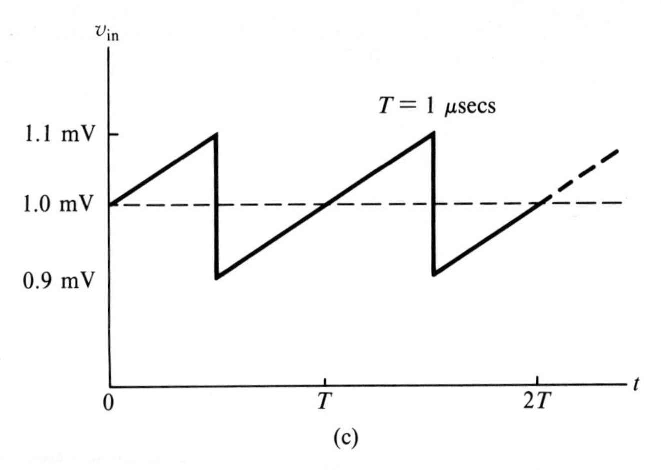

Figure P14.14

14.15 Find the approximate value of $i_c(t)$ after the switch is closed and the circuit is in the steady state in Fig. P.14.14(a) and (b).

14.16 Find the average power in the steady state value of $v_o(t)$ and compare it to the average power in $v_{in}(t)$.

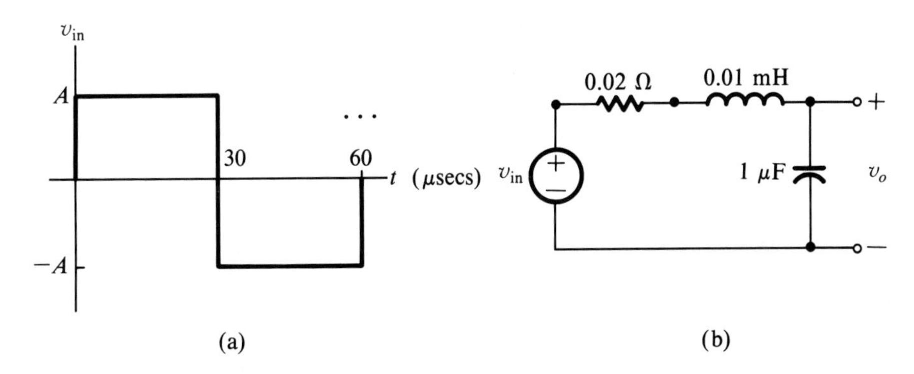

Figure P14.16

14.17 Obtain the cosine Fourier series coefficients of the signals given in Problem 14.3.

14.18 Draw the magnitude and power spectra for the complex Fourier series of $f(t)$ on the same scale. Discuss the difference in the shapes of the signals and their spectra. Assume that both signals have a period of 1 μsec.

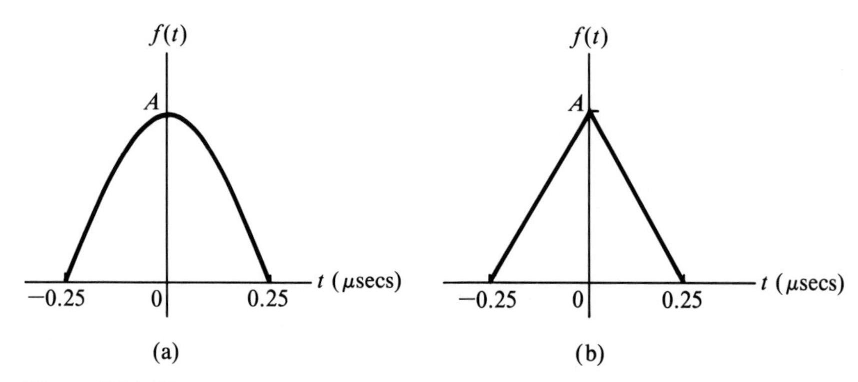

Figure P14.18

14.19 Find the cosine series coefficients for the signals in Problem 14.18.

14.20 Find the trigonometric series coefficients for the signals in Problem 14.18.

14.21 Draw the magnitude and power spectra for the signal shown below, and compare them to the spectra of the signal in Fig. P14.18(b).

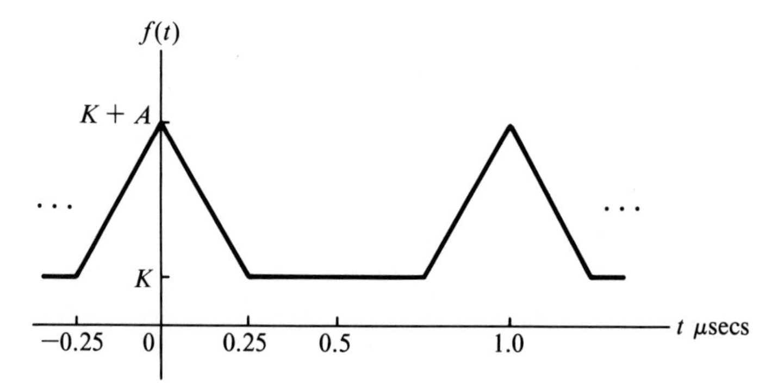

Figure P14.21

14.22 Obtain the complex Fourier series coefficients for $f(t)$.

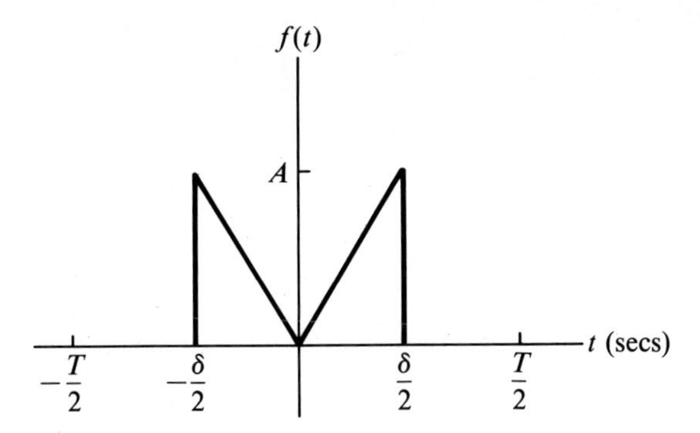

Figure P14.22

THE FOURIER TRANSFORM

INTRODUCTION

Many important physical signals are non-periodic and *cannot* be modelled by a Fourier series. This chapter will show that a non-periodic signal can be represented by a weighted *continuous sum* (or integral) of complex exponential signals, and will then use superposition to find the circuit's response to a non-periodic signal.

15.1 NON-PERIODIC SIGNALS: THE FOURIER TRANSFORM

The algorithm for calculating the Fourier coefficient $\mathbf{F}_n$ of a periodic signal [see (14.2)] may be impossible to evaluate if a signal is non-periodic because the integral could be meaningless as T becomes very large. So we'll need some other tools to help find the response of a circuit driven by a non-periodic signal.

A Fourier series represents a periodic signal by a discrete sum of suitably weighted phasor signals. The **Fourier transform** extends this key concept to non-periodic signals by representing them in terms of a continuous sum, or integral, of phasor signals. The Fourier integral (sum) of a signal $f(t)$ is defined as

$$f(t) = \frac{1}{2\pi}\int_{-\infty}^{\infty} \mathbf{F}(j\omega)e^{j\omega t}\,d\omega \tag{15.1}$$

where $\mathbf{F}(j\omega)$ is given by

$$\mathbf{F}(j\omega) = \int_{-\infty}^{\infty} f(t)e^{-j\omega t}\,dt \tag{15.2}$$

$\mathbf{F}(j\omega)$ is called the Fourier Transform of $f(t)$, and $f(t)$ and $\mathbf{F}(j\omega)$ are called a Fourier transform pair. The operation of taking the transform of a signal is denoted by $\mathfrak{F}\{\cdot\}$ and the transform itself is the function $\mathbf{F}(j\omega)$. The operation of taking the inverse transform to synthesize $f(t)$ is denoted by $\mathfrak{F}^{-1}\{\cdot\}$. Thus

$$\mathbf{F}(j\omega) = \mathfrak{F}\{f(t)\}$$

and

$$f(t) = \mathfrak{F}^{-1}\{\mathbf{F}(j\omega)\}$$

The pair formed by a time-domain signal and its frequency-domain transform will be symbolized by $f(t) \leftrightarrow \mathbf{F}(j\omega)$.

We will now explore the relationship between a pulse and a pulse train to reveal some notions about the Fourier transform, and develop the Fourier transform of a signal from the Fourier series of its periodic extension.

Given a signal $f(t)$ defined on an interval $(-T/2, T/2)$, its **periodic extension** is the signal $\tilde{f}(t)$ obtained from replicating the waveform of $f(t)$ along the time axis as shown in Fig. 15.1. Since $\tilde{f}(t)$ is periodic its Fourier series (assuming it exists) can be written as

$$\tilde{f}(t) = \sum_{-\infty}^{\infty} \tilde{\mathbf{F}}_n e^{jn\omega_o t}$$

where

$$\tilde{\mathbf{F}}_n = \frac{1}{T}\int_{-T/2}^{T/2} f(t)e^{-jn\omega_o t}\,dt$$

The periodic extension of the pulse function in Fig. 15.1 is a pulse train (see example 14.1) having the Fourier series defined by

$$\tilde{\mathbf{F}}_n = A\frac{\Delta}{T}\,\mathrm{Sa}\left(n\omega_o\frac{\Delta}{2}\right)$$

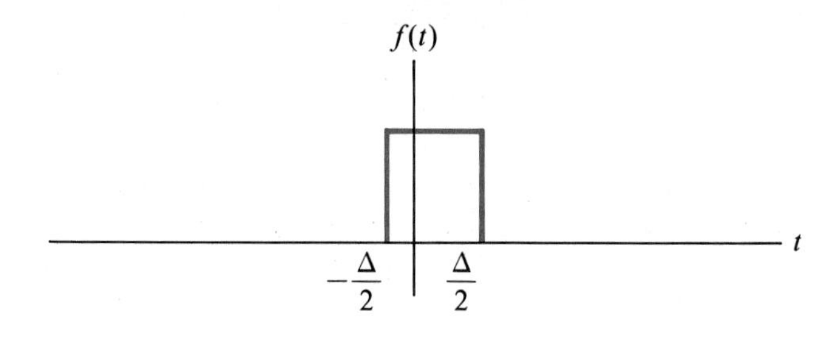

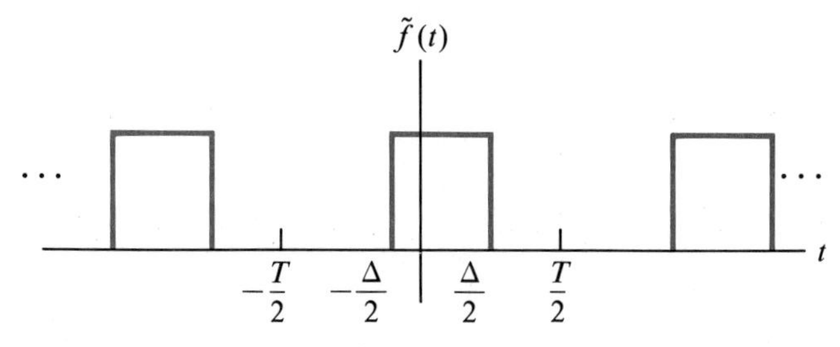

Figure 15.1 A rectangular pulse and its periodic extension.

Note that as $T \to \infty$ the signal $\tilde{f}(t)$ becomes $f(t)$, since

$$\lim_{T\to\infty} \tilde{f}(t) = f(t)$$

and the Fourier coefficients of $\tilde{f}(t)$ vanish:

$$\lim_{T\to\infty} \tilde{\mathbf{F}}_n = \lim_{T\to\infty} \quad A\frac{\Delta}{T}\,\mathrm{Sa}\left(n\omega_o\frac{\Delta}{2}\right) = 0$$

As T becomes large ω_o becomes small, causing the value of the sample function to become 1, so $\tilde{\mathbf{F}}_n$ approaches zero. The spectral components of $\tilde{f}(t)$ are closer together, but have small heights.

For any finite T the Fourier coefficients $\tilde{\mathbf{F}}_n$ still synthesize $\tilde{f}(t)$ but *not* the single pulse $f(t)$. Since T is increasing in size while $\tilde{\mathbf{F}}_n$ is decreasing, we are led to examine their product

$$T\tilde{\mathbf{F}}_n = A\Delta\,\mathrm{Sa}\left(n\omega_o\frac{\Delta}{2}\right)$$

Because the sample function has $\mathrm{Sa}(0) = 1$, and $\omega_o = 2\pi/T$, we discover that

$$\lim_{T\to\infty} T\tilde{\mathbf{F}}_n = A\Delta$$

Although $\tilde{\mathbf{F}}_n$ vanishes, the product of the Fourier coefficients and the period on which the signal is expanded does not vanish. We use this result to define the following function in the frequency domain:

$$\tilde{\mathbf{F}}(j\omega) = \begin{cases} T\tilde{\mathbf{F}}_n & \omega = n\omega_o,\ n = 0, 1, \pm 2, \ldots \\ 0 & \text{elsewhere} \end{cases}$$

and so the pulse train has

$$\tilde{\mathbf{F}}(j\omega) = \begin{cases} A\Delta \, \mathrm{Sa}\left(\omega\dfrac{\Delta}{2}\right) & \omega = n\omega_o,\ n = 0, \pm 2, \ \ldots \\ 0 & \text{elsewhere} \end{cases}$$

$\tilde{\mathbf{F}}(j\omega)$ as defined above consists of a scaled (by T) copy of the Fourier coefficients of the periodic extension signal $\tilde{f}(t)$ at the points ω that are multiples of the fundamental frequency.

In general $\tilde{\mathbf{F}}(j\omega)$ provides an alternate expression for the Fourier series of the periodic extension $\tilde{f}(t)$:

$$\tilde{f}(t) = \sum_{-\infty}^{\infty} \tilde{\mathbf{F}}_n e^{jn\omega_o t} = \sum_{-\infty}^{\infty} \frac{1}{T} T \tilde{\mathbf{F}}_n e^{jn\omega_o t} = \sum_{-\infty}^{\infty} \frac{\omega_o}{2\pi} \tilde{\mathbf{F}}(j\omega_n) e^{jn\omega_o t}$$

with $\omega_n = n\omega_o$. Therefore

$$\tilde{f}(t) = \sum_{-\infty}^{\infty} \frac{1}{2\pi} \tilde{\mathbf{F}}(j\omega_n) e^{j\omega_n t} \omega_o$$

Next we let $\Delta\omega_n = \omega_o$, so

$$\tilde{f}(t) = \sum_{-\infty}^{\infty} \frac{1}{2\pi} \tilde{\mathbf{F}}(j\omega_n) e^{j\omega_n t} \, \Delta\omega_n$$

The above sum defines $\tilde{f}(t)$ for any value of T. If $(1/2\pi)\tilde{\mathbf{F}}(jn\omega_o)e^{j\omega n_o t}$ is interpreted as a point along the curve defined by $(1/2\pi)\tilde{\mathbf{F}}(j\omega)e^{j\omega t}$ (as shown in Fig. 15.2) the sum is just an approximation to the area under the curve.

In the limit, as $T \to \infty$, $\tilde{f}(t) \to f(t)$, $\tilde{\mathbf{F}}(j\omega) \to \mathbf{F}(j\omega)$, $\Delta\omega \to d\omega$ and the sum becomes the integral or continuous sum

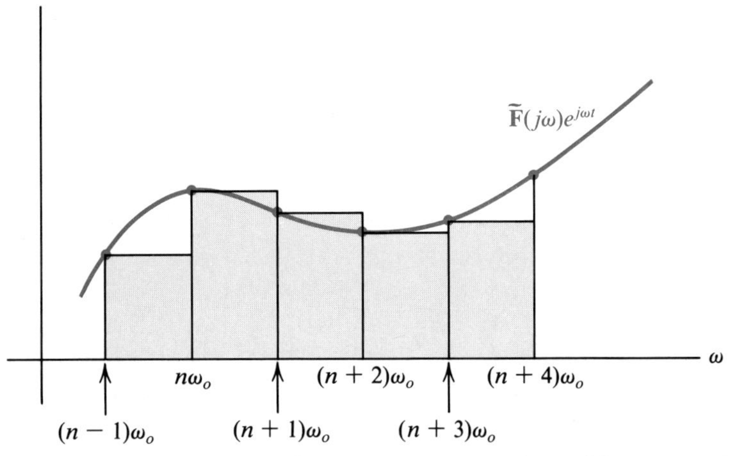

Figure 15.2 The area under a curve is approximated by a sum of the areas of the rectangles defined by values of the function at discrete points.

$$f(t) = \lim_{T\to\infty} \tilde{f}(t) = \lim_{T\to\infty} \frac{1}{2\pi} \sum_{-\infty}^{\infty} \tilde{\mathbf{F}}(j\omega_n)e^{j\omega_n t}\,\Delta\omega_n$$

$$= \lim_{\Delta\omega_n\to 0} \frac{1}{2\pi} \sum_{-\infty}^{\infty} \tilde{\mathbf{F}}(j\omega_n)e^{j\omega_n t}\,\Delta\omega_n$$

and

$$f(t) = \frac{1}{2\pi}\int_{-\infty}^{\infty} \mathbf{F}(j\omega)e^{j\omega t}\,d\omega$$

Example 15.1

Use the Fourier series of a pulse train to derive the Fourier transform of a single pulse.

Solution: For the rectangular pulse train $T\mathbf{F}_n = A\Delta\ \mathrm{Sa}(n\omega_o\Delta/2)$ and replacing $n\omega_o$ by ω gives

$$\mathbf{F}(j\omega) = \lim_{T\to\infty} T\mathbf{F}_n = \lim_{T\to\infty} A\Delta\ \mathrm{Sa}\left(n\omega_o\frac{\Delta}{2}\right)$$

$$= A\Delta\ \mathrm{Sa}\left(\omega\frac{\Delta}{2}\right)$$

The next example will support our conclusion that this expression is the Fourier transform of the rectangular pulse.

Example 15.2

Derive the Fourier transform of the rectangular pulse.

Solution: Recall that the pulse signal has

$$f(t) = \begin{cases} A & |t| \le \dfrac{\Delta}{2} \\ 0 & |t| > \dfrac{\Delta}{2} \end{cases}$$

Using (15.2) gives

$$\mathbf{F}(j\omega) = \int_{-\Delta/2}^{\Delta/2} Ae^{-j\omega t}\,dt = \frac{-A}{j\omega}e^{-j\omega t}\Big|_{-\Delta/2}^{\Delta/2}$$

$$= \frac{A}{j\omega}\left[e^{j\omega\Delta/2} - e^{-j\omega\Delta/2}\right]$$

When Euler's law is used to write the exponentials in terms of sines and cosines this becomes

$$\mathbf{F}(j\omega) = \frac{2A}{\omega} \sin \omega\frac{\Delta}{2}$$

Multiplying by $\frac{\Delta}{\Delta}$ puts $\mathbf{F}(j\omega)$ in the familiar (sin x) x form

$$\mathbf{F}(j\omega) = A\Delta \text{ Sa}\left(\omega\frac{\Delta}{2}\right)$$

The Fourier integral (series) in (15.1) synthesizes $f(t)$ from a continuous (discrete) sum of weighted phasor signals. In the Fourier integral, the phasor signal $e^{j\omega t}$ is weighted by $1/2\pi\, \mathbf{F}(j\omega)$.

15.2 AMPLITUDE SPECTRUM

The amplitude spectrum of $f(t)$ consists of a magnitude spectrum and a phase spectrum. The magnitude spectrum is the graph of $|\mathbf{F}(j\omega)|$ vs ω and the phase spectrum is the graph of $\measuredangle\mathbf{F}(j\omega)$ vs ω.

Example 15.3

The rectangular pulse has the magnitude and phase spectra shown in Fig. 15.3.

In engineering work the portion of the spectrum from the origin to the first spectral zero crossing is sometimes used to approximate the entire spectrum of the pulse signal. The first zero-crossing frequency $\hat{f}$ is defined by $\hat{\omega} = 2\pi/\Delta$ or, in units of Hz,

$$\boxed{\hat{f} = \frac{1}{\Delta}}$$

A convenient rule of thumb is that the spectral width is the reciprocal of the pulse width. A narrower pulse will have a wider spectrum. The rapid on-off-on sequence of the pulse requires higher frequency content than does a broader pulse. Note, too, that reducing the pulse width causes the value of the spectrum at $\omega = 0$ to decrease—meaning that a narrower pulse has relatively less signal density at DC.

15.3 EXISTENCE OF THE FOURIER TRANSFORM*

The Fourier transform of a function $f(t)$ is said to exist if $\mathbf{F}(j\omega)$ is defined for all ω. Formally, we mean that $|\mathbf{F}(j\omega)| < \infty$ for $-\infty < \omega < \infty$. The exis-

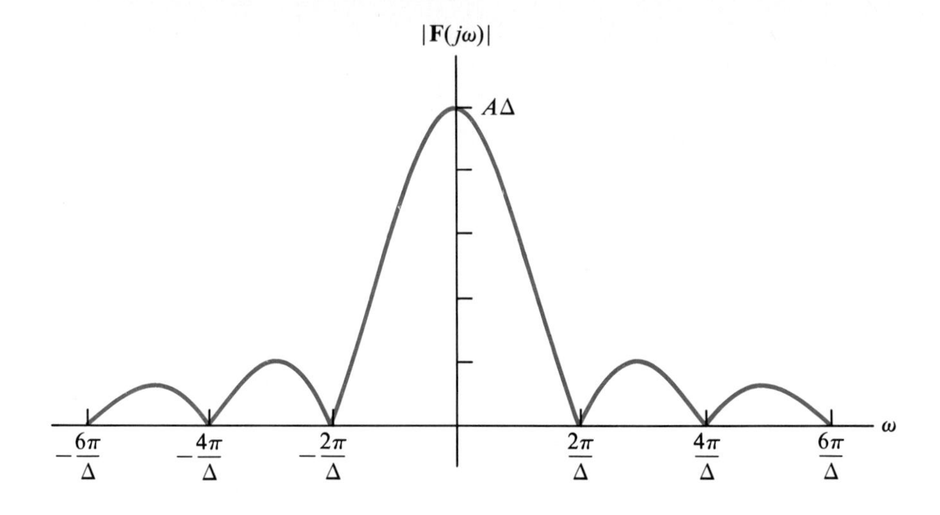

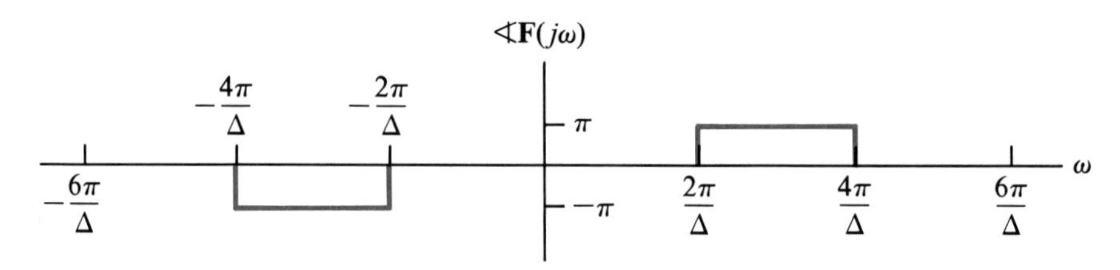

Figure 15.3 The magnitude and phase spectra of the rectangular pulse.

tence of the Fourier transform can be guaranteed if $f(t)$ meets a simple test. From the definition of the transform in (15.2) we can write

$$|\mathbf{F}(j\omega)| \leq \int_{-\infty}^{\infty} |f(t)|\, dt$$

since $|e^{j\omega t}| = 1$. We conclude that every absolutely integrable function has a Fourier transform. This condition only states that the function $\mathbf{F}(j\omega)$ exists for all ω, it does *not* guarantee that $f(t)$ can be synthesized from $\mathbf{F}(j\omega)$ in the Fourier integral of (15.1).

To ensure that the Fourier integral converges to $f(t)$ it is sufficient that $f(t)$ satisfy the Dirchlet conditions (See Chapter 14). For such functions the Fourier integral (15.1) converges to $f(t)$ at all points where $f(t)$ is continuous, and to $\frac{1}{2}[f(t^-) + f(t^+)]$ at points where $f(t)$ has a jump discontinuity.

The class of functions that satisfy the Dirchlet conditions is rather large for our purposes, but functions that do not satisfy the Dirchlet conditions have utility in engineering. By using impulses, we can extend Fourier transforms to include an even richer class of signals than those which satisfy the Dirchlet conditions.

Example 15.4

Find the Fourier transform of the decaying exponential $f(t) = e^{-\alpha t}u(t)$, with $\alpha > 0$.

Solution: Since

$$\int_{-\infty}^{\infty} |f(t)|\, dt = \int_{0}^{\infty} e^{-\alpha t}\, dt = \frac{1}{\alpha}$$

$f(t)$ is absolutely integrable and satisfies the Dirchlet conditions. Hence

$$\mathbf{F}(j\omega) = \int_{0}^{\infty} e^{-\alpha t} e^{-j\omega t}\, dt = \frac{-1}{\alpha + j\omega} e^{-(\alpha + j\omega)t} \Big|_{0}^{\infty}$$

If $\alpha > 0$ the exponential factor can be evaluated as

$$e^{-(\alpha+j\omega)t} \Big|_{0}^{\infty} = e^{-\alpha t} e^{-j\omega t} \Big|_{0}^{\infty} = \lim_{t\to\infty} e^{-\alpha t} e^{-j\omega t} - 1 = -1$$

so

$$\mathbf{F}(j\omega) = \frac{1}{\alpha + j\omega}$$

The magnitude and phase spectra of $f(t)$ are shown in Figure 15.4, where

$$|\mathbf{F}(j\omega)| = \frac{1}{\sqrt{\alpha^2 + \omega^2}} \qquad \sphericalangle \mathbf{F}(j\omega) = -\tan^{-1} \frac{\omega}{\alpha}$$

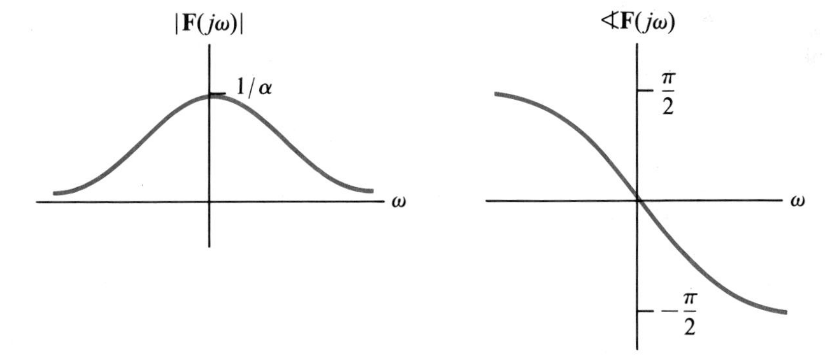

Figure 15.4 The magnitude and phase spectra of the exponential signal.

Skill Exercise 15.1

If $f(t) = e^{at}u(-t)$, with $a > 0$, find $\mathfrak{F}\{f(t)\}$.

Answer: $\mathbf{F}(j\omega) = \dfrac{1}{a - j\omega}$

15.3.1 Generalized Functions and Fourier Transforms

The impulse function and other generalized functions do not themselves exist, let alone meet the sufficiency conditions given for the existence of the Fourier transform. However, they play such an important role in engineering work that we associate with them a Fourier transform.

Example 15.5 (Time Impulse)

The impulse function does not satisfy the Dirchlet conditions because $\delta(t)$ is not even defined at $t = 0$. We saw in Chapter 6 that the impulse function can be viewed as a generalized function whose effect on a circuit can be obtained as the limiting behavior of the circuit when a rectangular pulse function is applied.

Recall that with the impulse $\delta(t)$ we associated the limit of the response of a circuit to the pulse signal $g_\Delta(t)$ (shown in Fig. 15.5) when $\Delta \to 0$. The area under the curve of $g_\Delta(t)$ is 1 for any finite but arbitrarily small Δ. We define the Fourier transform of the impulse function to be the limit of the Fourier transform of $g_\Delta(t)$ as $\Delta \to 0$:

$$\mathbf{F}(j\omega) = \lim_{\Delta\to 0} \frac{\Delta}{\Delta} \,\mathrm{Sa}\left(\omega\frac{\Delta}{2}\right) = \lim_{\Delta\to 0} \mathrm{Sa}\left(\omega\frac{\Delta}{2}\right) = 1$$

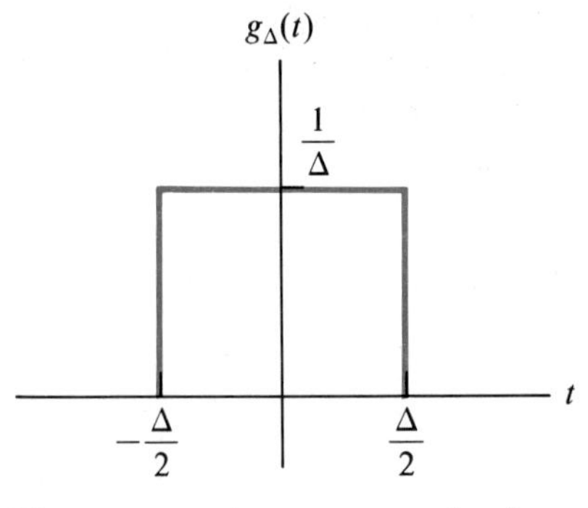

Figure 15.5 A unit-area pulse function.

The impulse function has the uniform magnitude spectrum shown in Fig. 15.6. All phasor signals are present in the same proportion. Using the sampling property (see Chapter 6) of the unit-impulse function in (15.2) gives

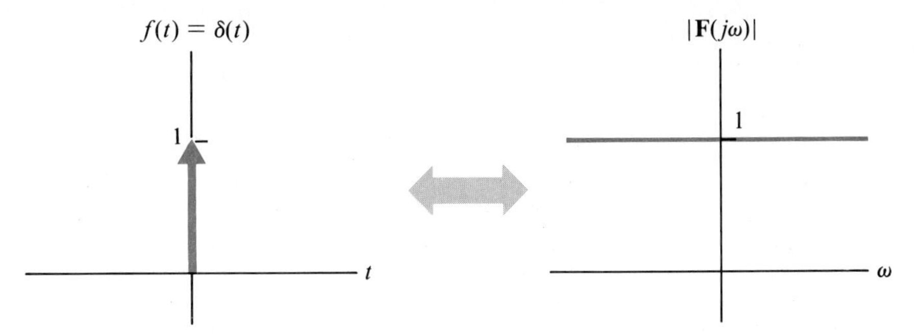

Figure 15.6 The unit-impulse function has a uniform (flat) magnitude spectrum.

$$\mathbf{F}(j\omega) = \int_{-\infty}^{\infty} \delta(t)e^{-j\omega t}\,dt = e^{-j\omega t}\Big|_{t=0} = 1$$

Figure 15.7 summarizes the "limit relationship" between the pulse functions, the generalized impulse function, and their spectra.

The impulse and its Fourier transform illustrate the fact that a generalized function can have a Fourier transform that exists everywhere in the frequency domain, even though the generalized function does not exist, in the strict sense, in the time domain.

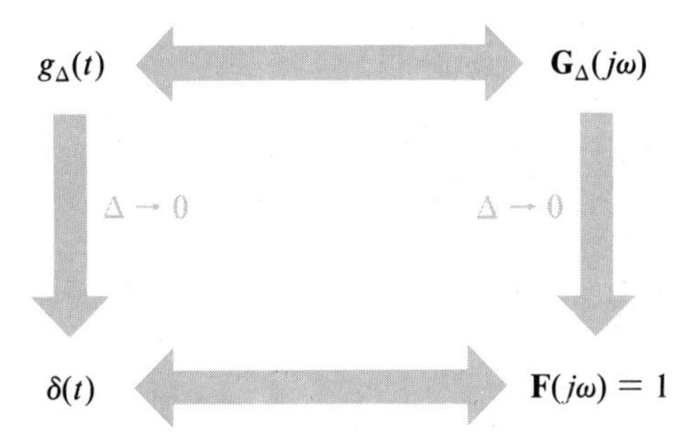

Figure 15.7 The impulse function is the limit of a sequence of "unit-area" rectangular pulses. Its Fourier transform is the limit of the Fourier transforms of the rectangular pulses.

Example 15.6 (Frequency Impulse)

We will now show that the Fourier transform of a constant signal is an impulse at the origin of the frequency domain: $\mathfrak{F}\{1\} \leftrightarrow 2\pi\delta(\omega)$. That is, a constant in the time domain has a frequency spectrum consisting of a single impulse at $\omega = 0$, as shown in Fig. 15.8.

For all t the sampling property of the impulse gives

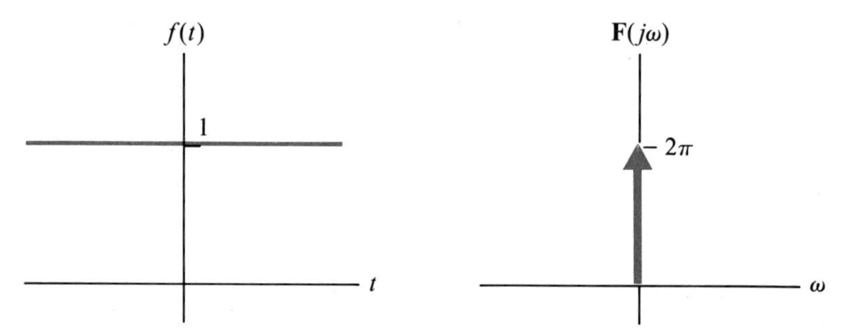

Figure 15.8 The Fourier transform of a constant signal is a weighted impulse at the orgin in the frequency domain.

$$\mathfrak{F}^{-1}\{2\pi\delta(\omega)\} = \frac{1}{2\pi}\int_{-\infty}^{\infty} 2\pi\delta(\omega)e^{j\omega t}\, d\omega = 1$$

This result has the following interpretation: A constant signal in the time domain can be considered to be the limit signal of a sequence of signals whose Fourier transforms have the impulse as their limit. Thus, a well-defined time-domain signal may have a generalized function as its Fourier transform. This relationship is illustrated by the next example.

Example 15.7

The function $f_\alpha(t) = e^{-\alpha|t|}$ with $\alpha > 0$ has the constant signal as its limit as $\alpha \rightarrow 0$. Each $f_\alpha(t)$ is absolutely integrable for $\alpha > 0$, with Fourier transform

$$\mathbf{F}\alpha(j\omega) = \frac{2\alpha}{\alpha^2 + \omega^2}$$

The sequence of Fourier transforms $\mathbf{F}_\alpha(j\omega)$ has the defining properties of an impulse function (in ω) as $\alpha \rightarrow 0$, since

i. The function vanishes everywhere but at the origin:

$$\lim_{\alpha \rightarrow 0} \mathbf{F}_\alpha(j\omega) = 0 \qquad \omega \neq 0$$

ii. The function is unbounded at the origin:

$$\lim_{\alpha \rightarrow 0} \mathbf{F}_\alpha(0) = \infty$$

iii. The functions all have the same area 2π under their curve.

$$\int_{-\infty}^{\infty} \mathbf{F}_\alpha(j\omega)\, d\omega = \int_{-\infty}^{\infty} \frac{2\alpha d\omega}{\alpha^2 + \omega^2} = 2\pi \qquad \text{for all } \alpha$$

Each Fourier transform $\mathbf{F}_\alpha(j\omega)$ in the sequence has a time-domain counterpart $f_\alpha(t)$ and we associate their time-domain limit with the limit of the sequence of transforms, as summarized in Fig. 15.9, i.e. $f(t) \leftrightarrow 2\pi\delta(\omega)$.

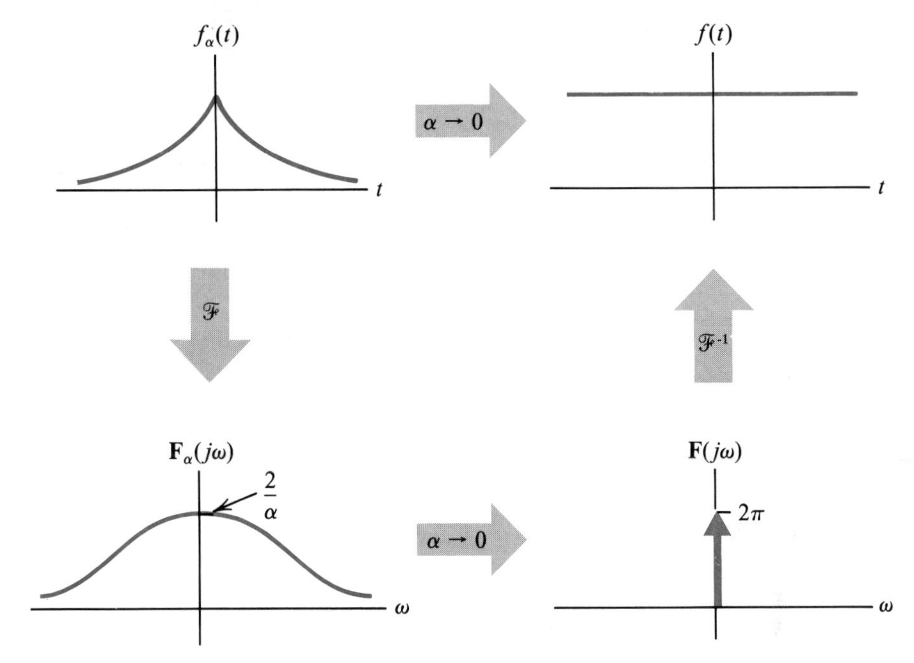

Figure 15.9 The Fourier transform of $f(t)$ is the limit of the Fourier transform of $f_\alpha(t)$ as $\alpha \to 0$.

15.3.2 Fourier Transforms "In the Limit"

A Fourier transform is said to exist "in the limit" when either it or its corresponding time-domain function is a generalized function. Strictly speaking, such transforms or their time-domain signals do not exist in the sense of (15.1). The impulse and constant signals are examples that we have already considered. The next example will demonstrate that the cosine function, which is *not* absolutely integrable, has a Fourier transform in the limit.

Example 15.8

The absolutely integrable function $f_\alpha(t) = e^{-\alpha|t|} \cos(\omega_o t)$, for $\alpha > 0$ defines the cosine function in the limit by

$$f(t) = \lim_{\alpha \to 0} e^{-\alpha|t|} \cos \omega_o t = \cos \omega_o t$$

When α becomes small in Fig. 15.10, the curve for $f_\alpha(t)$ approaches the curve for $f(t)$. The Fourier transform of $f_\alpha(t)$ can be determined directly from (15.2) as

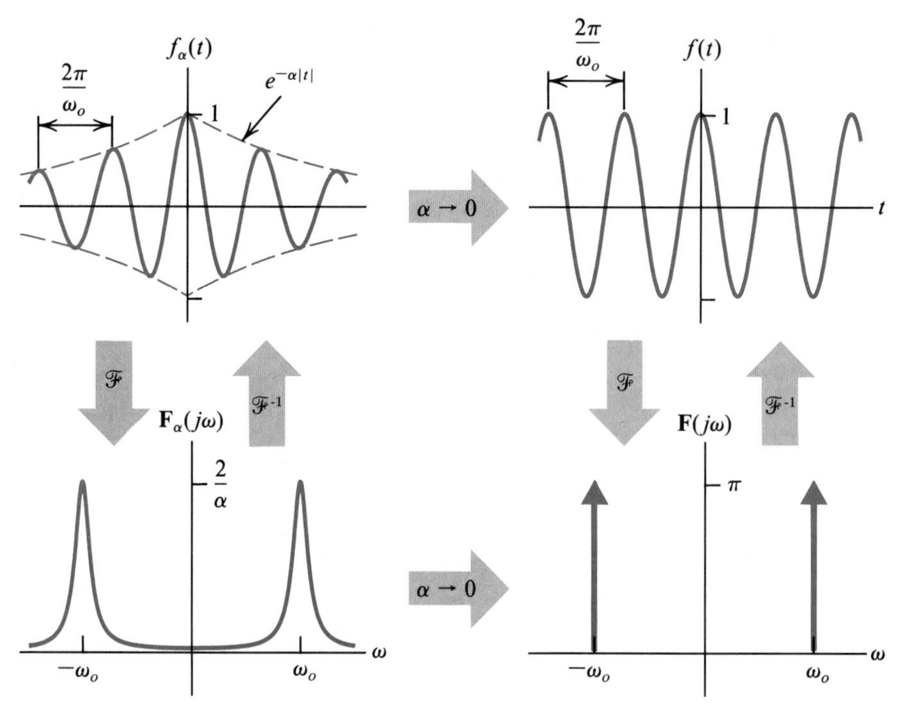

Figure 15.10 The Fourier transform of the periodic cosine signal is the limit of the Fourier transform of the damped cosine $f_\alpha(t)$ as $\alpha \to 0$.

$$\mathbf{F}_\alpha(j\omega) = \frac{\alpha}{\alpha^2 + (\omega + \omega_o)^2} + \frac{\alpha}{\alpha^2 + (\omega - \omega_o)^2}$$

(Verify this result.) As a check, we point out that letting $\omega_o \to 0$ gives the same $\mathbf{F}_\alpha(j\omega)$ as in Example 15.7. Now we will show that the sequence of Fourier transforms defined by $\mathbf{F}_\alpha(j\omega)$ defines a pair of impulse functions at $\omega = \pm\omega_o$ as $\alpha \to 0$. For simplicity, let

$$\hat{\mathbf{F}}_\alpha(j\omega) = \frac{\alpha}{\alpha^2 + (\omega - \omega_o)^2}$$

Then

i. $\hat{\mathbf{F}}_\alpha(j\omega)$ vanishes for $\omega \neq \omega_o$ as $\alpha \to 0$.
ii. $\hat{\mathbf{F}}_\alpha(j\omega)$ is unbounded at $\omega = \omega_o$ as $\alpha \to 0$, since $\lim_{\alpha \to 0} \mathbf{F}_\alpha(j\omega_o) = \infty$.
iii. The area under the curve of $\mathbf{F}_\alpha(j\omega)$ is constant for arbitrarily small but nonzero α because

$$\int_{-\infty}^{\infty} \frac{\alpha\, d\omega}{\alpha^2 + (\omega - \omega_o)^2} = \alpha\left(\frac{1}{2}\right)[\tan^{-1}(\alpha) - \tan^{-1}(-\infty)] = \pi$$

Likewise, as $\alpha \to 0$ the first term in the expression for $\mathbf{F}_\alpha(j\omega)$ satisfies the defining properties of an impulse function at $\omega = -\omega_o$. As $\alpha \to 0$, the sequence of Fourier transforms converges

$$\mathbf{F}_\alpha(j\omega) \to \mathbf{F}(j\omega) = \pi[\delta(\omega + \omega_o) + \delta(\omega-\omega_o)]$$

and we associate $\mathbf{F}(j\omega)$ with $\cos \omega_o t$ (the time-domain signal that is the limit of the sequence $f_\alpha(t)$ whose transforms converge to $\mathbf{F}(j\omega)$):

$$\mathfrak{F}\{\cos \omega_o t\} = \pi[\delta(\omega + \omega_o) + \delta(\omega - \omega_o)]$$

Figure 15.10 illustrates these relationships.

Notice that **the cosine function is not absolutely integrable,** that **its spectrum consists only of generalized functions,** and that **the impulses occur at $\pm\omega_o$, the frequency of the cosine**—which makes sense because $\mathbf{F}(j\omega)$ represents the frequency content of a signal.

Skill Exercise 15.2

Show that the signal $f(t) = \cos \omega_o t$ can be synthesized by using $\mathbf{F}(j\omega) = \pi[\delta(\omega + \omega_o) + \delta(\omega - \omega_o)]$ in (15.1).

Skill Exercise 15.3

Find the Fourier transform of $f(t) = \sin \omega_o t$.

Answer: $\mathfrak{F}\{\sin \omega_o t\} = j\pi[\delta(\omega + \omega_o) - \delta(\omega - \omega_o)]$

Example 15.9

The function Sgn (t) shown in Fig. 15.11(a) is not absolutely integrable. To find its Fourier transform we consider the sequence of absolutely integrable functions $f_\alpha(t)$ shown in Fig. 15.11(b). Note that $f_\alpha(t) \to$ Sgn (t)

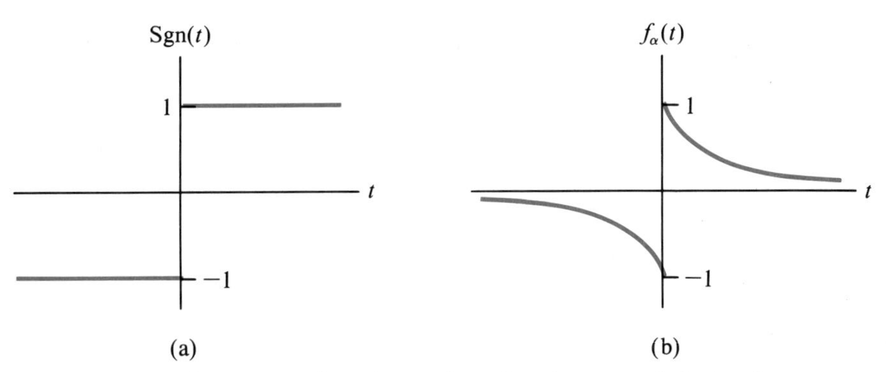

Figure 15.11 The signum function and a related exponential signal.

as $\alpha \to 0$. For $t \neq 0$ we have $f_\alpha(t) = e^{-\alpha t}$ if $t > 0$ and $f_\alpha(t) = -e^{\alpha t}$ if $t < 0$. Using (15.2) we get

$$\mathbf{F}_\alpha(j\omega) = -\int_{-\infty}^{0} e^{\alpha t} e^{-j\omega t}\,dt + \int_{0}^{\infty} e^{-\alpha t} e^{-j\omega t}\,dt$$

$$= \frac{-1}{\alpha - j\omega} e^{(\alpha - j\omega)t}\Big|_{-\infty}^{0} + \frac{-1}{\alpha + j\omega} e^{-(\alpha + j\omega)t}\Big|_{0}^{\infty}$$

For $\alpha > 0$ the evaluated terms vanish at $t = \pm\infty$, so

$$\mathbf{F}_\alpha(j\omega) = \frac{-1}{\alpha - j\omega} + \frac{1}{\alpha + j\omega} = \frac{-2j\omega}{\alpha^2 + \omega^2}$$

In the limit as $\alpha \to 0$, $\mathbf{F}_\alpha(j\omega) \to \mathbf{F}(j\omega)$ and

$$\mathbf{F}(j\omega) = \lim_{\alpha \to 0} \frac{-2j\omega}{\alpha^2 + \omega^2} = \frac{2}{j\omega} \text{ for } \omega \neq 0.$$

For $\omega = 0$ we note that $\mathbf{F}_\alpha(0) = 0$ for any $\alpha \neq 0$, so we assign the value $\mathbf{F}(j0) = 0$. Then

$$\mathbf{F}(j\omega) = \begin{cases} \dfrac{2}{j\omega} & \omega \neq 0 \\ 0 & \omega = 0 \end{cases}$$

15.4 FOURIER TRANSFORM PROPERTIES

15.4.1 Uniqueness

Distinct signals have distinct Fourier transforms. That is, if $f_1(t) \leftrightarrow \mathbf{F}_1(j\omega)$ and $f_2(t) \leftrightarrow \mathbf{F}_2(j\omega)$, then $f_1(t) = f_2(t) \leftrightarrow \mathbf{F}_1(j\omega) = \mathbf{F}_2(j\omega)$. This property lets us associate a Fourier transform and a time-domain signal, without requiring direct use of the Fourier integral. (Strictly speaking, this property does not apply to signals that differ only at discrete points.)

15.4.2 Symmetry

If $f(t)$ is real valued, then $\mathbf{F}(j\omega) = \mathbf{F}^*(-j\omega)$, $|\mathbf{F}(j\omega)| = |\mathbf{F}(-j\omega)|$ and $\sphericalangle\mathbf{F}(j\omega) = -\sphericalangle\mathbf{F}(j\omega)$. The proof of the symmetry property is left as an exercise.

15.4.3 Area

When $\mathbf{F}(j\omega)$ is finite at $\omega = 0$ the area under the curve of $f(t)$ is related to $\mathbf{F}(j\omega)$ by

$$\int_{-\infty}^{\infty} f(t)\, dt = \mathbf{F}(0) \tag{15.3}$$

Likewise, the "area" under the curve of $\mathbf{F}(j\omega)$ determines f(0):

$$\frac{1}{2\pi}\int_{-\infty}^{\infty} \mathbf{F}(j\omega)\, d\omega = f(0) \tag{15.4}$$

provided that $f(0)$ is defined. We prove (15.3) by evaluating (15.2) at $\omega = 0$, and 15.4 by evaluating (15.1) at $t = 0$. This property can be used to check your work.

Example 15.10

The rectangular pulse has an area given by $A\Delta$ and inspection shows that

$$\mathbf{F}(0) = A\Delta \text{ Sa}\left(\omega\frac{\Delta}{2}\right)\Bigg|_{\omega=0} = A\Delta$$

because $\text{Sa}(0) = 1$.

15.4.4 Linearity

The relationship between time domain signals and their Fourier transforms is linear. Thus, if

$$f_1(t) \longleftrightarrow \mathbf{F}_1(j\omega)$$

and

$$f_2(t) \longleftrightarrow \mathbf{F}_2(j\omega)$$

then

$$a_1 f_1(t) + a_2 f_2(t) \longleftrightarrow a_1\mathbf{F}_1(j\omega) + a_2\mathbf{F}_2(j\omega).$$

This property can be used to find the Fourier transform of a signal from the Fourier transform of other signals that were combined to form it. The proof of this property is left to the reader.

Example 15.11

The unit-step function $u(t)$ can be written as the sum of a constant (1/2) and 1/2 Sgn (t), as shown in Fig. 15.12. Therefore, by linearity

$$\mathfrak{F}\left\{u(t)\right\} = \mathfrak{F}\left\{\frac{1}{2}\right\} + \mathfrak{F}\left\{\frac{1}{2}\text{ Sgn }(t)\right\}$$

Using the results of Examples 15.6 and 15.10 gives

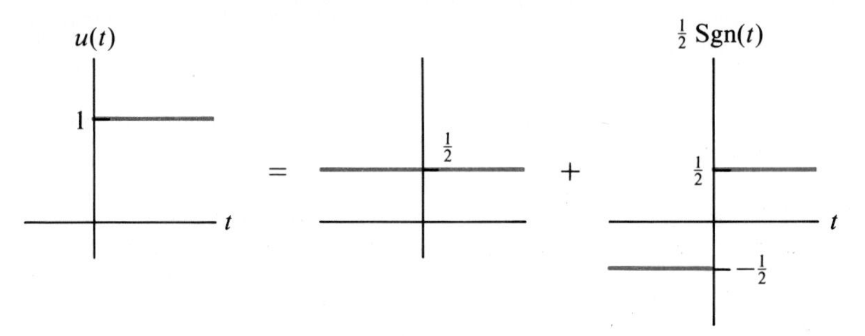

Figure 15.12 The unit-step function and its components.

$$\mathfrak{F}\left\{u(t)\right\} = \frac{1}{2}\left[2\pi\delta(\omega)\right] + \frac{1}{2}\left(\frac{2}{j\omega}\right)$$

and

$$\mathbf{F}(j\omega) = \pi\delta(\omega) + \frac{1}{j\omega} \tag{15.5}$$

In using this expression, remember that the $1/j\omega$ term only defines $\mathfrak{F}\{1/2 \text{ Sgn } t\}$ for $\omega \neq 0$ (see Example 15.10), and that $\mathfrak{F}\{1/2 \text{ Sgn } t\} = 0$ at $\omega = 0$. So it is possible to sensibly evaluate $\mathbf{F}(j\omega)$ in (15.7) at $\omega = 0$.

The impulse at the origin in the Fourier transform of the unit-step function accounts for the fact that the step has an average or DC value of 1/2 when averaged over the entire time axis using

$$f_{DC} = \lim_{T\to\infty} \frac{1}{2}\int_{-T}^{T} f(t)\, dt = \lim_{T\to\infty}\left[\frac{1}{2T}\, T\right] = \frac{1}{2}$$

The frequency-domain impulse in $\mathfrak{F}\{u(t)\}$ also indicates that the area under the time-domain curve of $u(t)$ is not bounded.

15.4.5 Time Shift

If a signal $g(t)$ is formed by shifting $f(t)$ on the time axis (i.e., $g(t) = f(t-\tau)$), the Fourier transform of $g(t)$ can be found directly from $\mathbf{F}(j\omega)$. If

$$f(t) \longleftrightarrow \mathbf{F}(j\omega)$$

then

$$f(t-\tau) \longleftrightarrow \mathbf{F}(j\omega)e^{-j\omega\tau}.$$

That is, the spectrum of the shifted signal is obtained by multiplying the original signal's spectrum by $e^{-j\omega\tau}$. (To prove this property, define $\beta = t - \tau$ and $d\beta = dt$. Then take $\mathfrak{F}\{f(t-\tau)\}$.)

Example 15.12

Find the Fourier transform of g(t) in Fig. 15.13.

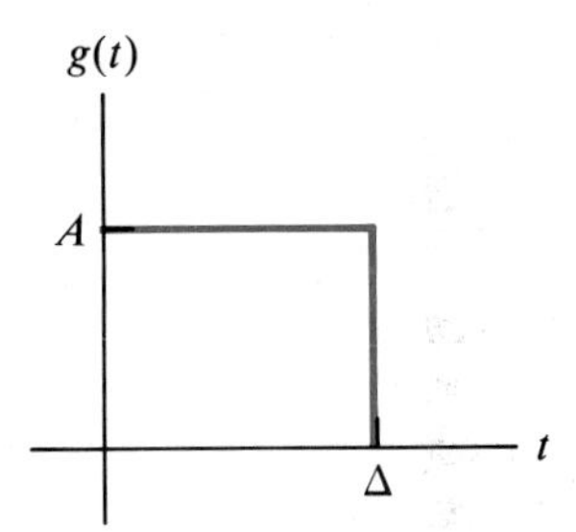

Figure 15.13 A delayed rectangular pulse.

Solution: Since g(t) is formed by delaying the pulse function by $\tau = \dfrac{\Delta}{2}$ we have

$$\mathbf{G}(j\omega) = e^{-j\omega\Delta/2}\,\mathrm{Sa}\left(\omega\frac{\Delta}{2}\right)$$

There are two important consequences of the time-shift property, which determine the magnitude and phase spectra of the shifted signal:

i. $|\mathbf{G}(j\omega)| = |\mathbf{F}(j\omega)|$

ii. $\sphericalangle\mathbf{G}(j\omega) = \sphericalangle\mathbf{F}(j\omega) - \omega\tau$

The first property states that time shifting a signal does not affect its magnitude spectrum. The second property states that time shifting corresponds to a linear phase-angle charge of a signal's spectral components.

When a signal is shifted on the time axis, its phase spectrum must be modified to create a proportional change in each phasor signal. Since phase angle and time translation are related by $\theta = -\omega\tau$ the required phase-angle change varies linearly with ω.

Example 15.13

Draw the phase spectrum of the delayed pulse signal in Example 15.13.

Solution: The phase spectrum of g(t) is obtained by combining the phase spectrum of the pulse signal with the phase angle of $e^{-j\omega\tau}$ as shown in Fig. 15.14.

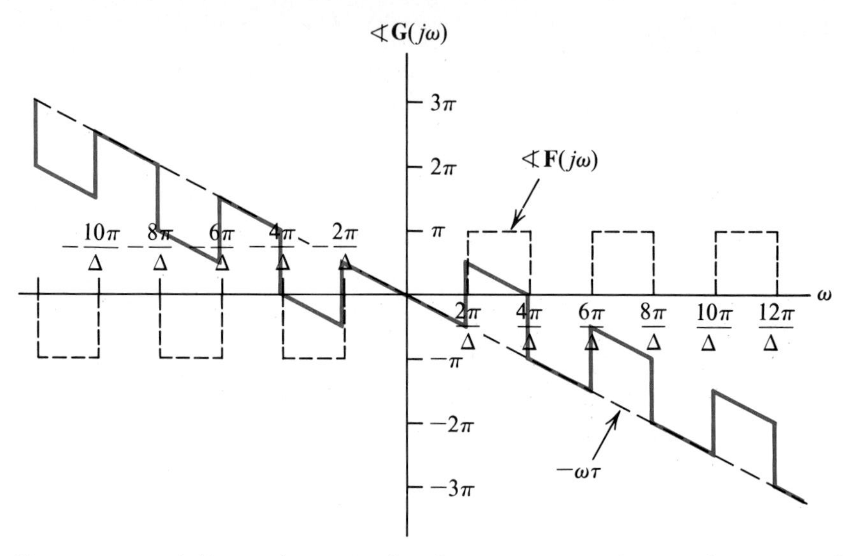

Figure 15.14 A linear change in the phase spectrum of a signal corresponds to a shift of its time domain waveform.

15.4.6 Frequency Shift (Modulation)

If $\mathfrak{F}\{f(t)\} = \mathbf{F}(j\omega)$ then $\mathfrak{F}\{f(t)e^{j\omega_o t}\} = \mathbf{F}[j(\omega - \omega_o)]$. Translation in frequency by ω_o corresponds to multiplication in the time domain by $e^{j\omega_o t}$. The proof is that

$$\mathfrak{F}\{f(t)e^{j\omega_o t}\} = \int_{-\infty}^{\infty} f(t)e^{j\omega_o t}e^{-j\omega t}\,dt$$

$$= \int_{-\infty}^{\infty} f(t)e^{-j(\omega-\omega_o)t}\,dt = \mathbf{F}[j(\omega - \omega_o)]$$

This property is also called the **modulation** property of the Fourier transform. It is used extensively in communication theory.

Example 15.14

Find the Fourier transform of $f(t) = e^{j\omega_o t}$.

Solution: Since $\mathfrak{F}\{1\} = 2\pi\delta(\omega)$, and $e^{j\omega_o t} = 1\,e^{j\omega_o t}$, we conclude that

$$\mathfrak{F}\{e^{j\omega_o t}\} = 2\pi\delta(\omega - \omega_o)$$

Note that $f(t)$ is a complex signal whose magnitude spectrum is not symmetric.

15.4.7 Derivative

Suppose that $f(t)$ has $\mathbf{F}(j\omega)$ as its Fourier transform. Then *if* df/dt has a Fourier transform it is given by

$$\mathfrak{F}\left\{\frac{df}{dt}\right\} = j\omega\mathbf{F}(j\omega) \tag{15.6}$$

provided that $\lim_{t\to\pm\infty} f(t) = 0$

Proof:

$$\mathfrak{F}\left\{\frac{df}{dt}\right\} = \int_{-\infty}^{\infty} \frac{df}{dt} e^{-j\omega t}\, dt$$

Using integration by parts, with $u = e^{-j\omega t}$ and $dv = \frac{df}{dt}dt$, we have

$$\int_a^b u\, dv = uv\Big|_a^b - \int_a^b v\, du$$

so

$$\mathfrak{F}\left\{\frac{df}{dt}\right\} = f(t)e^{-j\omega t}\Big|_{-\infty}^{\infty} - \int_{-\infty}^{\infty} f(t)(-j\omega e^{-j\omega t})\, dt$$

$$= f(t)e^{-j\omega t}\Big|_{-\infty}^{\infty} + j\omega\int_{-\infty}^{\infty} f(t)e^{-j\omega t}\, dt$$

$$= f(t)e^{-j\omega t}\Big|_{-\infty}^{\infty} + j\omega\mathbf{F}(j\omega)$$

Noting that $f(\pm\infty) = 0$ concludes the proof.

Example 15.15

In Example 15.7 we found the Fourier transform of the two-sided exponential

$$f(t) = e^{-\alpha|t|} \longleftrightarrow \mathbf{F}(j\omega) = \frac{2\alpha}{\alpha^2 + \omega^2}$$

with $\alpha > 0$. If

$$g(t) = \frac{d}{dt}e^{-\alpha|t|} = \begin{cases} \alpha e^{\alpha t} & t < 0 \\ -\alpha e^{-\alpha t} & t > 0 \end{cases}$$

as shown in Fig. 15.15, it can be verified that $g(t)$ is absolutely integrable. Therefore

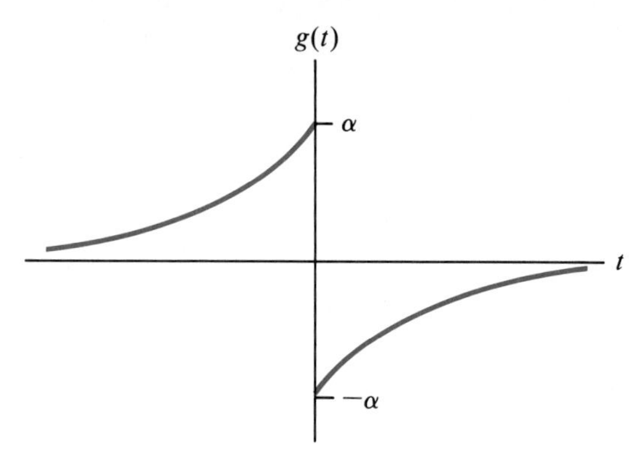

Figure 15.15

$$\mathbf{G}(j\omega) = j\omega\mathbf{F}(j\omega) = j\frac{2\alpha\omega}{\alpha^2 + \omega^2}$$

Generalized functions can be used with the derivative theorem to extend it to functions that do not satisfy the sufficient condition that $f(\pm\infty) = 0$.

Skill Exercise 15.4

Show that the Fourier transforms of $\sin \omega_o t$ and $\cos \omega_o t$ satisfy the derivative property, even though $f(\pm\infty) \neq 0$; that is, show that $\mathfrak{F}\left\{\frac{d}{dt} \sin \omega_o t\right\} = \mathfrak{F}\left\{\omega_o \cos \omega_o t\right\} = j\omega\mathfrak{F}\left\{\sin \omega_o t\right\}$

Care must be taken to *avoid misapplication of the derivative property*. A common mistake is to use it to obtain the Fourier transform of $f(t)$ indirectly from the Fourier transform of its derivative, by solving 15.5 for $\mathfrak{F}\{f(t)\}$ using

$$\mathfrak{F}\{f(t)\} = \frac{1}{j\omega}\mathfrak{F}\{\frac{df}{dt}\} \tag{15.7}$$

This division by $j\omega$ is only valid when $\omega \neq 0$. Consequently, at $\omega = 0$ we cannot use (15.7) to determine $\mathfrak{F}\{f(t)\}$ from $\mathfrak{F}\left\{\frac{df}{dt}\right\}$ [$\mathbf{F}(j0)$ cannot be found using (15.6)]. An example will illustrate the problem.

Example 15.16

The unit-step function and the impulse function are related by a derivative according to $\delta(t) = du(t)/dt$. Therefore

$$\mathfrak{F}\{\delta(t)\} = j\omega\mathfrak{F}\{u(t)\}$$

We might be led to *incorrectly* form the Fourier transform of $u(t)$ by taking $\mathfrak{F}\{u(t)\} = 1/j\omega$ but Example 15.12 showed that

$$\mathfrak{F}\{u(t)\} = \pi\delta(\omega) + \frac{1}{j\omega}$$

The derivative property only indicates how to form $\mathfrak{F}\left\{\frac{df}{dt}\right\}$ from $\mathfrak{F}\{f(t)\}$. Thus

$$\mathfrak{F}\{\delta(t)\} = j\omega\mathfrak{F}\{u(t)\} = j\omega\left[\pi\delta(\omega) + \frac{1}{j\omega}\right]$$
$$= j\pi\omega\delta(\omega) + 1 = 1$$

since $\omega\delta(\omega) = 0$.

Another way to appreciate the need for caution in using the derivative theorem is to realize that all functions that differ by a constant from $f(t)$ have the same derivative as $f(t)$. So

$$\frac{df}{dt} = \frac{d}{dt}[f(t) + K]$$

and

$$\mathfrak{F}\left\{\frac{df}{dt}\right\} = j\omega\mathfrak{F}[f(t) + K] = j\omega\mathfrak{F}\{f(t)\} + j\omega\mathfrak{F}\{K\}$$
$$= j\omega\mathbf{F}(j\omega) + j\omega K 2\pi\delta(\omega)$$

The spectral content of df/dt is not affected by K and K cannot be determined from $\mathfrak{F}\{df/dt\}$. In short, the derivative property determines the spectrum of the impulse from the spectrum of the step function, but not vice versa. The next property provides the missing link.

15.4.8 Integral

If

$$g(t) = \int_{-\infty}^{t} f(\alpha)\, d\alpha$$

then

$$\mathbf{G}(j\omega) = \pi\mathbf{F}(0)\delta(\omega) + \frac{1}{j\omega}\mathbf{F}(j\omega)$$

provided that $f(\pm\infty) = 0$, where $\mathbf{F}(0)$ is the area under the curve of $f(t)$

$$\mathbf{F}(0) = \int_{-\infty}^{\infty} f(\alpha)\, d\alpha$$

Example 15.17

If $f(t) = \delta(t)$ we know that $\mathbf{F}(j\omega) = 1$ and $\mathbf{F}(0) = 1$. Therefore, since

$$u(t) = \int_{-\infty}^{t} \delta(\alpha)\, d\alpha$$

and $\delta(\pm\infty) = 0$ we know that

$$\mathfrak{F}\{u(t)\} = \pi(1)\delta(\omega) + \frac{1}{j\omega}(1) = \pi\delta(\omega) + \frac{1}{j\omega}$$

Example 15.18

The integral of the rectangular pulse function is shown in Fig. 15.16. To derive its Fourier transform, we note that

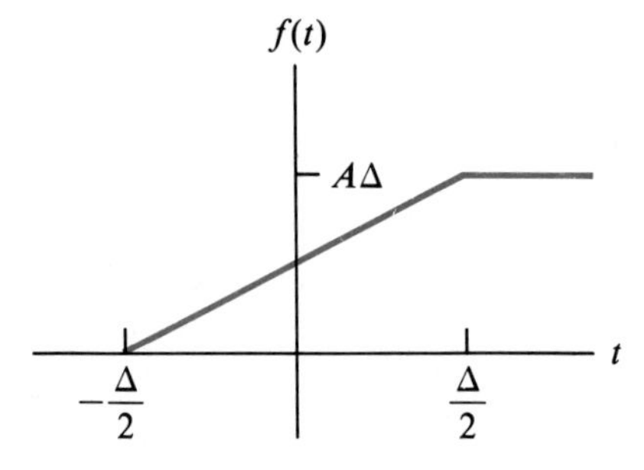

Figure 15.16 The integral of a symmetric pulse.

$$g(t) = \int_{-\infty}^{t} f(\alpha)\, d\alpha$$

where $f(t)$ is a pulse, and $\mathbf{F}(j\omega) = A\Delta\ \mathrm{Sa}\left(\omega\frac{\Delta}{2}\right)$. The integral property gives the transform of $g(t)$

$$\mathbf{G}(j\omega) = \frac{1}{j\omega}A\Delta\ \mathrm{Sa}\left(\omega\frac{\Delta}{2}\right) + \pi A\ \Delta\delta(\omega)$$

Example 15.19

Now we will use the integral property to find the Fourier transform of the triangular pulse signal shown in Fig. 15.17. We could, of course, obtain $\mathbf{F}(j\omega)$ directly by using (15.2) or indirectly by taking limits using the Fourier series coefficients of the related periodic extension of $f(t)$. Instead, we take the time derivative of $f(t)$

$$f'(t) = \begin{cases} \dfrac{2A}{\Delta} & -\dfrac{\Delta}{2} < t < 0 \\ -\dfrac{2A}{\Delta} & 0 < t < \dfrac{\Delta}{2} \\ 0 & |t| > \dfrac{\Delta}{2} \end{cases}$$

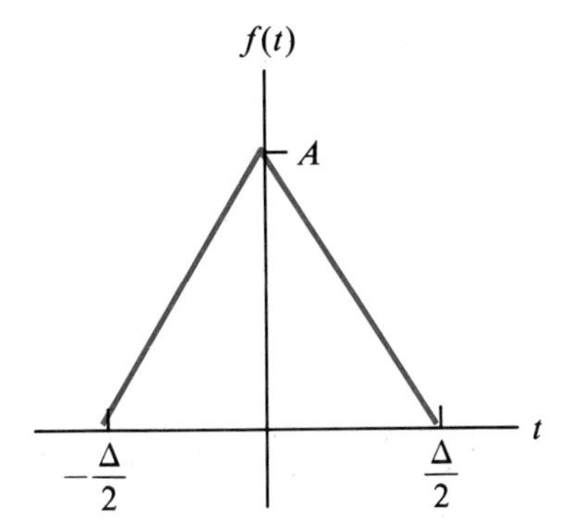

Figure 15.17 A triangular pulse.

The graph of $f'(t)$ is shown in Fig. 15.18, where we recognize that $f'(t)$ is the sum of two shifted rectangular pulse functions $f_1(t)$ and $f_2(t)$ whose graphs are shown in Fig. 15.19. Because $f'(t)$ is absolutely integrable its Fourier transform exists, and we also note that $f'(\pm\infty) = 0$. By the linearity property we get

$$\mathfrak{F}\{f'(t)\} = \mathfrak{F}\{f_1(t)\} + \mathfrak{F}\{f_2(t)\}$$

Also, because $f'(t)$ has zero area under its curve, the integral property gives

$$\mathbf{F}(j\omega) = \mathfrak{F}\{f(t)\} = \frac{1}{j\omega}[\mathfrak{F}\{f_1(t)\} + \mathfrak{F}\{f_2(t)\}]$$

The transforms of $f_1(t)$ and $f_2(t)$ are obtained directly from the transform of the pulse function and the shifting property, with adjustments made for the height and width of the pulses in Fig. 15.19. Noting that each pulse width is $\Delta/2$ with a height of $2A/\Delta$ or $-2A/\Delta$, and that the pulses are shifted by $\pm\ \Delta/4$ we have

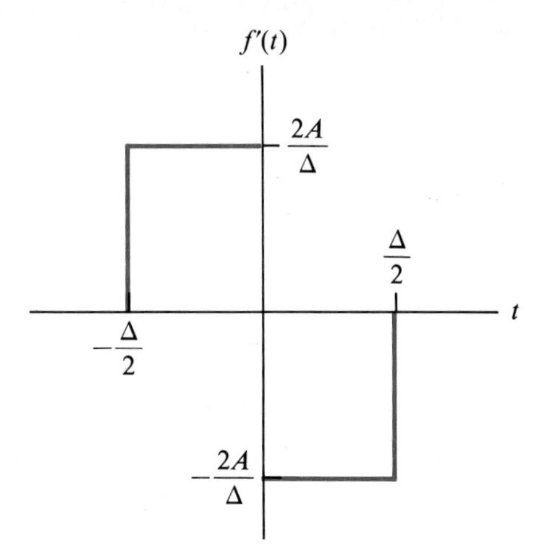

Figure 15.18 The derivative of the signal in Fig. 15.17.

$$f_1(t) \longleftrightarrow \frac{2A}{\Delta}\left(\frac{\Delta}{2}\right) \text{Sa}\left(\omega\frac{\Delta}{4}\right) e^{-j\omega(-\Delta/4)} = A \text{ Sa}\left(\omega\frac{\Delta}{4}\right) e^{j\omega\Delta/4}$$

and

$$f_2(t) \longleftrightarrow -\frac{2A}{\Delta}\left(\frac{\Delta}{2}\right) \text{Sa}\left(\omega\frac{\Delta}{4}\right) e^{-j\omega(\Delta/4)} = -A \text{ Sa}\left(\omega\frac{\Delta}{4}\right) e^{-j\omega\Delta/4}$$

Figure 15.20 shows the graph of $\mathfrak{F}\{f'(t)\}$ given by $f_1(t)$ and $f_2(t)$. Lastly, we find $\mathbf{F}(j\omega)$ as

$$\mathbf{F}(j\omega) = \frac{A\Delta}{2} \text{Sa}^2\left(\omega\frac{\Delta}{4}\right)$$

The graph of $|\mathbf{F}(j\omega)|$ for the triangular pulse is shown in Fig. 15.21. Note that the effect of the factor $1/j\omega$ is to reduce the high-frequency spec-

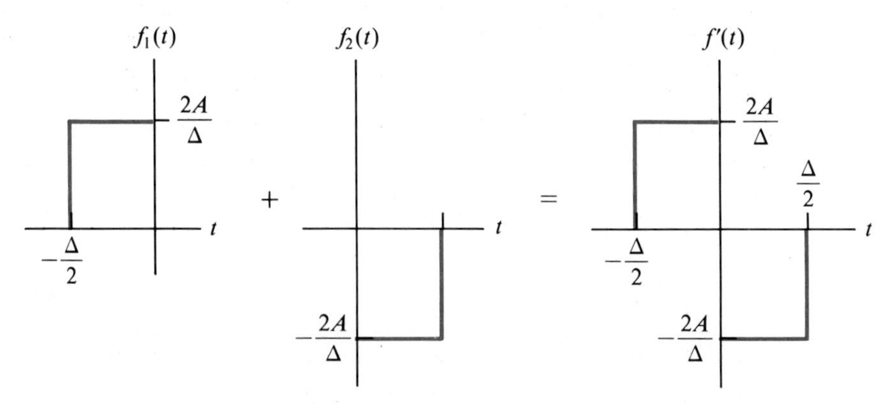

Figure 15.19 The sum of $f_1(t)$ and $f_2(t)$ forms $f'(t)$.

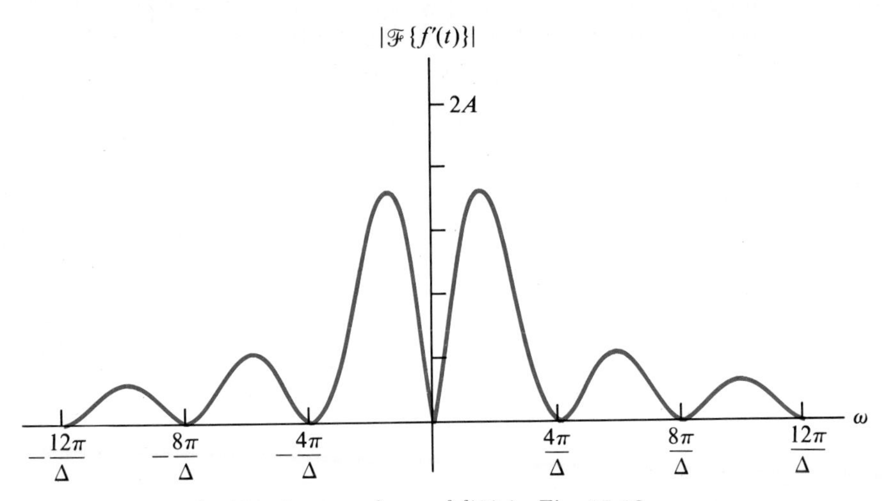

Figure 15.20 The Fourier transform of $f'(t)$ in Fig. 15.19.

tral content in $f(t)$ compared to $f(t)$ and to increase the low-frequency spectral content. As a check, we note that $\mathbf{F}(0) = A\Delta/2$ (the area under the curve of $f(t)$) and the value of $\mathfrak{F}\{f'(t)\}$ is 0 at $\omega = 0$—corresponding to the area under the curve of $f'(t)$.

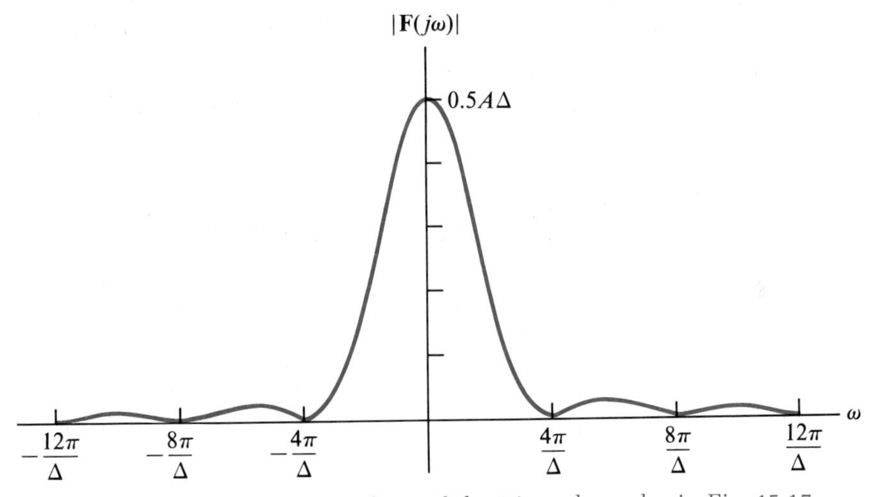

Figure 15.21 The Fourier transform of the triangular pulse in Fig. 15.17.

15.5 ENERGY SPECTRUM

The energy in a signal is defined by

$$W = \int_{-\infty}^{\infty} f^2(t)\, dt \tag{15.8}$$

where $f(t)$ can be considered to be analogous to a current dissipating power in a 1-Ω resistor. Next, we write (15.8) using (15.1)

$$W = \int_{-\infty}^{\infty} f(t)\left[\frac{1}{2\pi}\int_{-\infty}^{\infty} \mathbf{F}(j\omega)e^{j\omega t}\, d\omega\right]dt.$$

Interchanging the order of integration leads to

$$\begin{aligned} W &= \frac{1}{2\pi}\int_{-\infty}^{\infty} \mathbf{F}(j\omega)\left[\int_{-\infty}^{\infty} f(t)e^{j\omega t}\, dt\right]d\omega \\ &= \frac{1}{2\pi}\int_{-\infty}^{\infty} \mathbf{F}(j\omega)\mathbf{F}(-j\omega)\, d\omega \\ &= \frac{1}{2\pi}\int_{-\infty}^{\infty} |\mathbf{F}(j\omega)|^2\, d\omega \end{aligned}$$

Since $|\mathbf{F}(j\omega)|$ is an even function of ω

$$W = \frac{1}{\pi}\int_{0}^{\infty} |\mathbf{F}(j\omega)|^2\, d\omega$$

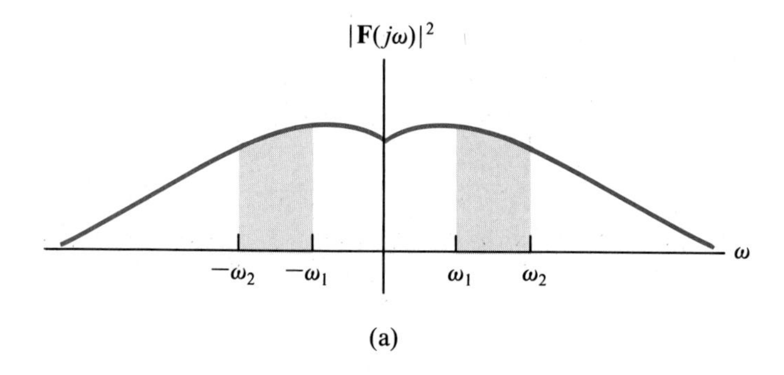

(a)

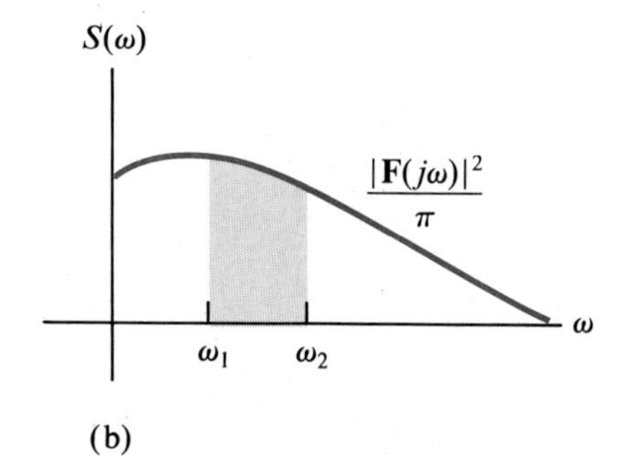

(b)

Figure 15.22 The energy content of a signal can be associated with its spectral content over a band of frequencies.

The "one-sided" energy spectrum, or the **energy spectral density** of $f(t)$ is defined to be

$$S(\omega) = \frac{1}{\pi} |\mathbf{F}(j\omega)|^2$$

When the energy in a signal is finite it can be associated with its frequency-domain spectral content by using $S(\omega)$ to obtain the energy contained in $f(t)$ in a given band of frequencies. Then

$$W_{12} = \int_{\omega_1}^{\omega_2} S(\omega)\, d\omega$$

is the energy in $f(t)$ between ω_1 and ω_2 (as shown in Fig. 15.22).

Example 15.20

Figure 15.23 shows the energy-spectral-density function of the rectangular pulse. Since $S(\omega)$ depends on the square of $|\mathbf{F}(j\omega)|$ the curve for $S(\omega)$ tends to have relatively smaller contribution from those frequencies where $|\mathbf{F}(j\omega)|$ is small, i.e. beyond the first zero crossing of $|\mathbf{F}(j\omega)|$.

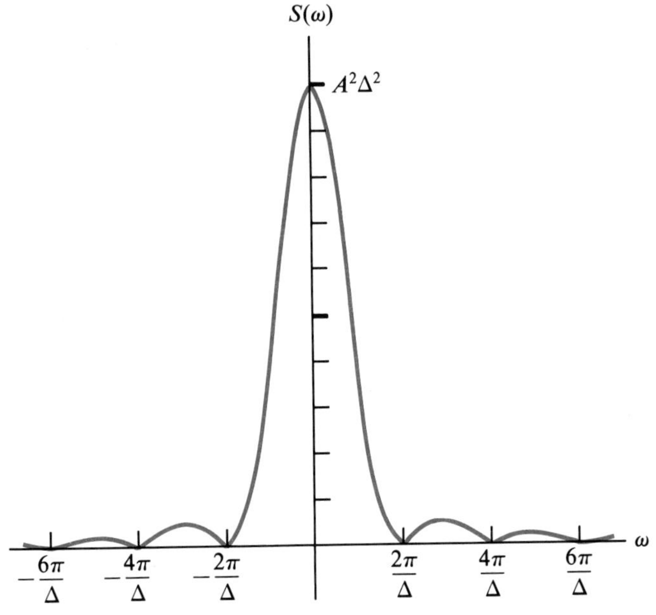

Figure 15.23 The energy-density spectrum $S(\omega)$ of the rectangular pulse.

TABLE 15.1 SELECTED FOURIER TRANSFORM PAIRS

$f(t)$	$F(j\omega)$
$\delta(t)$	1
$u(t)$	$\pi\delta(\omega) + \dfrac{1}{j\omega}$
1	$2\pi\delta(\omega)$
$e^{-at}u(t)$	$1/(a + j\omega)$
$\cos \omega_o t$	$\pi\delta(\omega - \omega_o) + \pi\delta(\omega + \omega_o)$
$\sin \omega_o t$	$j\pi[-\delta(\omega - \omega_o) + \delta(\omega + \omega_o)]$
$e^{-a\|t\|}$	$\dfrac{2a}{a^2 + \omega^2}$
$\cos \omega_o t u(t)$	$\dfrac{\pi}{2}\left[\delta(\omega - \omega_o) + \delta(\omega + \omega_o)\right] + \dfrac{j\omega}{\omega_o^2 - \omega^2}$
$\sin \omega_o t u(t)$	$j\dfrac{\pi}{2}\left[-\delta(\omega - \omega_o) + \delta(\omega + \omega_o)\right] + \dfrac{\omega_o}{\omega_o^2 - \omega^2}$

15.6 CIRCUIT ANALYSIS WITH FOURIER TRANSFORMS

Fourier transforms represent signals in terms of phasor signals. Now we will describe the input/output relationship governing the zero-state response (ZSR) of a circuit in terms of the Fourier transforms of its input and output signals.

Recall that the ZSR of a circuit is described by the input/output differential equation model

$$a_n\frac{d^ny}{dt^n} + \cdots + a_1\frac{dy}{dt} + a_0y = b_m\frac{d^mx}{dt^m} + \cdots + b_1\frac{dx}{dt} + b_0 \tag{15.9}$$

where $x(t)$, $y(t)$ and all of their derivatives are zero for $t < 0$. Taking the transform of both sides of (15.9) for $t > 0$ gives

$$a_n\mathfrak{F}\left\{\frac{d^ny}{dt^n}\right\} + \cdots + a_1\mathfrak{F}\left\{\frac{dy}{dt}\right\} + a_0\mathfrak{F}\{y(t)\} = b_m\mathfrak{F}\left\{\frac{d^mx}{dt^m}\right\} + \cdots + b_1\mathfrak{F}\left\{\frac{dx}{dt}\right\} + b_0\mathfrak{F}\{x(t)\}.$$

Using the derivative property, we have

$$\mathfrak{F}\left\{\frac{d^ny}{dt^n}\right\} = (j\omega)^n\mathfrak{F}\{y(t)\} = (j\omega)^n\mathbf{Y}(j\omega)$$

and likewise

$$\mathfrak{F}\left\{\frac{d^m x}{dt^m}\right\} = (j\omega)^m \mathfrak{F}\{x(t)\} = (j\omega)^m \mathbf{X}(j\omega)$$

Substituting such expressions into (15.9) to replace all of the derivative terms gives

$$a_n(j\omega)^n \mathbf{Y}(j\omega) + \cdots + a_0 \mathbf{Y}(j\omega) = b_m(j\omega)^m \mathbf{X}(j\omega) + \cdots + b_0 \mathbf{X}(j\omega)$$

and factoring both sides leaves

$$\mathbf{Y}(j\omega)[a_n(j\omega)^n + \cdots + a_1 j\omega + a_0] = \mathbf{X}(j\omega) \qquad (15.10)$$
$$[b_m(j\omega)^m + \cdots + b_1 j\omega + b_0]$$

Equation 15.10 is a frequency-domain expression of the input/output model of the circuit. The important and convenient feature of this form of the I/O model is that it is algebraic, thereby skipping the need to solve a differential equation. Solving (15.10) for $\mathbf{Y}(j\omega)$ gives

$$\mathbf{Y}(j\omega) = \frac{b_m(j\omega)^m + \cdots + b_0}{a_n(j\omega)^n + \cdots + a_0} \mathbf{X}(j\omega) \qquad (15.11)$$

and

$$\boxed{\mathbf{Y}(j\omega) = \mathbf{H}(j\omega)\mathbf{X}(j\omega)} \qquad (15.12)$$

The Fourier transform of the output signal is the product of the transfer function and the Fourier transform of the input signal. $\mathbf{Y}(j\omega)$ can be found for a given $\mathbf{X}(j\omega)$ and $\mathbf{H}(j\omega)$. The inverse transform provides $y(t)$. Recall that we began our study of dynamic circuits by learning that the particular solution to an exponential input is itself an exponential, whose amplitude is the product of the input amplitude and the transfer function evaluated at the frequency of the exponential. The Fourier transform synthesizes $x(t)$ as a continuous sum of phasor signals (exponentials) $\mathbf{X}(j\omega)e^{j\omega t}$ and $\mathbf{H}(j\omega)$ specifies the synthesis of the output to be the continuous sum of $\mathbf{H}(j\omega)\mathbf{X}(j\omega)e^{j\omega t}$. The conceptual model that we used for a single exponential signal has now been extended to handle any Fourier transformable signal! Indeed, the zero-state response of the circuit is the superposition of its response to the individual input exponentials, or

$$y(t) = \frac{1}{2\pi} \int_{-\infty}^{\infty} \mathbf{X}(j\omega)\mathbf{H}(j\omega)e^{j\omega t} d\omega \qquad (15.13)$$

The response defined by (15.13) is the zero-state response of the circuit because no explicit use has been made of the circuit's boundary conditions.

The time-domain and frequency-domain I/O relationships are depicted in Fig. 15.24, where $\mathbf{H}(j\omega)$ represents the circuit whose transfer

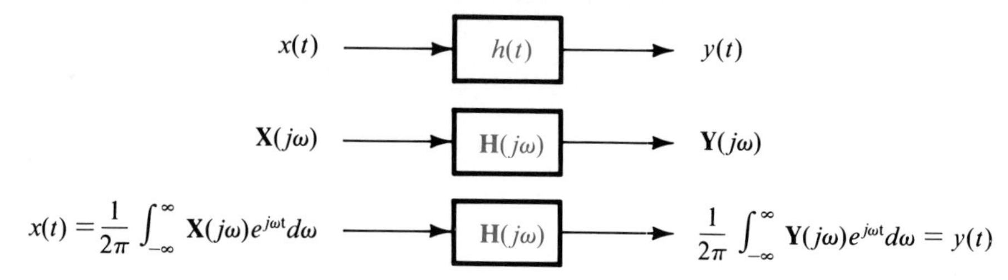

Figure 15.24 Time-domain and frequency-domain input/output relationships of a circuit.

function is $\mathbf{H}(j\omega)$. The transfer function determines the ratio of the output Fourier transform to the input Fourier transform.

$$\mathbf{H}(j\omega) = \frac{\mathbf{Y}(j\omega)}{\mathbf{X}(j\omega)}$$

A circuit's input/output differential equation is a **time-domain** model of its behavior; its transfer function provides a **spectral model**. The former does not in itself provide the algorithm for finding the response. On the other hand, Equation (15.12) and the Fourier integral together form the algorithm for determining the output spectrum and for constructing the output signal from the input spectrum and the transfer function. Fourier-signal synthesis assures us that the signal whose spectrum is formed from (15.12) satisfies the I/O differential equation *automatically*. The I/O magnitude and phase-angle relationships follow from (15.12):

$$|\mathbf{Y}(j\omega)| = |\mathbf{X}(j\omega)||\mathbf{H}(j\omega)|$$

and

$$\measuredangle\mathbf{Y}(j\omega) = \measuredangle\mathbf{X}(j\omega) + \measuredangle\mathbf{H}(j\omega)$$

The magnitude and phase spectra of the output signal are copies of the input spectra modified by the transfer function. This fact has very important consequences in circuit analysis and design because it allows the output signal to be determined if the input is known, and because it lets us choose circuit components to create a desirable shape for $\mathbf{H}(j\omega)$. Chapter 17 will elaborate on this to design filters—circuits that selectively modify whatever spectrum the input signal has.

Example 15.21

The *RC* circuit in Fig. 15.25 is called a lowpass filter because its magnitude spectrum attenuates high-frequency spectral components much more than it does those at low frequency. Find the ZSR of $y(t)$ to $x(t) = Au(t)$.

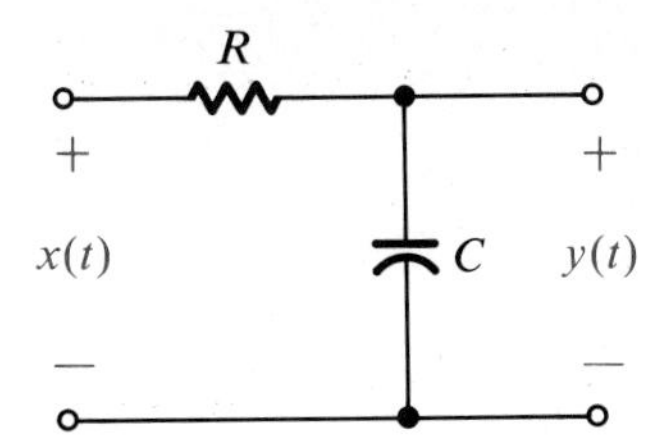

Figure 15.25 Circuit for Example 15.23.

Solution: For the given x(t)

$$\mathbf{X}(j\omega) = A\left[\pi\delta(\omega) + \frac{1}{j\omega}\right]$$

Since the filter transfer function is

$$\mathbf{H}(j\omega) = \frac{1}{1 + j\omega RC}$$

the output signal's Fourier transform must be

$$\mathbf{Y}(j\omega) = A\left[\pi\delta(\omega) + \frac{1}{j\omega}\right]\left[\frac{1}{1 + j\omega RC}\right]$$
$$= \frac{A\pi\delta(\omega)}{1 + j\omega RC} + \frac{A}{j\omega(1 + j\omega RC)}$$

The inverse transform of $\mathbf{Y}(j\omega)$ gives:

$$y(t) = \mathfrak{F}^{-1}\left\{\frac{A\pi\delta(\omega)}{1 + j\omega RC}\right\} + \mathfrak{F}^{-1}\left\{\frac{A}{j\omega(1 + j\omega RC)}\right\}$$

Taking the first term we have

$$\mathfrak{F}^{-1}\left\{\frac{A\pi\delta(\omega)}{1 + j\omega RC}\right\} = \frac{1}{2\pi}A\pi\int_{-\infty}^{\infty}\frac{\delta(\omega)e^{j\omega t}}{1 + j\omega RC}\,d\omega = \frac{A}{2}$$

The second term can be written in partial fraction expansion form as

$$\mathfrak{F}^{-1}\left\{\frac{A}{j\omega(1 + j\omega RC)}\right\} = \mathfrak{F}^{-1}\left\{\frac{A}{j\omega} + \frac{-RC\,A}{1 + j\omega RC}\right\}$$
$$= \mathfrak{F}^{-1}\left\{\frac{A}{j\omega}\right\} - RC\,\mathfrak{F}^{-1}\left\{\frac{A}{1 + j\omega RC}\right\}$$

$$= \frac{1}{2}A\ \mathrm{Sgn}\ (t) - \mathfrak{F}^{-1}\left\{\frac{A}{j\omega + \frac{1}{RC}}\right\}$$

$$= \frac{1}{2}A\ \mathrm{sgn}\ (t) - Ae^{-t/RC}u(t)$$

Combining all of the terms gives

$$y(t) = \frac{1}{2}A + \frac{1}{2}A\ \mathrm{Sgn}\ (t) - Ae^{-t/RC}u(t)$$

and

$$y(t) = \begin{cases} 0 & t \leq 0 \\ A(1 - e^{-t/RC})u(t) & t > 0 \end{cases}$$

This agrees with the step response found in Chapter 8.

It should be noted in the preceding example that:

1. All of the steps taken to find $y(t)$ were algebraic, and required no reference to the circuit's time-domain behavior.
2. The input signal was *not* an exponential signal.
3. The input signal could be represented by superposition (Fourier integral) of phasor signals.
4. The spectral components of the output signal could be determined from $\mathbf{H}(j\omega)$ and $\mathbf{X}(j\omega)$, (the spectral amplitude of $e^{j\omega t}$).
5. Although the output signal could be synthesized by means of the Fourier integral, it was obtained directly from the time-domain behavior of the component parts of $\mathbf{Y}(j\omega)$.

15.7 IMPULSE RESPONSE

The impulse response $h(t)$ of a circuit is obtained from (15.12) with $\mathbf{X}(j\omega) = 1$. In this case, as shown in Fig. 15.26, $\mathbf{Y}(j\omega) = \mathbf{H}(j\omega)$ and therefore $y(t) = h(t)$ where

$$\boxed{h(t) = \mathfrak{F}^{-1}\{\mathbf{H}(j\omega)\}}$$

A circuit's time-domain response to an impulse is given by the inverse transform of its transfer function.

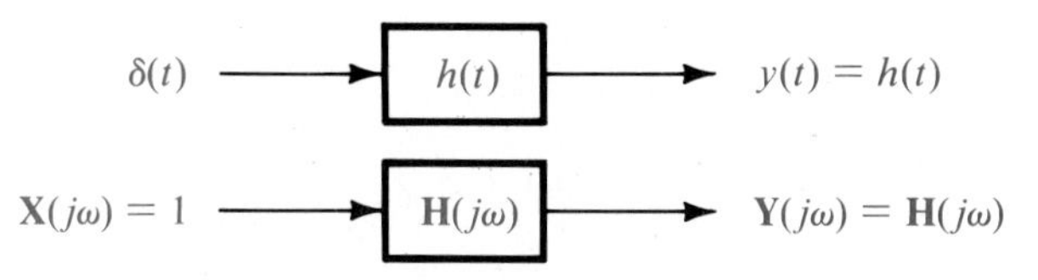

Figure 15.26 Time-domain and frequency-domain representations of a circuit's impulse response.

Example 15.22

The impulse response of the RC circuit in Example 15.22 is

$$h(t) = \mathfrak{F}^{-1}\left\{\frac{\frac{1}{RC}}{\frac{1}{RC} + j\omega}\right\} = \frac{1}{RC}e^{-t/RC}u(t)$$

This agrees with the result found in Chapter 9.

15.8 CONVOLUTION*

Frequency-domain tools find the response of a circuit directly (*without* its I/O DE). They have a counterpart in the time domain. Just as the frequency-domain method constructs the output *spectrum* directly, convolution in the time domain constructs the output *signal* directly—with no reliance on the frequency domain, or on exponential signals!

The convolution of two signals $x(t)$ and $h(t)$ denoted by $x(t) * h(t)$ is an operation that forms a signal $y(t)$ from the so-called **convolution integral**

$$y(t) = x(t) * h(t) = \int_{-\infty}^{\infty} x(\alpha)h(t-\alpha)\,d\alpha$$

Time-domain convolution states that

$$x(t) * h(t) \longleftrightarrow \mathbf{X}(j\omega)\mathbf{H}(j\omega)$$

If a signal is formed from the convolution of $x(t)$ and $h(t)$ its spectrum is the product of $\mathbf{X}(j\omega)$ and $\mathbf{H}(j\omega)$ or

Time-domain convolution is equivalent to spectral multiplication.

To prove this, let

$$\mathfrak{F}\{x(t) * h(t)\} = \int_{-\infty}^{\infty}\int_{-\infty}^{\infty} x(\alpha)h(t-\alpha)d\alpha e^{-j\omega t}\,dt$$

Interchanging the order of integration gives

$$x(t) * h(t) = \int_{-\infty}^{\infty}\int_{-\infty}^{\infty} h(t-\alpha)e^{-j\omega t}\,dt\,x(\alpha)\,d\alpha$$

Applying the shifting theorem to $h(t-\alpha)$ gives

$$\mathfrak{F}\{x(t) * h(t)\} = \int_{-\infty}^{\infty} \mathbf{H}(j\omega)e^{-j\omega\alpha}x(\alpha)\,d\alpha$$

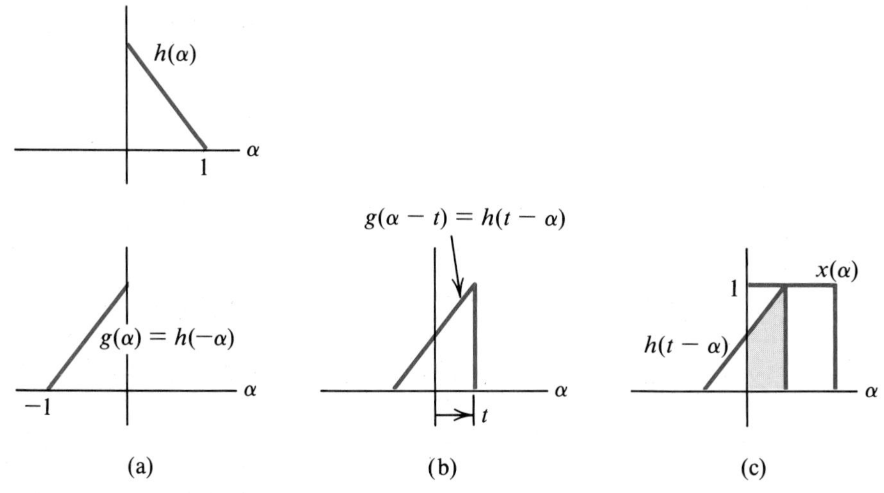

Figure 15.27 Convolution.

$$= \mathbf{H}(j\omega) \int_{-\infty}^{\infty} x(\alpha) e^{-j\omega\alpha} \, d\alpha = \mathbf{H}(j\omega)\mathbf{X}(j\omega)$$

Understanding the behavior of the function $h(t - \alpha)$ is the key to understanding convolution. If $g(\alpha) = h(-\alpha)$ the graph of $g(\alpha)$ vs α is the mirror image of the graph of $h(\alpha)$ [as shown in Fig. 15.27(a)]. Next we note that for a fixed t the graph of $g(\alpha - t)$ is a delayed copy of $g(\alpha)$ [see Fig. 15.27(b)]. The curve of $g(\alpha)$ is shifted to the right if $t > 0$. However, $g(\alpha - t) = h[-(\alpha - t)] = h(t - \alpha)$ so the graph of $h(t - \alpha)$ is actually a **shifted mirror image** of $h(\alpha)$. It is essential to note that the image translates to the right for $t > 0$. For a given t convolution calculates the area under the curve formed by the product of $x(\alpha)$ and $h(t - \alpha)$ [as shown in Fig. 15.27(c)]. As t varies, the area being integrated varies, with the resultant shape being the curve for $y(t)$.

Now we'll show that convolution provides a *direct* time-domain description of the input/output relationship of a circuit. If $x(t) = 0$ for $t < 0$ and if the circuit's impulse response is causal [$h(t) = 0$ for $t < 0$] the convolution integral is simplified. First, using $x(\alpha) = 0$ for $\alpha < 0$ gives

$$x(t) * h(t) = \int_{-\infty}^{\infty} x(\alpha)h(t - \alpha) \, d\alpha = \int_{0}^{\infty} x(\alpha)h(t - \alpha) \, d\alpha$$

Then using $h(t - \alpha) = 0$ for $t - \alpha < 0$ or $\alpha > t$ we get

$$\boxed{x(t) * h(t) = \int_{0}^{t} x(\alpha)h(t - \alpha) \, d\alpha}$$

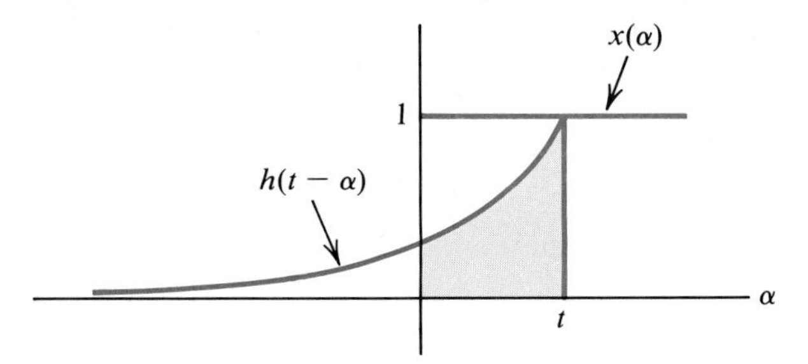

Figure 15.28 The reduced range of integration for the convolution integral.

The reduced range of integration is apparent in Fig. 15.28, which illustrates convolution when the circuit input is a step and its impulse response is a first-order exponential decay.

If $x(t)$ is the input signal to a circuit whose impulse response is $h(t)$ the output signal $y(t)$ can be computed directly from

$$y(t) = \int_0^t x(\alpha)h(t-\alpha)\,d\alpha$$

Implementing the convolution algorithm usually requires a digital computer, but has the attractive practical feature of eliminating the need to find the Fourier transform of the input signal.

Since convolution is an operation on a pair of waveforms, it can be performed on two spectra to form a third. In this case, the convolved Fourier transforms and their signals are related by

$$x(t) \cdot h(t) \longleftrightarrow \frac{1}{2\pi}\mathbf{X}(j\omega) * \mathbf{H}(j\omega)$$

or

$$x(t) \cdot h(t) \longleftrightarrow \frac{1}{2\pi}\int_{-\infty}^{\infty} \mathbf{X}(j\alpha)\mathbf{H}[j(\omega-\alpha)]\,d\alpha$$

The proof of this property follows the same steps taken for time-domain convolution.

Example 15.23

The signal $y(t) = \cos\omega_o t\, u(t)$ is a cosine that turns on at $t = 0$. Since it is the product of two signals whose spectra are known we take

$$\mathfrak{F}\{\cos\omega_o t\, u(t)\} = \mathfrak{F}\{\cos\omega_o t\} * \mathfrak{F}\{u(t)\}$$

$$= \frac{1}{2\pi}\left[\pi\delta(\omega - \omega_o) + \pi\delta(\omega + \omega_o)\right] * \left[\pi\delta(\omega) + \frac{1}{j\omega}\right]$$

$$= \frac{1}{2}\int_{-\infty}^{\infty}\left[\delta(\alpha - \omega_o) + \delta(\alpha + \omega_o)\right]\left[\pi\delta(\alpha - \omega) + \frac{1}{j(\alpha - \omega)}\right] d\alpha$$

$$= \frac{\pi}{2}\delta(\omega_o - \omega) - \frac{\pi}{2}\delta(\omega + \omega_o) + \frac{1}{2}\left[\frac{1}{j(\omega_o - \omega)} - \frac{1}{j(\omega_o + \omega)}\right]$$

$$= \frac{\pi}{2}\left[\delta(\omega - \omega_o) + \delta(\omega + \omega_o)\right] + \frac{j\omega}{\omega^2 - \omega_o^2}$$

Note that $\mathfrak{F}\{\cos \omega_o t\, u(t)\} \neq \mathfrak{F}\{\cos \omega_o t\}$. Why?

SUMMARY

For each frequency ω the Fourier transform of a signal $f(t)$ prescribes the relative amount of complex exponential that must be used to synthesize $f(t)$ from a continuous sum of exponentials, according to the Fourier integral algorithm. Fourier transforms represent non-periodic signals in terms of complex exponential signals, and are obtained by direct evaluation.

The Fourier transform of a circuit's zero-state output signal is the product of the Fourier transform of the input signal and the circuit's transfer function—the same algorithm used to calculate the exponential amplitude of the circuit's forced response to a single complex exponential input. The time-domain waveform of the output signal can be obtained from its Fourier transform by direct evaluation or, more simply, by using the uniqueness property of the transform and identifying it from a table of known Fourier transform pairs. The amplitude spectrum of a circuit's output signal can be compared to the spectrum of its input signal to understand of its frequency-domain characteristics. Conversely, the transfer function shows how a circuit will shape the spectral characteristics of the input signal to form the spectral characteristics of its output signal.

Time-domain convolution is equivalent to frequency-domain multiplication, and vice versa. The ZSR of a circuit can be found by convolving the input signal with the circuit's impulse response.

Problems - Chapter 15

15.1 Find the Fourier transform of $f(t)$.

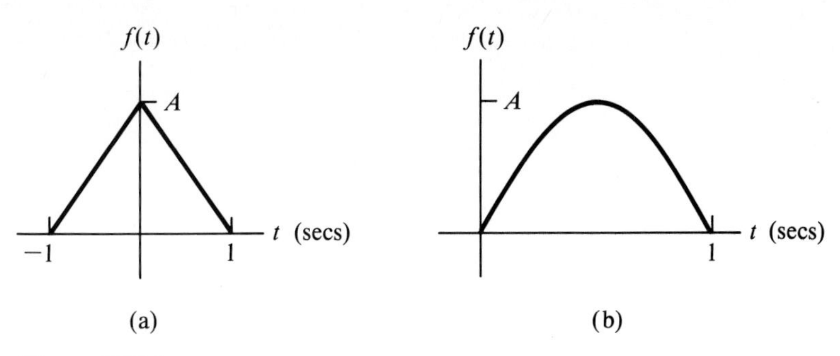

Figure P15.1

15.2 Draw the magnitude and phase spectra of the Fourier transforms found in Problem 15.1.

15.3 Use the integral or derivative properties of the Fourier transform to find the transforms of the signals shown below.

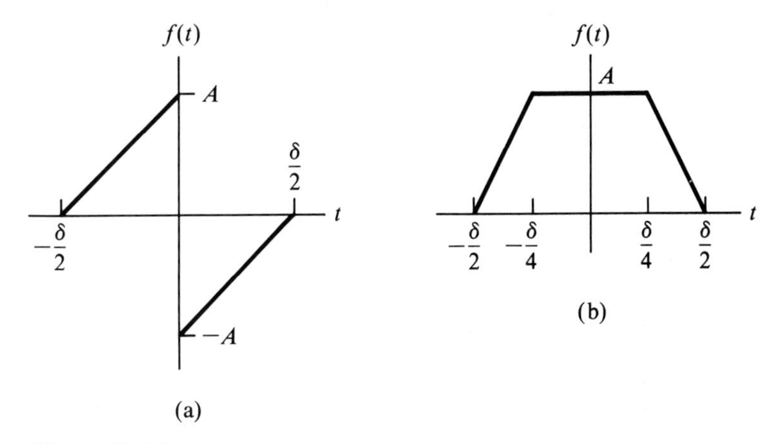

Figure P15.3

15.4 Find $f(t)$ for the $\mathbf{F}(j\omega)$ shown below.

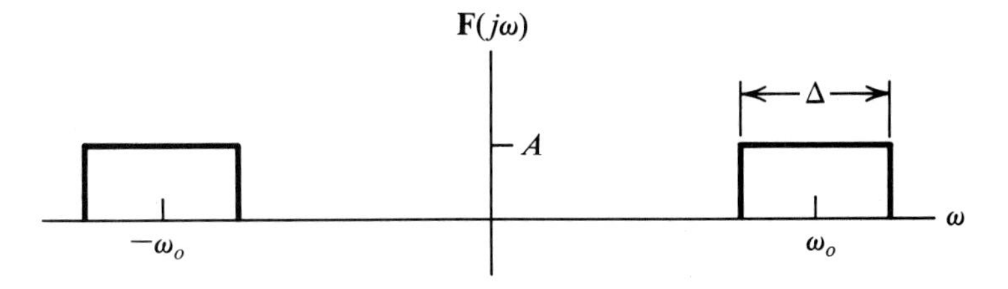

Figure P15.4

15.5 Find $\mathbf{F}(j\omega)$.

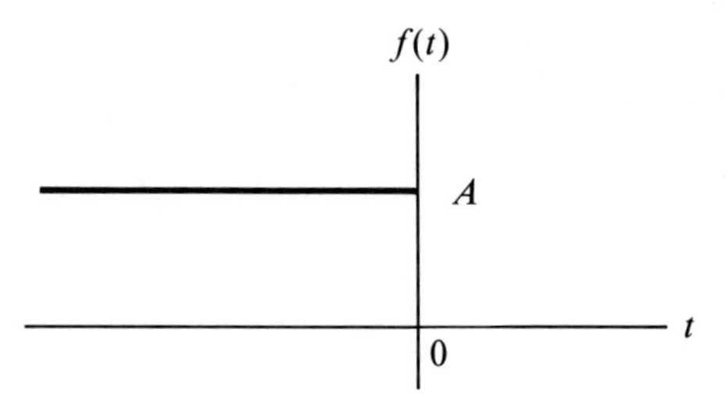

Figure P15.5

15.6 Obtain the Fourier transform of the signal $f(t) = 2e^{-0.5t}u(t)$. What is the total energy in $f(t)$?

15.7 Find the energy spectral density for the signals in Problem 15.1.

15.8 Find the energy spectral density of $f(t) = te^{-\alpha t}u(t)$ and use it to find the total energy in $f(t)$.

15.9 Obtain the Fourier transform of $f(t)$.

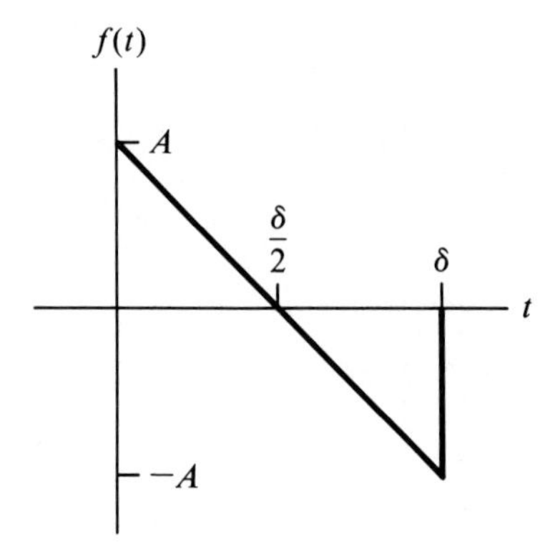

Figure P15.9

15.10 If $f(t) = 10e^{-(t-4)}u(t)$, find $\mathbf{F}(j\omega)$.

15.11 The input signal to the filter shown is $f(t)$.

a. Draw $|\mathbf{Y}(j\omega)|$.

b. Draw the energy spectral density for $y(t)$.

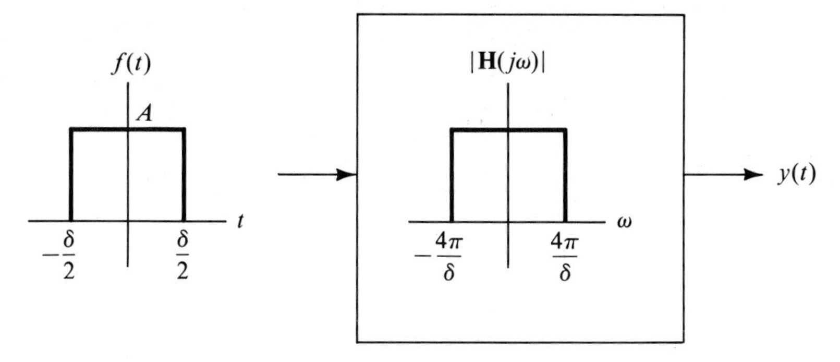

Figure 15.11

15.12 If $f(t)$ in Fig. P15.9 is the input to the circuit in Fig. P9.4 with $R_1 = 1\ \Omega$, $R_2 = 2\ \Omega$, $C = 0.25$ F, $A = 10$ and $\delta = 1$, find the energy spectral density of $v_o(t)$.

15.13 If a series RL circuit has $R = 1\ \Omega$ and $L = 4$ H find the impulse response of the resistor voltage.

15.14 Use the convolution property of the Fourier transform to find $\mathbf{F}(j\omega)$ for $f(t) = \sin \omega_c t\ u(t)$.

15.15 If a circuit has an impulse response given by $h(t) = e^{-t}\,u(t)$ use convolution to find its ZSR to $x(t) = u(t)$.

15.16 Use the convolution property of the Fourier transform to find the Fourier transform of the cosine pulse signal $f(t) = A \cos \pi/\delta t$ for $|t| \leq \delta/2$. Hint: consider the cosine pulse to be the product of a suitably chosen gate function and a cosine signal.

15.17 Using convolution, show that in general

$$f(t) \cos \omega_c t \leftrightarrow 1/2\ \mathbf{F}[j(\omega - \omega_c)] + 1/2\ \mathbf{F}[j(\omega + \omega_c)]$$

This property is known as the modulation property of the Fourier transform. It plays an important role in communication theory, where $\cos \omega_c t$ is a carrier signal whose amplitude is modulated by $f(t)$. This property determines the spectrum of the modulated carrier from the spectrum of the modulating signal. To gain some appreciation for its great practical significance note that it allows the spectrum of the modulating signal to be translated to an arbitrary carrier frequency. If their carrier frequencies are far enough apart, several AM radio stations can broadcast simultaneously in the same geographical area without interfering with each other.

15.18 Show that $x(t) * \delta(t) = x(t)$ and that $x(t - \beta) * \delta(t) = x(t - \beta)$.

15.19 If the ZSR of a circuit to a unit-step function is given by $y(t) = 10(1 - e^{-2t})$, find the I/O transfer function of the circuit.

15.20 Show that $x(t) * h(t) = h(t) * x(t)$ (convolution is commutative).

15.21 Using convolution, find the ZSR of the capacitor voltage to a unit-step input voltage in a series RC circuit.

15.22 Show that the Fourier transform of a periodic signal is

$$\mathbf{F}(j\omega) = \sum_{-\infty}^{\infty} \mathbf{F}_n \delta(\omega - n\omega_o)$$

where $\mathbf{F}_n$ is the Fourier coefficient of the signal.

15.23 Draw the amplitude spectrum of the Fourier transform of the pulse train in Fig. 14.2.

CIRCUIT ANALYSIS WITH LAPLACE TRANSFORMS

INTRODUCTION

The analytical tools developed in the previous chapters let us find a circuit's response to exponential signals, periodic signals, and Fourier-transformable signals. These tools have restricted use, and even the Fourier transform provides only the zero-state response (ZSR) of a circuit. Also, its convergence properties either limit its use, or require the use of impulse functions to describe the spectral components of a signal. The Laplace transform does not have these limitations.

16.1 THE LAPLACE TRANSFORM

A signal $f(t)$ is said to be *causal* if $f(t) = 0$ for $t < 0$. The **Laplace transform** of a causal signal is defined by

$$\mathbf{F}(s) = \int_{0^-}^{\infty} f(t)e^{-st}\,dt \tag{16.1}$$

and the **inverse Laplace transform** of $\mathbf{F}(s)$ can be used to form $f(t)$ according to

$$f(t) = \frac{1}{2\pi j}\int_{c-j\infty}^{c+j\infty} \mathbf{F}(s)e^{st}\,ds \tag{16.2}$$

The function $f(t)$ and its Laplace transform $\mathbf{F}(s)$ are said to form a Laplace transform pair, denoted by

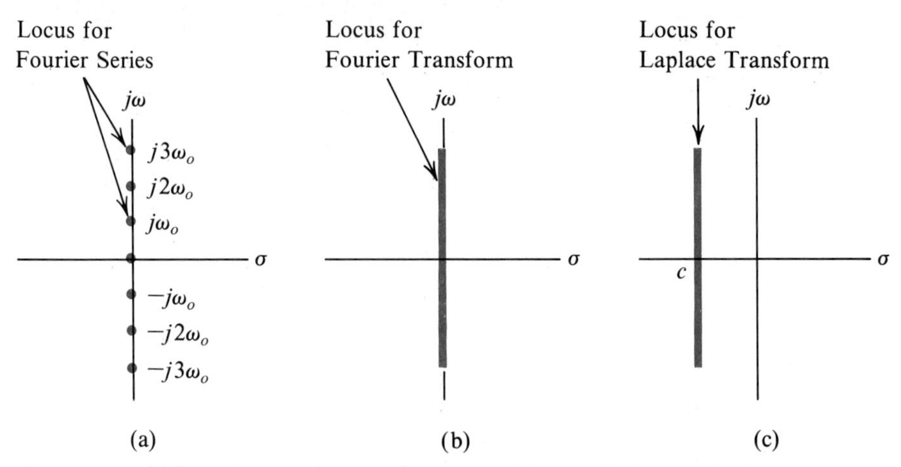

Figure 16.1 The s-domain locus of exponential signals for (a) the Fourier series, (b) the Fourier transform, and (c) the Laplace transform.

$$\mathbf{F}(s) = \mathbf{L}\{f(t)\} \text{ and } f(t) = \mathbf{L}^{-1}\{\mathbf{F}(s)\}$$

or, symbolically, by writing $f(t) \longleftrightarrow \mathbf{F}(s)$.

The Fourier series synthesizes a periodic signal from a **discrete sum** of phasor signals whose sinusoidal signal components have frequencies located at $n\omega_o$ on the j-axis of the complex plane. The Fourier transform synthesizes a non-periodic signal from a **continuous sum** of phasor signals with frequencies along the entire j-axis. The Laplace transform uses **damped or undamped** phasor signals of the form $\mathbf{F}(s)e^{st}$ and forms a continuous sum along the line in the complex plane from $c - j\infty$ to $c + j\infty$ for a specific value of the real number c. Figure 16.1 shows the locus of the phasor-signal frequencies used to synthesize signals for the Fourier series, the Fourier transform, and the Laplace transform. In general $\mathbf{F}(s)$ is a complex quantity; it defines the relative amount of the damped phasor signal e^{st} in $f(t)$.

16.1.1 Existence

A causal function $f(t)$ is said to be **exponentially bounded** if for some real number σ_c

$$\int_{0^-}^{\infty} |f(t)e^{-\sigma_c t}|\, dt < \infty$$

Every exponentially bounded function has a Laplace transform because

$$|\mathbf{F}(s)| = \left|\int_{0^-}^{\infty} f(t)e^{-st}\, dt\right| \le \int_{0}^{\infty} |f(t)e^{-st}|\, dt$$

$$\le \int_{0^-}^{\infty} |f(t)e^{-\sigma t}||e^{-j\omega t}|\, dt$$

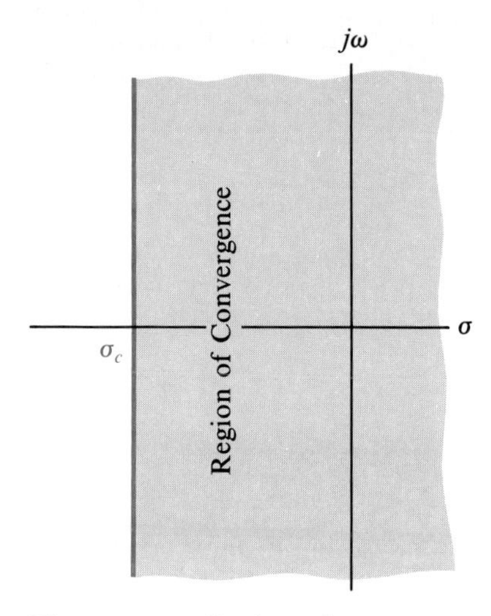

Figure 16.2 Region of convergence for the Laplace transform.

Since $|e^{-j\omega t}| = 1$ for any value of t we have

$$|\mathbf{F}(s)| \leq \int_{0^-}^{\infty} |f(t)e^{-\sigma t}|\, dt \tag{16.3}$$

If $f(t)$ is exponentially bounded there is some value of σ (denoted by σ_c) for which the right-hand side of (16.3) is bounded, so $|\mathbf{F}(s)| < \infty$ and $\mathbf{F}(s)$ exists. The region of convergence in the complex plane is defined by $\sigma > \sigma_c$ (as shown in Fig. 16.2).

The Laplace transform's use of damped phasors to synthesize signals has the advantage that transforms can be found for signals whose Fourier transform does not exist, because signals that are themselves not absolutely bounded can be exponentially bounded. This eliminates the need to define transforms "in the limit."

Example 16.1

The Laplace transform of the unit-step function is given by

$$\mathbf{F}(s) = \int_{0^-}^{\infty} 1\, e^{-st}\, dt = -\frac{1}{s}e^{-st}\bigg|_{0^-}^{\infty} = -\frac{1}{s}e^{-s\infty} + \frac{1}{s}$$

Noting that $e^{-s\infty} = e^{-(\sigma+j\omega)\infty} = e^{-\sigma\infty}e^{-j\omega\infty}$ and $e^{-\sigma\infty} = 0$ if $\sigma > 0$, we conclude that $\mathbf{F}(s) = 1/s$ for $\text{Re}\{s\} > 0$.

If the region of convergence of the Laplace transform includes the j-axis, the Fourier transform can be obtained from the Laplace transform by letting $s = j\omega$. In the case of the step function, the Laplace region of convergence does not include the j-axis, so an attempt to evaluate $\mathbf{F}(j\omega) = 1/j\omega$ as the Fourier transform of $u(t)$ is incorrect.

16.1.2 Uniqueness

No two distinct signals have the same Laplace transform and vice versa. That is, $f_1(t) = f_2(t) \leftrightarrow \mathbf{F}_1(s) = \mathbf{F}_2(s)$. This property lets us find a signal from a transform without resorting to (16.2). Thus, if $\mathbf{F}(s) = 1/s$ we know that $f(t) = u(t)$.

Example 16.2

Find the Laplace transform of the switched, damped phasor

$f(t) = e^{s_o t}u(t)$ where $s_o = \sigma_o + j\omega_o$.

Solution: $$\mathbf{F}(s) = \int_{0^-}^{\infty} e^{s_o t}e^{-st}\,dt = \frac{-1}{s - s_o}e^{-(s-s_o)t}\bigg|_{0^-}^{\infty}$$

If $\text{Re}\,\{s - s_o\} > 0$ the exponential term vanishes at the upper evaluation limit, because

$$\left|e^{-(s-s_o)t}\right| = \left|e^{-(\sigma+j\omega-s_o)t}\right| = e^{-(\sigma-\sigma_o)t}$$

If s is chosen such that $\text{Re}\,\{s - s_o\} = (\sigma - \sigma_o) > 0$ the exponent is a large negative number as t becomes large. Thus, the Laplace transform exists for $\sigma > \sigma_o$ and is given by

$$\mathbf{F}(s) = \frac{1}{s - s_o}$$

For example, if $f(t) = e^{-3t}u(t)$ then $\mathbf{F}(s) = 1/(s + 3)$.

Formally, $\mathbf{F}(s)$ is not restricted to arguments s such that $\sigma > \sigma_o$. The restriction is only needed to ensure convergence when (16.2) is used to recover $f(t)$ from $\mathbf{F}(s)$. We will not use (16.2).

The limiting behavior of the Laplace transform is usually straightforward, because it does not rely on generalized functions. (Recall that the Fourier transform of the step could not be obtained directly from the exponential.)

16.1.3 Linearity

If $f_1(t)$ and $f_2(t)$ have Laplace transforms $\mathbf{F}_1(s)$ and $\mathbf{F}_2(s)$ respectively, then for any constants a_1 and a_2

$$\mathbf{L}\{a_1f_1(t) + a_2f_2(t)\} = a_1\mathbf{F}_1(s) + a_2\mathbf{F}_2(s) \tag{16.4a}$$

and

$$\mathbf{L}^{-1}\{a_1\mathbf{F}_1(s) + a_2\mathbf{F}_2(s)\} = a_1 f_1(t) + a_2 f_2(t) \tag{16.4b}$$

The proof of (16.4) is left as an exercise.

Skill Exercise 16.1

a. Find the signal whose Laplace transform is given by $\mathbf{F}(s) = 10/s + 5/(s + 2)$.
b. Find the Laplace transform of $f(t) = (3 - 5e^{-t})u(t)$.

Answer: (a) $f(t) = 10u(t) + 5e^{-2t}u(t)$.
(b) $\mathbf{F}(s) = 3/s - 5/(s + 1)$.

Example 16.3

Find the Laplace transform of the impulse function.

Solution: $\mathbf{F}(s) = \int_{0^-}^{\infty} \delta(t)e^{-st}\, dt = 1$

The Laplace transform of the unit impulse function is a surface of constant height above the entire complex plane.

Skill Exercise 16.2

Find the Laplace transform of $f(t) = e^{-\alpha t}e^{j\beta t}u(t)$.

Answer: $\mathbf{F}(s) = 1/(s + \alpha - j\beta)$.

Example 16.4

If $f(t) = e^{-\alpha t} \cos \omega_d t\; u(t)$ find $\mathbf{F}(s)$.

Solution:
$$f(t) = e^{-\alpha t}\frac{e^{j\omega_d t} + e^{-j\omega_d t}}{2}$$
$$= \frac{1}{2}e^{(-\alpha + j\omega_d)t} + \frac{1}{2}e^{(-\alpha - j\omega_d)t}$$

Using the linearity property and the result of Skill Exercise 16.2 we get

$$\mathbf{F}(s) = \frac{1/2}{s - (-\alpha + j\omega_d)} + \frac{1/2}{s - (-\alpha - j\omega_d)}$$

and

$$\mathbf{F}(s) = \frac{s + \alpha}{(s + \alpha)^2 + \omega_d^2} = \frac{s + \alpha}{s^2 + 2\alpha s + \alpha^2 + \omega_d^2}$$

Skill Exercise 16.3

Find $\mathbf{L}\{\cos \omega_d t\, u(t)\}$.

Answer: $\mathbf{F}(s) = s/(s^2 + \omega_d^2)$.

16.1.4 Symmetry

If $f(t)$ is real-valued, then $\mathbf{F}(s^*) = \mathbf{F}^*(s)$. That is, $|\mathbf{F}(s^*)| = |\mathbf{F}^*(s)| = |\mathbf{F}(s)|$ and $\sphericalangle\mathbf{F}(s^*) = \sphericalangle\mathbf{F}^*(s) = -\sphericalangle\mathbf{F}(s)$.
The proof is

$$\mathbf{F}(s^*) = \int_{0^-}^{\infty} f(t)e^{-s^*t}\, dt = \int_{0^-}^{\infty} f(t)e^{-(\sigma - j\omega)t}\, dt$$
$$= \left[\int_{0^-}^{\infty} f(t)e^{-(\sigma + j\omega)t}\, dt\right]^* = \mathbf{F}^*(s)$$

16.1.5 Derivative

If $\mathbf{F}(s) = \mathbf{L}\{f(t)\}$, then $\mathbf{L}\{df/dt\} = s\mathbf{F}(s) - f(0^-)$, because

$$\mathbf{L}\left\{\frac{df}{dt}\right\} = \int_{0^-}^{\infty} \frac{df}{dt} e^{-st}\, dt$$

and integration by parts gives

$$\mathbf{L}\left\{\frac{df}{dt}\right\} = f(t)e^{-st}\Big|_{0^-}^{\infty} - \int_{0^-}^{\infty} f(t)(-s)e^{-st}\, dt$$
$$= -f(0^-) + s\int_{0^-}^{\infty} f(t)e^{-st}\, dt = s\mathbf{F}(s) - f(0^-)$$

Example 16.5

Obtain the Laplace transform of $f(t) = \sin \omega_d t\, u(t)$ from the Laplace transform of $g(t) = \cos \omega_d t\, u(t)$ by using the derivative property.

Solution: First we note that

$$\frac{d}{dt} \cos \omega_d t\, u(t) = -\omega_d \sin \omega_d t\, u(t) + \cos \omega_d t\, \delta(t)$$

But since $\cos \omega_d t\, \delta(t) = \delta(t)$ we can write

$$d/dt \cos \omega_d t\, u(t) = -\omega_d \{\sin \omega_d t\, u(t)\} + \delta(t)$$

Thus, the derivative property implies that

$$\mathbf{L}\left\{\frac{d}{dt}\cos\omega_d t\, u(t)\right\} = s\mathbf{L}\{\cos\omega_d t\, u(t)\} - 0$$
$$= -\omega_d\mathbf{L}\{\sin\omega_d t\, u(t)\} + 1$$

and since

$$\mathbf{L}\{\cos\omega_d t\, u(t)\} = \frac{s}{s^2 + \omega_d{}^2}$$

we have

$$\frac{s^2}{s^2 + \omega_d{}^2} = -\omega_d\mathbf{L}\{\sin\omega_d t\, u(t)\} + 1$$

and so

$$\mathbf{L}\{\sin\omega_d t\, u(t)\} = \frac{\omega_d}{s^2 + \omega_d{}^2}$$

The derivative property can be generalized to treat higher-order derivatives. In the case of the second derivative

$$\mathbf{L}\left\{\frac{d^2 f}{dt^2}\right\} = s\mathbf{L}\left\{\frac{df}{dt}\right\} - \left.\frac{df}{dt}\right|_{0^-} = s[s\mathbf{F}(s) - f(0^-)] - \frac{df}{dt}(0^-)$$

which gives

$$\mathbf{L}\left\{\frac{d^2 f}{dt^2}\right\} = s^2\mathbf{F}(s) - sf(0^-) - \frac{df}{dt}(0^-)$$

These same steps can be extended to show that

$$\mathbf{L}\left\{\frac{d^n}{dt^n}f(t)\right\} = s^n\mathbf{F}(s) - s^{n-1}f(0^-) - s^{n-2}\frac{df}{dt}(0^-) - \cdots - \frac{d^{n-1}f}{dt^{n-1}}(0^-)$$

16.1.6 Integration

If $f(t)$ has a Laplace transform $\mathbf{F}(s)$ the Laplace transform of the integral of $f(t)$ is

$$\mathbf{L}\left\{\int_{0^-}^{t} f(\alpha)\, d\alpha\right\} = \frac{1}{s}\mathbf{F}(s)$$

First note that

$$f(t) = \frac{d}{dt}\int_0^t f(\alpha)d\alpha$$

Therefore, the derivative property implies that

$$\mathbf{L}\{f(t)\} = \mathbf{L}\left\{\frac{d}{dt}\int_{0^-}^{t} f(\alpha)d\alpha\right\} = s\mathbf{L}\left\{\int_{0^-}^{t} f(\alpha)d\alpha\right\} - \int_{0^-}^{0^-} f(\alpha)d\alpha$$

and

$$\mathbf{L}\{f(t)\} = s\mathbf{L}\left\{\int_{0^-}^{t} f(\alpha)d\alpha\right\}$$

so

$$\mathbf{L}\left\{\int_{0^-}^{t} f(\alpha)d\alpha\right\} = \frac{1}{s}\mathbf{F}(s)$$

As an exercise, use integration by parts to demonstrate this equivalence.

The derivative and integral properties of the Laplace transform can be used to derive the transforms shown in Table 16.1 for derivatives and integrals of the step function.

TABLE 16.1 LAPLACE TRANSFORMS OF DERIVATIVES AND INTEGRALS OF THE UNIT STEP FUNCTION

$f(t)$	$F(s)$
$\delta^{(n)}(t)$	s^n
$\delta''(t)$	s^2
$\delta'(t)$	s
$\delta(t)$	1
$u(t)$	$\frac{1}{s}$
t	$\frac{1}{s^2}$
$\frac{1}{2}t^2$	$\frac{1}{s^3}$
$\frac{1}{n!}t^n$	$\frac{1}{s^{n+1}}$

16.1.7 Complex Frequency Shift

If $\mathbf{L}\{f(t)\} = \mathbf{F}(s)$ then

$$\mathbf{L}\{f(t)e^{-s_o t}\} = \mathbf{F}(s + s_o) \tag{16.5}$$

Multiplying $f(t)$ by $e^{-s_o t}$ has the same effect as shifting the origin of the transform by s_o. The proof is that

$$\mathbf{L}\{f(t)e^{-s_o t}\} = \int_{0^-}^{\infty} f(t)e^{-s_o t}e^{-st}\,dt$$

$$= \int_{0^-}^{\infty} f(t)e^{-(s+s_o)t}\,dt = \mathbf{F}(s + s_o)$$

This property is especially useful because it enables us to generate additional transforms. For example, from Table 16.1 we have

$$\mathbf{L}\left\{\frac{t^n}{n!}\right\} = \frac{1}{s^{n+1}}$$

Applying the shifting theorem gives

$$\mathbf{L}\left\{\frac{t^n e^{-s_o t}}{n!}\right\} = \frac{1}{(s + s_o)^{n+1}}$$

Table 16.2 can be created from Table 16.1.

TABLE 16.2

f(t)	F(*s*)
$e^{-s_o t}$	$\frac{1}{s + s_o}$
$te^{-s_o t}$	$\frac{1}{(s + s_o)^2}$
$\frac{t^2 e^{-s_o t}}{2}$	$\frac{1}{(s + s_o)^3}$
$\frac{t^n e^{-s_o t}}{n!}$	$\frac{1}{(s + s_o)^{n+1}}$

Example 16.6

Use the frequency-shift property to find the Laplace transform of $f(t) = e^{-\alpha t} \cos \omega_d t \, u(t)$

Solution: From Skill Exercise 16.3 we have

$$e^{-\alpha t} \cos \omega_d t \, u(t) \longleftrightarrow \frac{s + \alpha}{(s + \alpha)^2 + \omega_d^2}$$

16.1.8 Time Shift

If a signal $f(t)$ is translated on the time axis to form $f(t - \tau)$, the resulting signal has the Laplace transform given by

$$\mathbf{L}\{f(t - \tau)\} = e^{-s\tau}\mathbf{F}(s)$$

The proof is left to the reader.

Skill Exercise 16.4

If $g(t) = (t - 5)/5 \, u(t - 5)$ find $\mathbf{G}(s)$.

Answer: $\mathbf{G}(s) = 1/(5s^2)e^{-5s}$.

Example 16.7

If $\mathbf{F}(s) = \dfrac{5e^{-0.2s}}{s(s+4)}$ find $f(t)$.

Solution: $\mathbf{F}(s) = 5e^{-0.2s}\left[\dfrac{1}{s(s+4)}\right] = 5e^{-0.2s}\left[\dfrac{0.25}{s} - \dfrac{0.25}{s+4}\right]$

Because $0.25/s \longleftrightarrow 0.25u(t)$ and $-0.25/(s+4) \longleftrightarrow -0.25e^{-4t}u(t)$ we have, by the shifting theorem and the uniqueness property,

$$5e^{-0.2s}\left(\frac{0.25}{s}\right) \longleftrightarrow 5(0.25)u(t-0.2)$$

and

$$5e^{-0.2s}\left(\frac{-0.25}{s+4}\right) \longleftrightarrow 5(-0.25)e^{-4(t-0.2)}u(t-0.2)$$

so

$$f(t) = 1.25[1 - e^{-4(t-0.2)}]u(t-0.2)$$

16.2 INVERSE TRANSFORMS AND PARTIAL FRACTION EXPANSIONS

This section will present methods for finding $f(t)$ when $\mathbf{F}(s)$ is known, without using (16.2). It is typically the case in engineering work that $\mathbf{F}(s)$ is a ratio of polynomials, with

$$\mathbf{F}(s) = \frac{N(s)}{D(s)} = \frac{b_m s^m + b_{m-1}s^{m-1} + \cdots + b_1 s + b_0}{a_n s^n + a_{n-1}s^{n-1} + \cdots + a_1 s + a_0}$$

If $m \geq n$ the numerator polynomial can be divided by the denominator polynomial to form

$$\mathbf{F}(s) = Q(s) + \frac{R(s)}{D(s)}$$

where the quotient $Q(s)$ is a polynomial of degree $m - n$, with

$$Q(s) = q_{m-n}s^{m-n} + q_{m-n-1}s^{m-n-1} + \cdots + q_1 s + q_0$$

The time-domain signal whose Laplace transform is $Q(s)$ is obtained directly (from Table 16.2) as

$$q(t) = q_{m-n}\delta^{(m-n)}(t) + \cdots + q_1\delta'(t) + q_0\delta(t)$$

Because this part of $f(t)$ is always a sum of an impulse function and its derivatives, it can usually be written by inspection. Thus, we will focus our attention on ratios of polynomials in which the degree of the numerator polynomial is *less* than the degree of the denominator polynomial. Such functions are called **proper fractions.**

16.2.1 Case I: Simple Pole Factors

Suppose that $\mathbf{F}(s) = N(s)/D(s)$, with $n = \text{Deg } D(s)$ and $m = \text{Deg } N(s)$, and that $m < n$ (i.e., $\mathbf{F}(s)$ is a proper fraction). If $D(s)$ has distinct roots, we can write $\mathbf{F}(s)$ in factored form as

$$\mathbf{F}(s) = \frac{\prod_{j=1}^{m} (s - z_j)}{\prod_{k=1}^{n} (s - s_k)}$$

$$= \frac{K(s - z_1)(s - z_2) \cdots (s - z_m)}{(s - s_1)(s - s_2) \cdots (s - s_n)}$$

and $s_j \neq s_k$ for $j \neq k$.

The values of s for which $s = z_j$ are called the **zeros** of $\mathbf{F}(s)$ and the values of s for which $s = s_k$ are called the **poles** of $\mathbf{F}(s)$ because $\mathbf{F}(z_k) = 0$ and $\mathbf{F}(s_j) = \infty$. Since its denominator factors are distinct $\mathbf{F}(s)$ can be written in partial-fraction expansion form (PFE)

$$\mathbf{F}(s) = \sum_{i=1}^{n} \frac{K_i}{(s - s_i)}$$

The expansion coefficients are easily obtained by noting that

$$(s - s_j)\mathbf{F}(s) = K_j + \sum_{i \neq j} \frac{K_i(s - s_j)}{s - s_i}$$

If we evaluate both sides of this expression at $s = s_j$ all of the terms in the sum on the right side vanish, and we obtain

$$K_j = (s - s_j)\mathbf{F}(s)\big|_{s=s_j}$$

Example 16.8

Find the partial fraction expansion of

$$\mathbf{F}(s) = \frac{3}{s^2 + 7s + 10}$$

Solution: In factored form

$$\mathbf{F}(s) = \frac{3}{(s + 2)(s + 5)}$$

$\mathbf{F}(s)$ is a proper fraction; its PFE is

$$\mathbf{F}(s) = \frac{K_1}{s+2} + \frac{K_2}{s+5}$$

To find K_1 and K_2 take

$$K_1 = (s+2)\mathbf{F}(s)\Big|_{s=-2} = \frac{3}{s+5}\Big|_{s=-2} = 1$$

Likewise,

$$K_2 = (s+5)\mathbf{F}(s)\,|_{s=-5} = -1$$

Having determined K_1 and K_2 we form

$$\mathbf{F}(s) = \frac{1}{s+2} - \frac{1}{s+5}$$

Once the PFE of $\mathbf{F}(s)$ has been developed, the time-domain expression for $f(t)$ corresponding to $\mathbf{F}(s)$ is found by inverting the simple pole factors (using Tables 16.1 and 16.2)

$$f(t) = \mathbf{L}^{-1}\left\{\sum_{i=1}^{n} \frac{K_i}{s - s_i}\right\} = \sum_{i=1}^{n} \mathbf{L}^{-1}\left\{\frac{K_i}{s - s_i}\right\}$$

$$= \sum_{i=1}^{n} K_i e^{s_i t} u(t)$$

Skill Exercise 16.5

Find $f(t)$ if $\mathbf{F}(s) = \dfrac{12}{s^2 + 6s + 8}$

Answer: $f(t) = 6(e^{-2t} - e^{-4t})u(t)$

16.2.2 Case II: Repeated Pole Factors

A pole at $s = s_q$ is said to be **repeated** r times if the denominator of $\mathbf{F}(s)$ contains a factor of $(s - s_q)^r$. If the denominator of $\mathbf{F}(s)$ contains a repeated pole factor the simple poles are expanded in a partial fraction in the usual way. Any pole s_q that is repeated r times contributes r terms to the PFE in the form

$$\hat{\mathbf{F}}(s) = \frac{A_1}{s - s_q} + \frac{A_2}{(s - s_q)^2} + \cdots + \frac{A_r}{(s - s_q)^r} \qquad \textbf{(16.6)}$$

The time-domain expressions corresponding to these terms can be obtained from Table 16.2:

$$f(t) = \left[A_1 + A_2 t + A_3 \frac{t}{2!} + \cdots + A_r \frac{t^{r-1}}{(r-1)!}\right] e^{s_q t}$$

To find $A_1, A_2, \ldots, A_r$, note that $\mathbf{F}(s)$ can be written as

$$\mathbf{F}(s) = \sum_{\substack{\text{Simple} \\ \text{Poles}}} \frac{K_i}{s - s_i} + \hat{\mathbf{F}}(s)$$

and so

$$\mathbf{F}(s)(s - s_q)^r = (s - s_q)^r \left[\sum_{\substack{\text{Simple} \\ \text{Poles}}} \frac{K_i}{(s - s_i)}\right]$$
$$+ A_1(s - s_q)^{r-1} + A_2(s - s_q)^{r-2} + \cdots + A_{r-1}(s - s_q) + A_r$$

It is left as an exercise to show that

$$A_r = \mathbf{F}(s)(s - s_q)^r \big|_{s=s_q}$$

$$A_{r-1} = \left[\frac{d}{ds} \mathbf{F}(s)(s - s_q)^r\right]\bigg|_{s=s_q}$$

$$A_1 = \frac{1}{(r-1)!} \left[\frac{d^{r-1}}{ds^{r-1}} \mathbf{F}(s)(s - s_q)^r\right]\bigg|_{s=s_q}$$

Example 16.9

If $\mathbf{F}(s) = \dfrac{1}{(s+2)(s+4)^3}$ find $f(t)$.

Solution: The PFE of $\mathbf{F}(s)$ has

$$\mathbf{F}(s) = \frac{K_1}{(s+2)} + \frac{A_1}{(s+4)} + \frac{A_2}{(s+4)^2} + \frac{A_3}{(s+4)^3}$$

The unrepeated pole has $K_1 = 1/8$. Next, we form

$$(s+4)^3 \mathbf{F}(s) = \frac{1}{s+2}$$

Then

$$A_3 = \left(\frac{1}{s+2}\right)\bigg|_{s=-4} = -\frac{1}{2}$$

$$A_2 = \frac{d}{ds}\left(\frac{1}{s+2}\right)\bigg|_{s=-4} = -\frac{1}{4}$$

$$A_1 = 1/2 \frac{d^2}{ds^2}(s+4)^3\mathbf{F}(s)|_{s=-4}$$

$$= 1/2 \frac{d^2}{ds^2}\frac{1}{s+2}\bigg|_{s=-4} = -\frac{1}{8}$$

and the PFE of $\mathbf{F}(s)$ is

$$\mathbf{F}(s) = \frac{1/8}{s+2} + \frac{-1/8}{s+4} + \frac{-1/4}{(s+4)^2} + \frac{-1/2}{(s+4)^3}$$

and

$$f(t) = \left[\frac{1}{8}e^{-2t} - \left(\frac{1}{8} + \frac{1}{4}t + \frac{1}{4}t^2\right)e^{-4t}\right]u(t)$$

Skill Exercise 16.6

Find the PFE of $\mathbf{F}(s) = 1/[(s+4)(s+1)^2]$.

Answer: $\mathbf{F}(s) = \frac{1/9}{s+4} + \frac{-1/9}{s+1} + \frac{1/3}{(s+1)^2}$

16.2.3 Case III: Complex Pole Factors

If $\mathbf{F}(s)$ is restricted to have real coefficients, any complex pole in the denominator of $\mathbf{F}(s)$ will be accompanied by its complex-conjugate pole factor. The pair of terms due to the conjugate poles have expansion coefficients that can be combined, because the coefficients themselves must be a complex-conjugate pair (i.e., the coefficients K_1 and K_2 corresponding to the simple poles at $s = -\alpha \pm j\beta$ must be such that $K_1 = K_2{}^*$). To see this, suppose that

$$\mathbf{F}(s) = \frac{K_1}{s+\alpha-j\beta} + \frac{K_2}{s+\alpha+j\beta} + \hat{\mathbf{F}}(s)$$

where $\hat{\mathbf{F}}(s)$ accounts for all other poles of $\mathbf{F}(s)$. Next, we evaluate K_1 and K_2 in the usual way:

$$K_1 = (s+\alpha-j\beta)\,\mathbf{F}(s)|_{s=-\alpha+j\beta}$$

$$K_2 = (s+\alpha+j\beta)\,\mathbf{F}(s)|_{s=-\alpha-j\beta}$$

The denominator of $\mathbf{F}(s)$ can be written as $D(s)M(s)$ where $D(s) = (s+\alpha-j\beta)(s+\alpha+j\beta)$. Therefore

$$K_1 = \frac{N(s)}{(s+\alpha+j\beta)M(s)}\bigg|_{s=-\alpha+j\beta} = \frac{N(-\alpha+j\beta)}{2j\beta M(-\alpha+j\beta)}$$

$$K_2 = \frac{N(s)}{(s+\alpha-j\beta)M(s)}\bigg|_{s=-\alpha-j\beta} = \frac{N(-\alpha-j\beta)}{-2j\beta M(-\alpha-j\beta)}$$

Because $N(s)$ and $M(s)$ are restricted to real coefficients we must have

$$\frac{N(-\alpha + j\beta)}{M(-\alpha + j\beta)} = \left[\frac{N(-\alpha - j\beta)}{M(-\alpha - j\beta)}\right]^*$$

Likewise, $1/2j\beta = [1/(-2j\beta)]^*$. From these observations we conclude that K_1 and K_2 are a complex-conjugate pair, with $K_1^* = K_2$. Therefore, if we write the polar form

$$K_1 = |K_1|e^{j\psi_1} \text{ and } K_2 = K_1^* = |K_1|e^{-j\psi_1}$$

we can simplify the expansion of $\mathbf{F}(s)$ by combining terms:

$$\frac{K_1}{s + \alpha - j\beta} + \frac{K_2}{s + \alpha + j\beta} = \frac{|K_1|e^{j\psi_1}}{s + \alpha - j\beta} + \frac{|K_1|e^{-j\psi_1}}{s + \alpha + j\beta}$$

The corresponding time-domain signal is

$$\mathbf{L}^{-1}\left\{\frac{K_1}{s + \alpha - j\beta} + \frac{K_1^*}{s + \alpha + j\beta}\right\} = |K_1|e^{j\psi_1}e^{-(\alpha - j\beta)t} + |K_1|e^{-j\psi_1}e^{-(\alpha + j\beta)t}$$

$$= 2|K_1|e^{-\alpha t}\cos(\beta t + \psi_1) = 2|K_1|e^{-\alpha t}\sin(\beta t + \phi_1)$$

where $\phi_1 = \psi_1 + \pi/2$. This is the same time-domain expression developed in Chapter 11 for the natural response of an underdamped second-order circuit.

Example 16.10

Find $f(t)$ if $\mathbf{F}(s) = \dfrac{1}{s^2 + 2s + 5}$.

Solution: Factoring the denominator of $\mathbf{F}(s)$ leads to the PFE

$$\mathbf{F}(s) = \frac{K_1}{s + 1 - j2} + \frac{K_1^*}{s + 1 + j2}$$

and

$$K_1 = (s + 1 - j2)\mathbf{F}(s)\,\Big|_{s=-1+j2} = \frac{1}{s + 1 + j2}\,\Big|_{s=-1+j2} = \frac{1}{j4}$$

Likewise, $K_2 = -1/j4$ (note that $K_2 = K_1^*$). In polar form $K_1 = 0.25\angle -90^\circ$ and $K_2 = 0.25\angle 90^\circ$. Therefore

$$f(t) = 0.5e^{-t}\cos(2t - 90^\circ)u(t)$$
$$= f(t) = 0.5e^{-t}\sin 2t\, u(t)$$

Skill Exercise 16.7

Find $f(t)$ if $\mathbf{F}(s) = \dfrac{1}{(s+2)(s^2+2s+2)}$

Answer: $f(t) = [0.5e^{-2t} + 0.707e^{-t}\cos(t - 135°)]u(t)$

16.3 LAPLACE TRANSFORMS AND DIFFERENTIAL EQUATIONS

The Laplace transform provides an algebraic method for solving a differential equation. Recall that the I/O model for the initial-state response of a circuit is described by

$$a_n\frac{d^ny}{dt^n} + \cdots + a_1\frac{dy}{dt} + a_0y(t) = b_m\frac{d^mx}{dt^m} + \cdots + b_1\frac{dx}{dt} + b_0x(t) \quad (16.7)$$

for $t \geq 0$. Taking the Laplace transform of both sides of (16.7) gives

$$\mathbf{L}\{a_n\frac{d^ny}{dt^n} + \cdots + a_1\frac{dy}{dt} + a_0y(t)\} = \mathbf{L}\{b_m\frac{d^mx}{dt^m} + \cdots + b_1\frac{dx}{dt} + b_0x(t)\}$$

and by linearity

$$a_n\mathbf{L}\left\{\frac{d^ny}{dt^n}\right\} + \cdots + a_1\mathbf{L}\left\{\frac{dy}{dt}\right\} + a_0\mathbf{L}\{y(t)\} = b_m\mathbf{L}\left\{\frac{d^mx}{dt^m}\right\} + \cdots + b_1\mathbf{L}\left\{\frac{dx}{dt}\right\} + b_0\mathbf{L}\{x(t)\} \quad (16.8)$$

Using the derivative property of the transform, we write for the nth derivative

$$a_n\mathbf{L}\left\{\frac{d^ny}{dt^n}\right\} = a_n[s^n\mathbf{Y}(s) - s^{n-1}y(0^-) - s^{n-2}\frac{d}{dt}y(0^-) - \cdots - \frac{d^{n-1}}{dt^{n-1}}y(0^-)]$$

Each term can be written in a similar way on both sides of equation (16.8) to form

$$a_ns^n\mathbf{Y}(s) + a_{n-1}s^{n-1}\mathbf{Y}(s) + \cdots + a_0\mathbf{Y}(s) - \hat{B}(s) = b_ms^m\mathbf{X}(s) + b_{m-1}s^{m-1}\mathbf{X}(s) + \cdots + b_1s\mathbf{X}(s) + b_0\mathbf{X}(s) \quad (16.9)$$

$\hat{B}(s)$ includes all of the polynomial terms that involve initial conditions on

y and its derivatives. Remember that the initial-state response applies $x(t)$ for $t > 0$, for given initial conditions on y and its derivatives, so those terms do not appear on the right-hand side of (16.9). Next we group terms to get

$$\mathbf{Y}(s)[a_n s^n + \cdots + a_0] = \mathbf{X}(s)[b_m s^m + \cdots + b_0] + \hat{B}(s)$$

We then divide both sides by the characteristic polynomial from the left side to solve for the Laplace transform of the response:

$$\mathbf{Y}(s) = \frac{\mathbf{X}(s)[b_m s^m + \cdots + b_0]}{a_n s^n + \cdots + a_0} + \frac{\hat{B}(s)}{a_n s^n + \cdots + a_0}$$

Recognizing that the first expression includes the transfer function of the circuit we are finally able to write a more compact expression for the transform

$$\mathbf{Y}(s) = \mathbf{X}(s)\mathbf{H}(s) + \mathbf{B}_c(s) \tag{16.10}$$

where

$$\mathbf{B}_c(s) = \frac{\hat{B}(s)}{a_n s^n + \cdots + a_0}$$

If the initial conditions are all zero, the polynomial $\hat{B}(s)$ evaluates to zero.

Equation (16.10) gives an s-domain (complex frequency-domain) model for the input/output relationship of the circuit, and provides an algebraic method for obtaining the response $y(t)$ for a given $x(t)$.

Example 16.11

Find an expression for $\mathbf{Y}(s)$ in terms of $\mathbf{X}(s)$ and the boundary conditions for $y(t)$ if

$$\frac{d^2y}{dt^2} + 7\frac{dy}{dt} + 12y(t) = x(t)$$

Solution:

$$\mathbf{L}\left\{\frac{d^2y}{dt^2}\right\} = s\mathbf{L}\left\{\frac{dy}{dt}\right\} - \dot{y}(0^-)$$
$$= s[s\mathbf{L}\{y(t)\} - y(0^-)] - \dot{y}(0^-)$$
$$= s^2\mathbf{Y}(s) - sy(0^-) - \dot{y}(0^-)$$

Likewise,

$$\mathbf{L}\left\{\frac{dy}{dt}\right\} = s\mathbf{Y}(s) - y(0^-)$$

so

$$\mathbf{L}\left\{\frac{d^2y}{dt^2} + 7\frac{dy}{dt} + 12y(t)\right\} = \mathbf{L}\{x(t)\}$$

$$= s^2\mathbf{Y}(s) - sy(0^-) - \dot{y}(0^-)$$
$$+ 7s\mathbf{Y}(s) - 7y(0^-) + 12\mathbf{Y}(s)$$
$$= \mathbf{Y}(s)(s^2 + 7s + 12) - sy(0^-) - \dot{y}(0^-) - 7y(0^-)$$
$$= \mathbf{X}(s)$$

Rearranging gives

$$\mathbf{Y}(s) = \frac{\mathbf{X}(s)}{s^2 + 7s + 12} + \frac{sy(0^-) + \dot{y}(0^-) + 7y(0^-)}{s^2 + 7s + 12}$$

Example 16.12

Find the zero-state response of $y(t)$ if

$$\mathbf{H}(s) = \frac{1}{(s + 1)(s + 2)(s + 3)(s + 4)}$$

and $x(t) = u(t)$.

Solution: The Laplace transform method [using (16.10) with $\mathbf{B}_c(s) = 0$] allows us to write

$$\mathbf{Y}(s) = \frac{1}{s(s + 1)(s + 2)(s + 3)(s + 4)}$$
$$= \frac{1/24}{s} + \frac{-1/6}{s + 1} + \frac{1/4}{s + 2} + \frac{-1/6}{s + 3} + \frac{1/24}{s + 4}$$

which gives

$$y(t) = \left(\frac{1}{24} - \frac{1}{6}e^{-t} + \frac{1}{4}e^{-2t} - \frac{1}{6}e^{-3t} + \frac{1}{24}e^{-4t}\right)u(t)$$

This is somewhat easier than attempting to use the methods of Chapter 7, which would require solving four boundary-condition equations to find the coefficients of the natural solution of the differential equation. The Laplace-transform method directly evaluates the coefficients of the partial fraction expansion, instead of solving the system of simultaneous equations for the coefficients of the natural-solution terms.

16.3.1 Initial-State Response

In Chapter 8 we saw that the differential equation in (16.7) describes the initial state response of a circuit. Therefore, $\mathbf{Y}(s)$ given by (16.10) is the Laplace transform of the initial-state response of the circuit. It consists of the sum of a term due to the transform of the input and a term due to the initial conditions. In the time domain

$$y(t) = \mathbf{L}^{-1}\{\mathbf{X}(s)\mathbf{H}(s)\} + \mathbf{L}^{-1}\{\mathbf{B}_c(s)\} \qquad (16.11)$$

$$\updownarrow \qquad\qquad \updownarrow \qquad\qquad \updownarrow$$

ISR ZSR ZIR

The initial-state response defined by (16.11) is the sum of the zero-state response and the zero-input response. Thus the time-domain signals and their Laplace transforms satisfy the same response superposition property as we saw in Chapter 8, namely

$$\mathbf{Y}_{ISR}(s) = \mathbf{Y}_{ZSR}(s) + \mathbf{Y}_{ZIR}(s)$$

Example 16.13

If

$$\frac{d^2y}{dt^2} + 12\frac{dy}{dt} + 32y(t) = 2\frac{dx}{dt} + x(t)$$

and $y(0^-) = y_0$ and $\dot{y}(0^-) = \dot{y}_0$, find the ZIR, ZSR and ISR of $y(t)$ when $x(t) = tu(t)$, a unit-ramp signal.

Solution:

$$\mathbf{L}\left\{\frac{d^2y}{dt^2} + 12\frac{dy}{dt} + 32y(t)\right\} = \left[s\mathbf{L}\left\{\frac{dy}{dt}\right\} - \dot{y}_0\right]$$
$$+12[s\mathbf{L}\{y\} - y_0] + 32\mathbf{L}\{y\}$$
$$= (s^2\mathbf{L}\{y\} - s\dot{y}_0 - y_0) + 12(s\mathbf{L}\{y\} - y_0) + 32\mathbf{L}\{y\}$$
$$= \mathbf{Y}(s)(s^2 + 12s + 32) - sy_0 - \dot{y}_0 - 12y_0$$

On the right-hand side of the equation [with $x(t) = 0$ for $t \leq 0$] we get

$$\mathbf{L}\left\{2\frac{dx}{dt} + x(t)\right\} = 2s\mathbf{X}(s) + \mathbf{X}(s)$$

Therefore

$$\mathbf{Y}(s) = \frac{(2s + 1)\mathbf{X}(s)}{s^2 + 12s + 32} + \frac{sy_0 + \dot{y}_0 + 12y_0}{s^2 + 12s + 32}$$

Letting $\mathbf{X}(s) = 1/s^2$ (for the ramp signal) leads to

$$\mathbf{Y}(s) = \frac{(2s + 1)}{s^2(s + 4)(s + 8)} + \frac{sy_0 + \dot{y}_0 + 12y_0}{(s + 4)(s + 8)}$$

The first term corresponds to the ZSR and the second term corresponds to the ZIR:

$$y_{ZIR}(t) = \mathbf{L}^{-1}\left\{\frac{sy_0 + \dot{y}_0 + 12y_0}{(s + 4)(s + 8)}\right\}$$

$$y_{ZSR}(t) = \mathbf{L}^{-1}\left\{\frac{2s + 1}{s^2(s + 4)(s + 8)}\right\}$$

Taking partial-fraction expansions leads to

$$y_{ZIR}(t) = \mathbf{L}^{-1}\left\{\frac{\frac{1}{4}(-4y_0 + \dot{y}_0 + 12y_0)}{s+4} - \frac{\frac{1}{4}(-8y_0 + \dot{y}_0 + 12y_0)}{s+8}\right\}$$

$$= \left[\frac{1}{4}(8y_0 + \dot{y}_0)e^{-4t} - \frac{1}{4}(4y_0 + \dot{y}_0)e^{-8t}\right]u(t)$$

Likewise,

$$y_{ZSR}(t) = \mathbf{L}^{-1}\left\{\frac{13/256}{s} + \frac{1/32}{s^2} + \frac{-7/64}{s+4} + \frac{15/256}{s+8}\right\}$$

and

$$y_{ZSR}(t) = \left(\frac{13}{256} + \frac{t}{32} - \frac{7}{64}e^{-4t} + \frac{15}{256}e^{-8t}\right)u(t)$$

Then for specified values of y_0 and $\dot{y}_0$, $y(t) = y_{ZIR}(t) + y_{ZSR}(t)$.

In the preceding example we deliberately left the boundary conditions in generic, rather than numeric, terms so that the effect of each boundary condition could be seen before we evaluated the final result. This can be very useful when using a personal computer to simulate a circuit's response to a variety of initial conditions. Premature substitution of numeric values would conceal such information.

Example 16.14

If $\mathbf{H}(s) = 10/(s+2)$ and $x(t) = \ u(t)$ find $y(t)$ for $t \geq 0$ if $y(0^-) = 20$.

Solution: Before $y(t)$ can be found it will be necessary to find $\mathbf{Y}(s)$. Since the initial conditions do not correspond to the zero state, we *cannot* merely use $\mathbf{Y}(s) = \mathbf{X}(s)\mathbf{H}(s)$. Instead, we have to recover $\mathbf{Y}(s)$ from the I/O differential equation implied by $\mathbf{H}(s)$. Thus we must first obtain

$$\frac{dy}{dt} + 2y(t) = 10x(t)$$

Then, we take the Laplace transform of the equation:

$$s\mathbf{Y}(s) - y(0^-) + 2\mathbf{Y}(s) = 10\mathbf{X}(s)$$

and so

$$\mathbf{Y}(s) = \frac{10\mathbf{X}(s)}{s+2} + \frac{y(0^-)}{s+2} = \frac{10}{s(s+2)} + \frac{20}{s+2}$$

$$= \frac{5}{s} - \frac{5}{s+2} + \frac{20}{s+2}$$

Lastly, we take the inverse transform to get

$y(t) = (5 + 15e^{-2t})u(t)$

16.3.2 Zero-State Response

If the initial conditions on $y(t)$ are all set to 0 at $t = 0^-$ then $\mathbf{B}_c(s) = 0$ in (16.11) and $\mathbf{Y}(s)$ is the Laplace transform of the zero-state response of $y(t)$, with

$$\mathbf{Y}_{ZSR}(s) = \mathbf{X}(s)\mathbf{H}(s) \tag{16.12}$$

The Laplace transform of the zero-state response is the product of the Laplace transform of the input signal and the transfer function, as shown in Fig. 16.3.

The transfer functions $\mathbf{H}(s)$ govern the input/output relationship for the Laplace transform of signals because such signals are in effect the result of superposition of exponential signals. What is remarkable is the fact that $\mathbf{Y}(s)$ formed according to (16.12) automatically satisfies the boundary conditions for the ZSR.

Under the boundary conditions of the zero-state response

$$\mathbf{H}(s) = \frac{\mathbf{L}\{y_{ZSR}(t)\}}{\mathbf{L}\{x(t)\}}$$

and

$$\mathbf{H}(s) = \frac{\mathbf{Y}_{ZSR}(s)}{\mathbf{X}(s)}$$

$\mathbf{H}(s)$ determines the amplitude and phase shift of each damped phasor signal in $x(t)$ to synthesize $y_{ZSR}(t)$. *Avoid the trap* of thinking that $\mathbf{H}(s)$ always defines the ratio of $\mathbf{Y}(s)$ to $\mathbf{X}(s)$. (When doesn't it?)

In general, the poles of $\mathbf{Y}(s)$ include both the natural frequencies of the circuit and any pole factors that are in $\mathbf{X}(s)$. Therefore, the ZSR exhibits

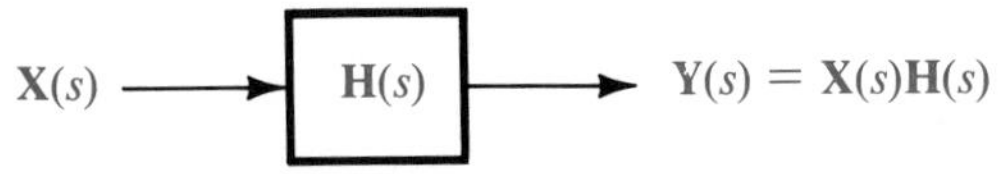

$\mathbf{X}(s) \longrightarrow$ [$\mathbf{H}(s)$] $\longrightarrow \mathbf{Y}(s) = \mathbf{X}(s)\mathbf{H}(s)$

Figure 16.3 A circuit's transfer function and the Laplace transform of its input signal determine its zero-state response.

behavior due to the natural frequencies of the circuit and behavior due to the poles of the input signal's Laplace transform. In contrast, the ZIR can only exhibit behavior due to the natural frequencies of the circuit.

Skill Exercise 16.8

If $\mathbf{H}(s) = \dfrac{1}{s^2 + 12s + 35}$ and if $x(t) = tu(t)$, find $y(t)$ if all initial conditions are 0.

Answer: $y(t) = \left(\dfrac{-12}{1225} + \dfrac{1}{35}t + \dfrac{1}{50}e^{-5t} - \dfrac{1}{98}e^{-7t}\right)u(t)$

When $x(t)$ is formed by adding two or more signals, the ZSR to $x(t)$ is the superposition of the zero-state response due to each signal. If $x(t) = \Sigma x_j(t)$ then $\mathbf{Y}(s) = \Sigma \mathbf{Y}_j(s)$ and so

$$\mathbf{Y}(s) = \Sigma \mathbf{H}(s)\mathbf{X}_j(s)$$

Skill Exercise 16.9

If $\mathbf{H}(s) = 1/(s + 1)$ find the ZSR of $y(t)$ when $x(t) = (1 + t)u(t)$.

Answer: $y(t) = tu(t)$. As an exercise, graph the components of $y(t)$ due to $u(t)$ and to $tu(t)$.

16.3.3 Zero-Input Response

If $x(t)$ is zero its Laplace transform is zero, and $y(t)$ in (16.8) is simply the zero-input response. So (16.10) becomes

$$\mathbf{Y}_{ZIR}(s) = \mathbf{B}_c(s) = \frac{\hat{B}(s)}{a_n s^n + \cdots + a_1 s + a_0} \tag{16.13}$$

In the time domain

$$y_{ZIR}(t) = \mathbf{L}^{-1}\left\{\frac{\hat{B}(s)}{a_n s^n + \cdots + a_1 s + a_0}\right\} \tag{16.14}$$

The exponential terms in the response of $y(t)$ will correspond to the natural frequencies of the circuit being modeled by the I/O equation. The numerator polynomial in (16.13) and (16.14) accounts for the initial conditions.

Skill Exercise 16.10

A circuit has $\mathbf{H}(s) = 2/(s^2 + 5s + 6)$. Find the ZIR of $y(t)$ when $y(0^-) = 5$ and $dy/dt(0^-) = 1$.

Answer: $y_{ZIR}(t) = (16e^{-2t} - 11e^{-3t})u(t)$

The transfer function of a circuit determines the particular solution of the I/O D.E. to a complex exponential input signal, provided that the complex frequency of the exponential is not a natural frequency of the circuit. If it is, the denominator of $\mathbf{H}(s)$ is zero, so $\mathbf{H}(s)$ is undefined, and we must resort to some other method to construct the particular solution. The Laplace-transform method circumvents this difficulty.

Example 16.15

Suppose that $\mathbf{H}(s) = 2/(s + 1)$ and that $x(t) = 10e^{-t}u(t)$. Then, if the initial conditions are zero

$$\mathbf{Y}(s) = \frac{20}{(s + 1)^2}$$

and from Table 16.2

$$y(t) = 20te^{-t}u(t)$$

16.3.4 Transient and Steady-State Response

The transient and steady-state responses of a circuit variable are related to its Laplace transform. In (16.10) the term $\mathbf{X}(s)\mathbf{H}(s)$ has poles due to $\mathbf{X}(s)$ and $\mathbf{H}(s)$ while the term $\mathbf{B}_c(s)$ has only poles due to $\mathbf{H}(s)$. Therefore, the poles can be grouped together in the partial-fraction expansion

$$y(t) = \left[\mathbf{L}^{-1}\{\mathbf{X}(s)\mathbf{H}(s)\} \Big|_{\substack{\text{Poles} \\ \text{of} \\ \mathbf{X}(s)}} \right] + \left[\mathbf{L}^{-1}\{\mathbf{X}(s)\mathbf{H}(s) + \mathbf{B}_c(s)\} \Big|_{\substack{\text{Poles} \\ \text{of} \\ \mathbf{H}(s)}} \right]$$

If all of the poles of $\mathbf{H}(s)$ are strictly in the left half-plane, the steady-state value of $y(t)$ is due entirely to the first term:

$$y_{ss}(t) = \lim_{t \to \infty} \mathbf{L}^{-1}\{\mathbf{X}(s)\mathbf{H}(s)\} \Big|_{\substack{\text{Poles} \\ \text{of} \\ \mathbf{X}(s)}}$$

y_{ss} will be non-zero if and only if $\mathbf{X}(s)$ has at least one pole on the j-axis or in the right half-plane. When the poles of $\mathbf{H}(s)$ are in the LHP, only the poles of the input signal determine the steady-state response. However, the transient response is determined by the poles of the circuit and the poles of the input signal.

16.4 IMPULSE RESPONSE

The impulse response $h(t)$ of a circuit is its ZSR to an impulse signal. Under those conditions, $x(t) = \delta(t)$ and $\mathbf{X}(s) = 1$. As a result, the impulse response is

$$h(t) = \mathbf{L}^{-1}\{\mathbf{H}(s)\}$$

16.5 CIRCUIT ANALYSIS WITH LAPLACE TRANSFORMS

The algebraic methods developed in Chapters 13–15 for phasors and Fourier transforms also apply to the Laplace transforms of all circuit variables. In particular, the Laplace transform of the branch voltages around a loop/mesh obeys KVL, and the Laplace transform of the branch currents entering a node satisfies KCL:

$$\Sigma \mathbf{V}_j(s) = 0 \text{ and } \Sigma \mathbf{I}_k(s) = 0$$

Likewise, the impedance and admittance of an individual circuit element define the ratios of the Laplace transforms of voltage and current at the element's terminals as

$$\mathbf{Z}(s) = \frac{\mathbf{V}(s)}{\mathbf{I}(s)} \text{ and } \mathbf{Y}(s) = \frac{\mathbf{I}(s)}{\mathbf{V}(s)}$$

provided that the circuit is initially relaxed (in the zero state). Thus, $\mathbf{Y}(s)$ **and** $\mathbf{Z}(s)$ **define the ratio of the Laplace transforms of the ZSR of** $v(t)$ **and** $i(t)$**, as well as the ratio of the complex exponential amplitude of the particular solution to a complex exponential input signal.** (See Chapters 9–11)

There are two basic approaches to using Laplace transforms to find the initial-state response or the zero-input response of a circuit. The first is to write the time-domain integro-differential equations describing the circuit, and then transform and solve the equations to obtain the transform of the response. Taking the inverse transform gives the time-domain solution. A second method for finding the ZIR or the ISR avoids the somewhat cumbersome step of writing time-domain equations altogether. Instead, it develops a model that directly describes relationships between the Laplace transforms of the circuit variables. This method creates a **Laplace model** for the circuit by replacing inductors and capacitors with their I/O Laplace-transform models.

First, consider the Laplace transform of the I/O equation for the voltage across a capacitor

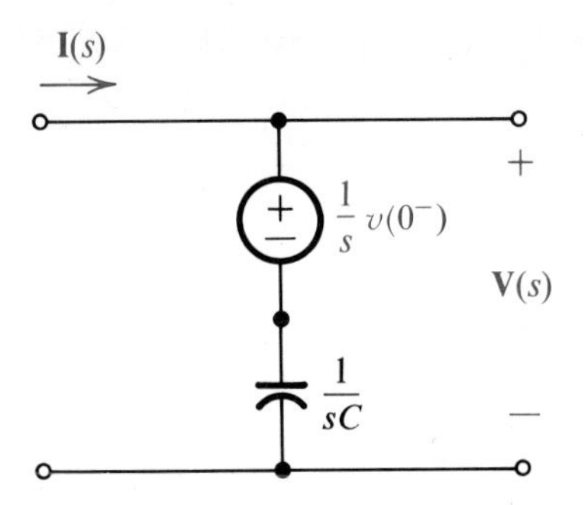

Figure 16.4 A series Laplace model of a capacitor.

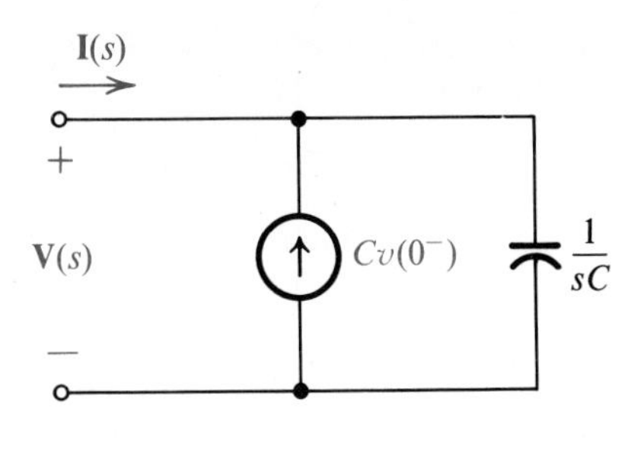

Figure 16.5 A parallel Laplace model of a capacitor.

$$\mathbf{L}\{v(t)\} = \mathbf{L}\left\{v(0^-) + \frac{1}{C}\int_{0^-}^{t} i(\alpha)d\alpha\right\} \tag{16.15}$$

The "transformed" circuit diagram in Fig. 16.4 represents equation (16.15) by a voltage source in series with an uncharged capacitor. The voltage source has the value of the Laplace transform of the time-domain effect of the initial stored energy, and the capacitor is labeled with its s-domain impedance.

An alternative Laplace model for the capacitor, shown in Fig. 16.5, can be obtained by solving for $\mathbf{I}(s)$:

$$\mathbf{I}(s) = sc\mathbf{V}(s) - Cv(0^-)$$

It consists of the s-domain impedance model of the capacitor connected in parallel with a current source whose value depends on the initial capacitor voltage.

The Laplace models for inductors are shown in Fig. 16.6(a), (b). Their derivation is left as an exercise.

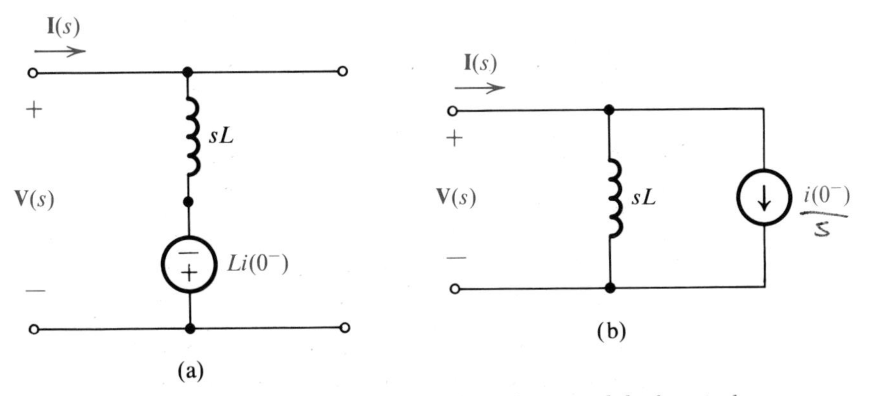

Figure 16.6 A (a) series, and (b) parallel Laplace model of an inductor.

Example 16.16

Find the ZIR of $v_o(t)$ in Fig. 16.7, with $v_o(0^-) = 10$ and $i_L(0^-) = 2$ A.

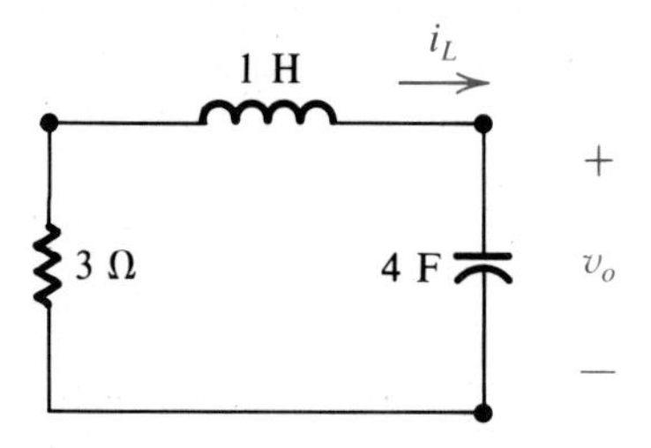

Figure 16.7

Solution: First, the circuit is replaced by the series Laplace models of the inductor and capacitor, as shown in Fig. 16.8. Next, we use voltage division to obtain an expression for the transformed output voltage, *being careful* to note that it is measured across the capacitor *and* the voltage source:

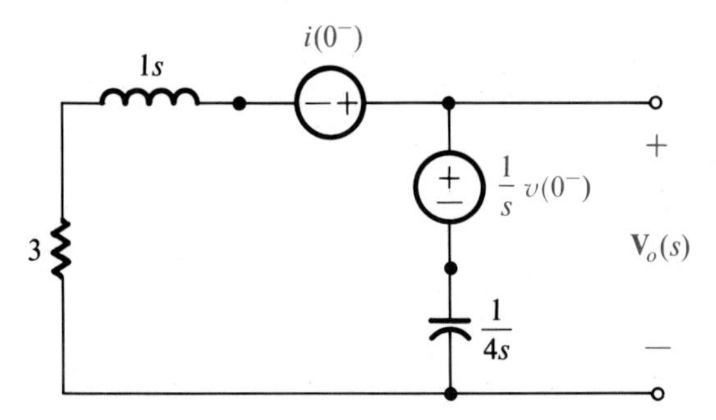

Figure 16.8

$$\mathbf{V}_o(s) = \frac{1}{s}v(0^-) + \frac{\frac{1}{4s}\left[i(0^-) - \frac{1}{s}v(0^-)\right]}{3 + s + \frac{1}{4s}}$$

and:

$$\mathbf{V}_o(s) = \frac{10}{s} + \frac{2 - \frac{10}{s}}{4s^2 + 12s + 1}$$

$$= \frac{10}{s} - \frac{10}{s} + \frac{10.48}{s + 0.858} - \frac{0.48}{s + 2.914}$$

Then

$v_o(t) = (10.48e^{-0.0858t} - 0.48e^{-2.914t})u(t)$

The waveform of $v_o(t)$ begins at a value of 10 volts and exhibits overdamped decay to zero in approximately 47 sec.

Example 16.17

If a series RC circuit with $R = 1\ \Omega$ and $C = 0.5$ F is driven by $v_{in}(t) = tu(t)$ and $v_o(0^-) = 5$ V, find $v_o(t)$, the capacitor voltage for $t \geq 0$.

Solution: The Laplace circuit model shown in Fig. 16.9 will be used to find the ISR of $v_o(t)$. Solving for $\mathbf{V}_o(s)$ gives

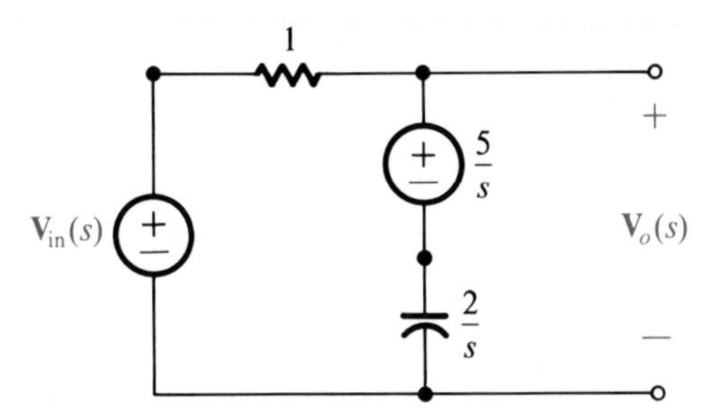

Figure 16.9

$$\mathbf{V}_o(s) = \frac{5}{s} + \frac{\mathbf{V}_{in}(s) - \frac{5}{s}}{1 + \frac{2}{s}} \times \frac{2}{s}$$

$$= \frac{2\mathbf{V}_{in}(s)}{s+2} + \frac{5}{s+2}$$

Note that one advantage of working in terms of a generic input signal $\mathbf{V}_{in}(s)$ is that the problem need not be completely re-solved for different source signals. The model is valid for an arbitrary source. This approach also separates the zero-input and zero-state components of the response.

Replacing $\mathbf{V}_{in}(s)$ by the Laplace transform of the ramp signal gives

$$\mathbf{V}_o(s) = \frac{-0.5}{s} + \frac{1}{s^2} + \frac{5.5}{s+2}$$

and

$$v_o(t) = (-0.5 + t + 5.5e^{-2t})u(t)$$

Figure 16.10 shows the graph of $v_o(t)$. The capacitor voltage initially discharges, because initially $v_o(t) > v_{in}(t)$ which causes current opposite the polarity of the capacitor, but eventually (when?) the capacitor follows the voltage due to the ramp signal.

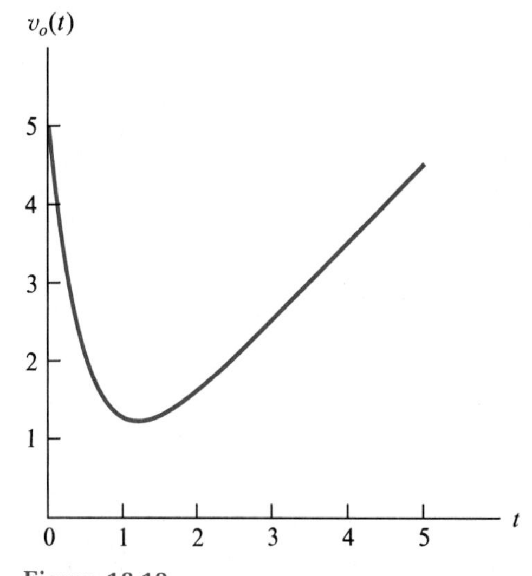

Figure 16.10

Example 16.18

Repeat Example 16.17 with $v_{in}(t) = 2u(t - 1)$.

Solution: $\mathbf{V}_{in}(s) = \dfrac{2}{s}e^{-1s}$

so

$$\mathbf{V}_o(s) = \frac{4}{s+2}\left(\frac{1}{s}e^{-1s}\right) + \frac{5}{s+2}$$

$$= 2e^{-1s}\left(\frac{1}{s} + \frac{-1}{s+2}\right) + \frac{5}{s+2}$$

and

$$v_o(t) = 2[1 - e^{-2(t-1)}]u(t - 1) + 5e^{-2t}u(t)$$

here the capacitor voltage decays for 1 second until the source voltage is turned on, as shown in Fig. 16.11.

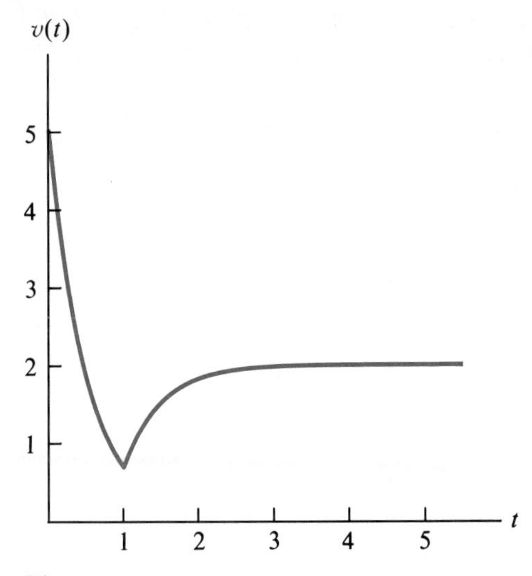

Figure 16.11

16.6 SOURCE SUPERPOSITION

When two or more sources simultaneously drive a circuit, the ZSR of the node voltages and branch currents will be the superposition of their response to the individual sources.

Example 16.19

The circuit in Fig. 16.12 has $v_{in1}(t) = 100 \sin \pi t/5u(t)$ and $v_{in2}(t) = tu(t)$. Find $v(t)$, assuming that $v(0^-) = 0$.

Solution: Writing KCL at the capacitor node leads to

$$\frac{\mathbf{V}_{in1}(s) - \mathbf{V}(s)}{5} + \frac{\mathbf{V}_{in2}(s) - \mathbf{V}(s)}{2} = 0.2s\mathbf{V}(s)$$

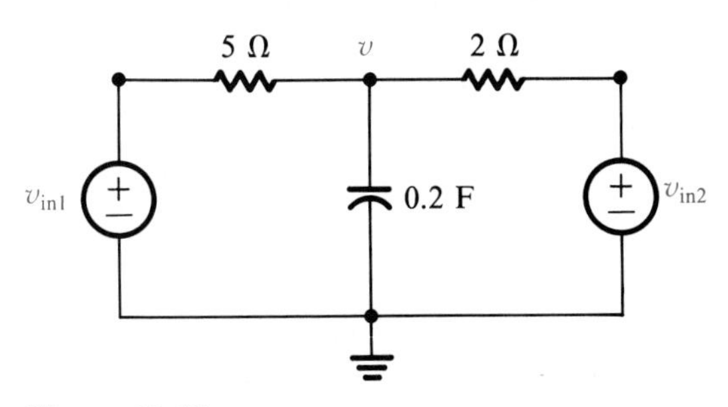

Figure 16.12

$$\mathbf{V}(s) = \frac{2/50}{s + 7/50}\mathbf{V}_{\text{in1}}(s) + \frac{1/10}{s + 7/50}\mathbf{V}_{\text{in2}}(s)$$

Because the Laplace transform of $v(t)$ depends on the Laplace transforms of $v_{\text{in1}}(t)$ and $v_{\text{in2}}(t)$, we know that $v(t)$ likewise depends on $v_{\text{in1}}(t)$ and $v_{\text{in2}}(t)$. Using

$$\mathbf{V}_{\text{in1}}(s) = \frac{20\pi}{s^2 + \pi^2/25} \text{ and } \mathbf{V}_{\text{in2}}(s) = \frac{1}{s^2}$$

we form $\mathbf{V}(s) = \mathbf{V}_1(s) + \mathbf{V}_2(s)$ where

$$\mathbf{V}_1(s) = \frac{6.07}{s + \dfrac{7}{50}} + \frac{3.106e^{j167.439°}}{s + \dfrac{j\pi}{5}} + \frac{3.106e^{-j167.439°}}{s - \dfrac{j\pi}{5}}$$

and

$$\mathbf{V}_2(s) = \frac{5.1}{s + \dfrac{7}{50}} + \frac{0.714}{s^2} + \frac{-5.1}{s}$$

In the time domain

$$v(t) = [11.17e^{-7t/50} - 5.1 + 0.71t + 6.2 \sin(\pi t/5 - 77.44°)]u(t)$$

The graph of $v(t)$ in Fig. 16.13 shows the result of superimposing the ramp-source response and the

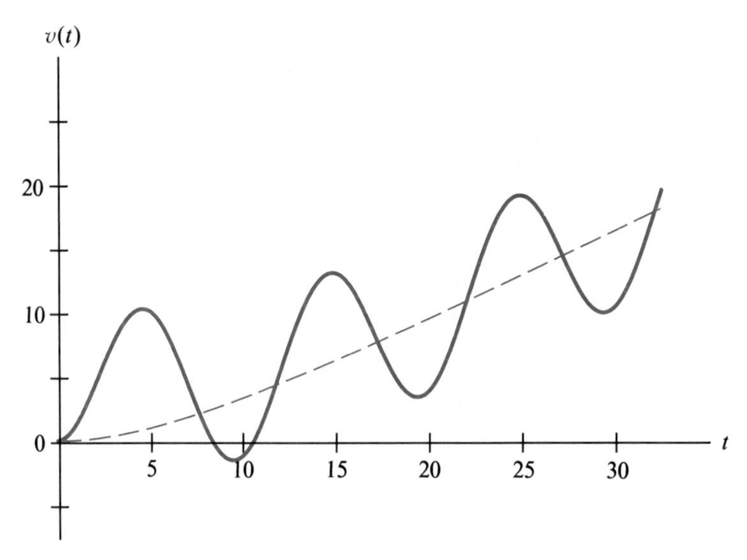

Figure 16.13

sinusoidal-source response. It also shows (dashed line) the response due to the ramp acting separately. The response due to both sources oscillates about a voltage that eventually grows linearly (the ramp response), with a constant steady-state amplitude to the oscillation.

16.7 THEVENIN AND NORTON EQUIVALENT CIRCUITS

The s-domain Thevenin equivalent of an *RLC* op amp circuit consists of a series connection of the Laplace transform of the open-circuit voltage measured at the terminals and the impedance measured at the terminals with all of the internal independent voltage and current sources turned off. The current source of the Norton circuit is the Laplace transform of the short-circuit current at the terminals.

Example 16.20

Find the Thevenin equivalent of the circuit in Fig. 16.14.

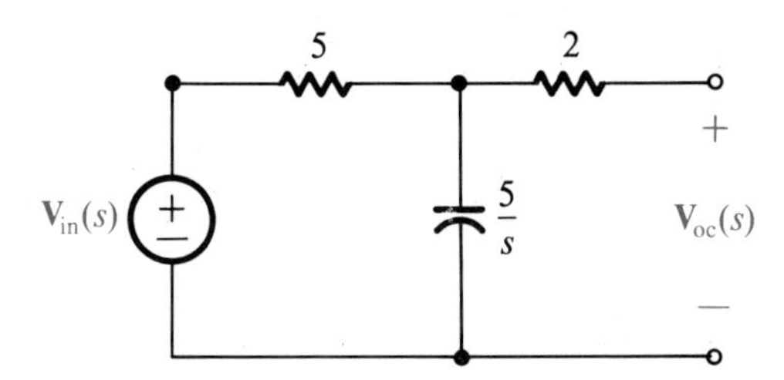

Figure 16.14

Solution: By voltage division

$$\mathbf{V}_{oc}(s) = \frac{\frac{5}{s}}{5 + \frac{5}{s}} \mathbf{V}_{in}(s) = \frac{\mathbf{V}_{in}(s)}{s + 1}$$

and

$$\mathbf{Z}_{th}(s) = 2 + \frac{5\left(\frac{5}{s}\right)}{5 + \frac{5}{s}} = 2 + \frac{5}{s + 1}$$

The Laplace transform of the KVL and KCL equations of a circuit can also be written and solved in matrix format.

Example 16.21

Write and solve the nodal equations for the Laplace transform of the ZSR of v_1 and v_2 in Fig. 16.15.

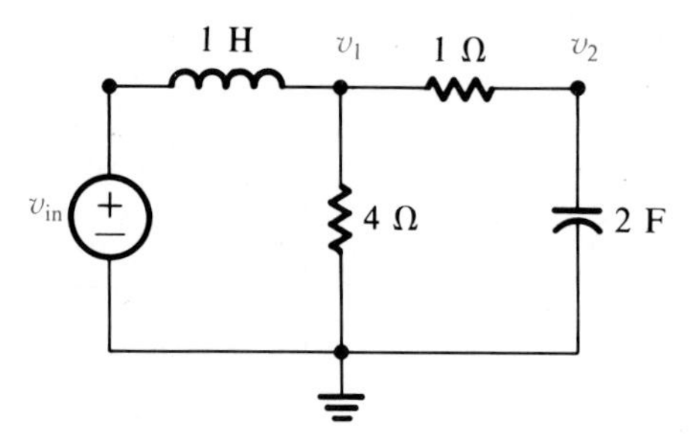

Figure 16.15

Solution:

$$\frac{\mathbf{V}_{in}(s) - \mathbf{V}_1(s)}{s} + \frac{\mathbf{V}_2(s) - \mathbf{V}_1(s)}{1} - \frac{\mathbf{V}_1(s)}{4} = 0$$

$$\frac{\mathbf{V}_1(s) - \mathbf{V}_2(s)}{1} - 2s\mathbf{V}_2(s) = 0$$

In matrix form

$$\begin{bmatrix} \frac{5}{4} + \frac{1}{s} & -1 \\ -1 & 1 + 2s \end{bmatrix} \begin{bmatrix} \mathbf{V}_1(s) \\ \mathbf{V}_2(s) \end{bmatrix} = \begin{bmatrix} \frac{1}{s} \\ 0 \end{bmatrix} [\mathbf{V}_{in}(s)]$$

Inverting the "admittance" matrix gives

$$\begin{bmatrix} \mathbf{V}_1(s) \\ \mathbf{V}_2(s) \end{bmatrix} = \frac{\begin{bmatrix} 1 + 2s & 1 \\ 1 & \frac{5}{4} + \frac{1}{s} \end{bmatrix} \begin{bmatrix} \frac{1}{s} \\ 0 \end{bmatrix}}{\left(\frac{5}{4} + \frac{1}{s}\right)(1 + 2s) - 1} [\mathbf{V}_{in}(s)]$$

So

$$\mathbf{V}_1(s) = \frac{(1 + 2s)\mathbf{V}_{in}(s)}{s\left[\left(\frac{5}{4} + \frac{1}{s}\right)(1 + 2s) - 1\right]}$$

$$\mathbf{V}_2(s) = \frac{\mathbf{V}_{in}(s)}{s\left[\left(\frac{5}{4} + \frac{1}{s}\right)(1 + 2s) - 1\right]}$$

For a given $v_{in}(t)$ the expressions for $\mathbf{V}_1(s)$ and $\mathbf{V}_2(s)$ can be expanded into partial fractions, and then inverted to obtain $v_1(t)$ and $v_2(t)$.

SUMMARY

The Laplace transform synthesizes a signal from a continuous sum of damped or undamped complex exponential signals. It provides a completely algebraic tool for finding the response of a circuit to any physical signal—periodic or non-periodic—without generalized functions. Its areas of use include circuit boundary conditions, circuits with repeated natural frequencies, and circuits excited by exponential sources whose frequency coincides with a natural frequency. The zero-input, zero-state, and initial-state responses of a circuit can be obtained directly from their Laplace transform in the s-domain, with no need to solve the differential equation in the time domain.

The transfer function of a circuit defines the ratio of the Laplace transform of its zero-state response to the Laplace transform of its input signal. It handles a circuit's nonzero initial conditions automatically by incorporating them in the model obtained from the Laplace transform of the input/output differential equation, or by including them in a Laplace model of the circuit. In the latter approach all steps can be taken in the s-domain—the boundary conditions on the energy storage elements do not have to be converted to boundary conditions on the output signal and its derivatives.

Problems -Chapter 16

16.1 Find $\mathbf{F}(s)$ for each $f(t)$.

a. $f(t) = 5e^{-3t}u(t)$

b. $f(t) = 10 \cos 4tu(t)$

c. $f(t) = 2e^{-5t} \cos (20\pi t - 30°)u(t)$

d. $f(t) = t^3e^{-8t}u(t)$

e. $f(t) = 5tu(t) + 2\frac{d}{dt}\delta(t)$

f. $f(t) = 10[u(t) - u(t - 2)]$

16.2 Find $f(t)$ for each $\mathbf{F}(s)$.

a. $\mathbf{F}(s) = 100/(s + 40)$

b. $\mathbf{F}(s) = 25/(s^2 + 10s + 21)$

c. $\mathbf{F}(s) = 4s/(s^2 + 4s + 104)$
d. $\mathbf{F}(s) = 12/(s + 5)^3$
e. $\mathbf{F}(s) = 200(s + 20)/(s^2 + 144)^2$
f. $\mathbf{F}(s) = 180(s + 15)/[(s + 3)(s + 12)(s + 30)]$
g. $\mathbf{F}(s) = 100/[(s + 2)^3(s + 5)(s + 1)]$
h. $\mathbf{F}(s) = s^3/[(s + 5)^2]$
i. $\mathbf{F}(s) = 1/[s(s + 1)^4]$
j. $\mathbf{F}(s) = 1/[(s + 3)(s + 2)^2]$
k. $\mathbf{F}(s) = 10/[(s + 2)^2] + 0.2e^{-0.15s}/(s^2 + 14)$
l. $\mathbf{F}(s) = (s + 10)/(s + 2)$

16.3 Find the Laplace transform of the signal shown in Fig. P6.17b.

16.4 Find the Laplace transform of the sine pulse $f(t) = A \sin 2\ t$ for $0 \leq t \leq \pi$ and $f(t) = 0$ elsewhere.

16.5 Find $\mathbf{F}(s)$ for $f(t)$ in Fig. P16.5.

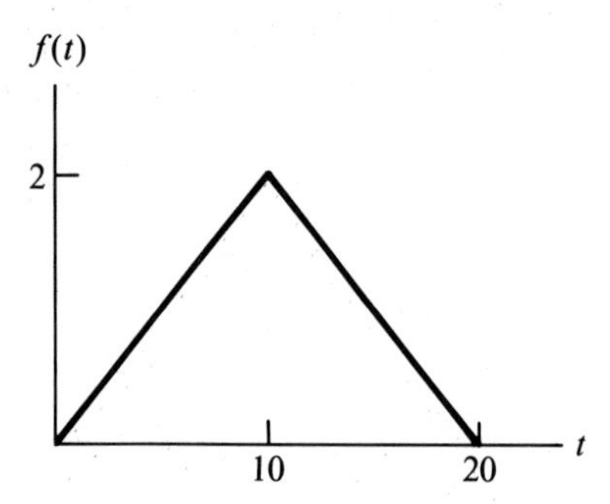

Figure P16.5

16.6 Show that the Laplace transform obeys the scaling property

$$\mathbf{L}\{f(at)\} = (1/a)\mathbf{F}(s/a)$$

for constant $a > 0$.

16.7 Find $f(t)$.

a. $\mathbf{F}(s) = \dfrac{e^{-0.2s}}{s + 10}$

b. $\mathbf{F}(s) = \dfrac{(s + 4)e^{-s}}{(s + 5)^3}$

16.8 Find the ZSR of the output signal $y(t)$ of a circuit having an impulse response

$$h(t) = (2e^{-5t} + 4e^{-3t})u(t)$$

and an input signal

$$x(t) = 10e^{-3t}u(t)$$

16.9 Find the ZSR of $y(t)$ of the circuit whose I/O transfer function is

$$\mathbf{H}(s) = (s + 5)/(s + 10)$$

when the input signal is

$$x(t) = 2te^{-10t}u(t)$$

16.10 Find the ISR of $y(t)$ when

$$\mathbf{H}(s) = \frac{s^2 + 6s + 8}{s^2 + 11s + 30}$$

and

$$x(t) = 20e^{-3t}u(t)$$
$$y(0^-) = 5$$
$$dy(0^-)/dt = -10$$

16.11 If a circuit is known to have a transfer function given by

$$\mathbf{H}(s) = \frac{s + 2}{s^2 + 5s + 4}$$

and if the zero-state response of the circuit has been observed to be

$$Y_{ZSR}(t) = (5/3e^{-t} - 20/3e^{-4t} + 5e^{-3t})u(t)$$

what must have been the input signal $x(t)$?

16.12 Find the ZIR of the circuit described by

$$\mathbf{H}(s) = \frac{s + 4}{s^2 + 4s + 3}$$

when $dy/dt = -5$ and $y = 2$ at $t = 0^+$.

16.13 If a circuit has a ZSR given by

$$y_{ZSR}(t) = 10(1 - e^{-2t} + 5te^{-2t})u(t)$$

when the input signal is

$$x(t) = 5u(t)$$

find the pole-zero pattern of the circuit's transfer function.

16.14 If a circuit has the input/output transfer function

$$\mathbf{H}(s) = \frac{1}{s^2}$$

find its ZSR output signal when its input signal is a unit-ramp signal.

16.15 Find the ZSR to a step input when a circuit is described by the I/O differential equation

$$\frac{d^3y}{dt^3} + 6\frac{d^2y}{dt^2} + 12\frac{dy}{dt} + 8y(t) = 5x(t)$$

16.16 Use Laplace transforms to solve, for $t \geq 0$,

a. $$2\frac{d^2y}{dt^2} + 300\frac{dy}{dt} + 10{,}000y(t) = 20\frac{dx}{dt} + 10x(t)$$

with $x(t) = e^{-200t}u(t)$ and $dy(0^-)/dt = 1$, $y(0^-) = 0$.

b. $$5\frac{d^2y}{dt^2} + 10\frac{dy}{dt} + 10y(t) = \frac{dx}{dt} + x(t)$$

with $x(t) = 2u(t)$ and $\frac{dy(0^-)}{dt} = 0$, $y(0^-) = 0$.

16.17 Find the impulse response of the circuits whose transfer functions are

a. $\mathbf{H}(s) = (s + 1)/(s + 2)$

b. $\mathbf{H}(s) = s/(s + 10)$

c. $\mathbf{H}(s) = s/(s + 25)^2$

d. $\mathbf{H}(s) = s/(s^2 + 81)$

e. $\mathbf{H}(s) = (s + 3)/[(s + 1)^2 + 4]$

16.18 Find the pole-zero pattern of the zero-state response of the output signal $\mathbf{Y}(s)$ when the circuit has an impulse response

$$h(t) = 10e^{-6t}u(t)$$

and

$$x(t) = 20 \cos t\, u(t).$$

16.19 The switch in Fig. P16.19 is opened after having been closed for a long time.

a. Find an expression for $\mathbf{I}(s) = \mathbf{L}\{i(t)\}$, for $t \geq 0$.

b. Find $i(t)$.

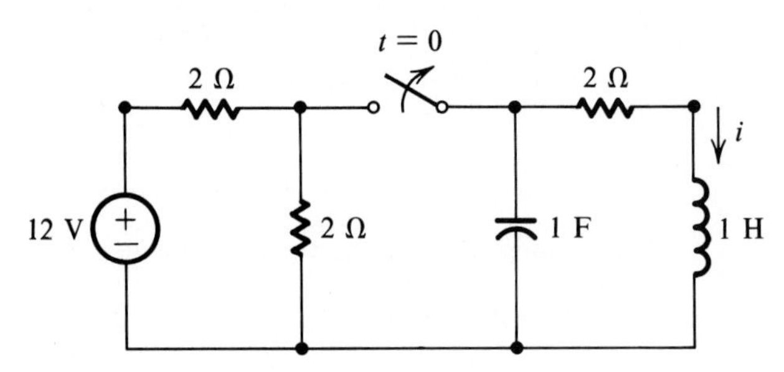

Figure P16.19

16.20 Find $\mathbf{V}_o(s) = \mathbf{L}\{v_o(t)\}$, for $t \geq 0$, when $i(0^-) = 2$ A, $v_o(0^-) = 5$ V, and $i_{in}(t) = u(t)$A.

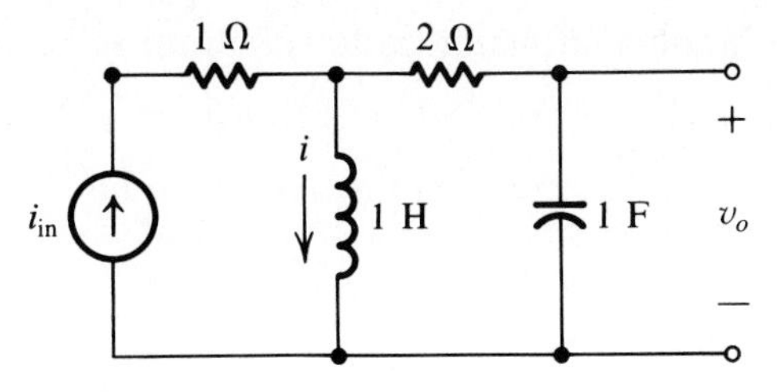

Figure P16.20

16.21 If $v_{in}(t) = tu(t)$ in Fig. P16.21, find the ZSR of $v_o(t)$.

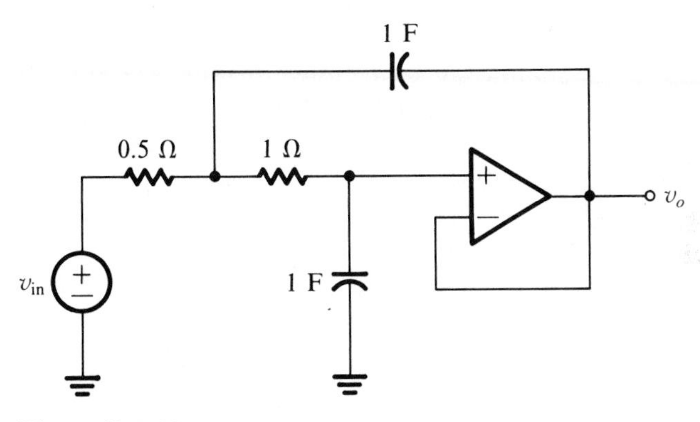

Figure P16.21

16.22 For the circuit in Fig. P16.22:

a. Find $\mathbf{H}(s) = \mathbf{V}_o(s)/\mathbf{V}_{in}(s)$.

b. Find $\mathbf{V}_o(s)$ if the circuit is initially relaxed and $v_{in}(t) = 2t^3e^{-2t}u(t)$.

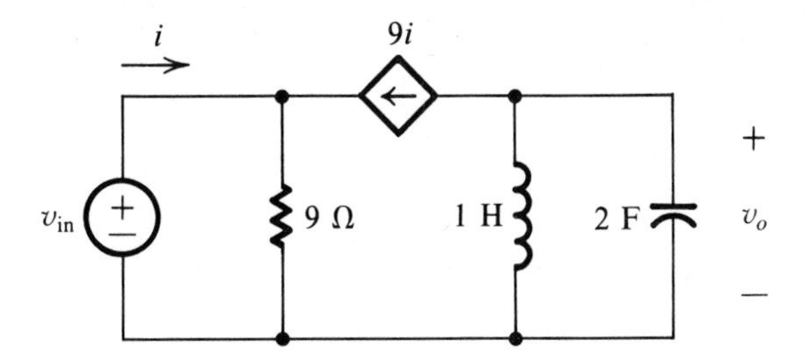

Figure P16.22

16.23 A circuit's ZSR to a unit-step input is

$$y(t) = (5e^{-2t} + 8e^{-10t})u(t)$$

a. Find its ZSR to a unit-ramp input.
b. Find its ZIR when $y(0^-) = 5$ and $\dot{y}(0^-) = 1$.
c. Find its ISR to the input $x(t) = 5e^{-t}u(t)$ for the same initial conditions as in b.

16.24 In Chapter 11 we saw how to obtain the boundary conditions of a circuit by inspection. We then used those conditions to solve the I/O differential equation. Alternatively, the solution could have been obtained by first using algebraic methods to find the circuit's transfer function. Taking the Laplace transform of the I/O equation would then automatically include the boundary conditions. For the circuit in Fig. P16.24, with $R = 2\ \Omega$, $L = 4$ H and $C = 1$ F:
a. Find $\mathbf{H}(s)$ and obtain the I/O DE.
b. Take the Laplace transform of the I/O equation and obtain an expression for $\mathbf{V}_o(s)$.
c. Find the ZSR of $v_o(t)$ to $u(t)$.
d. Find the ZIR to $i_L(0^-) = 2$ A and $v_o(0^-) = -5$ V.
e. Repeat c if $v_{in}(t) = 10r(t)$.

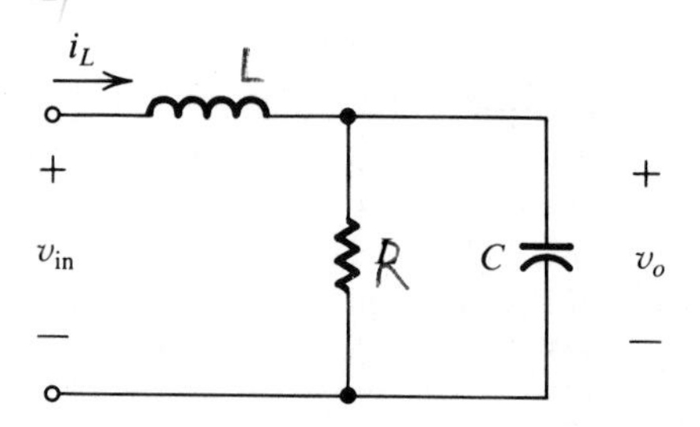

Figure P16.24

16.25 Use boundary-condition models (Laplace models) for the inductor and capacitor in Fig. P16.24, and obtain an expression for $\mathbf{V}_o(s)$.

16.26 A source is applied to a series RC circuit, with $v_{in}(t) = te^{-t/RC}u(t)$.
a. Find the ZSR of the capicitor voltage $v_o(t)$.
b. Find the ISR of $v_o(t)$ if $v_o(0^-) = 10$ V.

16.27 If a circuit has poles at $s = -1, -2, -3$ and -4, and has no zeroes, and if a unit-step input produces a signal $y(t)$ with $y(\infty) = 2$, find the circuit's ZSR to $u(t)$.

16.28 If a circuit has an I/O transfer function of

$$\mathbf{H}(s) = (s + a)/(s + b)$$

find its impulse response.

16.29 Find the unit-impulse response of the capacitor voltage in a series RLC circuit driven by a voltage source, when $R = 50\ \Omega$, $L = 250$ mH and $C = 400\ \mu$F.

16.30 If a series RL circuit is driven by a voltage source having the waveform in Fig. P16.30, find $i_L(t)$ for $t \geq 0$ when $i_L(0^-) = 5$ A, $R = 2\ \Omega$ and $L = 0.4$ H.

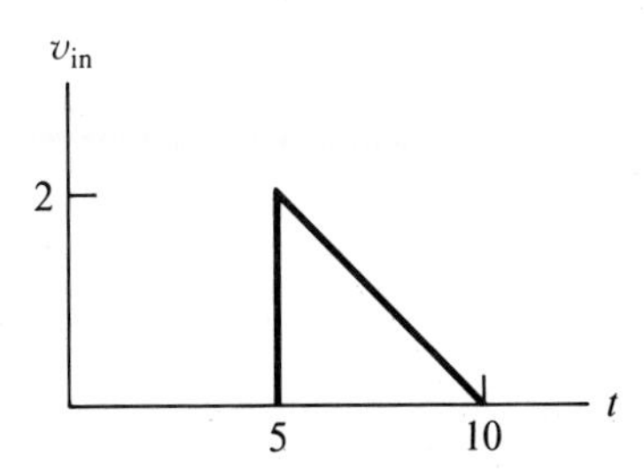

Figure P16.30

16.31 Using boundary-condition models for the inductor and capacitor in a series RLC circuit, find $\mathbf{I}(s)$ and solve for $i(t)$ for $t \geq 0$ when $i(0^-) = 2$ A, $v_c(0^-) = 5$ V and a source is applied with $v_{in}(t) = \sin t$. ($R = 6\ \Omega$, $L = 1$ H and $C = 0.5$F)

16.32 Using boundary-condition models for the inductor and the capacitor in a series RLC circuit, find the ZIR of $i(t)$ if $v_c(0^-) = -2$ V and $i(0^-) = 1$ A, with $R = 3\ \Omega$, $L = 0.25$ H and $C = 2$ F.

16.33 Find the ZSR of v_o to a unit step.

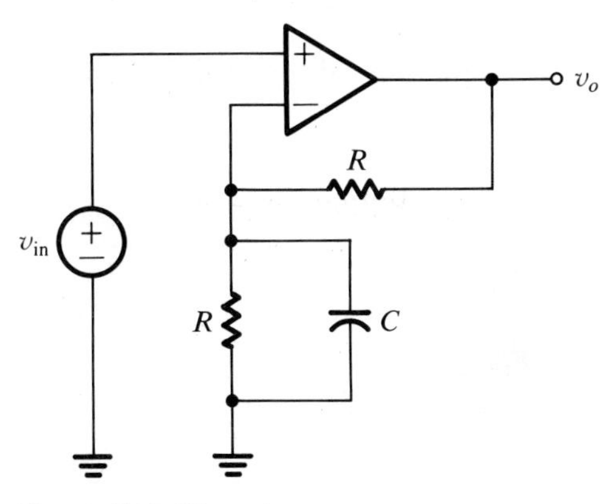

Figure P16.33

16.34 Write nodal equations for $\mathbf{V}_1(s)$ and $\mathbf{V}_2(s)$ and solve for $v_1(t)$ and $v_2(t)$ for $t \geq 0$ when $i_{in} = 10$ A and $v_{in} = 5$ V. Use $R_1 = 2\ \Omega$, $R_2 = 8\ \Omega$, $L = 0.25$ H and $C = 0.002$ F.

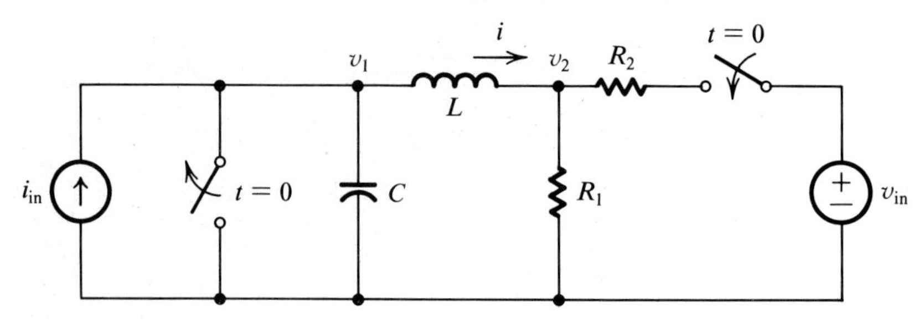

Figure P16.34

16.35 If a circuit has a "Laplace" Thevenin equivalent with $\mathbf{Z}(s) = (s + 5)/[(s + 10)(s + 20)]$ and $\mathbf{V}_{th}(s) = 10/s$, find the current $i_{in}(t)$ that would be supplied to the circuit by a source with $v_{in}(t) = r(t)$.

16.36 Show that $\mathbf{L}\{u(t)\}$ can be obtained from $\mathbf{L}\{e^{-at}u(t)\}$ in the limit as $a \rightarrow 0$, with $a > 0$. Compare the result with that obtained by taking the limits of Fourier transforms rather than Laplace transforms.

16.37 Derive $\mathbf{L}\{u(t)\}$ from $\mathbf{L}\{\cos \omega_d t\ u(t)\}$ in the limit as $\omega_d \rightarrow 0$.

16.38 Derive $\mathbf{L}\{\delta(t)\}$ from $\mathbf{L}\{u(t)\}$.

16.39 Derive $\mathbf{L}\{u(t)\}$ from $\mathbf{L}\{\delta(t)\}$.

16.40 Find $\mathbf{L}\{e^{-\alpha t} \sin \omega_d t\ u(t)\}$ with $\alpha > 0$.

16.41 An associate contends that the transfer function $\mathbf{H}(s)$ for a circuit can be found by driving the circuit with a step function and measuring the response. Prove or disprove this claim.

16.42 A circuit has an impulse response $h(t) = (4e^{-7t} - 3e^{-2t})u(t)$ Find the circuit's pole-zero pattern.

16.43 The circuit whose impulse response was given in the previous problem has a ZSR of $y(t) = 2e^{-7t} + 4e^{-2t} + 5e^{-4t}$. What must have been the circuit's input signal?

FREQUENCY RESPONSE AND AN INTRODUCTION TO FILTER DESIGN

INTRODUCTION

This chapter presents an introduction to analog filter design. The previous chapters showed how a circuit's transfer function defines a relationship between the frequency-domain (spectral) characteristics of any input signal and the frequency-domain characteristics of its output signal, as well as between the time-domain input signal and the time-domain output signal (convolution). In many engineering applications circuits are used to shape the spectral characteristics of a signal; these circuits are called *filters*.

For example, the commercial AM broadcasting system in the United States allows several stations to broadcast simultaneously under conditions assuring that, for all practical purposes, no two stations will have overlapping electrical signal spectra. A radio receiver must filter, or remove, input-signal spectral components due to other stations so that only the signal of the selected station is heard. How are such circuits designed?

17.1 FREQUENCY RESPONSE: MAGNITUDE AND PHASE CHARACTERISTICS

If $\mathbf{H}(s)$ is the transfer function of a circuit, the graph of $|\mathbf{H}(j\omega)|$ is called the **magnitude response** or **magnitude characteristic** of the circuit, and the

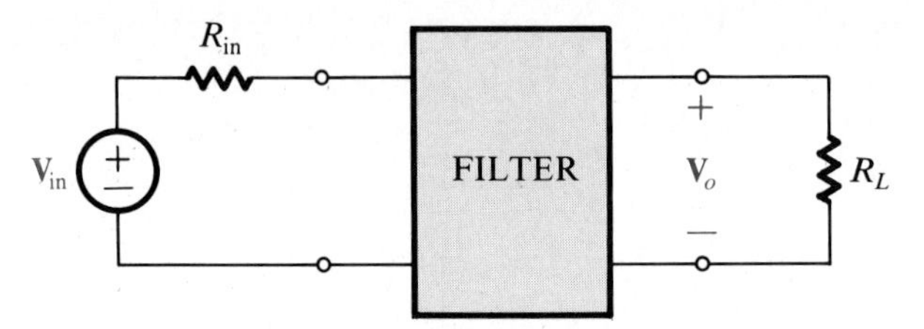

Figure 17.1 A circuit and filter configuration.

graph of $\theta(j\omega) = \measuredangle \mathbf{H}(j\omega)$ is called the **phase response** or **phase characteristic.** Designing a filter requires carefully choosing its component values, because they directly affect the shape of the magnitude and phase responses. (Design also involves cost tradeoffs between circuits that have similar performance but different physical implementations.)

Fig. 17.1 shows a typical filter configuration, where a filter is to be inserted between a given circuit, shown here by its Thevenin equivalent, and its load. The transfer function between the voltage source and the load voltage is given by

$$\mathbf{H}(s) = \frac{\mathbf{V}_o(s)}{\mathbf{V}_{in}(s)}$$

(We could use phasor amplitudes or complex amplitudes instead of Laplace transforms.) In filter design we are interested in the shapes of the graphs of $|\mathbf{H}(j\omega)|$ and $\measuredangle \mathbf{H}(j\omega)$.

The magnitude responses of ideal filters are shown in Fig. 17.2. These filters are ideal because it is physically impossible to design filters having the sharp corners shown. Nonetheless, ideal filters are useful for introducing concepts that have general application to real filters. A filter's **passband** is the range of frequencies for which a sinusoidal input signal produces a significant steady-state output signal. Conversely, a sinusoidal signal at a frequency in the **stopband** of a filter will not produce a significant output signal in comparison to a signal in the passband (assuming both have the same amplitude). In the ideal case, a signal in the stopband is attenuated to zero.

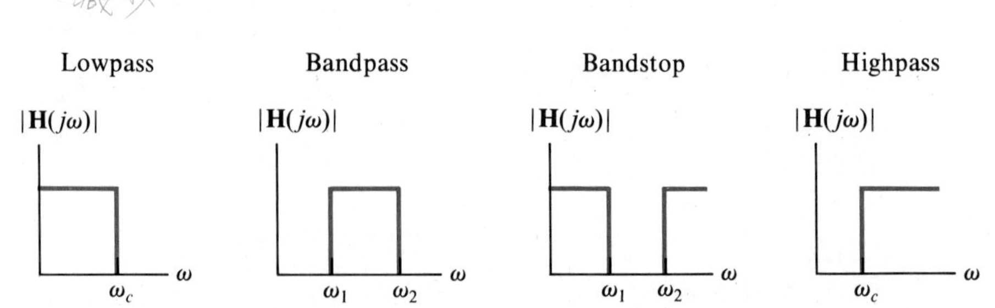

Figure 17.2 Ideal filter characteristics.

A **lowpass filter** passes sinusoids at frequencies below some cutoff frequency ω_c and attenuates sinusoids above the cutoff. Conversely, a **highpass filter** passes signals above its cutoff frequency and attenuates signals below the cutoff. **Bandpass filters** are *selective*—they pass signals in a passband between a lower cutoff frequency ω_1 and an upper cutoff frequency ω_2. Televisions and radios use bandpass filters to select the spectrum of the signal from a particular station. Notch filters selectively reject signals in a stopband. They can also be used to remove interference signals, such as 60 Hz hum.

A filter is said to pass its input signal $x(t)$ *without distortion* if its output, $y(t)$ is

$$y(t) = Kx(t - \tau)$$

for some value of delay τ. In this case, the waveform of $y(t)$ is just a scaled and delayed copy of the waveform of $x(t)$. When a filter passes $x(t)$ without distortion the input signal can be recovered from the output signal by descaling and advancing the waveform of $y(t)$. This simple principle has led to a vast market for high-fidelity recording and playback equipment that uses video tapes, records, and laser discs to make and re-create undistorted replicas of audio and video signals.

We will now develop technical specifications for a filter that guarantee that its time-domain output signal will be an undistorted copy of its input signal. If the output $y(t)$ is an undistorted copy of the input $x(t)$ then

$$\frac{1}{2\pi}\int_{-\infty}^{\infty} \mathbf{Y}(j\omega)e^{j\omega t}\,d\omega = K\frac{1}{2\pi}\int_{-\infty}^{\infty} \mathbf{X}(j\omega)e^{j\omega(t-\tau)}\,d\omega$$

Comparing integrands, we conclude that distortionless transmission of $x(t)$ requires that

$$\mathbf{Y}(j\omega) = K\mathbf{X}(j\omega)e^{-j\omega\tau}$$

or

$$\mathbf{H}(j\omega) = Ke^{-j\omega\tau}$$

The filter's magnitude and phase characteristics must be such that $|\mathbf{H}(j\omega)| = |K|$ and $\measuredangle\mathbf{H}(j\omega) = \measuredangle K - \omega\tau$. The filter must have a **flat magnitude response** and a **linear phase response.** Flat magnitude and linear phase characteristics cannot be realized over the entire j-axis. In practice, the spectrum of $x(t)$ is usually significant over only a relatively small range of frequencies. So a filter is designed to be distortionless over a finite range. An example is the case of high-fidelity stereo equipment, which usually achieves a close approximation to distortionless transmission for $0 \le f \le 20$ KHz.

Filter design involves two fundamental steps. First, we must choose a circuit whose generic frequency response is satisfactory. Then we must choose that filter's component values to realize acceptable passband and stopband characteristics. We'll first study generic filters having approximately ideal characteristics. Then we'll develop active and passive circuit realizations of their behavior.

17.2 FIRST-ORDER LOWPASS FILTER

The transfer function of a generic lowpass filter has

$$\mathbf{H}(s) = \frac{K\omega_c}{s + \omega_c} = \frac{K}{1 + s/\omega_c}$$

Its magnitude and phase responses are

$$|\mathbf{H}(j\omega)| = \frac{|K|\omega_c}{|j\omega + \omega_c|} = \frac{|K|}{\sqrt{1 + (\omega/\omega_c)^2}}$$

$$\measuredangle\mathbf{H}(j\omega) = \measuredangle K - \tan^{-1}(\omega/\omega_c)$$

The generic parameter ω_c is called the **cutoff frequency** of the filter. An important design problem is to choose component values to effect a desirable ω_c. First we must understand the relationship between ω_c and the filter's magnitude and phase responses.

Curves of $|\mathbf{H}(j\omega)|$ and $\measuredangle\mathbf{H}(j\omega)$ are shown in Fig. 17.3. The magnitude response begins at a value of $|K|$ when $\omega = 0$, and K is called the **DC gain** of the filter. As ω increases the magnitude response decreases asymptotically to zero. At $\omega = \omega_c$ the gain of the filter is $|\mathbf{H}(j\omega_c)| = 0.707\,|K|$.

The cutoff frequency of a lowpass filter is called its **half-power frequency** because an input sinusoid whose frequency is ω_c will produce a steady-state output sinusoid having one-half the average power of the input signal. At the cutoff frequency, $|\mathbf{H}(j\omega_c)|^2 = 1/2$ so $P_o = 1/2\ \mathbf{V}_o^2 = 1/2(1/2)\mathbf{V}_{in}^2 = 1/2\ P_{in}$. Sinusoidal signals at frequencies above a lowpass filter's cutoff frequency are considered insignificant in comparison to a signal having the same amplitude and lying in the passband.

17.2.1 Passive Realization

The generic first-order lowpass filter can be realized by placing a capacitor in parallel with the load, as in Fig. 17.4. At DC the capacitor

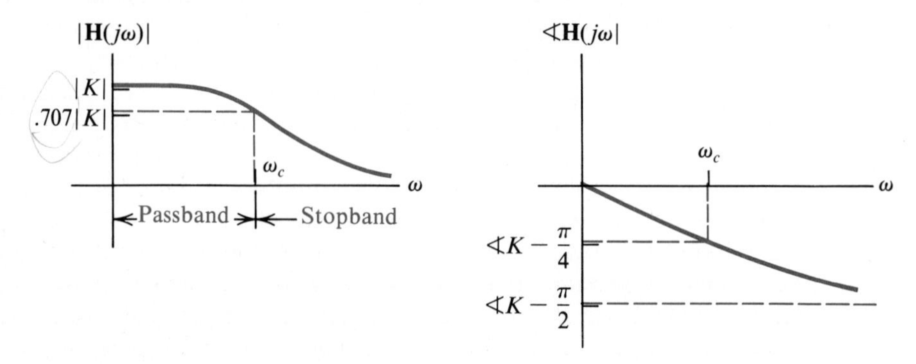

Figure 17.3 Lowpass filter frequency response.

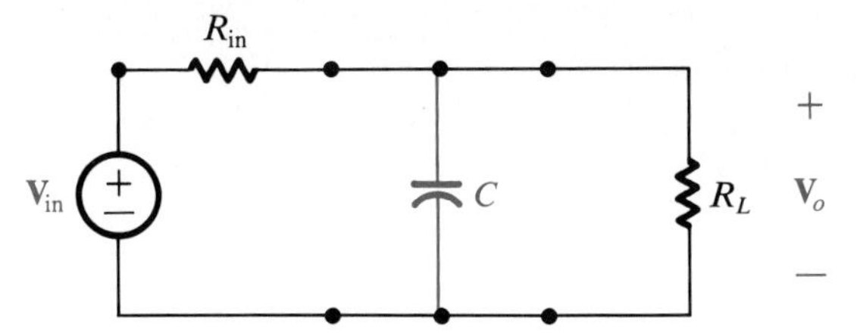

Figure 17.4 A passive lowpass filter realization.

behaves like an open circuit, and all of the current from the source is delivered to the load. At low frequencies the capacitor presents a high-impedance path to current, and at high frequencies it presents a low-impedance path. The output voltage will be reduced at high frequencies because relatively less current is in the load path.

The transfer function for the lowpass filter in Fig. 17.4 is

$$\mathbf{H}(s) = \frac{R_L/(R_{in} + R_L)}{1 + sR_LR_{in}C/(R_{in} + R_L)}$$

Comparing this expression with the generic transfer function gives $K = R_L/(R_L + R_{in})$ and $\omega_c = 1/RC$, where R is the parallel equivalent of R_L and R_{in}. The DC gain K is just the voltage division created by the two resistors, which agrees with our understanding that the capacitor current is zero at DC.

Skill Exercise 17.1

If the circuit in Fig. 17.4 has $R_L = 40$ kΩ and $R_{in} = 100$ Ω, choose C such that the circuit realizes a lowpass filter with a half-power frequency of 5 kHz.

Answer: $C = 0.319$ μF.

17.2.2 Lowpass Filter Pole-Zero Pattern Relationships

Fig. 17.5 shows the pole-zero pattern and the geometric relationship of a lowpass filter's pole and the vector drawn from the pole to a point on the j-axis at $s = j\hat{\omega}$. If $|\mathbf{H}(j\omega)|$ is calculated for ω in the range $0 \le \omega < \infty$ the vector drawn from the pole to the point $j\omega$ increases in length as ω increases. As a result the curve of $|\mathbf{H}(j\omega)|$ vs ω will decrease because the vector has its shortest length at $\omega = 0$. The filter's magnitude response at DC will be its maximum value. The vector from the filter's pole to a point on the j-axis swings from an angle of 0 at $\omega = 0$ to an angle of $\pi/2$ as $\omega \to \infty$. This angle is associated with the denominator factor of $\mathbf{H}(\cdot)$ so it is actually the negative of the change in the angle of $\mathbf{H}(\cdot)$, which swings from $\sphericalangle K$ to $\sphericalangle K - \pi/2$ in Fig. 17.3.

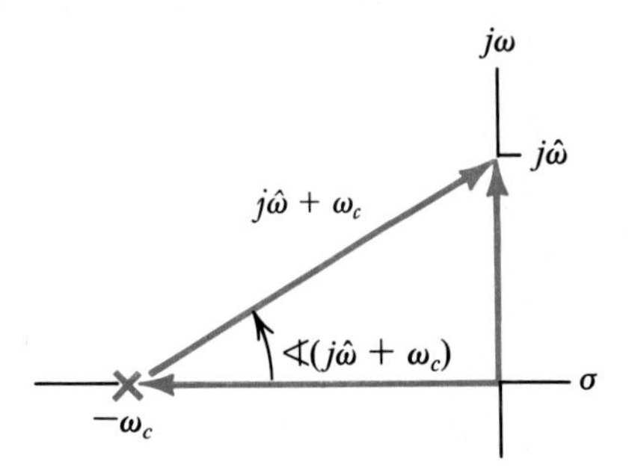

Figure 17.5 Lowpass filter pole-zero geometry for $s = j\omega$

The bandwidth of the first-order lowpass filter is directly related to its pole location, as shown in Fig. 17.6 where we juxtapose the pole-zero pattern and a rotated copy of the magnitude response. The length of the vector drawn from the pole to the point with $j\omega = j\omega_c$ has length $\sqrt{2}\omega_c$, so the transfer function will have value $|\mathbf{H}(j\omega_c)| = 0.707\,|K|$.

Figure 17.7 shows a three-dimensional plot of the lowpass filter's $|\mathbf{H}(s)|$ surface above the complex plane, with the graph of $|\mathbf{H}(j\omega)|$ defined by the surface's intersection with a vertical plane at $\sigma = 0$.

A filter's pole-zero pattern links its time-domain waveforms and its frequency-response characteristics. A high-bandwidth filter will have a quicker time response, in the sense that the transient response will decay faster, because its pole is located farther from the orgin. The time constant and bandwidth of the first-order lowpass filter are inversely related by

$$\omega_c = 1/\tau_c$$

A pulse signal will be transmitted through a lowpass filter if the filter bandwidth is much larger than the reciprocal of the pulsewidth.

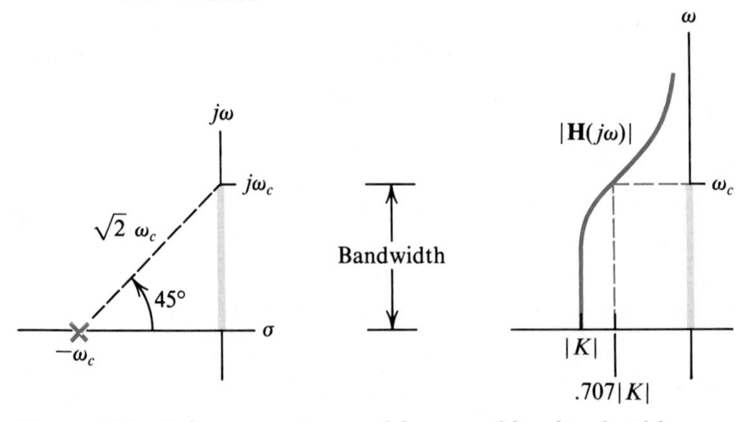

Figure 17.6 Pole-zero pattern and lowpass filter bandwidth.

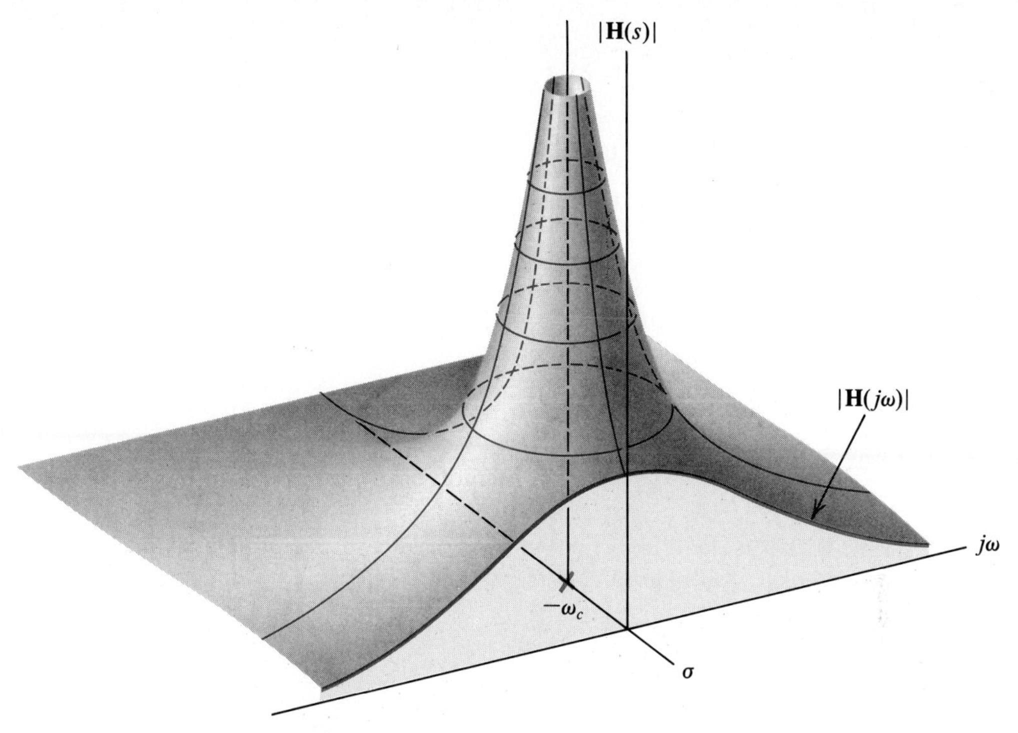

Figure 17.7 The intersection of a first-order lowpass filter's $|\mathbf{H}(s)|$ surface with the vertical plane at $\sigma = 0$ determines the filter response $|\mathbf{H}(j\omega)|$.

17.2.3 Lowpass Filter Design

A filter's **constraint equations** specify the value of its generic parameters as functions of its components (e.g., $\omega_c = 1/RC$). The constraint equations are developed by matching the coefficients of the transfer function of a circuit realization with the coefficients of the generic transfer function. In contrast, filter **design equations** specify component values that realize given performance parameters, such as K and ω_c. If R_{in} and R_L are given, the design equation specifying a choice of C that realizes a given ω_c is just

$$C = 1/(\omega_c R)$$

with $R = R_{in}R_L/(R_{in} + R_L)$. Design equations are obtained by solving the constraint equations for the component values.

Skill Exercise 17.2

The circuit in Fig. SE 17.2 realizes a generic first-order lowpass filter.

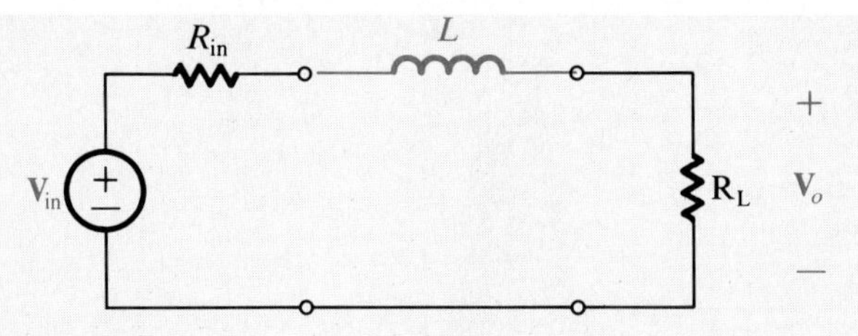

Figure SE17.2

Find the constraint equation and the design equation specifying the inductor value that realizes a given ω_c.

Answer:

$$\mathbf{H}(s) = \frac{R_L}{R_{in} + R_L} \frac{(R_{in} + R_L)/L}{s + (R_{in} + R_L)/L}$$

and $K = R_L/(R_{in} + R_L)$; $\omega_c = (R_{in} + R_L)/L$; $L = (R_{in} + R_L)/\omega_c$. An inductor in series with the load resistor behaves as a lowpass filter because it blocks current to the load at high frequency by having a relatively high impedance.

17.3 SECOND-ORDER LOWPASS FILTER

The cutoff and phase characteristics of a lowpass filter can be improved by increasing the complexity of the circuit. A generic model for a second-order lowpass filter is

$$\mathbf{H}(s) = \frac{K\omega_c^2}{s^2 + s\sqrt{2}\omega_c + \omega_c^2}$$

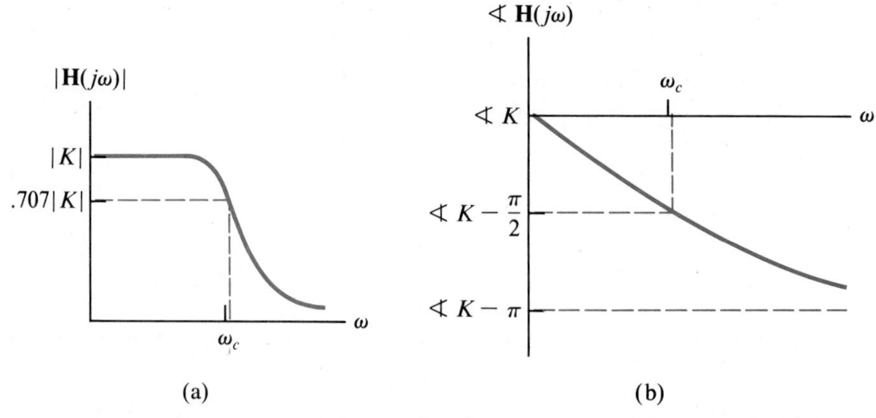

Figure 17.8 The (a) magnitude and (b) phase responses of a second-order lowpass filter.

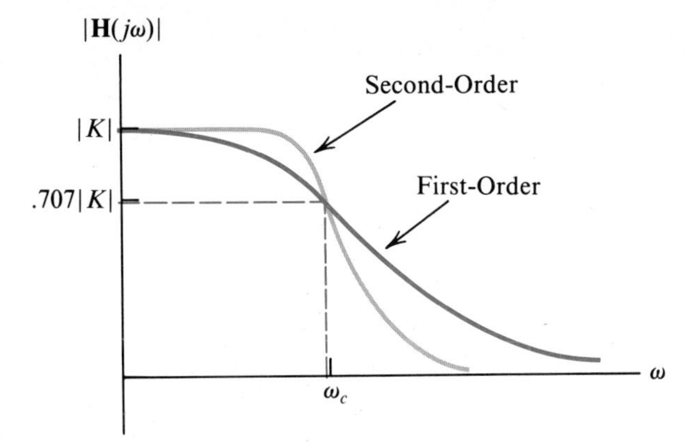

Figure 17.9 A comparison of the magnitude responses of first- and second-order lowpass filters.

The filter's DC gain is $|K|$ since $|\mathbf{H}(0)| = |K|$. Its magnitude and phase responses are shown in Fig. 17.8, where

$$|\mathbf{H}(j\omega)| = \frac{|K|}{\sqrt{1 + (\omega/\omega_c)^4}}$$

and

$$\measuredangle\mathbf{H}(j\omega) = \measuredangle K - \tan^{-1}\frac{\sqrt{2}\,\omega/\omega_c}{1 - (\omega/\omega_c)^2}$$

The second-order filter has $|\mathbf{H}(j\omega_c)| = 0.707$ and $|\mathbf{H}(j\omega_c)|^2 = 1/2$ so the generic parameter ω_c is in fact the half-power frequency of the filter. The phase response of the filter is approximately linear for $\omega < \omega_c$.

The performance improvement resulting from the increased complexity of the filter can be seen by comparing the first- and second-order magnitude characteristics in Fig. 17.9. The second-order filter has a flatter response for $\omega < \omega_c$ and it attenuates signals more sharply for $\omega > \omega_c$. The second-order phase characteristic in Fig. 17.10 is more linear than the first-order characteristic, leading to relatively less phase distortion of a signal in the passband.

17.3.1 Passive Realization

Placing a capacitor in parallel with the load in Fig. SE17.2 would create the second-order filter in Fig. 17.11. It combines two effects: the inductor blocks current to the load at high frequency, and the capacitor shunts current around the load. The circuit's transfer function is

$$\mathbf{H}(s) = \frac{\mathbf{V}_o}{\mathbf{V}_{in}} = \frac{1}{LC\left[s^2 + s\left(\dfrac{1}{CR_L} + \dfrac{R_{in}}{L}\right) + \dfrac{(1 + R_{in}/R_L)}{LC}\right]}$$

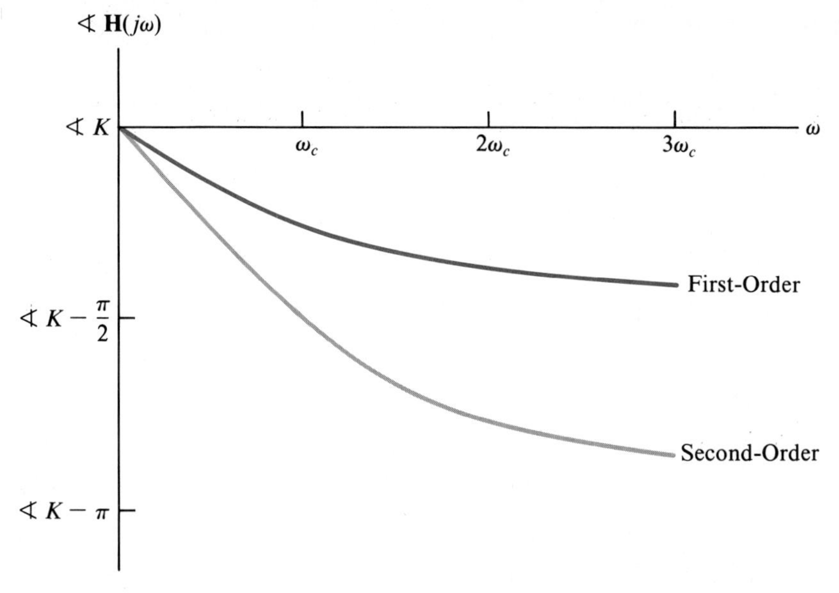

Figure 17.10 A comparison of the phase responses of first- and second-order lowpass filters.

Matching the coefficients of **H**(s) with those of the generic filter gives the constraint equations

$$\sqrt{2}\,\omega_c = \frac{R_{in}}{L} + \frac{1}{R_L C} \tag{17.1}$$

$$\omega_c^{\,2} = \frac{(1 + R_{in}/R_L)}{LC} \tag{17.2}$$

and $K\omega_c^{\,2} = 1/LC$, so $K = R_L/(R_{in} + R_L)$. The DC gain of the filter can also be obtained by inspection of the circuit with the inductor shorted and the capacitor open.

Since $K \le 1$ the filter's gain cannot exceed 1 and the filter always attenuates signals. (Designs with op amps can simultaneously amplify and filter a signal.) For a given R_{in} and R_L the value of K is fixed, so we solve (17.1) and (17.2) to obtain **design equations** for the values of L and C that realize a given K and ω_c from (17.2)

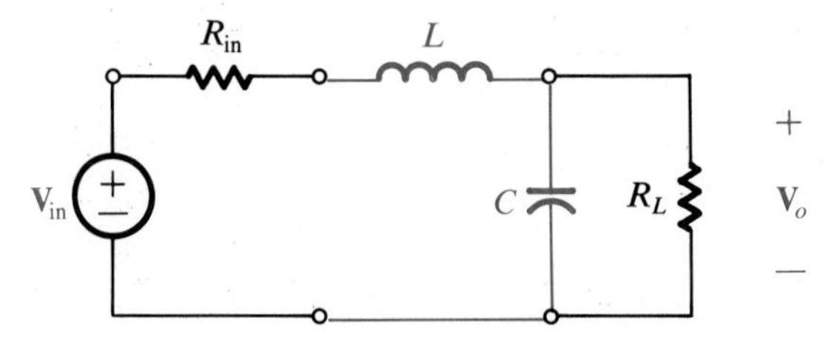

Figure 17.11 A passive realization of a second-order lowpass filter.

$$L = \frac{1 + R_{in}/R_L}{C\omega_c^2} \tag{17.3}$$

Substituting this value of L into (17.1) and rearranging gives a quadratic equation

$$\frac{R_{in}\omega_c^2 C^2}{1 + R_{in}/R_L} - \sqrt{2}\,\omega_c C + \frac{1}{R_L} = 0$$

Solving this equation specifies the capacitor for a given cutoff frequency. Using this value of C in (17.3) specifies L.

Skill Exercise 17.3

Find the design values of L and C to realize a second-order lowpass filter having $\omega_c = 2000\,\pi$ with $R_{in} = 10\ \Omega$ and $R_L = 100\ \Omega$. Also find the value of K.

Answer: $C = 23.6\ \mu F$ and $L = 1.18$ mH or $C = 1.18\ \mu F$ and $L = 23.6$ mH. $K = 0.91$.

17.3.2 Pole-Zero Pattern

The pole-zero pattern of the generic second-order lowpass filter shown in Fig. 17.12 has a complex pair of poles lying at the intersection

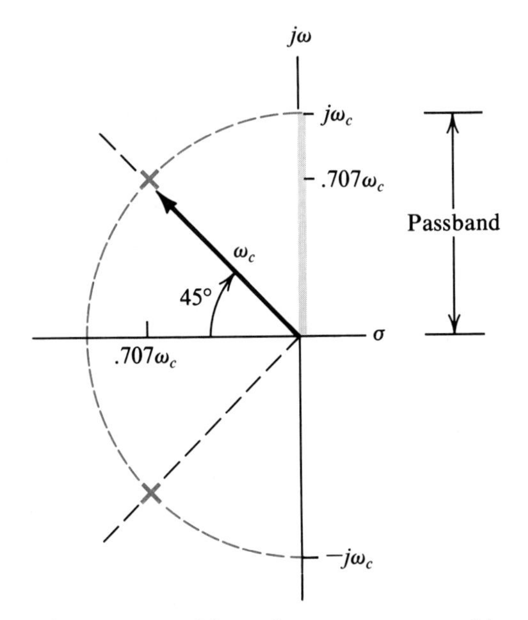

Figure 17.12 The pole-zero pattern and bandwidth of a generic second-order lowpass filter.

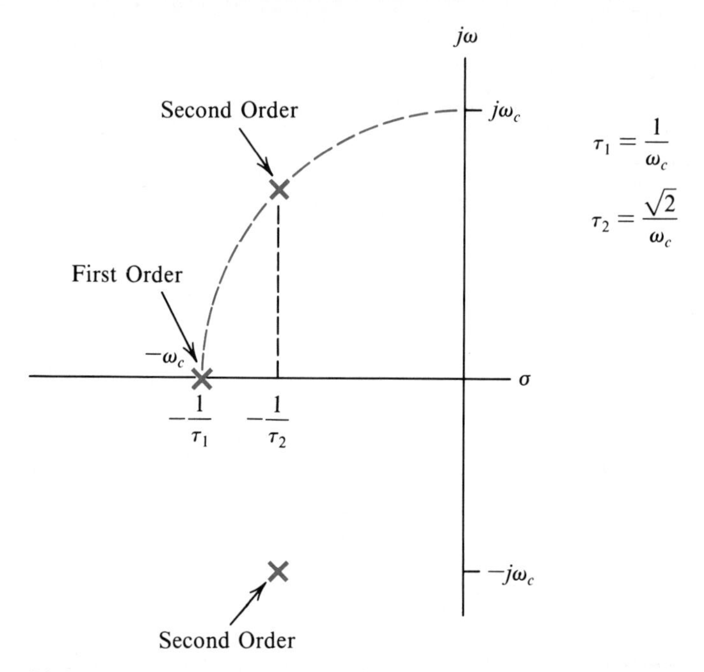

Figure 17.13 A comparison of pole-zero patterns for first- and second-order lowpass filters.

between a circle of radius ω_c and the 45° wedge (corresponding to a damping factor of 0.707), with s_1 and $s_2 = 0.707\ \omega_c(1 \pm j)$.

The bandwidth of the generic filter can be shown on the pole-zero pattern by locating the intersection of the circle of radius ω_c with the j-axis. The circle also passes through the poles. The time constant of the second-order lowpass filter is related to its bandwidth by $\tau = 1.414/\omega_c$. As in the first-order case, a second-order lowpass filter's bandwidth and time constant are inversely related.

Figure 17.13 shows an overlay of first- and second-order lowpass filter pole-zero patterns for the same bandwidth. The first-order filter has the shorter time constant because its pole is located with its real part at a greater distance from the orgin. Thus, using a second-order filter to get a sharper cutoff characteristic incurs the penalty of increasing the filter's time constant by 40%. The first-order filter will pass a time-domain pulse with less attenuation for the same bandwidth than the second-order filter will.

17.3.3 Active Lowpass Filter Design

Circuits using op amps can realize a second-order lowpass filter without inductors. This makes possible integrated circuit realizations of filters and eliminates the need for bulky, heavy, and expensive inductors—provided that the additional cost of supplying power to the op amp is offset by the other economies.

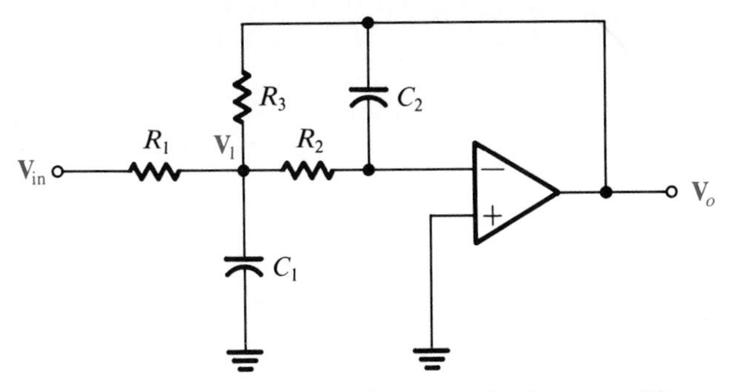

Figure 17.14 An active realization of a lowpass filter.

The active second-order lowpass filter in Fig. 17.14 has

$$\frac{\mathbf{V}_o}{\mathbf{V}_s} = \mathbf{H}(s) = \frac{\dfrac{-1}{R_1C_1R_2C_2}}{s^2 + \dfrac{1 + R_1/R_2 + R_1/R_3}{R_1C_1}s + \dfrac{1}{R_2C_2R_3C_1}}$$

Its constraint equations specify K and ω_c for given components

$$K = -R_3/R_1 \tag{17.4}$$

$$\sqrt{2}\omega_c = (1 + R_1/R_2 + R_1/R_3)/R_1C_1$$

and

$$\omega_c{}^2 = 1/(R_2C_2R_3C_1)$$

On the other hand, the components of the filter can be specified by choosing values for R_1, R_2, R_3, C_1 and C_2 that satisfy the three constraint equations given above for specified K and ω_c. If $|K|$ is given we cannot pick values for R_1 and R_3 independently since $R_3 = |K|R_1$. The two remaining constraint equations express the design relationship between R_1, R_2, C_1 and C_2. At this point the complexity of the design problem has been reduced, but the four remaining components cannot be uniquely specified from only two equations. One approach is to choose a component value, with the other component values following from the solution to the design constraints. If we fix the resistor ratio as $R_2/R_3 = \rho$, then $R_1/R_3 = 1/|K|$ and $R_1/R_2 = 1/(\rho|K|)$. With this additional condition the design constraints become

$$\sqrt{2}\omega_c = \frac{1}{R_1C_1}[\rho(|K| + 1) + 1]\frac{1}{\rho|K|} \tag{17.5}$$

and

$$\omega_c^2 = \frac{1}{\rho|K|^2R_1^2C_1C_2} \tag{17.6}$$

From (17.4) we get

$$C_1 = \frac{\rho(|K| + 1) + 1}{\rho|K|\sqrt{2}\omega_c R_1} \tag{17.7}$$

which, when substituted into (17.5) gives the value of C_2 as

$$C_2 = \frac{2\rho C_1}{[\rho(|K| + 1) + 1]^2}$$

A typical design might have $|K| = 10$, $R_1 = 10\ \text{k}\Omega$ and $R_3 = 100\ \text{k}\Omega$.

Example 17.1

Suppose that the signal $v_{in}(t) = 4 + 10 \sin 500\pi t + 2 \sin 600\pi t + 7 \sin 700\pi t + 5 \sin 1000\pi t$ is the input to a filter that must selectively reject certain components of $v_{in}(t)$ and produce an output signal that is very closely approximated by $v_o(t) = -(4 + 10 \sin 500\pi t)$. Specify the significant performance characteristics of the filter and determine the value of each of its components.

Solution: The filter must have a negative DC gain and must pass the DC and 500π rad/sec components of v_{in} while rejecting the components at 600π and above, so we select an active lowpass filter with $\omega_c = 500\pi$ and DC gain $K = -1$. Then, letting $R_1 = 1\ \text{k}\Omega$ and $\rho = 2$, we calculate $R_2 = 2\ \text{k}\Omega$, $R_3 = 1\ \text{k}\Omega$, $C_1 = 1.02\ \mu\text{F}$ and $C_2 = 0.163\ \mu\text{F}$.

17.3.4 Bode Plots

Since filters may operate over a wide range of input-signal frequencies, it is useful to introduce a unit of measure that effectively compresses the dynamic range of ω so that $|\mathbf{H}(j\omega)|$ can be plotted more conveniently. The unit of measure in common use is the **decibel (dB)** (after Alexander Graham Bell)—defined by

$$|\mathbf{H}(j\omega)|_{\text{dB}} = 20 \log_{10} |\mathbf{H}(j\omega)| = 20 \log_{10} |\mathbf{V}_{out}/\mathbf{V}_{in}|$$

Suppose that $\mathbf{V}_{out} = K\mathbf{V}_{in}$. Then $|\mathbf{H}(j\omega)|_{\text{dB}}$ is given in Table 17.1 for values of K ranging over six orders of magnitude.

TABLE 17.1 CHANGE IN $|H(j\omega)|_{DB}$ AS K VARIES OVER SIX ORDERS OF MAGNITUDE

K	$\|H(j\omega)\|_{DB}$
1000	60
100	40
10	20
1	0

TABLE 17.1 Continued

K	$\|H(j\omega)\|_{DB}$
0.1	−20
0.01	−40
0.001	−60

A difference of 20 dB between $\mathbf{V}_{out}$ and $\mathbf{V}_{in}$ corresponds to an order of magnitude increase between them. The decibel measure of $\mathbf{H}(\cdot)$ is convenient because it describes the *relative* values of $\mathbf{V}_{out}$ and $\mathbf{V}_{in}$ rather than their absolute values.

A graph of $|\mathbf{H}(j\omega)|_{dB}$ is called a **Bode plot** or a **Bode diagram** (after H. W. Bode). For the first-order lowpass filter with $K = 1$ we have

$$\mathbf{H}(j\omega) = \frac{1}{1 + j\omega/\omega_c}$$

so that

$$\begin{aligned}|\mathbf{H}(j\omega)|_{dB} &= 20 \log_{10} \left|\frac{1}{1 + j\omega/\omega_c}\right| \\ &= 20 \log_{10} \frac{1}{\sqrt{1 + (\omega/\omega_c)^2}} \\ &= -10 \log_{10} [1 + (\omega/\omega_c)^2]\end{aligned}$$

Before graphing $|\mathbf{H}(j\omega)|_{dB}$ we calculate its high- and low-frequency asymptotes. For $\omega \ll \omega_c$, $|\mathbf{H}(j\omega)|_{dB} \simeq 0$, and for $\omega \gg \omega_c$ $|\mathbf{H}(j\omega)|_{dB} \simeq -20 \log_{10} \omega/\omega_c$. The graph of the high-frequency asymptote will be linear if plotted on a semi-log scale instead of a linear scale. To see this, let

$$f(\omega) = -20 \log_{10} \omega/\omega_c = -20 \log_{10} \omega + 20 \log_{10} \omega_c$$

Next, we change variables, using $z = \log_{10} \omega$ to get

$$g(z) = -20z + 20 \log_{10} \omega_c$$

The new function $g(z)$ has a straight-line graph in z. A unit change in z causes $g(z)$ to change by -20 dB, and $g(\omega_c) = 0$. The graphs of $g(z)$ and $f(\omega)$ are shown plotted together in Fig. 17.15 for corresponding values of z and ω, with a linear scale for z and a semi-log scale for ω. Thus, the graph of the high-frequency asymptote of $|\mathbf{H}(j\omega)|_{dB}$ on semi-log paper will be linear.

Fig. 17.16 shows the low-frequency and high-frequency asymptotes (dashed lines) and the actual response for $|\mathbf{H}(j\omega)|_{dB}$. The asymptotes provide a quick sketch of the actual curve. They intersect at $\omega = \omega_c$ and for this reason ω_c is called the **corner frequency.** At the corner frequency the *actual* value of the magnitude response is $|\mathbf{H}(j\omega_c)|_{dB} = -3$ dB (i.e., the actual curve is 3 dB below the intersection of the asymptotes. The high-frequency asymptotic slope is -20 dB/decade, indicating that increasing

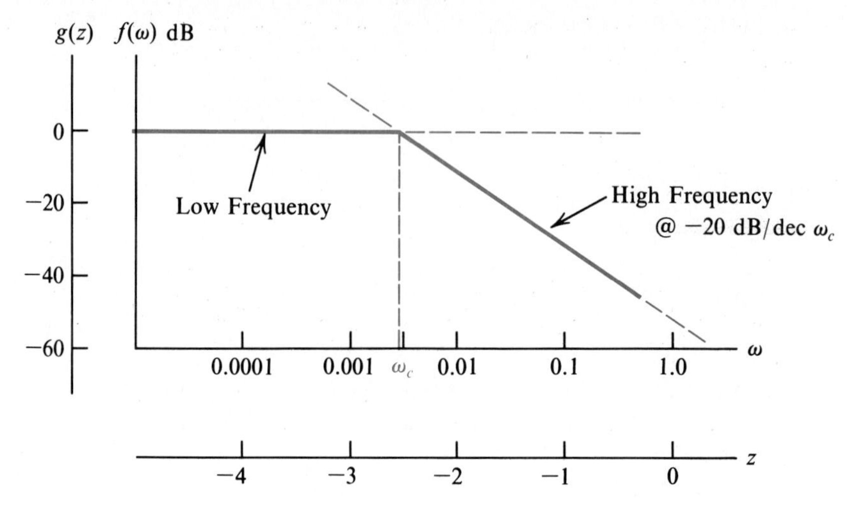

Figure 17.15 Lowpass filter frequency-response plots on linear and logarithmic scales.

the input-signal frequency by one decade causes the output signal to be attenuated by one order of magnitude compared to the input signal.

17.3.5 Second-Order Bode Plots

The Bode diagram for a generic second-order lowpass filter is obtained from

$$|\mathbf{H}(j\omega)| = \frac{|K|}{\sqrt{1 + (\omega/\omega_c)^4}}$$

so

$$|\mathbf{H}(j\omega)|_{\text{dB}} = 20 \log_{10}|K| - 20 \log_{10} \sqrt{1 + (\omega/\omega_c)^4}$$

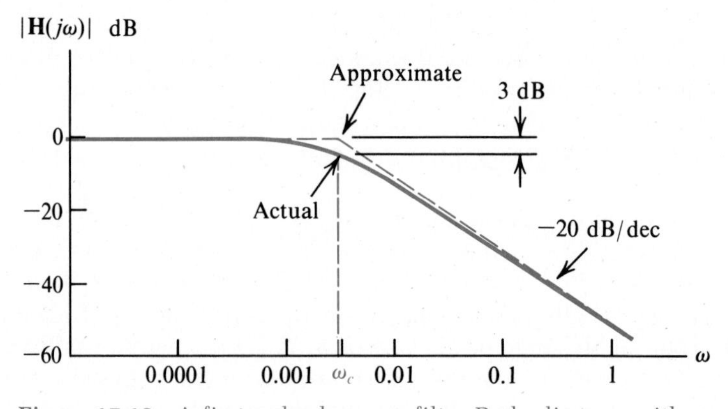

Figure 17.16 A first-order lowpass filter Bode diagram with asymptotes.

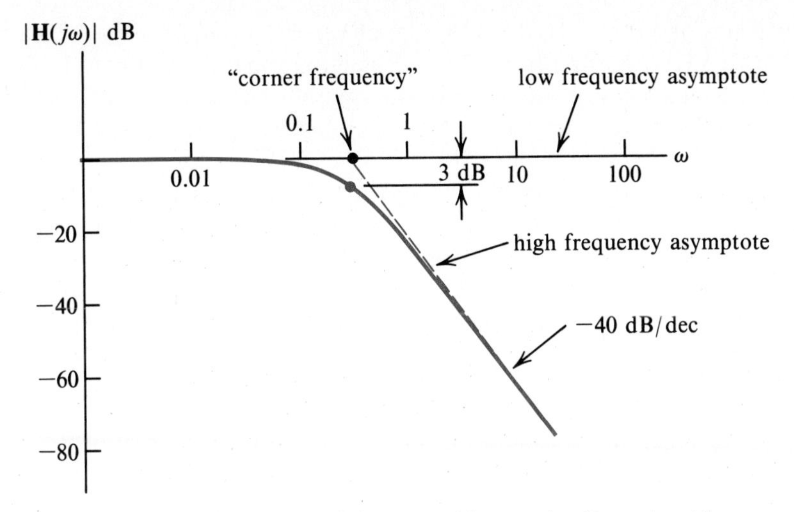

Figure 17.17 A second-order lowpass filter Bode diagram with asymptotes.

For $\omega \ll \omega_c$, $|\mathbf{H}(j\omega)|_{\text{dB}} \simeq 20 \log_{10} |K|$, and for $\omega \gg \omega_c$, $|\mathbf{H}(j\omega)|_{\text{dB}} \simeq 20 \log_{10} |K| - 40 \log_{10} \omega/\omega_c$. The low- and high-frequency asymptotes are shown plotted with $|\mathbf{H}(j\omega)|_{\text{dB}}$ on a semi-log scale in Fig. 17.17. They intersect at ω_c and the high-frequency asymptotic slope is −40 dB/decade. The actual curve of $|\mathbf{H}(j\omega)|_{\text{dB}}$ is 3 dB below the asymptote at the corner frequency.

17.4 HIGHPASS FILTER*

Highpass filters are used to place the passband *above* a cutoff frequency ω_c. One common application of a highpass filter is to remove the DC bias from signals in a multi-stage transistor amplifier. A simplified model of a stage of such an amplifier is shown in Fig. 17.18. The operating conditions of a transistor require that constant bias voltage be applied. When a transistor is used as an amplifier its output signal consists of the scaled sum of the applied input signal and the DC bias voltage. For example,

$$v_{\text{out}}(t) = V_{\text{DC}} + v_{\text{AC}}(t)$$

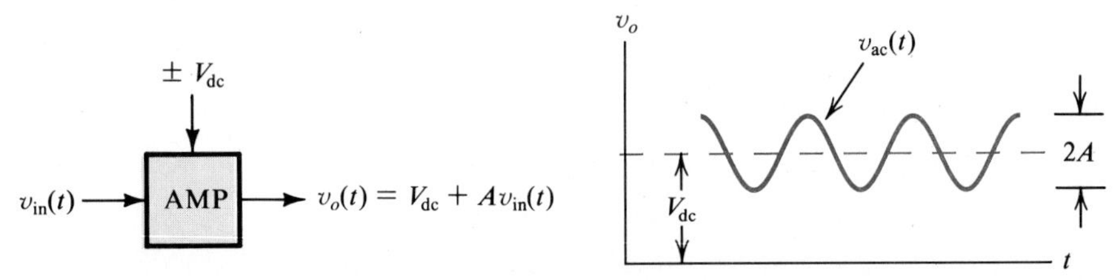

Figure 17.18 Amplifier output-signal bias.

where $v_{AC}(t) = Av_{in}(t)$. The DC signal V_{DC} is said to *bias* the AC signal (or displace it from the orgin). The bias signal must be removed in order for successive stages of the amplifier to operate properly. Otherwise, the signal levels at the input of successive stages would exceed the physical tolerances of the transistors.

In a multi-stage configuration, the second and subsequent stages of the amplifier each must have an unbiased signal as the input. A filter that passes $v_{AC}(t)$ but blocks V_{DC} must be inserted between the amplifier stages, so the filter's pole-zero pattern must have a zero at the orgin to block the DC signal component, but must not distort $v_{AC}(t)$.

A generic first-order highpass filter has a transfer function given by

$$\mathbf{H}(s) = \frac{Ks}{s + \omega_c}$$

with

$$|\mathbf{H}(j\omega)| = \frac{|K|}{\sqrt{1 + (\omega_c/\omega)^2}}$$

$$\sphericalangle \mathbf{H}(j\omega) = \sphericalangle K + \pi/2 - \tan^{-1} \frac{\omega}{\omega_c}$$

The parameter ω_c is the half-power frequency of the filter, since $|\mathbf{H}(j\omega_c)| = 0.707\,|K|$. The magnitude response shown in Fig. 17.19 has $|\mathbf{H}(0)| = 0$ and $|\mathbf{H}(\infty)| = |K|$. The parameter K is called the **high-frequency gain,** or the **AC gain** of the filter.

17.4.1 Passive Realization

There are two fundamental ways to create a first-order highpass filter. One is to block current to the load at low frequency; the other is to provide a path for current to bypass the load at low frequency. Placing a capacitor between the source and load in Fig. 17.20 blocks steady-state DC current

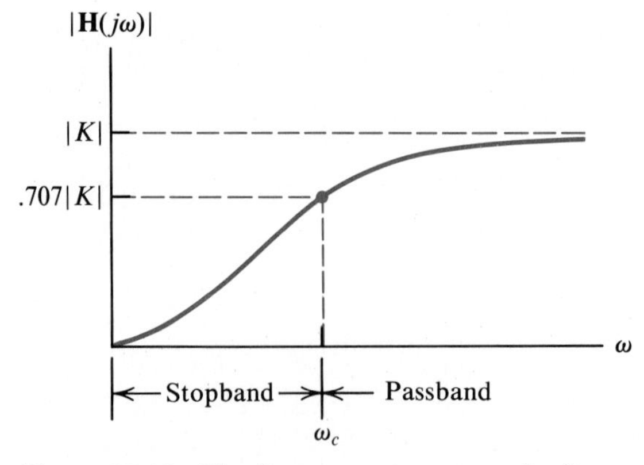

Figure 17.19 The frequency response of a first-order highpass filter.

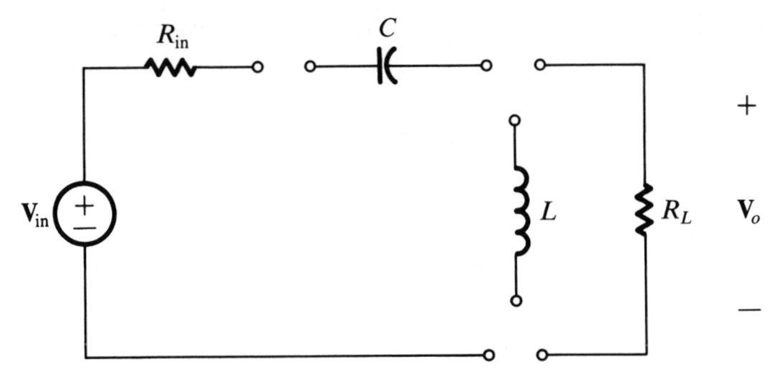

Figure 17.20 Realizations of a highpass filter.

from producing a steady-state component of load current to the load; placing an inductor in parallel with the load provides a shunt path for DC current. In either configuration, a DC source produces a steady-state output voltage of zero. The transfer function for the capacitor realization is

$$\mathbf{H}(s) = \frac{sR_L/(R_{in} + R_L)}{s + \dfrac{1}{C(R_{in} + R_L)}}$$

Skill Exercise 17.4

Find expressions for ω_c and K for the capacitor realization of a highpass filter.

Answer: $\omega_c = 1/(R_{in} + R_L)C$ and $K = R_L/(R_{in} + R_L)$.

17.4.2 Pole-Zero Pattern

The pole-zero pattern of the highpass filter has a zero at the orgin, corresponding to the filter's ability to block a DC signal. The dashed circle centered at the orgin and drawn through the pole in Fig. 17.21 intersects the *j*-axis at the cutoff frequency, and thereby fixes the stopband of the filter. The passband is infinite for the highpass filter; its bandwidth, as related to time constants, is the width of the stopband. (As an exercise, deduce the shape of the circuit's phase characteristic from Fig. 17.21.)

17.5 BANDPASS FILTERS

Bandpass filters selectively pass signals within a passband and reject signals at frequencies above and below the passband. They are used in television, radio, and communications receivers to eliminate unwanted signals and provide filtered (low-noise) inputs to amplifiers. Bandpass filters in

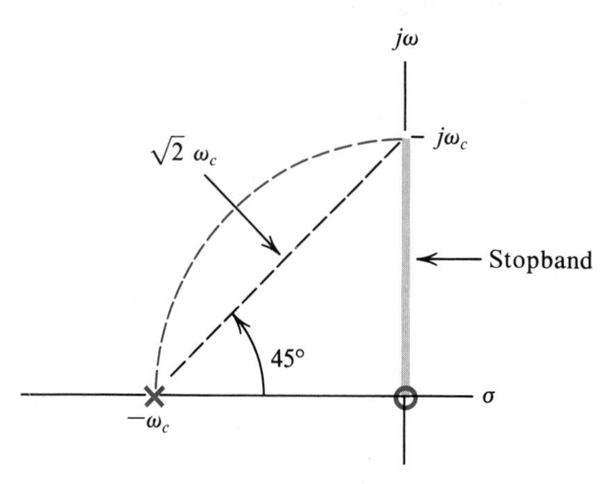

Figure 17.21 The pole-zero pattern and bandwidth of a generic first-order highpass filter.

telephones "decode" the composite signal tones to determine the dialed extension number.

17.5.1 Generic Filter Characteristics

The generic model for a bandpass filter has

$$\mathbf{H}(s) = \frac{Ks\omega_b}{s^2 + s\omega_b + {\omega_p}^2}$$

and

$$\mathbf{H}(j\omega) = \frac{jK\omega\omega_b}{{\omega_p}^2 - \omega^2 + j\omega\omega_b}$$

The magnitude and phase responses shown in Fig. 17.22 are obtained in the usual way

$$|\mathbf{H}(j\omega)| = \frac{|K|\,\omega\omega_b}{\sqrt{(\omega_p^2 - \omega^2)^2 + \omega^2\omega_b^2}}$$

and

$$\mathbf{H}(j\omega) = \begin{cases} \dfrac{\pi}{2} + \sphericalangle K - \tan^{-1}\dfrac{\omega\omega_b}{\omega_p^2 - \omega^2} & \omega \le \omega_p \\ -\dfrac{\pi}{2} + \sphericalangle K + \tan^{-1}\dfrac{\omega\omega_b}{\omega^2 - \omega_p^2} & \omega \ge \omega_p \end{cases}$$

The parameters of the generic transfer function have special significance: ω_b is called the **bandwidth** of the filter and ω_p is called the **peak frequency** or the **resonant frequency.** Since $|\mathbf{H}(j\omega_p)| = |K|$ the parameter K is called the **peak gain** or the **resonant gain** of the filter. A sinusoid at the resonant frequency is said to be **in resonance.** Independent values can be chosen for ω_b and ω_p to specify a variety of filter characteristics, and the process of "tuning" a filter refers to selection of circuit parameters to place ω_p at a desired value.

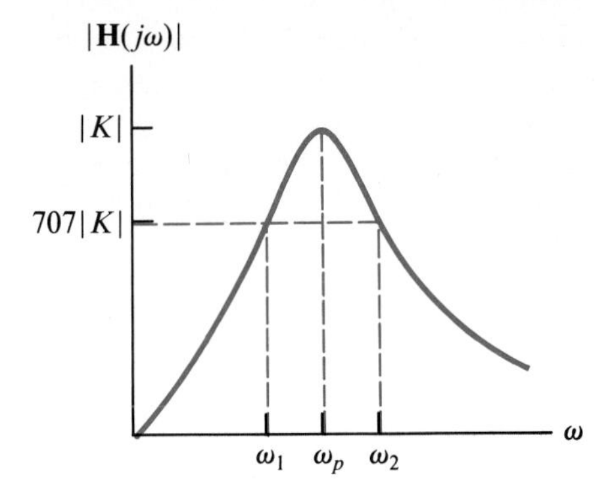

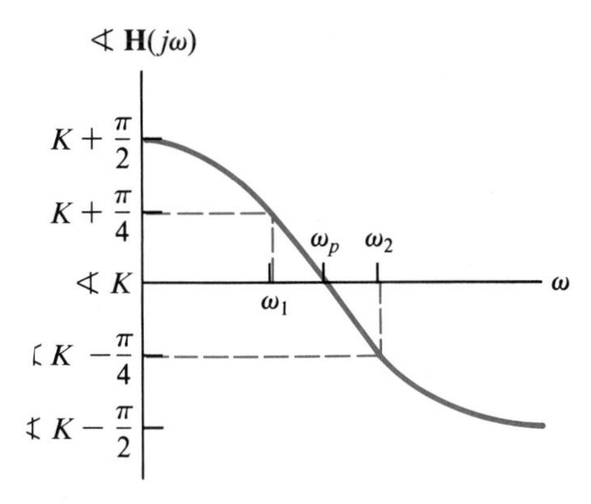

Figure 17.22 The magnitude and phase responses of a bandpass filter.

The *half-power frequencies* ω_1 and ω_2 of a bandpass filter are the frequencies at which $|\mathbf{H}(j\omega_1)| = |\mathbf{H}(j\omega_2)| = 0.707$, with $\omega_1 < \omega_2$. It can be shown that the peak frequency and the bandwidth are related to the half-power frequencies by

$$\omega_b = \omega_2 - \omega_1$$
$$\omega_p = \sqrt{\omega_1 \omega_2}$$

The bandwidth is the difference of the half-power frequencies, and the peak frequency is their geometric mean.

The **quality factor** Q of a bandpass filter is a measure of the filter's selectivity:

$$Q = \frac{\omega_p}{\omega_b}$$

and the filter's **fractional bandwidth** is defined by taking the reciprocal of Q ($1/Q = \omega_b/\omega_p$). A **high-Q** filter passes signals within a relatively narrow frequency band. The tuning circuit in a radio must have Q large enough to reject all but the station to which the tuner is set. Otherwise, more than one station will be heard at a time. High-Q filters are almost distortionless in the passband.

It is left to the reader to show that for a given ω_b and ω_p the half-power frequencies can be written as

$$\omega_2, \omega_1 = \omega_p \sqrt{1 + \frac{1}{4Q^2}} \pm \frac{\omega_b}{2}$$

The half-power frequencies are symmetrically located about the **center frequency** $\omega_c = \omega_p \sqrt{1 + 1/(4Q^2)}$ and since $\omega_c > \omega_p$ the curve for $|\mathbf{H}(j\omega)|$ is always skewed in the direction of increasing ω, especially for small values of Q.

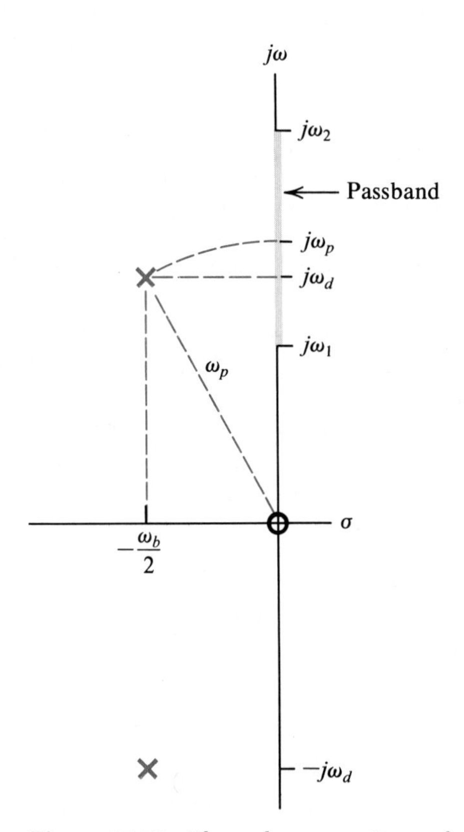

Figure 17.23 The pole-zero pattern of a generic bandpass filter.

17.5.2 Actual and Approximate High-Q Pole-Zero Patterns

When $Q > \frac{1}{2}$ the bandpass filter has a pair of complex poles at

$$s_1, s_2 = -\omega_b/2 \pm j\omega_p\sqrt{1 - 1/(4Q^2)}$$

These poles are associated with a damped oscillatory time-domain response. The pole-zero pattern in Fig. 17.23 shows the location of the half-power frequencies, the resonant frequency (undamped frequency of oscillation), the center frequency, and the damped frequency of oscillation. The complex poles lie on a circle of radius ω_p. The location of the resonant frequency is determined by the intersection of this pole-locus circle with the j-axis. The real part of the complex poles is exactly one-half of the filter bandwidth, so **selective bandpass filters will have their poles located relatively close to the j-axis.**

The step response of a bandpass filter with $Q > \frac{1}{2}$ will have a damped frequency of oscillation $\omega_d = \omega_p\sqrt{1 - 1/(4Q^2)}$ and a transient decay envelope with a time constant $\tau = 2/\omega_b$. This time-constant/bandwidth tradeoff implies that a low-bandwidth filter will have a lightly damped—i.e., highly oscillatory (ringing)—transient response. Dividing the transient-decay interval 4τ by the period of the damped oscillation, $2\pi/\omega_d$ and using $\omega_d \simeq \omega_p$ gives the approximate number of cycles in the transient: $N = 2/\pi\ \tau\omega_d \simeq 4\omega_p/(\pi\omega_b) \simeq 1.3\ Q$.

The pole-zero pattern of a high-Q filter can be drawn very quickly by noting in Fig. 17.24 that the half-power frequencies ω_1 and ω_2 are approximately determined by the intersection of a half-circle with the j-axis. The half-circle is centered at $j\omega_p$ and has a radius equal to $\omega_b/2$. It passes through the pole and its intersections with the j-axis give the approximate location of ω_1 and ω_2. The center frequency is approximated by $\omega_c \simeq \omega_p$ to give $\omega_2, \omega_1 \simeq \omega_p \pm \omega_b/2$.

The resonant-frequency response of the bandpass filter can be related to the geometry of its pole-zero pattern in Fig. 17.25. Since $\mathbf{H}(s)$ has a zero at the orgin $\mathbf{H}(j0) = 0$ and because there are more poles than zeros in the transfer function $\mathbf{H}(j\infty) = 0$. Beginning at $\omega = 0$ the vectors drawn to a point on the j-axis from the zero and the lower pole increase in length as ω increases. But the vector from the upper pole decreases in length until $\omega > \omega_d$. Since this vector is in the denominator of $\mathbf{H}(j\omega)$ the value of $|\mathbf{H}(j\omega)|$ becomes relatively large especially when Q is large, because the poles are very close to the j-axis. $|\mathbf{H}(j\omega)|$ has its maximum value above the point where the upper pole vector reaches its minimum length. The phase response also has a geometric basis, but we will not consider it here.

Example 17.2

If the input to a bandpass filter is $v_{in}(t) = 10 \sin{(\omega_p + \omega_b/2)t}$ find an approximate expression for the steady-state output $v_o(t)$.

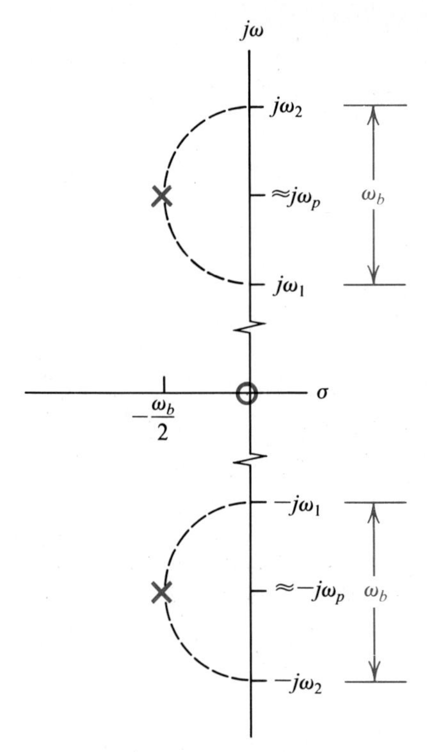

Figure 17.24 High-Q approximations for the upper and lower half-power frequencies.

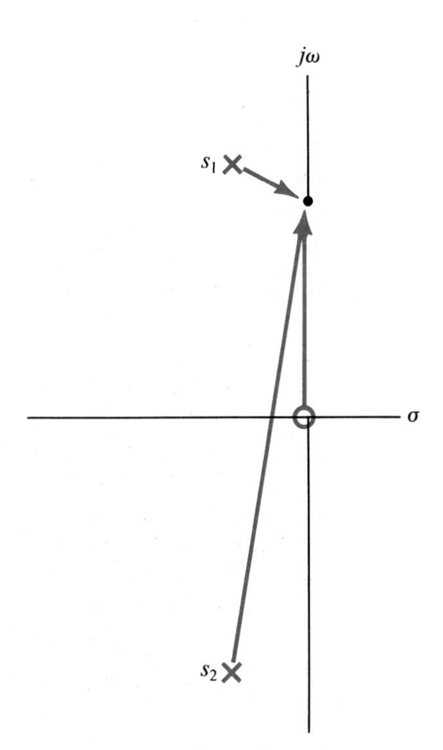

Figure 17.25 The pole and zero vectors of a bandpass filter near resonance.

Solution: Since the sinusoidal frequency is approximately the half-power frequency, the output signal $v_o(t) \simeq 7.07 \sin(\omega_2 t + 45°)$, where $\omega_2 \simeq \omega_p + \omega_b/2$.

17.5.3 Passive Bandpass Filter Design

There are two fundamental passive filter schemes that realize an overall generic bandpass filter characteristic. One is to place a resonant circuit in series with the load; the other is to place a resonant circuit in parallel with R_L. If the series impedance $\mathbf{Z}(j\omega)$ in Fig. 17.26(a) is relatively small for $\omega = \omega_p$ and large otherwise, the high current in R_L at ω_p will give rise to a bandpass-like resonant characteristic. Likewise, if $\mathbf{Y}(j\omega)$ in Fig. 17.26(b) is small for $\omega = \omega_p$ and large otherwise, the circuit will behave like a bandpass filter because relatively more current will be shunted to R_L at $\omega = \omega_p$.

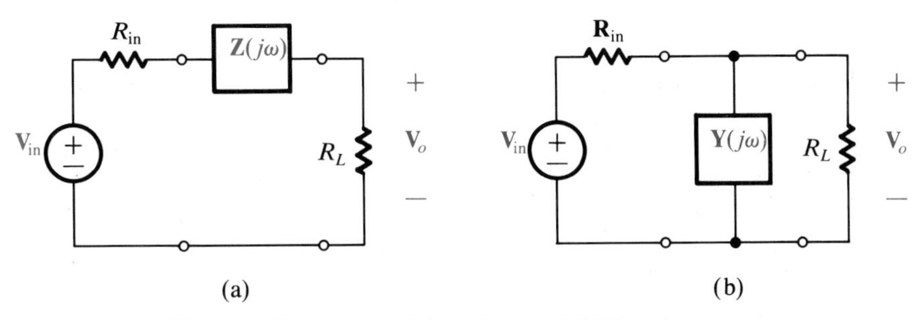

Figure 17.26 Filter configurations: (a) series, and (b) bandpass.

The series *LC* section shown in Fig. 17.27(a) has the desired characteristic of acting like a short circuit to current at its resonant frequency. Its

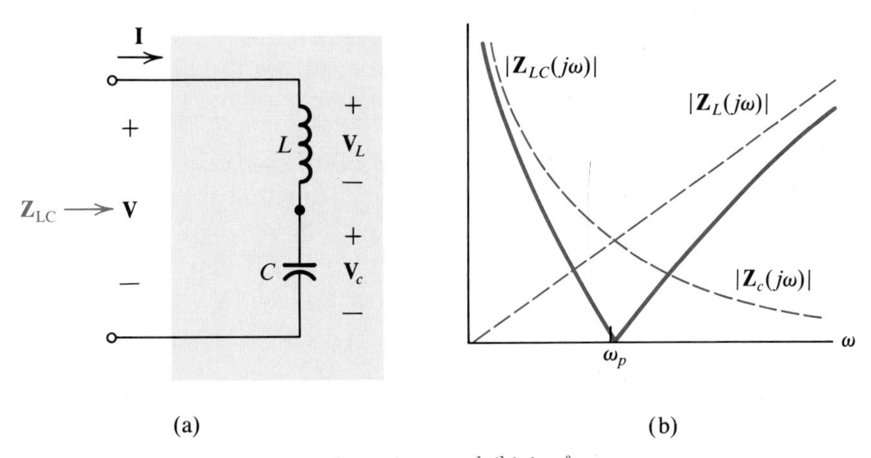

Figure 17.27 (a) A series *LC* section, and (b) its frequency response.

impedance is the sum of the impedances of its individual components: $\mathbf{Z}_{LC}(j\omega) = j[\omega L - 1/\omega C)]$. Figure 17.27(b) shows a graph of the two components of $\mathbf{Z}_{LC}$ (their sum being zero at the point where the curve for $|\mathbf{Z}_L(j\omega)|$ intersects the curve for $|\mathbf{Z}_C(j\omega)|$). At low frequencies $\mathbf{Z}_{LC}$ is dominated by $\mathbf{Z}_C$ and at high frequencies $\mathbf{Z}_L$ prevails. The point of intersection can be obtained analytically by solving $|\mathbf{Z}_{LC}(j\omega)| = 0$ to get $\omega_p = 1/\sqrt{LC}$. It is important to understand that this does *not* imply that $\mathbf{Z}_L = \mathbf{Z}_C = 0$ but only that the impedance measured across the series *pair* is zero. Failure to realize this could have serious consequences! Consider the voltage phasors at resonance

$$\mathbf{V}_L = \mathbf{I}\mathbf{Z}_L(j\omega_p) = j\sqrt{L/C}\,\mathbf{I}$$

and

$$\mathbf{V}_C = \mathbf{I}\mathbf{Z}_C(j\omega_p) = -j\sqrt{L/C}\,\mathbf{I}$$

While it's true that $\mathbf{V} = \mathbf{V}_L + \mathbf{V}_C = 0$, note that $\mathbf{V}_L \neq 0$ and $\mathbf{V}_C \neq 0$. This condition presents a potential safety hazard when the *LC* circuit is operated at or near its resonant frequency because even relatively small currents can produce dangerously large voltages across the reactive components.

Skill Exercise 17.5

If $L = 100$ mH and $C = 0.01\ \mu$F find the resonant frequency of the series *LC* section, and calculate $|\mathbf{V}_L|$, $|\mathbf{V}_C|$, $\measuredangle\mathbf{V}_L$, $\measuredangle\mathbf{V}_C$ and steady-state values of $v_L(t)$ and $v_c(t)$ of the section if $i(t) = 50 \sin \omega t$ mA.

Answer: $\omega_p = 31{,}633$ rad/sec, $|\mathbf{V}_L| = |\mathbf{V}_C| = 158$ volts, $\measuredangle\mathbf{V}_L = -\measuredangle\mathbf{V}_C = \pi/2$ and $v_L(t) = -v_c(t) = 158 \cos(31633t)$.

At $\omega = \omega_p$ the sinusoidal voltages corresponding to the phasors $\mathbf{V}_L$ and $\mathbf{V}_C$ in the series *LC* section are 180° out of phase and cancel each other when measured across the input terminals. Internally, $v_c(t) \neq 0$ and $v_L(t) \neq 0$ except at their zero crossings.

Placing the series *LC* section in series with the load creates the bandpass filter shown in Fig. 17.28. Its transfer function is

$$\mathbf{H}(s) = \frac{s(R_L/L)}{s^2 + s\dfrac{R_{in} + R_L}{L} + \dfrac{1}{LC}} = \frac{\mathbf{V}_o}{\mathbf{V}_{in}}$$

with constraint equations

$$K\omega_b = R_L/L \tag{17.8}$$

$$\omega_b = (R_{in} + R_L)/L \tag{17.9}$$

$$\omega_p{}^2 = 1/LC \tag{17.10}$$

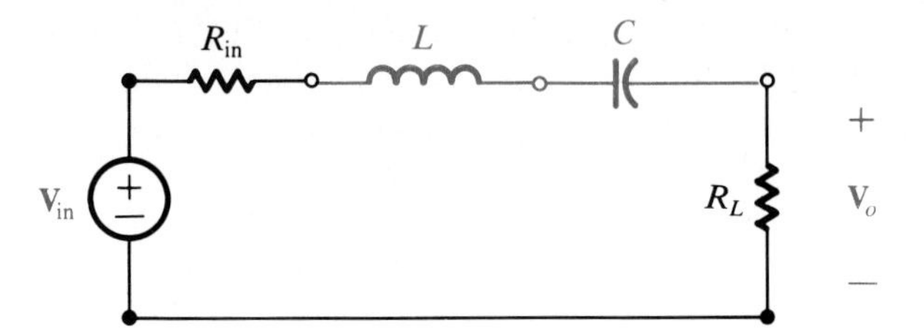

Figure 17.28 A bandpass filter with a series LC section.

Combining (17.8) and (17.9) leads to $K = R_L/(R_{in} + R_L)$. (Note that K could also be obtained by inspection with $\mathbf{V}_{LC} = 0$). The peak gain of the passive filter is fixed by R_{in} and R_L and cannot exceed one.

For a given R_{in}, R_L, ω_p and ω_b the **design equations** specifying the components of the filter follow from (17.8) and (17.9):

$$L = \frac{R_{in} + R_L}{\omega_b}$$

$$C = \frac{\omega_b}{\omega_p^2(R_{in} + R_L)}$$

A bandpass filter can be realized by connecting a parallel LC section across the load, as shown in Fig. 17.29. Understanding the behavior of the parallel LC section in Fig. 17.30 is the key to understanding the bandpass response of the filter. First, note that

$$\mathbf{Z}_{LC}(j\omega) = \frac{s/C}{s^2 + \dfrac{1}{LC}}$$

$\mathbf{Z}_{LC}$ has a pair of poles on the j-axis and a zero at the orgin. Its denominator has the form of a generic bandpass transfer function with $\omega_b = 0$ and $\omega_p = 1/\sqrt{LC}$. At the resonant frequency $\mathbf{Z}_{LC} = \infty$, and $\mathbf{Y}_{LC} = 0$ so for a given $\mathbf{V}_{LC}$ the circuit is open! In the filter configuration of Fig. 17.29 the LC section is in parallel with the load, so at resonance we have $\mathbf{V}_o/\mathbf{V}_{in} = R_L/(R_{in} + R_L)$.

Avoid the trap of thinking that $\mathbf{I}_{LC} = 0$ at $\omega = \omega_p$ implies that $\mathbf{I}_L = \mathbf{I}_C = 0$. To the contrary, at resonance

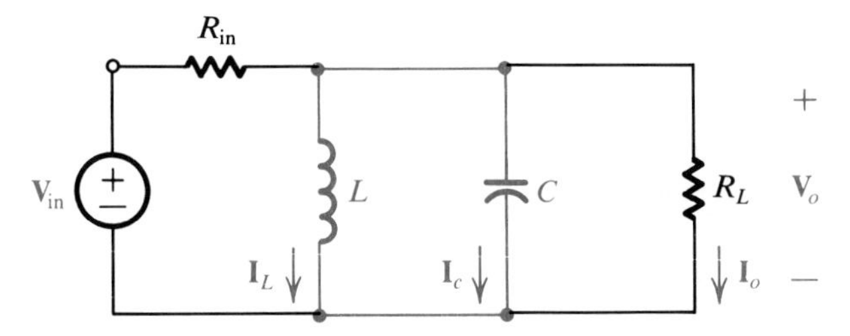

Figure 17.29 A bandpass filter with a parallel LC section.

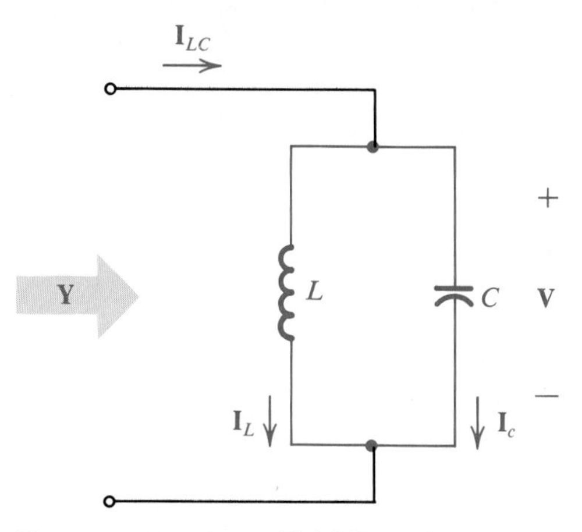

Figure 17.30 A parallel *LC* section.

$$\mathbf{I}_L = \mathbf{V}_o\mathbf{Y}_L = \mathbf{V}_o 1/(j\omega_p L)$$

and

$$\mathbf{I}_L = -j\mathbf{V}_o\sqrt{C/L}$$
$$\mathbf{I}_C = j\mathbf{V}_o\sqrt{C/L}$$

So the steady-state time-domain currents $i_c(t)$ and $i_L(t)$ have equal magnitude but are 180° out of phase with each other, corresponding to **current circulation** at the resonant frequency of the filter.

The parallel configuration of the bandpass filter cannot be realized exactly because physical inductors also have a small amount of resistance. Instead, the **lossy** bandpass filter in Fig. 17.31 is actually realized. The filter is called lossy because the inductor's resistor dissipates energy.

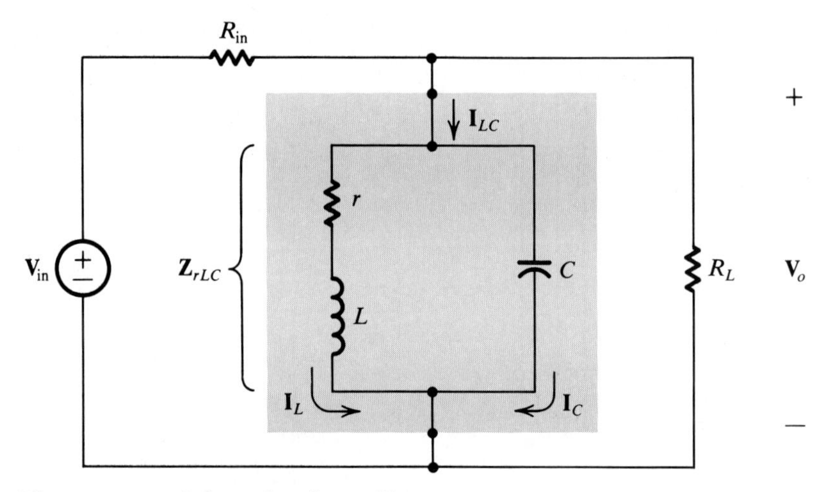

Figure 17.31 A lossy bandpass filter.

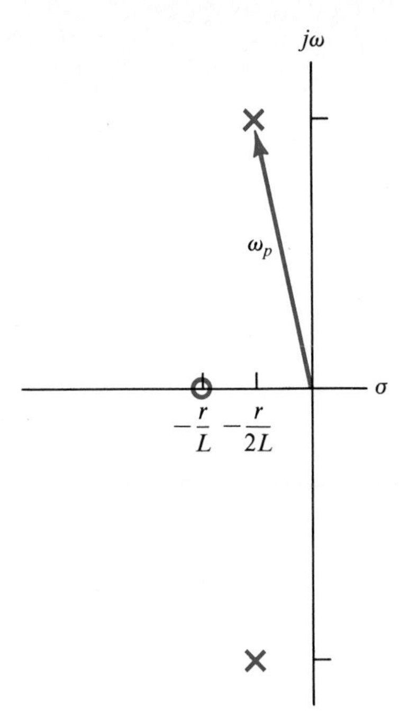

Figure 17.32 The pole-zero pattern of a lossy *LC* circuit.

The lossy *rLC* section has

$$\mathbf{Z}_{rLC}(s) = \frac{\dfrac{1}{C}\left(s + \dfrac{r}{L}\right)}{s^2 + s\dfrac{r}{L} + \dfrac{1}{LC}}$$

Instead of having a zero at the orgin $\mathbf{Z}_{rLC}(s)$ has a zero at $s = -r/L$ and its poles are displaced from the *j*-axis by $-r/2L$ as shown in Fig. 17.32.

The lossy *rLC* circuit cannot exhibit the same resonance condition as the lossless *LC* circuit because it cannot have $\mathbf{I}_{LC} = 0$ unless $r = 0$. The phasor diagrams in Fig. 17.33 show that if $\mathbf{I}_L = -\mathbf{I}_C$ the phasors of the voltage across the capacitor and the voltage across the lossy inductor can-

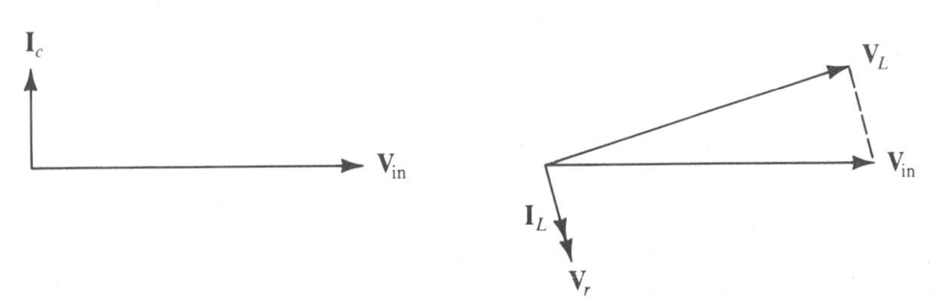

Figure 17.33 Phasor diagrams for the lossy *LC* section.

not align if $r \neq 0$. The phasor of the voltage across r and L is the sum of $\mathbf{V}_r$ and $\mathbf{V}_L$. It can align with $\mathbf{V}_{in}$ only if $\mathbf{V}_r = 0$, and this is possible if and only if $r = 0$.

Now let's return to the lossy bandpass filter in Fig. 17.31. Its transfer function has

$$\mathbf{H}(s) = \frac{\frac{1}{R_{in}C}\left(s + \frac{r}{L}\right)}{s^2 + s\left(\frac{r}{L} + \frac{R_{in} + R_L}{CR_{in}R_L}\right) + \left(\frac{1}{LC} + \frac{1}{C}\frac{R_{in} + R_L}{R_{in}R_L}\frac{r}{L}\right)}$$

If $r \neq 0$ the filter will have a DC response component to a step input signal, and the complex poles will be farther from the orgin and farther from the j-axis (ω_p, ω_d and ω_b all increase). (See Problem 17.30).

17.5.4 Active Bandpass Filter Design

The active bandpass filter shown in Fig. 17.34 is popular in applications where inductors are to be avoided, or where a broad range of pole-zero locations is desired. The physical operation of the circuit can be understood by realizing that $\mathbf{V}_o$ will be nonzero if and only if $\mathbf{I}_3$ is nonzero. At DC, C_2 blocks current to R_3, so $\mathbf{V}_o$ is zero. At high frequency C_1 and C_2 short $\mathbf{V}_o$ to ground. At intermediate frequencies current in R_3 causes $|\mathbf{V}_o|$ to be positive.

The circuit's transfer function is obtained by writing, combining, and simplifying KCL at $\mathbf{V}_1$ and at the input node

$$\mathbf{H}(s) = \frac{\frac{-1}{R_1C_1}s}{s^2 + s\frac{C_1 + C_2}{R_3C_1C_2} + \frac{R_1 + R_2}{R_1R_2R_3C_1C_2}} = \frac{\mathbf{V}_o}{\mathbf{V}_{in}}$$

Matching the coefficients of $\mathbf{H}(s)$ with those of the generic filter gives the design equations:

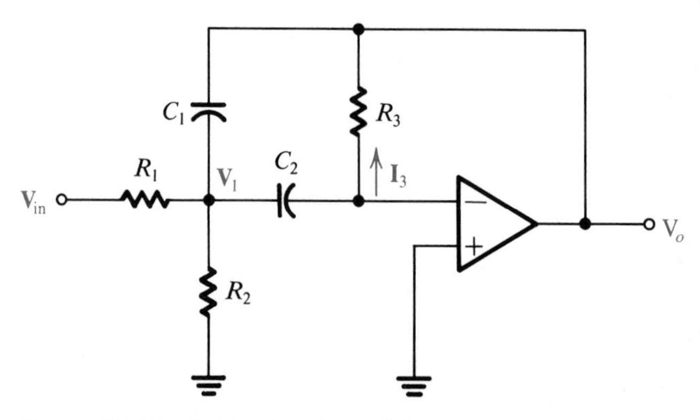

Figure 17.34 Active bandpass filter.

$$|K| = \frac{1}{\omega_b R_1 C_1} \tag{17.11}$$

$$\omega_b = \frac{C_1 + C_2}{R_3 C_1 C_2} \tag{17.12}$$

and

$$\omega^2{}_p = \frac{R_1 + R_2}{R_1 R_2 R_3 C_1 C_2} \tag{17.13}$$

To *choose* filter components to realize a given $|K|$, ω_b and ω_p we fix C_1 and let $C_2 = \beta C_2$ for a chosen value of β. This reduces the number of unspecified components to four. Then (17.12) and (17.11) define R_3 and R_1

$$R_3 = \frac{1 + \beta}{\beta \omega_b C_1}$$

$$R_1 = \frac{1}{\omega_b C_1 |K|}$$

Then we use these solutions for R_1 and R_3 to solve for R_2 in (17.13)

$$R_2 = \frac{\omega_b}{(1 + \beta)\omega_p^2 C_1 - \omega_b^2 C_1 |K|}$$

To ensure that R_2 is nonnegative we must require $\beta > |K|/Q^2 - 1$

Example 17.3

Design a filter whose steady-state output signal will be approximately $v_o(t) = -20 \sin 600\pi t$ when the input is $v_{in}(t) = 2 + 6 \sin 500\pi t + 10 \sin 600\pi t + \sin 700\pi t + 5 \sin 1000\pi t$. Specify the performance characteristics of the filter and determine the value of each circuit component.

Solution: An active bandpass filter with $k = -2$, $\omega_p = 600\pi$ and $Q = 10$ will ensure adequate attenuation of the unwanted signal components. If we arbitrarily choose $C_1 = 1\ \mu F$ and $\beta = 1$ the design values for the other components are: $R_1 = 2652.58\ \Omega$, $R_2 = 26.79\ \Omega$, $R_3 = 10610.33\ \Omega$ and $C_2 = 1\ \mu F$.

17.6 BANDSTOP FILTER*

Placing a series *LC* section in parallel with a load will create a complete bypass path for current and force the load voltage to zero at the resonant frequency. Likewise, placing a parallel *LC* section in series with the load in Fig. 17.35 blocks current to the load at the resonant frequency. Both circuits are examples of **bandstop filters.** Ideal bandstop filters reject sig-

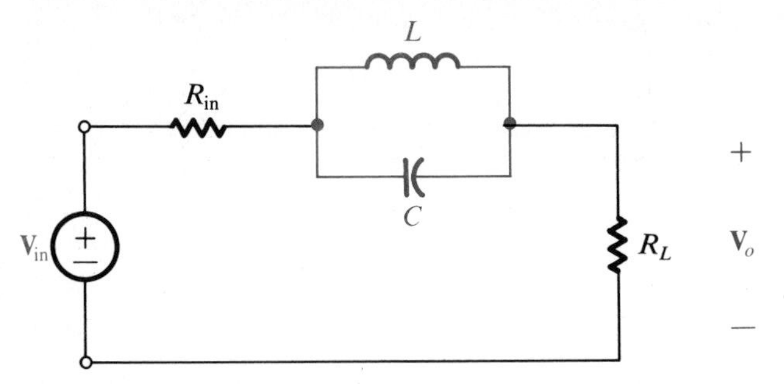

Figure 17.35 Bandstop filter with a parallel *LC* section.

nals at a specific frequency ω_R (the **rejection frequency**) and provide significant attenuation of sinusoidal signals whose frequency is within a stopband that includes ω_R. The width of the stopband is the bandwidth of the filter.

The generic transfer function of a second-order bandstop filter is

$$\mathbf{H}(s) = \frac{K(s^2 + \omega_R{}^2)}{s^2 + s\omega_b + \omega_R{}^2}$$

and Fig. 17.36 shows the magnitude and phase responses defined by

$$|\mathbf{H}(j\omega)| = \frac{|K|\,|\omega_R{}^2 - \omega^2|}{[(-\omega^2 + \omega_R{}^2)^2 + (\omega\omega_b)^2]^{1/2}}$$

$$\sphericalangle\mathbf{H}(j\omega) = \sphericalangle[K - \tan^{-1}\frac{\omega\omega_b}{\omega_R{}^2 - \omega^2} \qquad 0 \le \omega \le \infty.$$

The magnitude response has a notch at $\omega = \omega_R$ with $|\mathbf{H}(j\omega_R)| = 0$. We call ω_R the **frequency of complete rejection.** The filter is **flat** at DC and at high frequency.

The half-power frequencies of the bandstop filter are the frequencies ω_1 and ω_2 at which $|\mathbf{H}(j\omega_1)| = |H(j\omega_2)| = 0.707\ |K|$. It can be shown that $\omega_b = \omega_2 - \omega_1$ and $\omega_R = \sqrt{\omega_1\omega_2}$. The filter's quality factor is defined as $Q = \omega_R/\omega_b$. It provides a measure of the rejectivity of the filter, and $1/Q$ is the fractional bandwidth. The expressions for ω_R and ω_b can be combined to produce the half-power frequencies of the filter

$$\omega_2,\ \omega_1 = 1/2\sqrt{\omega_b{}^2 + 4\omega_R{}^2} \pm \omega_b/2$$

The filter's **center frequency** is defined to be the midpoint of the passband $\omega_c = 1/2\sqrt{\omega_b{}^2 + 4\omega_R{}^2}$ with the half-power frequencies being symmetrically located about ω_c. The center and half-power frequencies can also be written as

$$\omega_2,\ \omega_1 = \omega_R\sqrt{1 + 1/(4Q^2)} \pm \omega_R/(2Q)$$
$$\omega_c = \omega_R\sqrt{1 + 1/(4Q^2)}$$

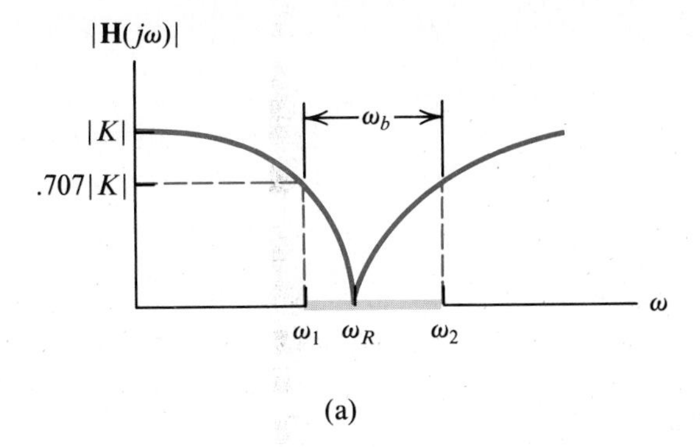

(a)

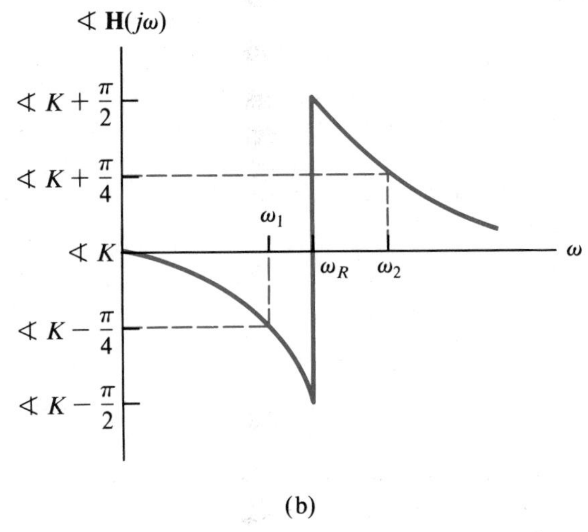

(b)

Figure 17.36 A bandstop filter: (a) magnitude response, (b) phase response.

Skill Exercise 17.6

Find K, ω_R, ω_b, Q, ω_c, ω_1 and ω_2 for a bandstop filter having

$$\mathbf{H}(s) = \frac{4s^2 + 57600\pi^2}{2s^2 + 40\pi s + 28800\pi^2}$$

Answer: $K = 2$, $\omega_R = 120\pi$, $\omega_b = 20\pi$, $Q = 6$, $\omega_c = 120.42\pi$, $\omega_1 = 130.42\pi$ and $\omega_2 = 110.42\pi$.

The circuit using the parallel *LC* section in Fig. 17.35 has

$$\mathbf{H}(s) = \frac{R_L}{R_{in} + R_L} \frac{s^2 + \dfrac{1}{LC}}{s^2 + s\dfrac{1}{(R_{in} + R_L)C} + \dfrac{1}{LC}}$$

with design constraints $K = R_L/(R_{in} + R_L)$, $\omega_b = 1/(R_{in} + R_L)C$ and $\omega_R = 1/\sqrt{LC}$. Notice that K can be obtained as either the high- or low-frequency gain of the circuit by inspection of Fig. 17.35. The rejection frequency can be tuned by selecting L and C, but since ω_b does not depend on L the filter can be tuned by choosing only L (without affecting the bandwidth). For a given R_{in} and R_L the design equations are $C = 1/\omega_b(R_{in} + R_L)$ and $L = \omega_b(R_{in} + R_L)/\omega_R{}^2$.

17.7 SCALED-FILTER DESIGN*

Filter design equations do not necessarily produce component values that are realistic or desirable. It is generally advisable to use large component values to minimize a circuit's sensitivity to thermal or manufacturing changes in those values. On the other hand, it is more convenient to work with small numbers when using the design equations. And sometimes it is desirable to be able to quickly convert a filter designed for one specification to work for another. Component scaling can meet new design specifications (without re-solving the design equations) or achieve realistic component values. There are two kinds of scaling: magnitude scaling and frequency scaling.

17.7.1 Magnitude Scaling

The impedances of individual R, L and C are given by $\mathbf{Z}_R = R$, $\mathbf{Z}_L = sL$ and $\mathbf{Z}_c = 1/sc$. If we make the following changes in the component values: $R \rightarrow K_m R$, $L \rightarrow K_m L$ and $C \rightarrow C/K_m$ we get new impedances defined by $\mathbf{Z}_{newR} = K_m \mathbf{Z}_R$, $\mathbf{Z}_{newL} = K_m \mathbf{Z}_L$ and $\mathbf{Z}_{newC} = K_m \mathbf{Z}_c$.

In general, the impedance of a circuit will be scaled by K_m if its components are scaled by

$$R_{new} = K_m R_{old}$$
$$L_{new} = K_m L_{old}$$
$$C_{new} = C_{old}/K_m$$

This form of scaling is called **magnitude scaling** because it scales the magnitude of the impedance by the same factor K_m at every frequency, so $|\mathbf{Z}_{new}(j\omega)| = K_m |\mathbf{Z}_{old}(j\omega)|$. This has the effect of changing the vertical scale of the graph of $|\mathbf{Z}_{old}(j\omega)|$ without changing the shape of the graph.

Magnitude scaling of impedances does not affect transfer functions that are dimensionless quantities (i.e., voltage ratios or current ratios).

Example 17.4.

If $\mathbf{Z}_1$ and $\mathbf{Z}_2$ are both scaled by K_m in Fig. 17.37, find the transfer function for the scaled circuit.

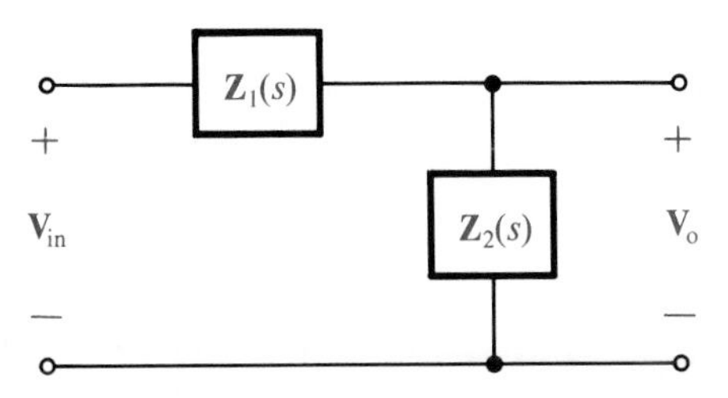

Figure 17.37 Voltage division across impedances.

Solution: The unscaled circuit has

$$\mathbf{H}(s) = \frac{\mathbf{Z}_2(s)}{\mathbf{Z}_1(s) + \mathbf{Z}_2(s)}$$

The magnitude-scaled circuit will have

$$\tilde{\mathbf{H}}(s) = \frac{K_m\mathbf{Z}_2(s)}{K_m\mathbf{Z}_1(s) + K_m\mathbf{Z}_2(s)} = \frac{\mathbf{Z}_2(s)}{\mathbf{Z}_1(s) + \mathbf{Z}_2(s)} = \mathbf{H}(s)$$

Since voltage transfer functions are not affected by magnitude scaling their components can be scaled without affecting the filter's performance characteristics.

17.7.2 Frequency Scaling

Magnitude scaling only affects the vertical scale of a graph of an impedance or admittance; it does not affect a voltage or current transfer function. Frequency scaling affects the horizontal scale of graphs of all four circuit functions. A transfer function is frequency scaled by forming $\mathbf{H}_{new}(s) = \mathbf{H}(s/K_f)$. The graph of the scaled function is either a compressed $(0 < K_f < 1)$ or stretched $(K_f > 1)$ copy of the original function. This is illustrated in Fig. 17.38 for a lowpass filter with $K > 1$. Thus, frequency scaling allows us to redesign the cutoff frequency of a filter.

In general, frequency scaling is accomplished by noting that for inductors $\mathbf{Z}_L(s) = sL$ and $\mathbf{Z}_L(s/K_f) = (s/K_f)L = s(L/K_f)$ and for capacitors $\mathbf{Z}_c(s) = 1/sC$ and $\mathbf{Z}_c(s/K_f) = K_f/sC = 1/s(C/K_f)$. Resistors do not depend on frequency. A circuit is **frequency scaled** by scaling each L and C according to

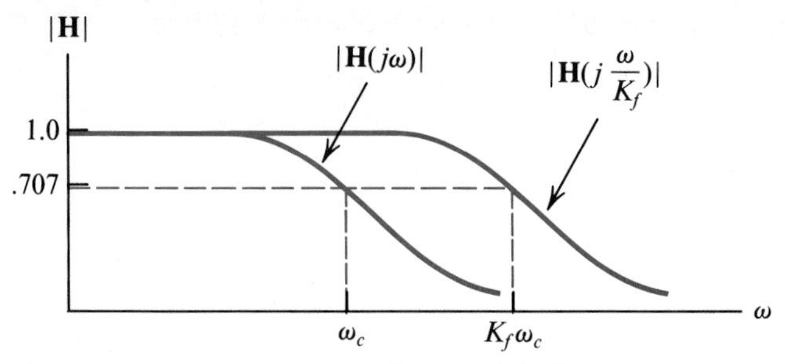

Figure 17.38 Frequency scaling with $|K_f| > 1$.

$$L_{new} = L_{old}/K_f$$
$$C_{new} = C_{old}/K_f$$
$$\omega_{new} = K_f \omega_{old}$$

Frequency scaling is equivalent to relabeling the horizontal axis of the filter's magnitude and phase characteristics.

Example 17.5

When the passive bandpass filter of Fig. 17.28 has $R_{in} = 10\ \Omega$, $R_L = 10\ \Omega$, $L = 31.8$ mH and $C = 0.795\ \mu$F its generic parameters are $\omega_b = 200\pi$, $\omega_p = 2000\pi$, $Q = 10$ and $K = 1/2$. If the circuit is frequency scaled with $K_f = 100$ then $L_{new} = 31.8/100$ mH $= 0.318$ mH, $C_{new} = 0.795/100\ \mu$F $= 7.95\ \mu$F, $\omega_{bnew} = 20{,}000\pi$, $\omega_{pnew} = 100{,}000\pi$, and $Q_{new} = (\omega_{p/\omega b})_{new} = (K_f\omega_p/K_f\omega_b)_{old} = Q_{old}$. Frequency scaling scales ω_b and ω_p by K_f but does not affect K or Q of the bandpass filter.

17.8 CASCADED FILTERS AND LOADING*

Higher-order filters can be formed by cascading sections of first- and second-order filters. If the transfer function of each filter stage in Fig. 17.39 is not affected by the stage attached to its output, the overall transfer function of the cascaded section is

$$\mathbf{H}(s) = \mathbf{H}_1(s)\mathbf{H}_2(s) \cdots \mathbf{H}_N(s) \qquad N \geq 2$$

Figure 17.39 Cascaded transfer functions.

If identical lowpass filters are cascaded, with the ith and every other stage having

$$|\mathbf{H}_i(s)| = \frac{|K|}{\sqrt{1 + (\omega/\omega_c)^2}}$$

the cascaded filter has

$$|\mathbf{H}(j\omega)| = |K|^N[1 + (\omega/\omega_c)^2]^{-N/2}$$

with DC gain of $|K|^N$. At the half-power frequency of the overall filter we have

$$|\mathbf{H}(j\tilde{\omega})| = |K|^N[1 + (\tilde{\omega}/\omega_c)^2]^{-N/2} = \frac{|K|^N}{\sqrt{2}}$$

With logarithms we can solve this equation for

$$\boxed{\tilde{\omega} = (2^{1/N} - 1)^{1/2}\omega_c}$$

The cutoff frequency of the overall filter is always below the cutoff frequency of a section and approaches 0 and N becomes large.

Example 17.6

Cascading two generic first-order lowpass filters forms a second-order filter having

$$\mathbf{H}(s) = \left(\frac{k\omega_c}{s + \omega_c}\right)^2 = \frac{K^2\omega_c{}^2}{s^2 + 2\omega_c s + \omega_c{}^2}$$

with a DC gain of K^2. The cascaded filter's cutoff frequency is given by

$$\tilde{\omega} = (2^{0.5} - 1)^{1/2}\omega_c = .644\omega_c$$

Its pole-zero pattern in Fig. 17.40 has a repeated pole on the real axis at $s = -\omega_c$. A generic second-order lowpass filter would have its poles located on the 45° wedge. Cascading two generic first-order lowpass filters has produced a non-generic second-order lowpass filter having a lower cutoff frequency than either stage.

A family of non-generic second-order lowpass filters is defined by

$$\mathbf{H}(s) = \frac{K\omega_c{}^2}{s^2 + 2\zeta\omega_c + \omega_c{}^2}$$

Figure 17.41 shows $|\mathbf{H}(j\omega)|$ for various ζ with $\omega_c = 1$ and $K = 1$. The filter of Fig. 17.40 is a special case of this filter with $\zeta = 1$, corresponding to a critically damped filter. The generic second-order lowpass filter has $\zeta =$ 0.707 and its magnitude response has a sharper cutoff for the same ω_c without having a peak at ω_c.

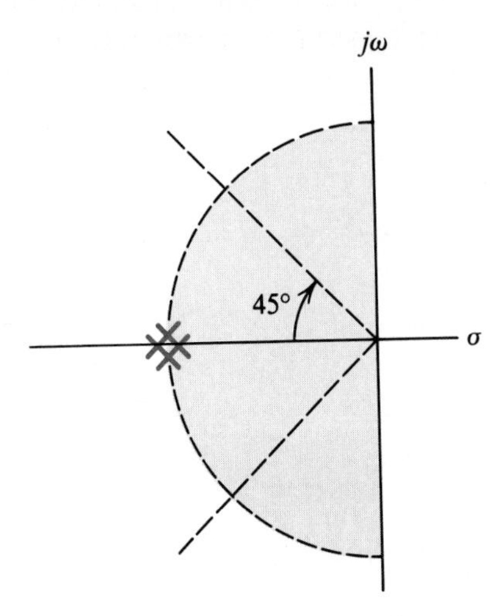

Figure 17.40 Non-generic lowpass filter pole-zero pattern.

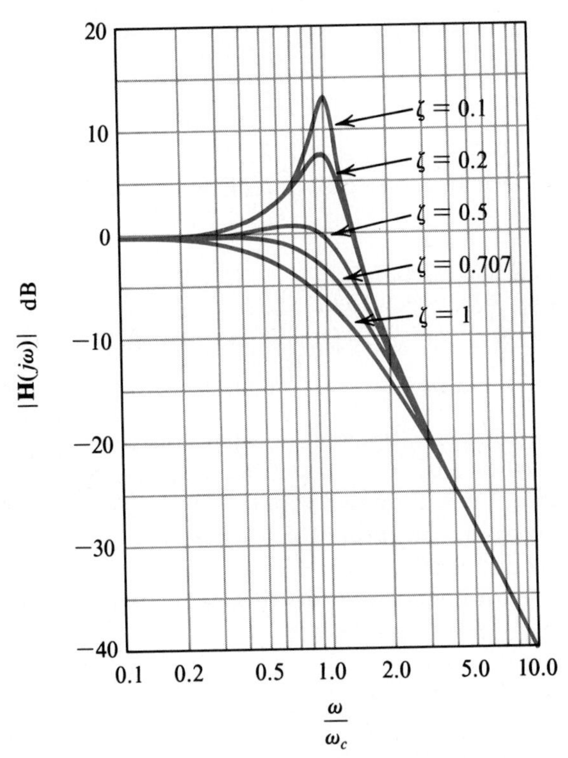

Figure 17.41 Second-order bandpass filter characteristics for various damping ratios.

When passive filters are cascaded a given stage will be affected by the addition of successive stages, unless op amps are used as buffers.

SUMMARY

A filter will pass a signal without distortion if the significant part of the signal's frequency spectrum lies in a region where the filter's magnitude response is flat and its phase response is linear. The frequency-response characteristics of lowpass, highpass, bandpass, and stopband filters were defined, and methods given to design their physical implementations. Filter design constraints specify the filter's performance in terms of its component values. Conversely, filter design equations specify component values that realize selected performance characteristics, such as bandwidth, gain, and the location of the half-power frequencies.

Bode plots of a filter's magnitude and phase responses are helpful when the band of frequencies of interest extends over several orders of magnitude. First-order lowpass filters are characterized by 20 dB/decade attenuation of sinusoidal signal amplitudes.

A filter's frequency and time-domain responses are related to the shape of its pole-zero pattern. The shape of the pole-zero pattern is related to the values of the circuit's components. Filters can be designed by selecting component values that lead to a pole-zero pattern with a suitable frequency response.

Problems - Chapter 17

17.1 Verify that the filter in Fig. P17.1 realizes a generic first-order lowpass filter and explain the physical basis for the shape of its magnitude response.

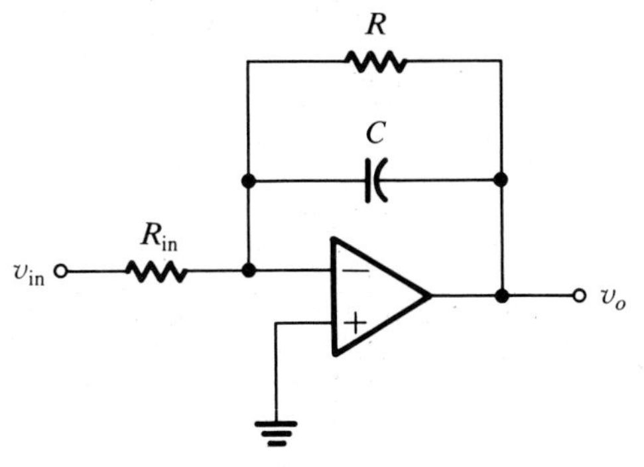

Figure P17.1

17.2 Design an active second-order lowpass filter with a DC gain of -2 and a cutoff frequency of 5 kHz, using the circuit in Fig. 17.14 with $R_1 = 15$ kΩ and $R_2 = 5\ R_3$.

17.3 If $f(t)$ is the periodic input signal to a first-order lowpass filter having DC gain $K = 3$ and cutoff frequency 10 Hz, find the approximate steady-state output signal if (a) $T = 1$ msec and (b) $T = 1$ nsec.

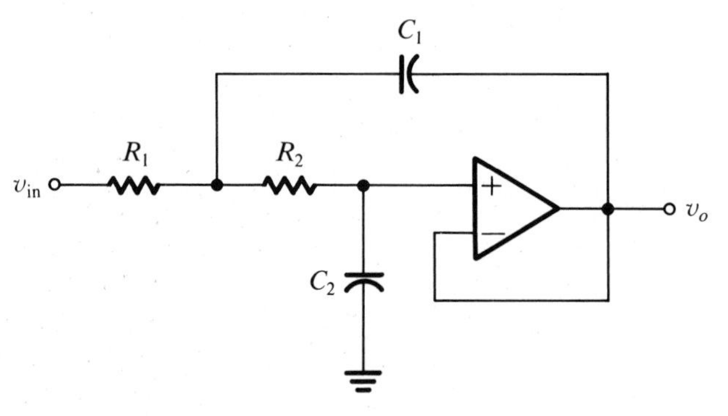

Figure P17.3

17.4 Find the DC gain and cutoff frequency of the lowpass filter described by

$$\mathbf{H}(s) = \frac{3}{6s + 2}$$

Draw the filter's Bode plot, including its asymptotes.

17.5 For the circuit in Fig. P17.5:

a. Give an explanation for the high- and low-frequency response of the filter.

b. Determine the constraint equations for K and ω_c.

c. Assuming that $R_1 = R_2 = R$, find the design equations that specify C_1 and C_2 for a given R, K and ω_c.

d. Design the filter to have a cutoff frequency of 5000π rad/sec if $R_1 = R_2 = 100\ \Omega$.

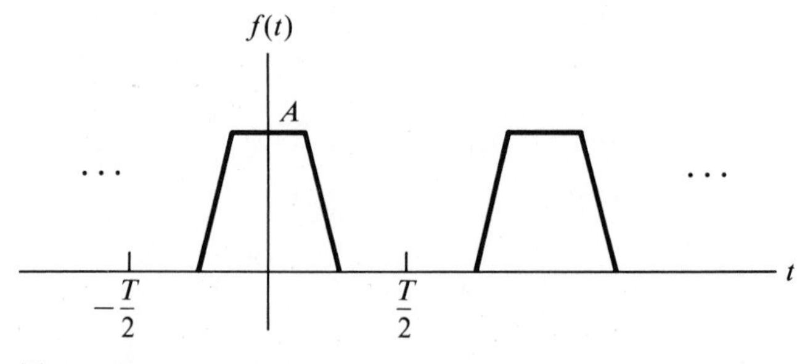

Figure P17.5

17.6 Find the input/output transfer function of the circuit in Fig. P17.6 with and without the buffer. Draw the pole-zero pattern and sketch the magnitude response of each filter. Compare them to the pole-zero pattern and magnitude response of a generic second-order filter having the same DC gain and cutoff frequency.

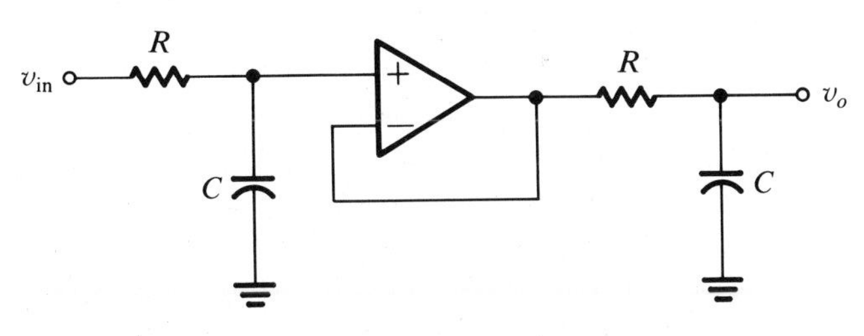

Figure P17.6

17.7 If $R_L = 20\ \text{k}\Omega$ and $C = 1\ \mu\text{F}$ in Fig. 17.4, find the range of values for R_{in} that specifies voltage sources for which the filter behaves as a lowpass filter (with cutoff frequency f_c such that $50\ \text{kHz} \leq f_c \leq 60\ \text{kHz}$).

17.8 A lowpass filter is described by

$$\mathbf{H}(s) = \frac{3 \times 10^{10}}{2s^2 + 6 \times 10^5 s + 18 \times 10^{10}}$$

a. Calculate the DC gain.
b. Calculate the bandwidth.
c. Draw the pole-zero pattern.

17.9 The pole-zero pattern of a second-order filter has poles at $s = -2\pi \times 10^5 \pm j\sqrt{2}\pi \times 10^5)$. The filter has a steady-state output of 2 V when the input is a step of height 8 V.

a. Determine the filter's cutoff frequency.
b. Calculate the average power in the filter's steady-state output signal when $v_{in}(t) = 10 \sin(2\pi \times 10^5 t + 45°)$.

17.10 If $v_{in}(t) = 10 \cos(2\pi \times 10^6 t)$ is the input signal to a filter whose phase characteristic has $\theta(j2\pi \times 10^6) = -2°$, find the time shift of the steady-state output signal relative to $v_{in}(t)$.

17.11 A first-order lowpass filter has been designed with a cutoff frequency of 22 kHz for use with a source having $R_{in} = 1000\ \Omega$ and a load having $R_L = 2000\ \Omega$. If temperature variations cause R_{in} and R_L to decrease by 5% of their values at the nominal design temperature, calculate the percentage change in the cutoff frequency of the filter.

17.12 The filter in Fig. P17.12 must completely attenuate a DC signal, and must attenuate a signal at 1000 Hz by 3 dB compared to its value at 10^8 Hz. It is also required that the steady-state output be $v_o(t) = 15 \sin(2\pi \times 10^6 t + 180°)$ when the input is $v_{in}(t) = 3 \sin(2\pi \times 10^6 t)$. Design the filter.

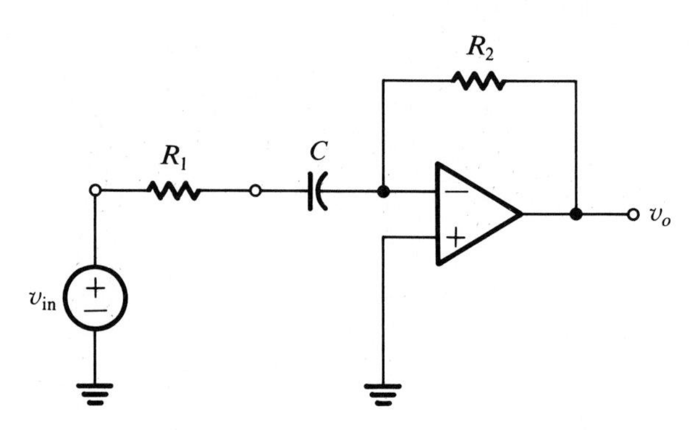

Figure P17.12

17.13 If the transfer function of a generic second-order highpass filter is

$$\mathbf{H}(s) = \frac{Ks^2}{s^2 + s\sqrt{2}\omega_c + \omega_c^2}$$

a. What is the filter gain at DC?
b. What is the high-frequency gain of the filter?
c. Prove that the generic parameter ω_c is the half-power frequency of the filter.
d. Draw the filter's pole-zero pattern.

17.14 Verify (physically and analytically) that the filter in Fig. P17.14 realizes a generic second-order highpass filter and find the constraint equations for K and ω_c for given values of R_{in}, R_L, L and C. (See the preceding problem.) For given R_{in} and R_L obtain the design equations for L and C.

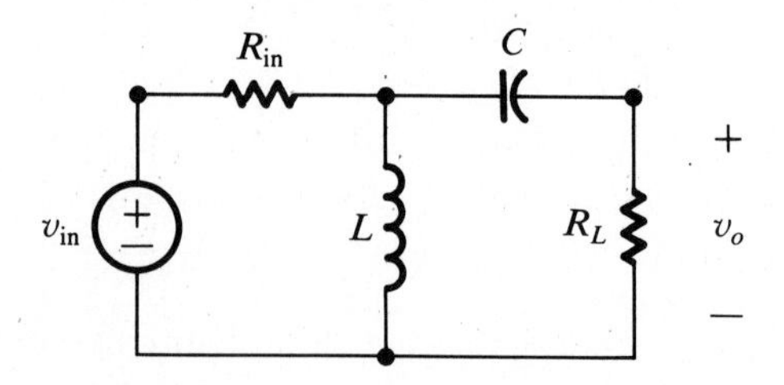

Figure P17.14

17.15 A generic second-order highpass filter (See Problem 17.13) has a cutoff frequency at 15 kHz and a high-frequency gain of 0.75.

a. Draw the pole-zero pattern of the filter.

b. Find the zero-state response of the filter to a unit-step input signal.

c. Find the steady-state response of the filter to $v_{in}(t) = 25 \sin(\pi \times 10^4 t)\, u(t)$.

17.16 Carefully draw the pole-zero patterns and graphs of $|\mathbf{H}(j\omega)|$ for the following filters:

a. $$\mathbf{H}(s) = \frac{100s}{s^2 + 100s + 10^4}$$

b. $$\mathbf{H}(s) = \frac{10s}{s^2 + 10s + 10^4}$$

c. $$\mathbf{H}(s) = \frac{s}{s^2 + s + 10^4}$$

d. Discuss the results obtained in *a*, *b* and *c*.

17.17 If a generic bandpass filter (with $Q > 1/2$) is initially at rest, find an expression for its output signal $v_o(t)$ when its input signal is given by

a. $v_{in}(t) = A \sin(\omega_p t)\, u(t)$.

b. $v_{in}(t) = A \sin(\omega_p t + \omega_b/2)\, u(t)$.

17.18 Design a parallel *RLC* bandpass filter to receive a radio signal having spectral components in a band of frequencies centered at a broadcasting station's carrier frequency of 1240 kHz. The filter must pass signal components having frequencies 10 kHz above and below the center frequency of the filter. Assume $R_{in} = R_L = 100\ \Omega$.

a. Specify the half-power frequencies.

b. Specify the peak resonant frequency of the filter.

c. Calculate Q for your design.

d. Draw the pole-zero pattern for your design.

e. Calculate the component values.

17.19 The circuit shown in Fig. P17.19 realizes a bandpass filter by cascading a generic lowpass filter with a buffered-input generic highpass filter. Obtain an expression for the location of the filter's peak (resonant) frequency, half-power frequencies, bandwidth, and center frequency.

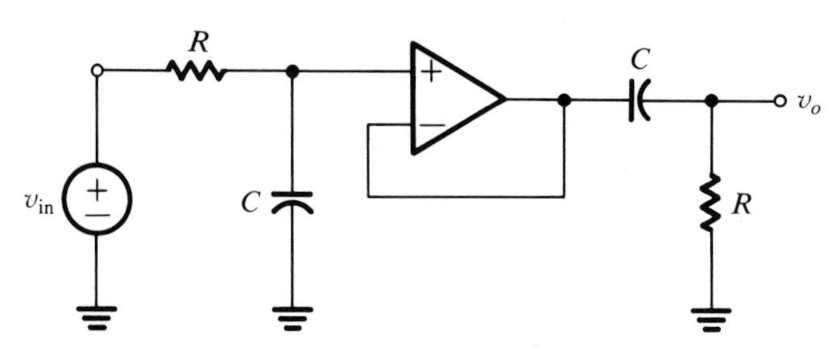

Figure P17.19

17.20 A filter has poles at $s = -100 \pm j$ and a zero at $s = 0$.
a. Find ω_p and ω_b.
b. Calculate ω_p, ω_b and Q.
c. Sketch $|\mathbf{H}(j\omega)|$ vs ω.

17.21 Find an expression for the poles of a generic bandstop filter and draw the pole-zero pattern, including the location of ω_R and ω_d.

17.22 Draw the high-Q approximation to the pole-zero pattern of a generic highpass filter.

17.23 Using the pole-zero pattern for a generic bandstop filter, develop a geometric explanation for the discontinuity in the filter's phase characteristic at the frequency ω_R.

17.24 A bandpass filter has $Q = 0.25$ and $K = -10$, with $\omega_p = 2\pi \times 10^3$ rad/sec. If its input signal is $v_{in}(t) = (1 + 0.3t)u(t)$ find $v_o(t)$.

17.25 A bandpass filter has half-power frequencies $f_1 = 5100$ Hz and $f_2 = 7100$ Hz.
a. Find the filter parameters ω_p and Q.
b. Draw the filter's pole-zero pattern.
c. Design an active realization of the filter with $K = -1$.

17.26 Calculate the approximate steady-state average power in the output signal of a notch filter if it is driven by a periodic, rectangular pulse train having an amplitude of 4 volts, pulse width of 0.25 μsec, and period of 1 μsec. The filter has a notch center frequency of $4\pi \times 10^6$ rad/sec and a Q of 100.

17.27 A notch filter has half-power frequencies at $\omega = 2500 \pm 500$ rad/sec.

a. Draw the pole-zero pattern.
b. Calculate the filter's Q.
c. Calculate the damped frequency of oscillation of the filter's zero-input response.

17.28 Design an active notch filter having $Q = 10$, $K = 5$ and $\omega_R = 2000\pi$ rad/sec. Use $R_1 = 50$ kΩ and $C_1 = 0.1\mu$F. What is the time constant of the filter's time-domain response to a step input signal?

17.29 If a voltage source with $R_{in} = 3\ \Omega$ drives a lossy parallel RLC circuit with $R = 160\ \Omega$, $L = 0.1$ mH and $C = 5$ F, compare the pole-zero patterns and plots of $|\mathbf{H}(j\omega)|$ when the inductor has $r = 0$ and $r = 2\ \Omega$.

17.30 The circuit shown in Fig. P17.30 approximately realizes a notch filter.

a. Find the design constraints that describe the location of the circuit's poles and zeros in the complex plane.
b. Draw the circuit's pole-zero pattern.
c. Find the DC and high-frequency gains of the filter.
d. Sketch the Bode plot of the filter's magnitude response function.
e. Explain why the filter cannot realize complete rejection at the frequency ω_R.

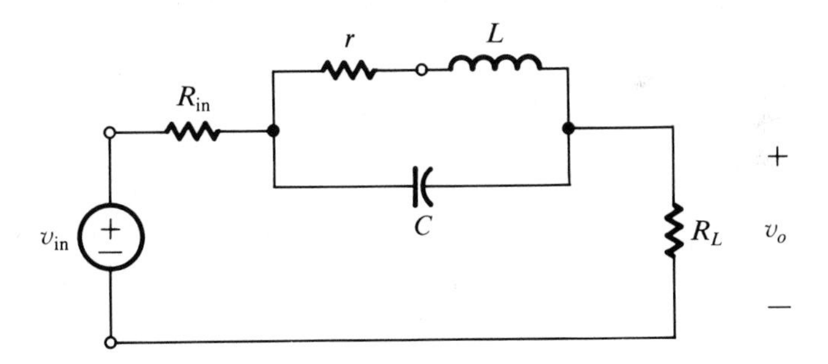

Figure P17.30

17.31 Find $\mathbf{H}(j\omega) = \mathbf{V}_o/\mathbf{V}_{in}$ and draw $|\mathbf{H}(j\omega)|$ and ang $\mathbf{H}(j\omega)$ vs ω for the circuit in Fig. P17.31. Calculate the circuit's DC and high-frequency gains.

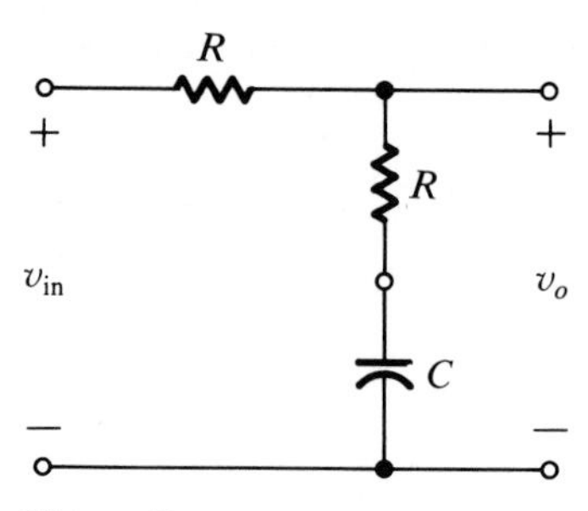

Figure P17.31

17.32 For the circuit in Fig. P17.32:

a. Find $\mathbf{H}(s) = \mathbf{V}_o(s)/\mathbf{V}_{in}(s)$.

b. Draw the pole-zero pattern

c. Draw $|\mathbf{H}(j\omega)|$ and ang $\mathbf{H}(j\omega)$ vs ω.

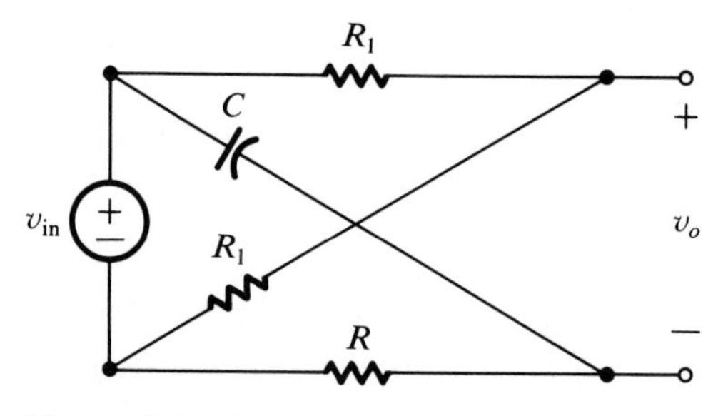

Figure P17.32

17.33 Repeat Problem 17.32 for the circuit in Fig. P17.33.

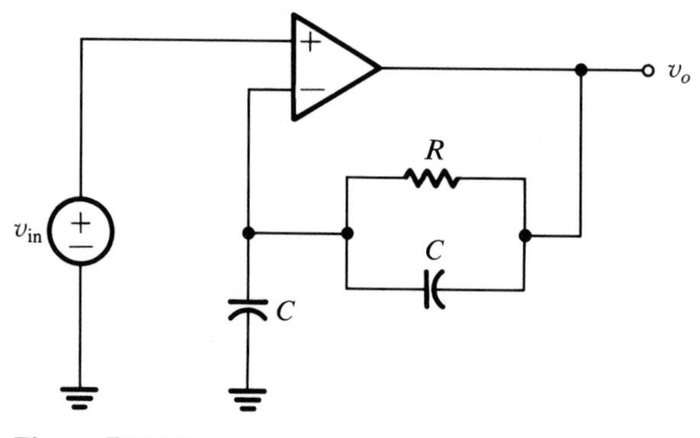

Figure P17.33

17.34 The bandpass circuit shown above in Fig. 17.34 has $R_1 = 10^{-3}\Omega$, $R_2 = 10^{-2}\Omega$, $R_3 = 5 \times 10^{-2}\Omega$, $C_1 = 10$ F and $C_2 = 40$ F.

a. Scale the components in the circuit to have $R_1 = 10$ kΩ, without changing ω_p and Q. Recalculate the values of K and ω_b and compare to their original values.

b. Rescale the filter of *a* to have $R_1 = 10$ kΩ and $\omega_p = 1180 \times 2\pi$ rad/sec, and recalculate K, Q, and ω_b of the scaled filter. List all component values.

MODELS FOR TWO PORT NETWORKS

INTRODUCTION

Thevenin and Norton models of circuits are limited to situations where a circuit has a single port, or pair of terminals, at which a source may be connected. This chapter will develop similar models for circuits that have two ports. Typically, one pair of terminals serves as an input port and the other pair serves as an output port. Two-port circuit models can be constructed from *external* voltage and/or current measurements at the circuit's driving ports, which can be very helpful when it is impossible or inconvenient to gain direct access to a circuit's internal components.

18.1 IMPEDANCE MODEL OF A CIRCUIT*

Assume that the circuit shown in Fig. 18.1 has exactly two ports, contains no internal independent sources, and has no other external connections between the ports. For the labeled currents and voltages shown, the impedance model or **z-model** of the circuit describes the relationship between the port voltages and the port currents by

$$\mathbf{V}_1(s) = \mathbf{I}_1(s)\mathbf{z}_{11}(s) + \mathbf{I}_2(s)\mathbf{z}_{12}(s) \tag{18.1a}$$

$$\mathbf{V}_2(s) = \mathbf{I}_1(s)\mathbf{z}_{21}(s) + \mathbf{I}_2(s)\mathbf{z}_{22}(s) \tag{18.1b}$$

Figure 18.1 Voltage and current polarities for a two-port circuit.

One way to visualize the model is to imagine that current sources are connected to the circuit as shown in Fig. 18.2 and that the port voltages are measured. The model expresses the fact that the port voltages measured are due to the superposition of the port voltages that would be caused by the current sources acting separately.

The impedance functions $\mathbf{z}_{11}(s)$, $\mathbf{z}_{12}(s)$, $\mathbf{z}_{21}(s)$ and $\mathbf{z}_{22}(s)$ are called the model parameters, or the **z-parameters** of the circuit. They can be obtained from external measurements by noting that

$$\mathbf{z}_{11}(s) = \left.\frac{\mathbf{V}_1(s)}{\mathbf{I}_1(s)}\right|_{\mathbf{I}_2(s) = 0}$$

$$\mathbf{z}_{12}(s) = \left.\frac{\mathbf{V}_1(s)}{\mathbf{I}_2(s)}\right|_{\mathbf{I}_1(s) = 0}$$

$$\mathbf{z}_{21}(s) = \left.\frac{\mathbf{V}_2(s)}{\mathbf{I}_1(s)}\right|_{\mathbf{I}_2(s) = 0}$$

$$\mathbf{z}_{22}(s) = \left.\frac{\mathbf{V}_2(s)}{\mathbf{I}_2(s)}\right|_{\mathbf{I}_1(s) = 0}$$

The z-parameters are also called **open-circuit** parameters because the driving-point impedance at a given port is calculated (or measured) under the condition that the other port is an open circuit, and the transfer impedance at a port is calculated under the condition that it is an open circuit (i.e., $\mathbf{z}_{12}$ is calculated with $i_1 = 0$).

When the internal configuration of a circuit is known, its z-parameters can be calculated by inspection.

Figure 18.2 A two-port circuit driven by current sources.

Example 18.1

The circuit shown in Fig. 18.3 has z-parameters

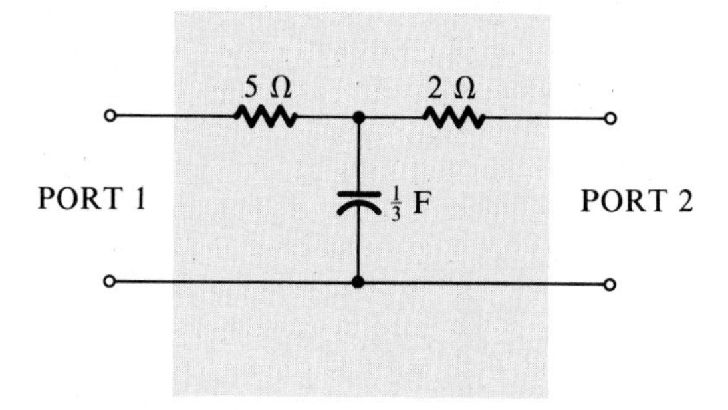

Figure 18.3

$$z_{11}(s) = 5 + \frac{3}{s} = \frac{5s + 3}{s}$$

$$z_{12}(s) = \frac{3}{s}$$

$$z_{21}(s) = \frac{3}{s}$$

$$z_{22}(s) = 2 + \frac{3}{s} = \frac{2s + 3}{s}$$

Note that if $i_1 = 0$ then v_1 is the voltage across the capacitor.

The z-parameter model can be written in matrix form as

$$\mathbf{V}(s) = \mathbf{Z}(s)\mathbf{I}(s) \tag{18.2}$$

where $\mathbf{V}(s)$ and $\mathbf{I}(s)$ are vectors and $\mathbf{Z}(s)$ is a matrix of the impedance parameters.

The impedance model specifies the port voltages $\mathbf{V}_1(s)$ and $\mathbf{V}_2(s)$ for given port currents $\mathbf{I}_1(s)$ and $\mathbf{I}_2(s)$. Alternatively, if $\mathbf{V}_1(s)$ and $\mathbf{V}_2(s)$ are known the driving-port currents can be identified from

$$\mathbf{I}(s) = \mathbf{Z}^{-1}(s)\mathbf{V}(s) \tag{18.3}$$

provided that $\mathbf{Z}(s)$ has an inverse. Once the z-parameters are known, the (ZSR) port-voltage behavior can be calculated without further reference to the internal components and their configuration.

Example 18.2

Find the ZSR of $v_1(t)$ and $v_2(t)$ in Fig. 18.3 when $i_1(t) = 5u(t)$ and $i_2(t) = 2e^{-t}u(t)$.

Solution: Since $\mathbf{I}_1(s) = 5/s$ and $\mathbf{I}_2(s) = 2/(s+1)$ we have

$$\mathbf{V}_1(s) = \frac{25}{s} + \frac{15}{s^2} + \frac{6}{s} + \frac{-6}{s+1}$$

and $v_1(t) = (31 + 15t - 6e^{-t})u(t)$. Similarly

$$\mathbf{V}_2(s) = \frac{6}{s} + \frac{15}{s^2} - \frac{2}{s+1}$$

and $v_2(t) = (6 + 15t - 2e^{-t})u(t)$. As a check on our work, note that $v_1(0^+) = 25$ and $v_2(0^+) = 4$, which agrees with the effect of $i_1(0^+)$ and $i_2(0^+)$ in the circuit.

No matter how complex a linear two-port circuit might be, the z-parameter model has the equivalent circuit shown in Fig. 18.4. Verify (18.1) using Fig. 18.4.

A two-port circuit is said to be **reciprocal** if $\mathbf{z}_{12}(s) = \mathbf{z}_{21}(s)$. Under this condition, the circuit can be represented by the generic "tee circuit" shown in Fig. 18.5. (Again, verify that (18.1) is satisfied for this circuit.) In this case the equivalent-circuit model does not require dependent sources.

When a two-port circuit is reciprocal a current source at one port may be exchanged with a voltmeter at the other port, without a change in the voltmeter's reading. So external measurements can be used to determine whether a circuit is reciprocal.

If $\mathbf{z}_{11}(s) = \mathbf{z}_{22}(s)$ a reciprocal two-port circuit is said to be **symmetric.** Symmetric, reciprocal circuits are characterized by measurement of only two, rather than four, impedance functions.

The z-parameters of a circuit are properties of the circuit itself; current sources need not actually be attached to the circuit's ports. The parameters just prescribe the ratio of v to i at the ports. In fact, the model can be used to determine the characteristics of the circuit, such as its input/output voltage gain, and its input impedance, when it has a load rather

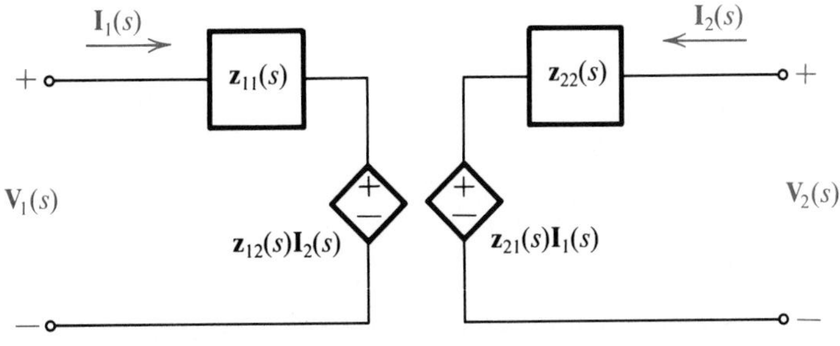

Figure 18.4 Equivalent circuit- 2 parameter model.

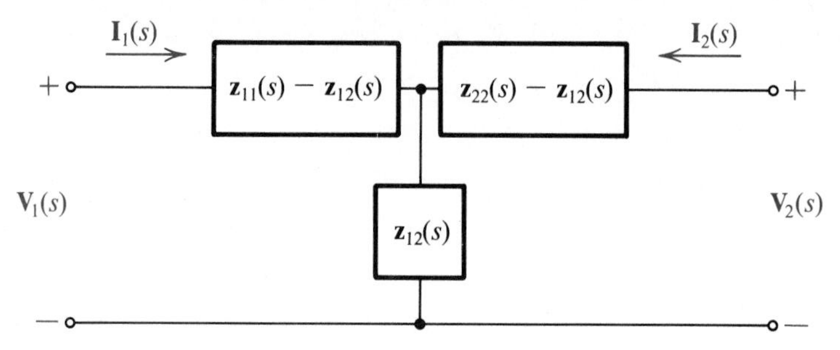

Figure 18.5 A "Tee" circuit.

than a source attached, as shown in Fig. 18.6. In this case the loaded circuit must have

$$\mathbf{I}_2(s) = -\frac{\mathbf{V}_2(s)}{\mathbf{Z}_L(s)}$$

Therefore, the z-parameter model can be simplified to

$$\mathbf{V}_1(s) = \mathbf{z}_{11}(s)\mathbf{I}_1(s) - \mathbf{z}_{12}(s)\frac{\mathbf{V}_2(s)}{\mathbf{Z}_L(s)} \tag{18.4a}$$

$$\mathbf{V}_2(s) = \mathbf{z}_{21}(s)\mathbf{I}_1(s) - \mathbf{z}_{22}(s)\frac{\mathbf{V}_2(s)}{\mathbf{Z}_L(s)} \tag{18.4b}$$

Next, to find the input/output voltage gain of the circuit we use (18.4b) to get

$$\mathbf{I}_1(s) = \frac{1}{\mathbf{z}_{21}(s)}\left(1 + \frac{\mathbf{z}_{22}(s)}{\mathbf{Z}_L(s)}\right)\mathbf{V}_2(s) \tag{18.5}$$

Substituting (18.5) into (18.4a) gives

$$\mathbf{V}_1(s) = \left[\frac{\mathbf{z}_{11}(s)}{\mathbf{z}_{21}(s)}\left(1 + \frac{\mathbf{z}_{22}(s)}{\mathbf{Z}_L(s)}\right) - \frac{\mathbf{z}_{12}(s)}{\mathbf{Z}_L(s)}\right]\mathbf{V}_2(s)$$

$$= \frac{\mathbf{z}_{11}\mathbf{Z}_L + \mathbf{z}_{11}\mathbf{z}_{22} - \mathbf{z}_{12}\mathbf{z}_{21}}{\mathbf{z}_{21}(s)\mathbf{Z}_L(s)}\mathbf{V}_2(s)$$

Therefore, the input/output voltage gain of the circuit is given by

Figure 18.6 A "loaded" two-port circuit.

$$\frac{\mathbf{V}_2(s)}{\mathbf{V}_1(s)} = \frac{\mathbf{z}_{21}(s)\mathbf{Z}_L(s)}{\mathbf{z}_{11}(s)\mathbf{Z}_L(s) + \mathbf{z}_{11}(s)\mathbf{z}_{22}(s) - \mathbf{z}_{12}(s)\mathbf{z}_{21}(s)}$$

If the circuit is unloaded ($\mathbf{Z}_L = \infty$) the voltage gain becomes

$$\left.\frac{\mathbf{V}_2(s)}{\mathbf{V}_1(s)}\right|_{\mathbf{Z}_L = \infty} = \frac{\mathbf{z}_{21}(s)}{\mathbf{z}_{11}(s)}$$

which is just a voltage divider. (What if $\mathbf{Z}_L = 0$?).

The input impedance of the loaded circuit can also be found from its z-parameters by solving (18.4b) for $\mathbf{V}_2(s)$

$$\mathbf{V}_2(s) = \frac{\mathbf{z}_{21}(s)\mathbf{Z}_L(s)}{\mathbf{z}_{22}(s) + \mathbf{Z}_L(s)}\mathbf{I}_1(s)$$

Substituting this expression into (18.4a) gives

$$\mathbf{V}_1(s) = \mathbf{z}_{11}(s)\mathbf{I}_1(s) - \frac{\mathbf{z}_{12}(s)}{\mathbf{Z}_L(s)}\frac{\mathbf{z}_{21}(s)\mathbf{Z}_L(s)}{\mathbf{z}_{22}(s) + \mathbf{Z}_L(s)}\mathbf{I}_1(s)$$

$$= \left(\mathbf{z}_{11}(s) - \frac{\mathbf{z}_{12}(s)\mathbf{z}_{21}(s)}{\mathbf{z}_{22}(s) + \mathbf{Z}_L(s)}\right)\mathbf{I}_1(s)$$

so

$$\mathbf{Z}_{in}(s) = \frac{\mathbf{V}_1(s)}{\mathbf{I}_1(s)} = \frac{\mathbf{z}_{11}(s)\mathbf{z}_{22}(s) - \mathbf{z}_{12}(s)\mathbf{z}_{21}(s) + \mathbf{z}_{11}(s)\mathbf{Z}_L(s)}{\mathbf{z}_{22}(s) + \mathbf{Z}_L(s)}$$

18.2 ADMITTANCE PARAMETERS*

The admittance-parameter model of a circuit relates its port currents to its port voltages according to

$$\mathbf{I}_1(s) = \mathbf{y}_{11}(s)\mathbf{V}_1(s) + \mathbf{y}_{12}(s)\mathbf{V}_2(s) \tag{18.6a}$$

$$\mathbf{I}_2(s) = \mathbf{y}_{21}(s)\mathbf{V}_1(s) + \mathbf{y}_{22}(s)\mathbf{V}_2(s) \tag{18.6b}$$

or in matrix format

$$\mathbf{I}(s) = \mathbf{Y}(s)\mathbf{V}(s) \tag{18.6c}$$

The admittance parameters are also called **short-circuit parameters** because they are measured with a port voltage set to zero

$$\mathbf{y}_{11}(s) = \left.\frac{\mathbf{I}_1(s)}{\mathbf{V}_1(s)}\right|_{\mathbf{V}_2(s)=0}$$

$$\mathbf{y}_{12}(s) = \left.\frac{\mathbf{I}_1(s)}{\mathbf{V}_2(s)}\right|_{\mathbf{V}_1(s)=0}$$

$$\mathbf{y}_{21}(s) = \left.\frac{\mathbf{I}_2(s)}{\mathbf{V}_1(s)}\right|_{\mathbf{V}_2(s)=0}$$

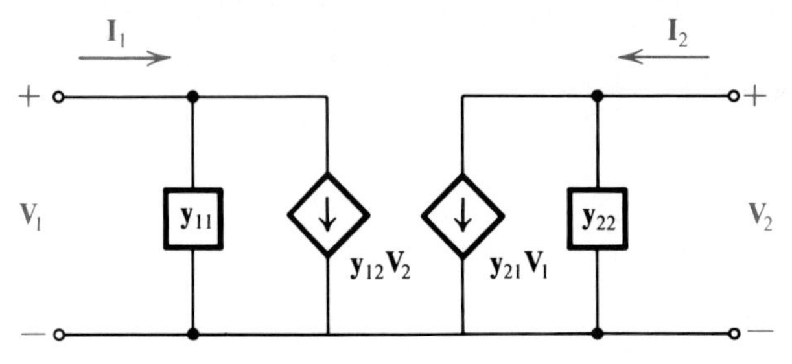

Figure 18.7 Equivalent circuit- Y parameter model.

$$\mathbf{y}_{22}(s) = \left.\frac{\mathbf{I}_2(s)}{\mathbf{V}_2(s)}\right|_{\mathbf{V}_1(s)=0}$$

The generic circuit model in Fig. 18.7 implements (18.6).

A relationship between the y- and z-parameters of a circuit can be obtained by comparing (18.3) and (18.6) to get $\mathbf{Z}^{-1} = \mathbf{Y}$ (provided that $\mathbf{Z}^{-1}$ exists). If $\mathbf{Z}^{-1}$ does not exist, then $\mathbf{Y}$ does not exist, in the sense that one or more of the y-parameters of the circuit is not defined.

Example 18.3

The resistive circuit in Fig. 18.8 has admittance parameters given by

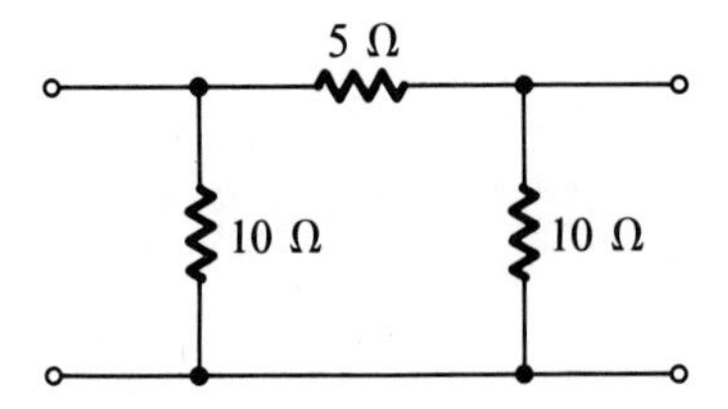

Figure 18.8 A symmetrical lattice circuit.

$$\mathbf{y}_{11} = \frac{1}{10} + \frac{1}{5} = 0.3 \text{ mhos}$$
$$\mathbf{y}_{12} = -1/5 \text{ mhos}$$
$$\mathbf{y}_{21} = -1/5 \text{ mhos}$$
$$\mathbf{y}_{22} = \frac{1}{10} + \frac{1}{5} = 0.3 \text{ mhos}$$

The circuit is symmetric and reciprocal.

Example 18.4

Find the y- and z-parameters of the circuit in Fig. 18.9.

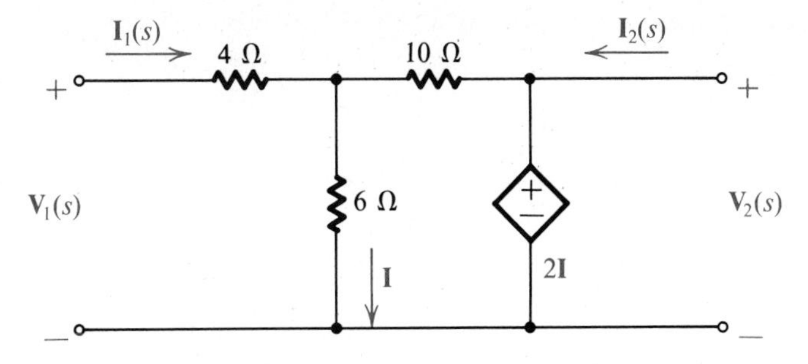

Figure 18.9

Solution: Before attempting to calculate the parameters of the circuit, we note that KVL with Laplace transforms gives

$$6\mathbf{I}(s) = 10[\mathbf{I}_1(s) - \mathbf{I}(s)] + 2\mathbf{I}(s)$$

which reduces to

$$\mathbf{I}(s) = 5/7\ \mathbf{I}_1\ (s).$$

Then, with $\mathbf{I}_2(s) = 0$

$$\mathbf{V}_1(s) = 4\mathbf{I}_1(s) + 6(5/7)\mathbf{I}_1(s) = 58/7\ \mathbf{I}_1\ (s)$$

so $\mathbf{z}_{11} = 58/7\ \Omega$. Similarly, with $\mathbf{I}_2(s) = 0$ again, we have

$$\mathbf{V}_2(s) = 2\mathbf{I}(s) = 2(5/7)\mathbf{I}_1(s)$$

so $\mathbf{z}_{21} = 10/7\ \Omega$. On the other hand, if $\mathbf{I}_1(s) = 0$ then $\mathbf{V}_2(s) = 0$ and $\mathbf{V}_1(s) = 0$ for any $\mathbf{I}_2(s)$ because the voltage source acts like a short circuit to the current $\mathbf{I}_2(s)$. That is, we must have

$$(10 + 6)\mathbf{I}(s) = 2\mathbf{I}(s)$$

so necessarily $\mathbf{I}(s) = 0$. Consequently

$$\mathbf{Z}(s) = \begin{bmatrix} 58/7 & 0 \\ 10/7 & 0 \end{bmatrix} \Omega$$

For this example $\mathbf{Z}^{-1}$ does not exist, and so $\mathbf{Y}(s)$ cannot be determined by inverting $\mathbf{Z}$. To understand the physical significance of this restriction, note that to calculate $\mathbf{y}_{11}$ we must have $\mathbf{V}_2(s) = 0$ or $\mathbf{I}(s) = 0$. This condition cannot be imposed on the circuit because $\mathbf{I}(s)$ cannot be forced to 0. Likewise, if $\mathbf{V}_1(s) = 0$ and $\mathbf{V}_2(s)$ is applied, we must have

$$\mathbf{V}_2(s) = 2\mathbf{I}(s) = 2\mathbf{V}_2(s)/31$$

which can only be satisfied by $\mathbf{V}_2(s) = 0$. We conclude that the y-parameter model of the circuit does not exist.

Skill Exercise 18.1

Find the y-parameters of the symmetrical lattice circuit shown in Fig. 18.10.

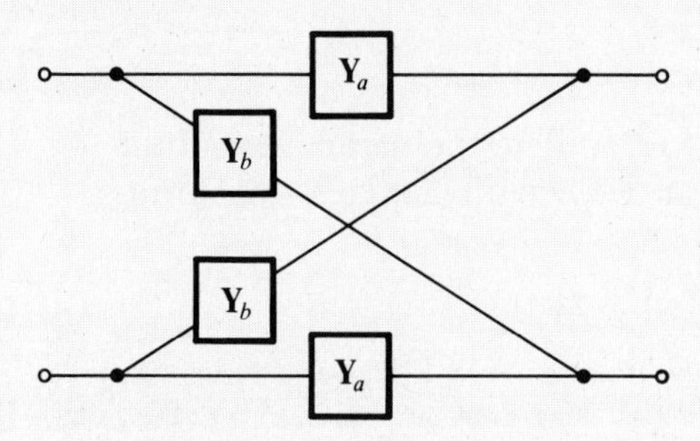

Figure 18.10

Answer: $\mathbf{y}_{11} = \mathbf{y}_{22} = 1/2(\mathbf{Y}_a + \mathbf{Y}_b)$ and $\mathbf{y}_{12} = \mathbf{y}_{21} = 1/2(\mathbf{Y}_a + \mathbf{Y}_b)$

Example 18.5

The feedback amplifier circuit in Fig. 18.11 has z-parameters but does not have y-parameters. To support this claim, we note that $\mathbf{z}_{11} = R_1$ and $\mathbf{z}_{21} = -R_2$. On the other hand $\mathbf{z}_{12} = 0$ and $\mathbf{z}_{22} = 0$ (because $i_1 = 0$ implies that $v_2 = 0$). Any current applied at the op amp output must return through the op amp's internal ground path, but not pass through R_2. Therefore, the impedance-parameter matrix is

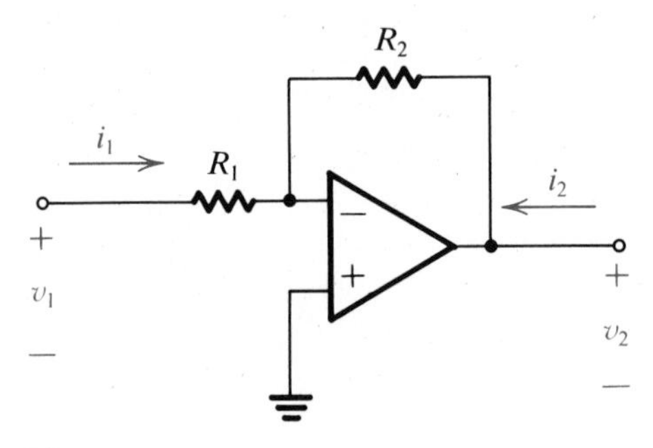

Figure 18.11

$$\mathbf{Z}(s) = \begin{bmatrix} R_1 & 0 \\ -R_2 & 0 \end{bmatrix}$$

Next, since $\mathbf{Z}^{-1}(s)$ does not exist $\mathbf{Y}(s)$ *does not exist*. This means that one or more of the components of $\mathbf{Y}(s)$ is not defined. This has physical significance. Note that we are prohibited from calculating $\mathbf{y}_{11}$ by taking the ratio $\mathbf{I}_1/\mathbf{V}_1$ with $\mathbf{V}_2 = 0$ because requiring $\mathbf{V}_2(s)$ to be zero while applying $\mathbf{V}_1(s)$ to create $\mathbf{I}_1(s)$ *violates* the restriction that an ideal voltage source may not be shorted; so $\mathbf{y}_{11}$ and $\mathbf{y}_{12}$ are not defined. Physically, such a procedure would make the op amp exceed its current rating by conducting infinite current, thereby destroying the device. If a voltage source is applied at the output port while the input port is grounded, we must have $i_1 = 0$. Otherwise, the voltage across the input terminals of the op amp itself would be nonzero. Consequently, no current may flow in R_2 and therefore $v_2 = 0$. The values of $\mathbf{y}_{21}$ and $\mathbf{y}_{22}$ are not defined either. In general, we may not calculate the two-port parameters of a circuit by creating a short across an ideal voltage source, or by opening the path of an ideal current source. To do so will lead to nonsense. For example, we might be led to mistakenly claim that $\mathbf{z}_{21} = -1/R_2$ and $\mathbf{z}_{22} = 1/R_2$ in the circuit in Fig. 18.11.

18.3 HYBRID PARAMETERS*

Another way to define a model for a two-port network is to describe how the input voltage and the output current depend on the input current and the output voltage. Since this model mixes current and voltage variables as inputs and outputs, it is called a **hybrid or *h*-parameter model.** The model equations are

$$\mathbf{V}_1(s) = \mathbf{h}_{11}(s)\mathbf{I}_1(s) + \mathbf{h}_{12}(s)\mathbf{V}_2(s) \tag{18.7a}$$

$$\mathbf{I}_2(s) = \mathbf{h}_{21}(s)\mathbf{I}_1(s) + \mathbf{h}_{22}(s)\mathbf{V}_2(s) \tag{18.7b}$$

The generic circuit shown in Fig. 18.12 implements the hybrid model of a resistive two-port network.

The external conditions that must be imposed on a circuit in order to calculate or measure its hybrid parameters follow from (18.7)(i.e., we calculate $\mathbf{h}_{11}$ and $\mathbf{h}_{21}$ with the output shorted, and we calculate $\mathbf{h}_{12}$ and $\mathbf{h}_{22}$ with the input open).

$$\mathbf{h}_{11}(s) = \left.\frac{\mathbf{V}_1(s)}{\mathbf{I}_1(s)}\right|_{\mathbf{V}_2(s)=0}$$

$$\mathbf{h}_{12}(s) = \left.\frac{\mathbf{V}_1(s)}{\mathbf{V}_2(s)}\right|_{\mathbf{I}_1(s)=0}$$

$$\mathbf{h}_{21}(s) = \left.\frac{\mathbf{I}_2(s)}{\mathbf{I}_1(s)}\right|_{\mathbf{V}_2(s)=0}$$

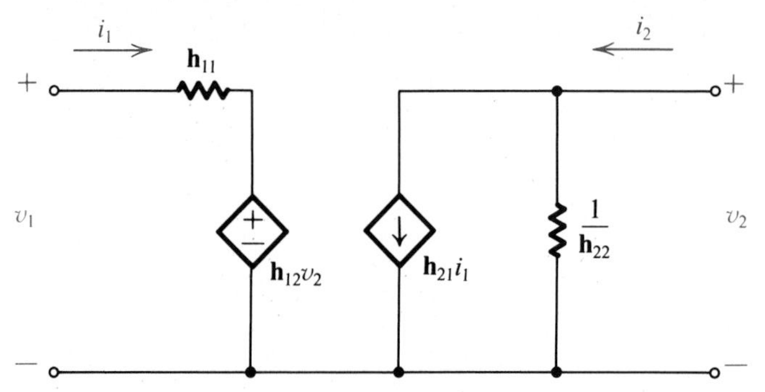

Figure 18.12 Equivalent circuit- h parameter model.

$$\mathbf{h}_{22}(s) = \left.\frac{\mathbf{I}_2(s)}{\mathbf{V}_2(s)}\right|_{\mathbf{I}_1(s)=0}$$

The parameter $\mathbf{h}_{11}$ is the input impedance of the circuit when the output is grounded, and $\mathbf{h}_{22}$ is the output admittance when the input is open. The parameters $\mathbf{h}_{12}$ and $\mathbf{h}_{21}$ have special significance since $\mathbf{h}_{12}$ is the reverse voltage gain of the circuit with the input open and $\mathbf{h}_{21}$ is the forward current gain with the output shorted.

The hybrid model is often used to describe the small-signal (linear) behavior of a transistor, and in that case the parameters are given special labels: $\mathbf{h}_{11} = \mathbf{h}_i$, $\mathbf{h}_{12} = \mathbf{h}_r$, $\mathbf{h}_{21} = \mathbf{h}_f$ and $\mathbf{h}_{22} = \mathbf{h}_o$.

Hybrid parameters are usually used to calculate the forward current gain of a two-port circuit when a load is connected at the output port. When the hybrid model is loaded by $\mathbf{Z}_L$ then $\mathbf{V}_2(s) = -\mathbf{I}_2(s)\mathbf{Z}_L$ so

$$\mathbf{I}_2(s) = \mathbf{h}_{21}(s)\mathbf{I}_1(s) - \mathbf{h}_{22}(s)\mathbf{I}_2(s)\mathbf{Z}_L(s)$$

and the ratio of the current is

$$\frac{\mathbf{I}_2(s)}{\mathbf{I}_1(s)} = \frac{\mathbf{h}_{21}(s)}{1 + \mathbf{h}_{22}(s)\mathbf{Z}_L(s)} \tag{18.8}$$

This defines the **current gain** between the input port and the load. The voltage gain will be developed next. Using (18.8) in (18.7a) to replace $\mathbf{I}_1(s)$ we write

$$\mathbf{V}_1(s) = \frac{\mathbf{h}_{11}(s)}{\mathbf{h}_{21}(s)}\left[1 + \mathbf{h}_{22}(s)\mathbf{Z}_L(s)\right]\mathbf{I}_2(s) + \mathbf{h}_{12}(s)\mathbf{V}_2(s)$$

But $\mathbf{I}_2(s) = -\mathbf{V}_2(s)/\mathbf{Z}_L$ so

$$\mathbf{V}_1(s) = \frac{-\mathbf{h}_{11}(s)}{\mathbf{h}_{21}(s)\mathbf{Z}_L(s)}\left[1 + \mathbf{h}_{22}(s)\mathbf{Z}_L(s)\right]\mathbf{V}_2(s) + \mathbf{h}_{12}(s)\mathbf{V}_2(s) \tag{18.9a}$$

$$= \frac{-\mathbf{h}_{11}(s)[1 + \mathbf{h}_{22}(s)\mathbf{Z}_L(s)] + \mathbf{h}_{12}(s)\mathbf{h}_{21}(s)\mathbf{Z}_L(s)}{\mathbf{h}_{21}(s)\mathbf{Z}_L(s)}\mathbf{V}_2(s) \tag{18.9b}$$

Rearranging (18.9b) creates the expression for the **voltage gain** under load conditions

$$\frac{\mathbf{V}_2(s)}{\mathbf{V}_1(s)} = \frac{-\mathbf{h}_{21}(s)\mathbf{Z}_L(s)}{\mathbf{Z}_L(s)[\mathbf{h}_{11}(s)\mathbf{h}_{22}(s) - \mathbf{h}_{12}(s)\mathbf{h}_{21}(s)] + \mathbf{h}_{11}(s)}$$
$$= \frac{-\mathbf{h}_f\mathbf{Z}_L}{\mathbf{Z}_L(s)[\mathbf{h}_i\mathbf{h}_o - \mathbf{h}_r\mathbf{h}_f] + \mathbf{h}_i}$$
$$= \frac{-\mathbf{h}_f}{\mathbf{h}_i\mathbf{h}_o - \mathbf{h}_r\mathbf{h}_f + \mathbf{h}_i/\mathbf{Z}_L}$$

Notice that the value of $\mathbf{h}_i$ determines how the loading by $\mathbf{Z}_L$ affects the I/O voltage gain, and that the maximum gain occurs when $\mathbf{Z}_L$ is large compared to $\mathbf{h}_i$.

Skill Exercise 18.2

Calculate the forward voltage gain of the hybrid model when the output is open circuited.

Answer:

$$\frac{\mathbf{V}_2(s)}{\mathbf{V}_1(s)} = \frac{-\mathbf{h}_{21}(s)}{\mathbf{h}_{11}(s)\mathbf{h}_{22}(s) - \mathbf{h}_{12}(s)\mathbf{h}_{21}(s)}$$
$$= \frac{-\mathbf{h}_f}{\mathbf{h}_i\mathbf{h}_o - \mathbf{h}_r\mathbf{h}_f}$$

Example 18.6

The common-emitter (bipolar) transistor amplifier shown in Fig. 18.13 has $\mathbf{h}_i = 2500\ \Omega$, $\mathbf{h}_r = 4 \times 10^{-4}$, $\mathbf{h}_f = 50$ and $\mathbf{h}_o = 10^{-5}$mhos. Calculate the forward voltage gain of the circuit with and without a resistive load of $\mathbf{R}_L = 5000\ \Omega$.

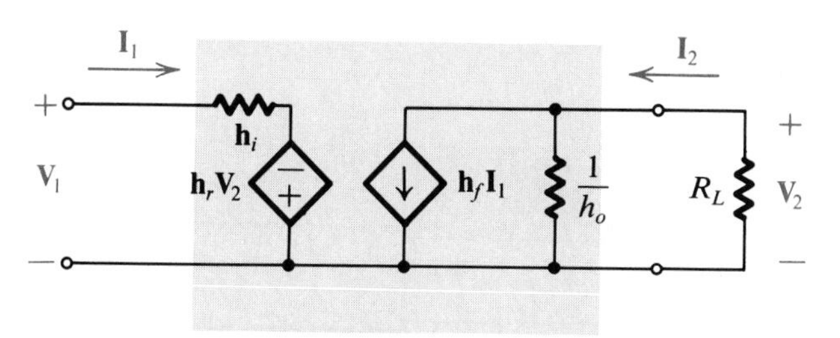

Figure 18.13 A loaded h-parameter model.

Solution: With the given parameters we have, with $R_L = 5000\ \Omega$,

$$\frac{\mathbf{V}_2(s)}{\mathbf{V}_1(s)} = \frac{-50}{2500(10^{-5}) - 4 \times 10^{-4}(5) + 2500/5000} = 99.01$$

If $\mathbf{R}_L = \infty$ then

$$\frac{\mathbf{V}_2(s)}{\mathbf{V}_1(s)} = \frac{-50}{2500(10^{-5}) - 4 \times 10^{-4}(50)} = -10000$$

Skill Exercise 18.3

If the 5000-Ω load resistor in Example 18.6 is to dissipate an average power of 160 milliwatts, what must be the effective value of $\mathbf{V}_1$ for a sinusoidal input voltage?

Answer: $\mathbf{V}_1 = 285.7$ mV rms

18.4 TRANSMISSION PARAMETERS*

Transmission parameters, or ABCD parameters, model a cascade circuit configuration in terms of its individual stage models. The ABCD model for a two-port circuit defines its input voltage and current in terms of its output voltage and current according to

$$\mathbf{V}_1(s) = \mathbf{A}(s)\mathbf{V}_2(s) - \mathbf{B}(s)\mathbf{I}_2(s) \tag{18.10a}$$

$$\mathbf{I}_1(s) = \mathbf{C}(s)\mathbf{V}_2(s) - \mathbf{D}(s)\mathbf{I}_2(s) \tag{18.10b}$$

In matrix form

$$\begin{bmatrix} \mathbf{V}_1(s) \\ \mathbf{I}_1(s) \end{bmatrix} = \begin{bmatrix} \mathbf{A}(s) & \mathbf{B}(s) \\ \mathbf{C}(s) & \mathbf{D}(s) \end{bmatrix} \begin{bmatrix} \mathbf{V}_2(s) \\ -\mathbf{I}_2(s) \end{bmatrix} \tag{18.11}$$

The model parameters are obtained from

$$\mathbf{A}(s) = \left.\frac{\mathbf{V}_1(s)}{\mathbf{V}_2(s)}\right|_{\mathbf{I}_2(s)=0}$$

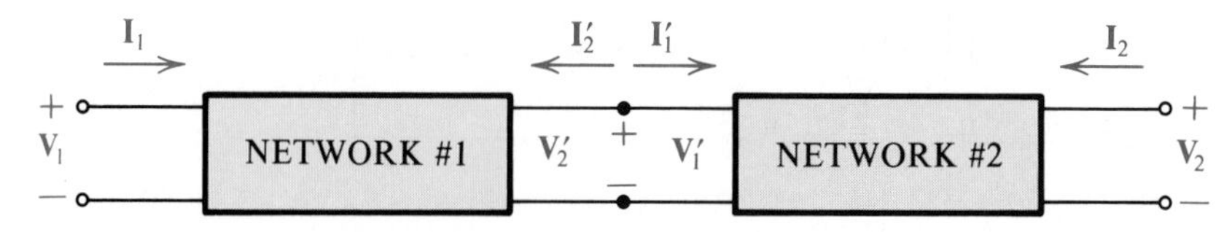

Figure 18.14 A pair of cascaded two-part circuits.

$$-\mathbf{B}(s) = \left.\frac{\mathbf{V}_1(s)}{\mathbf{I}_2(s)}\right|_{\mathbf{V}_2(s)=0}$$

$$\mathbf{C}(s) = \left.\frac{\mathbf{I}_1(s)}{\mathbf{V}_2(s)}\right|_{\mathbf{I}_2(s)=0}$$

$$-\mathbf{D}(s) = \left.\frac{\mathbf{I}_1(s)}{\mathbf{I}_2(s)}\right|_{\mathbf{V}_2(s)=0}$$

A circuit's ABCD parameters can also be obtained from its impedance parameters, since (18.1b) implies that

$$\mathbf{I}_1(s) = \frac{\mathbf{V}_2(s) - \mathbf{I}_2(s)\mathbf{z}_{22}(s)}{\mathbf{z}_{21}(s)} \tag{18.12a}$$

$$= \frac{1}{\mathbf{z}_{21}(s)}\mathbf{V}_2(s) - \frac{\mathbf{z}_{22}(s)}{\mathbf{z}_{21}(s)}\mathbf{I}_2(s) \tag{18.12b}$$

Using the expression in (18.1a) gives

$$\mathbf{V}_1(s) = \frac{\mathbf{z}_{11}(s)}{\mathbf{z}_{21}(s)}\mathbf{V}_2(s) - \left[-\mathbf{z}_{12}(s) + \frac{\mathbf{z}_{22}(s)}{\mathbf{z}_{21}(s)}\mathbf{z}_{11}(s)\right]\mathbf{I}_2(s) \tag{18.13}$$

Comparing (18.12b) and (18.13) to (18.10), we conclude that the ABCD parameters of the two-port network are defined by

$$\mathbf{A}(s) = \frac{\mathbf{z}_{11}(s)}{\mathbf{z}_{21}(s)}$$

$$\mathbf{B}(s) = \frac{\mathbf{z}_{11}(s)\mathbf{z}_{22}(s) - \mathbf{z}_{12}(s)\mathbf{z}_{21}(s)}{\mathbf{z}_{21}(s)}$$

$$\mathbf{C}(s) = \frac{1}{\mathbf{z}_{21}(s)}$$

$$\mathbf{D}(s) = \frac{\mathbf{z}_{22}(s)}{\mathbf{z}_{21}(s)}$$

provided that $\mathbf{z}_{21} \neq 0$.

The ABCD model is distinguished by the fact that it defines the values of the input variables in terms of the values of the output variables. To find the model parameters of the two cascaded stages in Fig. 18.14 note that the first stage has

$$\begin{bmatrix}\mathbf{V}_1(s)\\ \mathbf{I}_1(s)\end{bmatrix} = \begin{bmatrix}\mathbf{A}_1 & \mathbf{B}_1\\ \mathbf{C}_1 & \mathbf{D}_1\end{bmatrix}\begin{bmatrix}\mathbf{V}_2'(s)\\ -\mathbf{I}_2'(s)\end{bmatrix}$$

and for the second stage

$$\begin{bmatrix}\mathbf{V}_1'(s)\\ \mathbf{I}_1'(s)\end{bmatrix} = \begin{bmatrix}\mathbf{A}_2 & \mathbf{B}_2\\ \mathbf{C}_2 & \mathbf{D}_2\end{bmatrix}\begin{bmatrix}\mathbf{V}_2(s)\\ -\mathbf{I}_2(s)\end{bmatrix}$$

The cascade combination requires that $\mathbf{V}_1'(s) = \mathbf{V}_2'(s)$ and $\mathbf{I}_1'(s) = -\mathbf{I}_2'(s)$. Therefore

$$\begin{bmatrix} \mathbf{V}_1(s) \\ \mathbf{I}_1(s) \end{bmatrix} = \begin{bmatrix} \mathbf{A}_1 & \mathbf{B}_1 \\ \mathbf{C}_1 & \mathbf{D}_1 \end{bmatrix} \begin{bmatrix} \mathbf{A}_2 & \mathbf{B}_2 \\ \mathbf{C}_2 & \mathbf{D}_2 \end{bmatrix} \begin{bmatrix} \mathbf{V}_2(s) \\ -\mathbf{I}_2(s) \end{bmatrix}$$
$$= \begin{bmatrix} \mathbf{A} & \mathbf{B} \\ \mathbf{C} & \mathbf{D} \end{bmatrix} \begin{bmatrix} \mathbf{V}_2(s) \\ -\mathbf{I}_1(s) \end{bmatrix}$$

The transmission-parameter matrix of the cascaded network is equal to the product of the transmission matrices of the individual circuits, or

$$\begin{bmatrix} \mathbf{A} & \mathbf{B} \\ \mathbf{C} & \mathbf{D} \end{bmatrix} = \begin{bmatrix} \mathbf{A}_1 & \mathbf{B}_1 \\ \mathbf{C}_1 & \mathbf{D}_1 \end{bmatrix} \begin{bmatrix} \mathbf{A}_2 & \mathbf{B}_2 \\ \mathbf{C}_2 & \mathbf{D}_2 \end{bmatrix}$$

If the input voltage and current of a two-port network are known, the output voltage and current can be obtained by solving (18.12)

$$\begin{bmatrix} \mathbf{V}_2(s) \\ -\mathbf{I}_2(s) \end{bmatrix} = \begin{bmatrix} \mathbf{A}(s) & \mathbf{B}(s) \\ \mathbf{C}(s) & \mathbf{D}(s) \end{bmatrix}^{-1} \begin{bmatrix} \mathbf{V}_1(s) \\ \mathbf{I}_1(s) \end{bmatrix}$$

provided the inverse of the tranmission matrix exists. This result automatically includes the effect of the second stage loading the first stage. When a two-port network is loaded by $\mathbf{Z}_L$ the output current must satisfy the (Ohm's Law) constraint that $\mathbf{I}_2(s) = -\mathbf{V}_2(s)/\mathbf{Z}_L$ so

$$\mathbf{V}_1(s) = \mathbf{A}(s)\mathbf{V}_2(s) + \mathbf{B}(s)/\mathbf{Z}_L(s)\mathbf{V}_2(s)$$
$$= [\mathbf{A}(s) + \mathbf{B}(s)/\mathbf{Z}_L(s)]\mathbf{V}_2(s).$$

and the *loaded* voltage gain is

$$\frac{\mathbf{V}_2(s)}{\mathbf{V}_1(s)} = \frac{1}{\mathbf{A}(s) + \mathbf{B}(s)/\mathbf{Z}_L(s)} \tag{18.14}$$

Also, since

$$\mathbf{I}_1(s) = -\mathbf{I}_2(s)\mathbf{C}(s)\mathbf{Z}_L(s) - \mathbf{D}(s)\mathbf{I}_2(s)$$

the **forward current gain** is

$$\frac{\mathbf{I}_2(s)}{\mathbf{I}_1(s)} = \frac{-1}{\mathbf{C}(s)\mathbf{Z}_L(s) + \mathbf{D}(s)}$$

Note that if $\mathbf{Z}_L(s) = \infty$ then $\mathbf{I}_2(s) = 0$ for any $\mathbf{I}_1(s)$.

The impedance seen at the terminals of the input port of a loaded circuit can be obtained from its hybrid parameters according to

$$\frac{\mathbf{V}_1(s)}{\mathbf{I}_1(s)} = \frac{\mathbf{V}_1(s)}{\mathbf{V}_2(s)} \frac{\mathbf{V}_2(s)}{\mathbf{I}_1(s)}$$
$$= \frac{\mathbf{V}_1(s)}{\mathbf{V}_2(s)} \frac{[-\mathbf{I}_2(s)\mathbf{Z}_L(s)]}{\mathbf{I}_1(s)}$$
$$= \left[\mathbf{A}(s) + \frac{\mathbf{B}(s)}{\mathbf{Z}_L(s)}\right] \frac{\mathbf{Z}_L(s)}{\mathbf{C}(s)\mathbf{Z}_L(s) + \mathbf{D}(s)}$$

and

$$\mathbf{Z}_{\text{in}}(s) = \frac{\mathbf{V}(s)}{\mathbf{I}_1(s)} = \frac{\mathbf{A}(s)\mathbf{Z}_L(s) + \mathbf{B}(s)}{\mathbf{C}(s)\mathbf{Z}_L(s) + \mathbf{D}(s)}$$

Example 18.7

Derive the transmission parameters of the voltage-divider circuit in Fig. 18.15.

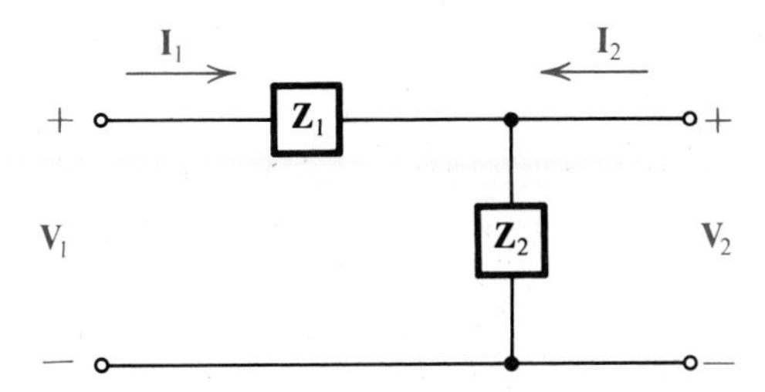

Figure 18.15 A voltage divider.

Solution: First, with the output circuit open

$$\mathbf{V}_2(s) = \frac{\mathbf{Z}_2}{\mathbf{Z}_1 + \mathbf{Z}_2}\mathbf{V}_1(s)$$

so

$$\mathbf{A}(s) = \left.\frac{\mathbf{V}_1(s)}{\mathbf{V}_2(s)}\right|_{\mathbf{I}_2=0} = 1 + \frac{1}{\mathbf{Z}_1} = 1 + \mathbf{Y}_1(s)$$

Likewise $\mathbf{V}_2(s) = \mathbf{I}_1(s)\mathbf{Z}_2(s)$ so

$$\mathbf{C}(s) = \left.\frac{\mathbf{I}_1(s)}{\mathbf{V}_2(s)}\right|_{\mathbf{I}_2=0} = \frac{1}{\mathbf{Z}_2(s)} = \mathbf{Y}_2(s)$$

Then we create the circuit shown in Fig. 18.16 by forcing $\mathbf{V}_2(s)$ to zero to give

$$\mathbf{I}_2(s) = -\mathbf{I}_1(s) = -\frac{\mathbf{V}_1(s)}{\mathbf{Z}_1(s)}$$

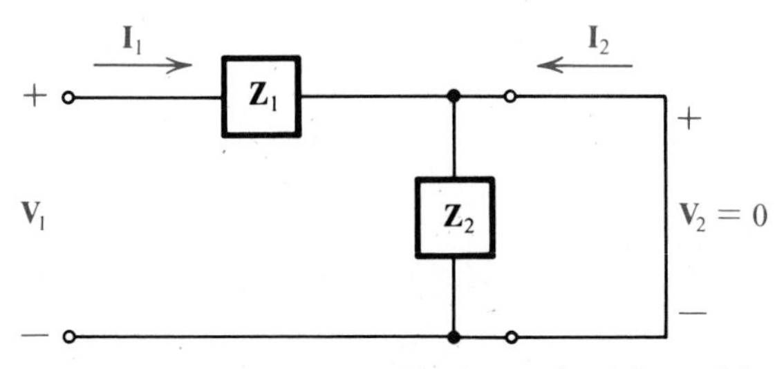

Figure 18.16 A voltage divider with a shorted load.

and so

$$\mathbf{B}(s) = -\left.\frac{\mathbf{V}_1(s)}{\mathbf{I}_2(s)}\right|_{\mathbf{V}_2(s)=0} = \mathbf{Z}_1(s)$$

The value of $\mathbf{D}(s)$ is obtained from

$$\mathbf{D}(s) = \left.\frac{\mathbf{I}_1(s)}{\mathbf{I}_2(s)}\right|_{\mathbf{V}_2=0} = 1$$

Therefore, the transmission-parameter matrix of the voltage divider is

$$\begin{bmatrix} \mathbf{A} & \mathbf{B} \\ \mathbf{C} & \mathbf{D} \end{bmatrix} = \begin{bmatrix} 1 + \mathbf{Y}_1(s) & \mathbf{Z}_1(s) \\ \mathbf{Y}_2(s) & 1 \end{bmatrix}$$

where $\mathbf{Y}_1 = 1/\mathbf{Z}_1$ and $\mathbf{Y}_2 = 1/\mathbf{Z}_2$.

Example 18.8

If two RC circuits are cascaded together without a buffer circuit separating them the input impedance of the combination is changed from that of a single stage. Using the results of Example 18.7 with the RC voltage divider in Fig. 18.17, we have

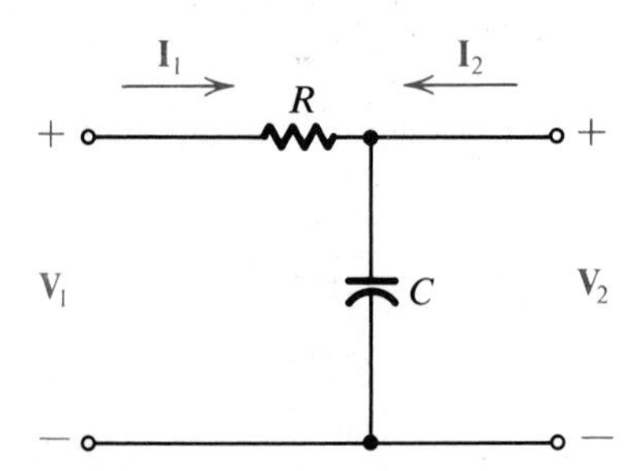

Figure 18.17

$$\begin{bmatrix} \mathbf{A} & \mathbf{B} \\ \mathbf{C} & \mathbf{D} \end{bmatrix} = \begin{bmatrix} (1 + sRC) & R \\ sC & 1 \end{bmatrix}$$

If two stages are cascaded together and connected to R_L the resulting circuit has

$$\begin{bmatrix} \mathbf{A} & \mathbf{B} \\ \mathbf{C} & \mathbf{D} \end{bmatrix} = \begin{bmatrix} (1 + sRC) & R \\ sC & 1 \end{bmatrix} \begin{bmatrix} (1 + sRC) & R \\ sC & 1 \end{bmatrix}$$

$$= \begin{bmatrix} (1 + sRC)^2 + sRC & sR^2C + 2R \\ sC(2 + sRC) & sRC + 1 \end{bmatrix}$$

The voltage gain [see (18.15)] of a resistive, loaded, two-stage RC circuit is

$$\frac{\mathbf{V}_2(s)}{\mathbf{V}_1(s)} = \frac{1}{(1 + sRC)^2 + sRC + (sR^2C + 2R)/R_L}$$

$$= \frac{1}{s^2R^2C^2 + s\left(3RC + \frac{R^2C}{R_L}\right) + \frac{2R}{R_L} + 1}$$

$$= \frac{1}{R^2C^2\left[s^2 + s\left(\frac{3}{RC} + \frac{1}{R_LC}\right) + \frac{2R + R_L}{R^2R_LC^2}\right]}$$

If the circuit is open (i.e., $R_L = \infty$)

$$\frac{\mathbf{V}_2(s)}{\mathbf{V}_1(s)} = \frac{1}{R^2C^2\left[s^2 + s\left(\frac{3}{RC}\right) + \frac{1}{R^2C^2}\right]}$$

Skill Exercise 18.4

Find the forward current gain of the loaded, cascaded RC network in Example 18.8.

Answer:

$$\frac{\mathbf{I}_2(s)}{\mathbf{I}_1(s)} = \frac{1}{RR_LC^2\left[s^2 + \frac{s(2R_L + R)}{RR_LC} + \frac{1}{RR_LC^2}\right]}$$

We have already seen how the transmission parameters and y-parameters of a two-port network can be derived from its z-parameters. In general, any of the parameter sets can be derived in terms of either of the others. The relationships are shown in Table 18.1.

TABLE 18.1 TWO-PORT PARAMETER CONVERSION TABLE

$$\mathbf{z}_{11} = \frac{\mathbf{y}_{22}}{\Delta\mathbf{y}} = \frac{\mathbf{A}}{\mathbf{C}} = \frac{\Delta\mathbf{h}}{\mathbf{h}_{22}} \qquad \mathbf{A} = \frac{\mathbf{z}_{11}}{\mathbf{z}_{21}} = \frac{-\mathbf{y}_{22}}{\mathbf{y}_{21}} = \frac{-\Delta\mathbf{h}}{\mathbf{h}_{21}}$$

$$\mathbf{z}_{12} = \frac{-\mathbf{y}_{12}}{\Delta\mathbf{y}} = \frac{\Delta}{\mathbf{C}} = \frac{\mathbf{h}_{12}}{\mathbf{h}_{22}} \qquad \mathbf{B} = \frac{-\Delta\mathbf{z}}{\mathbf{z}_{21}} = \frac{-1}{\mathbf{y}_{21}} = \frac{-\mathbf{h}_{11}}{\mathbf{h}_{21}}$$

$$\mathbf{z}_{21} = \frac{-\mathbf{y}_{21}}{\Delta\mathbf{y}} = \frac{1}{\mathbf{C}} = \frac{-\mathbf{h}_{21}}{\mathbf{h}_{22}} \qquad \mathbf{C} = \frac{1}{\mathbf{z}_{21}} = \frac{\Delta\mathbf{y}}{\mathbf{y}_{21}} = \frac{-\mathbf{h}_{22}}{\mathbf{h}_{21}}$$

$$\mathbf{z}_{22} = \frac{\mathbf{y}_{11}}{\Delta\mathbf{y}} = \frac{\mathbf{D}}{\mathbf{C}} = \frac{1}{\mathbf{h}_{22}} \qquad \mathbf{D} = \frac{\mathbf{z}_{22}}{\mathbf{z}_{21}} = \frac{-\mathbf{y}_{11}}{\mathbf{y}_{21}} = \frac{-1}{\mathbf{h}_{21}}$$

$$\mathbf{y}_{11} = \frac{\mathbf{z}_{22}}{\Delta\mathbf{z}} = \frac{\mathbf{D}}{\mathbf{B}} = \frac{1}{\mathbf{h}_{11}} \qquad \mathbf{h}_{11} = \frac{\Delta\mathbf{z}}{\mathbf{z}_{22}} = \frac{1}{\mathbf{y}_{11}} = \frac{\mathbf{B}}{\mathbf{D}}$$

$$\mathbf{y}_{12} = \frac{-\mathbf{z}_{12}}{\Delta\mathbf{z}} = \frac{-\Delta}{\mathbf{B}} = \frac{-\mathbf{h}_{12}}{\mathbf{h}_{11}} \qquad \mathbf{h}_{12} = \frac{\mathbf{z}_{12}}{\mathbf{z}_{22}} = \frac{-\mathbf{y}_{12}}{\mathbf{y}_{11}} = \frac{\Delta}{\mathbf{D}}$$

$$\mathbf{y}_{21} = \frac{-\mathbf{z}_{21}}{\Delta\mathbf{z}} = \frac{-1}{\mathbf{B}} = \frac{\mathbf{h}_{21}}{\mathbf{h}_{11}} \qquad \mathbf{h}_{21} = \frac{-\mathbf{z}_{21}}{\mathbf{z}_{22}} = \frac{\mathbf{y}_{21}}{\mathbf{y}_{11}} = \frac{-1}{\mathbf{D}}$$

TABLE 18.1 Continued

$\mathbf{y}_{22} = \dfrac{\mathbf{z}_{11}}{\Delta\mathbf{z}} = \dfrac{\mathbf{A}}{\mathbf{B}} = \dfrac{\Delta\mathbf{h}}{\mathbf{h}_{11}}$	$\mathbf{h}_{22} = \dfrac{1}{\mathbf{z}_{22}} = \dfrac{\Delta\mathbf{y}}{\mathbf{y}_{11}} = \dfrac{\mathbf{C}}{\mathbf{D}}$
$\Delta\mathbf{z} = \mathbf{z}_{11}\mathbf{z}_{22} - \mathbf{z}_{12}\mathbf{z}_{21}$	$\Delta\mathbf{h} = \mathbf{h}_{11}\mathbf{h}_{22} - \mathbf{h}_{12}\mathbf{h}_{21}$
$\Delta\mathbf{y} = \mathbf{y}_{11}\mathbf{y}_{22} - \mathbf{y}_{12}\mathbf{y}_{21}$	$\Delta = \mathbf{AD} - \mathbf{BC}$

SUMMARY

Two-port parameter models of circuits simplify circuit analysis by summarizing the internal detail of a circuit with parameters that describe its behavior at the input and output ports. Impedance parameters describe port voltages in terms of port currents, admittance parameters describe port currents in terms of port voltages, hybrid parameters describe the input voltage and output current in terms of the input current and output voltage, and transmission parameters describe how the input current and voltage depend on the output current and voltage.

Problems - Chapter 18

18.1 Find the z- and y-parameters of the circuit shown in Fig. P18.1.

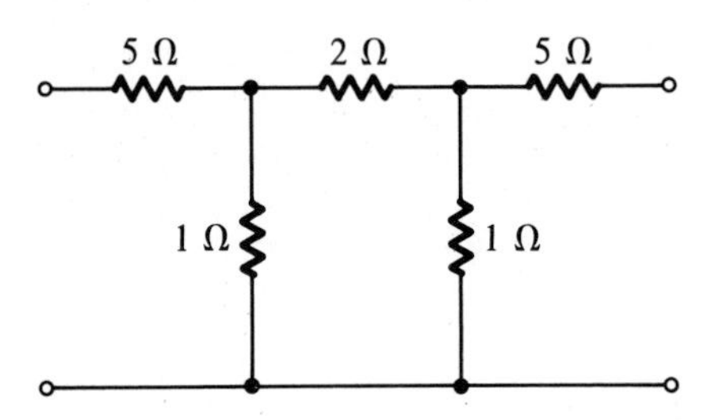

Figure P18.1

18.2 If a reciprocal two-port circuit has impedance parameters $\mathbf{z}_{11} = 10\ \Omega$, $\mathbf{z}_{22} = 20\ \Omega$ and $\mathbf{z}_{12} = 5\ \Omega$ find its admittance matrix.

18.3 Suppose that the input admittance and forward current gain of an unknown circuit are measured with the output terminals shorted, and are found to be 1/15 mhos and −2, respectively. Likewise, with the input circuit open, the reverse voltage gain and the output impedance are found to be 16 and 8 Ω. Find the impedance-parameter matrix of the circuit.

18.4 Calculate the ABCD parameters of the *RC* coupling circuit shown in Fig. P18.4. (Hint: Consider the circuit to be formed by cascading suitably chosen subcircuits.)

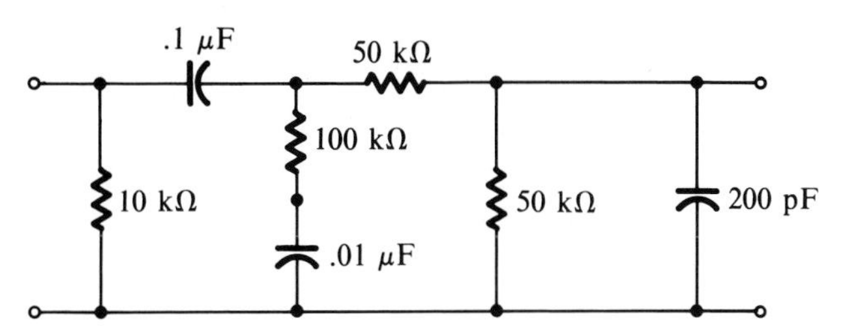

Figure P18.4

18.5 The circuit shown in Fig. P18.5 is used as a low-frequency model of the common-emitter configuration of a transistor.

a. Find the Thevenin impedance seen by v_{in}.

b. Find the transmission parameters of the circuit.

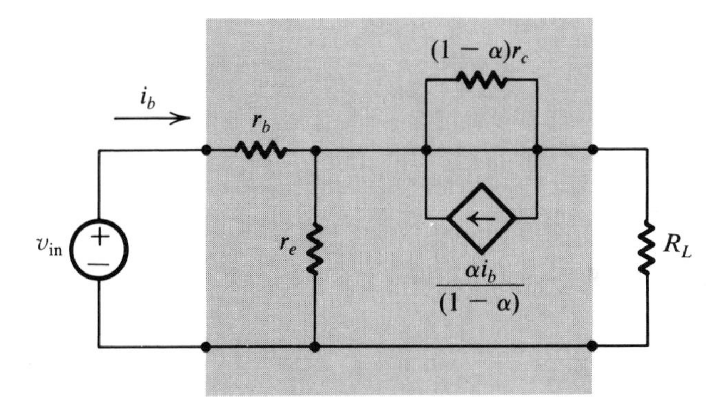

Figure P18.5

18.6 Find the impedance parameters of the bridge-T circuit shown in Fig. P18.6.

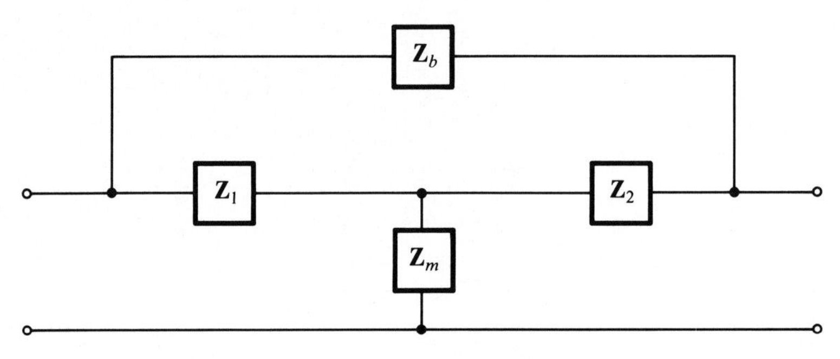

Figure P18.6

18.7 Find the impedance parameters of the symmetrical lattice circuit shown in Fig. 18.10.

18.8 Find the ZSR of v_1, v_2 and v_3 when $i_1(t) = 10 \sin 4t\, u(t)$ and $i_2(t) = 2 \cos t\, u(t)$.

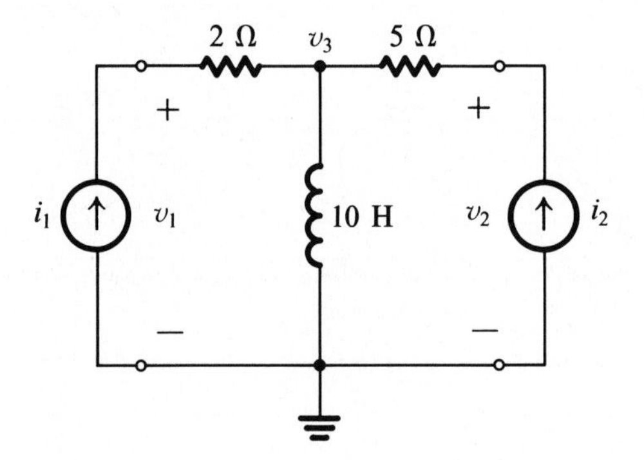

Figure P18.8

18.9 Find the admittance parameters of the controlled voltage source shown in Fig. P18.9

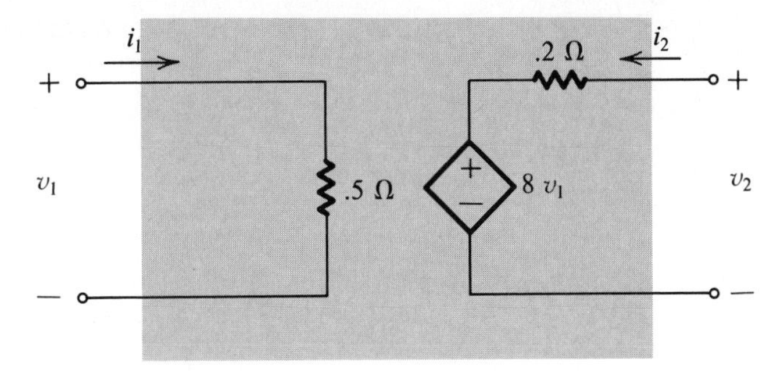

Figure P18.9

18.10 Two identical stages of a transistor amplifier are cascaded together as shown in Fig. P18.10. If the load requires an average power of 25 watts in steady state, find the required sinusoidal-source amplitude A, and find $i_3(t)$ and $v_3(t)$. Find the input impedance seen by the source.

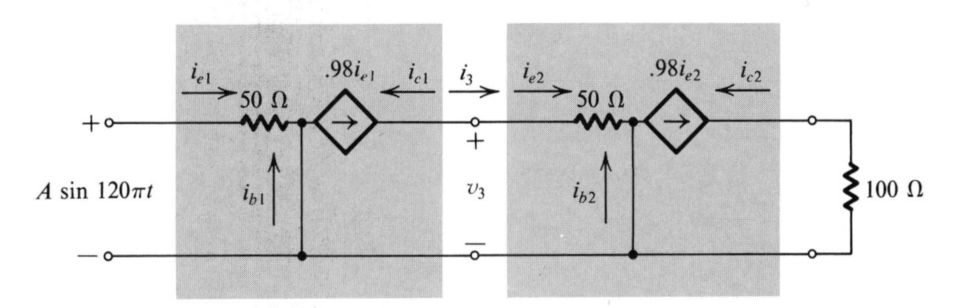

Figure P18.10

18.11 Find the ZSR of $v(t)$ when $v_{in}(t) = 10e^{-5t}u(t)$ and $i_{in}(t) = 20 \sin 5t\, u(t)$.

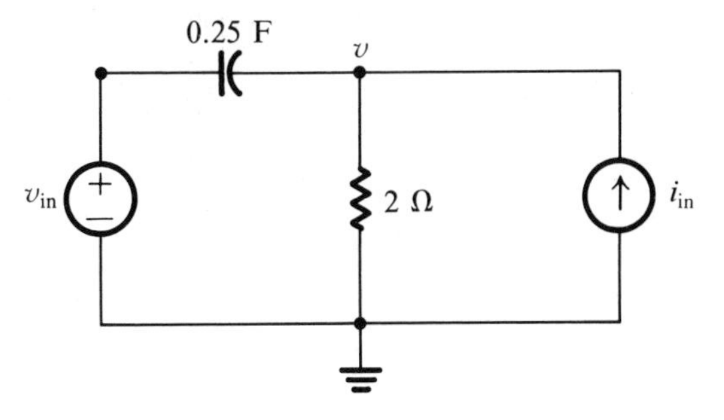

Figure P18.11

18.12 Explain why the y-parameter y_{21} is not defined for the op amp circuit in Fig. P18.11.

18.13 Find the input impedance of the loaded transistor circuit shown in Fig. P18.13.

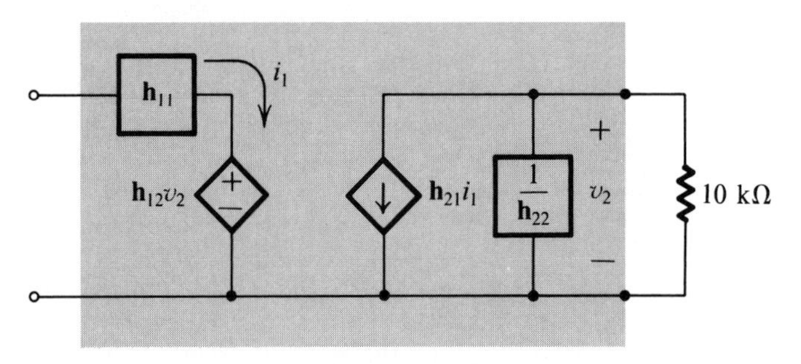

Figure P18.13

18.14 When an ideal voltage source is connected to the circuit in Problem 18.13, what impedance is seen by the load?

APPENDIX I: Matrices

A **matrix** of numbers is a rectangular array consisting of n rows and m columns of numbers. (In general, the numbers may be real or complex). **A, B, C** and **D** are matrices of real numbers:

$$\mathbf{A} = \begin{bmatrix} 2 & 7 & 1 \\ 5 & 3 & -8 \\ 9 & 0 & 2 \end{bmatrix} \qquad \mathbf{B} = \begin{bmatrix} 6 \\ 1 \end{bmatrix}$$

$$\mathbf{C} = [3 \quad 7 \quad -1] \qquad \mathbf{D} = \begin{bmatrix} -6 & 2 & -5 & 4 \\ 5 & 3 & -9 & 0 \end{bmatrix}$$

The label that identifies a matrix will always be written in bold uppercase notation (e.g., **A**). The numbers that form an array are called its elements. The element in the second row and third column of **D** is -9. Each element of an array is also labeled—with subscripted lowercase symbols corresponding to its position. For example, a_{ij} denotes the element in row i and column j of **A.** The first subscript denotes the row occupied by the element; the second subscript denotes its column. In the example given above $d_{23} = -9$ and $d_{24} = 0$. An array can also be represented by enclosing its element label in brackets; $[a_{ij}]$ denotes the same array as **A.** Similarly, $[\mathbf{A}]_{ij}$ denotes a_{ij}. For example $[\mathbf{D}]_{24} = 0$.

An array having n rows and m columns is said to be of order $n \times m$. The array **D** given above is of order 2×4. If $n = m$ the array is said to be square. The elements a_{ii} of a square array are said to form its **main diagonal.**

ADDITION AND SUBTRACTION OF ARRAYS

Matrices that have the same order may be added or subtracted on an element-wise basis. If a_{ij} and b_{ij} are the elements in the ith row and jth column of matrices **A** and **B** the **sum C** of **A** and **B** is formed by $\mathbf{C} = \mathbf{A} + \mathbf{B}$, where $c_{ij} = a_{ij} + b_{ij}$. Likewise, the **difference** of **A** and **B** is formed by $\mathbf{A} - \mathbf{B}$. For example

$$\mathbf{A} = \begin{bmatrix} 5 & 9 \\ 2 & -4 \end{bmatrix} \qquad \mathbf{B} = \begin{bmatrix} 7 & 1 \\ 6 & -5 \end{bmatrix}$$

$$\mathbf{A} + \mathbf{B} = \begin{bmatrix} (5+7) & (9+1) \\ (2+6) & (-4-5) \end{bmatrix} = \begin{bmatrix} 12 & 10 \\ 8 & -9 \end{bmatrix}$$

$$\mathbf{A} - \mathbf{B} = \begin{bmatrix} (5-7) & (9-1) \\ (2-6) & [-4-(-5)] \end{bmatrix} = \begin{bmatrix} -2 & 8 \\ -4 & 1 \end{bmatrix}$$

The reader can verify that the operations of **addition and subtraction of matrices commute.** That is, $\mathbf{A} + \mathbf{B} = \mathbf{B} + \mathbf{A}$ and $\mathbf{A} - \mathbf{B} = -\mathbf{B} + \mathbf{A}$. The $n \times n$ matrix **C** is called the **zero matrix** or the **null matrix** if $c_{ij} = 0$ for all i,j. Note that $\mathbf{A} + \mathbf{C} = \mathbf{A}$ if **C** is null. (When we write $\mathbf{A} + \mathbf{B}$ we assume that **A** and **B** have the same order.)

MULTIPLICATION OF MATRICES

If **A** has order $m \times n$ and **B** has order $n \times q$, the **right product** of **A** by **B** is the matrix $\mathbf{C} = \mathbf{AB}$ of order $m \times q$ formed by $c_{ij} = a_{i1}b_{1j} + a_{i2}b_{2j} + \ldots + a_{in}b_{nj}$ for $i = 1, \ldots, m$ and $j = 1, \ldots, q$. The elements in the ith row of **A** are multiplied term by term by the corresponding elements in the jth column of **B** and the resulting products are added. For **A** and **B** given above, we have

$$\mathbf{AB} = \begin{bmatrix} (5 \times 7 + 9 \times 6) & [5 \times 1 + 9 \times (-5)] \\ (2 \times 7 - 4 \times 6) & [2 \times 1 - 4 \times (-5)] \end{bmatrix} = \begin{bmatrix} 89 & -40 \\ -10 & 22 \end{bmatrix}$$

The matrix formed by multiplying the elements of a given matrix by a scalar has elements equal to the product of the scalar and the corresponding elements of the given matrix (i.e., if $\mathbf{C} = k\mathbf{A}$ then $c_{ij} = ka_{ij}$). For example

$$\mathbf{A} = \begin{bmatrix} 5 & 9 \\ 2 & -4 \end{bmatrix} \text{and } 3\mathbf{A} = \begin{bmatrix} 15 & 27 \\ 6 & -12 \end{bmatrix}$$

It is left as an exercise to verify that the operation of **multiplication does not commute.** That is, in general $\mathbf{AB} \neq \mathbf{BA}$.

THE INVERSE OF A MATRIX

The **transpose** of a matrix **A** is formed by interchanging its rows and columns. If $\mathbf{B} = \mathbf{A}^{\mathrm{T}}$ then $b_{ij} = a_{ji}$. For example

$$\mathbf{A} = \begin{bmatrix} 5 & 7 & 9 \\ \\ 2 & 3 & -4 \end{bmatrix} \text{and } \mathbf{A}^{\mathrm{T}} = \begin{bmatrix} 5 & 2 \\ 7 & 3 \\ 9 & -4 \end{bmatrix}$$

A matrix is said to be **symmetric** if $\mathbf{A} = \mathbf{A}^{\mathrm{T}}$.

The $n \times n$ matrix $\mathbf{I}$ is called an **identity matrix** if it has $[\mathbf{I}]_{ij} = 1$ for $i = j$, and $[\mathbf{I}]_{ij} = 0$ if $i \neq j$. The **inverse** of a square matrix $\mathbf{A}$, denoted by $\mathbf{A}^{-1}$, is the unique matrix having the property that $\mathbf{A}\mathbf{A}^{-1} = \mathbf{A}^{-1}\mathbf{A} = \mathbf{I}$, where $\mathbf{I}$ is the identity matrix having the same order as $\mathbf{A}$. Post- or pre-multiplying a matrix by its inverse creates an identity matrix.

If the inverse of a second-order square matrix exists, it is given by

$$\mathbf{A}^{-1} = 1/\Delta \begin{bmatrix} a_{22} & -a_{12} \\ -a_{21} & a_{11} \end{bmatrix}$$

where $\Delta = a_{11}a_{22} - a_{12}a_{21}$ is the **determinant** of the array of numbers forming $\mathbf{A}$. For example

$$\mathbf{A} = \begin{bmatrix} 5 & 9 \\ 2 & -4 \end{bmatrix} \text{and } \mathbf{A}^{-1} = -\frac{1}{38}\begin{bmatrix} -4 & -9 \\ -2 & 5 \end{bmatrix}$$

and the reader can verify that $\mathbf{A}\mathbf{A}^{-1} = \mathbf{I}$.

Given the square matrix $\mathbf{A} = [a_{ij}]$, the **cofactor** c_{ij} of element a_{ij} is given by

$$c_{ij} = (-1)^{i+j} \det \mathbf{A}_{ij}$$

where the reduced matrix $\mathbf{A}_{ij}$ is formed by eliminating the ith row and the jth column of $\mathbf{A}$. If

$$\mathbf{A} = \begin{bmatrix} 5 & 9 \\ 2 & -4 \end{bmatrix}$$

we get

$$\begin{aligned} c_{11} &= (-1)^{1+1} \det([-4]) = -4 \\ c_{12} &= (-1)^{1+2} \det([2]) = -2 \\ c_{21} &= (-1)^{2+1} \det([9]) = -9 \\ c_{22} &= (-1)^{2+2} \det([5]) = 5 \end{aligned}$$

The **inverse** of $\mathbf{A}$ is formed as

$$\mathbf{A}^{-1} = [c_{ij}]^{\mathrm{T}}\, 1/(\det \mathbf{A})$$

where the matrix $[c_{ij}]$ is formed with the cofactors defined above. Therefore,

$$\mathbf{A}^{-1} = \begin{bmatrix} -4 & -2 \\ -9 & 5 \end{bmatrix}^{\mathrm{T}} \left(-\frac{1}{38}\right) = -\frac{1}{38}\begin{bmatrix} -4 & -9 \\ -2 & 5 \end{bmatrix}$$

This agrees with the rule that the inverse of a 2×2 matrix is formed by (1) interchanging the elements on the main diagonal, (2) reversing the sign of the elements off the diagonal, and (3) dividing the result by the determinant of **A**. The more general rule must, however, be used to form the inverse of third- and higher-order matrices. Suppose

$$\mathbf{A} = \begin{bmatrix} 2 & -3 & -7 \\ 5 & 1 & 0 \\ -8 & 2 & 7 \end{bmatrix}$$

To find $\mathbf{A}^{-1}$ each of nine cofactors must be calculated. For example

$$c_{12} = (-1)^{1+2} \det \begin{bmatrix} 5 & 0 \\ -8 & 7 \end{bmatrix} = (-1)(35) = -35$$

Carrying out the remaining steps leads to

$$\mathbf{A}^{-1} = \begin{bmatrix} -1 & -1 & -1 \\ 5 & 6 & 5 \\ -\frac{18}{7} & -\frac{20}{7} & -\frac{17}{7} \end{bmatrix}$$

This result uses the fact that the determinant of a 3×3 array of numbers is given by

$$\det \mathbf{A} = a_{11}a_{22}a_{33} + a_{12}a_{23}a_{31} + a_{13}a_{21}a_{32} - a_{31}a_{22}a_{13} - a_{32}a_{23}a_{11} - a_{33}a_{21}a_{12}.$$

SOLUTION OF LINEAR ALGEBRAIC EQUATIONS

In circuits and many other engineering problems, matrices and vectors are used to compactly represent a system of linear algebraic equations. Suppose that $y_1, y_2, \ldots, y_n$ and $x_1, x_2, \ldots, x_n$ are related by

$$\begin{aligned} y_1 &= a_{11}x_1 + a_{12}x_2 + \cdots + a_{1n}x_n \\ y_2 &= a_{21}x_1 + a_{22}x_2 + \cdots + a_{2n}x_n \\ y_3 &= a_{31}x_1 + a_{32}x_2 + \cdots + a_{3n}x_n \\ &\;\vdots \\ y_n &= a_{n1}x_1 + a_{n2}x_2 + \cdots + a_{nn}x_n \end{aligned}$$

The equations describing these relationships can be written in matrix form as

$$\mathbf{y} = \mathbf{Ax}$$

where

$$\mathbf{y} = \begin{bmatrix} y_1 \\ y_2 \\ \cdot \\ \cdot \\ \cdot \\ y_n \end{bmatrix} \text{ and } \mathbf{x} = \begin{bmatrix} x_1 \\ x_2 \\ \cdot \\ \cdot \\ \cdot \\ x_n \end{bmatrix}$$

The arrays forming $\mathbf{x}$ and $\mathbf{y}$ are called **vectors.** A vector is an array consisting of a single column of numbers. The symbol for a vector is usually written in lowercase boldface notation.

If it exists, the unique solution of the algebraic equations defined by $\mathbf{y} = \mathbf{Ax}$ is formed by

$$\mathbf{x} = \mathbf{A}^{-1}\mathbf{y}$$

The proof is direct, since $\mathbf{A}^{-1}\mathbf{y} = \mathbf{A}^{-1}(\mathbf{Ax}) = (\mathbf{A}^{-1}\mathbf{A})\mathbf{x} = \mathbf{Ix} = \mathbf{x}$. In general, $\mathbf{A}^{-1}$ will exist if and only if det $\mathbf{A}$ is nonzero. As an example, suppose that

$$y_1 = 2x_1 + 5x_2$$
$$y_2 = -x_1 + x_2$$

In matrix form

$$\mathbf{y} = \begin{bmatrix} 2 & 5 \\ -1 & 1 \end{bmatrix} \mathbf{x}$$

Since det $\mathbf{A} = 7$, $\mathbf{A}^{-1}$ exists, and the unique solution to the set of equations is

$$\mathbf{x} = \frac{1}{7} \begin{bmatrix} 1 & -5 \\ 1 & 2 \end{bmatrix} \mathbf{y}$$

for a given value of $\mathbf{y}$. For example, if $\mathbf{y} = [2 \quad 5]^T$, then $\mathbf{x} = [-23/7 \quad 12/7]^T$.

The operation of matrix inversion has the property that

$$[\mathbf{AB}]^{-1} = \mathbf{B}^{-1}\mathbf{A}^{-1}.$$

The inverse of a product of matrices (each of whose inverse exists) is the product of the inverses in reverse order.

APPENDIX II: Complex Numbers

The **rectangular form** of a complex number z is the pair (x, y) of real numbers, where the first entry x is the **real part** of z (Re{z}) and the second entry is the **imaginary part** (Im{z}). Such a number is also written as $z = x + jy$, where j is used to distinguish the real and imaginary parts of the number (j's value will be shown to be the imaginary number defined by $\sqrt{-1}$). Example: $z = 3 + j4$.

A complex number can be represented by a point in the 2-dimensional plane of complex numbers, called the **complex plane** (as shown in Fig. A2.1). The horizontal axis is referred to as the **real axis,** and the vertical axis is referred to as the **imaginary axis.** Complex numbers have real and imaginary parts and can be thought of as points in this plane or, equivalently, a complex number can be thought of as a vector drawn from the origin to the point in the complex plane. The vector representation of a complex number also defines its magnitude and length.

MODULUS OF A COMPLEX NUMBER

The **modulus** of a complex number is its **magnitude,** (defined by $|z| = \sqrt{x^2 + y^2}$. This magnitude is the length of the vector that represents the number in the complex plane. (See Fig. A2.1) Example: $z = 3 + j4$, $|z| = 5$.

ANGLE OF A COMPLEX NUMBER

The angle of a complex number (ang z or $\measuredangle z$) is defined by the angle drawn counterclockwise from the positive horizontal axis to its vector.

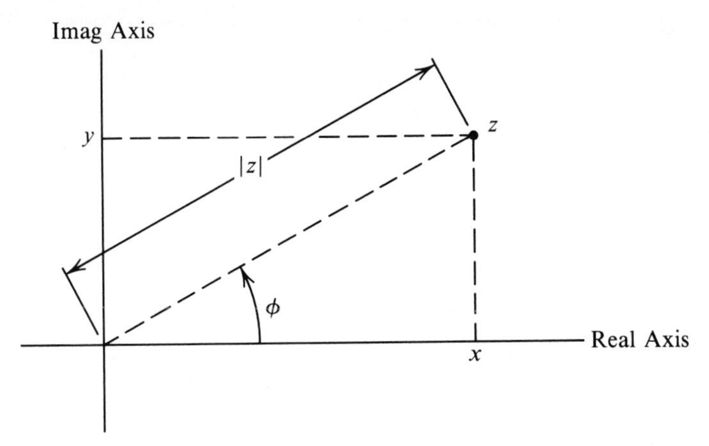

Figure A2.1

When z is in the first quadrant ang $z = \tan^{-1} y/x$. The expressions for $|z|$ and ang z can be used to construct the polar form of the number from its rectangular form. The polar form of a complex number is also written as $|z| = \angle \phi$ where ϕ = ang z. Example: $z = 3 + j4$, $|z| = 5$, ang $z = 51.3°$ and $z = 5 \angle 51.3°$.

EULER'S IDENTITY

Euler's identity, or law, states that

$$e^{j\phi} = \cos \phi + j \sin \phi$$

This result is used to express a complex number in exponential form as $z = |z|e^{j\phi}$ with ϕ = ang z because

$$|z|e^{j\phi} = |z|(\cos \phi + j \sin \phi) = x + jy$$

This expression also defines a conversion from the polar form to the rectangular form of a number, with

$$x = |z| \cos \phi$$

and

$$y = |z| \sin \phi.$$

ADDITION OF COMPLEX NUMBERS

Like real numbers, complex numbers obey a set of rules that defines a variety of operations that create other numbers. The sum of two complex numbers is the complex number formed by separately adding their real and imaginary parts. If $z_1 = (x_1, y_1)$ and $z_2 = (x_2, y_2)$, their sum is the number formed by

$$z_3 = z_1 + z_2 = [(x_1 + x_2), (y_1 + y_2)]$$

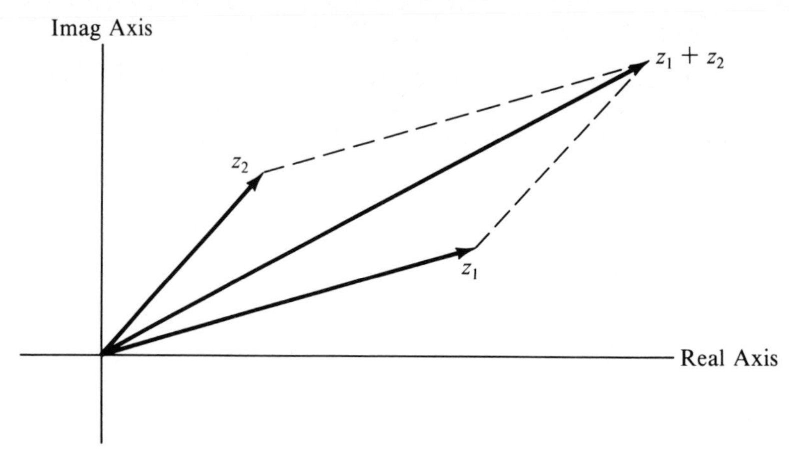

Figure A2.2

Note that

$$\mathrm{Re}\{z_1 + z_2\} = \mathrm{Re}\{z_1\} + \mathrm{Re}\{z_2\} = x_1 + x_2$$

and

$$\mathrm{Im}\{z_1 + z_2\} = \mathrm{Im}\{z_1\} + \mathrm{Im}\{z_2\} = y_1 + y_2$$

Example: $(3 + j4) + (7 - j2) = 10 + j2$
The vector representation of the sum of two complex numbers is displayed by the parallelogram shown in Fig. A2.2.

MULTIPLICATION BY A SCALAR

Multiplying a complex number $z = (x, y)$ by a scalar k forms the complex number $kz = (kx, ky)$. Multiplying z by k multiplies its real and imaginary parts by k.

PRODUCT AND QUOTIENT OF COMPLEX NUMBERS

The product of two complex numbers z_1 and z_2 is formed by

$$z_3 = z_1 z_2 = |z_1||z_2|e^{j(\phi_1 + \phi_2)} = |z_1||z_2| \underline{/ \phi_1 + \phi_2}$$

Example: $(5\underline{/30^\circ})(7\underline{/20^\circ}) = 35\underline{/50^\circ}$
Note that the magnitude of z_3 is the product of the magnitudes of z_1 and z_2 and the angle of z_3 is the sum of the angle of z_1 and the angle of z_2.
The quotient of z_1 and z_2 is defined by

$$z_3 = z_1/z_2 = |z_1|/|z_2|e^{j(\phi_1 - \phi_2)} = |z_1|/|z_2| \underline{/ \phi_1 - \phi_2}$$

Example: $(5\underline{/30^\circ})/(7\underline{/20^\circ}) = 5/7\underline{/-20^\circ}$
The magnitude of z_3 is the quotient of the magnitudes of z_1 and z_2 and the angle of z_3 is the difference of the angle of z_1 (numerator) and the angle of z_2 (denominator).

RECIPROCAL OF A COMPLEX NUMBER

The reciprocal of a complex number z is the complex number w such that $zw = 1$. Therefore $w = z^{-1}$ so $|w| = 1/|z|$ and ang $w = -$ang z.
Example: $z = 5\underline{/\ 30^\circ}$, $z^{-1} = 0.2\underline{/\ -30^\circ}$

ROOTS OF A COMPLEX NUMBER

The nth root of a complex number z is the number w such that $w^n = z$. Thus

$$|(z)^{-n}| = |z|^{-n}$$

and

$$\text{ang}\ (z)^{-n} = (1/n)\ \text{ang}\ z$$

The square root of z is given by

$$|\sqrt{z}| = \sqrt{|z|}$$

and

$$\text{ang}\ \sqrt{z} = 1/2\ \text{ang}\ z.\ \text{Therefore}$$

$$\sqrt{-1} = \sqrt{1\ \underline{/\ 180^\circ}} = 1\ \underline{/\ 90^\circ} = j$$

COMPLEX CONJUGATE

The conjugate z^* of a complex number z is the number with $|z^*| = |z|$ and ang $(z^*) = -$ang z. Note that $zz^* = |z|^2$, $z + z^* = 2\ \text{Re}\{z\}$ and $z - z^* = 2j\ \text{Im}\{z\}$.
Example: $z = 3 + j4$, $z^* = 3 - j4 = 5\ \underline{/\ -51.3^\circ}$

INDEX

Q

R

S

T